PLANT AND ANIMAL BIOLOGY
VOLUME II

PLANT AND ANIMAL BIOLOGY

VOLUME II

BY

A. E. VINES, B.Sc. (Hons.)

*Head of Science Department
and Principal Lecturer in Biology
The College of St. Mark and St. John, Chelsea*

AND

N. REES, B.Sc. (Hons.)

*Formerly Senior Lecturer in Biology
The College of St. Mark and St. John, Chelsea*

FOURTH EDITION

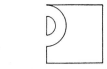

PITMAN PUBLISHING

First published 1959
Second edition 1964
Third edition 1968
Reprinted 1969
Reprinted 1970 (*twice*)
Reprinted 1971
Fourth edition 1972

SIR ISAAC PITMAN AND SONS LTD
Pitman House, Parker Street, Kingsway, London, WC2B 5PB
PO Box 6038, Portal Street, Nairobi, Kenya

SIR ISAAC PITMAN (AUST.) PTY LTD
Pitman House, 158 Bouverie Street, Carlton, Victoria 3053, Australia

PITMAN PUBLISHING COMPANY SA LTD
PO Box 11231, Johannesburg, South Africa

PITMAN PUBLISHING CORPORATION
6 East 43rd Street, New York, NY 10017, USA

SIR ISAAC PITMAN (CANADA) LTD
495 Wellington Street West, Toronto 135, Canada

THE COPP CLARK PUBLISHING COMPANY
517 Wellington Street West, Toronto 135, Canada

ISBN: 0 273 25223 2

Text set in 10/11 pt. Monotype Times New Roman, printed by letterpress,
and bound in Great Britain at The Pitman Press, Bath

G2—(T.1620/77)

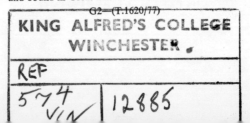

PREFACE TO THE FOURTH EDITION

It has been the aim, every four years or so, to try to bring this volume into line with any new developments in the biological field, including recent discoveries, changes in concepts, significantly different interpretations of older material and any other new trends that may have occurred in that period. It is certainly not before time that this latest revision work has been completed to catch up with the more important changes, since there have been a number of areas requiring alterations. Some have needed quite drastic changes in their treatment and there has been need to add at least two new topics for study, namely, endogenous rhythms and the genetic control of development.

The most extensive revision will be found in Chapters 2, 3, 6, 8, 9, 10, 12, 13, 16, 17 and 19.

Also, in view of the declared intention of this country to metricate and to adopt the use of SI units, this has been brought into effect throughout this fourth edition.

Lastly, fresh titles have been added to the reading lists and the examination questions have been augmented with more recent examples.

It is hoped that a useful purpose has been served in making these changes and that students will find the text of help to them in their immediate and possibly future studies.

We are grateful to all those readers who have contributed suggestions and we have tried to incorporate as many of these as possible.

A. E. V.
N. R.

Leigh-on-Sea
19th September, 1971

PREFACE TO THE FIRST EDITION

IN prefacing this textbook we can do little more than amplify some of the observations made in the preface to its companion.

Over the whole work we have tried to compile a reasonably complete introduction to the biological sciences, suitable for students who are preparing for examinations up to the General Certificate of Education, Advanced Level, and possibly to Scholarship standard. This will include members of Technical College classes, sixth forms of Grammar Schools and Colleges, students attending Preliminary Courses in Agriculture, Horticulture, Forestry, Dentistry, Medicine, Pharmacy and Veterinary Science, the National Certificate biologists, and trainee teachers.

We have explained our reasons for the integrated approach, and we hope it will be accorded some value since we should not like to think we had failed in one of our primary purposes of writing. Likewise, the emphasis given to the physical and chemical aspects of the subject has been mentioned. In this volume the student will encounter these far more frequently than in Volume I.

Also, as previously pointed out, our original conception of the order in which to present the topics was altered to facilitate a more convenient division of the work. It may be of some assistance to those who may use both volumes to be given a brief guide to their most advantageous use. We would suggest that the introductory chapters on cell structure and behaviour in Volume I, in conjunction with cell physiology and environmental conditions in the opening chapters of Volume II, be treated first. These may then be followed by a practical and theoretical study of the systematic and structural detail, together with the elementary physiology, of the plants and animals given in Volume I. The student should then be fitted for a clearer understanding of the problems with which organisms contend in their various environments, and be capable of a better comprehension of how the structures he has encountered are adapted to function in the variety of conditions. After the type studies, there should follow a treatment of the life processes as given in Chapters 7–17 of this book. The ultimate synthesis of his biological knowledge at this level can be achieved by reference to the final chapters. The same sequence can be applied to either botanical or zoological study. Consultation with a teacher will always produce equally good study programmes.

We must emphasize here the fact that in physiological and general biological study, quite as much as in morphological pursuits, practical work is of supreme importance, and the student is strongly advised to take as much care in this as he does in his sectioning, dissecting and other laboratory activities. For his guidance a good deal of practical work has been outlined in this volume. Some form of field work is essential.

With regard to the subject-matter, we have tried, as far as possible, to give recent information in the many topics discussed, but we admit that in this we must have failed in some cases. We have spent in the region of five years over the completion of the work, and in view of the fact that new interpretations and discoveries are being made with such great frequency, we have not always been able to keep up with them. Apologies are made in advance for any inaccuracies on this account. We hope we may have the opportunity of rectifying them at some future date.

Many of the illustrations to this volume are repetitions of those in its companion. They have been repeated, partly for their pertinence to the text, and partly for convenience of the reader who will not then have to search for them in Volume I. The remainder, as before, are simple line drawings which always bear some relation to the subject-matter.

Once more we acknowledge with gratitude all those who have assisted in the completion of our work. Our wives are thanked for their support and forebearance, and again we give special acknowledgement to Mrs. Rees for her very capable typing. From our publisher we have received nothing but kindly and helpful treatment and from the printer a very high standard of accuracy in reproduction. We know now that a "printer's error" is a rarity! Help with the proof reading by Miss P. Bishop, of Shoeburyness, was keenly appreciated, and to Mr. P. Lee, B.Sc., N.D.A., of Wye Agricultural College, we are indebted for helpful advice in the compilation of Chapters 10 and 11. To Mr. H. Miller, M.Sc., of the Municipal College, Southend-on-Sea, we are grateful for help relating to some matters of chemistry in this and Volume I. We also express our thanks to those who gave permission to use some of the illustrations to Chapter 20, and again to those who made loans of Examination Papers and to the various University Examination Boards for their kind permission to use the questions.

A. E. V.
N. R.

SOUTHEND-ON-SEA
28th March, 1959

CONTENTS

INTRODUCTION TO PHYSIOLOGY

In Chapter 1 of Volume I it was pointed out that the common characteristic of living things is that they consist of *protoplasm*. It was also stated that in the majority of cases, the protoplasm exists in small but definite units called *cells*. It was further indicated that in most multicellular plants and animals, the body is made up of specialized *tissues*, *organs*, and *organ systems* co-ordinated so that they function as a single entity, the *organism*. The organism is a self-sustaining system able to perform certain processes by which we recognize that it is alive. It has certain capabilities not shared by non-living systems. The most basic or fundamental of these are its powers of extracting from its surroundings energy and materials with which to maintain its existing composition and to synthesize more systems with the same properties.

As living systems have evolved so they have developed increasingly effective structures and mechanisms with which to perform these primary activities in a greater and greater variety of environmental conditions. Thus all organisms in their active state exhibit, to a more or less advanced degree, secondary functions that enable them to perform the primary ones more efficiently, so as to ensure survival for a sufficiently long period.

Even a cursory study of an active living thing will usually reveal that it makes manifest its property of being alive by the performance of what we sometimes call the "living functions" and biologists have, for convenience, attempted to study these individually. It must be remembered, however, that in the fully active organism they are almost always intricately interdependent for their continuous performance; failure on the part of a living thing to carry out any one of them at peak efficiency inevitably leads to loss of power to perform the others, culminating in death and disintegration of the once-living system. It should also be remembered that an organism may enter or be caused to enter a state of suspended activity, during which it is still alive in the sense that it retains its potential to perform the primary functions and so to manifest actively its living state once more when the conditions are favourable.

It is the consideration of all the "living functions" or "life processes", together with the nature of the structures developed for their performance, that forms the broad basis of the study of biology. Physiology is that part of the study that deals with the mechanisms by which the processes are carried out and physiologists generally refer to them under the headings given below.

1

Nutrition comprises all the activities associated with securing the materials necessary for the organism's continued survival. Nutritive substances serve a multitude of purposes, but broadly are needed for their energy content, for the building of additional protoplasm, for the manufacture of secretory substances which may be expended or stored, and for provision of offspring. There is considerable variation among living things both in nutritive requirements and in the methods of obtaining them. In the case of green plants, it is usual to include within the study of nutritional processes not only the means whereby basic materials are secured but also the mechanisms by which light energy is made available to convert these into more complex nutritive substances.

In *respiratory* processes, energy is made available to the protoplasm by the breakdown of high-energy compounds. Organisms which need free oxygen for the completion of the breakdown are termed aerobic; those which can respire without oxygen are anaerobic; some, such as yeast, can exist satisfactorily under either condition. The energy thus made available is used to do work of a variety of kinds.

Some of the energy is expended during activities involving *movement*. Movement in an organism can vary between the streaming of protoplasm within a cell and movement of the whole body from place to place. This latter type of movement is referred to as *locomotion*.

All living things *grow* and undergo stages of *development*. During its life, a unicellular organism increases the amount of its protoplasm. Multicellular organisms increase the number of cells as well as increasing the amount of protoplasm within them. All progress through different forms towards a fully functional maturity.

Any process leading to the elimination of unwanted or waste products is termed *excretory*. As a result of many of the activities of protoplasm, end-products may be formed which are of no further use to an organism, and indeed may be harmful. Also, an organism may absorb too much of a particular substance or may absorb substances with which it is quite unable to cope. In its correct use, the term excretion can be applied only to processes which rid the body of materials which have been part of the cell constituents. The elimination of unabsorbed food material is not excretion.

The ability of an organism to perceive and respond to changing conditions is called its *sensitivity* or *irritability*. Organisms normally respond to stimuli in such ways as to favour survival. There is great variation in sensitivity and also in type of response. To some stimuli, the response seems to be predetermined, to others, the organism may show a choice of responses. The power of choice culminates in man.

All the processes by which organisms increase their numbers and thus perpetuate their species, are called *reproductive*. No unit of protoplasm seems to be capable of perpetual existence without some form of rejuvenation. In some of the smallest and simplest creatures, the process manifests itself merely as fission into two or more parts, the division of the protoplasm apparently serving as sufficient stimulus for the continuance of life. In more complicated organisms, there is need for the fusion of protoplasmic material from two sources, normally called male and female, for the provision of offspring. Commonly, there is, besides this sexual reproduction, a phenomenon of propagation, whereby a single portion of an organism can give rise to a complete new one. Any such process is described as asexual reproduction. However reproduction is performed, all forms of living things produce offspring and thus strive to ensure the survival of their race. Indeed, with the exclusion of man, one may say that the ultimate objective of all organisms is to reproduce.

The performance of each of these functions involves a multiplicity of chemical and physical changes within the parts of an organism. All these parts must be interconnected and co-ordinated to keep the system balanced and unified, and whilst it is functioning, there is constant exchange of materials and energy between the organism and its environment. In a study of physiology, therefore, we must deal not only with those processes occurring in the separate cells, tissues and organs, but must investigate also the means whereby interconnexion and co-ordination of parts is achieved, and all the phenomena by which the plant or animal establishes relationships with its surroundings. Physiology embraces not only the specific processes of photosynthesis, digestion, absorption and assimilation, respiration, movement, growth and development, excretion, irritability and reproduction, but must include also the phenomena of osmosis, diffusion, evaporation; the absorption, transformation and external dissipation of energy; secretion and storage of materials, and the translocation of substances within the organism.

Generally, the term *metabolism* is used inclusively for all the chemical processes involved in maintaining life, and any such process in which substances are synthesized from simpler constituents is termed *anabolism*, whilst the reverse process is termed *catabolism*. In reality, the metabolic processes cannot be so sharply circumscribed. The modern concept is one in which the chemical make-up of an organism or any of its living parts is in a state of flux, involving cyclical changes which overlap and intersect one another at so many points that precise separation of one process from another cannot always be achieved. The organism is essentially an absorber of energy and materials and these are transformed and used in such ways that building and breakdown are continuous activities which may be going on in the same cell at the same time and involving the same substances. We may picture the metabolites, i.e. the substances taking part in metabolic processes, existing as interconnected reservoirs or pools, between which there may be flow from one to another in directions governed by many factors. Two of these are the levels of the materials in the pools, and some controlling influences along the interconnecting lines whereby the one substance is converted to another during transit. These latter are represented by the organic catalysts, the enzymes, without which no organism could metabolize at all; the enzymes themselves are influenced by many factors.

Not only is it difficult to separate one metabolic process from another, it is not easy to draw a line between the chemical processes and the physical, since in many cases the physical properties of protoplasm depend upon its metabolic activity. We must visualize the protoplasm as a physico-chemical system of such intricacy as to exhibit some characteristics which have as yet no clear parallels in the inorganic world and which cannot always be represented by man-made appliances. It has properties as yet inexplicable in terms of better-known events.

Nevertheless, our only approach to an elucidation of the physiological processes is through chemistry and physics, since we have no understanding of phenomena outside these to fall back upon. A special life-force may be postulated by which we may account for everything biological, but only when we understand what it is and how it operates, can it serve a satisfactory explanatory purpose. The student of physiology must therefore be equipped with a knowledge of physics and chemistry which will enable him to translate biological phenomena into a language which he can understand. Too great an emphasis cannot be placed on this requisite and in conjunction with the more purely biological study, we shall try to include in this text as much useful physics and chemistry concerning the organism and its environment as possible. The appendix, Matter and Energy in Relation to Living Things, may prove of assistance in this connexion to those whose studies of the physical sciences have been limited.

Since the characteristic of being alive resides in protoplasm, we can make no more appropriate approach to the subject of physiology than by gaining some appreciation of the nature and properties of this substance. This we can follow with study of the interchange of materials and energy between protoplasm and its surroundings and then study of the chemistry of plant and animal substances, including enzymes, and finally, before coming to grips with the physiology of the organism as a whole, we must appreciate the general characters of the environments in which it may be found. The next five chapters of this volume, in arrangement and content, are intended to provide this form of approach.

It has already been suggested that the broad basis of biological study is concerned with structure and function. In Vol. I much space was devoted to the former and it is assumed that the student at this stage will have had opportunity to digest its content. Where detailed knowledge of structure is required for proper understanding of function, reference to the appropriate chapter in Vol. I should be made.

Notes on Metrication and the Use of SI Units

It is now the intention of this country that in the near future all quantities and measurements shall be expressed in accordance with the metric system and that for all scientific and technological purposes a system of units known as the International System of Units (Le Système International d'Unités), abbreviated to SI units, shall be adopted. The latest revision of this volume has taken account of this and all quantitative statements conform to the principles agreed by the International Bureau of Weights and Measures (BIPM), published in Le Système International d' Unités, translated by the National Physical Laboratory, UK, and the National Bureau of Standards, USA, and published by HMSO in 1970.

These notes are intended to explain the principles upon which the International System of Units is based and to assist the reader in making any necessary adjustments that may be required.

There are three categories of SI units, namely *base, derived* and *supplementary.* The first is so-called because the units included form a group upon which the others are based. Derived units are clearly those formed from the base units by combining these to form products or quotients according to how they are connected one to another by multiplication or division. Such a resulting product/quotient may be given a special name, for example, the unit of force is called the newton (N) but is derived from a relationship between length × mass ÷ time. The third class of unit is supplementary in the sense that the units included in it have not yet been assigned to one or other of the two previously named categories. There are two units only included so far. They are the *radian*, the unit of plane angle, and the *steradian*, the unit of solid angle. Both are purely geometrical units. They can be regarded as either base or derived units according to preference.

There are six base units so far recognized with a seventh most likely to be officially accepted. They are:

unit of length: *metre* (m), equal to 1 650 763·73 wavelengths in vacuum of the radiation corresponding to the transition between the levels $2p_{10}$ and $5d_5$ of the krypton-86 atom;

unit of mass: *kilogram* (kg), equal to the mass of the international prototype of the kilogram made of platinum-iridium and kept by the

BIPM under specified conditions—this unit must not be used as a unit of weight or force;

unit of time: *second* (s), equal to the duration of 9 192 631 770 periods of the radiation corresponding to the transition between two hyperfine levels of the ground state of the caesium-133 atom—the duration 1/86 400 of the mean solar day is no longer accurate enough because the solar day varies irregularly;

unit of electric current: *ampere* (A), equal to that constant current which, if maintained in two straight parallel conductors of infinite length, of negligible circular cross-section, and placed one metre apart in vacuum, would produce between these conductors a force equal to 2×10^{-7} newton per metre of length;

unit of thermodynamic temperature: *kelvin* (K), equal to the fraction $1/273 \cdot 16$ of the thermodynamic temperature (T) of the triple point of water (*see* p. 1004)—note that the unit name is kelvin *not* degree kelvin and that it can be used to express an interval or a difference of temperature—note also that in addition to the thermodynamic temperature, use is made of Celsius temperature (t) given by $t = T - T_0$, where $T_0 = 273 \cdot 15$ K, (*see* p. 1003);

unit of luminous intensity: *candela* (cd), equal to the luminous intensity, in the perpendicular direction, of a surface of 1/600 000 square metre of a black body at a temperature of freezing platinum under a pressure of 101 325 newtons per square metre;

unit of amount of substance: *mole* (mol), equal to the amount of substance of a system which contains as many elementary entities as there are atoms in 0·012 kilogram of carbon 12—note that when the unit is used, the elementary entities such as atoms, molecules, ions, electrons, photons or other particles or groups of them must be specified. The unit must not be confused with the unit of mass. (At this date the mole has not received official recognition but is included here on the assumption that it will.)

There are rules for writing the units and symbols of these units. The unit name is always written with a small letter. The symbol for a unit is written as a capital letter only when the unit name is a derivative of a proper name, e.g. A (from Ampère), K (from Kelvin). A symbol does not change in the plural, i.e. 10 m not 10 m's, and is not completed with a full stop as abbreviations sometimes are.

Of the derived units there can be three kinds:

(i) those named and expressed in terms of base units only, e.g. volume, expressed as cubic metre with the symbol m^3;

(ii) those derived units given special names, having a distinctive symbol but expressible in terms of other units or base units, e.g. pressure, the unit of which is called the *pascal* (Pa), expressible as newton per square metre, $N\,m^{-2}$, or as base units $m^{-1}\,kg\,s^{-2}$: energy, quantity of heat, work, the unit for all of which is called the *joule* (J), expressible as newton metre, N m, or as base units $m^2\,kg\,s^{-2}$; (iii) those expressed by means of special names, e.g. moment of force, the unit of which is called the *metre newton* (N m), expressible in base units as $m^2\,kg\,s^{-2}$: surface tension, the unit of which is the *newton per metre*, $N\,m^{-1}$, expressible in base units as $kg\,s^{-2}$.

In writing these derived units, when a product is indicated, strict application of the rules includes a dot to indicate this, e.g. N·m, but it has been agreed that the dot need not appear when there is no possibility of confusion between a product and the symbol for another unit. In this text the dot has been dispensed with. When it is required to indicate that a derived unit is formed from others by division, there are three possibilities—use of the solidus or oblique stroke, as in N/m; the horizontal line, as in $\dfrac{N}{m}$; or negative powers, as in $N\,m^{-1}$. The last system is used in this text since difficulty can arise with the solidus when complicated expressions are required.

Some examples of commonly used derived units, their symbols and means of expression are given in the table on page 8.

Any units may be given as multiples or sub-multiples (by tens) and names and symbols of these decimal prefixes are:

$1\,000\,000\,000\,000 = 10^{12}$ tera T		$0{\cdot}1 = 10^{-1}$ deci d	
$1\,000\,000\,000 = 10^{9}$ giga G		$0{\cdot}01 = 10^{-2}$ centi c	
$1\,000\,000 = 10^{6}$ mega M		$0{\cdot}001 = 10^{-3}$ milli m	
$1\,000 = 10^{3}$ kilo k		$0{\cdot}000\,001 = 10^{-6}$ micro μ	
$100 = 10^{2}$ hecto h		$0{\cdot}000\,000\,001 = 10^{-9}$ nano n	
$10 = 10^{1}$ deca da		$0{\cdot}000\,000\,000\,001 = 10^{-12}$ pico p	
		$0{\cdot}000\,000\,000\,000\,001 = 10^{-15}$ femto f	
		$0{\cdot}000\,000\,000\,000\,000\,001 = 10^{-18}$ atto a	

It must be remembered that when an exponent is applied to a symbol with prefix, that exponent then applies to the multiple or submultiple of the unit concerned. For example, $1\,mm^3 = 10^{-9}\,m^3$ (not $10^{-3}\,m^3$), or one thousand millionth of a cubic metre not one thousandth of a cubic metre.

The fact that the base unit, kilogram, has a prefix is attributable to historical reasons. In this case, to avoid complications, the decimal multiples and submultiples of the unit of mass are attached to the word

Quantity described	Unit name	Symbol	Means of expression using base units only	Means of expression using derived unit special names
area	square metre	m^2	m^2	
volume	cubic metre	m^3	m^3	
speed	metre per second	$m\ s^{-1}$	$m\ s^{-1}$	
acceleration	metre per second per second	$m\ s^{-2}$	$m\ s^{-2}$	
density	kilogram per cubic metre	$kg\ m^{-3}$	$kg\ m^{-3}$	
concentration	mole per cubic metre	$mol\ m^{-3}$	$mol\ m^{-3}$	
luminance	candela per square metre	$cd\ m^{-2}$	$cd\ m^{-2}$	
force	newton	N	$m\ kg\ s^{-2}$	
pressure	pascal	Pa	$m^{-1}kg\ s^{-2}$	$N\ m^{-2}$
energy, work or quantity of heat	joule	J	$m^2kg\ s^{-2}$	$N\ m$
power	watt	W	$m^2kg\ s^{-3}$	$J\ s^{-1}$
quantity of electricity	coulomb	C	$s\ A$	$A\ s$
electric potential	volt	V	$m^2kg\ s^{-3}A^{-1}$	$W\ A^{-1}$
capacitance	farad	F	$m^{-2}kg^{-1}s^4A^2$	$C\ V^{-1}$
electric resistance	ohm	Ω	$m^2kg\ s^{-3}A^{-2}$	$V\ A^{-1}$
magnetic flux	weber	Wb	$m^2kg\ s^{-2}A^{-1}$	$V\ s$
inductance	henry	H	$m^2kg\ s^{-2}A^{-2}$	$Wb\ A^{-1}$
luminous flux	lumen	lm	$cd\ sr*$	
illuminance	lux	lx	$m^{-2}cd\ sr*$	
frequency	hertz	Hz	s^{-1}	
moment of force	metre newton	$N\ m$	$m^2kg\ s^{-2}$	
surface tension	newton per metre	$N\ m^{-1}$	$kg\ s^{-2}$	
heat capacity, entropy	joule per kelvin	$J\ K^{-1}$	$m^2kg\ s^{-2}\ K^{-1}$	

* Steradian is here used as a base unit

gram. The unit of mass is the kilogram, however, and its multiples and submultiples must be indicated by exponents attached to the symbol kg, e.g. kg^{-1}.

When a number is being written a dot is used to separate the integral part of it from its decimal part. Numbers may be separated into groups of three figures so that they may read more easily, but neither dot nor comma can be used between the groups of figures.

Whilst all kinds of quantities can and should be expressed in SI units, there are certain of them that have been conveniently expressed in other units, not SI, for a long time. It is considered acceptable that they should be allowed to continue in use. They are:

units of time: *minute* (min) = 60 s; *hour* (h) = 60 min = 3600 s; *day* (d) = 24 h = 86 400 s;

units of plane angle: *degree* (°) = $\pi/180$ rad; *minute* (') = $(1/60)° = \pi/10\ 800$ rad; *second* (") = $(1/60)' = \pi/648\ 000$ rad;

unit of volume: *litre* (l) = 1 dm³ = 10^{-3} m³ (note that the litre is here redefined as the *cubic decimetre*, the original definition, namely, the volume occupied by a mass of one kilogram of pure water, at its maximum density and at standard atmospheric pressure, being abrogated in 1964. This takes into account that the old litre and the cubic decimetre differ by twenty-eight parts in one million. To avoid confusion, litre should never be used when the results of high accuracy volume measurements are being expressed);

unit of mass: *tonne* (t) = 10^3 kg.

It was agreed in 1969 that certain other commonly used units could continue in use for a limited time. They include: *angstrom* (Å) = 0·1 nm = 10^{10} m: *hectare* (ha) = 1 hm² = 10^4 m²: *bar* (bar) = 0·1 MPa = 10^5 Pa: *standard atmosphere* (atm) = 101 325 Pa or N m⁻²: *curie* (Ci) = 3·7 × 10^{10} s⁻¹: *rontgen* (R) = 2·58 × 10^{-4} C kg⁻¹: *rad* (rd) = 10^{-2} J kg⁻¹. Certain units used in specialized fields can likewise be retained, for example, *electron volt* (eV) = 1·602 19 × 10^{-19} J approx. being the energy acquired by an electron in passing through a potential difference of one volt in vacuum. Others in this category are *unified atomic mass unit* (u) = to the fraction 1/12 of the mass of an atom of the nuclide ¹²C; 1 u = 1·660 53 × 10^{-27} kg approx; *astronomical unit* (AU) of distance = the length of the radius of the unperturbed circular orbit of a body of negligible mass moving round the sun with sidereal angular velocity of 0·017 202 098 950 radian per day of 86 400 ephemeris seconds; 1 AU = 149 600 × 10^6 m.

It is important to note that the unit of heat energy, the *calorie*, is to be abandoned and replaced by the *joule* (1 calorie = 4·186 8 J); the unit of length, the *micron* (μ), is also abolished, one thousandth of a millimetre should be expressed as 10^{-3} mm or 10^{-6} m.

CHAPTER 1

THE NATURE AND PROPERTIES OF PROTOPLASM

LITTLE knowledge of the essential nature of protoplasm can be discovered by examining it with the light microscope, since the instrument will not resolve particles below certain limits of size. Chemical and physical treatments interfere with its structure and often destroy its living nature, so that what is observed is artificial and requires interpretation in terms of the treatment. Nevertheless, by use of specialized instruments and new techniques, some of which are mentioned below, considerable progress has been made in the study of protoplasm in recent years. Without such means of investigation, the student will be unable to verify most of the information given in this chapter.

METHODS OF STUDYING PROTOPLASM

Apart from the use of a wide variety of microscopes, modern investigators use many physical and chemical techniques. The microscope is an instrument for giving an enlarged image of a small object. Chemical treatments emphasize particular parts of, or substances in, the protoplasm; physical treatments indicate certain of its properties.

Optical Methods

The light microscope is limited in its resolving power by the very nature of light itself. Two points cannot be seen separately if the wavelength of the viewing light (*see* p. 1008) is greater than twice the distance between the points. For yellow light of wavelength 600 nm (nanometres *see* p. 21) (0·0006 mm), the distance between the points for their resolution must not be less than 300 nm; for red light 400 nm, and for ultra-violet rays about 150 nm. Thus, the existence of particles smaller than 0·00035 mm cannot be demonstrated by a system of lenses using light. It must be noted here that the human eye cannot normally discriminate between objects less than 0·1 mm apart.

The ultra-violet microscope increases resolving power but necessitates the use of quartz lenses. Certain materials absorb ultra-violet light: the nucleic acids show strong absorption. Hence this instrument has been of great value in studying chromosome changes during mitosis. The technique has shown particularly valuable results in work on the movements of nucleic acids.

Dark field and ultramicroscopes are of value in detecting the presence and behaviour of small particles, rather than their size and shape. Light is directed on the material in a direction at right angles to the optical axis of the instrument. Small particles present reflect and diffract the light in various directions. Some of the light thus scattered will pass up the barrel to the observer's eye and the particles appear as bright specks in a dark field. In this way, particles of diameter no greater than 8 nm or even less, have been made visible. The motes in a sunbeam are familiar examples of the scattering phenomenon.

The phase contrast microscope was first described by Zernike. The optical system of this instrument makes it possible to distinguish between parts of a transparent specimen which have very slight differences of refractive index, something which is not possible with an ordinary light microscope. Slight variations in refractive index in parts of the specimen cause small differences in optical path of the transmitted light, and thus changes in phase of the light which passes through those parts. These "out-of-step" effects are quite invisible on an ordinary microscope since the eye can appreciate only differences in intensity, not differences in phase. In the phase contrast microscope, the phase effects are translated into variations in intensity. This is done by the introduction of a $\frac{1}{4}\lambda$ phase plate in the objective, used in conjunction with a special condenser, so that light passing directly through the specimen is made to interfere with that which is diffracted by it, so converting the phase changes into amplitude changes which are easily seen (*see* p. 1008). The major advantage of the instrument is that great contrast between parts of living protoplasm can be seen without the need for staining.

The polarizing microscope, as used with transparent objects, incorporates a Nicol prism, the polarizer, beneath the substage condenser and a similar prism, the analyser, in the eyepiece system. Thus the transmitted light is plane polarized before entering the object, and any effect which the object may have on the plane of polarization can be detected (*see* p. 1009). For example, any optically active substance, i.e. one which will rotate the plane of polarization, will show bright against a dark background, if the instrument was first set with the axes of polarizer and analyser at right angles, at which setting all light is eliminated before reaching the eye.

X-ray microscopes utilize rays of very short wavelength of the order of 0·1 nm and thus can be used to demonstrate objects less than 1 μm (micrometre) in diameter. An additional advantage is that X-ray photography will reveal internal structure and can be manipulated so that the living tissue is not killed. X-ray studies have revealed many important details of structure.

The Burch reflecting microscope uses mirrors for magnification instead of lenses. The instrument eliminates chromatic aberration: the rays of all colours will form a clearly-focused achromatic image. It can also provide a much greater working distance than glass lens systems can.

The electron microscope has revolutionized microscopy especially with respect to the sizes of objects that can be resolved. The wavelength of an electron beam is extremely small, being 0·005 nm or 5 picometres at a voltage of 60 kV. The lenses are represented by electromagnetic fields which can be manipulated to focus the beam of electrons. Aberration reduces clear definition to about 1·0 nm, but objects as small as this have been resolved. The outstanding disadvantage from the biological point of view is that only dead material can be examined, since the system is enclosed in a vacuum. A method of detecting fine detail of surface relief has been developed. It entails bombarding the specimen with atoms of a heavy metal such as gold or platinum. The result is a picture with all gradations of light and shade, corresponding to the surface relief of the specimen. By this means, many of the viruses can now be identified by their appearance.

Cinemicrography has enabled us to study continuous processes within cells. Microphotographs are taken at intervals on cine-film; projection shows in a few minutes a process which may have taken several hours. Thus, for example, the whole process of mitosis can be shown, immensely magnified and speeded up, on a screen.

Chemical and Physical Techniques

Many of the methods used involve the death of the protoplasm, and consequently great caution has to be exercised in interpretation. Some of the structures made visible by various processes are undoubtedly artifacts, caused by the particular type of treatment.

There are a number of dyes which can be used without apparent harm to the protoplasm. The process is known as *intra-vitam staining* and some dyes in common use are neutral red, janus green, methyl green and nile blue. Some of the dyes are indicators and thus acid or alkaline conditions can be determined; others stain specific constituents of a cell.

The examination of *fixed and stained sections* is a standard method in histology and cytology. The tissue is first killed by chemical fixatives which destroy the structure as little as possible. The material is cut into thin sections and suitably treated for mounting on glass slides. Delicate tissues may be embedded in molten paraffin wax, or frozen rapidly; either method preserves the relative positions of the parts. A revival of an old process in improved form is the Altmann-Gersh technique, whereby the tissue is suddenly frozen at the temperature of liquid air; the proteins of the protoplasm are thus fixed without becoming denatured. A very great variety of treatments has been developed for sections, smears, teased tissues and whole mounts; the use of double and triple staining gives contrasted colours to various structures in the cell.

Incineration of cells on mica strips destroys all the organic constituents leaving only mineral ash. Thus the presence of specific elements such as iron or calcium can be determined and their positions in the cell located.

Cell fractionation is the separation of the various cell components into "clean" portions. To achieve this, so that the parts that can be identified as discrete entities by the electron microscope can be determined chemically, it is first necessary to break up cell structure. This can be done by mincing, grinding or homogenizing the tissue using minute cutting blades revolving at high speed or subjecting it to vibrations of ultrasonic frequency. The disintegrated cell material is then treated by *centrifugation* at various speeds so that different components are "spun" out according to the gravitational forces applied. Certain cell components such as nuclei, chloroplasts, portions of membranes, mitochondria and lysosomes sediment at different rates and by choosing appropriate speeds and times of operation of the centrifuge these can usually be fairly cleanly separated. With an ultracentrifuge operating at up to one million revolutions per second, forces of fifty million times gravity can be achieved but small quantities only of material can be dealt with so it is used mainly to measure sedimentation rates. The whole operation is carried out in a vacuum chamber and under refrigeration. The sedimentation rates are converted to molecular masses and thus individual proteins can be selected out for analysis.

By means of *micromanipulation*, delicate operations can be performed on living cells. Very ingenious apparatus has been devised whereby tiny instruments can be mechanically moved through very small distances. Thus micro-dissection and micro-injection are performed. In some spectacular operations, nuclei have been removed; they have even been interchanged between two species of *Amoeba*.

Radioactive isotopes of many elements can be produced in an atomic pile (*see* p. 944). The isotopes emit rays and their presence and concentration can be detected by the Geiger-Muller counter or by their effect on photographic film or plate. The isotopes of carbon, oxygen, nitrogen and phosphorus have been particularly useful in tracing the course of various aspects of metabolism in the protoplasm. Examples of their use will be found in Chaps. 9 and 10.

The development of *chromatography* has simplified the identification of specific chemicals in cell extracts. Substances in solution vary in the rate at which they will move through an absorbent material such as filter-paper, or percolate through closely-packed columns of absorbent powders. There is a separation of the constituents of a mixed solution into fairly well-defined bands. The presence of uncoloured substances can be demonstrated by spraying the paper with indicators giving coloured products. Chromatography can also be used in conjunction with radioactive tracers which can be identified by the usual methods. A number of the intermediate products of photosynthesis have thus been established; separation of amino-acids has been achieved.

Tissue culture is the technique whereby cells are grown in specially prepared nutritive fluids. The activities of a cell from mitosis to maturity can be studied. The formation of dentine and enamel in teeth, the behaviour of chromosomes at mitosis, the growth and division of white corpuscles, are only a few of the sequences which have been carefully observed. It is interesting that this method of studying cells has provided considerable vindication of the older fixation and staining techniques, in that many of the structures tentatively postulated by the cytologists of an earlier day, have been confirmed by the modern exponents of tissue culture.

Other tools such as the spectrophotometer, flame photometer, the laser, respirometers, molecular sieves, electronic devices and computers are being used to study the structure and mode of functioning of proto-plasm. Each technique, in the hands of skilled investigators, is adding its increment of knowledge. The picture is not yet complete, but the main outlines are beginning to emerge. Protoplasm is not a homo-geneous substance. It exists mainly in membrane-bound units or cells, each cell being composed of a variety of clearly defined particulate structures or organelles such as chromosomes, mitochondria, ribosomes, Golgi bodies, centrosomes, plastids, and others, in a more or less uniform background of hyaloplasm (*see* Chap. 2). The structures and compositions of a number of the larger organelles have been elucidated and their roles in the chemistry of cells have been worked out. Little is known, however, of the hyaloplasmic substance beyond the fact that it possesses some invisible molecular organization in an aqueous medium. The system as a whole possesses the properties of substances in the colloidal state (*see* pp. 969 and 973).

THE CHEMICAL NATURE OF PROTOPLASM

The presence of various specific constituents in protoplasm can be demonstrated by micro-injection, incineration and spectrographic methods. Such experiments are not quantitative. Gross analysis of dead protoplasm gives the percentages of various constituents. An analysis of a myxomycete fungus, which consists of little but naked protoplasm, gives the following composition expressed as percentages of dry weight—

ORGANIC MATERIALS

(a) Water soluble		(b) Insoluble	
Monosaccharide sugars	. 14	Nucleo-protein . . .	. 32
Proteins	. 2	Free nucleic acid . .	. 3
Amino-acids . . .	. 24	Protein	. 1
Inorganic salts . . .	. 4	Lipoprotein . . .	. 5
		Neutral fats . . .	. 7
		Sterols	. 3
		Phosphatides . . .	. 1
		Polysaccharides and pigments, etc.	4

In the living protoplasm there is normally 85 to 90 per cent of water, so that all the above materials together represent only 10 to 15 per cent of the living substance. It must also be emphasized that the figures given above apply only to one particular kind of protoplasm, at one stage in its life, and under one set of environmental conditions. All three factors are subject to variation, and we are inevitably brought to the conclusion that there are many kinds or states of protoplasm.

In all the kinds of protoplasm, certain essential materials are always present. Water, to the extent of from 85 to 90 per cent, is an invariable constituent of active protoplasm. In special conditions, such as are found in dormant seeds and spores, the water content may fall as low as 10 per cent. The total protein will rarely exceed 10 per cent, and carbohydrate 2 per cent. Fats and fatty compounds will total less than 2 per cent. In specialized storage tissues, these figures will be considerably exceeded. There will always be a variable but small percentage of enzymes and intermediate metabolites. Finally, the protoplasm will contain inorganic salts in solution to the extent of 1 or 2 per cent. Generally the following ions are always present; potassium, calcium, magnesium, iron, phosphate, sulphate, bicarbonate and chloride. There will also be special elements associated with particular metabolic systems (see Chaps. 4, 10, 11).

Protoplasm is almost always slightly acid; in the cytoplasm very little removed from neutral, but in the nucleoplasm very definitely acid.

The protoplasm is a highly complex, heterogeneous colloidal system in a state of dynamic equilibrium, both physical and chemical. Nevertheless, over and above the multiplicity of changes constantly taking place, there is some overriding organization. Our widening knowledge of metabolism, of enzymes, and of genetics, is slowly yielding some clues as to the nature of this organization, but the whole pattern is far from completion.

THE ENERGY RELATIONS OF PROTOPLASM

The ceaseless chemical and physical activity of the protoplasm involves the performance of work (*see* p. 987). This entails a supply of energy, and ultimately this energy must come from the environment. There are two ways of obtaining this energy. Radiant energy may be absorbed, or compounds containing bond energy may be taken in. Broadly, green plants obtain energy by the former method, animals and fungi by the

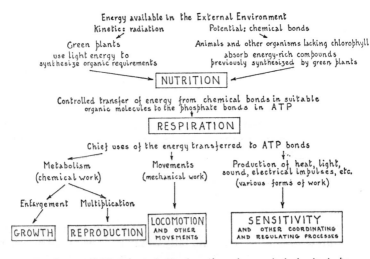

Fig. 1.1. Energy transformations in living things.

latter. The methods by which energy is absorbed are fully discussed in Chap. 10, and the quantization of energy is treated in the Appendix, pp. 986–992.

Within the protoplasm, the energy is utilized for chemical and physical work, and some is stored as potential energy in chemical compounds. Output of energy by the protoplasm is manifested as heat, as energy of motion, as chemical energy, and in some cases as sound, light and electricity (*see* Chap. 3). Fig. 1.1 shows schematically the broad outline of the energy transformations within living things. They are more fully treated in Chap. 3 and in the Appendix.

PROTOPLASM AND THE DIFFERENCES BETWEEN ORGANISMS

All living organisms are built up of units of protoplasm. In all, the materials and structure of the protoplasm are essentially the same; all perform their work in basically the same ways. The differences between organisms, even between those included in the same species, appear to lie in smaller or greater differences in the protoplasm, particularly in those parts of the living system we call the genes. Evidence of these differences lies chiefly in the ability or inability of particular organisms to break down or synthesize certain materials.

Enzyme Differences

The most outstanding chemical differences between organisms, and thus between their protoplasm, are the differences in the enzymes they produce. Careful study of bacterial enzymes has shown that even within a species, there are strains which show slight variation in the substrates they are able to attack. Classification of doubtful species of *Amoeba* has been achieved by comparing the protein-digesting enzymes they produce. Green plants possess the enzyme cellulase: in animals, this enzyme is confined to comparatively few groups: the trichonymphids, snails, earthworms, the mollusc *Teredo*, some insects and their larvae. A number of species of bacteria also produce this enzyme; some of these species are of the greatest service in the gut of herbivorous animals. Although many animals can manufacture the skeletal material chitin, no green plants appear to do so and only certain snails and a few insects can digest it. Albinos lack melanin pigment; they do not possess the enzyme tyrosinase which catalyses the conversion of the amino-acid tyrosin to melanin. This lack of tyrosinase has been linked with the absence of a specific gene. These are but a few of countless examples which show that a specific difference between organisms lies in their different enzymes, and hence ultimately with the genes which control their production (*see* p. 794).

Nucleic Acid Differences

The nucleic acids, together with the protein histone, are the fundamental constituents of chromosomes. The nucleic acids are themselves composed of chains of nucleotides, the number of links varying from 50 to 3000. Each nucleotide consists of three parts: one phosphoric acid unit, one ribose sugar unit, and a purine or pyrimidine base (*see* Chap. 4). The fact that the nucleotides are irregularly arranged on the chains gives rise to the possibility of an infinite variety of nucleic acids. The differences in length and arrangement of these nucleic acids may prove to be the ultimate facts which determine species. It is now known that the types of proteins synthesized in any cell depend on the linear arrangement of the bases adenine, thymine, guanine and cytosine. Each group of three determines the position of one particular amino-acid. Thus the sequence of these triplets determines the proteins synthesized (*see* pp. 112 and 794). It is noteworthy that while the amounts of these bases vary from species to species, they are constant for any one species irrespective of which individual or from which part of the body the material for examination is taken.

Differences in Kinds of Proteins

Analysis of the proteins of a wide range of organisms has shown that there is an infinite variety in the proportions and types of proteins present. In an experiment on the separation of the proteins in fish muscle by electrophoresis, it was found that the number of extractable proteins in various fishes ranged from six to twelve, and that there were great differences between species. Within a species, however, the numbers and types of proteins were so constant as to "finger-print" the species.

Differences in Volatile Amines in Plants

A great many plants produce volatile amines (*see* Chap. 4), some of which are responsible for the smell of flowers, such as trimethylamine, $(CH_3)_3N$, in hawthorn and rowan flowers. In an investigation of these volatile amines in more than one hundred different species, it was found that there was a chemical method of distinguishing species which agreed closely with the orthodox morphological classification.

Differences in Plant Seed Fats

Fats are all mixed glycerides (*see* Chap. 4). It is a matter of common knowledge that olive oil is different from linseed oil, castor oil or coco-nut oil. When the types and proportions of various fats present in many types of seeds were investigated, it was found that there was a basis for chemical classification which agreed closely with accepted botanical classification.

Serology in Plants

Carl Mez and his associates have attempted to establish relationships between plants by a serological method. A protein from a particular plant was injected into a rabbit. After a suitable time, the rabbit serum was extracted and mixed with a protein extract from a second plant. Precipitation indicated relationship. By examining the amount of precipitation caused by various dilutions of the rabbit serum, the nearness of the relationship between the two plants could be estimated. Extension of the work enabled Mez to construct a remarkable phylogenetic tree of the monocotyledons which showed general, but not exact conformity with existing morphological classification.

Serology in Animals

The methods of serology are well known nowadays in connexion with human blood groups (*see* Vol. I, Chap. 20). Here, it is interesting to note that if only the better known seven types of blood-grouping systems are considered, over one million different combinations are possible. It is obviously conceivable that blood-group identification may enable us to pin-point a single individual out of the whole world population. Serological methods have confirmed what comparative anatomy has also shown, that the anthropoid apes are more nearly related to man than are the monkeys. The technique has also been used in other groups of animals to determine their affinities. Serological studies of present world races have afforded interesting evidence of prehistoric migrations.

Protoplasmic Similarity and Difference

The above and many other equally striking lines of research lead us to the conclusion that every organism has its own protoplasmic order, the only possible exception being one-egg twins. The key pattern lies in the genes handed on from generation to generation (*see* p. 759). The genes determine the enzymes, the enzymes the metabolism, and the metabolism determines the physical form and physiological capability. Slight variation in the genic pattern will give slightly different individuals, large variations give widely different organisms. Between man and the bacterium lies an infinite series of finely graded differences. Yet all the multitudes of living organisms consist of protoplasm, all metabolize through the medium of enzymes; all are alive.

THE LIVING AND THE NON-LIVING

As the physiological processes going on in living things, referred to in the Introduction, are studied more fully, it will become abundantly clear that protoplasm, in any of its various forms, possesses properties not shared by any other known system of particles. Although some of its chemical components can be synthesized from inorganic sources, no single one of these, or collection of them artificially put together, shares its unique quality of being "alive," the integration of all its properties. This quality is not easy to define, although it is not too difficult to recognize when present. We recognize that a system is "alive" when it shows ability to carry out certain processes, for example, when it is able to nourish itself, to respire, to grow and develop, to excrete, to detect and respond to changes in conditions, to reproduce and sometimes to move. But none of these abilities, separately or together, constitutes being "alive," they are merely manifestations of the quality. We can use them to distinguish between living and non-living things, but by so doing we get no nearer to an understanding of the meaning of being "alive." We can approach an understanding by thinking more fundamentally and considering the energy and material relationships which known systems of particles have with their environments. In these terms, systems of particles fall into two main groups. On the one hand there are those which, under the physical and chemical forces of the universe, are tending always to disintegrate passively and to be forced towards a lower individual energy content. On the other hand there are systems of particles which, although subject to the same forces, are actively increasing their energy content and material substance by extracting and converting (not creating) energy and materials from their surroundings. Such systems are found only in protoplasm, the "living" substance. Thus a distinction between living and non-living things could be made in these more fundamental terms, that is, according to their possession or otherwise of the property of being able actively to obtain and make use of energy and materials from their surroundings, so to increase or at least maintain their own bulk and energy content. The quality of being "alive," with all its manifestations, could then be regarded as the expression of this property.

The living things which we encounter on this planet not only energize and materially sustain themselves to withstand the natural forces of disintegration and energy dissipation, but they do so in a perfectly self-regulated way. All the "manifestations of life" are contributory to this self-regulated "living" activity, the end-point of which is really the replication of the system so that the "living" material is perpetuated.

As has been stated, by this definition, protoplasm is the only living substance to be found on this planet and is alive only when it exists as a system of particles in its complex entirety. Any one of its separate components, from DNA molecules and proteins to water and inorganic ions, are separately as inanimate as the particles of a rock, when isolated from the system as a whole. There is, however, no reason to suppose that under different physical and chemical conditions, such as may be found on other bodies of the universe, equivalent living systems could not exist. Throughout the universe there may be many other forms of living systems, each fitted to energize, materially sustain and replicate itself in its own particular environment.

To indicate the opposite of "alive" we use the word "dead." It should be used only to describe a once-living protoplasmic system which has lost this quality. Dead protoplasm has lost the power to extract energy and materials from its surroundings and so to replicate itself. It will tend ultimately to disintegrate in the same way as any other non-living system, but this latter should not be described as dead, since it never was alive.

The question is often asked, "Is a virus particle alive?" In composition it is protein enclosing nucleic acid, and if these substances are taken separately then each is no more alive than any other individual organic substance. Together, as the whole virus particle, but held in a purely inorganic environment, they are still inanimate, for no virus can replicate under such conditions. The same virus particle inside a suitable living protoplasmic system, however, soon becomes a multitude of identical particles. Has the virus itself actively utilized energy and materials to bring about its own replication or has its presence merely caused an already living protoplasmic system, equipped with the necessary synthesizing machinery, to do the replicating work? If the second of these is correct, then the virus particle cannot be considered to be alive in the sense described above.

Between the presumably non-living particle system of the virus and the highest grades of living protoplasmic systems, such as are encountered in the clearly defined plants and animals, there exist intermediate conditions. The lowest grade of living system so far discovered exists in the extremely minute pleuropneumonia-like organisms (PPLO) (*see* Chap. 36, Vol. I). These possess lipoid-protein membranes, DNA in the form of a double helix, RNA in ribosome particles and protein substances, totalling about 1200 large molecules. They appear to be alive since they can replicate in a suitable non-living environment. The discovery of these minute living systems poses a question. How small are the smallest living organisms? Theoretical calculations suggest that the absolute limit lies in a unit of about 50 nm diameter, with a content of about 150 large molecules. The smallest PPLO so far discovered are only twice this diameter. They are, therefore, approaching the lower limit of size for living creatures, but others still smaller may yet be discovered.

TABLE OF MEASUREMENT OF LENGTH IN SI UNITS

1×10^{-18} metre	= 1 attometre	(am)
1×10^{-15} ,,	= 1 femtometre	(fm)
1×10^{-12} ,,	= 1 picometre	(pm)
1×10^{-9} ,,	= 1 nanometre	(nm)
1×10^{-6} ,,	= 1 micrometre	(μm)
1×10^{-3} ,,	= 1 millimetre	(mm)
1×10^{-2} ,,	= 1 centimetre	(cm)
1×10^{-1} ,,	= 1 decimetre	(dm)
basic unit ,,	= 1 metre	(m)
1×10 ,,	= 1 decametre	(dam)
1×10^2 ,,	= 1 hectometre	(hm)
1×10^3 ,,	= 1 kilometre	(km)
1×10^6 ,,	= 1 megametre	(Mm)
1×10^9 ,,	= 1 gigametre	(Gm)
1×10^{12} ,,	= 1 terametre	(Tm)

CHAPTER 2

THE FINE STRUCTURE OF CELLS

IN Chapter 1, Volume I, an account of the generalized structure of plant and animal cells has been given. This describes, in simple terms, what may be learnt of their composition using the light microscope, and from chemical analysis of crudely separated cell portions. With the invention of the electron microscope and skilled application of new biochemical techniques, a great deal more detail has been elucidated in the last few years. The work of cytologists, aimed at dissecting the cell down to its fundamental particles, and the work of the biochemists, aimed at elucidating the precise chemical activity of protoplasm, have blended to the extent that the chemical pathways by which protoplasm carries out the living processes can be related directly to its sub-microscopic structure. Between them, cell biologists have made great strides in explaining and relating, in molecular terms, the structure of protoplasm and the biochemical activities in which it is engaged.

It is true that within the wide range of plant and animal cells, there may be seen many different forms, constructed specifically for the performance of particular functions. Nevertheless, recent work has confirmed that practically all cells are built on a common fundamental structural pattern and carry out similar functional activities. For example, structurally, all are bound externally by a membrane which encloses the cytoplasm in which are included various "organelles," and a controlling nucleus; functionally, all are able to harness and transform energy.

It is the purpose of this chapter to give the reader a current understanding of cell structure, by outlining some of the more clearly established features of their sub-microscopic detail and the ways in which this is thought to be related to their activities. Diagrammatic representations of plant and animal cells, illustrating some of the structure which has been disclosed in electron micrographs are shown in Figs. 2.1 and 2.2. (See also the Plates between pp. 22 and 23.)

Cell Membranes

These include the *external cell membrane* or *plasma membrane* and a variety of intra-cellular membranes that by their nature and form and possibly the kind of material they enclose mark off from the background substance a range of inclusions or "organelles" mentioned above

The following electron micrographs illustrate the degree to which the electron microscope, especially at the higher magnifications, can reveal the finer structure of the different cell components—structure which is not observable under the light microscope. For example, the nuclear membrane is seen to be double; the endoplasmic reticulum and the Golgi body are seen to consist of canals which are often lined with small round bodies, and the nucleoli appear as aggregates of similar bodies; the mitochondria show a definite, highly ordered structure; in the chloroplasts the layered structure of the grana can be seen; the fine pores in the cell wall are also made visible.

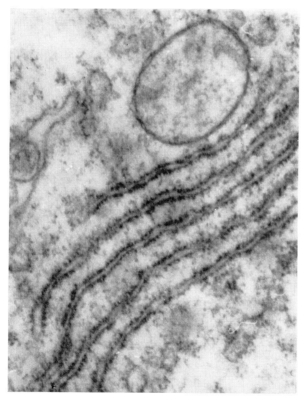

Plate 1. **Endoplasmic reticulum lined with ribosomes,** from the cytoplasm of an epidermal cell of the insect *Rhodnius prolixus* (*Hemiptera*) (× 44 000).

(*Courtesy, Dr. W. E. Beckel, Dept. of Zoology, University of Toronto*)

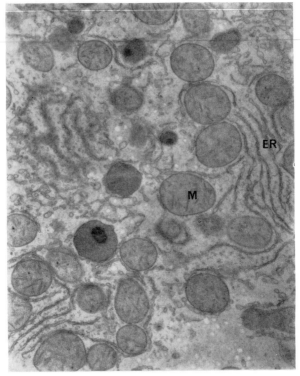

Plate 2. **A portion of the cytoplasm of a cell from the liver of a rat** (× 4000).
  ER, endoplasmic reticulum; *M*, mitochondrion.

Plate 3. **A portion of the cell nucleus of an insect,** *Rhodnius prolixus,* showing the nucleolus as an aggregate of ribosome-like particles (× 44 000).
  N, nucleoplasm; *Nu*, nucleolus; *NM*, nuclear membrane (double); *C*, cytoplasm.

Plate 4. **A Golgi body from a cell of the cotyledon of a bean,** *Phaseolus multiflorus* (× 100 000). The membranes are covered with ribosomes.
  C, canal; *V*, vesicle; *F*, fat body.

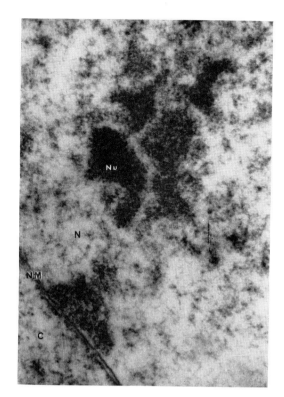

Plate 3

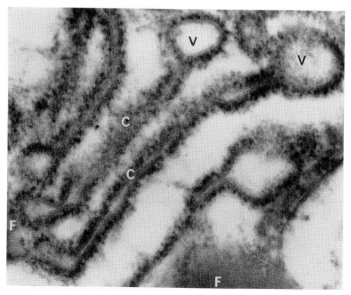

Plate 4

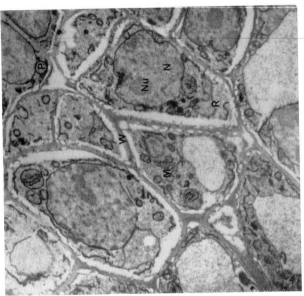

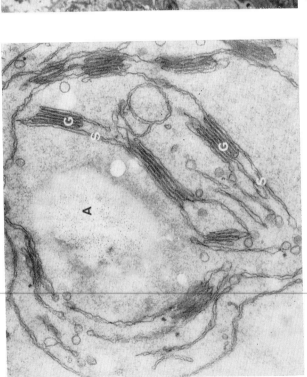

Plate 6. **Young cells from the stem meristem of a grass** (× 3300). Note the large nuclei in relation to the rest of the cell, characteristic of actively dividing meristematic tissue. In some cells the cytoplasm has shrunken away from the cell wall. *P*, plastid; *W*, cell wall; *Nu*, nucleolus; *N*, nucleus; *M*, mitochondria; *R*, endoplasmic reticulum.

Plate 5. **A green amyloplast from the stem of** *Pellionia daveauana* (× 3000). Amyloplasts are normally plastids in which starch is stored and become starch grains when mature. They are of interest because in them the various stages in the development of the typical chloroplast can be observed. *S*, loosely packed lamellae in a potential grana region; *G*, tightly packed lamellae in a grana region; *A*, starch granule.

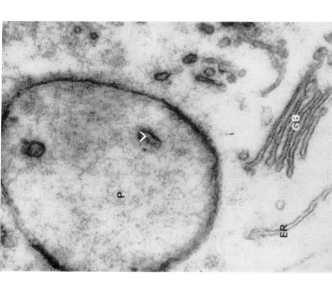

Plate 8. **A portion of the cytoplasm of a cell from the growing point of a grass,** showing a very young plastid with two vesicles that will later proliferate and flatten to form the characteristic lamellar structure of the chloroplast (× 50 000). A Golgi body is present in the cytoplasm. *P*, plastid; *GB*, golgi body; *V*, vesicle; *ER*, endoplasmic reticulum.

(*Courtesy, Dr. N. P. Badenhuizen, Dept. of Botany, University of Toronto*)

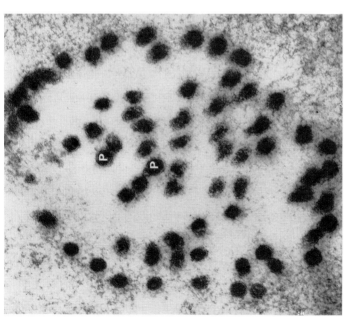

Plate 7. **A parallel section through the cell wall in the bottom of a pit, cutting a number of plasmodesmata at right angles** (× 20 000). From a cell in the growing point of a grass (*Cynodon*). *P*, plasmodesmata.

(*Courtesy, Dr. N. P. Badenhuizen, Dept. of Botany, University of Toronto*)

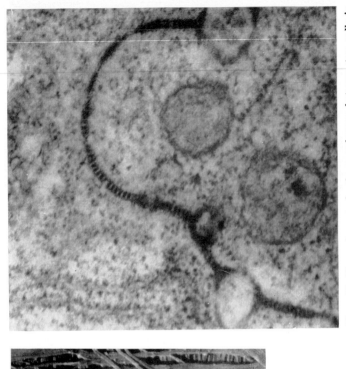

Plate 10. **Plasmodesmata in the membranes between two cells in the insect** *Rhodnius prolixus* (× 58 800).

(*Courtesy, Dr. W. E. Beckel, Dept. of Zoology, University of Toronto*)

Plate 9. Cellulose micelles from the cell walls of *Valonia* (× 30 000).

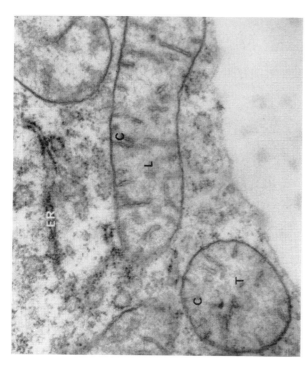

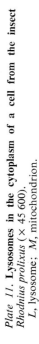

Plate 12. **Mitochondria, sectioned longitudinally and transversely, from a cell of the insect *Rhodnius prolixus* (× 63 000).**

ER, endoplasmic reticulum; *L*, mitochondrian sectioned longitudinally; *C*, cristae; *T*, mitochondrion sectioned transversely.

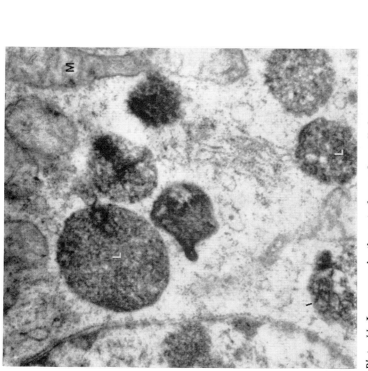

Plate 11. **Lysosomes in the cytoplasm of a cell from the insect** *Rhodnius prolixus* (× 45 600).

L, lysosome; *M*, mitochondrion.

(*Courtesy, Dr. W. E. Beckel, Dept. of Zoology, University of Toronto*)

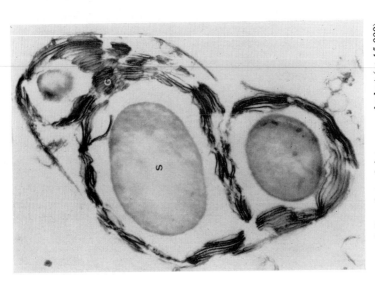

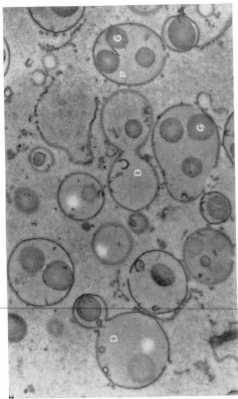

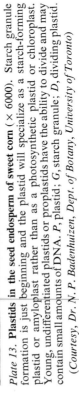

Plate 13. **Plastids in the seed endosperm of sweet corn** (× 6000). Starch granule formation is just beginning and the plastid will specialize as a starch-forming plastid or amyloplast rather than as a photosynthetic plastid or chloroplast. Young, undifferentiated plastids or proplastids have the ability to divide and may contain small amounts of DNA. P, plastid; G, starch granule; D, dividing plastid.

(Courtesy, Dr. N. P. Badenhuizen, Dept. of Botany, University of Toronto)

Plate 14. **A plastid from a cell in the cotyledon of a young pea, being transformed by the accumulation of starch into an amyloplast** (× 15 000). This type of plastid is known as a chloroamyloplast because in its early development it possesses the characteristic grana typical of the mature chloroplast. G, granum; S, starch.

(Courtesy, Dr. N. P. Badenhuizen, Dept. of Botany, University of Toronto)

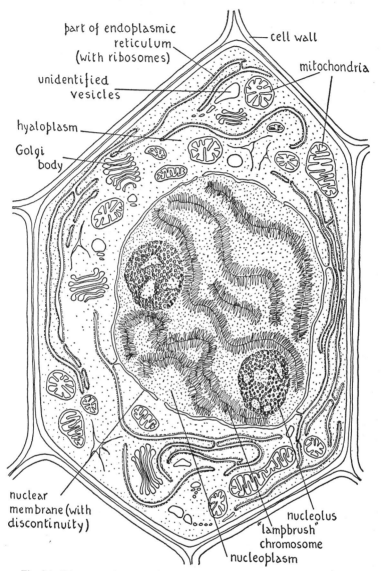

part of endoplasmic reticulum (with ribosomes)

cell wall

unidentified vesicles

mitochondria

hyaloplasm

Golgi body

nuclear membrane (with discontinuity)

nucleolus

"lambbrush" chromosome

nucleoplasm

Fig. 2.1. Diagrammatic representation of a young plant cell as seen with the electron microscope.

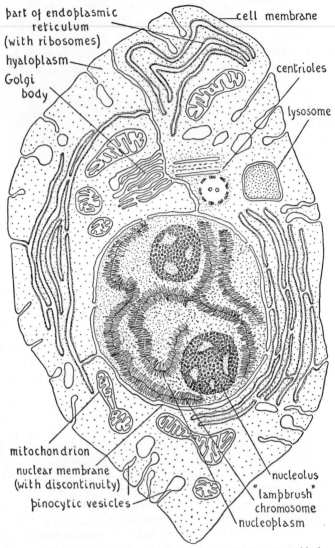

part of endoplasmic reticulum (with ribosomes)

cell membrane

hyaloplasm

Golgi body

centrioles

lysosome

mitochondrion

nuclear membrane (with discontinuity)

pinocytic vesicles

nucleolus

"lampbrush" chromosome

nucleoplasm

Fig. 2.2. Diagrammatic representation of an animal cell as seen with the electron microscope.

Among these, the *endoplasmic reticulum*, the *nuclear membrane*, the *Golgi apparatus, mitochondria, lysosomes* and other vesicles are commonly encountered, each with its particular kind of limiting membrane.

Owing to the extremely delicate nature of the external cell membrane, not even the electron microscope has been able to show with certainty all its structural properties. From information gained from permeability and surface tension studies, it has been postulated as a double protein and lipide layer, with a thickness in the neighbourhood of 8 nm (0·000008 mm) (*see* p. 21). In the best electron micrographs, the membranes of animals cells can be seen as two dense lines of about 2 nm each, separated by a lighter band of about 3·5 nm. This would correspond to the layers postulated above, and have approximately the same dimensions. It is sometimes referred to as "unit membrane," but not all the authorities agree that it is a universal external membrane structure nor does it necessarily represent the intracellular membrane form for some of the cell inclusions. A number of other complex arrangements of protein and lipide have been postulated and there is the possibility that some membranes are patchworks of differing molecular groupings. In some observed cases, the membrane has appeared unbroken and regular, in others, irregular with many invaginations. These invaginations could be the result of pinocytic activity of the cell (*see* p. 62), or could contain discharges from the cell. There seems to be no complete agreement as to whether some of the invaginations communicate directly with the internal reticulum, as is shown in Figs. 2.1 and 2.2, thus placing the deeper parts of the cell substance in straight communication with the external medium.

In some epithelial cells of animals there are thickenings of the external membrane, constituting *desmosomes*, at which points adjacent cells are caused to adhere more strongly, partly by the agency of minute fibrils of keratin that enter the thickened membrane regions from the surrounding cytoplasm. Plasma membranes of adjacent meristematic plant cells have been seen in continuity with one another, suggesting that at least in their early stages of development, adjacent masses of protoplasm in the plant are continuous with one another.

Permeability studies have also indicated that the plasma membrane may contain pores, i.e. is sieve-like, but so far no electron micrographs have shown the occurrence of gaps in the membrane structure. This may be because the pores, if they exist, are very small and widely separated or it is possible that they may be temporary only.

The tonoplasts of plant cells (not occurring in animal cells), which undoubtedly exercise control over movement of substances between the protoplast and the vacuole, show a structure comparable with that of the external membrane. Direct information on the nature of the tonoplast has come from studies with isolated vacuoles. A vacuole in this condition can be obtained if an isolated protoplast (without cell wall) is allowed to expand. The cytoplasm often ruptures and disintegrates, leaving the vacuole and tonoplast intact. Such isolated vacuoles show normal semipermeable properties in a sugar solution, that is, they have normal osmotic relations with the exterior, and show differentially permeable properties in sugar solutions containing fat-soluble substances. The fact that a tonoplast can be detached so readily from the protoplast may mean that there is no clear molecular bonding to the cytoplasm outside it.

Chemically, membranes are almost entirely lipides and proteins varying in type and quantity according to their origin and nature. The lipides are chiefly complex or compound lipides (*see* p. 91) such as *glycerophosphatides*, *sphingolipides* and the steroid, *cholesterol*. The protein substance of the membrane is partly structural but probably mostly enzymic in many cases, in which the membrane is acting as a surface for catalytic activity. All plasma-membranes appear to carry the enzyme Na-K-Mg-ATPase and a glycoprotein enzyme the sugar-protein product of which can form a mucus layer at the cell surface. Small quantities of sialic acid, phosphatidic acid, glycolipides and inorganic ions such as calcium are also present.

Membranes are replaced or newly constructed as required from their chemical components which are synthesized within the cell, for example, proteins at ribosomes, fatty acids at the membranes of the mitochondria. The construction sites of the different kinds of membranes vary according to the membrane. For instance, reticulum membrane is formed within the existing structure, the membranes surrounding lysosomes appear to originate in the Golgi apparatus, the membranes surrounding pinocytic vesicles and phagosomes, that is, vesicles containing ingested liquid and solid material respectively, are derived from the plasma membrane whilst new nuclear membrane substance originates in the endoplasmic reticulum. The inner membranes of mitochondria and chloroplasts are known to be formed only by existing membranes of the same kind and this must happen before similar new structures can be formed in the cytoplasm. A cell lacking some initial membrane material of these kinds cannot produce them *de novo*.

Further detail of other intra-cellular membranes or *cytomembranes* are given with the descriptions of the organelles with which they are associated.

Cytoplasm

From numerous studies it would appear that the cytoplasmic structure of both plant and animal cells is of similar nature. The electron microscope has shown that the cytoplasmic component of the cell is made up of a more or less homogeneous "ground substance" or *hyaloplasm*, in which may be found a variety of membrane-bound "organelles." Not even the electron microscope can show any structural detail of this ground substance although the structures of a number of the cytoplasmic granules have been clearly made out. The hyaloplasm must possess an, as yet, invisible molecular organization. So far nothing is known for certain what this might be, although there have been numerous postulates. One of these supposes a system of membrane-enclosed *microtubules* acting as a support system.

It appears from the study of flow in protoplasmic substance that the hyaloplasm has much in common with the physical state of matter known as *liquid crystal* or the *paracrystalline state*. The flowing substance is mostly protein and it is probable that when flow occurs the protein exists mainly as short, rod-like, macromolecules or micelles, each surrounded by a sheath of water molecules, forming thus a lyophilic colloidal system (*see* p. 971). Such micelles can line up parallel to one axis and slide freely over one another in that direction. Substances that take on the paracrystalline state are all composed of elongated molecules or groups of molecules and such structural units can take on one of several possible states of organization as distinct from the rigid spatial patterns taken up by molecules or atoms in true crystals. They can be arranged completely at random and show a truly liquid condition. They can be arranged so that the axes of all the molecules are parallel but the molecules themselves are irregularly spaced, like logs moving in a fast current. This is the *nemetic state*. A third possibility is known as the *smectic state* and is that in which the molecule axes are all parallel but the molecules have lost their freedom of movement in one of the three dimensions, that is, they form sets of regularly spaced parallel surfaces that can move over one another. This can be further complicated by the condition of a specific ordering of the molecules within a layer. The most complex arrangement is that in which the molecule axes are not only parallel but the centres of the molecules form a regular three-dimensional network corresponding to the solid crystalline state. In protoplasm, all these arrangements could occur and the importance of the ability of the protoplasm to take on the paracrystalline state is that it can then combine the fluid and diffusion properties of a liquid with an internal structure characteristic of a crystalline solid, that is, it can have a structure peculiar to itself whilst still able to take part in physical and chemical activities.

What has been shown very clearly is that, in both plant and animal cytoplasm, extending through the ground substance is a system of membrane-enclosed vesicles, forming a meshwork or reticulum. This

is now commonly known as the *endoplasmic reticulum* or *ergastoplasm*
although strictly speaking it is neither confined to the inner portions of
the cytoplasm nor does it necessarily create an intercommunicating
meshwork of vesicles. It seems to vary in its extensiveness according
to the type, state of activity and state of development of the cell. Mature
red blood cells show no signs of it. There are two forms of the endoplasmic reticulum recognizable, namely, *rough*, in which the vesicles or
tubules are commonly closely placed to one another and hence tend to
be flattened into *cisternae* and bear at their outer faces small particles
called *ribosomes* (*see* p. 29) and *smooth*, lacking the ribosomes and
not forming flattened cisternae. The two systems seem to be interconnected. The membranes which delimit this reticulate vesicular
system vary in different cells but, in general, have similar structural
properties to the external cell membrane and there are some indications
that the reticulum membranes are continuous with it. If this is so,
then the vesicular contents are really extensive continuations of the
external fluids, penetrating deeply into all parts of the cell. This
would increase enormously the surface area over which communication
between the cytoplasm and the external medium could occur. Physiological evidence has also accumulated which suggests that part of the
cytoplasm is freely available to externally applied solutions. The
volume of tissue into which external solute can penetrate by diffusion
cannot be explained unless a large part of the cytoplasm is freely
available to the solute. The volume of freely available tissue is called
the *apparent free space* and the volume not available is called the
apparent osmotic volume.

If there is such continuity between the deeper cytoplasm and the
outside of the cell then it would appear to extend to the nucleus since
it has been clearly shown that the membranes of the endoplasmic
reticulum are continuous with the double-layered nuclear envelope.

In some instances, in very young plant cells, membranes of the
endoplasmic reticulum have been located within the primary cell walls,
and in the views of some investigators the endoplasmic reticulum
membranes of adjacent cells may be continuous at some points. They
certainly appear to bridge the cell plate when this is being formed at the
end of cell division. This would mean that such intercellular connexions of the vesicular system would extend the contact from the
nucleus of one cell to the cytoplasmic matrix and nuclei of neighbouring
cells. There is no evidence that such intercellular contact is maintained
between the cells of mature plant tissues. There is a suggestion, however, that in some plants at least there exists a protoplasmic continuity
via plasmodesmata between cells, constituting what is then described
as the *symplasm*.

The tonoplasts of plant cells appear not to show continuity with the reticulum, although occasionally cytoplasmic vesicles may be attached to isolated vacuoles. The tonoplast apparently seals off the contents of the vacuole from the surrounding cytoplasm.

Within the ground substance, there are a variety of granular bodies or more correctly, membrane-limited zones or particles associated with membranes, generally referred to by names as *microbodies, ribosomes, mitochondria, lysosomes, Golgi bodies, centrosomes, blepharoplasts* (*basal bodies, kinetosomes*) and sometimes others according to the nature of the tissue.

Microbodies

These may be otherwise known as *peroxisomes* or *uricosomes* in animal cells and are very small rounded granules with a single-layered surface membrane found in a wide variety of plant and animal cell forms. Each is enzymic in nature, being composed of catalase and sometimes urate oxidase and possibly amino acid oxidase, indicating a probable role in uric acid and amino acid metabolism although the role of the catalase, an oxidation-reduction enzyme, in this connection is uncertain. Microbodies originate from the endoplasmic reticulum or the Golgi apparatus.

It should be noted that an older term, *microsome*, once used to describe any unidentified small granular inclusions of cytoplasm, is now used to designate small fragments of endoplasmic reticulum and Golgi body produced during cell homogenization and then appearing as artefacts during subsequent centrifugation.

Ribosomes

These appear as granular structures occurring in large numbers on the outer surfaces of the rough endoplasmic reticulum membranes. Each is composed of two rounded portions, one larger than the other arranged in "cottage loaf" form and consists of ribonuclear protein, RNA. It is known that the sequence of bases, adenine, guanine, thymine and cytosine, in the DNA molecules of the chromosomes encodes the information that fixes what kinds of protein shall be manufactured by the cell, and that this information is passed to the ribosomes via "messenger RNA" produced according to the coding on the chromosomes. Another kind, "transfer RNA," carries to the ribosome the right number of amino-acids, in the right sequence to fit the code, so that the ribosome can build a particular protein (*see* p. 795).

Mitochondria

Mitochondria from all sources seem to have a common basic form (*see* Fig. 2.3). Each is bounded by a double membrane, the outer smooth and regular, the inner involuted to form shelf-like or tubular processes or cristae, which penetrate deeply into the central fluid matrix. The detail of the membrane structure has been carefully resolved and fits the metabolic role of mitochondria which is primarily the operation of Krebs' cycle and oxidative phosphorylation in which ATP is formed (*see* p. 369). The enzymes necessary for this and some other activities, and at least fifty have been identified, are located on the membrane surfaces and grouped in such a way as to make the long series of steps at maximum efficiency, particularly electron transfer through the cytochromes (*see* p. 431). Some of the enzymes associated with the Krebs' cycle are linked to the outer membrane with others possibly in the space between the two layers whilst enzymes such as ATPase and the cytochromes are associated with the cristae inner membranes.

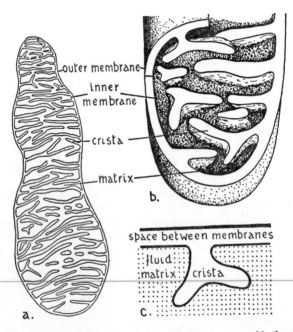

Fig. 2.3. Mitochondrion. (*a*) Natural appearance as seen with the electron microscope. (*b*) and (*c*) Diagrams to show the arrangement of membranes, cristae and matrix.

Mitochrondria are also concerned in the metabolism of fatty acids and the enzymes for extending fatty acid chains are located on the outer surface. There are also transphosphorylating enzymes positioned here as well.

A form of nucleic acid, unlike DNA in that the molecules are ring-formed, is attached to the boundary membrane and this suggests a kind of independence or self-controlling property of mitochondria, not shared by other cytoplasmic substance, as evidenced by their ability to repli-cate, separate into parts and fuse in pairs, as though they were not entirely under the direction of a nucleus.

Because the inner membrane can only originate from itself, a mito-chondrion must be formed from another one and from nowhere else in the cell.

Lysosomes

These have been described as occurring in some animal tissues, e.g. rat liver cells, but they have not been clearly shown in any plant cells. Where they occur they have been shown to contain the enzymes which control the breakdown of large molecules such as protein and fat, both of intra- and extra-cellular origin. There are several different forms of membrane-bound vesicles that constitute what may be called a *lysosome system*. *Primary lysosomes* contain the wide variety of digestive en-zymes, *heterophagosomes* enclose materials of extra-cellular origin for digestion and *autophagosomes* contain cell components to be degraded back into their simpler constituents. When the first meets either of the other two there is a coalescence of the vesicles to form a *secondary lysosome* in which the material to be digested is in contact with the digestive enzymes. As the final products of digestion are formed they are transferred through the membrane to the cytoplasm. Any residual undigested matter may be retained in the vesicle within the cell as a *post-lysosome* or may be discharged through the plasma membrane externally.

Primary lysosomes would clearly digest all the cell substance unless the investing membrane were present to prevent such *autolysis* or self-digestion, something that occurs when cells are badly crushed or otherwise damaged or die for some other reason. Because some sub-stances such as carbon tetrachloride, silica and some antibiotics can cause the investing membrane to break down, they can cause cell destruction via the lysosomes if ingested. Used primary lysosomes are replaced by the budding of vesicles from the Golgi apparatus, each containing enzymes formed at the ribosomes of the rough endoplasmic reticulum.

Heterophagosomes may be formed as a result of both phagocytic or pinocytic activity of the cell (*see* p. 62), the vesicle content being solid or a solution of macro-molecules accordingly. Autophagosomes are formed when the organelles or portion of cytoplasm to be degraded becomes invested by the appropriate smooth membrane.

Golgi bodies

Golgi bodies, once thought to be found in animal cells only, are now known to occur in plant cells also, at least, during their meristematic stages. In plant cells they appear to be more discrete and possibly simpler structures. In the animal cell, the electron microscope has shown that the Golgi bodies are vesicular regions of the cytoplasm surrounded by membranes lacking ribosomes but continuous with those of the endoplasmic reticulum. At their most complex, as for instance in secretory cells of pancreas, the Golgi apparatus is composed of a central vacuolar region surrounded by a complex meshwork of vesicles budding off membrane-bound packages of secretions and may be large enough to make up 5 to 10 per cent of the cell volume. Animal Golgi bodies are certainly associated with secretory activity of cells as well as the formation of primary lysosomes. In very young plant cells, each Golgi body, and there are usually very many, seems to be composed of a "pile" of about six plate-like, membrane-bound sacs or cisternae, associated with the edges of which are numerous small, more or less spherical vesicles (*see* Fig. 2.1). These Golgi bodies have been demonstrated in concentration in the region of cell-plate formation during cell division and it has been suggested that they may play a part in primary wall formation. In older plant cells, Golgi bodies of less distinctive structure have been seen. These show no sign of vesicle formation at the edges of the cisternae.

Centrosomes

Centrosomes or *centrioles*, in plants, found only in the cells of some algae and fungi, are very constant features of animal cells. What could be seen with a light microscope as a minute granular particle at the time of cell division, has been translated by the electron microscope into a pair of particles, each known as a centriole, and each being composed of a cylindrical body about 0·3–0·5 μm long and about 0·15 μm in diameter. The axes of the two paired centrioles always lie at right angles to one another. The wall of the cylinder, which surrounds two central strands, appears to be formed by nine groups of membrane-bound tubule-like bodies each group usually with three microtubules (*see*

Fig. 2.4). There is no doubt that these centrioles are very much concerned with spindle formation during nuclear division. Prior to chromosome activity, each member of the pair of centrioles divides and the two pairs migrate to opposite poles of the cell. Each cell, therefore, inherits one pair of centrioles from its parent and then duplicates the structure prior to further division. If this duplicating process is prevented, normal nuclear division in the cell will not occur. The fibrillar parts of the spindle and asters are thought to be constructed of similar microtubules.

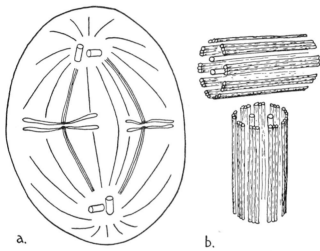

a.

b.

Fig. 2.4. Centrosome. (*a*) Positions of centrioles and spindle during nuclear division. (*b*) Diagram representing the structure of the centrioles.

Although no comparable structures have been seen in higher plant cells, the fact that spindle formation in plants seems to be controlled in a way comparable with that in animals, points to the existence of cytoplasmic particles of an equivalent nature.

Blepharoplasts or Kinetosomes

Otherwise known as *basal granules*, these have been seen at the roots of cilia and flagella in both plant and animal cells. Their structure closely corresponds with that of a centriole. It is presumed that such bodies are connected in some way with the formation of cilia and flagella and with their mode of functioning. Kinetosomes have the property of self-replication in common with centrioles; each duplicates every time that a new cilium or flagellum is formed.

Plastids

Plastids, by which is meant large or small, variously shaped cytoplasmic structures, are found only in plant cells and in the bodies of a few autotrophic protozoans. They are described as *leucoplasts* if colourless and *chromoplasts* if coloured. Of the latter, the chlorophyll-containing plastids or *chloroplasts* have been studied in much detail with the electron microscope. There appear to be three kinds within the plant kingdom. The *chromatophores* or *pigment-carriers* of bacteria and blue-green algae are each composed of a minute central core of bacteriochlorophyll and carotenoids. In the *lamellar chloroplasts* of the photosynthetic protozoa and the algae, the pigment is confined to dense parallel bands within the plastid membrane. In the complex chloroplasts of higher plants the chlorophyll is confined to an orderly arrangement of *grana* which are separated from one another by a network of fibrils within the *stroma* or background substance of the chloroplast (*see* Fig. 2.5). It is the chlorophyll molecules which supply the high-energy electrons under the influence of light, from which the energy is initially

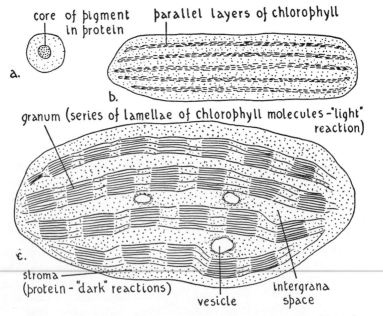

Fig. 2.5. Plastids. Diagrams representing: (*a*) pigment-carrier of bacteria and blue-green algae, (*b*) lamellar chloroplast of some green algae, (*c*) chloroplast of higher plant.

derived to do all the work of protoplasm. The "dark" synthesis reactions, involving the incorporation of more carbon from carbon dioxide into the system, are carried out in the stroma of the plastid.

Plastids originate as small proplastids. These enlarge to become first the leucoplasts. If these stay colourless and become centres of starch formation they are called *amyloplasts*. If chlorophyll develops then the proplastid becomes a chloroplast. Sometimes when amyloplasts are exposed to light they become converted to the chloroplast form and are known as green amyloplasts, of interest because they exhibit the stages of chloroplast formation.

In addition to the above well-defined cytoplasmic inclusions, others of far less easily distinguishable form have been reported from time to time. These include *annulate lamellae* and the microtubules of various disposition. The former is a membranous system originating from the nuclear membrane in some cells such as animal oocytes. Their structure closely resembles the porous nuclear membrane but no special function has as yet been attributed to them.

The Nucleus

The nuclear region, which may be more than two-thirds of the total volume of a very young cell, when examined by the electron microscope, is shown to be bounded by a clear nuclear membrane and to contain a background nuclear substance or *nucleoplasm*. In this are to be seen *nucleoli* and filaments of chromatin, the *chromosomes*.

NUCLEAR MEMBRANE

This is clearly a double layer, showing continuity with the endoplasmic reticulum at many points, and perforated with a number of circular openings measuring from 20–40 nm in diameter. These pores are thought to allow passage of large molecules, but there is not necessarily an absolute continuity between the nucleoplasm and the outside hyaloplasm, since the densities of these two components have been shown to be different from one another. It has been suggested that the continuity between the nuclear membrane and the endoplasmic reticulum membrane indicates that the former may play a part in the formation of the latter. In plant cells, during interphase, segments of the reticulum have been seen extending from the nucleus to the cell surface, and occasionally into the cytoplasm of adjacent cells. During the divisional phases, the whole system of membranes seems to break up and the fragments become scattered through the cell. Some stay aligned at the edge of the mitotic figure as it develops. When the nucleus reforms, there seems to be growth and fusion of the fragments to reconstitute the nuclear membrane. During this period the nuclear membrane substance must be duplicated and this is from the rough endoplasmic reticulum.

NUCLEOPLASM

This appears to have a consistency comparable with the cytoplasmic ground substance, but is not identical since the two have different densities. The detailed composition of the nucleoplasm has never been satisfactorily investigated but it is known to be basically proteinaceous.

CHROMOSOMES

The chromosomes, seen easily with the light microscope during nuclear division as short, spiralized, easily stainable bodies have no such appearance between these periods. The substance of each chromosome is extended and attenuated through the nucleoplasm. It has been calculated that if the DNA molecules of a single human cell nucleus were extended end to end, they would stretch to over a metre in length. Even with the electron microscope, no absolutely certain characteristic structure of a chromosome can be discerned. There is some indication that each consists of an *axial filament* which contains, at short intervals, pairs of thicker and denser swellings, the *chromomeres*. These appear to be densely spiralized parts of the chromosome axis (*see* Fig. 2.6). At each chromomere a pair of thin, matrix-encased filaments, probably despiralized parts of the chromosome axis, extend out and form a loop on each side of the chromosome and then return to it, entering the second chromomere of the pair. Maximum surface contact with other materials in the nucleus is thus made, and this is presumably where the chromosome pieces together the molecules of "messenger and carrier RNA" and replicates its own DNA structure at the appropriate times. The appearance of the chromosomes in this extended state has given rise to the name "lampbrush" chromosomes. The forms of the loops arising from the chromomeres are very variable but for any particular pair of chromomeres on a particular chromosome the form is constant. The chromosome axis, the chromomeres and the lateral loops consist of DNA, the sequence of bases in which determines the inheritable characteristics (*see* p. 792). The matrix enclosing the lateral loops is composed of RNA and protein. It should be understood that the true form of the extended chromosome has not yet been discerned clearly, partly because sections through such intricately tangled components are almost impossible to decipher.

Centromeres or *kinetochores*, seen clearly as unstainable portions on dividing chromosomes at points of spindle attachment, have constant positions on the same chromosome. Even the electron microscope, as

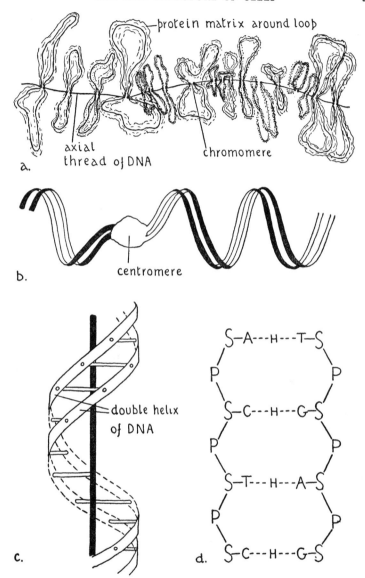

Fig. 2.6. Chromosomes. Diagrams representing: (a) "lampbrush" chromosome (part only), (b) spiralized double strand with centromere, (c) double helix of DNA, (d) same straightened to show the relative positions of sugar (S), phosphate (P) and bases (A, T, C, G).

yet, has failed to yield any definite information as to the structure of these bodies.

The chromosome axis is not regarded as being a single structure but as being made up of a number of identifiable fibrillar units called *chromonemata*. It has been postulated that the average somatic chromosome in plant cells has as many as 64 chromonemata and that in the giant salivary gland chromosomes of some insects there may be over a thousand.

NUCLEOLI

These are spherical bodies, lacking any definite boundary membrane, seen under the electron microscope to be packed with minute granules similar to the ribosomes of cytoplasm. They are very rich in RNA and appear to be the active centres of protein and RNA synthesis. It has been postulated that there is one nucleolus for each haploid set of chromosomes.

The "working" nucleus undoubtedly controls the metabolic activity of the cell, and there is good evidence to suppose that the "master plan" of the cell's activity is contained in the DNA molecules. These direct the manufacture of the enzyme proteins which ultimately control all chemical activity (*see* p. 794).

The Plant Cell Wall

This has two kinds of components, a discontinuous *microfibrillar portion* embedded in a continuous *amorphous matrix*. The microfibrils are comparatively inert and rigid and their number and the way in which they are oriented and interlinked in the matrix is mainly responsible for the final form of the wall. The matrix substances do partly contribute to the wall strength and firmness, but they are regarded as being more concerned in control of wall extensibility and hence in cell growth, since they are more inclined to change and extension. Most of what follows has been discovered from studies of primary walls of young cells.

The chemical compositions of the two classes of wall components have been worked out in some detail. The microfibrillar element is usually of *cellulose*, and seems to be the same in both primary and secondary walls of higher plants, but *chitin* replaces the cellulose in fibrils of many fungal cell walls. The formless background substance is composed chiefly of polysaccharides which are generally referred to simply as *hemicelluloses* and *pectic substances*.

Cellulose, which may make up as much as 25–35 per cent of dry walls of higher plants, occurs naturally as microfibrils 10–20 nm wide. Each microfibril consists of a core of long parallel polymers of chiefly β-1,4 linked D-glucose molecules (*see* Fig. 2.7), surrounded and interrupted by regions of substance of less precise molecular structure and

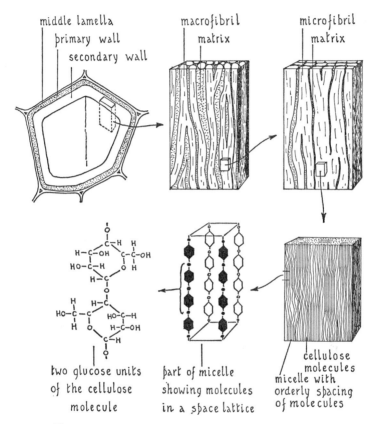

middle lamella
primary wall
secondary wall

macrofibril
matrix

microfibril
matrix

two glucose units
of the cellulose
molecule

part of micelle
showing molecules
in a space lattice

cellulose
molecules
micelle with
orderly spacing
of molecules

Fig. 2.7. Series to show cell wall structure in increasing detail.

A macro-fibrillar structure of cellulose, embedded in a matrix, is composed of smaller, microfibrillar strands. Each of these is formed of cellulose molecules, some of which show, at the micellar portions, an orderly lattice arrangement of the molecules.

containing some non-glucose molecules. It is possible that some hemicellulose of the matrix becomes adsorbed on to the cellulose core. The chitinous microfibrils of fungal cells consist of polymers of N-acetyl glucosamine. There have been reports of other types of fibrillar polysaccharides in the algae, but little is known about them.

The preponderating substance of the matrix is hemicellulose. This binds easily to cellulose and the two are often very difficult to separate. The term, hemicellulose, covers a heterogeneous group of compounds including D-xylose, L-arabinose, D-galactose, D-glucose, D-mannose and D-glucuronic acid residues. Generally, in any one wall, the matrix seems to be composed chiefly of one or two predominating substances, so that four relatively distinct forms of compounds can be separated. These are the xylans; the mannans and glucomannans; the galactans and arabogalactans; and the glucans. The xylans and arabogalactans show a branched molecular structure; the mannans and glucans are found commonly in fungal cell walls but not in higher plants. Very detailed analyses of the cell walls of oat coleoptiles show the presence of all the above component substances with the hemicellulose occurring as several different fractions.

The pectic substances are formed basically from the straight chain polymerization of α-1,4 linked D-galacturonic acid molecules which are commonly partially esterified with methyl groups.

There have been many reports of the isolation of nitrogenous substances from primary cell walls. These are considered to be proteinaceous and could be due to the presence of plasmodesmata and/or contamination of the wall substance under test with cytoplasm. Very well washed walls still show 2·5–5·0 per cent protein, however, and this may represent the presence of enzymes bound in the wall, e.g. pectin methyl esterase. Another suggestion to account for the presence of protein is that this and other non-cellulosic substances occur in the cementing material between the cells. This is considered not to be very probable.

Lipide substances, such as waxes, have been reported in some primary walls but these are probably part of cuticular layers. Electron microscope preparations treated with osmic acid do not show lipides as any essential part of the wall structure.

It should be noted that in all fresh primary cell walls there is always a good deal of water, as much as 60 per cent in young walls.

Between the walls of adjacent cells, the *middle lamella* is generally considered to be an amorphous layer of pectates, but the evidence for this is not entirely conclusive. It is possible that the middle lamellar region between two cells is only a cellulose-free region of the wall matrix, rather than a separate concentration of pectates. In this case, there is then no specialized cementing layer between adjacent cells.

A simple analogy between cell wall structure and reinforced concrete can be drawn. The concrete is represented by the matrix and the microfibrils of cellulose represent the reinforcing steel strands. Strength is gained from interactions between the individual microfibrils, between the matrix substances themselves, and between both these components. Such interactions, however, must be loose enough to allow for wall expansion and deposition of more substance. Apart from the fact that auxins are in some way connected with the "plasticization" of young growing walls, the nature of the interactions, and the changes which occur in them between wall components, is not yet understood.

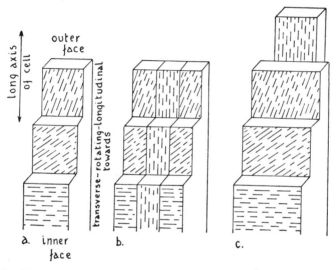

Fig. 2.8. Arrangement of cellulose microfibrils in plant primary cell walls. (a) Multinet structure. (b) "Integral" ribs of longitudinal microfibrils. (c) "Outer" ribs of longitudinal microfibrils.

The arrangement of the microfibrils, with respect to the longitudinal axis of the cell, in primary walls has been very carefully worked out in a number of cases (*see* Fig. 2.8). There appear to be two fairly distinct patterns of microfibril orientation, known as the *multinet arrangement* and the *longitudinal arrangement*. The former varies between an apparently disorganized meshwork of microfibrils, as in young onion cells, to a more organized pattern of fibril arrangement which is predominantly transverse, as in young oat coleoptile cells. In this case, during elongation, a pattern of fibrils is formed in which, on the inner surface, the fibrils are more or less transversely orientated, gradually increasing in their angle to the transverse direction until, on the outer surface, the fibrils are approaching the longitudinal direction. There is no layering in this system, the microfibrils at varying distances from the inner surface are all interwoven but gradually assume a wider and wider rotation from the transverse direction. This form of multinet structure seems very common but does not occur in root hairs. In this case, the meshwork of microfibrils generally lacks distinct orientation in the primary wall, but when the secondary wall is formed on the inside of this, its microfibrils are longitudinally placed. It is not known what determines the initial orientation of the microfibrils but it is known that they usually lie parallel to certain *microtubules*, elongated cylindrical structures of diameter 23 to 27 nm, found in the outside layers of the cytoplasm. Treatment of young cells with colchicine disrupts both the microtubule and microfibril arrangement. Thus there may be some direct connection between them.

The walls of many parenchymatous cells and young conifer tracheids show parts of the wall where longitudinal arrangements of microfibrils occur between regions of multinet structure and appear to be of two kinds. The so-called "integral ribs" are those in which all the microfibrils throughout the thickness of that part of the wall are longitudinally placed; the so-called "outer ribs" are separate groupings of longitudinally orientated fibrils, outside the multinet structure. Longitudinal microfibrils also occur in the walls of collenchyma cells. The young walls have the multinet structure but, with age, distinct layers of longitudinal microfibrils are formed at points around the circumference of the cell. In the cases of thickened primary walls, such as occur in outer epidermal cells, guard cells and some sub-epidermal and cortical cells, there are also longitudinal microfibrils.

To generalize, it seems that in most primary walls the patterns of arrangement of microfibrils are variations of the same basic form. There are two distinct microfibrillar components. These are transverse fibrils, which as a result of elongation of the cell (during which outer layers of these originally transverse microfibrils become twisted further and further towards the longitudinal direction) form the characteristic multinet structure, and longitudinal microfibrils which are laid down in clearly organized layers. Different types of primary walls are formed by differences in the relative proportions of these components. In some cases there may be no longitudinal component at all, in others such as collenchyma, they may predominate, whilst in ordinary parenchyma the condition is intermediate.

There is some variation in the way in which primary cell walls originate and develop during cell elongation. Some walls, such as those of oat coleoptiles, are known to expand uniformly over the whole length and new wall material is deposited continuously over the whole wall surface. In other cases, such as root hairs, cambial initials, tracheids and the phloem and xylem fibres, there is deposition of new materials and elongation at localized points only.

The possible mechanisms of cell expansion have been very fully explored but, so far, beyond certain generalizations, little has been discovered by which the process can be explained in detail. There are three physiological processes known to be involved when a cell increases in size; water enters the cell, new wall material is deposited and the wall increases in area. The regulation of the process could be through any one of these. There is no doubt that turgor pressure acts as the driving force in increasing the cell volume, and hence the surface area of the wall, but there is no evidence that this pressure is controlled by the cell. The evidence is that water movement into young cells is purely an osmotic phenomenon and not an "active," controlled occurrence. Thus there is no direct regulation of increase in size by control of osmosis, so that control must be through one of the other two agencies. The general concensus of opinion at present is that the walls become more "plastic" in response to the presence of auxins (probably coupled with gibberellins), that turgor pressure "stretches" the existing structure, and that growth of the cell wall is finally achieved by the deposition of new wall material in the extended structure. The chemical nature of the plasticizing effect of growth substances is not understood but could be a change in the nature of the pectin substances of the wall matrix.

CHAPTER 3

THE RELATIONS BETWEEN CELLS AND
THEIR SURROUNDINGS

PROTOPLASM is a colloidal system, its particles being dispersed in a continuous aqueous phase (*see* p. 969). It has been pointed out that such a living system is normally restricted to units of microscopic dimensions, and that each unit has an outer limit or boundary termed the plasma membrane. This membrane is of different construction from the rest of the protoplast; it maintains the protoplast as a single mass, and prevents its dispersion into the surrounding medium. If the plasma membrane is destroyed, the rest of the protoplasm quickly disperses. Each protoplast is not, however, completely isolated from its immediate environment by this external boundary; its own dispersion medium, water, is continuous with the water outside it. In order for the protoplast to continue its activity, there must be constant interchange of substances with the aqueous surroundings. In the following account we shall consider chiefly the interchange of materials between the protoplast and its environment and, in general, only the simplest case will be considered—that of an isolated protoplast not adjacent to, or connected with any other.

WATER RELATIONSHIPS OF THE CELL

A normal mature plant cell, such as that of onion epidermis or green alga filament, shows the protoplast contiguous with the cell wall at all points, with a large central vacuole. If such a cell is placed in a concentrated solution of common salt or sugar, certain changes soon become apparent. The protoplast loses its contact with the cell wall at one or more points and the volume of the vacuole decreases. With continued immersion in such a strong solution, the whole protoplast is soon shrunk away from the wall and the vacuole is considerably reduced. The cell is said to be *plasmolysed* (*see* Fig. 3.1)[1]. If the cell is not kept too long in the concentrated solution, complete recovery can be brought about by transferring the cell to distilled water. The

---

[1] Plasmolysis of plant cells may take a number of forms, often indicating that the external membrane of the protoplast is not merely adjacent to the cell wall and separate from it, but actually within the spaces between the microfibrils of cellulose. It is probably too simple a view to regard the plant cell protoplast as being pressed against the cell wall like the tube in a tyre.

vacuole now increases in volume slowly, and the protoplast occupies its original position against the cell wall. The cell is now *deplasmolysed*.

The same type of phenomenon can be demonstrated with animal cells. If red blood corpuscles are mounted in a similar strong solution, they shrivel and become wrinkled. They are said to be *crenated*. The decrease in volume indicates that they have lost something to the surrounding solution. If similar fresh red corpuscles are mounted in distilled water, they swell rapidly and soon disrupt, dispersing their

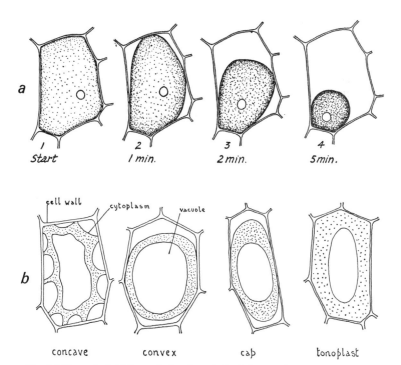

Fig. 3.1. Plasmolysis (*a*) Stages in the plasmolysis of pigmented cells of *Tradescantia* leaf epidermis in strong salt solution. (*b*) Types of plasmolysis.

material into the water. Obviously something is being absorbed from the surrounding medium in quantity sufficient to create an internal pressure which disperses the cell. In this latter condition, they are said to be *haemolysed*. But if fresh corpuscles are mounted in 0·6 per cent sodium chloride solution, neither crenation nor haemolysis results.

In each of these cases, it is apparent that changes in the concentration of the external solution affect the volume of the fluid within the cell, by causing a flow into or out of the cell according to whether the concentration is above or below a certain value. At this critical value, no net flow occurs in either direction.

Osmosis

To find an explanation of such experiments with plant and animal cells, we have recourse to certain well-known physical phenomena.

If two solutions of cane sugar (or salt) of different concentrations are separated by a membrane which will allow free passage of both water molecules and solute molecules (or ions), then after sufficient time, the two solutions will be of the same concentration, which will be intermediate between the two initial values. The two solutions will have become completely mixed, as if one had been shaken up with the other. The mixing is brought about simply as a result of the random movement of the individual solvent and solute particles. There is no difference in the movement of the particles and there is no justification for a distinction, but the movement of solute particles through the membrane is called *diffusion* and the movement of the solvent particles is called *osmosis*.

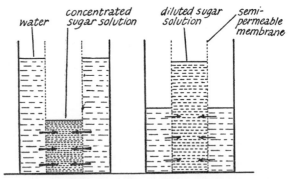

Fig. 3.2. Osmosis through a semi-permeable membrane.

A membrane which has pores too small to allow the passage of the solute molecules, but large enough to permit the passage of the solvent molecules is called a *semi-permeable membrane*. One type can be made artificially by filling a porous pot with a dilute solution of copper sulphate, and immersing it in an outer vessel containing a weak solution of potassium ferrocyanide. The two solutions pass into the walls of the pot and, where they meet, they form a gelatinous precipitate of copper ferrocyanide, filling up the pores of the pot.

The process of osmosis can be readily observed if such a pot is partly filled with a strong solution of sugar and then placed in a container of distilled water. The volume of the water will decrease, and the volume of the sugar solution will increase (*see* Fig. 3.2). The osmosis is due to excess diffusion of solvent only, through the semi-permeable membrane into a solution, caused by a lowering of the diffusion potential of the solvent in the solution by the solute. It can be shown that the excess movement of water particles in one direction through the membrane can continue against an opposing force. If a heavy piston is placed on the sugar solution it will be lifted until the hydrostatic pressure created in the solution by the weight of the piston is equal to the apparent force causing solvent movement into the solution. The pressure which must be applied to a solution in order just to prevent movement of solvent into it through a membrane is known as its *osmotic pressure*. Note that the osmotic pressure of a solution comes into existence only when the solution is separated from solvent by a suitable membrane. It can be defined as *the excess hydrostatic pressure which must be applied to a solution in order to make the diffusion potential of solvent in the solution equal to that of the pure solvent at the same temperature.* Osmotic pressure can be measured with an *osmometer*, a simple form of which is shown in Fig. 3.3.

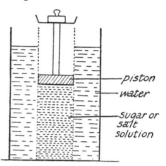

Fig. 3.3. A form of osmometer.

The Physical Laws of Osmosis

The osmotic pressures of dilute solutions of non-electrolytes were first investigated by Pfeffer in 1877. Van't Hoff pointed out in 1885 that the measurements made showed that dilute solutions obey the gas laws. His conclusions may be summarized thus—

1. The osmotic pressure of a solution is proportional to its molecular concentration: to the number of solute molecules per unit volume of solution.

2. The osmotic pressure is proportional to the absolute temperature.

3. The osmotic pressure produced by n molecules of solute in a volume V of solution is the same as the pressure which would be exerted by n molecules of a gas, occupying a volume V at the same temperature. These laws are equivalent to the three fundamental gas laws of Boyle, Charles and Avogadro. If one mole (*see* p. 6) of a solute, say cane sugar, is dissolved in water and the solution made up to

22·4 dm³ (22·4 litres—*see* p. 9), theoretically its osmotic pressure at 0°C is 1 standard atmosphere (101 325 N m⁻²—*see* p. 9). Alternatively, one mole of the same solute in 1 dm³ is theoretically capable of exerting an osmotic pressure of 22·4 atm. In practice, the osmotic pressure exerted by a molar solution (*see* p. 971) of cane sugar is about 25 atm. Thus the osmotic pressure of an ideal solution, at a given temperature can be defined as equal to the gas pressure which the solute would exert if it were present as a gas in a volume equal to the volume of the solution at that temperature.

Since electrolytes dissociate in solution (*see* p. 963), the ions acting individually, their osmotic pressures exceed those of non-electrolytes at the same molecular concentrations. In calculating the osmotic pressure of an electrolyte its dissociation factor for that particular temperature must be known.

So far we have considered the osmotic pressure of a single solution. When two solutions of unequal concentration are separated by a semi-permeable membrane, there is an excess movement of solvent from the weaker (*hypotonic*) to the stronger (*hypertonic*). The force causing movement is equal to the difference between the osmotic pressures of the two solutions. When two solutions have equal osmotic pressures they are said to be *isotonic*. The osmotic pressure of a solution containing more than one solute is equal to the total of the osmotic pressures exerted by each according to its particle concentration. Solutions of very high concentration do not obey the laws of osmosis perfectly due to interference between particles.

These physical facts of osmosis can now be applied to explain plasmolysis, deplasmolysis, crenation and haemolysis.

Plasmolysis and Related Phenomena

When a plant cell is immersed in a strong sugar or salt solution, the cell wall is completely permeable to both solvent and solute. Thus the solution comes into direct contact with the protoplast. Inside the vacuole is a comparatively weak solution of numerous osmotically-active substances. Such are the conditions under which there is a flow of water from the vacuole and possibly from the protoplasm also, into the external solution. Some part or all of the protoplast has acted as a semi-permeable membrane. Thus, plasmolysis is caused by the flow of water from the solution of lower osmotic pressure in the vacuole, to the solution of higher osmotic pressure outside the plasma membrane. If the plasmolysed cell is now placed in distilled water, deplasmolysis occurs because there is a solution with no osmotic pressure outside the protoplast, and a solution of high osmotic pressure in the vacuole.

Similar reasoning will explain crenation and haemolysis of red blood corpuscles. The different behaviour of the red cells is due to a basic difference in structure from the plant cells. The rapid inflow of water into the corpuscles is resisted only by the protoplasm and the delicate outer membrane, thus water continues to enter until the cell disperses. In plant cells, during deplasmolysis, the enlarging protoplast is forced against the cell wall. Further enlargement is restricted by the limited elasticity of the wall. When fully stretched, it exerts an inward force on the protoplast, and when this force is equal to the force by which the protoplast still tends to take in water, equilibrium is reached and the cell expands no further. Thus plant cells cannot be disrupted by these means; this is a considerable advantage, which animal cells do not normally possess. When a red blood cell is placed in a 0·6 per cent NaCl solution or a 5·5 per cent sucrose solution, it remains unchanged because these solutions are isotonic with its protoplasm. Thus animals have to maintain body fluids in such a condition that their tonicity does not interfere with the working of the protoplasm of the cells; small aquatic organisms must be isotonic with the water around them or make provision to regulate the flow of water. Even slight alterations in the concentration of extracellular fluids can lead to disaster for some animals. Most can osmoregulate (*see* Chap. 6).

Forces Affecting Movement of Water Into and Out of Cells

The total force by which a cell is able to take in water from its surroundings is known as its *diffusion pressure deficit* (*DPD*) which is defined as the difference between the diffusion pressure (*see* p. 247) of the water in an aqueous system and the diffusion pressure of pure water at the same temperature and pressure. It should now be obvious that this *DPD* at any particular time is equal to the osmotic pressure difference between the internal and external liquids minus the pressure exerted by the cell wall (if any) in resisting enlargement. This pressure exerted by the cell wall is known as *wall pressure* or *turgor pressure* often used synonymously but differing in the fact that they are acting in opposite directions to cause the same effect, that is to express water from the cell. Whereas wall pressure describes the tendency of the cell wall to shrink and so squeeze water from the vacuole, turgor pressure indicates a hydrostatic pressure within the cell above atmospheric pressure also tending to cause water to leave the vacuole. Using *DPD* for diffusion pressure deficit, *OP* for osmotic pressure, and *TP* for turgor pressure, we thus have—

$$DPD = (OP \text{ inside cell} - OP \text{ outside cell}) - TP.$$

This *DPD* may be negative and the cell will lose water when

$$OP \text{ outside cell} > OP \text{ inside cell}.$$

For a plant cell placed in distilled water, we have

$$DPD = (OP \text{ inside cell} - 0) - TP$$

or

$$DPD = OP \text{ inside cell} - TP$$

since the *OP* of distilled water is zero.

There are two extreme values for the *DPD* of a plant cell. When the *DPD* is at its maximum, the cell is said to be *flaccid*; this occurs when *TP* is zero, and then:

$$DPD = (OP \text{ inside cell} - OP \text{ outside cell}).$$

When the *DPD* is zero, the cell is said to be *turgid*; this occurs when the *OP* difference between the intra- and extracellular fluids is exactly equal to the *TP*. This would be the case when the protoplast had gained enough water to stretch the wall to its limit. For an animal cell placed in distilled water, we have

$$DPD = (OP \text{ inside cell} - OP \text{ outside cell}) - TP$$

$$= (OP \text{ inside cell} - OP \text{ outside cell}) - 0.$$

It is not usual to describe animal cells as being turgid or flaccid, since they rarely vary from an equilibrium. Fig. 3.4 shows the changing conditions in cells under the circumstances described, whilst the diagram in Fig. 3.5 indicates the changes in osmotic pressure of sap, turgor pressure and diffusion pressure deficit which occur in a plant cell during its change from a flaccid to a turgid condition in distilled water.

In the foregoing account, it has been assumed that the plasma membrane (or some part or all of the protoplasm) acts as a perfect semi-permeable membrane in all solutions. This is not strictly true, since solutes do enter the protoplasm and thus interfere with the osmotic equilibria which would otherwise be set up if the protoplasm acted as a perfect semi-permeable membrane. The penetration of most solutes under natural conditions is slow, but can be demonstrated by using solutes which enter almost as quickly as water. If an algal cell is placed in a weak solution of glycerine, it will plasmolyse in a few minutes. If, however, it is left in the solution, it will begin to deplasmolyse, and will not take long to return to its original condition. Initially the glycerine solution is hypertonic to the cell sap and thus the cell loses water rapidly. As the glycerine diffuses slowly into the cell, the internal osmotic pressure rises and the external falls. This continues until the cell sap is hypertonic to the external solution, and thus water passes inwards until the cell is deplasmolysed. Similar phenomena may be demonstrated with certain animal cells.

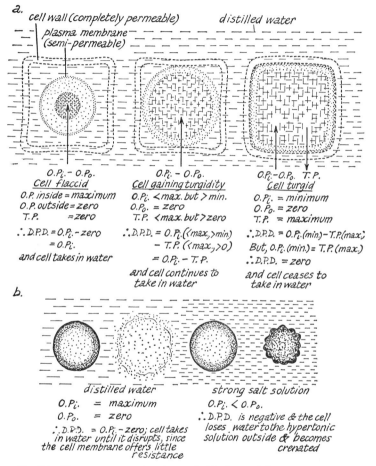

Fig. 3.4. Diagrams representing changes in (*a*) plant and (*b*) animal cells under different osmotic conditions.

Water Movement in a Biological System—Elementary Thermodynamical Treatment

On p. 999 reference is made to the fact that the capacity of a system for doing useful work is given by G, the Gibbs free energy of the system, although this is not its total energy. Now, it is the case that each component of a system contributes a share to the value of *G*. Thus a system such as that of a cell immersed in an aqueous solution or pure water will have its free energy contributed to by the water molecules and any particles such as ions, atoms, molecules dispersed through this solvent phase. The amount of free energy subscribed to the total by a mole of each component is called the *chemical potential* of that component, usually designated by the symbol, *μ*. It is also known as the *partial molar free energy* of the component. It is the difference in chemical potentials of a component in different parts of a system that determines the movement of that component between those parts.

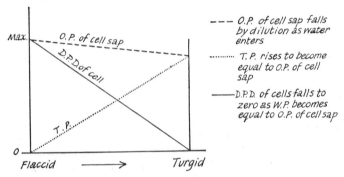

Fig. 3.5. Diagram to show changes in sap osmotic pressure, turgor pressure and cell diffusion pressure as a plant cell changes from flaccid to turgid in distilled water.

It is in fact the case that the driving force tending to move any component from one point to another in a system is the negative gradient of the chemical potential provided that no other components are moving. This is expressed as $F = -\mathrm{d}\mu/\mathrm{d}x$ where F is the driving force per mole of the component, $\mathrm{d}\mu$ is an infinitesimally small difference in chemical potentials and $\mathrm{d}x$ is the infinitesimally small distance over which this is measured.

In the system described the chemical potential of water will be the contribution made per mole of water to the free energy, *G*, of the system. This chemical potential of water within such a plant cell system will not be the same as that for a standard pure free water system. This is so because the water in a cell system is subjected to certain forces that do not exist in a pure water system. For example, there are solute particles present exerting forces on the water molecules, there may be a hydrostatic pressure above standard atmospheric squeezing the water molecules together causing greater repulsive forces between them and there may be various solid—liquid (or gas—liquid) interfaces subjecting the water molecules to special forces not encountered in a standard pure water system. All these factors will tend to affect the chemical potential of water in a cell system and this change from the standard pure water system can be expressed as the *water potential* in the given system. It is usually denoted by the symbol Ψ (psi) and it has as its units energy per unit volume (J cm^{-3}) or force per unit area, i.e. pressure (N m^{-2}). It is the resultant of three components: (i) P, the pressure in excess of atmospheric (101 325 N m^{-2}) exerted on the system, this having a positive effect by compressing the particles and giving them an increased free energy; (ii) π (pi), resulting from the presence of dissolved solutes in the system, tending to lower the chemical potential of the water and hence acting negatively on the water potential (π is really equivalent to the osmotic pressure of the solution which for dilute, ideal solutions is given by $RTCs$ where Cs is the concentration of the solutes in the system); (iii) τ (tau), the so-called "matric potential," reflecting the interactions of water molecules at interfaces, acting as a negative contribution to water potential because it arises from extra attractive forces to them.

Thus,

$$\Psi = P - \pi - \tau.$$

When dealing with the water potential of vacuolar fluid in a cell it is usual to remove τ from this equation on the grounds that the matric potential is negligible, that is, there are few or none of the macro-molecules that could create interface conditions. Hence for the vacuolar solution, the water potential is given by

$$\Psi = P - \pi.$$

It is the value of this vacuolar water potential with respect to the equivalent quantity for fluid external to the cell membrane that determines whether water will pass into or out of the vacuole. If there is a lower water potential within the cell than outside it, water will tend to flow from outside to inside and vice versa.

It should be possible to compare this newer concept, based on energies, with the older interpretation based on pressures as given on p. 49. By the latter, movement of water between vacuolar sap and external fluid is dependent on three pressures, the osmotic pressures of the fluids inside (OP_i) and outside (OP_e) and the wall pressure or turgor pressure (TP) exerted by the stretched wall. The resultant of the two forces ($OP_i - OP_e$) and TP determines whether water flows or not and in which direction. This resultant force, known as diffusion pressure deficit (DPD) was defined on p. 49 as the pressure by which the diffusion pressure of a solution was less than that of pure water at the same temperature and pressure. The comparison between the two methods of interpretation of the conditions is made possible by the recognition of the fact that DPD and water potential are exactly equivalent entities. In fact water potential describes the same property of a system as does DPD and they are numerically equal. The only difference is that they are opposite in sign, i.e. $\Psi = -DPD$. It should now be possible to recognize that the statements $DPD = OP - TP$ and $\Psi = P - \pi - \tau$ are equivalent, since P is really the same as turgor pressure, π is really osmotic pressure and τ is negligible in the system described. Thus $-DPD = -OP + TP$ or $DPD = OP - TP$ which is the same as $-\Psi = \pi - P$.

Water will tend to be taken up by a cell so long as the water potential of the vacuole, Ψ_i, is more negative (lower) than the water potential of the surrounding solution, Ψ_e. If water does pass into a cell, the vacuolar fluid is diluted and this causes an increase in the value of π for the vacuole, meaning that it becomes less negative or more positive. But as more water passes into the cell the value of P for the vacuole also increases or becomes more positive. Thus the value for Ψ_i becomes greater (more and more positive or less and less negative) until it becomes equal to Ψ_e at which point water ceases a net flow inwards. If the cell is immersed in pure water for which Ψ_e is zero, water will be taken up until Ψ_i becomes zero at which point the cell is fully turgid. When this happens $\pi = -P$ or $OP = TP$. If the cell is immersed in a solution for which $\Psi_e < 0$ there will be a cessation of net water flow before the cell reaches its maximum turgidity. At this point, $\pi - \Psi_e = -P$.

The water relationships of whole plants can be described using similar thermodynamical explanations of water movement within the complex system but at this level of study such treatment is not necessary.

Osmotic Relationships between Cells and Their Surroundings

Under natural conditions, some kind of osmotic equilibrium between the cell and its surroundings is achieved; there are several ways of attaining this. In the case of organisms living in fresh-water, the cell is hypertonic, and tends to take in water continuously by osmosis. When its requirements are satisfied, it must either resist further intake, or eliminate excess water. In the case of most plants, the wall is stout enough to exert a wall pressure which stops further intake when the cell is fully turgid. In the case of animals, and some plants which have no cell wall or only a delicate one, the difficulty is overcome by the active expulsion of water in bulk. In unicellular organisms the special cytoplasmic structures which accomplish this are called contractile vacuoles; they are osmoregulatory devices. In higher animals the excretory organs also osmoregulate (*see* Chaps. 6 and 15).

Provided that water is plentiful, plants rooted in the soil, exist under the same conditions as immersed plants, with the major difference that the parts in the atmosphere must inevitably lose water by evaporation. There cannot, therefore, be osmotic equilibrium between the plant and its surroundings except in a saturated atmosphere. Normally, the plant will always be taking in water through the roots and losing it to the air. This has no ill effect as long as water is plentiful and the plant cells are hypertonic to the soil solution.

Plants and animals which live in the sea, in many cases have their body fluids isotonic with the water. This obviates the necessity for continuous osmoregulation. The teleost (bony) fishes with a tendency for body fluids to become hypertonic to the sea water, relieve the condition by excreting salts from the gills. Elasmobranch (cartilaginous) fishes have the opposite condition, the body fluids tending to become hypotonic to the sea water, therefore they retain urea in the blood, to raise its concentration (*see* Chap. 6).

In terrestrial animals, even the most highly evolved, the body fluids closely approximate to sea water in composition and osmotic activity. This is adduced as an argument in favour of a marine ancestry for land animals. Their cells are isotonic with such a body fluid and must be maintained in this condition. It is one of the reasons why mineral salts form an essential part of the diet. Drastic dilution or increased concentration of body fluids can have serious effects on metabolism, and some animal diseases can be traced to this cause.

Water Relationships Inexplicable by Osmosis

Investigations into plasmolysis of plant cells, have shown that for some tissues the *DPD* exerted by a cell when just plasmolysed is often several atmospheres in excess of the osmotic pressure of the cell sap. On the classical theory, at plasmolysis, the *DPD* of the cell should be exactly equal to the osmotic pressure of the cell sap. The cells which exhibited this discrepancy were, in general, those of actively growing tissues, and it is suggested that the increased rate of metabolic activity influences the rate of water intake. It is also noteworthy that slowing down of respiratory activity by the lowering of temperature, also causes a fall in the rate of water intake; respiratory inhibitors also produce the same effect. There is justification then, in some cases at least, to speak of the "active secretion" of water by a cell into itself, the secretion being under metabolic control. In this connexion, the meaning of the word "secretion," differs from its normal usage; the "active secretion"

3

explanation disguises the fact that our knowledge of osmotic phenomena is inadequate to explain all cases of water relationship.

However, there are good reasons why any "active" movement of water into most cells is unlikely. In the first place, water is a major constituent of cells, with higher concentration than any other, and secondly, the permeability of cell membranes to water is many times greater than the permeabilities of other usual components. Taken together, these two conditions would make it necessary for there to be great expenditure of energy on the part of the cell to secrete water actively into itself against a chemical potential gradient.

THE EXCHANGE OF DISSOLVED SUBSTANCES BETWEEN CELLS AND THEIR SURROUNDINGS

It can easily be demonstrated that certain solutes will pass into and out of cells. The transmission may be inwards only, outwards only, or both at the same time; it may be rapid or slow according to the conditions prevailing and the solutes used.

Penetration into the Living Cell

Penetration of substances in solution from the external medium can be clearly observed in numerous cases by the effects which the solutes have on the protoplast. A motile cell of an alga or a protozoon is used preferably; such a cell will continue to indicate that it is still alive while the demonstration is in progress. If such a cell is immersed in a solution of an intra-vitam, non-toxic dye, such as janus green or neutral red, it can be seen that the protoplast will be coloured by the penetrating dye. This method is frequently used to demonstrate the presence of a substance in the cell for which the selected dye has a special affinity. Thus the neutral red vacuoles of animal cells may be indicated by this method. The same technique may be employed to follow the subsequent migrations of a group of embryonic cells.

Some substances kill cells as a consequence of their penetration. Ferric chloride will precipitate tannins in cells and produce a blue-black coloration. Iodine passes into cells and demonstrates the presence of starch. Many fixatives precipitate or denature proteins; these effects can be observed. Chloroform will destroy the protoplasmic structure and cause leakage of cell contents.

The penetration of radioactive isotopes (*see* p. 944) makes them a powerful tool in cell physiology. Radioactive carbon, administered as CO_2 in aqueous solution, can be located in specific cells by Geiger counters or by photographic methods; this technique has shed new light on the process of photosynthesis. Similarly, radioactive phosphorus, combined in phosphate, has aided the study of respiration.

Passage of Solutes Out of Living Cells

Gases emitted by plant and animal cells may be collected. Rapidly photosynthesizing plants give out oxygen; this may be conveniently collected from aquatic plants. Carbon dioxide may be collected over mercury from respiring seeds. Bromo-thymol blue may be used as an indicator to show that respiring roots in water give out carbon dioxide. The colour of the indicator changes from blue to yellow as the acidity of the solution increases (*see* p. 67).

There are many naturally-occurring exudates from cells, indicating the outward passage of solutes. The nectar produced by flowers is a strong solution of soluble sugars which must have been produced within the cells. Sweat and urine both contain a wide variety of soluble compounds, all of which have passed out from within particular cells. The digestive juices of the gut, of fungi, and of bacteria, all contain soluble substances exuded for particular purposes. The list of such substances could be extended indefinitely, and would cover a wide variety of organic and inorganic compounds.

The Method of Entry and Exit of Solutes

Considerable investigation has been made into the manner in which solutes are transmitted between the protoplast and its surroundings. It was at one time considered to be a straightforward diffusion mechanism obeying the normal diffusion laws applicable to aqueous solutions.

Diffusion

Diffusion in a solution is the migration of solute from a higher to a lower concentration, and is the result of the random movement of molecules or ions. Its measurement concerns the amount of dissolved substance which passes from one region of the solution to another in a given time, and this clearly depends on the difference in concentrations of the dissolved substance in the two regions considered. *The amount of solute diffusing through unit cross-section of area is directly proportional to the concentration gradient across this section.* This is Fick's law of diffusion. The concentration gradient is arbitrarily defined as the difference in concentrations at two points C_1 and C_2, divided by the distance d between them. C_1 and C_2 are expressed in moles per cubic decimetre ($mol\ dm^{-3}$) and d in centimetres. When this gradient is unity, the amount of solute transferred across $1\ cm^2$ in one second is known as the *diffusion coefficient* of the solute (flux of water or material is given in mol cm^2 s^{-1}). The quantity is difficult to measure since the concentration gradient changes as diffusion proceeds.

It has now become clear that the passage of solutes into or out of cells is not explicable by diffusion alone. Often, such passage is so slow that the rate is less than one-millionth of the rate calculated from a knowledge of the diffusion coefficient of the solute in water. In such cases the cell must offer considerable resistance to the movement. In yet other cases, the rate of passage is considerably greater than the theoretical value; here the cell is accelerating the movement. Furthermore, the rate of entry or exit of solutes varies with the nature of the cell, its age, and its rate of metabolism. No general conclusions can be drawn from any individual experiment. Any measurement of cell permeability can apply only to the particular cells used, under the particular conditions of the experiment.

Permeability of Cells

The movement of dissolved substances through cellulose walls seems to be due solely to diffusion. The water-filled spaces between the units of cellulose offer no resistance to the passage of even very large molecules. But when such cell walls are impregnated with lignin, cutin or suberin, the permeability is greatly reduced and they may even become completely watertight and air-tight.

The usual intercellular substances between animal cells are completely permeable and do not inhibit the passage of solutes.

Thus the effective control of entry and exit of soluble materials must lie in the protoplasm itself. There is considerable evidence that the plasma membrane controls the permeability of the cell to solutes, and in vacuolated plant cells the tonoplast exercises control over exchanges between the cytoplasm and the vacuole. There appear to be different mechanisms controlling the passage of non-electrolytes and electrolytes.

Passage of Non-electrolytes. There are indications that the two main factors which influence the rate of penetration of non-electrolytes are their *fat-solubility* and the *sizes of their molecules*. Studies with plant cells and with red corpuscles indicate that fat-solubility is a considerable factor. Work on a green alga has shown that the greater the fat-solubility of a substance, the faster it will penetrate the protoplast. For example, methyl alcohol with a high fat-solubility penetrates thousands of times faster than sucrose with a very low fat-solubility. There is also evidence that the size of the particle is a factor determining rate of entry. Smaller molecules often penetrate much more rapidly than would be predicted from their fat-solubilities, though this appears to be a factor only where groups of homologous compounds are considered. In such groups, as for example, the acid amides, or carboxylic acids, speed of entry increases as size of molecule decreases. If the penetration of two compounds of somewhat similar molecular size is compared, then fat-solubility is the overriding factor. Thus if propionamide, relative molecular mass 73, is compared with glycerol, relative molecular mass 92, with regard to rate of entry into the ox red corpuscle, the former penetrates four hundred times as rapidly as the latter; the greater fat-solubility of propionamide accounts for this.

Transmission of Electrolytes. The passage of electrolytes into and out of cells appears to be a much more complex process than the passage of non-electrolytes. In the first place, because the electrolyte is ionized, movement of separate ions takes place, and these do not necessarily move at the same rate or even in the same direction. Secondly, though the ions are comparatively much smaller than the molecules of non-electrolytes, their rate of penetration through cell membranes is much slower. This could be due to the electric charge on the ion causing the attraction of water dipoles around it (*see* p. 962). Consequently, the energy necessary for the particle to penetrate the membrane is of a much higher order. This, however, is not true for all ions and all cells; the anions Cl^- and HCO_3^- enter red corpuscles very rapidly, and H^+ and OH^- ions seem to penetrate practically all cells with ease.

ACCUMULATION OF IONS AGAINST A DIFFUSION GRADIENT. A notable feature of both plant and animal cells in relation to absorption of electrolytes, is the fact that intake of a particular ion will continue when its concentration within the cell is many times greater than that of the same ion in the external solution. This phenomenon is known as the power of *accumulation*; it is true of all types of cells with regard to particular ions. The table on p. 60 indicates the ratio of internal to external concentration for some ions, as measured by analyses of the cell sap, and of the external solution.

Erythrocytes of many animals contain concentrations of K^+ many times greater than its concentration in the blood plasma. Such accumulation implies continued movement of the ions against a diffusion gradient, and this necessitates the expenditure of energy by the cell. It has been shown that the cell cannot perform this work against a diffusion gradient unless it is vigorously metabolizing. Experiments on root systems and portions of potato and carrot tissue show that the cells will not accumulate ions from the surrounding culture solution, unless oxygen is bubbled through it. When the respiration rate falls, accumulation is slowed down or prevented altogether. This may be shown by reducing the supply of oxygen, or by the presence of specific respiratory inhibitors such as hydrogen cyanide, hydrogen sulphide or carbon monoxide, in the solution. The amount of respirable food material in the cells is also a factor affecting accumulation. When there is little food material present, cells can only accumulate ions for a short time. Thus it is certain that accumulation

THE FRESHWATER ALGA, *Nitella clavata*

Ion	Relative concentration in cell	Relative concentration in pond water	Ratio of internal/external concentration
K+ . .	54·3	0·051	1,065
Na+ . .	10·0	0·217	46
Ca++ . .	10·2	0·775	13
Cl− . .	90·8	0·903	100

THE MARINE ALGA, *Valonia macrophysa*

Ion	Relative concentration in cell	Relative concentration in sea water	Ratio of internal/external concentration
K+ . .	0·50	0·012	42
Na+ . .	0·09	0·498	0·18
Ca++ . .	0·0017	0·012	0·14
Cl− . .	0·597	0·580	1·0

is related to the rate of respiration, but there is no clear evidence as to how the energy is utilized for this work.

THE SELECTIVE POWER OF THE CELLS. Examination of the tables for *Nitella* and *Valonia* shows that very large numbers of ions of K+ have been absorbed, but the amounts of Na+ and Ca++ are considerably less. This is an indication of the *selective ability* of these cells. Red corpuscles of many animals show the same phenomenon, and here the concentration of Na+ within the cells is always less than that in the blood plasma.

It is generally the case that monovalent cations such as K+ are accumulated at a more rapid rate than divalent cations such as Ca++ or Mg++, or polyvalent cations. Similarly, the anions Cl−, Br− and NO_3^-, tend to accumulate much more readily than $SO_4^=$.

In addition to this power of selection between ions of similar charge, cells take up unequal amounts of cation and anion of a single salt. Experiments with wheat roots and with red corpuscles show that different amounts of cations and anions are absorbed from a given solution. With wheat roots, it has been shown that for a solution of common salt, the Cl⁻ is taken up four times as rapidly as the Na⁺, and strangely enough, with a solution of caesium chloride, the Cs⁺ is taken up eight times as fast as the Cl⁻.

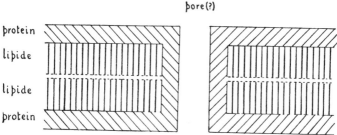

Fig. 3.6. Diagram to represent cell membrane structure.

Membrane Structure. Many speculations as to the nature of cell membranes have been made. Until the electron microscope became available, all such hypothetical structures were based on permeability studies and allowed for observations made during these. Since fat-soluble substances seem to penetrate comparatively easily it was thought that the membrane must be at least partly lipide. Permeability properties in relation to charged particles were accounted for by the assumption that the membrane possessed electrically charged areas. A porous or sieve-like nature accounted for the passage of comparatively large water-soluble particles. The low surface tension and elastic properties, which would exclude a wholly lipide constitution, were accounted for by the assumption of a protein component. Danielli proposed a structure represented diagrammatically in Fig. 3.6.

In this, the membrane was supposed to consist of a double layer of fat molecules, covered inside and out by a meshwork of protein. At intervals, pores in the system were assumed to occur. The total thickness was calculated as being about 8 nm and hence beyond resolution by the light microscope. Electron micrographs show the membrane to be a double layer of about 7·5–10 nm thickness, but so far no pores have been seen. The tonoplast of the vacuolated plant cell seems to have a comparable structure.

Membrane Potential. By connecting across a very sensitive voltmeter, two electrodes, one inserted through the cell membrane, the other in an external fluid medium, it is possible to show that there is usually a difference in electrical potential energy between the inside and the outside of the cell. In some animal cells this registers as about 60 to 80 millivolts. It is not a special condition of the cell but is its normal state so long as it is actively alive and is known as the *resting potential* of the cell membrane. It is due to unequal distribution of positively and negatively charged ions within and without the cell. The normal condition of the animal cell surface appears to be one in which there is a high concentration of Na⁺ and Cl⁻ ions and a low concentration of K⁺ ions in the external medium with the opposite condition inside the cell. The imbalance of ion distribution does not extend all through the external fluid or even through all the cytoplasm of the cell but is restricted to comparatively thin layers immediately at the membrane. The potential difference does not remain constant unless the cell remains alive. There is in fact a steady movement of Na^+ ions inwards that have to be "pumped" out if this is to be so. In general terms, the same conditions seem to apply to plant cells, there being imbalance between the same three ions inside and outside cells.

Phagocytosis and Pinocytosis. *Phagocytosis* or "eating by cells" has been observed and studied for a long time, but intake of liquid drops of surrounding fluid, that is, *pinocytosis* or "drinking by cells," although reported in 1931, was not studied seriously until recently. It has now become clear that material can be taken into cells of many kinds as fluid droplets. The essential feature of the phenomenon is that a portion of the membrane is invaginated into the cell and with it passes a small quantity of the external fluid. When the invagination reaches a great enough depth, its inner margin forms a fluid-filled vesicle which is then cut off from the external membrane to migrate to the cell interior. The process does not seem to proceed all the time, the cells appear to rest between periods of activity. Pinocytosis by cells can be induced by suitable solutions and is an active, energy-consuming process.

THE IMPORTANCE OF THE IONIC CONSTITUTION OF THE MEDIUM. The permeability of the cell to ions, to water, and to non-electrolytes is considerably affected by the ionic constitution of the external medium. Experiments with sea-urchin eggs show that water is absorbed from a solution of a non-electrolyte more rapidly than it is from sea-water. Addition of Ca^{++} ions brings the permeability back to normal, whereas addition of Ca^{++} and Na^+ ions is much less effective in reducing the water-permeability, although the same amounts of Ca^{++} may be present in both cases. The Na^+ ions are said to *antagonize* the action of the Ca^{++} ions; this antagonism between univalent and divalent ions is of common occurrence.

When sections of fresh beetroot tissue are placed in distilled water, there is a very slow escape of the red pigment from the vacuoles. If a little NaCl is added to the water, the pigment escapes rapidly, showing that the permeability of the cell membranes to the pigment has been greatly increased. But if sections are placed in a mixed solution of NaCl and $CaCl_2$, again the escape of pigment is barely noticeable. Here again, we have the antagonism of Na^+ and Ca^{++} affecting permeability.

Cells can exhibit normal permeability only when in a physiologically balanced solution containing the correct proportions of monovalent and divalent ions. Na^+, K^+ and Ca^{++} seem to be the most important ions in this respect. Hence it is important that the solutions used for tissue culture and hydroponics, i.e. growing plants in culture solution instead of in soil, must not only contain all the essential materials, but must have them in the correct amounts. Then osmotic relationships will be normal, and the cells will be able to exhibit their normal permeability to the dissolved substances.

Variations in temperature and acidity of the external medium are known to affect the permeability of cell membranes. Both these conditions affect the rate of metabolism, and it is probably for this reason that they affect permeability. Narcotics such as ether, chloroform and alcohol generally increase permeability. Beetroot cells exude their red pigment rapidly in chloroform vapour. This suggests that cell membranes are damaged by the narcotic; in the case of chloroform which is a fat solvent, there may be serious damage to the structure of the membranes and hence their permeability will be increased.

Some General Conclusions

The facts which are known about permeability, especially of electrolytes, cannot be explained by any known physical system. Accumulation, selective absorption, antagonism and speed of movement are all difficult of explanation. The workers most closely associated with permeability research can, at present, offer only hypothetical accounts. This is very understandable, since a complete explanation will involve accurate and detailed knowledge of the structure and activities of protoplasm.

Because of the lesser difficulty of experimenting with large cells of algae such as *Chara*, *Nitella*, *Valonia* and *Hydrodictyon spp.* most knowledge about movements of ions through plant cell membranes has been gained using these as subjects of study. Certain fundamental features seem to have been disclosed. For example, the two membranes, the outer cell membrane and the tonoplast, both exhibit an "ion pumping" character. Sodium seems to be pumped out and potassium and chloride pumped in, although there are differences in the permeability of the two membranes to these ions. The tonoplast appears to show the greater permeability. Pumps moving cations appear to be powered by ATP bonds. Movement of chloride ion on the other hand is not dependent on ATP, depending instead on an electron transfer mechanism to supply energy, see p. 432.

With regard to animal cells in general, there seems to be some similarity to plants between the movements of cations into and out of cells; sodium is "pumped" out against a potential difference across the cell membrane and potassium is accumulated inside much above its external concentration.

It is not within the scope of this book to discuss the various hypotheses put forward to account for the peculiarities of cell permeability.

However, most experts would agree that the passage of solutes into and out of cells is an "active" process. It is generally agreed that electrolytes are "carried" across the cell membrane and the protoplasm. Various carriers have been postulated; they include the respiratory substance cytochrome, organic phosphate compounds, protein molecules, and protoplasmic inclusions such as the mitochondria. Whatever the ultimate mechanism may prove to be, it is certainly an "active" process and not mere diffusion.

pH RELATIONSHIPS

It has long been known that conditions of acidity and alkalinity have important effects on cell metabolism.

The Meaning of pH

The degree of acidity or alkalinity of any solution depends upon the proportion of free hydrogen ions, H^+, to free hydroxyl ions, OH^-, in it (see p. 963). We say that an acid solution is one containing an excess of H^+ ions, and an alkaline solution is one containing an excess of OH^- ions. A neutral solution contains equal numbers of each. If the properties of diluted solutions of hydrochloric acid and acetic acid are compared, it is found that the former possesses very strongly all the characteristics associated with acids, whereas the latter possesses them to a much more limited degree. On the other hand, if each of these solutions contains one mole of hydrogen ions (36·5 g of HCl and 60 g of CH_3COOH), both will require the same amount of alkali for neutralization, e.g. 40 g of NaOH, and are therefore apparently equally acidic. The reason for the apparent contradiction is to be found in the different degrees to which the two acids dissociate into ions. The hydrochloric acid is almost completely dissociated into H^+ and Cl^- ions, the equilibrium in the equation being largely to the right—

$$HCl \rightleftharpoons H^+ + Cl^-$$

but the acetic acid is very weakly dissociated and the equilibrium lies largely to the left—

$$CH_3COOH \rightleftharpoons H^+ + CH_3COO^-$$

Since the characteristic properties of an acid depend on the amount of hydrogen ions present, the stronger powers of the hydrochloric acid in this respect are understandable. The equality of the neutralizing powers is due to the fact that the acetic acid exists in equilibrium with the hydrogen and acetate ions, so that as fast as hydrogen ions are neutralized by alkali, more acetic acid dissociates in order to maintain the equilibrium. This supplies more hydrogen ions. Neutralization by an alkali is a measure of two things, the number of hydrogen ions in the solution and the potential source of them. The full potential of a substance for releasing hydrogen ions when in solution, is known as the *total acidity* of the solution, and solutions of different acids containing equal quantities of ionizable hydrogen thus possess the same total acidity determinable by titration. Investigation has shown that it is not usually the total acidity of a solution which affects biological processes, but instead, the actual *concentration of hydrogen ions*, or in other words, the amount of dissociation of the solutes. It is desirable that this concept of acidity should be expressed in a convenient way and this is done from the following reasoning.

Pure water has a low but measurable electrical conductivity, showing that its molecules are to some extent dissociated. It may be regarded as an extremely weak acid and base, but represents true neutrality since the number of hydrogen ions must exactly equal the number of hydroxyl ions.

$$H_2O \rightleftharpoons H^+ + OH^-$$

It has been shown that when hydrogen and hydroxyl ions are present in the same solution, whether acid, alkaline or neutral, at 25°C the product of the concentration of these ions, i.e. $H^+ \times OH^-$, is a constant, 10^{-14} mol^2 dm^{-6}. Temperature affects ionization, and at 40°C this product for water is $10^{-13.42}$. In a neutral solution such as pure water, where the hydrogen ions are equal in number to the hydroxyl ions, these ions will each be present in a concentration of 10^{-7} mol dm^{-3}, since $10^{-7} \times 10^{-7} = 10^{-14}$. When a solution is made alkaline by the addition of hydroxyl ions, the hydrogen ions will decrease as the hydroxyl ions increase, but the product $H^+ \times OH^-$ will remain constant at 10^{-14} mol^2 dm^{-6}. Clearly then, as a solution becomes more alkaline, the hydrogen ions decrease in number, but they will always be present, even in very strongly alkaline solutions. It is possible therefore, to express alkalinity as well as acidity in terms of the hydrogen ion concentration of the solution. For this purpose, we could say that a given solution contains 10^{-x} mol dm^{-3} of hydrogen ions, but this is very cumbersome since x may have any value from zero to -14 and we would often find values such as $10^{-6.71}$ as quoted above for pure water at 40°C. Instead, we use a symbol pH meaning the logarithm of $1/H^+$, or pH $= -\log H^+$. More simply, pH *is merely the negative logarithm of the hydrogen ion concentration*. An example will help in elucidating this—

H^+ concentration of pure water at 25°C $= 10^{-7}$ mol dm^{-3}.

$\log_{10}$ of $\qquad 10^{-7} = -7$

But $\qquad\qquad$ pH $= -\log_{10} H^+$ concentration $= -(-7)$

Therefore $\qquad$ pH $= 7$

The pH of any neutral solution will always be given by the figure 7, since for any aqueous solution the product $H^+ \times OH^- = 10^{-14}$. The pH of an acid will be given by a figure less than 7 and that of an alkali by a figure greater than 7. This is obviously more convenient than the other method. Thus, the acidity of saliva may be written pH 6·9 instead of hydrogen ion concentration = $10^{-6.9}$ mol dm^{-3}.

Buffering

Drastic changes in pH can be resisted by means of *buffers*. Those of most importance in the biological field resist pH changes when acids are added to the solution which contains the buffers. Salts of weak acids, such as sodium acetate and sodium phosphate are typical of this class. There is a much smaller increase in pH when strong acids are added to their solutions, than would occur in their absence. In natural conditions the salts of weak acids such as malic and citric, phosphates, carbonates and some proteins, are all effective buffers. In mammalian blood, buffering is achieved principally by the interaction of sodium bicarbonate in the plasma and haemoglobin in the erythrocytes. Large amounts of acid can be transported in the blood and finally excreted, without interfering with the slight alkaline reaction which is the normal condition. Serious disturbance of this condition results in the quick death of the individual. Indeed, the extreme importance of buffers in the body fluids of both plants and animals cannot be over-emphasized.

In living tissues, buffering is usually effected by the presence of "buffer pairs," composed of a mixture of (1) a weak acid (one which does not strongly ionize) and (2) a salt of this acid with a strong base. We may take as a simple example to illustrate the mode of action, the mixture of acetic acid, CH_3COOH, and its sodium salt, CH_3COONa. This mixture in solution may be represented—

$$[CH_3COOH \rightleftharpoons CH_3COO^- + H^+] + [CH_3COONa \rightleftharpoons CH_3COO^- + Na^+]$$

When strong acid, e.g. $H^+ + Cl^-$, is added, the excess hydrogen ions so introduced combine with acetate ions to form the weakly ionized acetic acid. This prevents any drastic change in hydrogen ion concentration. When strong alkali, e.g. $Na^+ + OH^-$, is added, the excess hydroxyl ions combine with hydrogen ions to form water.

$$CH_3COO^- + Na^+ + H^+ + Cl^- \rightarrow Na^+ + Cl^- + CH_3COOH$$
$$CH_3COO^- + H^+ + Na^+ + OH^- \rightarrow CH_3COO^- + Na^+ + H_2O$$

In plants, buffering of the cell sap is achieved by various methods. The salts of malic, citric and phosphoric acids are commonly present. Often, there are colloidal surfaces which adsorb hydrogen ions.

Calcium ions which combine with free acids such as oxalic to form insoluble salts can also be regarded as buffering agents. All these buffers may be absorbed from the outside medium or produced within the cell as a result of metabolism. It is possible that respiration, by producing CO_2, and photosynthesis by removing CO_2, may play a part in resisting changes in the pH of green plants.

Determination of pH

The pH of a solution may be determined accurately by electrical methods, and for coloured solutions these are the only methods

Indicator	Colour range	pH range
Thymol blue (acid range) . .	red–yellow	1·2–2·8
Brom-phenol blue . . .	yellow–purple	3·0–4·6
Brom-cresol green . . .	green–blue	3·8–5·4
Methyl red	red–yellow	4·2–6·3
Brom-cresol purple . . .	yellow–purple	5·2–6·8
Brom-thymol blue . . .	yellow–blue	6·0–7·6
Phenol red	yellow–red	6·8–8·4
Thymol blue (alkaline range) .	yellow–blue	8·0–9·6

possible. pH of colourless solutions may be estimated fairly accurately by means of coloured indicators. Various series of indicators which cover a wide pH range can easily be obtained. One such series of synthetic indicators is tabulated above.

The rough pH value is first found by means of a Universal Indicator which is a mixture of indicators, so designed that its colour changes approximate to the colours of the spectrum as the pH changes from 3 to 11.

pH	Colour of indicator	pH	Colour of indicator
3	pink	7–7·5	greenish-yellow
4	red	8	green
5	orange-red	9	greenish-blue
6	orange-yellow	10	violet
6·5	yellow	11	reddish-violet

After a rough determination, one of the specific indicators from the list above can be selected. By its use in conjunction with a colour chart or disc, estimates to 0·1 pH can be made with some accuracy. For rough work such as soil-testing, special Universal Indicator outfits may be obtained.

Some Effects of pH Changes on Living Organisms

It is certain that pH changes in intracellular or extracellular fluids have important physiological effects, though the precise mechanisms are not always clearly understood. The important effects of the pH of the soil solution and of fresh waters will be discussed in Chapter 6 and the constancy of pH in the sea will be stressed.

The pH of body fluids, such as blood, has marked effects on the cells and tissues. Mammalian blood must always be maintained at the slightly alkaline level of pH 7·4. Serious deviation is fatal. Plants are somewhat more tolerant to pH changes, but for each species there is an optimum pH at which it grows best. For the higher plants, this is on the acid side of neutral, ranging around pH 6·7. Growing fungi are fairly tolerant, but their spores normally germinate only within a narrow range. Bacteria are very sensitive to pH changes.

Enzyme activity is correlated with the pH of the medium in which the enzymes act. Hence, in animals, the digestive juices contain substances which correct the acidity of the food mass. For every enzyme there is an optimum pH at which it works most efficiently; for many enzymes, moderate change of pH will reverse a chemical reaction. Such reversals are sometimes essential and hence there has to be precise local adjustment.

There may be many reasons why cells are so sensitive to pH changes. Two will be mentioned here. It is known that the permeability of cell membranes is affected by the pH of the medium; this will obviously interfere with intake and output and will therefore affect metabolism. Also we have the fact that since proteins are amphoteric electrolytes (*see* p. 94), they will be iso-electric at certain pH values. For any particular protein, the iso-electric point is reached when the number of cations is equal to the number of anions. The properties of a protein vary according to whether it is at the iso-electric point, near it, or far removed from it. At the iso-electric point, proteins are least soluble and are frequently precipitated. Hence pH changes will affect the nature as well as the properties of the fundamental constituents of the living protoplasm.

ENERGY RELATIONSHIPS

A living organism can be compared with a machine in that it converts one form of energy into another and performs work. Energy cannot be produced, but only converted; all machines must be supplied with energy, i.e. be fuelled, before they can perform work, and the living cell is no exception. If regular supplies of energy in acceptable form are not obtained, the cell ceases to function and must be considered dead. The majority of living cells obtain potential energy locked up in complex chemical compounds such as carbohydrates, fats and proteins. This energy is converted by carefully-graded small steps during respiration and, as energy of ATP molecules, is used to do work. Some may be translated into heat, light, sound or electricity, and some is used to perform chemical and mechanical work. In all cases, the source of the energy is originally external to the cell and must be captured. The only energy sources which can be tapped by living organisms are those of light and chemical bond energy, and it is on the initial absorption of these by two types of autotrophic organisms that all life depends. The absorption and transformation of light energy is the prerogative of pigmented plants and certain bacteria; transformation of external chemical energy is restricted to certain bacteria.

Absorption of Light Energy

Pigmented plants have the power of synthesizing special colouring materials, the chlorophylls and other compounds, which perform the special function of absorbing light energy. Such plants are said to be *photosynthetic*. Of the total light falling on a plant cell containing chlorophyll, only about 1 per cent is converted to use by the cell. Certain algae may double this efficiency. The remaining 99 per cent of the light is reflected, transmitted, or absorbed by other parts of the cell. Only the light absorbed by the special pigmented compounds is used to initiate the synthetic processes of the cell. Light absorption by green plants and the photosynthetic processes are dealt with in Chap. 10. The nature of light is treated more fully on p. 1005.

Absorption of Chemical Energy

Chemosynthetic organisms derive their energy from exothermic oxidation processes initiated by themselves. They can live on inorganic materials alone, using CO_2 or some form of carbonate as their sole source of carbon.

Among the better-known examples are the nitrifying bacteria, *Nitrosomonas* and *Nitrobacter*. The former oxidizes ammonia to produce nitrite, whilst the latter oxidizes nitrite to nitrate. The iron bacteria are said to oxidize ferrous iron to ferric, with the precipitation of ferric hydroxide. Thus it is that natural waters in which these oxidations have occurred are red in colour. These bacteria, together with others which perform similar activities, have been discussed in Vol. I, Chap. 36, and are mentioned again in Chap. 10 of this volume.

The energy thus made available to the organism by one set of chemical reactions, is utilized to initiate the synthesis of carbohydrate, fat and protein by another.

Animals, fungi and some bacteria, are in the last analysis dependent upon the autotrophic pigmented organisms for their supply of energy. They absorb high-energy compounds already synthesized by the auto-trophic organisms. The ultimate sources of all the energy utilized for life, are the sunlight and the chemical reactions induced by the chemo-synthetic bacteria. In connexion with the absorption of compounds containing potential energy, heterotrophes have developed a variety of methods, and even of modes of life such as parasitism, saprophytism and symbiosis.

Utilization and Dissipation of Energy

However obtained, every living cell contains substances capable of releasing energy when broken down. Should the sources of energy fail, the cell will lose the power to perform its functions and will be dead. The ways in which energy is released and work done, are described in Chaps. 11 and 12. Here, we shall survey briefly the ways in which energy is dissipated into the environment.

Loss of Energy as Heat

It is very easily recognized that warm-blooded animals dissipate heat. They maintain body temperatures which are normally higher than that of the environment. To do this, they must generate heat continuously. This is effected by the conversion of some of the potential energy of carbohydrates, fats and proteins into heat energy. The high-energy compounds are broken down into low-energy compounds such as CO_2, H_2O and NH_3. The liberation of heat is an inevitable concomitant of any energy transformation, and thus all living cells must generate some heat. In the warm-blooded animals, a greater proportion of the energy appears as heat. Experiments with respiring seeds show that there is considerable loss of energy as heat. A hand thrust into a compost heap or a pile of grass-cuttings will convince anyone of the ability of bacteria and fungi to generate heat. Damp haystacks and grain cargoes can take fire spontaneously as a result of heat production by these organisms. The following table indicates the quantities of heat released by three specific organisms. Expressed as kilojoules per kilogram of body weight we have the ratio, tortoise:dog:man = 1:4·5:3 approx.

		Body weight	Kilojoules in 24 hours
Giant tortoise . .	cold-blooded	117 kg	3680kJ
Dog	warm-blooded	11·5 kg	1634kJ
Man	warm-blooded	109 kg	10752kJ

Loss of Energy as Light

The phenomenon of light emission by living creatures is known as *bioluminescence*. The production of light as a result of chemical reactions is quite familiar in the form of flame. It represents the return to their basal energy state of highly excited atoms and molecules. Usually, this emission of light occurs only at high temperatures and when the reacting substances are gaseous. The high temperature is necessary to activate the molecules sufficiently; the gaseous state enables the molecules to emit their energy as light before losing it in collisions with other molecules. It follows that if any activated molecule can retain its energy for a sufficient time, it can produce light without any appreciable heat emission, even when the reactants are in solution. Such phenomena as the phospho- or chemiluminescence of phosphorus and the production of cold flames are well known.

Bioluminescence has aroused interest and comment for centuries; it was mentioned several times by Aristotle. Well-known examples of organisms that emit light are the glow-worm and the firefly. The "phosphorescence" of the sea, especially in tropical waters, is due to organisms such as the protozoan *Noctiluca*. The eerie glow of rotting meat, fish and vegetable matter is due to the light-emitting properties of the bacteria and fungi which cause the decomposition. Many deep-sea fish produce light. In all the cases which have been investigated, the light emission can be traced to the oxidation of a substance called luciferin, by the agency of a luciferase enzyme system. The process is therefore chemiluminescence and not one of phosphorescence (*see* p. 1015). Luciferin consists of three substances at least; adenosine triphosphate, bivalent ions of Mg^{++}, Mn^{++} or Co^{++}, and a complex material containing organic phosphate.

The purpose or value of light emission is not always explicable. In some fishes, it undoubtedly attracts prey, while in others it may be protective, tending to blind an assailant. The light is certainly a mating signal in fireflies, and in some marine worms. But in bacteria, fungi and certain other organisms, it is difficult to conceive of any advantage in giving out light. The tube-living annelid worm, *Chaetopterus*, which never leaves its tube, likewise dissipates energy as light, without apparent purpose.

Loss of Energy as Sound and Other Vibrations

The radiation of sound waves is confined to the animal kingdom. It is produced by vibrations of parts of the body, the vibrations being maintained by muscular activity. The noises of mammals, from the squeaking of the mouse to the roar of the lion, are produced by the vibration of the vocal cords in the larynx; the singing of birds is produced by the vibration of the semilunar membrane in the syrinx. The sound-radar device employed by bats in avoidance of obstacles, makes use of vibrations which produce sounds inaudible to human ears. These are all sound emissions usefully employed, but it must be remembered that whenever an organism moves, it must impart some vibrational energy to its surroundings. Most of these vibrations do not fall within the range of sound. The buzz of the bee, and the high-pitched whine of the mosquito are examples of audible vibration, produced by the rapid movement of the wings.

Movements of aquatic animals impart vibrations to the water, and many organisms are equipped with apparatus to detect such oscillations. Fishes have evolved special sense organs for detecting these low-frequency vibrations; together they form the lateral line system. The fish in a shoal are able to keep together by means of this system. The nearness of an inanimate obstacle may be judged by detection of the reflected vibrations; the same principle is used by man in making depth measurements by "echo-sounding." The ear itself is a modified lateral line organ.

Energy dissipated in movement of the body comes from muscular contraction associated with the movement of all or part of an animal's body; it involves the expenditure of large amounts of energy. The work done is the movement of the body or parts of it against a force (see p. 987). Details of the processes involved in muscle contraction will be found in Chap. 12.

Loss of Energy as Electricity

The transformation of chemical energy into electrical energy by the use of cells and batteries has its counterpart in living organisms. The conduction of impulses in the nervous systems of animals has been studied at great length and is known to be an electrical phenomenon. It will be discussed in Chap. 8. The electric eel, *Electrophorus*, the electric ray, *Torpedo*, and the electric catfish, *Malopterus*, have been objects exciting curiosity for centuries. Their electric organs are used for killing prey and for warding off attacks by enemies. Considerable power can be generated; one large specimen of *Torpedo occidentalis* has been credited with the development of an output of 6 kW. It is known that small discharges are made intermittently by these fish when they are cruising about in search of food; the discharges may be used to locate prey.

INTERNAL USE OF ENERGY

Energy that is not lost to the environment is used by the organism for maintaining metabolic activity and in keeping the protoplasm as a functional physico-chemical system (*see* p. 4). Provision must also be made, in the form of potential energy endowment, for the next generation and thus reserves of carbohydrate, protein and fat must be deposited in the egg, seed or spore. In many cases, it is also necessary to store supplies of potential energy as a reserve for periods when energy intake is low. The predominant materials used are carbohydrate and fat.

We would remind the reader that all organisms have evolved regulating mechanisms by which they have some degree of control over their energy exchanges, their material exchanges and over their pH. In general, the more highly evolved an organism is, the greater is its power of regulation.

CHAPTER 4

THE CHEMICAL NATURE AND IMPORTANCE OF THE COMMONER PLANT AND ANIMAL SUBSTANCES

THE greater part of any living organism is made up of water containing dissolved salts, but the most vital fraction of its composition is the smaller proportion of organic matter. This consists mainly of three groups of compounds: the *carbohydrates*, *lipides* and *proteins*. All three are of frequent occurrence as pure substances, but for the most part, they exist combined in various ways to form very large molecules. The study of carbohydrates, lipides, proteins, their derivatives, and the changes which they undergo, forms a large part of the branch of biology known as biochemistry. Here, we shall deal briefly with the more important and widely known biochemical substances. It should be noted that for the most part they are common to both plants and animals and frequently play exactly equivalent roles in the lives of these two forms of organisms.

THE CARBOHYDRATES AND THEIR DERIVATIVES

Carbohydrates are usually defined as compounds of carbon, hydrogen and oxygen, the latter two elements being in the proportion of two to one. This is not an accurate definition since it would have to include substances such as formaldehyde (CH_2O) and acetic acid ($C_2H_4O_2$) that are not carbohydrates and exclude rhamnose ($C_6H_{12}O_5$) which is. The best, but vague, definition is to include as carbohydrates all those substances that are simple sugars and compounds formed by combinations of these. The sugar can then be defined as below. Carbohydrates form important food storage compounds (starch, glycogen, sucrose, inulin), structural materials (cellulose, lignin, chitin) and the principal respiratory substrates (hexose sugars). Transport of hexoses is one of the chief ways in which carbon compounds are circulated in the body of an organism. In addition, since the primary product of photosynthesis is carbohydrate, it forms the source, directly or indirectly, of all the organic compounds of which a green plant is built.

The basic carbohydrate molecule, the sugar, is a derivative of a *polyhydroxy alcohol*. An alcohol, chemically, is a *paraffin* in which one hydrogen atom has been replaced by the univalent hydroxyl (—OH) group. A paraffin is an aliphatic or chain compound of carbon

74

and hydrogen in which the carbon atoms are linked by single bonds to the adjacent carbon atoms—

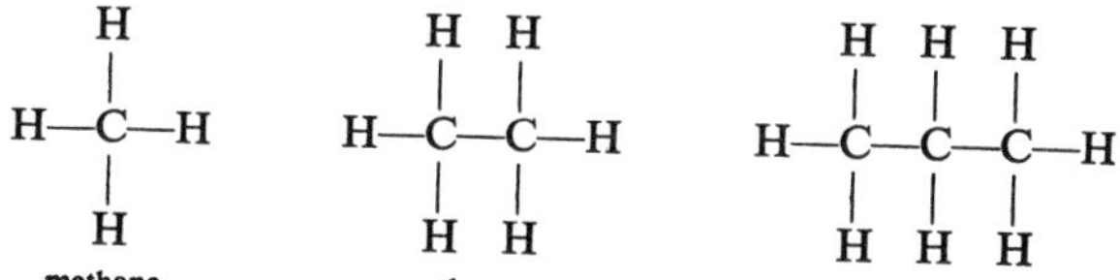

methane   ethane   propane

Thus methane may give rise to methyl alcohol, ethane to ethyl alcohol, propyl alcohol and so on as follows—

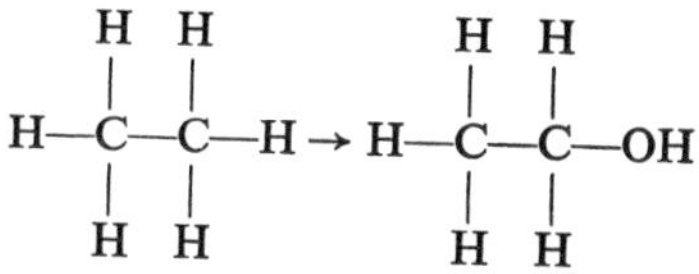

ethane   ethyl alcohol

A polyhydroxy alcohol is one which contains more than one alcohol group in the hydrocarbon molecule. The simplest are glycol and glycerol.

The simplest of the sugars is glycerose (glyceric aldehyde), derived from glycerol.

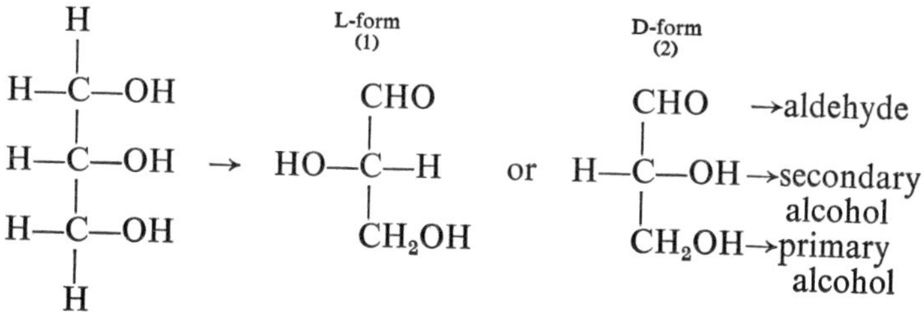

glycerol   two forms of glycerose (isomers)

It possesses 3 carbon atoms and is therefore known as a *triose* sugar. It shows an asymmetry about the central carbon atom, possessing an aldehyde on one side and a primary alcohol on the other. Further, the structure of the secondary alcohol is reversible, hence the sugar will be "optically active" and may be laevo- or dextro-rotatory according to whether it rotates the plane of polarization of plane-polarized light to the left or to the right (*see* p. 1009 and below). Sugars containing more than three carbon atoms are known as *tetroses* (4C), *pentoses* (5C), *hexoses* (6C) and so on up to *decoses* (10C). These are formed by the addition of more secondary alcohol groups and, according to the orientation of the parts of the alcohol groups, so isomers are possible. A further characteristic of the carbohydrate molecule is that it contains either an aldehyde (—CHO) or ketone (—CO) group. Typical examples are glucose which is a pentahydroxy alcohol with an aldehyde group, and fructose which is a pentahydroxy alcohol with a ketone group.

$$\text{Glucose} = CH_2OH.CHOH.CHOH.CHOH.CHOH.CHO$$

$$\text{Fructose} = CH_2OH.CO.CHOH.CHOH.CHOH.CH_2OH$$

The complex carbohydrates are built up of pure or mixed aggregates of these or similar units by a process known as *condensation* (*see* p. 78). All are either reducing sugars or give rise to reducing sugars after enzyme or acid *hydrolysis* (*see* p. 78–87).

Compounds which rotate the plane of polarization in a clockwise direction, as viewed through a polarimeter, are called *dextro-rotatory*; those that rotate the plane in an anti-clockwise direction are called *laevo-rotatory*. In the formulae on p. 75, (1) is L-glycerose and (2) is D-glycerose. Each is the mirror-image or *optical stereoisomer* of the other. This fact may not be obvious from the plane representation above, but if correct solid models are constructed, the reader will find that they cannot be superimposed.

Every optically active substance possesses a *specific rotation* which depends upon the concentration of the substance in solution in grams per cubic centimetre, the length of the path travelled by the light through the substance in decimetres, the wavelength of the light used and the temperature.

There is some confusion as to the method of indicating the dextro- and laevo-sugars and their parentage, but it is now becoming usual to use (+) for dextro-rotatory, and (−) for laevo-rotatory. Sugars derived from D-glycerose should carry the prefix D, and from L-glycerose, the prefix L. Thus D(−)-fructose is a laevo-rotatory sugar derived from dextro-glycerose. α, β and other Greek letters are used to denote stereoisomers which are not optical stereoisomers. For instance, αD(+)-glucose is a stereoisomer of βD(+)-glucose, but not an optical stereoisomer.

The true carbohydrates are all *saccharides*, being sugars or various combinations of them. A single sugar is a *monosaccharide*, two sugars make a *disaccharide*, and so on. A number of compounds included in the following tabular classification are not true saccharides, but they are either saccharide derivatives or consist mainly of saccharides,

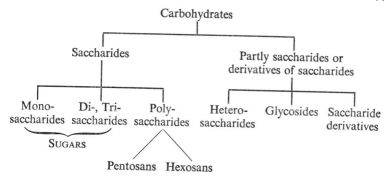

Sugars

The sugars, which include the mono- and disaccharides, are all soluble in water and have a sweet taste. Those with potentially active aldehyde or ketone groups are the reducing sugars, e.g. glucose, galactose, fructose, maltose and lactose. Sugars without the potentially active reducing groups are the non-reducing sugars, e.g. sucrose and trehalose.

The monosaccharides have the general formula $C_nH_{2n}O_n$ and all are reducing sugars. They range from monose to decose, the first true sugar being triose. The most widely occurring are pentoses, $C_5H_{10}O_5$, and hexoses, $C_6H_{12}O_6$. Pentoses never occur free, but only in combination with other groupings. L-Arabinose is found in gums, pectins and mucilages; D-xylose is found in straw, canes and husks of seeds; D-ribose occurs in one kind of nucleic acid, and a derivative, deoxy-D-ribose, in another. The perspective formulae of these two substances are given below.

The initial letters, RNA and DNA, used to represent the two different kinds of nucleic acid, indicate which form of the pentose sugar occurs in the molecule.

The prefix "2-deoxy" is derived from the fact that the oxygen atom lost, was attached to carbon atom 2.

Hexoses are the most important free sugars, D-glucose and D-fructose being the most common.

The disaccharides are *condensation* products of two molecules of hexose. Their general formula is $C_{12}H_{22}O_{11}$. They consist of two hexose residues, united by an oxygen bridge, one molecule of water having been eliminated—

$$(C_6H_{11}O_5)\text{—}O\text{—}(C_6H_{11}O_5)$$

The general equation for such a condensation is—

$$C_6H_{12}O_6 + C_6H_{12}O_6 \underset{\text{hydrolysis}}{\overset{\text{condensation}}{\rightleftharpoons}} C_{12}H_{22}O_{11} + H_2O$$

the reverse direction representing *hydrolysis*.

In the reducing disaccharides, one of the hexose residues retains its aldehyde or ketone group as an intact unit as a reducing group; the other hexose residue is linked by its own reducing group. Hence they are not such effective reducing agents as the monosaccharides. In the non-reducing disaccharides, both hexose residues are united through their reducing groups. Both types of linkage are shown below. It is thought that the aldehyde and ketone reducing groups are reactive only in the chain form of the hexose molecules (*see* p. 79).

Lactose: a reducing disaccharide

glucose residue

oxygen bridge

galactose residue

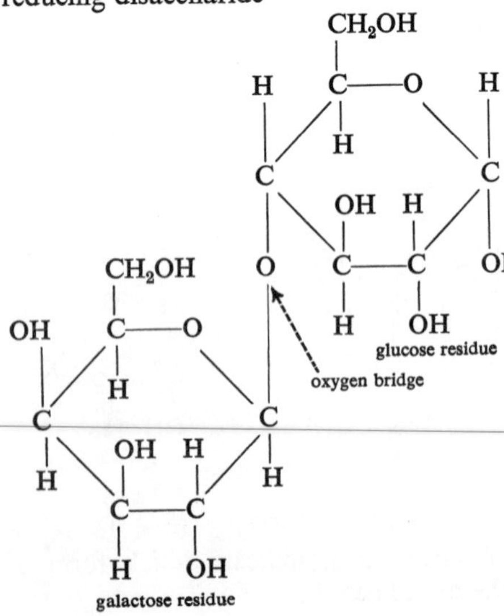

Sucrose: a non-reducing disaccharide

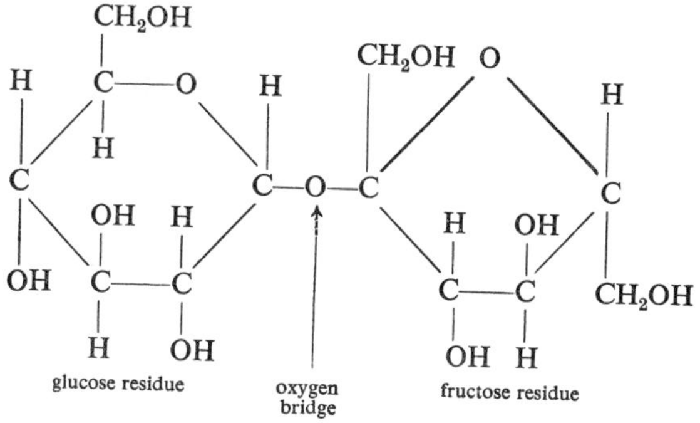

glucose residue   oxygen bridge   fructose residue

There are a number of naturally-occurring disaccharides, of which sucrose, maltose and lactose are the chief.

Monosaccharides

Glucose

D-Glucose, also known as dextrose or grape-sugar, occurs in all living cells, being especially plentiful in plant juices, and in the blood and tissue fluids of animals. It is the chief end-product of carbohydrate digestion in the gut. The α form rotates the plane of polarization of plane-polarized light $+ 113°$, and the β form, $+ 19°$. A fresh solution of glucose is an equilibrium mixture of both, with a specific rotation of $+ 52\cdot2°$. α-Glucose in solution loses rotating power and β-glucose gains it. The two forms are interconvertible until an equilibrium mixture is produced. When one type changes to the other, there is a mutarotation form with a straight-chain configuration.

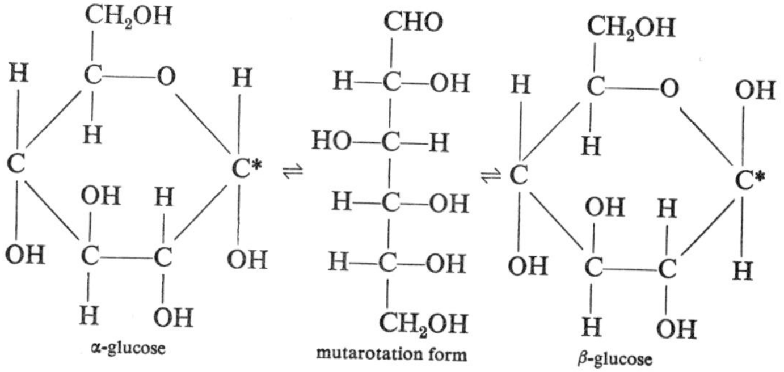

α-glucose   mutarotation form β-glucose

The change from α to β form occurs by a reversal of the —H and —OH groupings attached to the asymmetrical carbon atom marked with an asterisk. Because of its potentially active aldehyde (—CHO) group, glucose is called an aldose sugar.

Fructose

D-Fructose, also known as laevulose or fruit sugar, is found in fruits and in honey. It is one of the products of hydrolysis of sucrose, and the sole product of hydrolysis of inulin. It is laevo-rotatory (− 92°). The perspective and straight-chain formulae are given below—

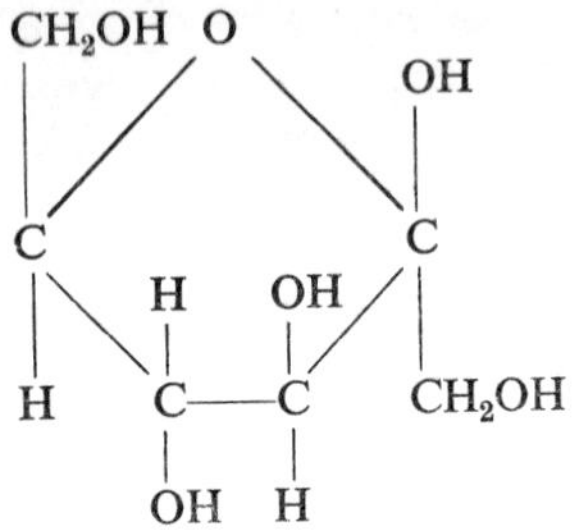

α-D-fructose 5-ring form

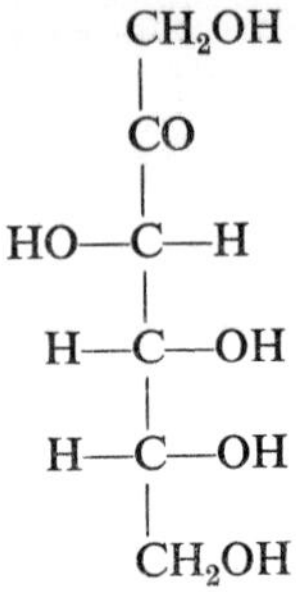

α-D-fructose straight chain form

Possessing the potentially active ketone, (—CO) group, fructose is called a ketose sugar.

Other Monosaccharides

D-Galactose occurs in lactose or milk sugar, and is released together with glucose when lactose is hydrolysed. D-Mannose occurs in certain glycosides; it is prepared industrially by the hydrolysis of mannan from the ivory nut (Tagua palm). Monosaccharides of the whole series monose to decose are known, though not all of them occur naturally.

Disaccharides

Sucrose

Sucrose, cane-sugar or beet-sugar is probably present in all green plants. Commercially, it is extracted from sugar-beet, sugar-cane and certain maples. It is a non-reducing, dextro-rotatory sugar, the angle of rotation being + 66·5°. It is readily hydrolysed by dilute acids, or by the enzyme invertase (sucrase), the products being equimolecular proportions of glucose and fructose.

Maltose

Maltose or malt-sugar, is produced by germinating barley; it does not occur widely in the free state. It is hydrolysed by dilute acids and by the enzyme maltase, to glucose. Maltose is a dextro-rotatory reducing sugar, with a rotating power of $+ 137\cdot5°$.

Lactose

Commonly known as milk-sugar, this occurs in the milk of all mammals, the percentage present ranging from 2 per cent in the rabbit to 7 per cent in the elephant. Human milk averages 6 per cent. It is doubtful whether lactose occurs in plants. Hydrolysis by dilute acids or by the enzyme lactase gives equimolecular proportions of glucose and galactose. Lactose is a reducing, dextro-rotatory sugar, the angle of rotation being $+ 52\cdot5°$.

Cellobiose

This sugar does not occur free. It is obtained when cellulose is acetylated and hydrolysed. Cellobiose is a stereoisomer of maltose.

The table below, shows the products of hydrolysis of the four principal disaccharides.

Disaccharide	Hydrolysed by	Products
sucrose . .	dilute acids and sucrase	glucose and fructose
maltose . .	dilute acids and maltase	glucose
lactose . .	dilute acids and lactase	glucose and galactose
cellobiose . .	dilute acids and cellobiase	glucose

Trisaccharides and Tetrasaccharides

Five trisaccharides, $C_{18}H_{32}O_{16}$ are known; raffinose is the most widely distributed. It is found in sugar-beet, in cereals and in some fungi. When completely hydrolysed, it yields equimolecular proportions of glucose, fructose and galactose. One tetrasaccharide, stachyose, is important. It occurs in many leguminous seeds, in the roots of the white dead-nettle, and tubers of *Stachys tubifera* (Chinese artichoke). Complete hydrolysis of stachyose gives two molecules of galactose, one of glucose and one of fructose.

Polysaccharides

These are compounds of high molecular weight formed by the condensation of a large number of monosaccharide units. Pentoses are condensed to form pentosans, and hexoses to form hexosans—

$$n(C_5H_{10}O_5) - nH_2O \rightleftharpoons (C_5H_8O_4)_n$$
$$n(C_6H_{12}O_6) - nH_2O \rightleftharpoons (C_6H_{10}O_5)_n$$

The hexosans are by far the more important. All the polysaccharides are either insoluble in water or form colloidal solutions; all are non-reducing and do not possess a sweet taste. Most of them can be identified by the coloured products they form with iodine solution. Starch, inulin and glycogen, are important carbohydrate food reserves, while cellulose and hemicellulose are structural materials in plants.

Hexosans

Starch

Starch can be extracted as a white powder which forms a sol with hot water; this sets to a gel on cooling. When heated to about 200°C, starch is partially converted into a mixture of dextrins. By boiling with dilute acids, it is hydrolysed via dextrins to maltose, and finally to glucose, showing that it consists mainly of glucose units. However, hydrolysis is never quite complete. A starch granule consists of a core of amylose enclosed by a rind of amylopectin all within a membrane-bound amyloplast. The amylose can be hydrolysed completely to glucose; the amylopectin is a phosphocarbohydrate and has to be dephosphatized before hydrolysis can take place. The enzyme phosphatase will accomplish this. The reactions which occur in living creatures with amylose are—

$$\underset{\text{amylose}}{(C_6H_{10}O_5)_n} \xrightarrow{\text{diastase}} \underset{\text{dextrins}}{(C_6H_{10}O_5)_n} \xrightarrow{\text{diastase}} \underset{\text{maltose}}{C_{12}H_{22}O_{11}} \underset{\text{glucose}}{\overset{\text{maltase}}{\rightleftharpoons} C_6H_{12}O_6}$$

Starch occurs only in plants, in the form of granules which have characteristic structures for various plants. It is an important storage product in seeds, especially those of cereals and legumes. Many roots, stems, tubers and corms store starch, providing important articles of human food such as potatoes, tapioca, sago and arrowroot. Starch occurs as a product of photosynthesis in the green leaves of most plants. Typical amylose molecules consist of helical chains of 300–400 glucose residues with molecular weights of the order of 60 000. The type of linkage is shown below—

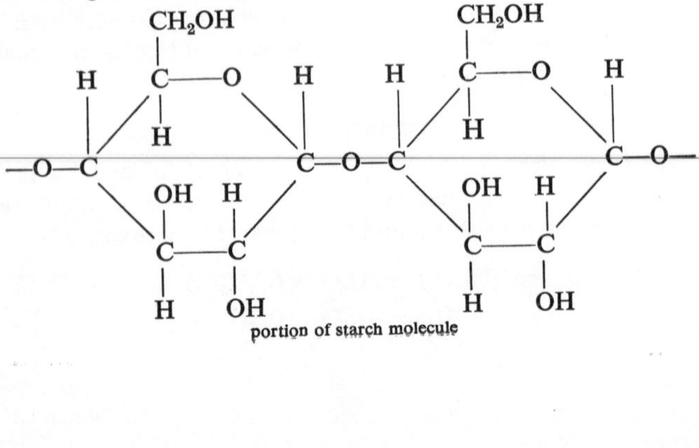

portion of starch molecule

Inulin

This is another storage polysaccharide. It occurs in the roots and tubers of many Compositae, especially dahlia, dandelion and chicory, and is present in small quantities in many monocotyledons. It can be extracted as a white powder soluble in warm water; the solution is slightly reducing, laevo-rotatory, with specific rotation of $-40°$. Hydrolysis by dilute acids or by the specific enzyme inulase, produces fructose only. Dahlia tubers contain 42 per cent dry weight of inulin, and attempts have been made to cultivate them for their inulin content, and hence to produce the sweet fructose.

Glycogen

Sometimes called "animal starch," glycogen is a condensation product of glucose units. Hydrolysis by dilute acids or by diastase, proceeds via dextrins and maltose to glucose. It can readily be extracted from minced fresh liver and purified as a white, soluble powder. It is non-reducing and dextro-rotatory, the angle of rotation being $+196·6°$. Glycogen is an important food storage substance in the liver and muscles of vertebrates; it is found also in many invertebrates. Of plants, only the Cyanophyta, some fungi, and maize seeds have been found to contain it. Its synthesis in the animal from glucose, takes place in several stages, involving the formation of hexose phosphates, and finally the separation of glycogen from the phosphate groups.

Cellulose

Cellulose is an important structural material in green plants. The walls of all young green plant cells are largely made of it. In fully differentiated cells, lignin, cutin, suberin or hemicellulose, may be associated with it during wall formation. It is insoluble in water, and chemically very inert: both properties would be expected from a consideration of its functions in the plant. It dissolves in cold concentrated sulphuric acid; if this solution is diluted with water and then boiled, complete hydrolysis into glucose units takes place. In living plants the hydrolysis probably takes place in two stages.

$$\underset{\text{cellulose}}{(C_6H_{10}O_5)_n} \underset{\text{cytase}}{\rightleftharpoons} \underset{\text{cellobiose}}{C_{12}H_{22}O_{11}} \underset{\text{cellobiase}}{\rightleftharpoons} \underset{\text{glucose}}{C_6H_{12}O_6}$$

Practically pure cellulose can be obtained by alkaline washing of cotton-wool. It has a very high molecular weight and consists of large numbers of glucose units, linked in alternate fashion, as shown on p. 84.

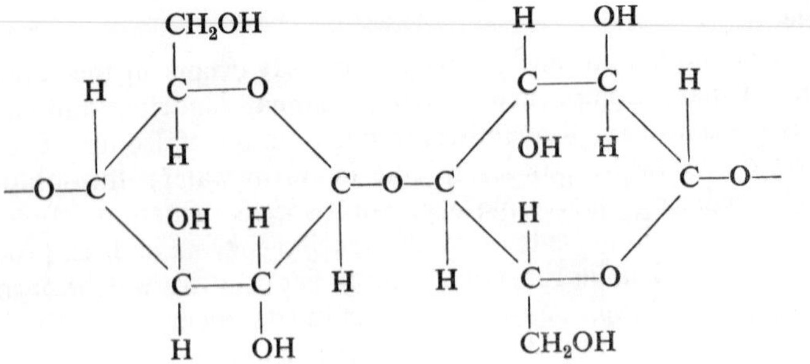

Structure of portion of a cellulose molecule

Hemicelluloses

These differ from cellulose in that they are easily hydrolysed by boiling with dilute acids; the products of hydrolysis are mixtures of hexoses, pentoses and uronic acids. They form part of cell walls and may be food storage substances in certain seeds such as those of the date, coffee, lupin, pea and bean. During germination they yield sugars by hydrolysis due to the enzyme hemicellulase.

Pentosans

Pentosans are condensation products of pentose units, the two most important being xylan and araban. Xylan occurs in straw, bran, canes and especially in the wood of deciduous trees; on hydrolysis it yields the pentose, xylose. Araban occurs in gums and mucilages; its hydrolysis gives arabinose, $(C_5H_8O_4)_n + nH_2O \rightleftharpoons n(C_5H_{10}O_5)$.

Heterosaccharides

These are not strictly carbohydrates but are grouped with them, since on hydrolysis, part of the yield consists of hexoses, pentoses or both. Usually they also yield glucuronic or galacturonic acids, derived respectively from glucose and galactose. The group includes pectins, mucilages, gums, chitin and hyaluronic acid.

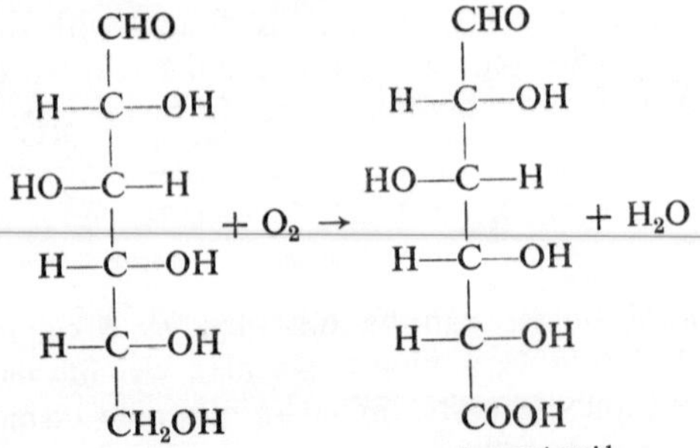

glucose   glucuronic acid

Pectin

Pectin is an important constituent of the middle lamella in green plants. Soluble pectin is present in many fruits, e.g. oranges, red currants and strawberries. Setting of jams is due to precipitation of pectin as a gel, in the presence of acid and sugar. Various pectins consist of chains of eight or more galacturonic acid units.

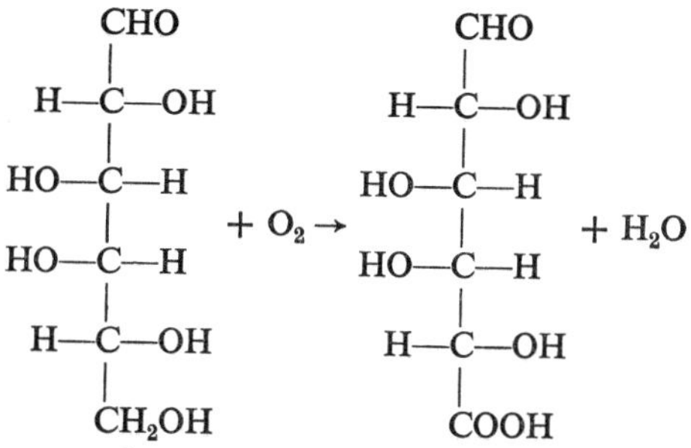

$$
\begin{array}{c}
\text{CHO} \\
| \\
\text{H—C—OH} \\
| \\
\text{HO—C—H} \\
| \\
\text{HO—C—H} \\
| \\
\text{H—C—OH} \\
| \\
\text{CH}_2\text{OH} \\
\text{galactose}
\end{array}
\ + O_2 \rightarrow
\begin{array}{c}
\text{CHO} \\
| \\
\text{H—C—OH} \\
| \\
\text{HO—C—H} \\
| \\
\text{HO—C—H} \\
| \\
\text{H—C—OH} \\
| \\
\text{COOH} \\
\text{galacturonic acid}
\end{array}
\ + H_2O
$$

Mucilages

These occur widely in plants and have the characteristic property of absorbing water and swelling. They increase water-holding capacity in many xerophytes, help imbibition of water in seeds, and often play a part in spore and seed dispersal. Acid hydrolysis produces hexoses, pentoses and uronic acids. Agar-agar, from sea-weeds, contains also organic sulphate; it is widely used as a culture medium for bacteria.

Gums

Gums are usually produced as abnormalities due to injury of the plant. On acid hydrolysis, they yield pentoses, hexoses and uronic acids. Gum arabic (*Acacia*) and gum tragacanth (*Astragalus*) are commercially important.

Chitin

This is an important structural material in many animal groups, e.g. exoskeleton of arthropods and cuticle of annelids. Hydrolysis yields acetic acid and glucosamine.

Hyaluronic Acid

This is composed of acetyl-glucose-amine and glucuronic acid units. It is a major intercellular substance in many soft tissues of animals especially epithelia and glands. It is also present in synovial fluid, vitreous humour and the jelly of the umbilical cord.

Glycosides

The glycosides form a very heterogeneous collection of substances which are widely distributed in plants. On hydrolysis they all give sugars, together with non-sugar or aglucone residues. When the sugar is glucose, the parent substance is called a glucoside.

Cyanophoric glycosides yield glucose, prussic acid (HCN) and benzaldehyde (C_6H_5CHO). They are present in the leaves of many plants such as cherry laurel, hawthorn, gooseberry and currant. Collectors of insects will be familiar with the use of shredded cherry laurel leaves for killing their specimens in the killing-bottle: the lethal substance is prussic acid. Probably the best-known cyanophoric glycoside is amygdalin from bitter almonds.

Mustard-oil glycosides are found in the seeds of many Cruciferae; black mustard, white mustard, watercress and wallflower, all contain them. On hydrolysis, they give glucose and various mustard oils.

Most of the saponin glycosides are toxic; they have the property of forming soapy colloidal solutions with water. Hydrolysis gives sugars and sapogenins. A number of important drugs are saponin glycosides. Digitalin from the foxglove, strophanthin, a tropical arrow-poison from various Apocynaceae, and cymasin from Indian hemp, are well-known examples. Sarsaparilla, for flavouring drinks, is extracted from the roots of *Smilax*.

There is a large group of aromatic glycosides which yield sugars and various aromatic compounds on hydrolysis. They are found in most green plants. The dyes, madder and indigo, the red, blue and violet sap-soluble anthocyanin pigments, are all glycosides. The anthoxanthin glycosides are all practically colourless; their presence has been shown in many white flowers.

Glycosides may confer advantages upon the plant which produces them. Those in leaves, may protect the plant from the ravages of herbivorous animals; those in seeds may help to preserve them until ripe and ready for dispersal; glycosides in bark can restrict fungus attacks. The coloured glycosides protect from excessive insolation; in flowers they attract pollinating insects; in fruits they attract animal dispersal agents.

Saccharide Derivatives

In this group is a heterogeneous collection of alcohols, acids, amines and esters derived from the sugars.

A number of polyhydric alcohols related to the sugars occur in plants. Common examples are mannitol, $C_6H_8(OH)_6$ which occurs in the sap of various trees and in fungi, and sorbitol, an isomer of mannitol, found in the fruits of many Rosaceae. The most important of these alcohols is glycerol, a trihydric alcohol corresponding to glycerose. Glycerol is an essential constituent of the fats of both plants and animals.

$$
\begin{array}{ccc}
CH_2OH & & CH_2OH \\
| & & | \\
CHOH & + H_2 \rightarrow & CHOH \\
| & & | \\
CHO & & CH_2OH \\
\text{glycerose} & & \text{glycerol}
\end{array}
$$

The acids derived from sugars, the uronic acids, have already been mentioned. They do not occur free, but are products of the hydrolysis of gums, mucilages and hemicelluloses.

Amine derivatives of the sugars are formed by replacing an —OH group by the amino-group —NH_2. Glucosamine,

$$CH_2OH.(CHOH)_3.CH.NH_2.CHO$$

is obtained by the hydrolysis of chitin. Chondrosamine, an isomer of glucosamine, results from the hydrolysis of chondroitic acid from cartilage.

Saccharide esters are formed by the reaction of sugars with acids, the most important being the hexose phosphates. They are formed during the early stages of respiration with a carbohydrate substrate, and are extremely important substances.

$$
\underset{\substack{\text{starch or}\\\text{glycogen}}}{(C_6H_{10}O_5)_n} + \underset{\substack{\text{phosphoric}\\\text{acid}}}{nH_3PO_4} \rightleftharpoons \underset{\substack{\text{glucose}\\\text{monophosphate}}}{nC_6H_{11}O_5.H_2PO_4}
$$

Glucose monophosphate is then converted to its isomer, fructose monophosphate, then another phosphate group is added, to produce fructose diphosphate—

$$
\underset{\text{fructose monophosphate}}{C_6H_{11}O_5.H_2PO_4} + H_3PO_4 \rightleftharpoons \underset{\text{fructose diphosphate}}{C_6H_{10}O_4(H_2PO_4)_2} + H_2O
$$

The manner in which this complex respiratory process continues will be explained fully in Chap. 11.

THE LIPIDES

The lipides are esters of the higher aliphatic alcohols. Esters are compounds formed by the reaction of an aliphatic acid with an alcohol, a simple example being the formation of ethyl acetate.

$$
\underset{\text{acetic acid}}{CH_3COOH} + \underset{\text{ethyl alcohol}}{C_2H_5OH} \rightleftharpoons \underset{\text{ethyl acetate}}{CH_3COO.C_2H_5} + H_2O
$$

All the lipides are insoluble in water, but soluble in a number of organic solvents such as ether, chloroform and hot alcohol. The chief concentrations of lipide material are found in the adipose tissues of

4

animals and in oily seeds. In addition, lipides form an integral part of the protoplasm, and are thus present in every living cell.

An outline classification of the lipides is given below—

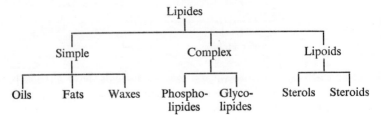

The lipoids, though not true lipides, are included for convenience here. They resemble fats in certain physical properties, and are extracted with fat solvents. The carotenoid pigments (*see* p. 101) are sometimes included with the lipides.

Simple Lipides

These include the oils and fats which are esters of glycerol and are known as glycerides. The waxes are esters of alcohols higher than glycerol.

Fatty Acid Series

Four series of fatty acids form esters with glycerol. Members of the *stearic* series have the general formula $C_nH_{2n+1}COOH$. They all contain an even number of carbon atoms, and each higher member contains two more carbon atoms than the preceding one.

STEARIC SERIES

Acid	Formula	Natural occurrence
Butyric .	$CH_3(CH_2)_2 COOH$	Butter
Caproic .	$CH_3(CH_2)_4 COOH$	Coco-nut and palm oils
Caprylic .	$CH_3(CH_2)_6 COOH$	Coco-nut and palm oils
Capric . .	$CH_3(CH_2)_8 COOH$	Coco-nut and palm oils
Lauric . .	$CH_3(CH_2)_{10}COOH$	Coco-nut, palm, laurel oils
Myristic .	$CH_3(CH_2)_{12}COOH$	Coco-nut, palm, myrtle oils
Palmitic .	$CH_3(CH_2)_{14}COOH$	Most animal fats and plant oils
Stearic . .	$CH_3(CH_2)_{16}COOH$	Most animal fats and plant oils
Arachidic .	$CH_3(CH_2)_{18}COOH$	Peanut and Macassar oils
Behenic . .	$CH_3(CH_2)_{20}COOH$	Ben oil
Lignoceric .	$CH_3(CH_2)_{22}COOH$	Pea-nut oil and Carnaüba wax
Cerotic . .	$CH_3(CH_2)_{34}COOH$	Beeswax, poppywax and Carnaüba wax

All these fatty acids of the stearic series are straight-chain saturated compounds, e.g. caproic acid—

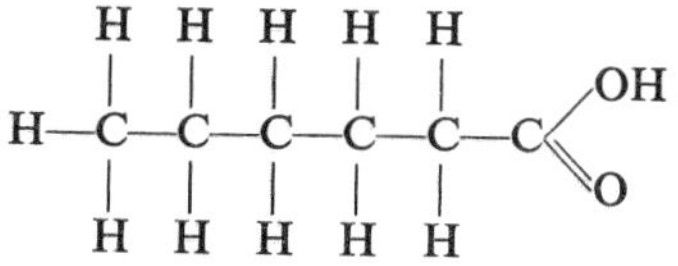

caproic acid

Acids of the *oleic* series have the general formula $C_nH_{2n-1}COOH$. They are unsaturated, with one double bond. It has become conventional to indicate the position of the double bond thus, $\Delta^{9:10}$. In this case for oleic acid, it indicates that the double bond is between the ninth and tenth carbon atoms, numbering from the right. Thus oleic acid, $C_{17}H_{33}COOH$, has the structure—

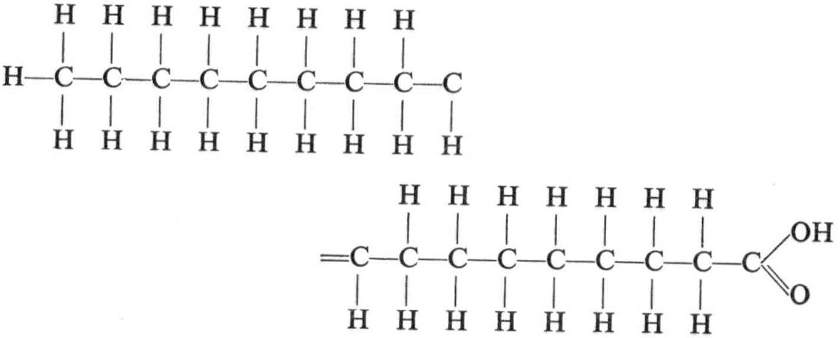

oleic acid

Oleic acid is the most widely distributed of all the fatty acids, being present in most fats and oils. Erucic acid, $C_{22}H_{43}COOH$, $\Delta^{13:14}$, is found in mustard oil, and petroselinic acid, $C_{17}H_{33}COOH$, $\Delta^{6:7}$, occurs in many oils from Umbelliferae seeds.

The *linoleic* series consists of a number of isomers of linoleic acid, $C_{17}H_{31}COOH$, $\Delta^{9:10 \text{ and } 12:13}$. The isomers vary in the position of the double bonds. Linoleic acids are present in linseed and cotton-seed oils. These oils are known as drying oils, since they harden to form a film, and hence they are used as the basis of paints and varnishes.

The *linolenic* series contains the unsaturated linolenic acid $C_{17}H_{29}COOH$, and its isomers, with three double bonds. It is found in all the drying oils and especially in linseed oil.

Natural Fats and Oils

The natural fats and oils are mixtures of glycerides of the fatty acids. Oils have a greater proportion of unsaturated acids and are liquid at 20°C; fats have a greater proportion of saturated acids and are solid at 20°C. Three molecules of fatty acid combine with one molecule of glycerol to form a non-polar structure (*see* p. 961).

$$\begin{array}{ccc}
\text{CH}_2\text{OH} & & \text{CH}_2\text{COO}.\text{C}_{17}\text{H}_{35} \\
| & & | \\
\text{CHOH} + 3\text{C}_{17}\text{H}_{35}.\text{COOH} & \rightleftharpoons & \text{CHCOO}.\text{C}_{17}\text{H}_{35} + 3\text{H}_2\text{O} \\
| & \text{stearic acid} & | \\
\text{CH}_2\text{OH} & & \text{CH}_2\text{COO}.\text{C}_{17}\text{H}_{35} \\
\text{glycerol} & & \text{tristearin}
\end{array}$$

Simple glycerides such as tristearin, tripalmitin and triolein are formed by the condensation of three identical aliphatic acid radicals with one molecule of glycerol. In the mixed glycerides, different radicals combine with one molecule of glycerol. Hydrolysis of fats and oils with dilute sulphuric acid or by the enzyme lipase gives the fatty acids and glycerol again. The metabolism of fats is more fully described on pp. 383 and 417.

Fats and oils of animals vary greatly in composition, depending largely on the types of fatty acids in the diet. Human fat contains a high percentage of oleic acid, and is liquid at body temperature, 37°C. Butter is a mixture of glycerides of butyric, oleic, palmitic, stearic, lauric, myristic, arachidic and other fatty acids. Lard contains the glycerides of oleic, stearic and linoleic acids. Beef and mutton fat consist principally of mixed glycerides of stearic, palmitic and oleic acids.

Plant fats and oils have a fairly constant composition. Palm oils consist mainly of palmitic, oleic, and small proportions of stearic and linoleic glycerides. Palm kernel oils contain mainly lauric, myristic and small fractions of caprylic, capric, palmitic, stearic, oleic and linoleic glycerides. Coco-nut oil contains mainly di-lauro-myristic glyceride. Of the drying oils, linseed oil consists principally of mixed glycerides of linoleic and linolenic acids. Cotton-seed oil glycerides contain chiefly linoleic and oleic acids. Soya bean oil has a large proportion of oleic and linoleic glycerides.

Waxes

Waxes are esters of the higher fatty acids with monohydric alcohols of high molecular weight. They are insoluble in water and difficult to hydrolyse; these properties explain their value as protective layers on leaves, stems, fruits, on the fur of animals and the integument of insects. Lanolin, from sheep's wool is used as a medium for application of external drugs, beeswax is used for polishes, Carnaüba wax for varnishes, candles and polishes, and spermaceti for candles. Spermaceti is composed principally of the palmitate of cetyl alcohol, $C_{16}H_{33}OH$; beeswax consists largely of the palmitate of myricyl alcohol, $C_{30}H_{61}OH$.

Complex Lipides

These occur in every living cell and are extracted by fat solvents. The chief phospholipides are *lecithins* and *cephalins*. Lecithins are widely distributed and are plentiful in bone-marrow, nervous tissue and embryonic tissue. On hydrolysis they yield glycerol, fatty acids, phosphoric acid and the nitrogenous base *choline*. Cephalins are plentiful in the brain; on hydrolysis they yield the same substances as lecithins, except that the base is *colamine*. Both phospholipides are present in egg-yolk. *Glycolipides* are abundant in brain and in nerves where they form the major part of the myelin sheath. On hydrolysis, they yield a fatty acid, galactose, and the nitrogenous base, *sphingosine*.

Steroids

When the fatty components of tissues are extracted, and the solution hydrolysed with NaOH, the true lipides are saponified and the residues remaining in the solution are *steroids*. All are based on the common skeleton shown below, and in the body, all are synthesized from

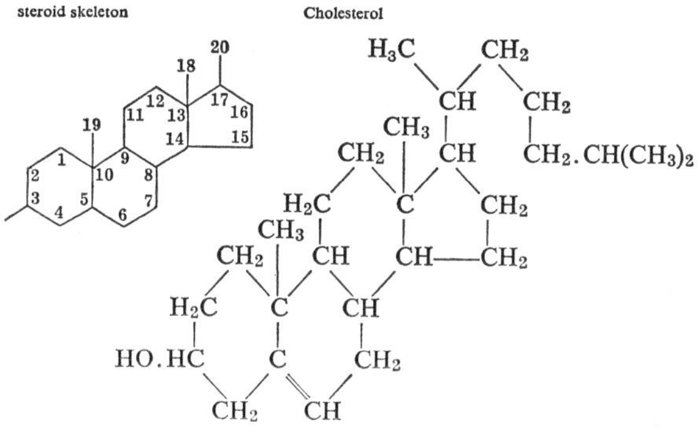

steroid skeleton   Cholesterol

cholesterol. Steroids with an OH group in position 3, are monohydroxy alcohols and were formerly called sterols. The remainder usually have an O at position 3 and were called steroids. It is now usual to use the latter name for both kinds.

Cholesterol is a very stable, white crystalline substance often esterified in the body to form a waxy material. It is obtainable from the diet, for example in egg-yolk, butter, milk and liver, but the body can also synthesize it from carbohydrates and proteins. Cholesterol is abundant in the brain, the adrenal cortex, and the sex glands; it is present in small quantities in all cells of animals, in blood and in the walls of blood-vessels. It is said to be strongly associated with some vascular diseases such as thrombosis and "hardening" of the arteries.

Important steroids are vitamin D and its precursors (*see* p. 389), the bile acids, *glycocholic* and *taurocholic* acids and their derivatives, the sex hormones of the ovary and testis, and the hormones of the adrenal cortex (*see* pp. 283–287). The moulting hormones, *ecdysones*, of insects and crustacea are also steroids. The main steroids of plants are *sitosterol* and *stigmasterol*; both are plentiful in embryos in seeds. *Ergosterol*, pro-vitamin D, has been found in many plants and animals.

PROTEINS

Proteins are nitrogenous compounds formed by the condensation of large numbers of amino-acids. On complete hydrolysis, the amino-acids are separated again. The proteins are the most fundamental constituents of protoplasm; the enzymes and chromosome framework also, are of protein nature. All proteins differ in the numbers and types of amino-acids they contain.

Amino-acids

About twenty-five amino-acids are known to enter into the structure of various proteins. All are white, crystalline, soluble in water, and with the sole exception of the simplest member, glycine, all are optically active. They all contain the amino-group, —NH_2, which has replaced a hydrogen atom in an aliphatic acid. Thus, from acetic acid, the amino-acid glycine is formed—

$$CH_3COOH \rightarrow \underset{\underset{NH_2}{|}}{CH_2COOH}$$

The amino-acids are amphoteric substances, having basic properties by virtue of the amino group, and acidic properties due to the carboxyl group, —COOH. They ionize in aqueous solution; those with one amino and one carboxyl group will give neutral solutions; those with more carboxyl than amino groups will form acid solutions, and those with more amino than carboxyl groups will give basic solutions.

Amino-acids are essentially derived from the fatty acids; some contain, in addition, the benzene ring, the pyrrolidine ring, the iminazole ring or the indole ring.

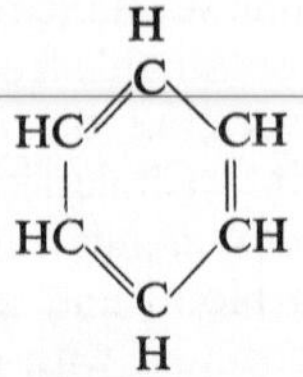

benzene ring 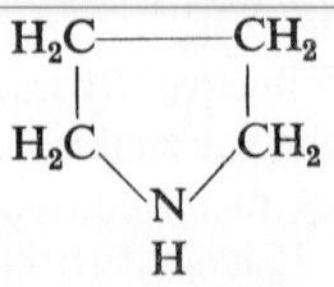   pyrrolidine ring

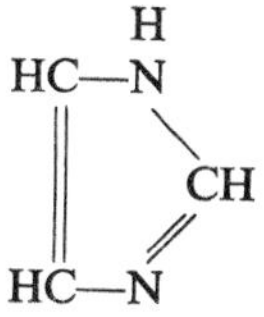

iminazole ring

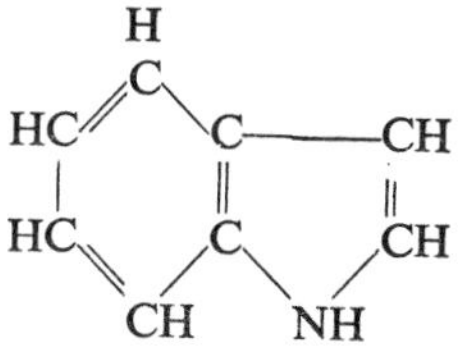

indole ring

The following list of the amino-acids is based on the fatty acids from which they are derived—

Fatty acid	Amino-acid		Derivative name
Acetic acid			
$CH_3.COOH$	Glycine	(G)	Aminoacetic acid
Propionic acid			
$CH_3.CH_2.COOH$	Alanine	(A)	α-Aminopropionic acid
	Serine	(S)	β-Hydroxyalanine
	Cysteine	(C)	β-Thioalanine
	Cystine		β-Dicysteine
	Phenylalanine	(P)	β-Phenylalanine
	Tyrosine	(T)	p-Hydroxyphenylalanine
	Iodogorgoic acid		3:5-Di-iodotyrosine
	Tryptophane		β-Indolealanine
	Histidine	(H)	β-Iminazolealanine
Butyric acid			
$CH_3(CH_2)_2COOH$	Butyrine		α-Aminobutyric acid
	Threonine	(Th)	β-Hydroxybutyrine
	Methionine		γ-Methylthiolbutyrine
Valeric acid			
$CH_3(CH_2)_3COOH$	Ornithine		α-δ-Diaminovaleric acid
	Arginine	(Ar)	α-Amino-δ-guanidino valeric acid
	Citrulline		α-Amino-δ-carbamido valeric acid
isoValeric acid			
$(CH_3)_2.CH.CH_2.COOH$	Valine	(V)	α-Aminoisovaleric acid
Caproic acid			
$CH_3(CH_2)_4COOH$	Lysine	(L)	α-ε-Diaminocaproic acid
	Leucine	(Le)	α-Aminoisocaproic acid
	isoLeucine	(iLe)	γ-Methyl valine
Succinic acid			
$COOH(CH_2)_2COOH$	Aspartic acid		Aminosuccinic acid
	Asparagine	(Asp)	β-Aminosuccinamide
Glutaric acid			
$COOH(CH_2)_3COOH$	Glutamic acid	(Gl)	α-Aminoglutaric acid
	Hydroxyglutamic acid		β-Hydroxyglutamic acid

There are two imino-acids derived from pyrrolidine—

	Proline	(Pr)	α-Carboxypyrrolidine
	Hydroxyproline		β-Hydroxyproline

The primary purpose of the amino-acids is to provide the groups necessary for the synthesis of proteins. In green plants, the requisite amino-acids are elaborated by the plant itself. Heterotrophic organisms cannot synthesize all the amino-acids they need, thus some must be supplied in the diet. Ten amino-acids are generally essential in the diet of the higher animals. They are histidine, lysine, tyrosine, phenylalanine, tryptophane, methionine, leucine, *iso*leucine, valine and threonine. Arginine appears to be a constituent of all proteins, and in some organisms at least, is essential for growth and for the production of urea in the ornithine cycle (*see* Chap. 15). From phenylalanine, the hormones adrenaline and thyroxine, and from tyrosine the pigment melanin are produced. Tryptophane provides the pyrrole groups for haemoglobin. Methionine provides sulphur and methyl groups necessary for metabolism; deficiency of methyl groups leads to the condition known as amethylosis, which is characterized by degeneration of the kidneys, spleen and thymus gland. Lysine is largely responsible for the base-neutralizing power of proteins by reason of its free NH_2 group.

The General Characteristics of Proteins

Determinations of the amino-acid composition of most proteins, are still somewhat approximate. To indicate the large numbers of amino-acids involved, a few are given here: gliadin of wheat, 240; albumin of hen's egg, 370; serum globulin of human blood, 736; haemoglobin of the horse, 541; myosin of rabbit muscle, 780. Some proteins have considerably more amino-acids: a toxic protein from the bacterium *Clostridium botulinum* contains at least 7 944.

The general characteristics of proteins are summarized below—

1. *Colloidal Nature.* Although many proteins can be obtained in the crystalline state, in living cells they are all colloidal (*see* p. 969).

2. *Amphoteric Properties.* Consisting of amino-acids, the proteins have their amphoteric properties. In solution, they are usually charged positively or negatively, and thus will move towards the cathode or anode under the influence of an electric current. Owing to their possession of some facility for internal interchange of H^+ ions, at a certain pH, the protein molecules will carry no net charge; this is the *isoelectric point* (I.E.P.) and at this pH they are referred to as *zwitterions*. At the isoelectric point, proteins have properties different from those that they possess above or below it. The following properties are all at a minimum value at the I.E.P.—solubility (hence there is great tendency for the protein to precipitate or coagulate), stability as emulsoid colloids, osmotic pressure, swelling by imbibition of water, viscosity and acid- and base-binding properties. Mixtures of proteins can be analysed

by electrophoresis, each protein in the mixture having a characteristic speed of movement which depends on its size, shape and charge.

3. *Large Size of Molecules.* Accurate determinations of the molecular weights of most proteins have not yet been accomplished. Methods based on osmotic pressure, and rate of sedimentation have given closely approximating results. Gliadin of wheat is about 30 000, egg albumin about 40 000, haemoglobin of man over 60 000, the enzyme urease nearly 500 000: the molecular weights of some of the virus proteins are said to run into millions.

4. *Specificity of Proteins.* The proteins of even closely related animals and plants differ. The phenomena observed in serology depend upon this specificity of proteins (*see* Chap. 1). The enzymes are proteins which show this specificity to a marked degree; they are specific in the substrates they affect, and in the organisms which elaborate them.

5. *Denaturation.* Alteration of the nature of a protein can be effected by a variety of methods. Heating will denature many proteins: thus, boiling an egg denatures the albumin. Various enzymes destroy the structure by hydrolysing proteins to amino-acids. Many chemical reagents will cause irreversible precipitation; such reagents are commonly used in the fixation of tissues for histological purposes. Exposure to ultra-violet rays, and often mere shaking, will both denature some proteins.

6. *Hydrolysis of Proteins.* Complete hydrolysis of proteins yields mixtures of amino-acids and usually some ammonia. The modern technique of paper chromatography has proved to be extremely useful in separating the amino-acids in such mixtures. In life, hydrolysis occurs by enzyme action; in the test-tube, strong HCl or H_2SO_4, or extracted enzymes, give the same results.

The Structure of Protein Molecules

A protein molecule consists of one or more chains, each made of a large number of amino-acid residues which are joined together by the *peptide linkage* $-CO.NH-$. The type of linkage is illustrated here in the formation of the dipeptide glycylglycine from two molecules of glycine. In a similar manner, a large number of amino-acids are condensed to form

$$H_2N.CH_2.CO|OH + H|HN.CH_2COOH \rightleftharpoons$$

$$H_2N.CH_2.CO.NH.CH_2COOH + H_2O$$

a polypeptide. The example below illustrates the sequence, using CH—R to denote an amino-acid residue. The arrangement of amino-acids in sequence in such chains is known as the *primary structure* of

$$H_2N—CH—\overset{\overset{\textstyle O}{\|}}{C}—N—CH—\overset{\overset{\textstyle O}{\|}}{C}—N—CH \ldots \text{etc.} \ldots COOH$$

$$\underset{R_1}{|} \qquad \underset{H}{|} \; \underset{R_2}{|} \qquad \underset{H}{|} \; \underset{R_3}{|}$$

proteins. Several polypeptides may be bonded laterally by a S—S linkage between two cysteine residues, or such a S—S linkage on one chain may cause a kink or a loop. The primary structure of

P—V—Asp—Gl—H—Le—C—Gl—S—H—Le—V—Gl—A—Le—T—Le—V—C—G—Gl—Ar—G—P—P—T—Th—Pr—L—Th

G—iLe—V—Gl—Gl—C—C—Th—S—iLe—C—S—Le—T—Gl—Le—Gl—Asp—T—C—Asp

For abbreviations see p. 93

human insulin is shown above. There are two peptide chains of 30 and 21 amino-acids respectively.

X-ray study of proteins has shown that the peptide chains are not in the extended condition as shown above but are folded in various ways. This folding is known as the *secondary structure* of proteins and in many cases it is a helix with the coils maintained in position by hydrogen bonding between the H of a HN group and the O of a CO group at fixed intervals along the spiral. (*See* Fig. 4.1.)

Such a helix would provide thin and fairly rigid rods which would be compatible with the structure of fibrous proteins such as myosin of muscle, collagen of connective tissue, or silk. But many proteins have molecules which are roughly globular in shape. In these cases, the helix is loose and folded at some points and tightly coiled at others. The various foldings of the helix constitute the *tertiary structure* of proteins. Kendrew has shown this type of structure for the myoglobin of muscle (*see* Fig. 4.2 (*a*)). It is now known that the haemoglobin molecule consists of four globular groups each like that of myosin, all pressed together (*see* Fig. 4.2 (*b*)). This arrangement of the peptide chains is the *quaternary structure* of a protein molecule.

The fibrous proteins have the helical secondary structure but the tertiary structure consists of elongated not folded units, and the quaternary structure contains aggregates of such units, some joined end to end and some laterally.

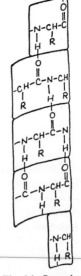

Fig. 4.1. Secondary structure of protein; the helix shown as if written on a spiral tape.

Classification of Proteins

Proteins are divided into three major groups. The simple proteins do not contain a prosthetic (non-protein) group; conjugated proteins contain one or more prosthetic groups; derived proteins are obtained by denaturation or cleavage of simple or conjugated proteins.

Simple Proteins

These include protamines, histones, prolamins (gliadins), glutelins, scleroproteins, albumins and globulins. Protamines are the simplest, and have been found only in the sperm of fishes. Salmin, from salmon sperm has a molecular weight of 2855; it contains 14 arginine, 1

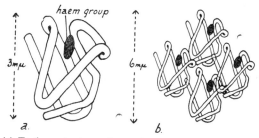

Fig. 4.2. (a) Tertiary structure of protein; the myoglobin molecule (after J. C. Kendrew). (b) Quaternary structure of protein; diagrammatic representation of the haemoglobin molecule.

valine, 3 proline and 3 serine units. Globin of haemoglobin is a common histone; the histones contain a somewhat greater variety of amino-acids than the protamines. Cell nuclei contain nucleohistones. The prolamins are found only in plants; they include gliadin from wheat, zein from maize, and hordein from barley. Glutelins are also solely vegetable proteins; the glutelin of wheat is the binding agent in dough and flour pastes. Scleroproteins are found only in animals; collagens of various kinds occur in cartilage, bone and white fibres; elastins occur in yellow elastic tissue, and keratins in epidermal structures such as hair, nails, horn and feathers. The scleroproteins are the most stable of all proteins; hence their use in protective and supporting structures. Albumins are common in both plants and animals. Ovalbumin occurs in egg-white, leucosin in wheat, lactalbumin in milk, and serum albumin in blood. Globulins are usually found associated with albumins. Ovoglobulin occurs in eggs, lactglobulin in milk, serum globulin in blood, legumin in peas and beans, and edestin in hemp seed.

Conjugated Proteins

These include phosphoproteins, nucleoproteins, glycoproteins, chromoproteins, zymoproteins and hormoproteins. Phosphoproteins contain phosphoric acid. The caseinogen of mammalian milk and the vitellin of egg-yolk are important members. In the nucleoproteins, the prosthetic group is nucleic acid which is allied to protamines or histones. Nucleoproteins form chromatin, and the virus proteins consist largely of them. Glycoproteins have polysaccharides as prosthetic groups; they occur in egg-white, in blood serum and in thyroid secretion. Chromoproteins contain a pigment. Important chromoproteins are the respiratory pigments haemoglobin, chlorocruorin (green, from some marine worms), and haemocyanin in the Crustacea. Two algal photosynthetic pigments, phycocyanin and phycoerythrin belong to this group. The cytochromes, important in cell oxidation systems, and the yellow flavoprotein respiratory catalysts are also chromoproteins. The zymoproteins are the enzymes. Hormoproteins include some of the hormones, especially insulin, secretin, prolactin and parathormone.

Derived Proteins

These are denatured or cleavage products of the original proteins. Denaturation by heat, by acids and by fixatives, causes physical changes which are usually irreversible. Cleavage of proteins by enzyme action gives successively smaller fractions, proceeding in this sequence: metaproteins, proteoses, peptones and peptides.

THE PIGMENTS OF LIVING ORGANISMS

The great majority of living organisms produce coloured substances which serve a variety of purposes. Some are light-sensitive, some protective from excessive insolation, some serve for attraction, camouflage or mimicry. Green is predominantly associated with plants, and red with animals. The main pigments may be considered in the following groups: linear tetrapyrroles, cyclic tetrapyrroles, carotenoids, flavins, melanins, pteridines, anthocyanins and anthoxanthins.

Linear Tetrapyrroles

These are based on the five-membered heterocyclic pyrrole ring. Four pyrrole units in a chain form the linear tetrapyrrole skeleton.

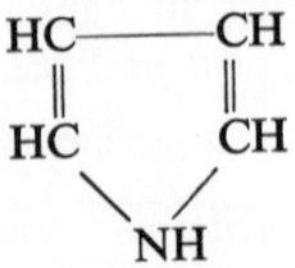

pyrrole ring

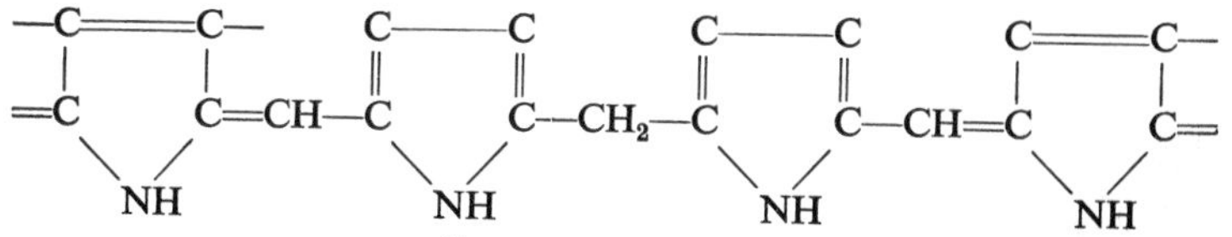

```
—C===C—        C——C        C——C        C===C—
 |    |        ‖    ‖        ‖    ‖        ‖    |
=C    C=CH—C   C—CH₂—C   C—CH=C        C=
  \  /     \  /     \  /     \  /
  NH       NH       NH       NH
```

linear tetrapyrrole skeleton

The pigments derived from this skeleton are waste products produced in the liver from haemoglobin of effete red corpuscles. The best-known are the yellow bilirubin, $C_{33}H_{36}O_6N_4$, and its green oxidation product biliverdin $C_{33}H_{34}O_6N_4$. The urochrome which colours the urine is probably a derivative of bilirubin, removed from the blood by the kidneys after re-absorption in the intestine.

Cyclic Tetrapyrroles or Porphyrins

The porphyrin pigments are widely distributed, and are extremely important substances. They are all based on a parent porphin structure which consists essentially of four pyrrole rings, numbered 1, 2, 3, 4 in the diagram on p. 100. The porphyrin pigments include the haemo-chromes, the cytochromes and the chlorophylls.

The Haemochromes

These are chromoproteins in which the prosthetic groups are either iron-containing haem, or copper-containing thiopeptide. The principal iron-containing pigments are the red haemoglobin of vertebrate blood, the red erythrocruorin found in many annelids and molluscs, and the green chlorocruorin of certain polychaete worms. The blue haemo-cyanin of many molluscs and arthropods contains copper. The red haemoerythrin of gephyrean annelids contains iron, but the prosthetic group is unknown. All these respiratory pigments form oxy-derivatives by the molecular union of oxygen with the metal atom; they serve for transport of oxygen in the body. The structure of haemoglobin is given on p. 100; the others vary from it in minor ways.

The Cytochromes

The cytochromes are chromoproteins, and also haem derivatives containing iron, combined with the protein globin. They are probably present in all living cells, acting as electron transporters. There appear to be several kinds, always present in very small amounts, but of exceedingly great importance. There is a considerable degree of cor-respondence between the amount of cytochrome in a cell and the respiratory activity. The continued activity of cytochromes depends on the action of cytochrome a_3, otherwise known as cytochrome oxidase. For more detail of the cytochrome system see p. 431.

The Chlorophylls

The metal in this case is magnesium and there is no protein in the molecule. The major constituent, apart from the porphyrin skeleton, is the phytol radical of phytyl alcohol, $C_{20}H_{39}OH$. Chlorophyll *a* is $C_{55}H_{72}O_5N_4Mg$, and chlorophyll *b*, $C_{55}H_{72}O_6N_4Mg$. The prime importance of the chlorophylls will be stressed later. The structure of chlorophyll *a* is given on p. 101.

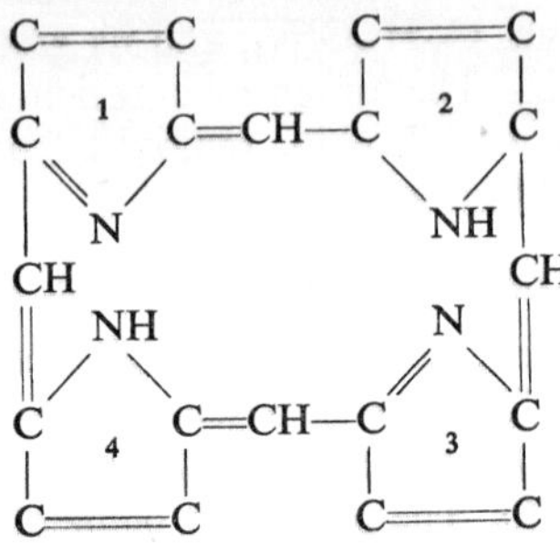

parent porphin structure

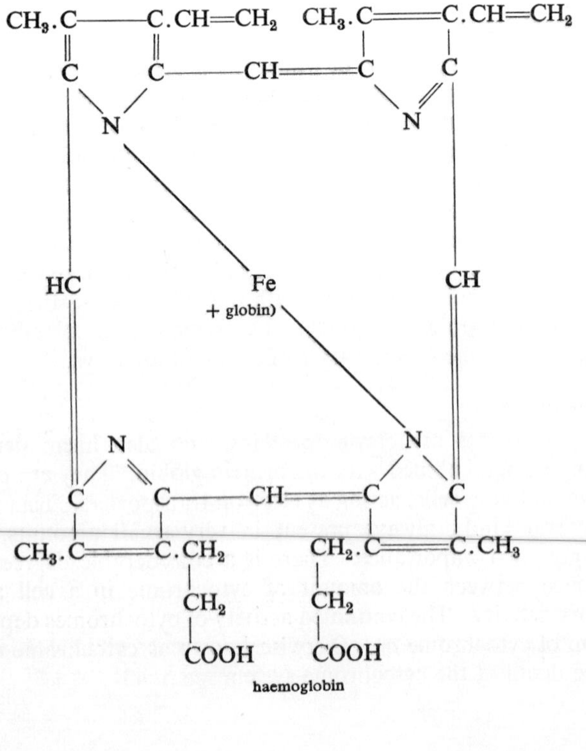

haemoglobin

The basic similarity in structure between the haemochromes, cytochromes and chlorophylls has often given rise to speculations as to a biochemical evolutionary link between plants and animals. In view of the universal presence of cytochromes, it seems feasible to think that they were evolved first; this possession of the haem nucleus would make it a comparatively easy step to the haemochromes; chlorophyll might also possibly have been developed from haem.

Carotenoids

These are fat-soluble pigments widely distributed in small amounts in plants and animals. They include the carotenes and the xanthophylls.

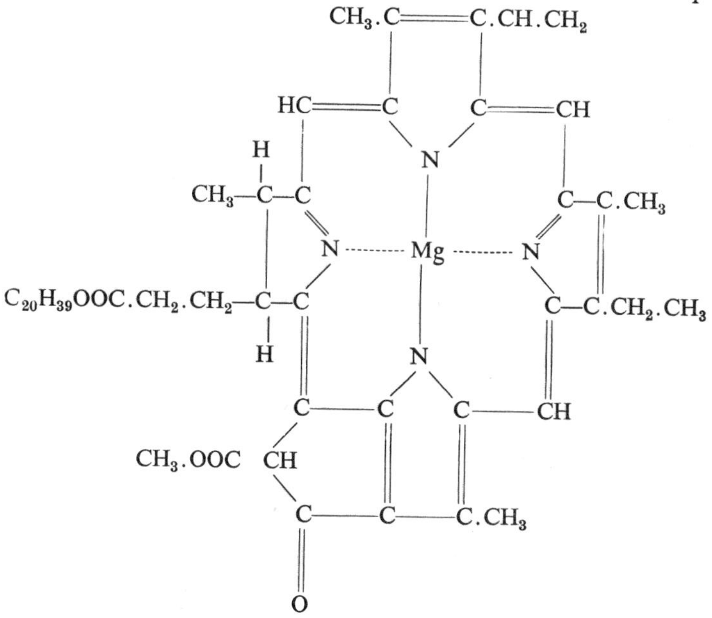

structure of the chlorophyll *a* molecule

Carotenes

The carotenes are all isomeric hydrocarbon oils with a formula $C_{40}H_{56}$. They are probably polymers of isoprene—

$$CH_2{=}C(CH_3){-}CH{=}CH_2$$

with a linear central arrangement and two cyclic end-groups. The three best known are α-carotene, β-carotene and lycopene. The α- and β-carotenes are present in green plant tissue and probably play a small part in photosynthesis, but are considerably less efficient than chlorophyll. In some plants they appear to be essential for phototropism. Lycopene is the colouring material of tomatoes and certain other fruits. It is worthy of note that hydrolysis of β-carotene yields two molecules of vitamin A.

$$C_{40}H_{56} + 2H_2O \rightarrow 2C_{20}H_{30}O$$

The structure of the vitamin A molecule is exactly half of the β-carotene molecule plus one molecule of water. Only one molecule of vitamin A is yielded by hydrolysis of α-carotene. Vitamin A has not been found in plants; animals make it from the carotene in the diet, and hence carotene is sometimes known as pro-vitamin A.

Xanthophylls

These are hydroxy, ketone, aldehyde or carboxylic acid derivatives of carotenes. The commonest is lutein, found in green leaves. Many crustacea possess a chromoprotein colouring the carapace. It has as its prosthetic group, the xanthophyll astaxanthine. Various colours are produced, including red, purple, blue and green. Boiling hydrolyses the chromoprotein, giving a characteristic xanthophyll red to crabs and lobsters. The same pigment is found in goldfish and in echinoderms.

The colours of most yellow and orange flowers, of some fruits and seeds, and of yellow leaves, are due to carotenoid pigments in chromoplasts. Animals possess carotenoids in the chromatophores of the skin; coupled with melanins, they act in producing colour changes. Dispersion of the carotenoids coupled with concentration of the melanins, will give the animal a lighter colour; the reverse will give a darker colour. The colours of butter, egg-yolk, yellow bird plumage, and straw-coloured blood serum, are all due to carotenoids.

Flavins

The flavins are yellow nitrogenous pigments soluble in water, the solution having a greenish-yellow fluorescence. Free flavins occur in many tissues of higher animals; all are probably identical with riboflavin, vitamin B_2. Riboflavin is a complex molecule consisting of D-ribose allied to dimethyl alloxazine; it is probably universally

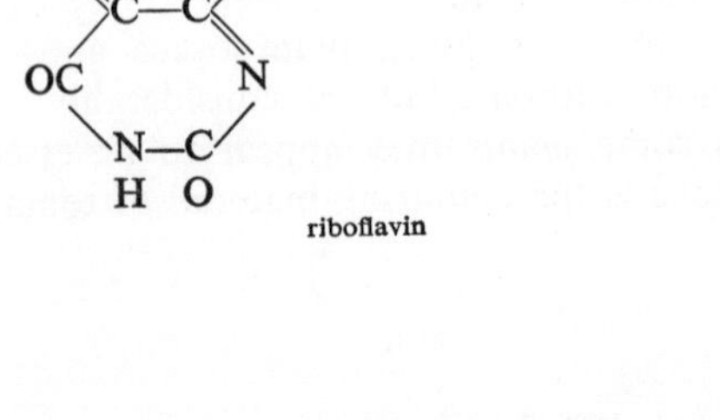

riboflavin

distributed in small amounts in all plant and animal cells. It is certainly essential in the diet of animals, and is present in all green leaves.

Several riboflavin nucleotides and proteins occur in the higher animals and plants and in yeast. They are all enzymes or co-enzymes concerned with cell oxidation systems. They accept hydrogen readily, and thus oxidize the substrate (*see* p. 429). They are colourless in the reduced condition, their yellow colour being restored by oxidation.

Melanins

These are widespread pigments ranging in colour from black, through brown to yellow. They are all produced from the amino-acid tyrosine, through the medium of the enzyme tyrosinase. The melanins are of special importance in animals, where they are concentrated as brownish-black granules in special cells called melanophores, normally situated immediately beneath the epidermis. They are very stable pigments, giving the colour to skin, hair, the choroid of the eye, the ink of the cuttle-fish, the seedlings of some plants, roots of beet and cabbage, and tubers of potato and *Dahlia*. Dispersion or concentration of the pigment, often in conjunction with other types of chromatophore, causes colour change in many animals. In cephalopods, there is nervous control of melanophores via muscle fibres attached to them. In vertebrates, control is mainly hormonal; in the frog the specific hormones are *B*-substance, causing darkening, and *w*-substance causing blanching. Both are produced by the pituitary gland. Hormones from the eye-stalk appear to initiate the change in crustaceans. In the great majority of cases investigated, light is the stimulus. Albinism, caused by extreme lack of melanin pigments, is due to lack of the enzyme tyrosinase. The value of control of pigmentation lies in protective coloration, and protection from excessive light. Great concentrations of melanin produce black, by which practically all the light is absorbed.

Pteridines

All these pigments are derivatives of the pteridine double ring. The pterins are predominantly insect pigments, though they are also found in vertebrates, and in many plant tissues. Leucopterin was obtained in 1894 from wings of the cabbage-white butterfly. Xanthopterin has been obtained from the lemon butterfly and from the yellow bands on the wasp.

Pteroic acid is an essential substance for growth and for feather-formation in birds. The vitamin, folic acid, is a polypeptide of glutamic acid units, with pteroic acid as the prosthetic group.

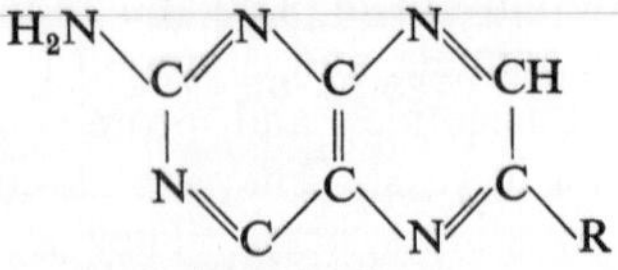

pteridine double ring

for simple pterins, R = OH.

for pteroic acid, R = CH_2 . NH . C_6H_4 . COOH.

Anthocyanins and Anthoxanthins

The anthocyanins are all red, blue or violet pigments which occur in solution in the cell sap of many flowers, fruits, stems and leaves. They are all glycosides of the sugars glucose, galactose and rhamnose. Hydrolysis of anthocyanins gives sugars, an anthocyanidin, and sometimes other residues. The anthocyanidins all contain the benzopyrilium nucleus. There are a very large number of different anthocyanins, all derived from the three anthocyanidins, pelargonidin, delphinidin, and cyanidin.

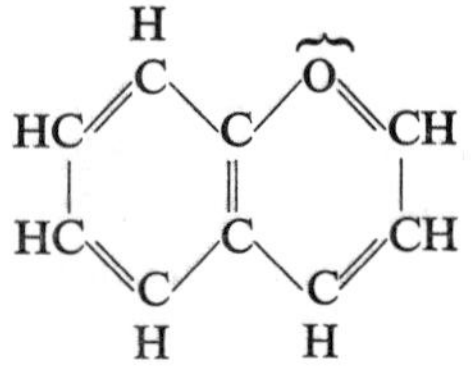

benzopyrilium nucleus

The anthoxanthins are usually almost colourless; their presence can be shown in many white and some yellow flowers. They are mainly glycoside derivatives of the benzopyrone nucleus.

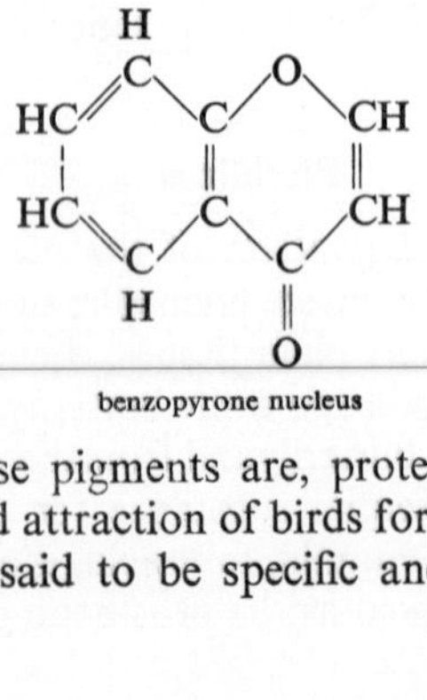

benzopyrone nucleus

The functions of these pigments are, protection from excess light, attraction of insects, and attraction of birds for fruit and seed dispersal. The various types are said to be specific and even to differ among varieties of a species.

ORGANIC ACIDS

There are many important organic acids which often exist free in organisms, as compared with the monobasic fatty acids which normally exist in combination with glycerol to form fats and oils. The more important of these acids are shown in three groups on below.

Dibasic non-hydroxy acids	Hydroxy acids
Oxalic, $(COOH)_2$	Glycollic, $CH_2(OH)COOH$
Malonic, $CH_2(COOH)_2$	Lactic, $CH_3.CH(OH)COOH$
Succinic, $(CH_2)_2(COOH)_2$	Malic, $CH(OH)CH_2(COOH)_2$
Glutaric, $(CH_2)_3(COOH)_2$	Tartaric, $(CHOH)_2(COOH)_2$
Adipic, $(CH_2)_4(COOH)_2$	Citric, $(CH_2)_2C(OH)(COOH)_3$
Fumaric, $(CH)_2(COOH)_2$	
cisAconitic, $CH_2.CH.C(COOH)_3$	

Keto acids

Pyruvic, $CH_3.CO.COOH$
Oxaloacetic, $CH_2.CO(COOH)_2$
α-Oxoglutaric$(CH_2)_2.CO(COOH)_2$
Oxalosuccinic, $CH_2.CH.CO(COOH)_3$

The main acids causing sourness in unripe fruits are succinic, glycollic, malic, tartaric and citric. During respiration, some of the acid is oxidized, and since the acids vary in the temperatures at which they are oxidized, we have a partial explanation of the difficulty of ripening certain fruits in the colder climates. Malic acid is oxidized at fairly low temperatures, tartaric acid requires a higher temperature, and citric higher still. Hence, apples containing malic acid as the prevalent acid, ripen in the temperate zones; grapes, with tartaric acid, need a more genial climate; the orange and grapefruit, in which citric is the prevalent acid, need still higher temperatures.

Probably all these organic acids play a part in regulating the pH of the cell sap. Oxalic acid is responsible for the removal of excess base as calcium, sodium and potassium oxalates; the insoluble calcium oxalate is deposited in many plants as crystals.

Some of these acids play a very important part in the respiratory carboxylic acid cycle or Krebs' cycle, a full account of which is given in Chap. 11.

Pyruvic acid occupies a key position in metabolism. It is the starting point of the carboxylic acid cycle, and is an important intermediate in anabolic and catabolic processes in both plants and animals. In the organism, it may be derived from carbohydrate by respiratory breakdown, the last reaction being the dephosphorylation of phosphopyruvic acid. Alternatively, it may arise from the amino-acid alanine by deamination—

$$CH_2.CO.PO_3.H_2.COOH + H_2O \rightleftharpoons CH_3.CO.COOH + H_3PO_4$$

phosphopyruvic acid   pyruvic acid   phosphoric acid

$$CH_3.CH.NH_2.COOH + O \rightleftharpoons CH_3.CO.COOH + NH_3$$

alanine   pyruvic acid   ammonia

Succinic and glutaric acids have particular importance in synthesis of proteins. Succinic acid may be obtained by deamination of aspartic acid, and glutaric acid by deamination of glutamic acid. It has been shown by tracer experiments that aspartic acid and glutamic acid are the two primary nitrogenous compounds synthesized in plants.

ESSENTIAL OILS AND MISCELLANEOUS COMPOUNDS

The essential oils are mixtures, but usually one predominates and gives its characteristic features to any particular oil. Some are derived from benzene and some from cyclohexane.

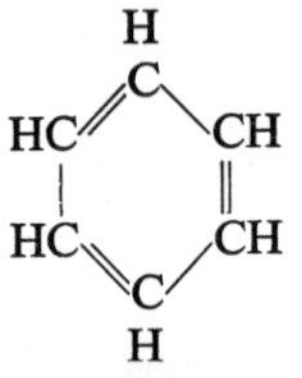

benzene 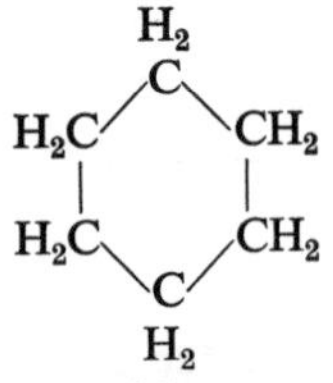   cyclohexane

The two main groups are the terpenes, derived from cyclohexane, and the camphors, which are derivatives of the terpenes. Some of the commoner oils with their constituents are listed below—

Common name	Occurrence	Essential oil constituents
Oil of turpentine . .	Coniferous trees	Pinene, limonene, dipentene
Eucalyptus oil . . .	*Eucalyptus spp.*	Pinene, cineole
Aniseed oil . . .	*Pimpinella anisum*	90% anethole
Cinnamon oil . . .	Bark of *Cinnamonum spp.*	Eugenol (mainly)
Attar of roses . . .	*Rosa damascena*	Geraniol and citronellol
Oil of wintergreen . .	*Gaultheria spp.*	Methyl salicylate

Many oils of importance are produced by plants in the family Labiatae. They include the oils of lavender, peppermint, spearmint, pennyroyal, marjoram, thyme, sage, basil and rosemary. The essential oils have different rôles in various plants. Some undoubtedly attract insects, some seem to protect against fungus attack. In leaves, they possibly play some part in regulating transpiration, and in preventing too much diurnal variation in internal temperature. Some oils seem to be merely end- or by-products of metabolism.

Rubber

Rubber and gutta-percha are the coagulated latex, respectively of *Hevea spp.*, and of the family Sapotaceae of Malay. They are hydrocarbons with the formula $(C_5H_8)_n$. All appear to be polymers or mixtures of polymers of isoprene, C_5H_8. It is thought probable that the number of units n, is about 11 000. The structure of isoprene is $CH_2{=}C(CH_3){-}CH{=}CH_2$.

Resins and Balsams

These are not to be confused with the plant gums and mucilages, which are heterosaccharides. Both resins and balsams are mixtures of substances of high molecular weight. The main constituents are resin acids, esters and terpenes. The most important of these is abietic acid, $C_{19}H_{29}COOH$, composed mainly of four isoprene units. Well-known resins are: rosin from pine exudates, dammar from several Malayan trees, dragon's blood from the Rattan palm, the fossil amber from *Pinites succinifer*, copal from several fossil deposits and from *Hymenoea spp.* of South America, and myrrh from *Commiphora spp.* The best-known balsams are gum benzoin from *Styrax spp.*, frankincense from *Boswellia spp.*, and Canada balsam from *Abies balsamea*.

Rubber, resin and balsam may protect against insects and fungi, especially by their exudation at wounds; on the other hand, they may be merely by-products of metabolism.

Tannins and Lignin

Tannins are very complex aromatic compounds. Some have been shown to be glucosides, containing glucose combined with various phenols or hydroxyacids. The most important tannin is gallotannin, a glucoside composed of one molecule of glucose and five molecules of digallic acid, $C_{14}H_{10}O_9$. It occurs in many types of galls, especially in those on the oak. The commercial tannins are mixtures obtained from the bark of various trees. They form leather from hide by precipitation of the proteins. In the plant, the tannins may give protection against fungi; it has also been suggested that they play a part in pigment formation.

Lignin is a colloidal thermoplastic polymer derived from coniferyl alcohol. There is no clear chemical definition of lignin, its structure varying between species. Cell walls of certain plant tissues such as the vessels and tracheids of xylem, and sclerenchyma fibres and sclereids, become impregnated with lignin at maturity. The wood of trees is composed of about 25 per cent lignin.

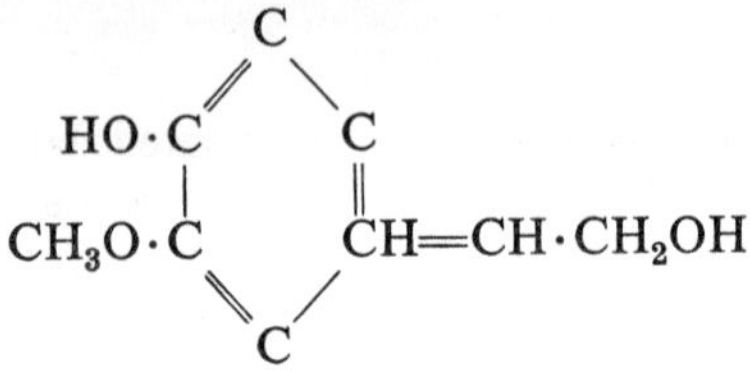

coniferyl alcohol

ALKALOIDS

All the alkaloids are derived from four parent heterocyclic ring structures: pyridine, pyrrole, tropane and quinoline. All are poisonous

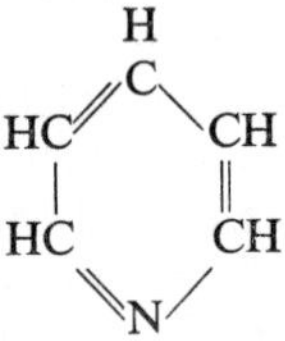

pyridine ring

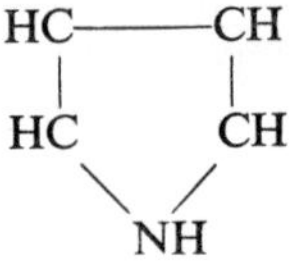

pyrrole ring

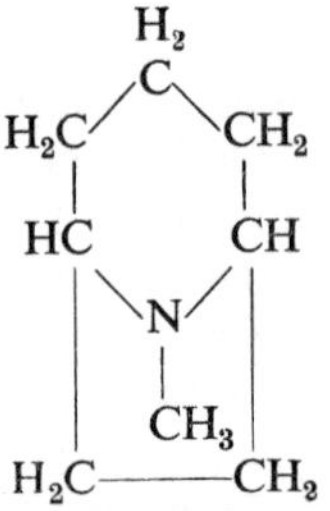

tropane ring

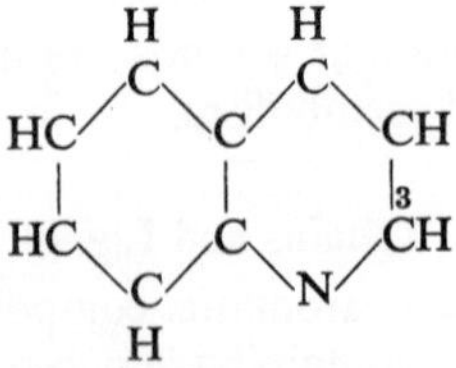

quinoline ring

and have been extensively used for lethal and medicinal purposes since ancient times. They rarely occur free, but usually as salts of malic, citric and succinic acids. Comparatively few plant families contain alkaloids: the Solanaceae, Ranunculaceae and Papaveraceae are outstanding.

The pyridine alkaloids include nicotine from *Nicotiana tabacum*, and piperidine from *Piper nigrum*. The main pyrrole alkaloids are pyrrolidine from carrot leaves and stachydrine from *Stachys spp.* Tropane alkaloids form an important group associated chiefly with the Solanaceae. Hyoscyamine is extracted from *Atropa belladonna*, atropine and hyoscine from *Datura spp.* and *Hyoscyamus spp.* The coca plant, *Erythroxylon*, produces cocaine. The quinoline group includes the quinines from the bark of *Cinchona spp.*, and strychnine, curine and curarine from *Strychnos spp.* The isoquinoline alkaloids have the nitrogen atom in the position marked (3). The most important are opium, morphine and codeine from *Papaver somniferum*, and colchicine from *Colchicum autumnale*. Colchicine has been used with some success to induce polyploidy in plants.

Some of these alkaloids appear to be of value in protecting the plant from herbivorous animals; in some plants they may take part in nitrogen metabolism; possibly the majority are end-products of nitrogen metabolism stored in plant organs, such as leaves, fruits and flowers, which will be eventually discarded.

AMINES AND AMIDES

Amines are the simplest organic compounds of nitrogen and may be regarded as derivatives of ammonia, in which the hydrogen atoms may individually be replaced by organic groupings. Thus, three methyl-amines are formed by replacing respectively one, two and three atoms of hydrogen by the methyl radical, $-CH_3$. These three amines are monomethylamine CH_3NH_2, dimethylamine $(CH_3)_2NH$, and trimethyl-amine $(CH_3)_3N$. Amines occur widely in plants, and their various proportions are said to be characteristic of the species. Putrescine, produced by decaying animal protein, is tetramethylene diamine, $NH_2.(CH_2)_4.NH_2$. It is derived from the amino-acid ornithine. Guanidine, $NH{=}C(NH_2)_2$, is a constituent of the amino-acids arginine and canavanine. It has not been found free in animals but has been detected in sugar-beet and in *Vicia* seedlings.

A number of betaines, derived from amino-acids by the complete methylation of the nitrogen atom, are known in plants, and one, thioneine, has been found in mammalian blood. Betaine, $C_5H_{11}O_2N$, is found in some Solanaceae and Chenopodiaceae, and trigonelline in legume seeds, potato and dahlia tubers. It is noteworthy that the acid amide, nicotinamide, one of the B vitamins, and a component of several respiratory enzymes, is derived from trigonelline. The betaines seem to play a part in transfer of methyl groups and significantly are most plentiful in young leaves and in growing-points.

Amides are derivatives of carboxylic acids, by replacement of the

—OH group with the amino group —NH_2. Thus from acetic acid CH_3COOH, acetamide, $CH_3CO.NH_2$ is formed. The most important amides in plants are asparagine and glutamine, derived respectively from aspartic acid and glutamic acid. Both asparagine and glutamine

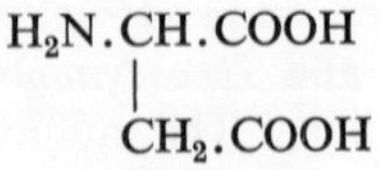

$$H_2N.CH.COOH$$
$$|$$
$$CH_2.COOH$$
aspartic acid

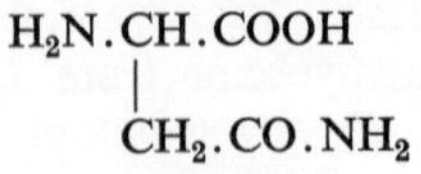

$$H_2N.CH.COOH$$
$$|$$
$$CH_2.CO.NH_2$$
asparagine

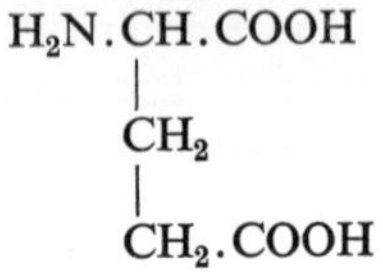

$$H_2N.CH.COOH$$
$$|$$
$$CH_2$$
$$|$$
$$CH_2.COOH$$
glutamic acid

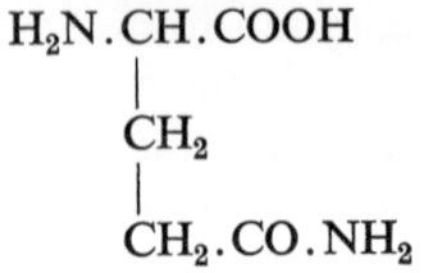

$$H_2N.CH.COOH$$
$$|$$
$$CH_2$$
$$|$$
$$CH_2.CO.NH_2$$
glutamine

seem to provide storage for ammonia, awaiting its utilization for amino-acid synthesis. It is believed that some of the urea produced in the liver, is derived from glutamine and asparagine. Urea may be regarded as a diamide of carbonic acid. Its importance in animal

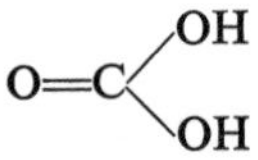

$$O=C\begin{cases}OH\\OH\end{cases}$$
carbonic acid

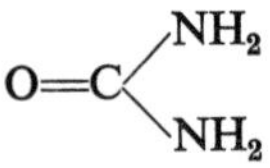

$$O=C\begin{cases}NH_2\\NH_2\end{cases}$$
urea

excretion will be discussed in Chap. 15. Urea is present in small quantities in many plants, derived from the amino-acid arginine.

PURINES, PYRIMIDINES AND THEIR DERIVATIVES

A number of very important compounds which exist in both plants and animals are derived from the pyrimidine and purine nuclei. Both occur free in nature, and are liberated by digestion of nucleoproteins.

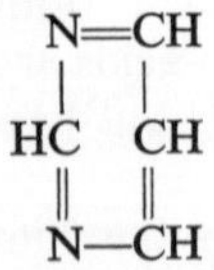

pyrimidine nucleus

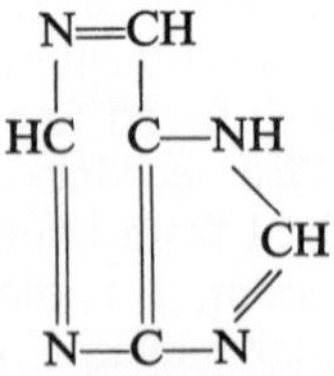

purine nucleus

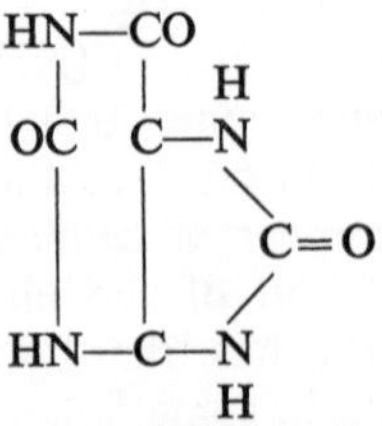

uric acid

The purines adenine, guanine, xanthine, hypoxanthine, and the pyrimidines cytosine, uracil and thymine are found free and as nucleosides. Adenine, guanine, cytosine and thymine occur in the DNA of plants and animals. Adenine, guanine, cytosine and uracil occur in the RNA of plants and animals.

Purine and Pyrimidine Derivatives

Uric acid, $C_5H_4N_4O_3$, is an oxidized purine. It is the chief nitrogenous excretory product in birds, many reptiles, and some invertebrates.

The purines adenine and guanine, and the pyrimidines cytosine, thymine and uracil, when allied to a ribose sugar or deoxyribose sugar, form important nucleosides. With phosphoric acid, each nucleoside forms an ester. These esters are the nucleotides, which are exceedingly important in cell metabolism, and in the synthesis of nucleic acids. Thus, the sequence in the synthesis of the nucleic acids is—

(1) Purine or pyrimidine + ribose or desoxyribose sugar = nucleoside
(2) Nucleoside + phosphoric acid = nucleotide
(3) Polymerized nucleotides = nucleic acid

Several important nucleotides are derived from adenosine (adenine nucleoside). They are, adenosine triphosphate (ATP), adenosine diphosphate (ADP), and adenosine monophosphate or adenylic acid. ATP is the principal phosphorylating agent in cells. After donating

$$ATP = adenine—ribose—O—\overset{\displaystyle OH}{\underset{\displaystyle O}{P}}—O\sim\overset{\displaystyle OH}{\underset{\displaystyle O}{P}}—O\sim\overset{\displaystyle OH}{\underset{\displaystyle O}{P}}—OH$$

one phosphoric acid group, the ATP is degraded to ADP. Finally, after yielding a second phosphoric acid group, the ADP becomes adenylic acid. In the degradation of ATP to ADP and of ADP to adenylic acid, considerable energy is released. The symbol $\sim$ denotes an energy-rich bond in the formula above (see p. 957).

The ATP is rebuilt in two stages by the addition of phosphoric acid groups. Other derivatives of adenosine, nicotinamide adenine dinucleotide (NAD)[1] and nicotinamide adenine dinucleotide phosphate (NADP)[1] are known as co-enzymes 1 and 2.

The nucleotides adenylic acid, guanylic acid, cytidylic acid, uridylic acid and thymic acid, are polymerized in various proportions to form nucleic acids.

Thymonucleic acid, which has been most examined, contains the bases adenine, guanine, cytosine and thymine. In 1953, Watson and Crick suggested that DNA molecules consisted of double helices united by H bonds (see p. 956) as below, with the bases paired thus: adenine—H—thymine and cytosine—H—guanine. Linear order and left and

[1] These substances were formerly known as diphosphopyridine nucleotide (DPN) and triphosphopyridine nucleotide (TPN).

right may vary, but the pairing is always the same. X-ray analysis has shown that this structure is correct. Chemical analysis has confirmed that the proportions of adenine to thymine are always 1 : 1 and so are those of guanine and cytosine.

It is now certain that the chromosomes consist of these double helices and that mitosis comprises the splitting into halves along the

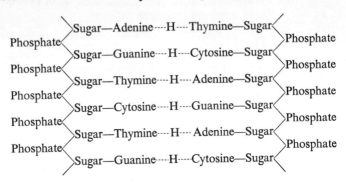

line of H bonds. Replication occurs by the addition to each helix of the correct nucleotides from the cytoplasm. Thus genetic identity is maintained in the daughter cells. As the phosphate and sugar remain unchanged, it is only the pattern of the bases that can vary. It is now clear that a gene consists of a specific sequence of bases along many links of a DNA molecule.

Synthesis of Proteins

The proteins synthesized in any cell depend on the linear sequence of adenine, guanine, cytosine and thymine in the DNA chains of the chromosomes (*see* p. 792). Corresponding to the DNA sequence, a messenger code of RNA is dispatched into the cytoplasm (*see* p. 794). This RNA contains, in various sequences, the bases adenine, guanine, cytosine and uracil instead of thymine. The amino-acids are aligned in chains,

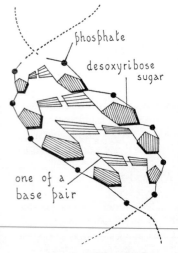

Fig. 4.3. The DNA molecule

each amino-acid fitting a particular triplet or *codon* of bases. For most amino-acids the triplets are known (*see* p. 797). An amino acid can be represented by more than one codon since there are 64 possible codons and only 20 amino acids. Although many of the details of protein synthesis still await investigation, enough is known to state with certainty that the DNA pattern in the chromosomes determines all the cell proteins, including the enzymes.

CREATINE, PHOSPHAGEN AND CREATININE

Creatine is an invariable constituent of vertebrate muscle, averaging 0·3 to 0·4 per cent of the tissue. The greatest concentration is found in voluntary muscle and the least in involuntary muscle. Vertebrate animals, especially carnivores, obtain some creatine in the diet, but all, and particularly herbivores, can synthesize it. The first stage of synthesis which takes place in the kidney, is the formation of glycocyamine from glycine and alanine. Then, in the liver, the transfer of a methyl group from methionine to glycocyamine, completes the synthesis.

In the resting muscle, creatine is combined with phosphoric acid to form creatine phosphate (phosphagen). When the muscle contracts, phosphagen is split into creatine and phosphoric acid, with the release of approximately 42 kJ mol^{-1} of phosphoric acid released. This is the energy responsible for the actual muscle contraction through the medium of ATP (see Chap. 12). The energy necessary for resynthesis of the phosphagen is provided by carbohydrate breakdown (glycolysis). Thus in contraction we have—

$$phosphagen \rightarrow creatine + H_3PO_4 + 42kJ$$

In recovery, we have—

$$creatine + H_3PO_4 + 42kJ \rightarrow phosphagen$$

In invertebrate muscle, arginine phosphate takes the place of creatine phosphate. Further detail of muscle contraction is given in Chap. 12.

Creatinine is a constant and characteristic constituent of the urine of all mammals. It appears to be solely a waste product, being filtered from the blood by the kidney tubules. It is the anhydride of creatine. Irrespective of the diet, there appears to be constant synthesis of

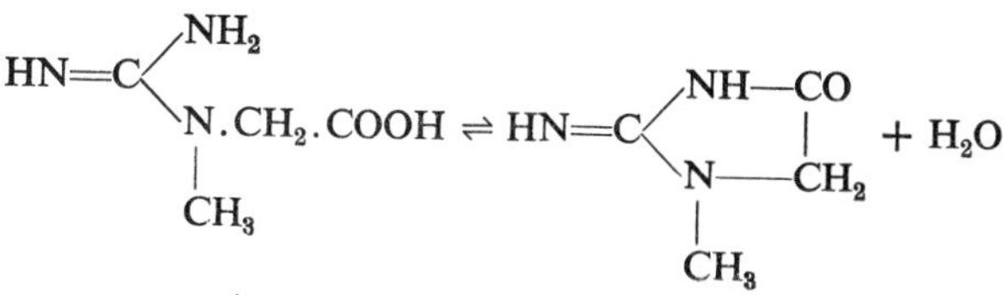

creatine   creatinine

creatine; as there is no appreciable storage, the excess may be excreted as creatinine.

HORMONES

The endocrine glands and the functions of their hormones are discussed in Chap. 8. Here, the chemical nature of hormones is briefly indicated. The great majority are either polypeptides or steroids.

The hormones derived from the hypothalamus by neurosecretion (*see* p. 213) are *oxytocin* and *vasopressin*; both are octapeptides. The releasing factors transmitted from the hypothalamus to the pars anterior of the pituitary body via the hypothalamo-hypophysial portal system are polypeptides (*see* p. 95). Also of a similar nature, secreted in the pars anterior, are *somatotropin, prolactin* and the *adrenocorticotropic* hormone, while those which stimulate the Graafian follicles, the corpora lutea and the interstitial tissue of the testes are glycoproteins. From the pars anterior, the two *melanocyte-stimulating* hormones, α and β, are polypeptides, as are *insulin* and *glucagon* from the pancreas, and also the neurosecretory hormones of insects and crustaceans.

Steroids include the *oestrogens* and *progesterone* from the ovaries, *testosterone* from the testes, and the hormones of the adrenal cortex which are *androgens, oestrogens, glucocorticoids* and *mineralocorticoids*. The moulting hormones, *ecdysones*, of insects and crustaceans are also steroids.

Thyroxin is derived from the amino-acid tyrosin, which is first iodinated to form di-iodotyrosin. Two molecules of this are condensed with the loss of one alanine side-chain to form the hormone which is tetraiodothyronin or thyroxin (*see* below).

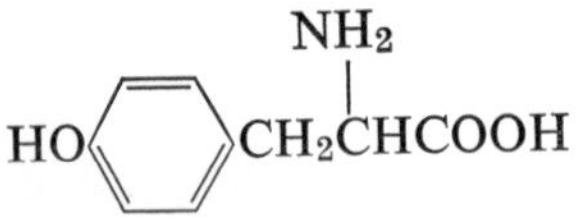

tyrosin

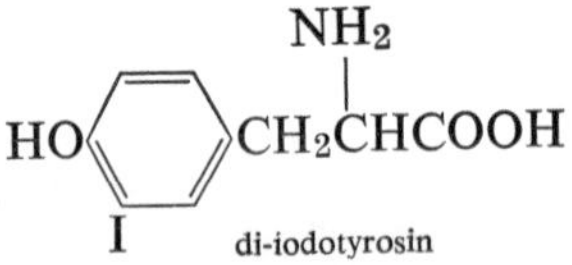

di-iodotyrosin

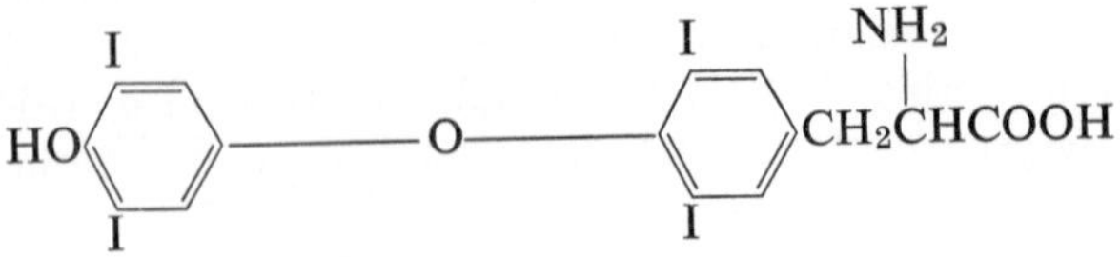

tetraiodothyronin

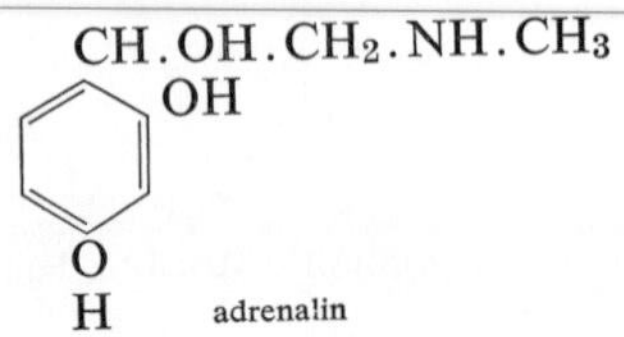

adrenalin

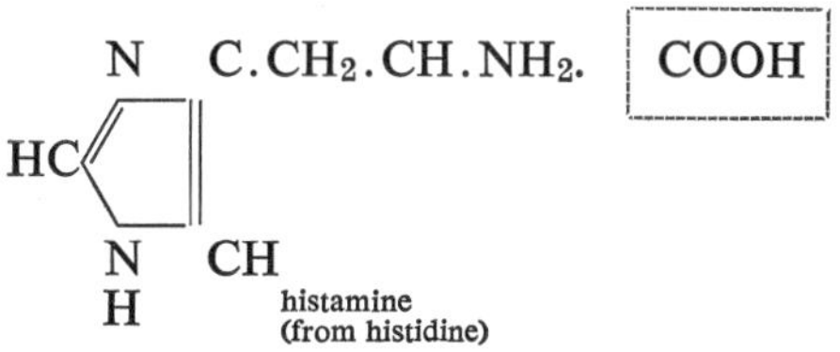

$$N \quad C.CH_2.CH.NH_2. \quad \boxed{COOH}$$

HC

$$N \quad CH$$
H

histamine
(from histidine)

Adrenalin and *nor-adrenalin*, from the medulla of the suprarenal glands, are both derived from a dihydroxybenzene nucleus with a methylamine and secondary alcohol group as a side-chain in positon 1. Nor-adrenalin lacks the terminal methyl group, CH_3. *Histamine*, a local hormone released by damaged tissue, is derived from the amino-acid, histidine, by removal of the acidic group, COOH. *Acetylcholine*, liberated at nerve-endings of somatic motor and parasympathetic nerve fibres, is synthesized from the nitrogenous base, choline, and acetic acid.

$$N(CH_3)_3(CH_2OH)_2 + CH_3COOH \xrightarrow{\text{choline acetylase}}$$
$$(CH_3)_3 N.OH.CH_2CH_2O.CO.CH_3 + H_2O.$$

Of the hormones secreted by the gut wall, only *secretin*, a proteose, is known with certainty. The pineal hormone, *melatonin*, is probably a polypeptide.

PHYTOHORMONES AND AUXINS

Plants produce hormones which control growth and certain other physiological functions. They are known as *phytohormones*. There is considerable evidence that the growth hormone is indole-3-acetic acid (*see* formula on p. 116). The term *auxin* is applied to any substance which will promote cell enlargement, and, therefore, it includes natural hormones as well as substances from other sources. Auxin can be extracted from plant parts where there is active growth, and in particular, from apical meristems. Substances having similar effects have been obtained from urine, malt extract and cereal germ oils, and many have been synthesized. The concentration of an auxin is estimated in *Avena units* from its ability to effect curvature or growth in oat coleoptiles. Auxins have become very important in agricultural practice. Other plant hormones include the giberellins and cytokinins. An account of these will be found on pp. 269–278.

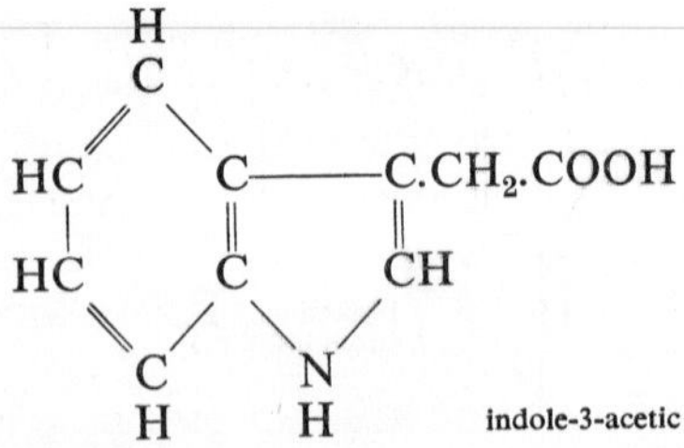

$$\text{indole-3-acetic acid.}$$

The molecular structure diagram shows:

$$
\begin{array}{c}
\overset{H}{C} \\
HC \diagdown \quad C——C.CH_2.COOH \\
HC \diagup \quad C \qquad CH \\
\underset{H}{C} \quad \underset{H}{N}
\end{array}
$$

indole-3-acetic acid.

ELEMENTS OF BIOLOGICAL IMPORTANCE

Thirty-nine elements are utilized by living organisms; of these, eighteen are invariably present. The greater part of the body weight of all creatures is made up of carbon, hydrogen, oxygen, nitrogen and phosphorus. Elements always present, but as less than 1 per cent of body weight, are sodium, potassium, magnesium, sulphur, chlorine, calcium and iron. Trace elements less than 0·05 per cent are boron, fluorine, silicon, manganese, copper and iodine. Twenty-one other elements appear to be necessary for certain organisms. They are aluminium, arsenic, bromine, beryllium, barium, chromium, copper, cadmium, caesium, germanium, lithium, lead, molybdenum, nickel, rubidium, strontium, silver, tin, titanium, vanadium and zinc. A brief indication of the major reasons for the importance of these elements now follows.

Carbon is the characteristic element of all organic substances. It forms about 20 per cent of the weight of most organisms, occurring largely in the essential proteins, lipides and carbohydrates. The carbon dioxide of the atmosphere is the source of all organic carbon compounds. *Hydrogen* is a component of water, which makes up 70 to 80 per cent of the weight of all organisms. H^+ is the most electro-positive of ions and is present in all aqueous solutions. Hydrogen is a constituent of all carbohydrates, proteins, lipides and their derivatives. The peroxide, H_2O_2, plays an important part in cell oxidation systems, and ammonia is important in nitrogen metabolism. *Oxygen* constitutes about two-thirds of the body-weight of plants and animals. It is an essential constituent of carbohydrates, lipides and most proteins. It is necessary for respiration, either directly, or indirectly, in all forms of life; aerobic organisms need a continuous supply of free oxygen. A number of prominent biologists have stated that most, if not all, the oxygen of the atmosphere is derived from photosynthesis in green plants. *Nitrogen* is present in all amino-acids, and hence in all proteins. It is also a constituent of amines, amides, purines, pyrimidines, porphyrins and many other substances. Urea and uric acid are the principal forms of nitrogenous excretion. *Phosphorus* is an essential constituent of all living cells, for the lipoid-protein structures and for nucleic acids; phosphorus compounds play a vital part in all carbohydrate metabolism. Creatine phosphate provides the initial energy for contraction of vertebrate muscle, arginine phosphate for invertebrate muscle. Calcium phosphate is the major constituent of bony skeletons; phosphates form esters for the conveyance of sugars and glycerol from the gut to the blood-stream; as phosphate esters, lipides are conveyed from the blood to the tissues. Phosphates also play an important part in buffering tissue fluids.

Sodium provides the chief circulating cation, Na^+, in body fluids, and thus provides most of the cations to balance the anions. With K^+ and Mg^+, it is important in maintaining a state of physiological excitability in the tissues. *Potassium* provides the chief metallic ion, K^+, inside cells; with Na^+ and Mg^+, it helps to maintain tissue excitability. It is essential for cell division and for protein synthesis; in green plants it is necessary for chlorophyll synthesis. Potassium is a growth factor in plants and animals, and is very important in the functioning of kidneys. *Magnesium* helps to maintain tissue excitability, and is a constituent of the chlorophyll molecule. Skeletons of many marine organisms are rich in $MgCO_3$; it is present in small quantities in skeletons, muscles and nerves of the higher animals. Mg^+ ions are activators of the phosphate enzymes. *Sulphur* is universally distributed in proteins, being present in the amino-acids cystine, cysteine and methionine. The dead layers of skin, feathers and hair contain about 20 per cent sulphur. It is essential for protein synthesis in plants, being taken up from the soil as $SO_4^=$ ions. *Chlorine* passes easily through the plasma membrane as Cl^-; hence it is an important agent in absorption, distribution and excretion of other radicals. NaCl is the principal electrolyte in body fluids, and hence is very important in osmotic relationships. The ease of transfer of Cl^- ions is of value in buffering. The HCl secreted in the gastric juice is necessary for the action of the enzyme pepsin. *Calcium* comprises 10 per cent of bony skeletons, and is present as the pectate in green plant cell walls. $CaCO_3$ forms the major part of the exoskeletons of the crustaceans, and of the shells of molluscs. Ca^{++} ions serve to counteract the state of excitability in tissues set up by Na^+, K^+ and Mg^+. Ca^{++} ions are also essential for muscle contraction, for blood coagulation, and for the precipitation of the casein of milk in mammalian stomachs. *Iron* is a constituent of the respiratory pigments haemoglobin, haemoerythrin and chlorocruorin, and also of the cytochromes within cells. The enzymes catalase and peroxidase contain iron.

All animal and plant tissues contain *boron* in very low concentrations. Its exact function is uncertain, but it appears to be concerned with carbohydrate transport. *Fluorine* is always present in minute quantities. Its general functions are unknown, but as CaF_2 it makes an important contribution to the hardness of skeletons and is a constituent of the enamel of teeth. *Silicon* is found in the higher plants as part of the cell wall substance. It is a microconstituent of animal cells, and is especially evident in connective tissue. Silica constitutes the skeletal material of many marine unicellular organisms, and the siliceous sponges have a skeleton of silica spicules. *Manganese* is a growth factor and is essential for bone development in animals. It is an activator of certain enzymes, especially phosphorylase, phosphatase, and dipeptidase. *Copper* is an invariable constituent of plants and animals. The respiratory pigment haemocyanin contains copper in the molecule, and it is present in certain feather pigments. It aids in haemoglobin synthesis in the higher animals. Some of the oxidase enzymes are copper-protein compounds. In minute traces, it is a growth factor in both plants and animals. *Iodine* is a microconstituent of all organisms, being especially plentiful in seaweeds. Iodogorgoic acid and its derivative thyroxine, contain iodine. It has also been identified in the skeletons of certain sponges and corals.

Elements not Invariably Present

These twenty-one elements are only present in certain groups, as far as is known. In many cases, their functions have not been determined.

Aluminium is found in all plants and in some animals. *Arsenic* is widely distributed in minute amounts. *Bromine* occurs in many marine plants and animals, especially in sea-anemones. *Beryllium* is present in a great many plants. *Barium* has been found in plants, especially marine, and in a few of the higher animals. *Chromium* occurs in a few mammals. *Cobalt* is common in plants and animals; its absence from the diet causes the disease "bush sickness" in sheep and cattle in Australia. *Cadmium* is always present in certain molluscs. *Caesium* is a microconstituent of certain algae, and of the shells of many molluscs. *Germanium* has been found in some algae and in mammalian blood. *Lithium* occurs in many plants, and in a few marine animals. *Lead* is commonly present in chordate animals, and in many plants, especially Gramineae. Most plant and animal tissues contain *molybdenum*, in minute amounts. *Nickel* is widely distributed in plants, especially in green leaves, and in the higher animals, especially in the pancreas. Some workers have claimed that nickel is an essential element. *Rubidium* has been reported to be a constant constituent of human tissues; many lower plants and animals contain it. *Strontium* is generally present in marine organisms; certain radiolarians construct a skeleton of strontium sulphate. *Silver* is common in fungi and in some marine molluscs. *Tin* is present in most mammalian tissues. *Titanium* has been found in many plants and in some higher animals. *Vanadium* is an essential constituent of the blood of sea-squirts and sea-cucumbers. It is also present in the wood of many trees, and in the liver of many animals. It appears to be concerned with the oxidation of phospholipides. *Zinc* is probably of universal occurrence. It has been shown to be a growth factor in plants and in some rodents. The mammalian enzyme, carbonic anhydrase, contains zinc.

The importance of many of the microconstituents is very great, as shown by work done on cobalt, manganese, boron and iodine. The presence or absence of one particular element can play a large part in determining the flora and fauna of an area.

TESTS FOR BIOCHEMICAL SUBSTANCES

Many of the materials for testing can be obtained directly from plant and animal tissues by the expression of their juices, but this invariably results in complicated mixtures of compounds which in the hands of the beginner can produce confusing results. For the purposes of the tests outlined in this and later sections, it is more profitable to use relatively pure substances which are readily obtainable as chemical reagents. Methods of demonstrating and identifying the more important secretory and storage substances as they occur in tissues are given in Chap. 14.

Carbohydrates

α-Naphthol Test (Molisch's test) for All Carbohydrates

To a small quantity of the carbohydrate solution in water add a few drops of alcoholic α-naphthol solution. Down the side of the test-tube pour slowly about 2 cm^3 of concentrated sulphuric acid. This can be done with a thistle funnel if necessary. Without shaking, hold the test-tube vertical and examine the interface of the two liquids. The presence of any carbohydrate is denoted by the formation of a purple ring at the interface. The colour is due to the reaction between the α-naphthol and furfural or its hydroxymethyl derivative, these being produced from the carbohydrate by the action of the sulphuric acid.

Tests for Sugars

To Distinguish Monosaccharides from Disaccharides. Boil about 5 cm^3 of Barfoed's reagent (copper acetate in acetic acid) in a test-tube. To this, add slowly a small quantity of the sugar solution. If the sugar is a monosaccharide, a red precipitate of cuprous oxide will appear. If the sugar is a disaccharide, no colour change is apparent unless the mixture is further boiled for a period, during which time hydrolysis may

5

occur and the monosaccharide reaction become apparent. The precipitation of the cuprous oxide in the one case and not in the other is due to the differences in reducing power of the two types of sugar (*see* p. 78).

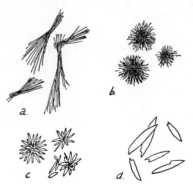

To Distinguish Reducing from Non-reducing Sugars. To the sugar solution add equal quantities of Fehling's solutions A (copper sulphate in water) and B (sodium potassium tartrate and sodium hydroxide in water). Boil. A reducing sugar causes the precipitation of red cuprous oxide showing that the cupric oxide has been reduced.

Fig. 4.4. Osazone-crystals of (*a*) glucose, (*b*) lactose, (*c*) maltose, (*d*) galactose.

A similar but more sensitive test can be performed using Benedict's solution (copper sulphate, sodium citrate and sodium carbonate). Add a small quantity of the solution to the sugar solution and boil vigorously for several minutes. Allow to stand. The appearance of a red or yellow precipitate indicates the presence of reducing sugars.

Tests for Specific Sugars. The ability to form *osazones* with phenylhydrazine is a property of some sugars. The osazones are yellow, crystalline compounds each with a definite melting point, but they can more readily be recognized by their form in most cases.

To about 10 cm³ of the sugar solution under test, 1 cm³ of glacial acetic acid, about 1 g of phenylhydrazine hydrochloride and 2 g of solid sodium acetate are added. The mixture is heated in a water bath and boiled for about half an hour and then allowed to cool slowly. In the case of glucose, the osazone will appear while the solution is still hot, but for most, the cooling must be complete and slowly accomplished. When the crystals are formed, mount some in the solution on a microscope slide and examine. Compare the crystals with those in Fig. 4.4. Glucosazone crystals are in bundles of needles as are those of mannosazone and fructosazone but these three sugars can be identified by other tests. Maltosazone crystals are lance-shaped in outline and appear in aggregates of rosette form. Lactosazone crystals are small needle-crystals appearing in spherical masses. Galactosazone crystals resemble broken dagger-blades and tend to appear separately. Sucrose does not form an osazone.

The ketonic sugars and those yielding ketonic sugars on hydrolysis, e.g. fructose and sucrose, can be identified by the application of

Seliwanoff's test. Mix equal volumes of concentrated hydrochloric acid and water and add a few crystals of resorcinol. When these have dissolved, add some sugar solution and heat. A red colour indicates fructose or sucrose.

Pentoses, e.g. arabinose, yield green compounds with Bial's reagent. This is composed of 0·3 g orcinol in 100 cm³ concentrated hychochloric acid to which 5 drops of 10 per cent ferric chloride solution have been added. Add the sugar to some of the solution and boil. A green colour appears. Fructose produces a red colour in the same reagent. Pentoses also give pink coloration with benzidine. To the sugar solution add a little 4 per cent benzidine solution in glacial acetic acid. Boil vigorously. Cool and add a little water. A pink to red colour is produced.

Tests for Polysaccharides

Hexosans. None of the polysaccharides will reduce Barfoed's, Fehling's or Benedict's solutions unless hydrolysed by boiling with dilute acids or other means. Addition of iodine to cold starch solution gives a blue-black coloration which disappears when the mixture is heated and reappears on cooling. A reddish-brown colour is produced when iodine is added to glycogen and this also occurs only when the mixture is cold. Inulin can be distinguished from these two polysaccharides by failure to give a colour reaction with iodine and by the fact that it gives a positive reaction to Seliwanoff's test when hydrolysed by mineral acid.

Cellulose in the pure form (filter paper is a good material), will give a blue colour with iodine if the cellulose has previously been treated with strong sulphuric acid. It will give a similar coloration directly with Schultze's solution (chlor-zinc-iodide). Hemicelluloses cannot be demonstrated easily by any general microchemical test, but some will give a blue-violet colour with iodine. Sections of a date stone, softened in 50 per cent glycerine and mounted in iodine, may be used to demonstrate this characteristic. The hemicellulose occurs in the heavily thickened cell walls.

Pentosans. The pentosans can be demonstrated by the fact that the pentose sugars, from which they are composed, readily yield furfural, which gives a bright red colour with aniline acetate. Boil a mixture of gum arabic in water and concentrated hydrochloric acid for some minutes and then place at the mouth of the test-tube a piece of filter paper wetted with aniline acetate. A bright red coloration of the paper will appear.

Demonstration of Heterosaccharides

Pectic compounds in the middle lamellae of plant tissues may be demonstrated by staining with ruthenium red, when they take on a dark red colour. Mucilage may be shown to be present in the walls of the epidermis of linseed testa. If the seeds are soaked in water, they become thickly covered with a slippery, transparent coat. If sections of a dry seed are mounted in strong glycerine, the mucilage is not apparent, but if the glycerine is replaced with water, the epidermal cell walls swell considerably and the cuticle is pushed off as the mucilage absorbs the water. If a section is stained with dilute $CuSO_4$, the mucilage takes on a pale blue colour.

There are no simple specific tests for chitin and allied substances which are found in arthropod exo-skeletons and in the cell walls of some fungi.

Demonstration of Glycosides

Amygdalin, from bitter almonds, can be recognized from the presence of one of its hydrolysis compounds, hydrogen cyanide or prussic acid. If some bitter almonds are ground with sand and water and the extract filtered and placed in a water bath at 35°C for half an hour, the amygdalin which it contains will be hydrolysed owing to the presence of emulsin (a glycosidase enzyme). Hydrogen cyanide can be detected by its ability to turn moist sodium picrate paper brown. Other hydrolysis products are glucose and benzaldehyde, the latter being recognizable by its smell. Salicin, a glycoside in the twigs of willows and poplars can be similarly extracted and demonstrated. When hydrolysed under the influence of emulsin, it yields saligenin which can be recognized from the deep purple colour it gives with ferric chloride. Aesculin can be obtained from horse-chestnut bark pounded with glacial acetic acid. If the solution is made alkaline with potassium hydroxide, a blue fluorescence is very apparent.

Demonstration of Saccharide Derivatives

Glycerol is a sugar alcohol and in common with the other polyhydric alcohols can be demonstrated by the following test. To a 5 per cent solution of borax add two or three drops of phenolphthalein and then a small quantity of 20 per cent aqueous solution of glycerol until the pink colour disappears. Boil. The pink colour will return if too much glycerol has not been used. Although other polyhydric alcohols such as mannitol and sorbitol give the same reaction, none is so distinct as glycerol. The presence of glycerol can also be shown if it is heated with powdered potassium hydrogen sulphate, upon which the very pungent smelling acrolein is developed. Glycerol is also shown to be present if on the addition of copper sulphate and a few drops of potassium hydroxide a deep blue colour develops without any precipitation. The amine derivative of glucose, glucosamine, resembles glucose in properties and reactions. The osazones are identical.

Lipides and Lipines

No great purpose is served in attempting to distinguish precisely between all the fatty and oily substances. All can be demonstrated by several general tests as below. Olive oil may be used as the reagent. The fats are all easily soluble in carbon bisulphide, ether, benzene and chloroform, but not in water, with which they form emulsions on vigorous shaking. The formation of an emulsion with water is the basis of a simple fat test. To about 2 cm^3 of ethyl alcohol add a little of the suspected fatty substance and shake well. Allow any solids to settle and then decant the alcohol into a clean test-tube. Add about 2 cm^3 of cold water. Mix and look for cloudiness which indicates the presence of a fat that is forming an emulsion with the water. The fat and water components of the emulsion can be delayed in their separation if a little bile salts powder or a few drops of sodium hydroxide are added before the shaking. All the lipides produce a "grease-spot" if dripped on to filter paper. Similarly, all become blackened in colour when in contact with osmic acid (osmium tetroxide). The fats readily take up the dyes Sudan III and Sudan IV (Scharlach Red). If the dye is added to olive oil in a test-tube and well shaken and then washed with alcohol or water by further shaking with one of these, on standing, it will be seen that the oil retains the dye in both cases. Note that castor oil is soluble in alcohol and should not be used for this demonstration. Alcoholic solution of alkannin will also dye fatty substances pink but some may take several hours to take up the stain. The waxy substances such as cutin, suberin, lanolin and beeswax all give similar responses to the tests outlined above. The first two can be found in epidermal and peridermal tissues of plants and sections of suitable material (holly leaf for cutin and cork for suberin) can be treated on a microscope slide and examined (see Chap. 14). The lipine, lecithin, can be distinguished from the lipides by its insolubility in acetone. If a small quantity of lecithin is dissolved in chloroform and acetone added to the solution, the lecithin will be precipitated. A lipide treated in similar fashion would remain in solution.

Proteins and Derivatives

All the proteins show reactions characteristic of the amino-acids of which they are composed and many of these can be demonstrated by qualitative colour tests with a variety of reagents. Dried blood albumin can be used in the following reactions.

The xanthoproteic test indicates the presence of tyrosine. To a small quantity of protein solution add nitric acid and warm gently. A yellow colour is produced. Cool and add a few drops of ammonia. A bright orange colour appears. The colouring is due to the formation of nitro-compounds by nitration of the benzene nucleus.

Tyrosine can also be demonstrated by the use of Millon's reagent (mercuric nitrate and nitrite). Heat the protein solution to which a few drops of Millon's reagent have been added and a deep red colour quickly appears. This is due to the presence of a phenolic group in the amino-acid.

The presence of tryptophane can be shown by adding glyoxylic acid (or glacial acetic acid which has been standing in the light) to a small quantity of the solution and then allowing concentrated sulphuric acid to run gently down the side of the test-tube (there must be no shaking). At the junction of the liquids a violet ring appears due to a reaction between the glyoxylic acid and the amino-group of tryptophane. This test is sometimes known as Adamkiewicz's test.

Cystine contains sulphur in the molecule and can be demonstrated accordingly. Add to the solution lead acetate and then sufficient sodium hydroxide to redissolve the precipitate. Boil. A dark brown precipitate of lead sulphide is developed.

All proteins respond positively to the biuret test. This is due to the presence of —NH—CO— groups in the molecule. To a portion of the protein solution add a little sodium hydroxide solution and then drop by drop 1 per cent copper sulphate solution, shaking at each drop. A violet colour is produced but this can be masked by the copper sulphate if this is added too rapidly. The test takes its name from the substance biuret, which can be obtained by heating urea.

The presence of a carbohydrate in conjunction with a protein molecule can be demonstrated by Molisch's α-naphthol test (*see* p. 119).

Metaproteins, proteoses and peptones, also consisting of amino-acids, will give positive results with any of the amino-acid tests outlined above if those amino-acids are present. Pure peptone can be distinguished from a protein by the fact that a rose-pink colour is produced in the biuret test.

Miscellaneous Compounds

Ethereal Oils, e.g. Turpentine, Eucalyptus Oil, Lavender Oil

These are stained by osmic acid, Sudan III and alkannin, but can be distinguished from lipides by being readily soluble in alcohol, being very volatile and by giving a yellow coloration with hydrochloric acid vapour. They give an intense emerald green colour when in contact with a saturated solution of copper acetate.

Resins and Balsams, e.g. Rosin, Dammar, Canada Balsam

These are closely related to the ethereal oils (terpenes) and give similar reactions to those above.

Tannins, e.g. Tannic Acid

The most characteristic reaction is the dark blue-black colour developed in the presence of ferric chloride. They will also develop a green colour (sometimes blue) in association with copper acetate and can therefore be confused with ethereal oils when this test is applied to plant sections.

Lignin

This is best demonstrated in wood where it is associated with cellulose. In aniline sulphate or chloride a deep yellow colour is developed. If the wood is dipped in alcoholic phloroglucin solution until the stain is taken up, and then transferred to strong hydrochloric acid, a deep red colour develops. A drop of acidified phloroglucin on newspaper will quickly indicate its wood-pulp origin. Filter paper (cellulose) will not give the same red colour.

CHAPTER 5

THE NATURE AND PROPERTIES OF ENZYMES

THE word "enzyme" is derived from the Greek "en zyme," meaning "in yeast." The action of yeasts in converting sugar solutions to alcohol, is the earliest recorded example of enzyme action.

All the vital activities of cells are controlled by enzyme mechanisms. Every organism produces a great variety of enzymes, but no organism can produce them all. Every aspect of metabolism is dependent upon their catalytic action; digestion of food, synthesis of storage and structural materials, contraction of muscle, respiration, light emission, production of special substances such as chitin, hormones, snake venom, pigments; these, and all other chemical changes performed by living organisms, come under their control. Some idea of their total range of activity is provided by the fact that all remains of living organisms down to the last vestige, are disintegrated by the enzymic activity of micro-organisms. The action of all enzymes is catalytic. The principal properties of catalysts are outlined below—

1. They cause increase in the velocity of a chemical reaction (*see* p. 980), without being themselves used up.

2. A very small amount of catalyst can effect the transformation of an indefinitely large amount of reactant.

3. The presence of the catalyst does not alter the nature or proportions of the final product.

4. The catalyst does not add any energy to the system.

5. The activity of the catalyst varies with temperature, pH, pressure, and with the proportions of the various reactants.

6. The reaction catalysed is reversible.

The names of enzymes are most commonly formed by the addition of -ase to the main part of the name of the substrate on which they act. For example, amylase acts on amylum (starch), lipase acts on lipides (fats) and maltase acts on maltose. There are some enzyme names in use that were not made originally in this way but are commonly retained because they have been applied for a long time, e.g. ptyalin, pepsin, erepsin. Others cannot be named in this way because some substances can be acted upon by more than one enzyme and confusion might arise. A dipeptide could be acted upon by any of three enzymes. One, in fact called a dipeptidase, can split it into its constituent amino acids, another, called a deaminase, can remove an amino group whilst a third

126

can remove a carboxyl group and is called a decarboxylase. These last two enzymes are named according to their catalytic activity and not by their substrate, which is the same for both.

CLASSIFICATION OF ENZYMES

Enzymes are classified according to the type of reaction they catalyse. It should be pointed out that there is no system of classification which is completely satisfactory. The table opposite shows the main groups and sub-groups but does not pretend to be exhaustive.

Hydrolysing Enzymes

These catalyse decomposition of a substrate by hydrolysis, attacking specific linkages. Conversely, they catalyse synthesis of complex substances by condensation, forging certain specific linkages.

(1) Hydrolysing enzymes	(2) Oxidation-reduction enzymes	(3) Enzymes attacking the C–C linkage	(4) Miscellaneous, not in groups (1) (2) or (3)
Esterases Carbohydrases Proteases Aminases	Dehydrogenases Oxidases Peroxidases Catalase	Carboxylases Decarboxylases	Isomerases Coagulating enzymes, etc.

Esterases

These enzymes attack organic esters, splitting them into two groups, one of which is normally an acid. *Lipases*, such as *steapsin*, split fats into aliphatic acids and glycerol. *Phosphatases* hydrolyse phosphoric esters such as hexosephosphates, glycerophosphates and nucleotides, producing phosphoric acid and a base.

Phosphorylases catalyse the reversible conversion of starch or glycogen, in the presence of phosphate, into glucose phosphate. Examples of the action of some esterases are given below—

$$\begin{array}{l} CH_2O.OC.C_{17}H_{33} \\ | \\ CHO.OC.C_{17}H_{33} + 3H_2O \rightleftharpoons \\ | \\ CH_2O.OC.C_{17}H_{33} \end{array} \xrightarrow{steapsin} \begin{array}{l} CH_2.OH \\ | \\ CH.OH + 3C_{17}H_{33}COOH \\ | \\ CH_2.OH \end{array}$$

fat (tri-olein)   water   glycerol   oleic acid

$$\underset{\substack{\text{hexose phosphate}}}{C_6H_{11}O_5 . H_2PO_4} + \underset{\substack{\text{water}}}{H_2O} \overset{\text{phosphatase}}{\rightleftharpoons} \underset{\substack{\text{hexose}}}{C_6H_{12}O_6} + \underset{\substack{\text{phosphoric acid}}}{H_3PO_4}$$

$$\text{nucleotide} + \text{water} \overset{\text{nucleotidase}}{\rightleftharpoons} \text{nucleoside} + \text{phosphoric acid}$$

$$\text{pectin} + \text{water} \overset{\text{pectase}}{\rightleftharpoons} \text{pectic acid} + \text{methyl alcohol}$$

$$\underset{\substack{\text{starch}}}{(C_6H_{10}O_5)_n} + \underset{\substack{\text{phosphoric acid}}}{nH_3PO_4} \overset{\text{phosphorylase}}{\rightleftharpoons} \underset{\substack{\text{glucose phosphate}}}{nC_6H_{11}O_5 . OPO_3H_2}$$

Carbohydrases

All the carbohydrases catalyse various stages in the hydrolysis of higher carbohydrates to simple sugars. Some of the names formerly used in this class for single enzymes, really denote a group. Thus *amylase* (*diastase*) is a general name for many enzymes such as *ptyalin* of saliva, and the *amylopsin* of the pancreas.

$$\underset{\substack{\text{starch}}}{2(C_6H_{10}O_5)_n} + \underset{\substack{\text{water}}}{nH_2O} \overset{\text{amylase}}{\rightarrow} \underset{\substack{\text{disaccharide}}}{n(C_{12}H_{22}O_{11})}$$

$$\underset{\substack{\text{inulin}}}{(C_6H_{10}O_5)_n} + \underset{\substack{\text{water}}}{nH_2O} \overset{\text{inulase}}{\rightleftharpoons} \underset{\substack{\text{fructose}}}{n(C_6H_{12}O_6)}$$

$$\underset{\substack{\text{cellulose}}}{2(C_6H_{10}O_5)_n} + \underset{\substack{\text{water}}}{nH_2O} \overset{\text{cellulase (cytase)}}{\rightleftharpoons} \underset{\substack{\text{cellobiose}}}{n(C_{12}H_{22}O_{11})}$$

A number of enzymes hydrolyse various disaccharides to monosaccharides. The type reaction is—

$$\underset{\substack{\text{disaccharide}}}{C_{12}H_{22}O_{11}} + \underset{\substack{\text{water}}}{H_2O} \overset{\text{enzyme}}{\rightleftharpoons} \underset{\substack{\text{monosaccharide}}}{2(C_6H_{12}O_6)}$$

Thus *maltase* hydrolyses maltose to glucose; *sucrase* (*invertase*) hydrolyses sucrose to glucose and fructose; *cellobiase*, cellobiose to glucose; and *lactase*, lactose to glucose and galactose. *Glucosidases*, which are plentiful in plants, catalyse the conversion of glucosides to glucose and aglucones.

Proteases

These are enzymes which attack the peptide link —CO—NH—, in proteins. *Endopeptidases* split the protein molecules into smaller units, proteoses, peptones and peptides. Thus *pepsin* and *trypsin* of animals split proteins to peptones. Many endopeptidases such as *papain* (melon tree) and *ficin* (fig tree), are found in plants. A number of similar enzymes have been discovered within animal cells. They are known as *cathepsin* I, II, III and IV, according to the substances which activate them.

Exopeptidases attack terminal —CO—NH— linkages, thus splitting off amino-acids. Each exopeptidase, such as *erepsin*, contains three component enzymes; an *amino-peptidase* which attacks peptides containing a free amino group; a *carboxyl-peptidase* which attacks peptides containing a free carboxyl group, and a *dipeptidase* attacking peptides which contain free carboxyl and amino groups.

Aminases

The *aminases* are hydrolysing enzymes which attack the —C—NH— and —C—NH_2— linkages in non-protein compounds. The —NH or —NH_2 group is replaced by —OH, and ammonia is liberated. These enzymes are very important in purine metabolism, and in the production of urea in those animals which excrete it. *Adenase* deaminates adenine with the formation of hypoxanthine and ammonia. *Guanase* similarly deaminates guanine, forming xanthine and ammonia. *Arginase* hydrolyses arginine to ornithine and urea; it plays an indispensable part in the ornithine cycle whereby urea is produced in animals. *Urease* is common in plants, but is not normally found in animals. It attacks urea, with the formation of ammonia and carbon dioxide—

$$\overset{\text{urease}}{CO(NH_2)_2 + H_2O \rightleftharpoons 2NH_3 + CO_2}$$

$$\underset{\text{urea}}{} \quad \underset{\text{water}}{} \quad \underset{\text{ammonia}}{} \quad \underset{\substack{\text{carbon}\\\text{dioxide}}}{}$$

Asparaginase, widely distributed in plants, attacks the amide asparagine, with the production of aspartic acid and ammonia—

$$\overset{\text{asparaginase}}{\underset{\text{asparagine}}{H_2N.CO.CH_2.CH.NH_2.COOH} + \underset{\text{water}}{H_2O} \rightleftharpoons}$$

$$\underset{\text{aspartic acid}}{COOH.CH_2.CH.NH_2.COOH} + \underset{\text{ammonia}}{NH_3}$$

Glutaminase similarly hydrolyses glutamine with the formation of glutamic acid and ammonia. This enzyme occurs commonly in plants and in the higher animals.

Oxidation-reduction Enzymes

The oxidation of a substrate in a living cell almost always involves the reduction of another substance (*see* p. 429). Let A be the substrate and B the other substance. If B donates oxygen to A, then B is reduced while A is oxidized.

$$A + BO \rightarrow AO + B$$

sub-   oxygen oxidized reduced
strate   donor substrate donor

Another method of oxidation commonly occurs when the substrate donates hydrogen to a hydrogen acceptor.

$$AH_2 + B \rightarrow A + BH_2$$

sub-   hydrogen oxidized reduced
strate   acceptor substrate acceptor

Both types of reaction are very common.

The principal oxidation-reduction enzymes are the *dehydrogenases*, the *oxidases*, the *peroxidases* and *catalase*.

Dehydrogenases

These catalyse the reaction involving transfer of hydrogen from a substrate to a hydrogen acceptor. A large number of these enzymes have been isolated from yeast and from various animal tissues. *Succinic dehydrogenase* plays a part in Krebs' cycle, catalysing the oxidation of succinic acid to fumaric acid (*see* p. 439).

succinic dehydrogenase
$$HOOC.CH_2.CH_2.COOH + B \rightleftharpoons HOOC.CH.CH.COOH + BH_2$$

succinic acid   hydrogen   fumaric acid   reduced
            acceptor   acceptor

Lactic dehydrogenase, which occurs commonly in plant and animal tissues activates the conversion of lactic acid into pyruvic acid; the reverse reaction is important in muscles.

lactic dehydrogenase
$$CH_3.CHOH.COOH + B \rightleftharpoons CH_3.CO.COOH + BH_2$$

lactic acid   hydrogen   pyruvic acid   reduced
         acceptor   acceptor

Triose phosphate dehydrogenase is one of a series of enzymes catalysing the breakdown of glucose to pyruvic acid. It activates the change of triose phosphate to phosphoglyceric acid.

$$\underset{\substack{| \\ CHOH \\ | \\ CH_2O.PO_3.H_2}}{CHO} + H_2O + \text{co-enzyme 1} + \text{phosphate} \underset{\text{dehydrogenase}}{\overset{\text{triose phosphate}}{\rightleftharpoons}} \underset{\substack{| \\ CHOH \\ | \\ CH_2O.PO_3H_2}}{\overset{COO.PO_3H_2}{}} + \quad \text{reduced co-enzyme 1}$$

Ethyl alcohol dehydrogenase in certain bacteria, catalyses the oxidation of ethyl alcohol to acetaldehyde. Co-enzyme 1 is the hydrogen acceptor as in the previous dehydrogenation. The reverse reaction is important in fermentation by yeast, when pyruvic acid is converted into acetaldehyde, then acetaldehyde into ethyl alcohol.

$$\underset{\text{ethyl alcohol}}{CH_3.CH_2.OH} + \underset{\text{hydrogen acceptor}}{\text{co-enzyme 1}} \overset{\text{ethyl alcohol dehydrogenase}}{\rightleftharpoons} \underset{\text{acetaldehyde}}{CH_3CHO} + \underset{\text{reduced acceptor}}{\text{reduced co-enzyme 1}}$$

Malic acid dehydrogenase activates the dehydrogenation of malic acid to oxaloacetic acid; this is another stage in Krebs' cycle—

$$\underset{\text{malic acid}}{HOOC.CH_2.CHOH.COOH} + \underset{\text{hydrogen acceptor}}{\text{co-enzyme 1}} \overset{\text{malic acid dehydrogenase}}{\rightleftharpoons} \underset{\text{oxaloacetic acid}}{COOH.CH_2.CO.COOH}$$
$$+ \underset{\text{reduced acceptor}}{\text{reduced co-enzyme 1}}$$

Oxidases

Oxidases are unable to utilize hydrogen acceptors, but they catalyse oxidations in the presence of free oxygen. *Amino-acid oxidases* are widespread in animals. The amino-acid is converted into the corresponding keto-acid and ammonia is evolved.

$$\underset{\text{amino-acid}}{R.CH.NH_2.COOH} + \underset{\text{oxygen}}{O} \overset{\text{amino-acid oxidase}}{\rightleftharpoons} \underset{\text{keto acid}}{R.CO.COOH} + \underset{\text{ammonia}}{NH_3}$$

Uric acid oxidase is important in the liver and kidney of many mammals. It catalyses the conversion of uric acid into allantoin, which is the principal end-product of purine metabolism in many mammals, especially herbivores.

Xanthine oxidase, present in the liver of many birds, mammals and plants, catalyses the oxidation of the purines xanthine and hypoxanthine, to uric acid. The presence of oxidases in plant tissues is demonstrated when they discolour on exposure to air. Polyphenolic substances in the cell are turned brown by oxidation in the presence of an oxidase and molecular oxygen. Broad bean tissue quickly turns reddish and later black due to the oxidation of tyrosine by tyrosinase. The so-called *cytochrome oxidase*, known to be responsible for the oxidation of the cytochrome series at the expense of oxygen and once thought of as an oxidase enzyme, is in fact a form of cytochrome a_3 (*see* p. 431).

Peroxidases

These enzymes act on hydrogen peroxide or organic peroxides only in the presence of a substrate which acts as an oxygen acceptor. The oxygen is liberated in an active state and transferred to the substrate. Peroxidases are found in all the higher plants and in certain mammalian tissues, especially spleen and lung. The type reaction is—

$$\underset{\substack{\text{oxygen} \\ \text{acceptor}}}{A} + \underset{\substack{\text{hydrogen} \\ \text{peroxide}}}{H_2O_2} \overset{\text{peroxidase}}{\rightleftharpoons} \underset{\substack{\text{oxidized} \\ \text{acceptor}}}{AO} + \underset{\text{water}}{H_2O}$$

Some plant juices contain substances which are easily oxidized to peroxides in the presence of free oxygen. These organic peroxides form important substrates for peroxidases and the atomic oxygen liberated is available for oxidation of other substances in the cell. Plants which produce peroxidases but not oxidases do not discolour when cut, but will do so if a few drops of hydrogen peroxide are added to the cut surface. Horseradish is a good example.

Catalase

This important enzyme is present in all aerobic tissues. It catalyses the breakdown of hydrogen peroxide into water and molecular oxygen, and is not dependent upon the presence of an oxygen acceptor.

$$\overset{\text{catalase}}{2H_2O_2 \rightarrow 2H_2O + O_2}$$

Hydrogen peroxide is very toxic to protoplasm; its production in cells occurs when the hydrogen acceptor in the oxidation of a substrate by dehydrogenation is molecular oxygen.

$$\underset{\text{substrate}}{AH_2} + \underset{\substack{\text{hydrogen} \\ \text{acceptor}}}{O_2} \overset{\text{dehydrogenase}}{\rightarrow} \underset{\substack{\text{reduced} \\ \text{substrate}}}{A} + \underset{\substack{\text{oxidized} \\ \text{acceptor}}}{H_2O_2}$$

The ubiquitous presence of catalase, where free oxygen is available, is thus a necessity. It can be demonstrated by dropping hydrogen peroxide on cut plant or animal tissue, when effervescence with liberation of oxygen occurs.

Enzymes Attacking the C—C Linkage

Enzymes which catalyse the splitting of C—C linkages are sometimes called *desmolases*. They are important enzymes forming parts of complex systems, especially associated with respiratory processes. The two principal types are carboxylases and decarboxylases.

Carboxylases

These activate the decarboxylation of ketonic acids such as pyruvic and oxaloacetic. They are ultimately concerned with the liberation of carbon dioxide as a result of cell respiration, and also with the assimilation of carbon dioxide by green plants. Hence they are extremely important substances. *Pyruvic carboxylase* decarboxylates pyruvic acid to acetaldehyde and carbon dioxide—

$$\underset{\text{pyruvic acid}}{CH_3.CO.COOH} \overset{\text{carboxylase}}{\rightleftharpoons} \underset{\text{acetaldehyde}}{CH_3.CHO} + \underset{\substack{\text{carbon} \\ \text{dioxide}}}{CO_2}$$

Oxaloacetic carboxylase converts oxaloacetic acid to pyruvic acid with the evolution of carbon dioxide

$$\underset{\text{oxaloacetic acid}}{HOOC.CO.CH_2.COOH} \overset{\text{carboxylase}}{\rightleftharpoons} \underset{\text{pyruvic acid}}{CH_3.CO.COOH} + \underset{\substack{\text{carbon} \\ \text{dioxide}}}{CO_2}$$

Some carboxylases work only in the presence of a co-carboxylase. This is thiamine pyrophosphate, a derivative of vitamin B_1. The discussion of Krebs' cycle in Chap. 11 will show several examples of the action of carboxylases.

Decarboxylases

Occurring principally in animal tissues, decarboxylases convert various amino-acids into the corresponding amines with evolution of carbon dioxide. The principal amino-acids attacked are tyrosine, histidine and tryptophane. The type reaction is—

$$\underset{\text{amino-acid}}{R.CH(NH_2)COOH} \overset{\text{decarboxylase}}{\rightleftharpoons} \underset{\text{amine}}{R.CH_2.NH_2} + \underset{\substack{\text{carbon} \\ \text{dioxide}}}{CO_2}$$

Miscellaneous Enzymes

A number of enzymes do not fall into the three major groups already specified. *Isomerases* catalyse the conversion of one isomer into another. Some of these are important in respiratory processes. *Phosphoglucomutase* converts glucose-1-phosphate into glucose-6-phosphate; the phosphate group being transferred from position 1 to position 6. It is found in animal tissues and in yeast, and is vitally necessary both for breakdown and synthesis of sugars. *Phosphohexo-isomerase* converts glucose-6-phosphate into fructose-6-phosphate, an essential stage in respiratory breakdown of glucose.

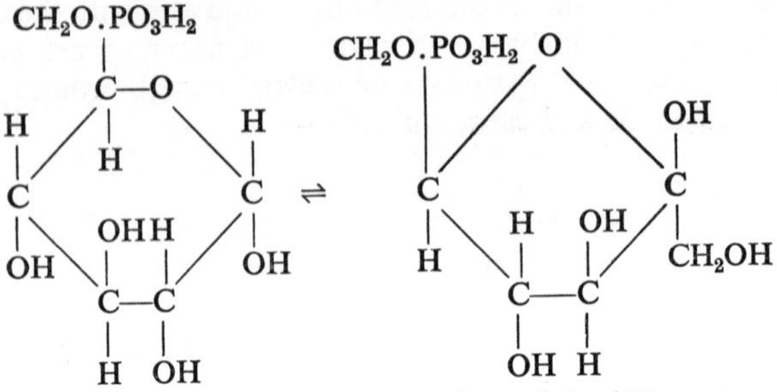

glucose-6-phosphate   fructose-6-phosphate

Phosphoglyceromutase catalyses the change of 3-phosphoglyceric acid into 2-phosphoglyceric acid, another stage in the catabolism or synthesis of glucose.

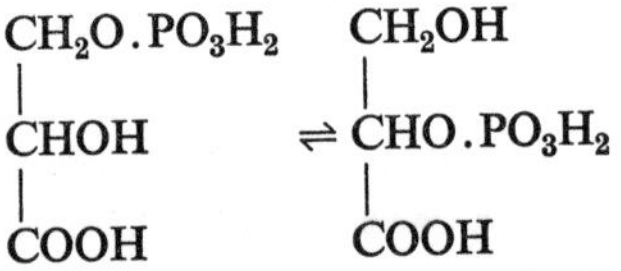

$$CH_2O.PO_3H_2 \quad\quad CH_2OH$$
$$| \quad\quad\quad\quad\quad\quad |$$
$$CHOH \quad \rightleftharpoons \quad CHO.PO_3H_2$$
$$| \quad\quad\quad\quad\quad\quad |$$
$$COOH \quad\quad\quad\quad COOH$$

3-phosphoglyceric   2-phosphoglyceric
acid   acid

Zymase, first extracted from yeast by Buchner in 1903, was at first thought to be a single enzyme catalysing the breakdown of glucose to ethyl alcohol and carbon dioxide, both end-products of fermentation.

$$\overset{\text{zymase}}{C_6H_{12}O_6 \rightarrow 2C_2H_5OH + 2CO_2}$$

glucose   ethyl alcohol   carbon
                            dioxide

Zymase is now known to be a complex mixture of at least fourteen enzymes, requiring also the presence of a number of co-enzymes and metallic ions for complete activity. The main stages of this fermentation process will be described in Chap. 11.

There are several well-known coagulative enzymes. *Rennin*, produced in the gastric pits of mammalian stomachs, catalyses the precipitation of the caseinogen of milk as insoluble casein. In the presence of calcium ions, the casein is converted into calcium paracaseinate. *Thrombase* plays a part in the clotting of blood, precipitating the soluble fibrinogen as insoluble fibrin.

Carbonic anhydrase splits carbonic acid into water and carbon dioxide. It is ultimately responsible for the rate at which carbon dioxide leaves the pulmonary circulation. The enzyme is present mainly in the red blood corpuscles, but is also found in small quantities in pancreas, in brain tissue, and in the mucous membrane of the stomach.

The enzyme *luciferase* which is found in all luminescent organisms appears to activate the substrate luciferin, though it is suggested by some workers that for certain cases the enzyme itself may be the light-emitting molecule

It should be pointed out that the majority of the examples cited here are individual enzymes catalysing individual reactions. Most enzymes, however, play small but important parts in complete systems.

PROSTHETIC GROUPS, CO-ENZYMES, ACTIVATORS, INHIBITORS, AND ANTI-ENZYMES

The great majority of enzymes need certain adjuncts without which their activity is not possible. The adjuncts are variously known as prosthetic groups, co-enzymes and activators. Conversely, the action of enzymes is stopped or prevented by specific inhibitors and by anti-enzymes.

Prosthetic Groups and Co-enzymes

These are non-protein groups permanently or temporarily attached to the zymoproteins. Without such an accessory or co-enzyme, the protein is incomplete and inactive. The protein component has been called the apoenzyme, and when the co-enzyme is present, the two together constitute the active holoenzyme.

In some cases, the co-enzyme is firmly attached to the zymoprotein, as for example peroxidase, which is quite active without the necessity of a second substance. In this case the prosthetic group or co-enzyme is an iron porphyrin. The same state of affairs exists in the case of catalase. In other examples, the prosthetic group is highly labile and is easily detachable from the apoenzyme. Several such co-enzymes have been isolated, particularly those which function as hydrogen carriers in oxidation-reduction reactions. These include co-enzymes 1 and 2 or nicotinamide adenine dinucleotide (NAD) and nicotinamide adenine dinucleotide phosphate (NADP) working with certain dehydrogenases in the metabolism of carbohydrate. The co-enzyme accepts hydrogen from the substrate and becomes reduced. It then passes the hydrogen to another hydrogen acceptor, so becoming oxidized and ready once more to co-operate with its dehydrogenase. Co-carboxylase (thiamine pyrophosphate) is known to act as co-enzyme to more than one apoenzyme. It acts as a co-enzyme to a carboxylase in the reaction pyruvic acid to acetaldehyde in the production of alcohol by yeast. Co-enzyme *A* is concerned in a number of reactions in which the transfer of acyl groups is involved. It contains an —SH (thiol) group and it is here that the acyl group is accepted. Glutathione acts as co-enzyme to glyoxalase which catalyses the conversion of methyl glyoxal to lactic acid. Co-enzyme R, otherwise biotin or vitamin H, assists in the conversion of pyruvic acid and carbon dioxide into oxaloacetic acid by yeast. No prosthetic groups have ever been associated with the proteolytic enzymes.

Activators

A substance which increases the activity of a holoenzyme is referred to as an activator. The absence of the activator may merely retard the action of the catalyst or it may completely prevent it. Activators are frequently inorganic ions, although some organic substances may act as such. Activating ions include Ca^{++} for thrombase, Cl^- for ptyalin and Mg^{++} for phosphatase of blood plasma. The kinases, enterokinase and thrombokinase, are enzymes which themselves function as activators.

It is not always clear what part the activator plays. Sometimes it simply removes the effect of an inhibitor, in other cases it may take some part in the reaction to be catalysed. In yet other instances, no clear function can be traced.

Inhibitors and Anti-enzymes

Inhibitors may be considered as substances which prevent the normal action of enzymes without actually destroying them. Salts of mercury, silver and gold are notable examples. Cyanides inhibit the activity of oxidases. Toluene is used as one of the most convenient enzyme preservatives because it has no effect on them at all. A substance such as formaldehyde, which is highly reactive, will not merely inhibit enzymes but destroy them.

An anti-enzyme is a substance, formed in the serum of an animal as a result of repeated injections of an enzyme into its blood, which prevents the normal action of the enzyme. Anti-urease, anti-pepsin, anti-glyoxalase and anti-trypsin are all known to exist. It is possible that the alimentary epithelium is not digested by the gut secretions because the membrane contains suitable anti-enzymes.

PROPERTIES OF ENZYMES

Enzymes possess the general properties of catalysts. Some of these properties need modification in their application to the enzymes, and in addition, there are certain special properties of enzymes which do not ordinarily apply to inorganic catalysts.

Specificity

Each enzyme either attacks a single specific substance or a particular atomic arrangement. For instance, lactic dehydrogenase will only catalyse the change of lactic acid to pyruvic acid, and cytochrome oxidase will cause oxidation of reduced cytochrome alone. On the other hand, some proteases will cause hydrolysis of any substance having the —CO—NH— linkage, while other proteases will only attack this linkage if there are certain special groups on either side of it. This specificity implies very exact organization of enzyme molecules, and of the charge distribution on their surfaces. Also, since the majority of enzymes play parts in complex chains of reactions it follows that there must be meticulous arrangement in space, so that either the substrate molecules are moved on from enzyme to enzyme, or the enzyme molecules themselves move from substrate to substrate.

Reversibility

Though the majority of enzymes catalyse a particular reaction in one direction, they can also catalyse the reverse reaction. The conditions which determine which reaction shall proceed, are principally the pH of the medium and the concentrations of the reactants. Some of the major enzyme systems, for example the glycolysis system, may be working in one part of an organism to catalyse anabolism, and at the same time, in another part of the same organism to catalyse catabolism. Indeed, both synthesis and breakdown of the same product, may take place in the same cell simultaneously.

Effect of the Amount of Enzyme Present

Since one enzyme molecule can deal only with one substrate molecule at one particular instant of time, it follows that the rate of conversion of a substrate will depend upon the amount of enzyme present, provided other conditions are favourable. The conversion of 200 000 molecules of hydrogen peroxide per second by pure extracted catalase represents an average result of experiments with relatively large quantities. Such quantities of enzyme and substrate are not so immediately available in the living cells. It has been estimated that the average conversion of substrate in an organism is 100 molecules per second per enzyme molecule. The temperature, pH, and concentrations of enzyme, substrate and end-products, are probably rarely combined to permit optimum activity. Indeed, such a condition of optimum activity for one particular enzyme would be quite useless in the organization of the cell, since each enzyme is only part of a complex system.

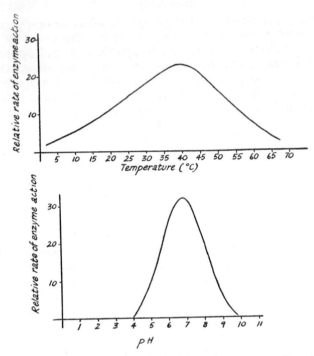

Fig. 5.1. Graphs representing the relative rates of pancreatic amylase action with variations in temperature and pH.

Effect of Temperature

The rate of chemical reactions is approximately doubled for every 10°C rise of temperature. This applies to enzyme-catalysed reactions at low temperatures, but above 40°C there is a rapid decrease in the rate of reaction owing to the denaturation of the enzyme protein. The optimum temperature for most enzymes is 35–40°C. All enzymes are destroyed at 100°C.

Effect of pH

Each enzyme-catalysed reaction proceeds most rapidly at a particular pH, and for most, the optimum is about pH 7. Some enzymes, however, require particular acid or alkaline conditions. The following table gives the optimum pH conditions for some enzymes.

Enzyme	Source	pH
Pepsin	Mammal stomach	1·5 to 2·5
Peroxidase	Plant roots	2·5 to 3·0
Zymase complex	Yeast	4·5 to 6·5
Amylase	Mammal pancreas	6·0 to 7·0
Trypsin	Mammal pancreas	8·0 to 9·0
Phosphatase	Mammal cartilage	9·0 to 9·2

THE NATURE OF ENZYMES AND THEIR MODE OF ACTION

Enzymes are soluble colloidal proteins produced only by living cells; they possess highly specific catalytic powers and are destroyed by moist heat at 100°C. All enzymes have high relative molecular masses; some examples are pepsin, 35 000, amylase (barley), 54 000, phosphorylase (muscle), 400 000, and urease (plants), 480 000. It is possible that all the protein in a cell either has, or is capable of, enzymic activity. Many enzymes need the presence of an attached non-protein or prosthetic group. Others require a co-enzyme, not permanently attached, and others still, require unattached activators.

Numerous enzymes have been extracted and isolated in pure crystalline form. These pure extracts are capable of a considerably greater degree of activity *in vitro* than they exhibit *in vivo*. It has been calculated that one molecule of the enzyme catalase can effect the decomposition of 200,000 molecules of hydrogen peroxide per second, at 0°C.

Production of Enzymes

Enzymes are produced by the protoplasm; like other proteins, their synthesis depends primarily on the DNA code (*see* p. 794). The sequences of bases in DNA determine the various RNA codes which are sent into the cytoplasm. On each RNA chain, the linear order of various triplets out of adenine, guanine, cytosine and uracil determines the kinds and sequences of amino-acids which will unite to form a protein. Most of the experimental work associating enzyme production with DNA has been carried out with fungi and bacteria. In experiments on mould fungi, mutations have been caused by X-rays. Organisms possessing these various mutations invariably showed some lack of synthetic power, indicating lack of a specific enzyme. The same type of result has been obtained with bacteria. The non-pigmented condition of albinism has been shown to be due to lack of the enzyme tyrosinase; this enzyme catalyses the formation of melanin from the amino-acid tyrosin; the absence of the normal gene means that there is no production of tyrosinase.

It may be mentioned here that all the cells of an organism do not produce all the enzymes of that organism. The process of mitosis indicates that all the somatic cells have an identical set of chromosomes. Nevertheless, it is only certain cells which produce certain enzymes. Pepsinogen is produced in mammals only by some cells of the gastric mucosa; carbonic anhydrase is found in red corpuscles, sparsely in brain, gastric mucosa and pancreas, but nowhere else; luciferase is restricted to special parts of fireflies and glow-worms; cholinesterase is distributed thinly throughout the muscle of an electric eel, but is present in large quantities in the electric organs. Many such examples could be cited for any species. It is now known that besides the chromosomal genes, cells possess self-replicating cytoplasmic determinants which certainly contain RNA, and in some cases, DNA also. It is possible that these determinants act partly together with and partly independent of the nuclear genes in producing the enzyme complex of any particular cell. Although the chromosomal genes are identical in every cell of one body, the cytoplasmic determinants may be unequally shared at a somatic cell division (see p. 786).

At present, the broad general picture with regard to enzyme production is that the gene-complex determines the enzyme pattern; the specificity of the enzymes determines both structure and function of the organism. Environmental factors undoubtedly play a great part since almost always the substrates for enzyme action, and often prosthetic groups, co-enzymes, activators and inhibitors, will be partly or wholly derived from the environment.

The Site of Enzyme Action

The majority of enzymes act within the cell, catalysing the multifarious changes which together constitute metabolism. In some organisms, there is extracellular digestion, the enzymes responsible being produced in inactive form and secreted either directly on the substrate, or into a lumen through which the substrate has to pass. Most of these extracellular enzymes need activators, which may themselves be enzymes. Thus, trypsinogen from the pancreas is inactive in the gut without the presence of enterokinase from the intestinal juice. In other cases activation is merely due to a suitable pH.

The sites of action of some of the intracellular enzymes have been determined. The enzymes which break down glycogen to pyruvic acid appear to be generally distributed throughout the cytoplasm, whereas those necessary to degrade the pyruvic acid to carbon dioxide and water, are located on the mitochondria. The demonstration of Krebs' cycle, requires not only the substrates and the specific enzymes, but also the presence of granules from the cytoplasm.

Mode of Action of Enzymes

In the economy of the cell, every substance that is metabolized passes through a sequence of chemical reactions, each involving relatively small alteration in chemical structure, each stage being catalysed by a specific enzyme. This procedure by short, easy steps avoids large energy exchanges and the inevitable dissipation of much energy as heat.

The absorption spectrum of an enzyme changes when the enzyme is active. This is taken as evidence that there is actual, though temporary, combination of the substrate molecules with the enzyme. The change then takes place in the substrate molecule; it is discarded by the enzyme, and a fresh substrate molecule is attacked.

It is often suggested that the specificity of enzymes may be explained by the lock and key analogy. The substrate is the lock, and one particular enzyme is the only key which fits it exactly and is therefore able to unlock it. There is no doubt that the actual physical configuration of enzyme and substrate is the deciding factor.

DEMONSTRATION OF ENZYME ACTION

Many enzymes can be obtained in the dry refined state: it is preferable to use these with reasonably pure substrates to demonstrate any action clearly. For most rapid action, experiments should take place at a temperature of 30–35°C, in an incubating oven. Note that with any such experiment, a control using boiled enzyme should be set up.

Hydrolysis of a Fat by Lipase

Shake up a little castor oil with 50 per cent alcohol and test it with litmus paper. The reaction should be neutral. Add a little lipase and shake vigorously. After some hours test again with litmus paper. The mixture shows an acid reaction due to the formation of fatty acid. Glycerine can be shown to be present by the addition of a little copper sulphate and a few drops of potassium hydroxide, when a blue colour is developed without a precipitate.

Hydrolysis of Sucrose by Invertase (Sucrase)

To a small quantity of sucrose solution, add one drop of invertase concentrate. After a few minutes, test for reducing sugars by Fehling's method. Their presence indicates the hydrolysis of the sucrose. With a polarimeter (saccharimeter), the activity of the enzyme can be shown by the rotation of the plane of polarization of light as the hydrolysis proceeds. Sucrose is strongly dextro-rotatory; as fructose is formed, its laevo-rotatory activity will cause the plane to be rotated less and less to the right. By the end of the hydrolysis, the glucose-fructose mixture will show slight laevo-rotation (*see* p. 76). The hydrolysis of sucrose can also be carried out by boiling with dilute hydrochloric acid.

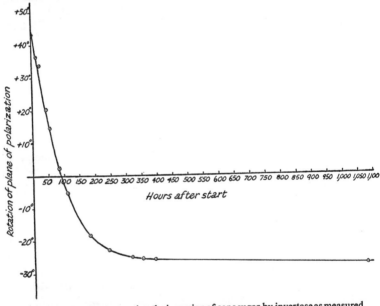

Fig. 5.2. Graph representing the inversion of cane sugar by invertase as measured by the rotation of the plane of polarization of light transmitted through a solution containing 20 per cent sugar and one drop of invertase concentrate at least 5 years old (temperature, 18°C).

Hydrolysis of Starch by Amylase

This can be quickly demonstrated, using a starch solution by the method employed above, but the demonstration is much more instructive if the following procedure is adopted.

Make up a 1 per cent starch solution by pouring boiling water on a cold starch paste. Allow to cool. Set out about a dozen watch glasses on a white background and in each place two drops of dilute iodine in potassium iodide solution. Take some of the starch solution in one test-tube and a little amylase shaken up in water in another. Into the first watch glass run ten drops of starch solution from a pipette and note the deep blue colour produced by the iodine already there. At a measured zero time add the enzyme to the starch solution and shake vigorously. At measured intervals of time from zero, add to each watchglass in turn, five drops of the starch and enzyme mixture. Suitable time intervals would be 15 s, 30 s, 45 s, 1 min, 1½ min, 2 min, 3 min, 5 min, 7 min and 10 min. As each watchglass is treated, note the colour reaction with iodine. As the enzyme hydrolyses the starch through dextrins to maltose, the blue becomes more purple, gradually getting fainter until no colour reaction is given at all. A test of the mixture left in the tube will show reducing sugars present. If the enzyme ptyalin is used directly from saliva, the hydrolysis is usually quite rapid and the times for transfer of the mixture may need to be changed. Dilution of the saliva with about its own volume of distilled water will slow the enzyme action.

Hydrolysis of Protein by Proteinases

Make up a suspension of coagulated egg albumin as follows. Stir egg-white in four times its own volume of water until there is an even consistency throughout. Heat in a water bath at about 60°C for 15–20 min to coagulate the albumin. Strain through muslin to remove any lumps. The strained liquid will be cloudy white. Dilute this in about ten times its own volume of water before use. To a quantity of the albumin suspension add a small quantity of 1 per cent pepsin powder in water. To the mixture must be added hydrochloric acid in sufficient quantity so that the pH is adjusted to allow the enzyme to work. Equal volumes of the albumin suspension and 0·4 per cent hydrochloric acid will acidify the mixture for this purpose. Incubate the tube for an hour or two and examine. The cloudiness will have disappeared due to the hydrolysis of the protein to polypeptides.

Demonstration of the Properties of Enzymes

Any of the preceding experiments can be adapted to illustrate one or more of the enzyme properties.

With each one, controls using boiled enzymes should be used. In each case, the enzyme is destroyed and its characteristic activity is not apparent.

The effect of temperature on the rate of reaction can be demonstrated roughly by employing controls at 0°C (refrigerator or ice-pack) and at room temperature. If means of regulating temperature at all values between 0°C and 40°C by about 10°C steps are available, the hydrolysis of starch against time demonstration can be used to compare the rate of enzyme activity at different temperatures.

The same experiment can be adapted to illustrate the effect of quantity of enzyme. It can be repeated using graded dilutions of saliva, note being taken in each case of the earliest time at which no colour is given with iodine.

The effect of pH is readily discernible in the case of pepsin. If a control tube is set up in which the albumin has been made slightly alkaline with dilute sodium carbonate, no hydrolysis of the protein will occur. The pepsin can be compared with trypsin with respect to its optimum pH.

Specificity of enzymes can be demonstrated by a further set of controls in all the experiments in which carbohydrases, proteinases and lipase are compared in their action on all the substrates mentioned.

Activity of Oxidizing and Reducing Enzymes

By using methylene blue as a hydrogen acceptor, the action of dehydrogenases in yeast extract (zymin) can be accomplished. Place about 1 cm³ of zymin suspension in a test-tube and add about the same quantity of 15 per cent glucose solution. Fill the tube with a very dilute (pale blue) solution of methylene blue. Cork to exclude all air. Leave in a warm place and examine later. The mixture will gradually lose its colour owing to the reduction of methylene blue to its leuco (colourless) state as the glucose is oxidized by the activity of the zymin. If, when the solution is colourless, the tube is uncorked and shaken with air, the methylene blue will regain its colour as it becomes oxidized.

An oxidase enzyme system can be demonstrated in many plant tissues by the use of freshly made guaiacum solution in alcohol. The guaiacum turns blue when dropped on the cut surfaces of the tissues. This is due to the oxidation of guaiaconic acid to guaiacum blue by direct oxidizing enzymes in the tissues.

Peroxidase can be shown to be present if the guaiacum turns blue only after hydrogen peroxide is added. This is clearly demonstrated by horseradish tissues.

Catalase in tissues can be demonstrated by its ability to form gaseous oxygen from hydrogen peroxide. Add to a test-tube nearly full of 20 vol. hydrogen peroxide some dried ground potato tissue. Quickly invert the mouth of the test-tube under water to trap the gas evolved. Test with a glowing splinter.

Activity of Miscellaneous Enzymes

The action of rennin can be demonstrated by adding a little to some fresh milk when the soluble caseinogen is converted to the insoluble calcium salt of casein which is then precipitated. If the milk is treated with sodium citrate or sodium oxalate before the rennin is added, the clotting does not occur, showing that the enzyme works only in the presence of calcium ions.

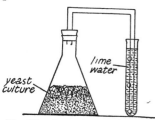

Fig. 5.3. Apparatus to show the evolution of carbon dioxide during sugar fermentation by yeast.

Urease action on urea can be detected by the evolution of ammonia and carbon dioxide when the two are brought into contact. Soya bean flour contains the enzyme and can conveniently be used.

To demonstrate the activity of the zymase enzyme system, it is convenient to ferment sugar with either brewer's or baker's yeast (*Saccharomyces cerevisiae*). The apparatus shown in Fig. 5.3 will suit the purpose. In the flask place about 300 cm³ of 15 per cent sucrose solution to which has been added 0·6 g potassium phosphate, 0·06 g calcium phosphate, 0·06 g magnesium sulphate and 3·0 g ammonium tartrate. This constitutes Pasteur's yeast culture solution but the additions to the sugar are not absolutely necessary for short working periods by the yeast. Add the yeast and shake up. Place the whole apparatus in a warm place and observe periodically. The contents of the flask will appear frothy in a few hours and will remain so until all the sugar is fermented. The gas evolved can be shown to be carbon dioxide by its action on the lime water. The culture solution can be shown to contain ethyl alcohol both by its smell and by the iodoform test. This last is performed by taking some of the clear liquid in the flask after all frothing has ceased, heating it and adding slowly to it a little iodine in potassium hydroxide solution until a yellow crystalline precipitate appears. This is iodoform and has the characteristic "hospital" smell. Under the microscope the crystals appear as yellow, hexagonal plates.

Further experiments involving the use of enzymes in animal digestive secretions are given in Chap. 10.

CHAPTER 6

THE NATURE OF THE ENVIRONMENTS OF LIVING THINGS: ADAPTATION: HOMEOSTASIS AND ITS MAINTENANCE: ENDOGENOUS RHYTHMS

THE environment of an organism is the aggregate of all the conditions, physical, chemical and biotic, under which it exists. We speak of an external environment when reference is made to the conditions imposed by the physical and chemical nature of the surroundings, such as temperature, light, presence or absence of particular substances and the presence or absence of other living things. We may also refer to an internal environment when we mean the conditions created inside the organism by its own metabolism, reactions to external changes and other activities. In the case of the former, all the factors are variable to a greater or lesser extent and over these external conditions organisms, apart from man, have little or no control; all are faced with the problem of adjustment to ceaseless fluctuations. The success of a particular organism, measured in terms of its numbers and the wideness of their dispersal through the biosphere, is very much dependent upon its ability to make the appropriate adjustments quickly enough. In the case of the internal environment, each kind of living thing has developed a pattern of conditions peculiar to itself, fitted to the broad nature of the external environment but generally subject to at least some measure of control. Hence, within the organism, there tends to be a state of *homeostasis* or constancy of conditions, maintained by specific regulatory processes. Good examples of such processes are osmo-regulation and the maintenance of constant body temperature by some animals.

The homeostatic animal can be likened to a self-regulating machine, that is, it constitutes a *cybernetic system. Cybernetics* is the study of control under guidance of internal information. The most important feature of a cybernetic system is that it can detect changes in its working conditions and formulate information that will lead to corrective responses when the changes threaten to interfere with the proper working of the mechanism.

All activities of living things are the expressions of literally thousands of processes being carried out at the same time within the cells. All these processes can compete and conflict with one another and yet in the normal organism they are arranged in such a way that they work together to produce a harmoniously working system. The achievement of this is based on what is called *feed-back control*, as distinct from control exercised through the rigid following of a predetermined sequence of events, each of which sets the nature and extent of the one following it. The former allows for the correction of errors, the latter does not. A system operating under feed-back control is thus more flexible and less liable to damage. A good example of a system controlled by rigid means is the driving of the wheels of a car by the engine. The piston moves a given distance and its rod turns the crankshaft which then turns the transmission shaft. An error in the movement of any part cannot be corrected for or smoothed out anywhere in the whole system and the final product, if there is one, must include the effects of all the errors in the system. An example of a feed-back control mechanism which represses a reaction by the product of the reaction is the bimetallic strip thermostat, the action of the strip being to make or break an electric circuit as its temperature changes, thus allowing more or less heat to be generated in the circuit according to its own temperature. The temperature of the surroundings of the strip is thus controlled in an *error-actuated way*, that is, an error in the temperature of the strip above or below its set point actuates the mechanism to correct it. In this case, the feed-back is called *negative feed-back* or *corrective feed-back*, since an error towards a condition away from the set point tends to diminish the effect causing it. The opposite kind of feed-back in which an error in one direction actuates the mechanism to increase the effect causing it is called *positive feed-back* or *runaway feed-back*. Such a system obviously has no stability and will tend ultimately to destroy itself.

Homeostatic systems are cybernetic systems operated by negative feed-back. The idea of the living organism being such a system can be traced back to Claude Bernard (1813–1878) who wrote of the chemical constancy of the body fluids of animals and how the constancy of this internal environment was essential to their survival as living organisms.

Every organism, during its life-time, inhabits more than one external type of environment. All began life in comparatively watery surroundings; even the most thoroughgoing terrestrial animals and plants have not been able to free themselves from this. After this aquatic phase, there are fixed or free-living stages in water or on the land. For many creatures, the likelihood of reaching favourable surroundings depends on mere chance. Thus, the spores of bacteria and fungi, the seeds of flowering plants, and the eggs of many marine animals, are produced in enormous numbers; these very numbers reduce the odds against survival. By this means and many others, all organisms have evolved ways of enhancing their offsprings' chances of success. Eggs, seeds and spores are produced at the most favourable season; dispersal devices ensure scattering; the butterfly lays its eggs on the food of the larva; the bees and other social insects nurture their young carefully; the female dogfish hides her egg-cases among sea-weeds; the mammal incubates her young, and nourishes them after birth. For organisms with the power of locomotion, there is some degree of choice of conditions. The herrings follow the plankton; herbivores seek less-grazed pastures; during the Antarctic winter, the whales leave the edge of the pack-ice for the shores of the southern continents; the little chlamydomonas swims to zones with suitable sunlight. Every mobile creature seeks the most suitable conditions.

After it has exhausted the food material bestowed on it by the parent, the organism is dependent on its surroundings for all its needs. Apart from material necessities such as nutrients, oxygen and water, it must obtain energy. Then there is a limited temperature range for every organism, outside which it cannot exist for its normal life period. The physical nature of the surroundings provides every organism with a certain measure of support and, if it is motile, there is always some degree of resistance which renders locomotion possible. Aquatic organisms are very sensitive to pH changes and many small organisms perish from slight changes in the acidity or alkalinity of the water.

To survive successfully and thus to reproduce, every creature has to "fit in" with its external environment; if it fails to do this, then it will cease to exist. This "fitting-in" is known as *adaptation*. Careful consideration of any animal or plant will reveal many adaptations which are strikingly "fitting." Such consideration can easily lead to the conclusion that the environment makes some impression on the organism, moulding it to fit in better with its surroundings. This was once the basis of a theory advanced by Lamarck (*see* Chap. 21).

But rarely will surroundings remain absolutely constant in all characteristics so that no matter how well adapted an organism may be to the broad conditions it may still need to make small adjustments to its own internal state in order to counteract the effects of the external fluctuations. For survival it must, to some extent, make itself independent of these minor changes or nullify their effects. When an organism achieves such independence by making adjustments to its internal condition, we say it is capable of *regulating* itself. Regulation is much more a property of animals than it is of plants.

SPATIAL DISTRIBUTION OF LIVING ORGANISMS

Life, as we know it, exists on this planet in the presence of a suitable temperature range, light, free oxygen and carbon dioxide, water, various inorganic ions, and pressure above a certain minimum.

Geographers divide the Earth into three zones; the atmosphere, the lithosphere and the hydrosphere. The atmosphere is not the normal or permanent habitat of any living organism, though a great many creatures use it for passage. Birds, insects, seeds and spores of various kinds may spend comparatively long periods in the air, but they cannot live there permanently. The lithosphere, which is the soil and rocky crust of the Earth, is thickly populated, but not to any considerable depth. In the hydrosphere, life is abundant, particularly in the seas.

To the best of our knowledge, life exists in a comparatively thin layer of atmosphere, hydrosphere and lithosphere. The thickness of this layer compared with the volume of the Earth, is considerably less than that of the skin to the apple.

Types of External Environment

In the small zone inhabited by living creatures, there is an infinite variety of different types of environment. Broadly, we may distinguish three major divisions; the *sea, fresh-water* and the *land*. Each is populated by a host of beings which have their own particular and smaller environments. Then, there are all degrees of gradation between these three major divisions. There is teeming life on the shore and in the intertidal zone, where sea and land meet and merge; there is a characteristic population of brackish waters where fresh and salt water mingle. There are special associations where land and fresh-water meet, as on the sides of lakes and banks of rivers; there is a specialized fauna of the abyssal ocean depths.

A host of creatures have become adapted for life in an *organic environment*. Saprophytes have their habitat in non-living organic material. Another large assemblage, the parasites, though unwelcome guests, make their homes in or on the bodies of other creatures.

THE SEA

The sea is the home of a plant and animal population of extreme diversity and a wide range of complexity. It is admirably suited to the development and maintenance of life; indeed, many biologists believe that life originated in the shallow coastal waters. The oceans and seas cover an area of about 363×10^6 km^2, which is approximately two-thirds of the earth's surface. The total volume of water is fifteen times greater than the total volume of the land above sea-level. The average depth of the water is estimated to be about 3840 m which is five times the average height of the land above sea-level. The greatest depth known at present is over 10 000 m in the Philippine trench. Depths below 6000 metres are termed abyssal; they constitute 1 per cent of the total area.

Physical Characteristics of the Seas and Their Biological Importance

The important physical characteristics of the seas as far as living organisms are concerned are buoyancy, pressure, movements of the water, the extent of light penetration, and osmotic pressure.

Buoyancy

The buoyancy of sea-water depends upon its relative density (*see* p. 808), which will vary with salinity and depth. The relative density at the surface is 1·0222 with a salinity of 34·32 parts per 1000; with a salinity of 34·6, the R.D. is 1·0277. At a depth of 10 000 m, with salinity 34·6, the R.D. is 1·0722. Nowhere in the open sea is there any considerable difference in buoyancy and because of its high value, marine organisms need little in the way of supporting structures; most of their weight is taken by the water. Thus creatures of exquisite fragility and delicate shape are possible. Locomotion also is simplified since the water offers the requisite resistance and the pull of gravity is considerably lessened. Vertical rise and fall in the water may be brought about by slight changes in the relative density of a body and often these changes are effected by secretion or withdrawal of gases. In many marine organisms there are precise adjustments to particular depths.

Pressure

The atmospheric pressure at sea-level is approximately 101 325 Nm^{-2}; there is an increase in pressure of this amount for every 10 m of descent in sea-water. Thus at 5500 m, there is the enormous pressure of 58 $MN\ m^{-2}$. This, however, has no effect on creatures living at these depths, since the pressure acts equally in all directions and is the same within the creature as it is outside. Nevertheless, the rapid changes of pressure with depth debar rapid ascent or descent. Thus the majority of marine organisms are restricted to a comparatively narrow vertical range.

The Movements of Sea-water

The waters of the sea are never still but are subject to constant and thorough mixing. If the surface water is moved away, more water wells up from beneath. There is horizontal movement caused by the wind, and vertical movement due to convection currents. Besides these, there are certain massive movements of huge quantities of water.

In the polar regions, the colder denser water on the surface sinks and flows slowly along the ocean floor to the equator to replace the warmer and less dense surface water which tends to flow north and south. Because of the earth's rotation on its axis, water or air moving towards the equator from the polar regions is deflected to the west. The NE. and SE. trade winds enhance this westward trend of the water, and hence we have the westward flowing N. and S. equatorial currents. The presence of the land masses diverts these currents, so that in the Atlantic ocean we have the northward-flowing Gulf stream, and the southward-flowing Brazil current. Similarly, in the Pacific, there are the northerly Kuro-Siwo current, and the southerly East Australia current. This by no means exhausts the tale of mass movements of water, but it will suffice to show that there is constant and considerable interchange.

Tides are due to the gravitational attraction of other bodies in space. The effects of all except the sun and moon are too small to be noticed, and owing to its comparative proximity, the moon has by far the greater effect. The hemisphere facing the moon will be more strongly attracted and hence at various places, the tide will be somewhere between half and full. The other hemisphere, more remote from the moon, will have tides between half and low. Once a fortnight, the earth, sun and moon will be in line, and the combined attraction of the sun and moon will cause the high spring tides. Also, once a fortnight, the earth, sun and moon will be at right angles, thus causing the low neap tides. The tides cause considerable movement and mixing of the water, especially in coastal areas.

Waves are caused by the action of the wind on the surface of the water. When the wind subsides, the waves continue as swells, which may travel for enormous distances on the open ocean. Careful measurement has shown that waves may reach a height of fifty feet during violent storms.

There are two main biological effects of these varied and incessant movements of the water. In the first place, the mixing tends to equalize temperature and salinity, and thus most marine species have a wide geographical range. Secondly, the water moves the drifting population, or plankton. Thus animals which feed on the plankton lead a migratory existence. Accurate observation of plankton movements is of immense importance to the fishing industry. The tides can cause considerable havoc to life in the shallow coastal waters. These neritic species have become strongly adapted to this environment, and nowhere do we find more striking adaptations than in intertidal organisms.

Temperature

The high specific heat capacity (*see* p. 1005) of sea-water (about $3.9 \text{ kJ kg}^{-1} \text{ K}^{-1}$), its bulk, and the continual mixing, ensure that the temperature is far more equable than it is on land. Nowhere in the surface waters is there an annual range of more than 10°C. In the colder seas, the daily variation is about 0.1°C, and in the warmer seas 0.4°C.

6

The highest sea temperature is over 30°C in the Red Sea; the lowest is — 1·5°C in the Antarctic. For comparison, the highest land temperatures are found in Central Asia where they may exceed 50°C, and the lowest in Antarctica with — 40°C. At least 80 per cent of the sea-water has a temperature less than 5°C. Since the maximum density of the water occurs at about 4°C then it is obvious that there will be comparatively little vertical variation in temperature. Except in the Polar regions, however, there is a slight decrease with depth.

Thus there are vast areas of water over which equable temperatures are found. This is reflected in the equally vast geographical range of most marine organisms. In addition, extremely fragile creatures can exist without the need for elaborate structures to protect them from wide changes in temperature.

Light Penetration

Photosynthesis is of fundamental importance to life in the sea, as well as to life on the land. The diatoms of the plankton form the base of the various food chains, as do the green plants on land. Owing to differential penetration of sunlight wavelengths (*see* p. 1008), the red and yellow are wholly absorbed at 50 m depth while most of the blue and violet rays penetrate deeper. At 180 m, there is insufficient light for plant life. Beyond this, darkness quickly ensues and prevails to the ocean bed. The diatoms flourish in the surface waters down to a depth of 80 m.

This restriction of plant life to a comparatively narrow surface layer makes it necessary for herbivorous organisms to be somewhat confined in their vertical range. Carnivores which feed on herbivores will be similarly restricted.

Osmotic Pressure

The accompanying table shows that osmotic pressure (*see* p. 47) is correlated with salinity.

Salinity: parts per 1000 . . .	10	20	30	40
Osmotic pressure in atmospheres . .	6·4	13·0	19·7	26·6

Since the salinity in the open sea varies only between 34 and 37, there is comparatively little difference encountered in external osmotic pressure except in particular cases. Migrant fish like the salmon and eel, on moving into fresh-water, become 10 to 15 per cent heavier, gaining water by endosmosis. On the other hand, a frog placed in sea-water soon becomes 20 per cent lighter. In coastal and estuarine waters the salinity is subject to changes and creatures in those areas have evolved special regulating mechanisms to meet the changing osmotic pressure. In the open sea, many organisms have evolved an internal solution which is almost always isotonic with the sea-water, and hence there is normally no difference between internal and external osmotic pressures.

The Chemical Characteristics of the Sea-water and Their Biological Importance

Chemically, the sea may be regarded as a nutrient medium. It contains ample carbon dioxide and the mineral salts necessary for food synthesis in chlorophyll-containing organisms. Plant life flourishes and consequently animal life is abundant. Oxygen, necessary for respiration, is present to saturation, and, what is perhaps a most vital point, the pH is remarkably constant.

Salinity

Sea-water is a dilute solution of all the elements which occur in the earth's crust, In such a dilute solution, practically all the salts are highly ionized (*see* p. 963). The following table shows the principal ions, expressed as percentages of the total ions—

Cations		Anions	
Na^+	30·4	Cl^-	55·2
Mg^{++}	3·7	$SO_4^=$	7·7
Ca^{++}	1 2	Br^-	0·19
K^+	1·1	$CO_3^=$	
Sr^{++}	0·04	HCO_3^-	0·35

Though the total salinity shows slight variations, an important characteristic of sea-water is that the proportions of the various ions are constant. Hence for salinity determinations, only one ion need be assessed; the rest will be in direct proportion. The chloride ion is the one usually determined, by precipitation with standardized silver nitrate solution. The residue left after evaporating sea-water gives the following salts in parts per 1000 of the water.

NaCl	27·21	K_2SO_4	0·86
$MgCl_2$	3·81	$CaCO_3$	1·02
$MgSO_4$	1·66	$MgBr_2$	0·08
$CaSO_4$	1·26		

The total of dissolved salts averages 35 parts per 1000, and varies only between 34 and 37 except in coastal and estuarine waters.

It will be noticed that nitrate and phosphate ions are in short supply; this is because these ions are vitally important in living bodies. Most of the nitrate and phosphate of the sea are present in the living bodies, and only seasonally is there any appreciable quantity present in the water. Magnesium, also very important to green plants, is abundant. Two salts are important for the secretion of skeletal materials; they are, for the animals, calcium carbonate, and for the green diatoms, silicates.

Dissolved Gases

The average gaseous content of one litre of surface water at 10°C with salinity 35 parts per 1000 is given in the table below—,

Total dissolved gas 18·7 cm³	N_2 12 cm³	O_2 6·4 cm³	CO_2 0·3 cm³

The oxygen is 34 per cent of the total dissolved gas, and the carbon dioxide 1·6 per cent, an interesting comparison with the atmosphere in which the oxygen percentage is 21 and carbon dioxide 0·03. The sea-water is generally saturated with oxygen; in the colder seas and the deep water there are 7 to 8 cm³ per dm³. The lowest oxygen content is 2 to 3 cm³ per dm³ in the tropical waters.

Hydrogen Ion Concentration

The hydrogen ion concentration of the sea-water averages 10^{-8} mol dm^{-3}, often stated more simply as pH 8 (*see* Chap. 3). Thus it is somewhat alkaline. Being an excellent buffer solution, it is resistant to large pH changes. Aquatic organisms are very sensitive to pH changes and are rapidly killed if the water becomes too acid or too alkaline. Large changes of pH never occur under natural conditions in the open sea.

Life in the Seas

The immensity of space and the dense and practically homogeneous medium are reflected in the wide geographical range of most marine species. The solution is a nutritive fluid, especially for the plants, which form the base of the food pyramid. The salts in the water are practically of the same kinds and concentrations as those in the body fluids of marine creatures, hence they will have the same osmotic pressure as the sea-water and there will be little or no expenditure of energy to keep the body fluids at constant concentration. The transparency permits of the life of plants to a considerable depth; for some distance below this zone, there will be a steady rain of slowly-descending dead bodies which provide food for scavenging organisms. These dead remains will, if not consumed, be fully decayed long before the abyssal depths are reached. In these remote dark regions only scavenging and carnivorous life is possible, and the strange inhabitants must eat each other or dead remains.

The two essential gases, oxygen and carbon dioxide, are in plentiful supply. The pH is constant, and buffered against violent change.

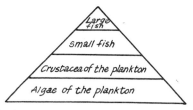

Fig. 6.1. Food chain (and pyramid of numbers) for life in the seas.

Over enormous expanses and to enormous depths, there is little variation of temperature.

These are the conditions, almost ideal, under which the creatures of the sea exist. For the purpose of this brief outline, the multifarious forms of marine life may be divided into three groups. These are, the *plankton* (Greek, wandering), the *nekton* (Greek, swimming), and the *benthos* (Greek, deep).

The Plankton

This is a very diverse assemblage of organisms, none of which are fixed to any substratum. Having little or no power of locomotion, they drift helplessly in incredible numbers. By far the most important organisms are the diatoms, which are microscopic plants with beautiful siliceous skeletons. They flourish down to 80 m and then diminish down to 180 m, below which the light is insufficient. The diatoms and other minute green plants form the marine pastures, which are assiduously grazed by a host of herbivorous animals; these in turn, fall prey to the carnivores.

The plankton animals range from huge jelly-fishes to minute protozoans. Of all the animals, probably the crustaceans predominate; copepods form the major part of the food of such fishes as the herring and mackerel; euphausiids feed the whalebone whales. Annelid worms, and the arrow-worm, *Sagitta*, are plentiful in the plankton.

Of the coelenterates, the large and small jelly-fish and the sea-gooseberries (Ctenophora) are best represented. Not many species of molluscs are plankton animals, but in some zones, the peculiar pteropods or sea-butterflies, are very plentiful. In warmer waters the sea-squirts or tunicates are common. Transitory members of the plankton are the larvae, of which enormous numbers are found drifting in the surface waters at certain seasons. Few survive to metamorphose into adult forms.

There is a distinct periodicity in the plankton population especially in the cooler waters. In spring, there is a great outburst of the minute green plants, then in early summer, hordes of larvae hatch from their eggs. By late summer, there is a decline in the plant life owing to the exhaustion of salts and of silica for their skeletons and the grazing by the animals. In early autumn, there is another flourish of diatoms with about 75 per cent of the spring abundance. Many of the larvae then settle down; later in autumn, the light becomes too weak and the winter paucity of numbers sets in.

The hazards of this drifting life may be gauged by the fecundity of some of the species with plankton larvae. The American oyster produces 1 800 000 eggs at one spawning, and the cod, 5 000 000. The sea-slug, *Doris*, lays 600 000 eggs in one spiral of jelly; the sea-hare, *Tethys*, crowns all their efforts by producing 500 000 000 eggs per annum.

The Nekton

This consists entirely of animals which swim powerfully enough to render them independent of tides and currents. Fishes predominate. Those which live on or near the sea-bed are known as *demersal* fish. They include round-fish, such as cod, haddock, hake and dogfish, and flat-fish which lie on the bottom. The best-known flat-fish are plaice, sole, turbot, brill, skate and ray. Apart from the dogfish, skate and ray, these demersal fish lay huge numbers of eggs which float to the surface and the fry add to the spring plankton. The adults remain on the sea-bed feeding on a variety of worms, molluscs and crustacea of the benthos. The *pelagic* fish swim near the surface, often congregating in shoals such as those of herring and mackerel, and feeding on the plankton. The eggs lie on the bottom; the fry rise slowly to the surface and join the plankton.

Certain mammals, secondarily marine, are important members of the nekton. The Cetacea include the whales, porpoises and dolphins. Of the carnivores, the seals and walruses are marine. In the phylum Mollusca, the cephalopods are marine; they include the octopus, cuttle-fish, squid, the argonaut and the pearly nautilus. The present

Reptilia are poorly represented; there are only the turtles and sea-snakes. None of these larger animals exist in excessively large numbers, but they make enormous depredations on the other inhabitants.

The Benthos

The benthos consists of an intensely rich and very diverse flora and fauna, especially characteristic of the *neritic* zone, which extends from the shore to the 180 m line. Beyond this, the population thins out rapidly, mainly because little organic detritus reaches the sea-bed. The creatures of the benthos creep upon, burrow into, or are fixed to, the bottom. Apart from the tidal areas fringing the coasts, the neritic zone is relatively calm and still, light penetrates the entire depth, and the water is enriched by salts from the land. There is, however, little fluctuation in salinity, density or temperature except in the regions near to the coasts.

All the great invertebrate phyla are well-represented. To deal with the fixed forms first; of the coelenterates there are the sea-anemones, the sea-fans, dead-man's fingers, and corals: sponges (Porifera) are particularly common, and vast beds of the sea-mats (Polyzoa) cover large areas. Lamp-shells (Brachiopoda) and sea-squirts (Tunicata) are plentiful. Of the echinoderms, there are countless, graceful sea-lilies. Plants are relatively sparse, except in the great coastal seaweed zones where they are very abundant. There are plentiful calcareous algae, especially on coral reefs. Most of these fixed forms require a firm substratum.

Burrowing forms thrive in the soft mud. Bristle-worms and tube-living worms exist in profusion; molluscs are plentiful; crustaceans and sea-cucumbers (Echinodermata) are common.

Moving about among the weird forests are the crabs and lobsters, the starfishes, sea-urchins and brittle-stars. The great majority of all benthos organisms are characterized by having plankton larvae in very large numbers. The vast majority perish; the survivors are widely dispersed.

The Abyssal Depths

Even in the greatest depths there is some life. The deepest net hauls, from about 3000 m, reveal a strange and grotesque sample of fish, crustaceans, sponges, sea-lilies, sea-squirts, sea-pens and segmented worms. Many of the creatures are of exquisite beauty and great fragility. Most of the benthos forms show adaptations for lifting their bodies clear of the deep and almost impalpable ooze. The sea-lilies and sea-pens are balanced on long, delicate stalks; the sea-urchins have excessively long spines; the crustacea, very long legs.

For such structures to exist, there must be almost profound calmness of the water. It is a cold, still and silent world, its total blackness illumined occasionally by the light emitted by some of the denizens. There is no plant life, but some organic detritus diminishing with depth; the inhabitants maintain an uneasy equilibrium by eating each other or scavenging.

FRESH-WATER

There is an infinite variety of types of fresh-water areas. In extent, there are all gradations from ephemeral puddles to Lake Superior (81 000 km²), from temporary trickles after rain to the Amazon river (6400 km). Consequent upon extent, geographic locality and depth, there is wide variation in salinity, pH, temperature fluctuation, and permanence. Because of these variations, it is not possible to describe a generalized fresh-water environment. Broadly, we may distinguish three types; they are, lakes, ponds and streams.

A lake has a surface area large enough to be considerably affected by winds; some lakes thus show notable tidal effects. In depth there are all stages from a few feet to abyssal depth. In the majority, however, light penetrates to the bottom and thus plant life is almost everywhere possible. Ponds are small bodies of water characterized by relative stillness; the areas are too small for wind to have any great effect. They are normally stagnant, well-supplied with vegetation and vary greatly in permanence. Streams have continuous flow of water in one direction. Their character changes considerably from source to outlet and they show gradual change in their flora and fauna. A typical sequence will be: spring, rill, brook, rivulet, river, estuary, sea.

In addition to these three main categories, there are many very specialized fresh-water areas such as hot springs, sulphur springs, bog pools, alkaline pools, and a variety of subterranean waters.

In spite of the heterogeneity of bodies of fresh-water, there are certain broad characteristics which distinguish them from the sea.

Physical Characteristics of Fresh-waters and Their Biological Importance

The same general characteristics affect organisms in any type of water. They are, buoyancy, pressure, movement, temperature, light penetration and osmotic pressure. In addition to these, there is, in fresh-water, another important property utilized by certain organisms, that of surface tension.

Buoyancy

Owing to the generally lower density of fresh-water compared with that of the sea, there is slightly less buoyancy, but the difference is extremely small. The weight of bodies is considerably reduced and there is easy locomotion as the pull of gravity is lessened. As in the sea, there is little need for heavy supporting structures.

Pressure

Again, this depends on the depth, and in any case, it has little effect on living creatures owing to the pervasive properties of the water. Pressure within and outside an organism is adjusted to equality except for sudden ascent and descent.

Movements of the Water

In ponds, there is no appreciable tidal movement, and mixing of the water is due only to convection currents, the kinetic movement of water molecules, and the movements of the living organisms. In large lakes there is considerable movement by tidal effects, largely limited to the surface layers. The greater depths tend to be very still. The constant movement in streams presents special problems to the inhabitants.

Temperature

Even though small bodies of water show considerable fluctuation of temperature, there is always less than occurs in the atmosphere. Even in a small pond, the temperature at the bottom, in winter, is 4°C. Long-continued freezing is necessary before even a shallow pond becomes solid ice. Thus though the surface may be coated with ice many organisms survive in the deeper warmer water. High temperature is a greater hazard than low temperature to the inhabitants of small ponds. Complete drying-up may be frequent; organisms survive by producing resistant resting-stages with an extraordinary capacity for longevity in this state. Larger bodies of water, like the sea, are subject to less fluctuation.

Light Penetration

As with the sea, there are differences of transparency with seasonal change, with the abundance of micro-organisms and with the amount of matter in suspension. There is, however, considerably greater variation than is found in the sea. In a clear lake, the penetration is often over 180 m, below which there is total darkness. In turbid rivers, light may penetrate only a few feet. In general there is less vertical scope for green plants in fresh-water.

Osmotic Pressure

The osmotic pressure of fresh-water is normally always less than that of the sea, and in some cases is practically negligible. This is reflected in the animals of fresh-water habitats, where osmoregulatory organs such as contractile vacuoles and large kidneys have been evolved.

Surface Tension

This property of water whereby its surface acts as an elastic skin is important to some fresh-water creatures. There are organisms such as the pond-skaters and spring-tails which glide over the surface without wetting themselves. The larvae of mosquitoes and gnats, of water-scorpions and water-boatmen, can remain suspended from this skin. Water-snails can frequently be seen travelling along the under-surface.

Chemical Characteristics of Fresh-water and Their Biological Importance

As in the sea, the density of animal population depends on the numbers of green plants. In turn, the green plants depend on the amount of carbon dioxide and essential salts in solution. In nature, it is the essential salt in minimal supply which is the limiting factor; in fresh-waters this is usually phosphate. All the aerobic organisms require adequate supplies of oxygen. Finally, any association of plants and animals in a particular habitat can tolerate only a narrow range of pH variation.

Salinity

In general, fresh-waters always show a considerably smaller salt content than the sea. A comparison of water from the North Atlantic

Dissolved Substance	Parts per 1000	
	North Atlantic	River Thames
NaCl	27·7	0·015
$MgCl_2$	3·4	–
$MgSO_4$	2·2	–
$CaSO_4$	1·4	0·045
KCl	0·7	0·01
$CaCO_3$	0·04	0·17
$MgCO_3$	–	0·02
SiO_2	0·015	0·009
K_2SO_4	–	0·002
Organic matter	–	0·04
Total	35·455	0·311

and from the River Thames above the tidal zone shows that salt content of the sea-water is more than one hundred times that of the river-water.

It will be noted that no nitrate or phosphate is quoted in either table. Both are present in such small quantities that they are usually estimated in parts per million. The spring outburst of plant life practically exhausts both; they are restored in autumn and winter by decay of plant and animal material. Thus for both salts there is a winter maximum. The silica necessary for the valves of diatoms will fall markedly in spring, rise in summer when crustacea feeding on the diatoms reject the valves, fall again with the autumn burst of diatoms, and rise to a maximum in the winter. Carbonate in solution diminishes rapidly in spring with the rise of photosynthetic activity and often insoluble $CaCO_3$ is deposited on submerged objects. In winter there is slow solution of CO_2 from the atmosphere and from decaying organisms; the water reaches its peak of acidity and more carbonate comes into solution as bicarbonate.

Dissolved Gases

In the surface layers, the dissolved oxygen and carbon dioxide will tend to be in equilibrium with the atmosphere and thus there is usually saturation in fairly pure water. At a little distance below the surface, and to a greater extent in deeper layers, the balance is considerably upset by living organisms and by the decay of organic matter. When green plants are plentiful, and temperature and light suitable, there is a rise in the oxygen content and decrease in the carbon dioxide. Often on sunny days, there is supersaturation, and bubbles of oxygen can be seen on submerged plants, and rising to the surface. In the colder months, with fewer green plants, the oxygen content is low, and the carbon dioxide rises to a maximum.

Hydrogen Ion Concentration

In conjunction with changes in salinity and gaseous content, there are marked changes in hydrogen-ion concentration. In general, the pH of natural fresh-waters varies with the amount of dissolved carbon dioxide. When green plants are most active in photosynthesis, the pH is high, and when they are least active, the pH is low. The type of rock substrate, and the amount of organic matter undergoing decay, sometimes have important effects. In waters overlying chalk or limestone, the pH averages about 7·8, but on sunny summer days, this may rise to pH 9. At the opposite extreme, in ponds overlying peat, the pH averages about 5·3, and may sometimes be as low as pH 4. Slightly alkaline waters, of pH 7·1 to 7·2 tend to support the greatest variety of organisms. Very few living creatures can tolerate a pH value below 4·7 or above 8·5.

Life in Fresh-waters

Even in a very large body of fresh-water, there is never the relative physical and chemical constancy which is found in the seas. Thus in fresh-water life, we find many widely-differing but characteristic communities. As with the oceans, we can find plankton, nekton and benthos populations. In this brief account only some general features of the flora and fauna will be indicated.

The most striking feature of fresh-water flora is the extent to which the Spermatophyta (seed-bearing plants) is represented. There is often a flourishing marginal flora in which the sedges, reeds and rushes are prominent. The true aquatics are rooted in mud or floating. Of the rooted types, some have their leaves and flowers above the water; examples are the water-lilies (*Nymphaea* and *Nuphar*), the pond-weeds (*Potamogeton*), and arrow-heads (*Sagittaria*). Totally submerged plants are represented by the Canadian pond-weed (*Elodea*), and the water-milfoil (*Myriophyllum*). There is a distinctive floating flora, of which the most common plants are the duckweed (*Lemna*), the frog-bit (*Hydrocharis*), the water-soldier (*Stratiotes*), and the insectivorous bladder-wort (*Utricularia*).

The green algae flourish in fresh-water. Large numbers of diatoms and desmids are often present, together with the flagellates such as *Euglena* and *Chlamydomonas*. There are several British species of the beautiful colonial *Volvox*. Most prevalent of all the green algae are the filamentous forms, among which *Spirogyra*, *Ulothrix*, *Oedogonium* and *Vaucheria* are common, and have been described in Vol. I. The peculiar stoneworts (Charales) are represented by about thirty British species. Fungi are represented in fresh-water, the best-known being the Saprolegniales, various members of which often form moulds on fishes. There are no British wholly aquatic ferns, and but a few mosses of the genus *Fontinalis*. There is a characteristic bacterial population which includes the peculiar sulphur and iron bacteria.

Except for the Echinodermata and Brachiopoda, which are exclusively marine, all the great animal phyla have fresh-water members. The Protozoa is well-represented, many species of Rhizopoda, Flagellata and Ciliophora being found. There are only two species of fresh-water sponge. Of the Coelenterata, there are several species of *Hydra*, and the uncommon *Microhydra ryderi* which has a medusoid as well as a hydroid form. The Polyzoa are poorly represented, with only a few species. Of the Annelida, the Polychaetae are exclusively marine, but there are several fresh-water species of Oligochaetae and Hirudinea (leeches).

The Arthropoda of fresh-waters constitute about 75 per cent of the total number of animal species. *Daphnia, Cyclops* and *Cypris* abound in some ponds, and fresh-water shrimps are common. *Argulus*, the fish-louse, is a well-known ectoparasite which causes serious loss in trout hatcheries. Of the larger Crustacea, the Decapoda, only the crayfish is represented. Perhaps the greatest difference between fresh-water and oceanic fauna is the large insect population. The insects are essentially terrestrial, none are marine, and certain of them are only secondarily aquatic. The larval stages of many develop in water; larvae of dragon-flies, may-flies, gnats, mosquitoes and caddis-flies are familiar examples. A number of adult insects have also become adapted to life in the water; the water-cricket, pond-skater, water-scorpion, water-boatman, and water-beetle are all plentiful. Of the Arachnida, there are the water-spiders and water-mites.

Free-living flatworms are common, and some parasitic flukes and tapeworms have aquatic hosts. The rotifers are almost entirely fresh-water, and are often present in enormous numbers. Mollusca of the two classes Gastropoda and Lamellibranchiata are fairly well represented. There are about thirty-six British species of aquatic gastropods; they include the pond-snails and fresh-water limpets. The Lamellibranchiata are represented by the fresh-water cockles and mussels.

Of chordate animals, the fishes are most numerous. The eel and salmon are migrants from the ocean. There are two British species of lamprey (Cyclostomata). Most amphibians have aquatic larvae and those of the frog, toad and newt are familiar sights in spring. There are no British aquatic reptiles. The birds are all terrestrial, but many are closely associated with water, common examples being the swan, duck, goose, kingfisher and moorhen. A few mammals, the otter, water-vole, and water-shrew, have become secondarily adapted to an aquatic life.

Almost any stretch of fresh-water is different from any other. All have characteristic living communities, only to be described after close study.

THE LAND

The land is the part of the Earth's crust not covered by water; it occupies about one-third of the total surface area. The areas not covered by water have changed their outlines many times; some areas have repeatedly been covered by and lifted above water, which could only settle in new depressions formed by colossal upheavals which elevated one crust area and depressed another. Such massive earth movements are still taking place. The original unbroken rocks have been reduced to fragments which have been transported by wind, water and moving ice. In most parts of the earth, the fragments have been compressed to form sedimentary rocks, but there are areas where the original igneous rocks still appear on the surface. Over great areas, however, the rock particles form a superficial layer of varying thickness; this layer forms the framework of the soil. The soil, formed from rock by complex processes collectively called "weathering," together with the atmosphere directly above it, forms the habitat of an extremely rich and diverse flora and fauna. This terrestrial environment differs from the aquatic in three major ways. These are: the relative scarcity of water, the existence of different gaseous relationships, and the lack of body support afforded by a continuous watery medium.

The Constituents of Soil

All terrestrial organisms are directly or indirectly dependent on green plants, whose existence is controlled by soil conditions. Soil is of very complex composition, and the range of living organisms which it will support, varies according to its nature. A soil capable of supporting healthy organisms of the higher grades is regarded as fertile, and is found only in the uppermost layers of the earth's crust; such soils have the most complex structure. The influence of the effective weathering agencies is concentrated on this upper region; the underlying strata are considerably less affected.

Any soil possesses a *profile*. This means that it changes in appearance and constitution from the surface downwards.

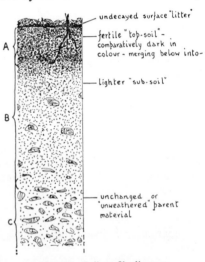

Fig. 6.2. Soil profile diagram.

A soil profile may be examined at the side of a trench excavated to a sufficient depth. Fig. 6.2 illustrates the profile that might be shown by a brown earth soil somewhere in Great Britain (*see* p. 180 for soil types).

A profile most commonly exhibits three zones or strata from ground level downwards, known as the *horizons A, B,* and *C.* Whilst they tend to merge into one another, each layer possesses sufficiently distinctive properties to make it separately recognizable. The surface layer or horizon *A* is commonly referred to as the "top soil." It contains the bulk of the soil population as well as their dead remains and this tends to darken the soil in comparison with deeper layers due to the formation of humus. It is the zone of greatest biological significance and is the truly "fertile" soil fraction. Where vegetation covers it, there is constant incorporation of matter and the withdrawal of the materials required for plant growth. It is the home of countless animals.

Overlying this mineral-organic amalgam there may be a layer of undecayed organic "litter" and in some places, such as moorlands or coniferous forest areas where decomposition is slow, this may be seen as separate layers above the true horizon *A.* According to the stages of degradation of this litter there may be seen two or more horizons *O.* Where this undecayed vegetation builds up in thickness a peaty over-layer may be produced.

Below the top-soil comes the "sub-soil" or horizon *B.* Although there is usually no sharp dividing line, this stratum is commonly less dark than the soil above it. If the area is well drained then it may be partly enriched by materials washed into it from above and contain a higher fraction of clay. This tends to create a less dark zone deficient in clay material at the lower limit of the horizon *A* and this band between *A* and *B,* if it stands out clearly, is sometimes referred to as horizon A_2. Where soil drainage is poor so that the water table is quite high for at least several months of the year, horizon *B* may show irregularities in colour, often showing a greyish, reddish or yellowish mottling due to the changing conditions of iron compounds contained in the soil as the air content varies.

Below horizon *B* occurs the relatively unaltered parent material from which the mineral portion of the soil above was derived. This is horizon *C,* said to be "unweathered" material and may even be solid rock. The nature of a soil profile will depend upon many factors, the chief being whether it is a sedentary or transported soil. Sedentary soils, formed *in situ,* depend upon the nature of the parent rock. Transported soils depend upon the nature of the transported material, as well as the kind of material on which they have been deposited. Both types of soil are considerably affected by rainfall and temperature variation.

In any natural soil the profile will have built up or evolved over many undisturbed years and there will be arrived at a climactic condition of development governed partly by the general climate and partly by the more local physiographic factors. In areas where horticulture and agriculture are practised, however, the natural state of horizon *A* is constantly being disturbed and this inevitably leads to changes in the physical and chemical states of this layer.

For purposes of further description reference will be confined to the fertile horizon *A*. Such a soil will have five main constituents which are: the *mineral skeleton*, the *organic material*, the *aqueous solution*, the *soil atmosphere* and the *living population*. Variation in the proportions and nature of these gives rise to a large number of soil types. Each constituent is closely interrelated with the others, and variation in any one, will affect all.

The Mineral Skeleton

This consists of the various particles derived from rock by weathering, which is the total effect of physical disintegration and chemical change. The principal physical agents are: the action of frost; the differences in thermal expansion between adjacent fragments; the abrasion of rocks or particles when moving over one another; grinding under moving ice-sheets; solvent action in removing the cementing substances between particles; organisms causing fissures; grinding-up of ingested particles (by earthworms). Chemical decomposition occurs by the solvent action of water and by the action of dissolved substances such as carbon dioxide, oxygen, acids and alkalis. Where the weathering is due solely to inorganic agencies, a "weathering-crust" is formed. This differs considerably from true soil which is found only in the presence of living organisms. The weathering-crust corresponds to horizon *C*, and the true soil-crust to horizon *A*, with *B* the intermediate condition. The change from a weathering-crust to a true soil-crust begins with the activities of the early colonizers, mosses, lichens, fungi and algae, which prepare the soil for the higher forms of life.

The mineral particles may be grouped physically on the basis of size. The following table shows an international classification.

Soil material	Mean diameter of particles	Soil material	Mean diameter of particles
Gravel	Above 2·0 mm	Fine sand	0·25–0·1 mm
Fine gravel	2·0–1·0 mm	Very fine sand	0·1 –0·05 mm
Coarse sand	1·0–0·5 mm	Silt	0·05–0·002 mm
Medium sand	0·5–0·25 mm	Clay	Below 0·002 mm

The proportions of these various sizes of particles have a profound bearing on the properties of the soil.

The chemistry of the soil particles is very difficult to elucidate, but certain salient facts have been established. Ninety-eight per cent of the soil-crust is composed of only eight chemical elements which are, in order of decreasing abundance: oxygen, silicon, aluminium, iron, calcium, magnesium, potassium and sodium. All the other natural chemical elements are contained in the remaining 2 per cent. Combined oxygen forms 50 per cent of the crust, and silicon 25 per cent. Aluminium is the most abundant metal, with iron second. Oxygen, silicon, aluminium and iron are rarely free, but occur mainly as complex alumino-silicates or ferro-silicates of calcium, magnesium, potassium and sodium. Smaller quantities occur as oxides and carbonates.

The sand and silt fractions consist mainly of quartz (SiO_2), with smaller quantities of felspars ($KAlSi_3O_8$, $NaAlSi_3O_8$) and micas ($H_2KAl_3(SiO_4)_3$). There are smaller quantities of other minerals such as apatite ($CaF_23Ca_3(PO_4)_2$), haematite (Fe_2O_3), magnetite (Fe_3O_4), and limonite ($FeO(OH)nH_2O$). Most of these are highly resistant to weathering and contribute little to the fluctuations of soil chemistry.

The clay particles consist of two main types, both being essentially alumino-silicates. Each has, in its minute particles, a characteristic crystalline structure. The smallest particles consist of *montmorillonite* clay in which the crystalline sheets are held together by weak binding forces which are easily broken. Hence the particles readily separate into pieces too small to be resolved by the light microscope. *Kaolinite* clays consist of slightly larger particles in which the crystal lattices are more strongly bound. In both cases, the particles are colloidal. They present large surfaces to the surrounding medium; water can enter between the crystal lattices, thus hydrating and swelling the particles. There are often excesses of negative charge at various points on the particles, and there, cations may be taken up from the soil solution and bound to the clay. This property is of the utmost importance to the plants inhabiting the land.

The Organic Material

This is derived from the decay of dead bodies of organisms, and by-products of their metabolism. The decay is due to the activities of bacteria and fungi, and all stages from fresh material to fully-decayed black or brown *humus* may be present. Undecomposed material is called *litter*; from litter to humus involves a whole series of chemical changes which have proved extremely difficult to elucidate. Humus is a fairly stable amorphous material, though it is not a single chemical compound. Treatment with sodium hydroxide separates an insoluble fraction called *humin*. If the soluble fraction is now treated with acid, a portion known as *humic acid* is precipitated. The fraction still in solution is termed *fulvic acid*. Humin is not a single compound; it contains carbon, hydrogen, oxygen, nitrogen, often phosphorus and sulphur, and sometimes various other substances. Humic acid is considered by some investigators to be a pure organic acid with the formula $C_{60}H_{52}O_{24}(COOH)_4$. The composition of fulvic acid is not known.

The nature and composition of the humus depends on the types of microbes present in any particular soil. Two extreme forms are known as *mor* humus and *mull* humus. Mor is formed in acid conditions where the calcium content is low; the pH ranges from 3 to 6·5. It has not passed through the gut of an earthworm; the soil is too acid for animals which ingest the organic matter. The decay process which results in mor humus is initiated by fungi. Mull is formed in soils where the calcium content is higher, in the pH range 4·5 to 8. It has probably passed through the alimentary canal of an animal at least once. For the formation of both mor and mull, the soil must be reasonably well-drained. Where drainage is bad, aerobic organisms do not flourish, and the remains of plants and animals form litter which is largely undecomposed. This is *peat*; it often forms a thick layer over the underlying mineral matter.

The humus is, like clay, in the colloidal state; this property has an important bearing on the soil population. Some of the organic matter is inseparably bound up with clay, forming the *clay-humus complex*.

The Aqueous Solution

Except under extremely arid conditions, the soil is always at least moist. The rain-water is slightly acidified by carbon dioxide dissolved during its descent. It acts as a solvent for materials of the mineral skeleton and the decaying organic substances. The principal ions in solution are potassium, calcium, magnesium and iron from the mineral skeleton, and nitrate, phosphate and sulphate from the decaying organic matter. There are also dissolved gases, principally nitrogen, oxygen and carbon dioxide, and complex organic substances derived from excretion or decay. Thus the soil solution is extremely complex, and variable from place to place, according to the nature of the other constituents.

It is customary to consider the liquid water under three categories. *Gravitational water* is that which will percolate through the soil and run away under the influence of gravity. *Capillary water* is that which lies in the pore-spaces, either completely filling them, or held on the surfaces of the particles. *Hygroscopic water* is held by adsorption and imbibition by the colloidal particles. The gravitational water will eventually reach an underground level which must rise and fall according to the amount of precipitation as rain. This changeable level of water in the soil is known as the *water-table*. From it, water ascends by capillarity into the pore-spaces, and it is this capillary water which is most easily available to plants. The depth of the water table is of the utmost importance to the living organisms which inhabit the soil. Water vapour can be regarded as a fourth soil water state.

The Soil Atmosphere

This occupies the pore spaces which are not filled with water. It is continuous with the atmosphere above the soil and there is free interchange between the two. Nevertheless their composition is slightly different on the average, as the following table shows—

	% by volume of the main constituents	
	Soil atmosphere	Atmosphere above soil
Oxygen	20·3 to 20·8	20·95
Nitrogen	78·0	78·09
Carbon dioxide . . .	0·15 to 0·65	0·03

The soil atmosphere tends to be richer in carbon dioxide and poorer in oxygen than the overlying air due to the respiratory exchanges of soil organisms. In a richly populated soil, the oxygen content will decrease slightly because of its use for respiration and from the lack of photosynthetic organisms to replenish it. The oxygen will therefore only be restored by diffusion from the atmosphere above. Again, because of the lack of photosynthetic organisms, the carbon dioxide evolved during respiration will tend to accumulate. Its greater density will retard the rate of its diffusion as compared with that of oxygen and nitrogen. The composition of the soil atmosphere then, at any particular place or time, will depend upon the rates of evolution of carbon dioxide, and its replacement by oxygen. The carbon dioxide content increases with depth, and the oxygen content decreases; this may be very marked in badly-drained soils in wet periods.

It must be remembered that a certain proportion of the total gas present in soil is in solution, and it is in solution that gases are absorbed by living organisms. Aeration of the soil is effected mainly by the rising and falling of the water level and by the burrowing of animals.

The Living Population of the Soil

The soil is the natural habitat of a host of living creatures. They form a delicately-balanced system of interdependence, and but for the activities of one another, most would cease to exist. Though smaller forms predominate, there is a great range of size from bacteria to tree-roots, from protozoa to earthworms. Without them, decay would never take place, and vital chemical elements would not be put back into circulation. It is the microbial forms which make the soil a suitable habitat for the larger and more advanced organisms.

It cannot be too strongly emphasized that *the soil is a living complex;* misunderstanding of its real nature has led to tragic consequences now apparent in various parts of the world. Some account of the denizens will be found on p. 177–9.

Properties of Soil of Biological Importance

We have outlined the origin and nature of soil, and it has been emphasized that variation in the nature and proportions of the constituents produces soils with widely different properties. Some account of the properties of soil will now be given.

Soil Texture

The texture of a soil is an expression of the sizes of the particles and hence of the spaces and channels which lie between the particles. Texture affects drainage, aeration, capillarity, water absorption, nature of the soil solution and ease of tillage. Soils may be placed in three groups according to their proportions of the various sizes of particles—

1. Coarse-textured soils—sands and sandy loams.
2. Medium-textured soils—loams and silt loams.
3. Fine-textured soils—clay loams and clays.

The approximate percentage constituents of each of these types of soil are given in the table below—

Class of soil	% coarse sand	% fine sand	% silt	% clay
Sand . . .	80	10	5	5
Sandy loam . .	65	20	5	10
Loam . . .	20	25	35 to 50	5 to 20
Silt loam . .	15	15	50	20
Clay loam . .	5	15	30	50
Clay . . .	1	9	25	65

The texture of a particular sample of soil may be estimated by sedimentation in a tall glass jar or by washing the soil through graded sieves. *Hardy's index* gives a mathematical expression for texture, based on the percentage of water the soil will hold at the *sticky-point*. This point is reached when the soil is wet enough to be plastic in the hands without sticking to them. The more clay in a soil, the higher is the index, with a maximum of about 55.

$$\text{Hardy's texture index} = \text{percentage of water at the sticky point} - \frac{\text{sand content}}{5}$$

Soil Structure

By soil structure, we mean the arrangement of the particles. They may occur as independent units as in pure sand, or as aggregates as in pure clay. Normally mixtures of both are present. In coarse-textured soils, the particles exist and function mainly as independent units. In fine-textured soils, groups of particles are bound together either as a dense continuous mass such as a *clod* of clay or as porous clusters of particles called soil *crumbs*. Because of its greater pore space and surface area, the crumb structure makes for greater fertility. The differences in structure are due to the amount of colloidal material present. Where there is no colloidal material, the soil will be pure sand. When colloids are present they may be of mineral or organic nature, or a mixture of both. Soil organisms affect the formation of crumbs by producing sticky gums and mucilages.

If a soil is rendered structureless by incineration or by prolonged washing, it becomes compacted and its pore space is reduced. The soil atmosphere almost disappears and drainage is practically nil. Puddled clay is a familiar example of this condition. Saturated clay soils form large sticky clods; the colloidal particles with their negative charges repel each other and will not form small aggregates. By the addition of suitable electrolytes, some of the negative charges will be neutralized and the particles will tend to aggregate. The clay is then said to be *flocculated*. Addition of lime or calcium carbonate to a sticky clay soil always helps to break up the clods by causing flocculation. The precise mechanism of the process is not clearly understood.

Porosity of Soil

The pore-space of a soil is the volume not occupied by solid material; it varies considerably with texture and structure. The finer the texture, the smaller will be the individual spaces, but total pore-space will be little affected as long as the soil is not compacted by pressure. Pore-space is increased by flocculation, or by the addition of organic matter which will leave spaces as it decays. The pore-space of dried fertile soils varies between 35 per cent and 50 per cent of their total volume.

The Soil as a Source of Nutrients

As far as their nutritional requirements are concerned, soil organisms may be divided into autotrophes and heterotrophes. Apart from the universal need for water, autotrophes will require only certain mineral substances, and heterotrophes will need a wide range of organic substances. To be fertile, a soil must provide for the needs of both groups.

The predominant minerals in soil are silicon and aluminium; neither contributes directly to the nutrition of plants and animals. Of the essential nutrient substances, we have to distinguish carefully between the available and non-available portions. Autotrophic plants require nitrogen, sulphur, phosphorus, potassium, calcium, magnesium and iron, together with minute quantities of boron, manganese, zinc, copper and sometimes other elements. These are not absorbed as elements or compounds but as ions, and therefore they must be present in the soil solution in that state. Though a soil may contain all the necessary ions, they may not be in sufficient quantity in the solution, because certain of them are attracted and held on the colloidal clay particles. The residual negative charges on these particles will attract cations and thus remove them from solution. Thus important cations such as potassium, magnesium and calcium may be totally absent from the soil solution. However, since these cations are not held equally strongly on the colloidal clay particles, it is possible to cause their release by *base exchange*. The calcium ion is held more strongly than the magnesium and the magnesium ion is held more strongly than the potassium. We can easily saturate the soil solution with calcium ions and thus free potassium, magnesium or other ions.

A similar system of anion exchange is postulated, but is not well understood. It is believed that anions may replace hydroxyl ions on the crystal lattices and thus alter the structure of the clay particles. However, the speed with which some anions such as nitrate are leached from a soil, shows that all anions are not adsorbed.

Thus for successful growth of autotrophic plants the following ions must be present in the soil solution: nitrate or nitrite or ammonium, sulphate, phosphate, potassium, calcium, magnesium and iron, together with ions of the trace elements. Some of these will be derived from the mineral skeleton itself, but the organic matter is a more important source, especially for nitrogen, sulphur and phosphorus. The dead remains of plants and animals are utilized as food by heterotrophic organisms and thus their decomposition is effected.

The wide variety of heterotrophic organisms in soil will require a correspondingly wide range of nutritive substances. Their foodstuffs are determined by the range of their enzymes. In a fertile soil, almost every conceivable natural organic compound will be present. All will be digested and will provide the carbohydrates, fats, proteins, etc., required by the heterotrophes.

The chief factors affecting nutrient availability are texture and organic content. In natural soils, there is a continuous cycle of extraction by plants and return, since no crop is being removed. In cultivated soils where crops are continually removed, the soil will become impoverished. The restoration of fertility is one of the major problems of today. From the time of Liebig, in 1840, the practice of artificially adding various mineral elements has grown to enormous dimensions. It is not, however, the whole solution, since it leads to continuous depletion of the organic matter, and eventually to a structureless soil.

Water Availability

The total water content of a soil gives very little indication of the amount available to living organisms. The soil may be regarded as a reservoir which is full when the water retained is the maximum possible

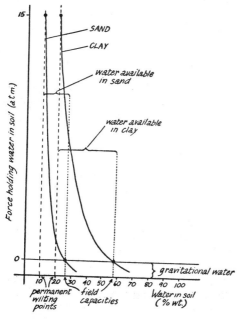

Fig. 6.3. Curves showing the water availability in sandy and clayey soils.

after gravitational loss is complete. In this condition, the soil is said to be at *field capacity*. On the other hand, when organisms can no longer exert forces great enough to pull the water from the soil particles, the soil is at the *permanent wilting point*. At this point, rooted plants begin to wilt and will not recover unless water is added to the soil. The quantity of water available to plants is sometimes expressed as the *pF* of the soil and is clearly not the total water in the soil that can be measured by a dry weight method. *pF* is defined as the logarithm of the pressure (in centimetres of water) with which the water in the soil is in equilibrium. There is no straightforward way of measuring this quantity and it is not often quoted.

If tests are carried out to find the force with which a soil holds water as its water content falls, curves as shown in Fig. 6.3 will be obtained. The smoothness of these curves indicates that there is no sharp dividing line between gravitational, capillary or hygroscopic water. The curves show that the water-holding capacity and water availability vary according to the texture of the soil. At field-capacity, clay holds more than twice as much water as sand. At permanent wilting-point, reached in both cases when the force holding the water is about fifteen atmospheres, sand has yielded only 10 per cent of its weight as water, while clay has yielded 35 per cent of its weight. At this permanent wilting-point, the sand retains only 10 per cent of its weight of water, while clay retains almost twice as much. The difference is due to the greater adsorptive and imbibitional powers of the clay particles. It should be realized that if humus is present, its colloidal particles will hold water in the same way as clay.

Many of the more lowly soil organisms need films of water in which to move, but it is probable that most of them could exist in a saturated atmosphere. The great majority imbibe sufficient water with their food materials.

Soil pH

The pH of a soil depends upon the extent to which the clay fraction contains hydrogen ions. It has been shown previously that clay particles carry residual negative charges and that these charges attract and hold basic cations such as calcium, potassium and magnesium. Two conditions create a shortage of these cations in the soil solution. They are leaching, and absorption by living organisms. Therefore, where leaching is rapid or organisms plentiful, the negative charges on clay particles are saturated by hydrogen ions. These ions come from carbonic acid or from acids produced by decay of organic material. Such clay particles in aqueous suspension give acidic properties to the soil solution around them; on the other hand, if saturated with basic cations, the clay colloids give alkaline properties to the solution in the same way as all salts of strong bases and weak acids are alkaline in solution.

The pH value has a considerable bearing on the types of organisms which can flourish in any particular soil. In natural soils, pH is usually between 4·5, which is strongly acid, and 8·3 which is fairly alkaline. The average pH of British soils is about 5. Hydrogen ion concentration may be artificially controlled by the addition of CaO, $Ca(OH)_2$ or $CaCO_3$ in the case of acid soils. Strongly alkaline soils are rare and can be corrected by addition of salts of weak bases and strong acids, such as sulphate of ammonia, $(NH_4)_2SO_4$, ferric chloride, $FeCl_3$, and calcium sulphate, $CaSO_4$.

Soil Temperature

Solar radiation (see p. 1006) is the principal source of soil heat, though some heat is produced by the decay of animal and vegetable material. The most important factor affecting soil temperature, apart from solar radiation, is the quantity of water present. A high specific heat capacity makes rise of temperature of water slow; thus wet soils will be colder than dry soils. In general, all the factors affecting water content will affect temperature. Fine-textured soils tend to hold more water than coarse-textured soils, and hence temperature rise will be slow.

When atmospheric temperature is high, soil temperature diminishes from the surface downwards, and the atmosphere will lose heat to the soil. When atmospheric temperature is low, soil temperature diminishes from the lower layers upwards, and the soil will lose heat to the air. However, fluctuations in soil temperature always lag considerably behind those of the atmosphere. They are more pronounced in summer than in winter, and more marked on bare soils than on soils covered with vegetation. A plant layer acts as an insulator, retarding both absorption and radiation.

Control of soil temperature is largely a matter of water control. Drainage has two important effects; it lessens heat loss by evaporation, and it lowers the average specific heat of the soil. Hence good drainage conduces to a higher soil temperature.

Soil temperatures affect the soil organisms because of the effect of temperature on the rate of metabolism. Rise of temperature in spring brings about a great awakening into activity after winter dormancy; fall of temperature in the autumn brings another decline. Seeds of all higher plants, spores of lower plants, encysted animals, all have a critical minimum temperature below which they will not be stirred into activity. As in other characteristics, each species is individual in this matter. Green plants are considerably influenced by the temperature of the atmosphere, as well as by the temperature of the soil.

Soil Colour

The colour of soils is due to the interaction of three main factors; humus, iron compounds and other minerals. Where humus is plentiful, the colour ranges from brown to black. The humus particles adhering

COMPARISON OF SANDY AND CLAYEY SOILS

Property	Sandy soil	Clayey soil
Texture	Coarse. Large percentage above 0·02 mm. Known as light soils.	Fine. Large percentage below 0·002 mm. Known as heavy soils.
Structure	Often structureless; very little crumb formation; no clodding.	Clods easily; large, sticky masses when wet; baked hard and cracked when dry; no crumb formation.
Porosity	Pore spaces comparatively large. Easily aerated since drainage rapid.	Pore spaces generally very small. Not easily aerated since water is strongly held in the small pore spaces.
Water-holding capacity	Low. Little water held by capillarity; drainage rapid; no waterlogging.	High. Considerable capillary and hygroscopic water; drainage poor; waterlogging common.
Temperature	Warm. When wet, total water too low to have much effect on average specific heat and the water effects rapid conduction. Heat more slowly when dry since air is a poorer conductor than water.	Cold. When wet, total water is high and raises average specific heat. Heat more rapidly when well-drained because average specific heat is lowered, but enough water present to conduct heat more readily.
Nutrients	Low retention of ions, since very small particles are absent. Rapid leaching. Base status very low.	Can be high: colloidal particles hold cations and some anions preventing rapid leaching. Base status usually high.
pH	Seldom acid since hydrogen ions are not retained in large numbers. Tends to be neutral.	Can be very acid since hydrogen ions instead of metallic cations held on particles. Can be made neutral or alkaline by displacing H^+ by Ca^{++} or similar ion.
Ease of tillage	Lightness and good drainage make for ease of tillage. Fine tilth readily obtained because particles separate easily.	Difficult. Heaviness and poor drainage make for stickiness. Fine tilth difficult to attain because of clodding and compacting. Flocculation and addition of humus will improve this condition.

to the surfaces of mineral fragments tend to obliterate or subdue their colour. In soils of low humus content, the colour is considerably lighter. Where drainage and therefore aeration are poor, reduced forms of iron and anaerobic decay of organic matter tend to bring about a grey coloration. Drainage and aeration of such soils causes oxidation, and hydrated oxides of iron give red to yellow shades. Such soils are low in humus; if humus is added, they quickly become brown or black.

Soil colour has an important influence on temperature since dark soils absorb more solar radiation than lighter-coloured soils. The benefit is somewhat diluted if the dark colour is due entirely to humus, since the higher moisture content tends to make the soil cold. Dark soils warm up more quickly in spring and germination is more rapid than in the cold light-coloured clays.

Properties of Loams

Because of the more even mixture of small and large particles, with their contrasted characteristics and properties, loams possess most of the good qualities of both types, with the poor qualities considerably subdued. Arable farmers and gardeners strive to obtain a loamy condition in their soil.

Marl

Marl is a soil consisting of a mixture of a calcareous material such as limestone with white argillaceous clay. The calcium carbonate component is often derived from shells of marine or fresh-water organisms, and the true marls are formed where these shelled creatures were once former inhabitants, and their shells impregnated a marine ooze. It is, however, common to refer to any soil rich in calcium carbonate as marl whether the preponderant particles are sand or clay. Sand marls will improve the condition of clayey soils, and clay marls will improve sandy soils.

Life in the Soil

The soil is the natural habitat for countless hordes of plants and animals of many species, and mainly belonging to the lower groups. Each species forms a part of a delicately balanced system of interdependence. The numbers of such organisms and their presence or absence depends on the nature and properties of the soil, of which they themselves form an integral part.

Bacteria are the most numerous of all soil organisms; they include some very important groups such as those concerned with decay and with nitrogen fixation. In well-cultivated soil, their numbers may reach the fantastic figure of 10^9 per g of soil. This bacterial population is subject to considerable fluctuation as conditions vary. Changes of temperature, moisture content, pH, organic material, and the activities of predatory protozoa, all affect the bacteria. They range from heterotrophic organisms using dead organic matter as a source of food, to purely autotrophic species which live photosynthetically or chemosynthetically. No comprehensive classification of soil bacteria has yet been completed.

Accompanying the bacteria are numerous bacteriophages which exist parasitically on the bacteria. They are able to bring about lysis of the bacterial wall, and use the bacterial protoplasm as a source of nutriment.

Actinomycetes are well represented; they are minute filamentous forms, considered by some authorities to be bacteria. They are predominantly heterotrophic and can use a wide range of carbon compounds such as cellulose and hemi-cellulose, or nitrogenous substances such as proteins, as sources of food supply. Thus they help in the decomposition of plant and animal material. Actinomycetes are said to be responsible for the earthy smell of freshly-turned soil, since they emit earthy or musty odours when sporing.

Fungi may be the dominant group of the soil flora in acid conditions where few bacteria can exist, though there is a considerable fungus population in most soils. The majority of soil fungi are microscopic, though when there is plentiful organic matter, the larger species are well represented. Phycomycetes predominate and include representatives of many common genera such as *Mucor, Pythium* and *Rhizopus*. The Ascomycetes include *Eurotium, Penicillium, Fusarium,* and many yeast-like forms. The Basidiomycetes are chiefly of the mushroom type (Hymenomycetes), or the puff-ball type (Gasteromycetes), the latter being more common in woodland soils. Most soil fungi are saprophytes, feeding on organic material; they play an important part in decay. Resting stages of many parasitic fungi may be found, awaiting a satisfactory host. The fungi are generally aerobic and are most common near the soil surface.

Soil algae include members of the Cyanophyta (blue-green), Xanthophyta (yellow-green), Bacillariophyta (diatoms), or the Chlorophyta (green). They are typically unicellular, colonial or filamentous species with mucilaginous walls, which help to retain moisture. The great majority are found at the surface, where light is available, though their presence has been demonstrated below the surface, where some forms seem to be able to live heterotrophically. The numbers of algae are very variable; counts have been made ranging from 100 000 to 3 000 000 per g of soil.

The higher plants have few, if any, representatives living entirely in a soil habitat. Certain of their organs function in the soil and have considerable effects upon its nature and properties. The chief of such organs are roots and rhizoids, together with subterranean resting and reproductive structures such as rhizomes, corms, bulbs and tubers.

Protozoa are almost always present; they consist of amoeboid, flagellate and ciliate forms. They thrive only in moist soil since their movement depends on the presence of water-films. In most cases, they can survive occasional drier conditions by encystment. The great majority are predators, feeding almost exclusively on bacteria; a few such as species of *Euglena*, can live autotrophically. There is enormous variation in the numbers of protozoans; counts vary between 300 and 300 000 per g of soil.

Nematode worms are very common, but little is known of the exact conditions under which they live. The forms most studied are parasites which attack plants. In arable land, their numbers may reach 10^6 per square metre, but in pasture land their numbers may exceed this figure by twentyfold.

Annelid worms form over half the weight of soil animals. All are called earthworms; there are about 25 British species, the best known being *Allolobophora longa*, *A. nocturna*, *A. caliginosa*, *A. chlorotica*, *Lumbricus terrestris*, *L. rubellus*, and *Eisenia rosea*. Compared with the enormous *Megascolides australis*, which may reach 3 or 4 m long, our native earthworms are very small, rarely exceeding 15 to 20 cm. They all feed on a variety of organic material. Numbers of earthworms vary according to locality; counts of 1.2×10^6 to 2.4×10^6 per hectare in forest soil, and up to 7×10^6 in pasture, have been recorded.

The arthropods are well represented in the soil. Acarine mites are numerous; they feed on decaying litter and fungal mycelium. Millipedes feed similarly, and also attack plant roots, while centipedes eat any type of smaller animal. Insects and their larvae are exceedingly numerous, feeding on decaying material or on living plant tissues. The total number of species would constitute a formidable list. Most students will, however, be familiar with the wire-worm, the leather-jacket and some of the soil beetles. Crustacea are represented by the wood-lice which feed exclusively on decaying material. The numbers and distribution of molluscs are but little known. Slugs and snails are mainly surface feeders, nourishing themselves with fresh vegetation or with fallen leaves; one species of slug preys on earthworms.

Flatworms and rotifers are commonly to be found in soil but little is known about their precise activities or the density of distribution.

Vertebrates include the burrowing animals such as mice, voles, shrews and moles. They feed on a variety of plants and animals;

the voracious appetite of the mole for earthworms and insect larvae is noteworthy. Rabbits, foxes, badgers and occasionally birds such as the sand-martin, may burrow into the soil, but apart from affecting the soil conditions locally, they are not of great significance in soil biology.

It may be pointed out here that the term "terrestrial" as applied to organisms may sometimes be misleading. It is applied both to creatures which live entirely in the soil and to those which use it as a surface for locomotion. It also connotes soil micro-organisms, many of which are essentially aquatic.

World Classification of Soils

Soils have been classified in a number of ways, one of which is in terms of their geological origins. Since it has been realized that soils owe their individuality more to the environment under which they develop and to a lesser extent to their parental material, however, it has been found more useful to define them by the number, arrangement, form and composition of their horizons. The great world soil groups include *tundra soils, desert soils, podsols, brown forest soils* (*brown earths,*) *laterites, chernozems* (*black earths*),

Soil group	Type of vegetation	Profile and depth	pH	Drainage	Occurrence and origin
Brown earth	Deciduous forest	Dark and humus-rich above. Uniform lighter colour in subsoil. Varies in depth—may be as much as 0·5 m.	Slightly acid. Never base-saturated. Mull humus.	Rarely completely free.	Occurs under deciduous woods in temperate climates. Often over clay in G.B.
Podsol	Heathland. Coniferous forest.	Horizons sharply contrasted. Surface, humus-rich above grey, leached layer (A). Clear, dark band rich in iron compounds and some humus. (B). Light brown below down to parent rock at about 0·5 m. (C). Horizon B may be cemented into a "pan."	Strongly acid. Base deficient. Mor humus.	Very free to excessive. Causes lower part of horizon A to be deficient in bases, iron and other compounds.	Developed on sandstones under moderately heavy rainfall in temperate climates.
Rendzina	Calcicolous (chalk-loving)	Upper horizon dark but may have whitish flecks. Grades into lighter colour down to parent rock. Shallow.	Neutral or alkaline. Base saturated.	Free.	Developed on chalk, carboniferous limestone or Great Oolite.
Fen or Alkaline peat.	Typical fen.	Very dark, sometimes down to considerable depths.	Alkaline. Contains basic substances.	Poor. Waterlogging leads to anaerobic conditions.	Developed in regions of continuous water logging. 65% or more undecayed organic material.
Acid peat	Typical peat bog.	As above.	Acid. Bases absent.	As above.	
Gley or glei.	Marshland.	Mottled rusty-brown above. Greyish green or blue in the gley horizon below.	Acid.	Impeded above, waterlogged below.	Developed alongside rivers and streams where drainage poor.

prairie, *steppe*, and *savannah* soils, *mountain* and *high plateau soils*, *oases soils* and *tropical and mangrove swamp soils; rendzinas* (chalk based). *Loess* (wind borne), *alluvial* (water borne), *gley* and *bog soils* are recognized as existing but of more local origin. The main natural soils of Great Britain, with their chief characteristics, are shown on p. 180.

THE ATMOSPHERE

The atmosphere is the gaseous layer which surrounds the earth and fills any otherwise empty interstices in its surface. It has a thickness of several hundred miles, becoming more and more attenuated as distance from the earth's surface increases. No living creatures normally exist above the first 6000 m. The atmosphere is composed of a mixture of gases and vapours. In the lower layers these are so thoroughly mixed that there is very little variation in composition as far as the permanent constituents are concerned. Air denotes both the dry mixture of gases and the same mixture with the presence of water vapour. The composition of dry air, in percentages by volume is given below—

Nitrogen	78·09	Helium	0·000524
Oxygen	20·95	Krypton	0·0001
Argon	0·93	Xenon	0·000008
Carbon Dioxide	0·03	Ozone	0·000001 (variable)
Neon	0·0018	Radon	6×10^{-18} (near ground)

Water vapour is present in the atmosphere near the ground in varying amounts dependent on climatic conditions. The percentage by volume rarely exceeds 4 per cent.

Properties of the Atmosphere of Biological Importance

A great many organisms exist in constant contact with the atmosphere. Its chemical composition and its physical characteristics present conditions with which their structure and functions must be in harmony. Those factors of the atmosphere which influence organisms in contact with it, are outlined below. It should be noted that no organisms have so far been described which can live wholly in and on the gases of the atmosphere.

Humidity or Moisture Content

Atmospheric water vapour results from evaporation (*see* p. 968) from water surfaces. The total amount present in any particular volume of air depends upon its temperature. The higher the temperature, the more water vapour can be held; the maximum at any particular temperature is said to saturate the atmosphere at that temperature.

If such a saturated atmosphere is cooled, its capacity for water vapour will be lessened, and the excess is precipitated as liquid. There is a natural cycle of evaporation and precipitation.

With the exception of a few isolated cases such as plants specially adapted to absorb water hygroscopically, the atmosphere cannot be a reliable source of liquid water for living creatures. Of much greater importance is the drying-out which they must undergo when in contact with an atmosphere below its saturation point. Their protoplasm in the normal active state is a source from which water may be evaporated. This renders the protoplasm of terrestrial organisms exceedingly vulnerable, and they have either evolved devices to prevent or restrict desiccation, or are limited to very humid localities. In any event, whether protected or not, they must all have access to liquid water to replenish losses which must inevitably occur in one way or another.

In plants, the development of almost waterproof coverings such as cuticle and corky bark is very widespread. Land plants such as some of the delicate mosses, which have no such protection, lose water rapidly in dry air and die in a few minutes; consequently they are restricted to damp regions. Terrestrial animals protect themselves and their eggs with waterproof coverings for similar reasons. Chitin and keratin are examples of waterproofing materials so used. The reproductive structures of plants such as seeds and spores may have to withstand prolonged drying conditions. Their protoplasm, however, contains as little as 10 per cent by weight of water, so that unless the air is absolutely dry, they cannot be desiccated much further. Many plants have developed special mechanisms for seed and spore dispersal which are associated with atmospheric humidity. The peristome of the moss capsule, the annulus of the fern sporangium, and the capsules of many flowering plants respond to changes in humidity by effecting movements designed to bring about the most advantageous dispersal of their contents.

The Atmosphere as a Source of Nutrients

No known organism is able to make use of the atmospheric gases alone as its source of food. Indeed, with the important exceptions of green plants which use the carbon dioxide as their source of carbon and a few bacteria and other micro-organisms which can utilize the free nitrogen, the constituents of the air play no part in the nutrition of terrestrial organisms.

Efficient absorption of carbon dioxide by green plants can only be achieved if the atmosphere is in direct contact with wet cell surface where the gas may go into solution and diffuse into the cells. But the presentation of unprotected tissues to the atmosphere would make the plant extremely vulnerable to desiccation, and hence the more successful land plants have become adapted so that their soft, wet, internal tissues are adequately ventilated through openings in the waterproof outer coverings. There are regulated stomatal openings in the cuticularized parts, and lenticels in the cork-covered organs which serve for this purpose. Such chinks in the waterproof covering may lead to excessive water loss by the plant, but this is a necessary evil. The abundance of the earth's vegetation indicates that it has never been of great importance except in those areas where either water supply is low or evaporation excessive. Even in those localities the land plants have evolved devices which overcome the danger.

Respiratory Relationships

Most organisms are aerobic and thus can only release energy efficiently by using oxygen for the oxidation of energy-rich compounds. The atmosphere is a rich source of oxygen and the problem which terrestrial organisms face is how to absorb it rapidly enough. The gas must be absorbed in solution, and all plants and animals which use it, must present a wet surface in which to absorb it. The land plants possess stomata and lenticels through which gases can diffuse to the tissues.

The gills and highly vascular skins of aquatic organisms are too vulnerable to desiccation on land; the successful land animals must have evolved from aquatic types with potentialities for developing respiratory surfaces more fitted to the drier environment. These may take the form of lungs as in vertebrates, the lung-books of spiders, and the tubular tracheae of insects. All communicate with the atmosphere by small external openings. The internal respiratory surfaces are thin, moist, well supplied with blood and of large area. Such special cases as the respiratory skins of earthworms can only function permanently if the animals can take in enough water to manufacture the wet slime with which the skins are coated. The skin of the frog, though not its only respiratory organ, is of a like nature. Such animals cannot exist for long in a drying atmosphere without repeated intake of water. The same consideration applies to the lunged and tracheate animals, but to a lesser degree, since with every exhalation, water vapour will be lost. The internal and protected respiratory surfaces reduce the degree of loss.

Temperature

The temperature of the atmosphere is its most variable characteristic. It varies with height above sea-level, with distance from the equator,

7

and also diurnally and seasonally. Altitude variation is relatively constant and averages 10°C fall for every mile above the earth's surface, At 10 000 m, known as the limit of the troposphere, the temperature is about − 50°C. From about 20 000 m, the temperature begins to rise. Variation with distance from the equator is recognized and expressed by dividing the earth into zones of latitude; they are the torrid zone, the temperate zones, and the frigid zones. The latitude lines do not coincide with the isothermal lines, which are plotted from yearly average temperatures, but there is rough agreement.

Exact locality within the zones has a considerable effect on atmospheric temperature. Coastal regions show less variation than centres of land masses at the same latitude. In general terms, the torrid zone shows a fairly uniform average day temperature of about 27°C, with a small annual range of about 6°C. The highest day temperature ever recorded was 58°C in Tunisia. The average night temperature is much lower according to locality and can be as low as freezing-point. Seasonal variation in the tropics is small.

The temperate zones show considerably greater variation, and the annual range exceeds the daily range. For the coastal areas, the annual range is about 8°C, and for the interiors of continents over 50°C. The distance from the equator causes the seasons to be well-marked. The average daily range of temperature rarely exceeds 6°C.

In the frigid zones, the temperatures are seldom above freezing point. In the Arctic regions, the temperature may range between − 40°C in midwinter to 0°C in midsummer. Less is known about temperature variation in Antarctic regions but in general, the temperature is lower.

Temperature affects most activities of living creatures. Metabolism cannot continue below freezing-point, nor for long above 38°C. Within this range, the rate of metabolism is approximately doubled for every 10°C rise in temperature. It is little wonder that the earth's surface shows such widely-differing scales of inhabitants as the lush flora and fauna of the tropics, where metabolic rates are high, and the sparse vegetation and animal life of the tundra, where metabolic rates may often be at a standstill. Water availability and temperature together account for the distribution of plants and animals over the earth's land surface; each climate, local or zonal, has its own peculiar flora and fauna. There is an altitude distribution which parallels the latitude distribution; even in the tropics, nothing can survive for long above the snow line of the mountains.

The only organisms which can exist with any degree of independence of temperature variation are those which have developed some means of keeping their tissues at a constant temperature. These are the homoiothermous (warm-blooded) birds and mammals.

Atmospheric Pressure

At sea-level, the average atmospheric pressure is about 101 kN m^{-2}. This corresponds to the weight of a column of mercury 760 mm high. The SI unit of pressure is the newton per square metre but in the case of atmospheric pressure this is often expressed in other ways. The *bar* is equal to 10^5 N m^{-2} and the *millibar* (mbar) one thousandth of this. The standard atmosphere (*see* p. 9) is equal to 101 325 N m^{-2} which is 1013·25 millibars. In some cases a pressure may be measured in terms of mm Hg, i.e. millimetres of mercury. Variations in pressure occur from moment to moment; the extremes recorded are 1075 mbar in Siberia in 1877, and 887 mbar in the Pacific Ocean in 1927. Pressure decreases with height thus—

Height (kilometres)	0	2	4	6	8	10	12	15
Pressure (millibars)	1013	795	615	470	355	260	190	120

Variations in pressure at any one altitude are not significant enough to affect living organisms seriously. Pressure changes with altitude are, however, sufficiently great to limit the activities of animals. With serious reduction of pressure in a short space of time, bleeding occurs from delicate vascular membranes.

Buoyancy

When the density of the air (0·001293 g per cm^3 at N.T.P.) is compared with the density of plant and animal bodies (about 1 g per cm^3), it is obvious that the air can offer practically no support. The earth's gravitational pull will always bind land organisms to its surface, unless they have special aerodynamic adaptations to flight, such as are found in insects and birds, and a few mammals. The land plant is less seriously handicapped than the animal; being sedentary, it has only to make provision for lifting and supporting its aerial organs. This it can do by maintaining a turgid condition of its living cells, and by the development of specially-strengthened tissues to carry the weight of the others.

The land animal must be motile to succeed in its hunt for food. Thus supporting structures have been developed which maintain support, and yet are flexible enough to allow movement. The three essential methods are turgidity of tissue, as in the earthworms, exoskeletons as in arthropods, and endoskeletons as in vertebrates. The two latter groups have developed jointed limbs. Animals which have conquered the air to the extent of transporting themselves without ground contact are the insects, birds, reptiles (mainly extinct) and a few mammals; these have developed wings, which when extended under proper control, make use of the lifting power of the air. Such animals are truly capable of active flight.

Many plants make use of passive flight for dispersal of spores and seeds. Of fruits, the parachute of the dandelion, the plumes of *Clematis*, and the wings of sycamore, are well-known examples. The seeds of many plants are winged; those of *Zanonia macrocarpa* are perfect gliders. The pollen grains of some plants have adaptations to reduce their density; the air-bladders on the pollen grains of many conifers are good examples. Similarly some animals, notably the spiders, may produce gossamer which keeps them afloat in the air and they may travel for miles in a slight breeze.

Wind (Currents)

There are occasional cases of destruction of plants and animals by high winds. Deformation of plants is much more common. Normally, however, air currents have little effect on land organisms except in three respects. Many plants rely on air currents, coupled with special adaptations, to distribute their reproductive structures. The turbulence of the air keeps the gases in a mixture of constant composition; otherwise there might be local concentrations of one constituent. Carbon dioxide concentration in and just above the soil surface would become too great for normal metabolism of many soil and surface creatures. Lastly, moving air increases the rate of evaporation of water from any exposed surface. Thus, plants living in situations which are very exposed to wind, have special adaptations to reduce the rate of evaporation.

Optical Properties of the Atmosphere

Clean air transmits radiation from the sun through the atmosphere's full depth of several hundred miles, to the surface of the earth. Since living organisms are confined to a narrow zone of atmosphere in contact with the earth's surface, there will be no great variation in the radiations received at any point. There is a wide range of radiations, but the most important biologically extend from 300 to 5300 nm, that is, from the ultra-violet through the visible spectrum to the infra-red. Radiant heat is essential for the maintenance of suitable temperatures, and light is the source of energy for photosynthesis. Ultra-violet radiation in small doses is harmless to large multicellular organisms; indeed in some cases it is beneficial; ergosterol in the skin of some animals is converted to vitamin D. These same ultra-violet rays can, however, be lethal to small micro-organisms without adequate protection. Death in such cases is due to coagulation of the protoplasm. The higher organisms are protected from this fate by the development of cuticle, skin, or pigment; often there is a combination of two or three of these. They prevent the ultra-violet rays from penetrating to too great a depth.

ORGANIC ENVIRONMENTS

Besides the oceans, fresh-water and the land, there is a fourth major type of environment, which we may term organic. For parasites and symbionts, it is a living organic environment, and for saprophytes, it is a non-living organic material.

The Living Organic Environment

It is probably true to say that the great majority of living organisms are afflicted with parasites, which range in size from animal viruses to large tapeworms, from plant-viruses to the huge *Rafflesia*. Endoparasites inhabit the tissues or fluids of their hosts. They thus obtain food, shelter, and considerable physical and chemical constancy in the conditions of their environment. Their special problem, for which many solutions have been evolved, is that of transfer of their offspring from host to host. Ectoparasites are attached to the outside of other living organisms; they are subject, therefore, to the same environmental variations as the host, and have to make the necessary adjustments. With symbionts, there is a very delicate balance between the two partners. Here again, the problem of dispersal with provision for successful continuance of the partnership, has been solved in various ways.

The Non-living Organic Environment

Saprophytes inhabit non-living organic material from which they derive both food and shelter. The types of saprophyte present depend on the nature and location of the substrate. The saprophytes are bacteria and fungi; they are subject to the environmental conditions of the substrate, and of the location in which it happens to be. Saprophytes are characterized by effective dispersal mechanisms and by resistant spores.

ADAPTATION

It was stated earlier in this chapter that for survival, plants and animals must "fit in" with their environments; they show adaptations which may be morphological and physiological. It is opportune here to survey broadly the nature of these adaptations. Details of adaptations in individual cases are described for the type genera in Vol. I.

In so far as they may be related to the main conditions prevailing in the two most contrasting environments, aquatic and terrestrial, adaptive features may be analysed as follows: features concerned with water relationships; with gaseous exchanges; with nutritional requirements; with body support; with locomotion; with stimuli perception; with the nature of the reproductive process, reproductive structures and mechanisms for the dispersal of these. If we except those organisms which have become secondarily aquatic, e.g. some angiosperms and mammals, we find that the plants and animals best adapted to the terrestrial life show broad correlation in their main structural and functional characteristics. A similar but less obvious correlation is found in the plants and animals best adapted for an aquatic life.

Plants which are wholly immersed in water need make no special provision for conserving it. Every cell is bathed in it and there is no risk of desiccation unless, as in some fresh-water environments, the water evaporates more quickly than it is replenished by rain. Under permanently wet conditions, there is no survival advantage in the development of organs specially associated with water absorption from a fixed source, nor in developing waterproof surface tissues. By contrast, the land plant must ensure continued intake of water from the only constant source, the soil, and must make provision for its conservation as far as that is compatible with other requirements (gaseous exchange). In the terrestrial plant, we see therefore, a well-developed root or rhizoid system through which water supplies can be maintained, and cutinized or suberized epidermal layers of cells which prevent excessive water loss to the atmosphere. Inhabitants of temporary pools and ditches overcome the risk of extinction during dry periods by developing highly drought-resisting reproductive structures which can serve the dual purpose of maintaining the species through times of water shortage and of being easily dispersed in a dry atmosphere, e.g. many fresh-water algae.

Gaseous exchanges between the tissues of a plant and its surroundings, whether these be water or air, are fundamentally the same, and adequate provision must be made for them to occur. For the aquatic plants, direct diffusion of dissolved gases into and out of cells can occur wherever there is a wet cell surface. For the terrestrial plant, the same applies if the wet cell surfaces are in contact with the atmosphere. But in this latter case, the exposure of wet cells to drying air involves the danger of excessive water loss. The true land plant is adapted to meet its requirements by including perforations, the stomata and lenticels, in its otherwise nearly waterproof covering.

All green plants need mineral salts in addition to water and carbon dioxide for nutritive purposes. In the aquatic environment, these are more or less evenly distributed throughout the medium. To obtain them, there is no need for special absorbing organs; any part of the plant may take them in, without the plant locating itself in any special position. On the land, however, as in the case of water, the soil is the only source of supply. The terrestrial plant must therefore anchor itself permanently to this source and present an adequate surface over which the nutrients may be absorbed. A root system serves both purposes. Another factor concerned in plant nutrition is light. Maximum light is obtained by aquatics if they are in shallow water or float near the surface. On land, a plant will obtain maximum light only if it overshadows its competitors and spreads its photosynthesizing tissue over a large surface area. The erect, branching habit coupled with the thin flat leaf are adaptations of the land plant to meet this requirement.

Gravity (see p. 936) affects all organisms, but in a liquid medium, the downward force is counteracted by upthrust due to displaced medium to a much greater extent than is the case in air. The effect of the upthrust is dependent upon the density of the body substance, but since in most cases this is not far removed from 1 g per c.c., in water a body is supported almost to the extent that it is weightless. In air, the same body would need to support itself to become elevated above the ground. To be adapted to such conditions, the land plant must necessarily make provision for supporting its own weight. Specialized mechanical tissues such as collenchyma and sclerenchyma, suitably distributed throughout the body, serve the purpose. Aquatic plants do not require them and seldom are there developed tissues of comparable function. Instead, the plant body can reach such extremes of fragility as seen in *Volvox* or the filamentous algae, without damage, as long as they are not too big. The same structure on land would collapse under its own weight.

Comparatively few plants make locomotive movements and those that do are microscopic and confined to water. Most motile plants move by means of flagella, but a few, such as diatoms, make use of protoplasmic streaming mechanisms. Adaptation of the body for locomotive purposes is shown only by these plants and in general they are comparable with very lowly animals in this respect.

The stimuli to which plants are known to respond do not vary with the nature of the environment and thus there are no adaptive features concerned with sensitivity which are specially associated with either aquatic or terrestrial conditions. But the same is not true of the kind of reproductive process adopted by the most successful plant inhabitants of the two environments, nor is it true of the reproductive structures and dispersal mechanisms which they exhibit. The coming together of gametes in water is obviously facilitated if at least one of them is motile and can seek out its partner. On land, a similar process is only fully efficient in wet conditions, which obtain in comparatively few places.

The most highly evolved land plants, the angiosperms, have become adapted to the dry conditions by evolving a fertilization mechanism involving air-borne microspores and a siphonogamous transfer of the male gametes. The mechanisms concerned with dispersal of reproductive structures in the two contrasting types of plant show clear adaptations to environment. In water, the opening of reproductive organs to release their contents is best achieved when the organ absorbs water, swells and bursts. The contents disperse most widely if they are motile. On land, efficient dehiscing mechanisms are dependent upon a high degree of desiccation of the organ concerned, which is so constructed as to split and fold itself open when it shrinks. Small, light and easily air-borne reproductive structures make for the widest dispersal.

Within the plant kingdom may be found forms which illustrate in their adaptive features the gradual transition of plants from the aquatic to the terrestrial environment. Commencing with the most primitive aquatic types and advancing to the fully adapted land plant, descriptions of selected examples will be found in Vol. I.

Animals show adaptive features corresponding in significance to those described above for plants. In aquatic environments, shortage of water is not a hazard but in contradistinction with the condition in plants, there is, in the majority of cases, the necessity of maintaining the body fluids at a narrowly limited osmotic pressure and ionic concentration. The problem differs for marine and fresh-water species. Marine species may be adapted in one of three ways. They may, as in the case of most invertebrates, maintain a body fluid which is both isotonic with the sea-water and ionically balanced with it. The bony fishes maintain a body fluid hypotonic to the medium by drinking the sea-water and excreting the excess salts. In the third group, which includes the cartilaginous fishes, the blood is kept rich in dissolved substances and is hypertonic to the sea-water, the condition being maintained by excretion of excess water (*see* p. 207). The body fluids of fresh-water animals are always hypertonic to the medium and the animals show osmoregulatory adaptations by the possession of contractile vacuoles in the more lowly forms and dual-purpose excretory organs in the higher. On land, the water problem is nearly always one of conservation and most animals have become adapted by developing relatively impervious outer coverings. Some land animals are able to exist on the chemically-combined water in the diet, but most must have continued access to liquid water to replenish the losses by evaporation and other ways. The excretion of nearly solid urine is a physiological adaptation to the dry conditions; it occurs in a wide variety of animals from insects to reptiles and birds.

Respiratory exchanges must occur somewhere at the body surface,

In water, the respiratory surfaces of lower organisms are those of each cell, but in more highly organized creatures, a delicate epithelium is exposed to the medium, and through this, the gases are diffused and transported around the body to supply each of the more internally disposed cells. This surface needs to be extensive and delicate, and thus gills, and a suitable means of keeping the water flowing over them, have been evolved to meet the requirements. On land, the gill structure could not function for long, because of the rapid drying of the exposed tissues. Many land animals have evolved the lung. This is adequately protected from drying by being internally placed but is also adequately ventilated by a muscular mechanism which keeps moist air moving in and out of it. The more highly adapted land animals possess a perfected double blood circulation, which coupled with the lungs, maintains very adequate gaseous exchange between the tissues and the atmosphere. By contrast, the aquatic animals with gills have evolved only the single blood circulation. Other types of respiratory surface, equally adapted to the land conditions, are to be found among the invertebrates. Insects show the highly intricate system of tubes, the tracheae, which serve to carry the atmosphere to all parts of the body. Spiders have evolved "lung books," and land molluscs the "pulmonary sacs."

Nearly all animals show adaptations in their nutritional processes, particularly in relation to the nature of the diet, of obtaining and ingesting it. Feeding mechanisms, and the structures evolved to operate them, show a very wide range of adaptive features, all of which are not necessarily correlated in any way with aquatic or terrestrial environments (see Chap. 10).

The remarks previously made concerning the greater need for body support on land as compared with water, apply equally well to animals as to plants. Stiffening structures are not essential to animals living in water as may be witnessed by the many successful but fragile forms of the Coelenterata, Platyhelminthes, Nematoda and Annelida, but an animal cannot be regarded as fully adapted to the land unless it possesses some means of supporting its own weight. Exoskeletons or endoskeletons, which have been evolved by many groups of animals, perform this function, but this is certainly not their only role or perhaps even the primary one in many cases. The mode of life of an animal, whether aquatic or otherwise, necessitates locomotive activity, and in most instances, the skeleton is developed in such a way as to form attachments for muscle, which, by its contractility, uses the jointed skeletal parts as a system of levers to produce movement of the whole body over a solid substratum. Such a system is employed by all the mobile land animals. In water, there may be movement through the medium as well as over the solid substratum. The action of swimming does not necessitate an endoskeleton, as long as some muscle attachment is provided which will allow waves of contraction to pass alternately down the right and left sides of the body, thereby imparting forward motion by the thrust exerted against the water. Such locomotion is achieved by some aquatic invertebrates which do not possess any hardened skeletal parts at all.

Mechanisms for stimulus perception are strongly developed in most animals. The chief stimuli to which animals respond are provided by variations in light intensity and direction, chemical conditions, in the nature of vibrations transmitted by the medium, the pressure of the medium and in physical contact with other objects. Some animals are better adapted to perceive these changes than others. In general, the higher forms show increasing ability in stimulus perception and have evolved more and more efficient organs for the purpose. In broad terms, the organs of special sense are functionally similar in both aquatic and terrestrial forms, but in most cases, there are numerous minor adaptive features to be found in them according to an animal's exact mode of life and particular habitat. For example, in nocturnal animals, the eyes are better adapted, both in structure and in operation, to dimly-lit conditions, than are the eyes of diurnal forms. The olfactory organs of animals vary in perfection of performance depending on whether the animal "hunts with its nose." In structure, the auditory organ of a fully adapted land animal is suitable for detecting sound vibrations of medium wavelength conducted through the atmosphere; that of a fish is more able to detect long wave vibrations transmitted by the water. Vertebrates on land, except in the cases of some birds, are not specially adapted to meet changing atmospheric pressure conditions. Normal changes are so small as to have little effect on their bodies or activities. Birds which may dive through many hundreds of feet can adjust internal and external pressures so that no harm is suffered during rapid descents. In water, pressure varies very quickly with depth, and although most aquatics live within comparatively narrow depth ranges, fish in particular are adapted to widening this range by eliminating the worst effects of rapidly changing pressures. These effects depend on the quantity of dissolved gas in the blood. The swim-bladder is an adaptative feature by which the fish can adjust its blood gases according to depth, by secreting gas into it or absorbing gas from it, as the pressure decreases or increases. Some fishes are so sensitive that they react to pressure changes of the order of a fraction of a millimetre.

In most animal groups, reproduction is by a sexual process only, and the majority of species are bisexual. The mingling of the gametes from both sexes is essential to the efficiency of the process, and whereas gametic union may be very largely influenced by chance in some cases, the majority of animals make very adequate provision for successful fertilization. In their mating behaviour and in the time and manner of liberation of their gametes, many creatures show adaptive features strongly favouring successful union of the gametes. This is particularly so of the land animals in which the unprotected gametes could not exist long enough outside the bodies of the parents to achieve their purpose. In such cases, the male gametes are transferred to the inside of the female animal, where they can fulfil their function protected from the effects of the external environment. In aquatic animals, the egg, fertilized or otherwise, is in the majority of cases shed by the female and, frequently unprotected, it continues its development completely at the mercy of the environmental conditions. A few aquatic animals lay protected eggs or the eggs may complete development inside the female parent, but neither condition is the general case. On land, the reverse is true. Eggs deposited by most invertebrate and vertebrate animals are covered externally to protect them from desiccation. The cleidoic eggs of reptiles and birds are the highest forms. But the mammal is adapted to the land conditions to an even greater extent in that eggs are not deposited at all but develop into embryos with tissue attachment to the parent, so that they are removed entirely from any direct external environmental influences.

Within each one of these major adaptive features in both plants and animals may be found countless variations in detail among the species. Each has its own minor adaptations to fit it to its own particular set of conditions and rarely in nature do plants and animals succeed outside their normal habitats.

Particularly interesting are the adaptations shown by organisms to an environment which they have secondarily invaded. Such are the aquatic angiosperms and the marine mammals. These forms have retained all the major features characteristic of their classes but exhibit many secondary adaptations necessary to their success in the new environment.

ADJUSTMENT TO ENVIRONMENTAL CHANGES: REGULATION AND SOME MECHANISMS LEADING TO HOMEOSTASIS

It was stated at the beginning of this chapter, that no environment remains absolutely constant for any considerable time. Many living things, particularly animals, are able to make chemical and physical adjustments to their bodies which counteract the effects of the fluctuating environmental conditions. All such adjustments are effected by *regulatory mechanisms*. The main effect of such homeostatic control processes is to give the organism some degree of independence of its surroundings and thus of the conditions which are thrust upon it. In general, the more lowly animals and plants do not possess any great powers of control and cannot tolerate more than slight variations in the environmental conditions; the more highly evolved types have developed regulatory powers to varying degrees, and are less dependent on unchanging conditions; for these, wider ranges of habitat are possible. The height of such power is seen in man, who can indeed be said to create his own external environment around himself.

The more obvious and important environmental fluctuations which are likely to occur are those in temperature, light intensity, osmotic conditions and hydrogen ion concentration. The mechanisms by which animals make the necessary bodily adjustments to meet these changing factors in order to prevent their worst effects, are briefly discussed below. By reason of their form and mode of life, plants generally are not capable of such finely balanced processes. They do, of course, make responses to changing external conditions and these together with comparable responses made by animals are described in Chap. 16.

Regulation of Body Temperature

Only the birds and mammals possess the power to regulate the body temperature so that it is kept at a constant value irrespective of the environment temperature. Their body temperature is, as it were, thermostatically controlled and we say they are *homoiothermic*. All other animals are, by contrast, *poikilothermic* and their body temperatures fluctuate with those of their surroundings. Because this is so, their metabolic rate shows the same trend of variation, causing them to be sluggish and inactive when it is cold and more lively and active when it is warm.

Homoiothermic animals control the body temperature in two main ways, through regulation of the actual amount of heat produced in the body and by control of the amount of heat lost from the body surface. Fluctuations in environmental temperature are perceived by the body in two ways. Firstly, sensitive nerve-endings in the skin, the *organs of Ruffini* and the *bulbs of Krause* (*see* Vol. I, Chap. 20), are specialized for the perception of temperature changes. Stimulation of these minute sense-organs leads to reflex control, through the autonomic system, of heat loss. Secondly, changes in the temperature gradient between the environment and the blood will affect blood temperature; changes in the blood temperature stimulate the temperature-sensitive centre in the brain, and this sets into action a number of mechanisms which affect heat release in the body.

Heat is released in the body by respiratory activity, most of which takes place with a carbohydrate substrate, but there is always some respiration with fat and protein substrates (*see* p. 443–4). Broadly speaking, about 55 per cent of the energy released is in the form of heat; most of this heat is released from the muscles (33 per cent), the liver (13 per cent) and the remainder in small amounts from all the other tissues of the body. The heat is equated approximately throughout the body by transport in the blood and tissue fluids. Heat is lost from the body by radiation, convection, evaporation from the surface and from breathing passages, and there is some small loss in defaecation and urination. The delicate balance between heat production and heat loss results in homoiothermy.

Respiratory activity, assuming a sufficient supply of substrate, and of oxygen, is controlled by a number of hormones; thyroxin, adrenalin, glucocorticoids, insulin, glucagon and somatotrophin all play a part in the control process under different conditions. The effects of these hormones are described in the next section, on the regulation of blood sugar. Two effects may be noted here in connection with heat release. At low temperatures, there is greater secretion of adrenalin, leading to greater output of glucose from the liver into the blood, thus providing more respiratory substrate for all the cells of the body. In connection with thyroxin, its presence affects uncoupling of oxidation from phosphorylation so that excess respiratory substrate can release more energy as heat, and less in making ATP molecules. Certainly, excess of thyroxin leads to increase in basal metabolic rate and loss of weight, whereas deficiency of thyroxin leads to decrease in basal metabolic rate and increase in deposited fat.

In cold conditions, any increase in muscular activity will cause greater heat release; shivering is reflex shaking of muscles; human beings rub their hands, stamp their feet or spend a few minutes in physical exercise for the same purpose. On the other hand, in hot conditions, the siesta in the shade cuts out all muscular activity except that of the heart and certain visceral muscles such as those of the gut and blood-vessels, and the muscles involved in breathing.

Regulation of Heat Loss

The main source of loss of heat is evaporation of sweat from the surface; most of the latent heat necessary for vapourization is taken from the body. Control of the amount of sweat will therefore help to control heat loss; the greater the amount of sweat excreted by the sudoriparous glands, the greater will be the loss of heat and vice versa. The amount of sweat is increased by a greater supply of blood to the surface blood-vessels; this is effected by reflex dilation of the vessels or *vasodilation*. Decrease of blood is effected by *vasoconstriction*.

Temperature changes perceived by the skin sensory receptors initiate reflex impulses which proceed via vasodilator and vasoconstrictor centres in the brain to the muscle of blood-vessels in the dermis, the ultimate response being dilation or constriction, especially of the precapillary arterioles. The nerve fibres concerned with vasodilation belong to the parasympathetic system and they liberate acetylcholine at their terminations in the muscle; those fibres concerned with vasoconstriction belong to the sympathetic system and liberate adrenalin. Thus the ultimate control of the unstriated muscle in the blood vessels is hormonal.

In hot conditions, mammals utilize more rapid and shallower breathing as a means of increasing the amount of evaporation from the respiratory tract and the lungs. The shallow breathing counteracts the higher rate so that oxygen intake and hence respiratory activity will not be increased. This rapid breathing, coupled with the open mouth and the lolling tongue can be seen in a dog on a hot day.

In addition to the major mechanisms of regulating body temperature, many mammals can erect the hair by reflex use of the arrectores pili muscles (see Vol. I, Chap. 29). This creates a thicker layer of still air immediately next to the skin, thus acting as insulation in cold conditions. Birds have a similar mechanism for erecting the filoplumes and down feathers (see Vol. I, Chap. 28).

Poikilothermic animals possess none of these mechanisms and are thus at a disadvantage when the surrounding temperature becomes too high or too low. If such animals live where the air temperatures are above 50°C for long periods during the day, they must shelter by burrowing or by some other means, otherwise they will die from over-heating. But it is possible that some do show a slight degree of thermoregulation through the activity of the pigment cells which occur in their skins. These may serve to control heat absorption and radiation under differing atmospheric temperature conditions. In some reptiles, a high body temperature (40°C) leads to a blanching of the colour and presumably decreased heat absorption. Conversely, a low temperature (5°C) causes darkening and presumably increased absorption. The colour changes are brought about by concentration (to lighten) and dispersion (to darken) of pigment in the chromatophores (see Chap. 16).

When poikilothermic animals are kept at comparatively low temperatures, they become exceedingly sluggish and show no normal activity whatsoever. The majority of these animals, and some which are homoiothermic, relapse into the condition known as *hibernation* if, during the winter, they are unable to migrate to warmer areas. They cease to move or feed and their metabolic rate drops to nearly zero. Hibernating mammals include insect-eating bats, hedgehogs and many rodents; of the carnivores, only the arctic bears and the racoon-dog hibernate. At low temperatures, usually a little above freezing point, these animals fall into the characteristic state of torpidity, but there is no clear-cut relationship between temperature and the state of the animal; hibernation is not due to low temperature alone. Such an interpretation would imply that a fall in environmental temperature to a certain level, would cause the animal to lose its ability to regulate its body temperature, i.e. from being homoiothermic it would become poikilothermic. That this explanation is too simple is clearly indicated by other known facts concerning hibernation. For example, the behaviour of an animal, after the torpid state has initially been reached, depends upon other factors besides low temperature. Some hibernating mammals can be aroused both by increases and decreases in the low temperature of their surroundings. Such is the marmot of North America. Further, the depth of coma reached by different animals is quite variable. By November, the dormouse has built up a large quantity of fatty tissue in its body and made a nest in which food material is stored. It retires to this nest and sinks into so torpid a state that it gives the impression of being dead. Its breathing is imperceptible; it is cold, and its body is rigidly rolled into a ball. Its comatose state is very profound. It can be aroused if the temperature is raised slowly, but is killed if exposure to high temperature is too sudden. All hibernating mammals do not sink into so deep a torpor or stay in the state as long. The pipistrelle bat shows an intermittent hibernating behaviour.

Lack of food could be an important factor in the hibernating habits of mammals. Preparatory to sinking into the resting state, most of them accumulate fat in the tissues; the fat is used during the long cold period. Some animals accumulate stores of food in the nest and make use of them during short bursts of activity between the longer comatose periods. Further, it is known that animals of the same species at different latitudes or different altitudes tend to commence hibernating at different times which can be correlated with the disappearance of suitable food. Obviously all the hormones concerned with the regulation of metabolism (see pp. 279–289) play an essential part in the process of hibernation.

Hibernation is the rule for those reptiles and amphibians which live in the cold and temperate latitudes. The body temperature cannot in any case be kept above that of the surroundings, and food would be scarce in winter. Snakes, lizards and tortoises either burrow or retire to rock crevices for their "winter sleep." The salamanders, toads, newts and most frogs hibernate in dry places but the common frog, *Rana temporaria*, prefers mud at the bottom of a pond. Fish in general, do not hibernate, but there are instances of some such as the carp, burrowing into mud and remaining there until the coldest periods are over.

Many invertebrates hibernate. Insects vary in the stage of their life history at which the hibernation period occurs. Some caterpillars hibernate immediately upon hatching, e.g. the silver-washed fritillary butterfly. The eggs hatch in August but the larvae do not commence activity until the following spring. The larvae of the pearl-bordered fritillary undergo first a period of feeding when they hatch in June, then in July retire to the undersides of the leaves of the food plant, dog-violet, where they remain inactive until the next spring, when they commence feeding again. Some caterpillars hibernate at the end of the larval period, burrowing into the ground in winter and pupating in the following spring. The large tortoise-shell butterfly hibernates in the imago state, very soon after its emergence from the pupa. The fact that some lepidopterans which do not feed during the imago period are fully active in the winter, points to the fact that food supply may be a controlling factor in inducing the condition in other animals. Such moths seem to be unaffected by a lowering of temperature and retain their vitality even when the temperature is quite low.

Land snails can be found hibernating in clusters in sheltered crevices, under stones or litter or buried in the ground. They are completely withdrawn into the shell and the entrance is sealed off by a membranous disc. The period lasts from late autumn until spring. The slug burrows, contracts into a ball and secretes a slimy covering. Water snails may burrow into the mud although some, like *Planorbis*, may remain active even under ice. Spiders do not seem to react to low temperatures in any corresponding way at all, beyond seeking cover. This may also indicate a controlling influence of food supply, since spiders are able to last many months without feeding and thus do not need to enter a torpid state when food is short.

Although possibly not connected in any way with the ability or inability to control body temperature, the condition of *aestivation* can be mentioned here since it has many resemblances to hibernation. It is the state which an animal is said to enter when subjected to long periods of heat or drought and may be described as a condition of suspended animation comparable with that of a hibernating animal. It occurs in every group of vertebrates except birds, and in many of the invertebrates. Feeding, movement, respiration and secretion are almost at a standstill. In some animals, it coincides with the hottest and driest periods of the year and can be automatic behaviour, irrespective of temperature, moisture or food supply. In the case of some aquatic animals such as fishes, the condition is forced upon them by the drying up of the water. Some fish, crocodiles and alligators will bury themselves in the mud and not revive until the water returns. When its habitat dries, the water tortoise, *Clemmys leprosa*, wedges itself into a rock cranny and stays in a state of torpor for months. It has been reported that even frogs can achieve the same state in sun-baked clay. The African lung-fish, *Protopterus*, and the South American *Lepidosiren*, are excellent examples of fish which are forced to aestivate. They burrow into the mud, secrete a covering of mucus which is open over the lips to allow air to pass to the lungs, and then remain completely inactive until the water channels are once more filled by the rains. Stored fat and some of the muscular tissue of the tail disappear during the period of aestivation. There are many examples of similar behaviour among representatives of the invertebrates, particularly the molluscs.

Regulation of Blood Glucose

The tissues of the body derive their energy almost entirely from the oxidation of glucose via the EMP pathway and Krebs' cycle (*see* Chap. 11). The glucose is supplied from the blood, which, in mammals, is regulated to contain approximately 0.1 g per 100 cm^3. After a meal, even though considerable amounts of carbohydrate may be taken, the blood glucose does not rise above 0.12 g per 100 cm^3, because absorbed hexoses are rapidly removed from the blood, converted into glycogen, and stored. In a healthy man weighing 70 kg, there will be about 500 g of glycogen; 60 per cent in the muscles, 35 per cent in the liver, and small quantities in the brain. The glycogen in the liver is the only store which is normally converted into glucose and released into the blood, and this interconversion,

$$\text{liver glycogen} \rightleftharpoons \text{blood glucose}$$

is under the control of hormones, so that the level of blood glucose remains fairly constant.

The blood is continually losing glucose to the tissues from the capillary circulation; blood in a venule contains 2 to 3 per cent less glucose than blood in an arteriole. Any excess of glucose above the glycogen storage maximum is converted into fat in the liver and stored in the various adipose regions. Hence, excess carbohydrate in the diet will lead to increase of weight by fat storage.

Blood glucose is continuously supplemented from liver glycogen plus small quantities from protein and fat breakdown, which also takes place in the liver. The delicate balance between gain and loss of blood glucose is maintained by the action of a number of hormones. The mode of action of hormones is discussed in Chap. 8.

Thyroxin affects the basal metabolic rate, which is the rate of energy release necessary to maintain the body in a resting condition. It is particularly concerned with slow adjustment of this rate to fluctuations of external temperature and to the varying growth rate at different times in the life cycle. Adrenalin causes rapid mobilization of glucose from the liver under conditions of stress, low temperature, or hypoglycaemia. Its effects are short-lived as it is rapidly destroyed in the tissues. The glucocorticoid hormones (see p. 283), from the cortex of the suprarenal glands, also affect carbohydrate metabolism by increasing the rate of conversion of fat and protein breakdown products to carbohydrate (see p. 416).

The most important regulators of the glucose-glycogen balance are, however, the hormones from the pancreas, insulin and glucagon. The cells of the islets of Langerhans are directly sensitive to the concentration of blood sugar and, in the presence of any excess, insulin is secreted. This has the effect of facilitating the conversion of blood glucose into glycogen in the liver, so tending to reduce the concentration in the blood to normality. The cells of the islets secrete glucagon in response to stimulation by somatotropin from the pituitary gland (see p. 280). Glucagon stimulates the breakdown of liver glycogen to glucose which is secreted into the blood, thus counteracting the effect of insulin.

These hormones together form a complex dynamic system which tends to ensure that there is always sufficient, but not excess, blood sugar to supply the respiratory needs of all the cells in the body.

Deficiency of insulin entails a shift in the glucose-glycogen balance, so that there is little conversion of glucose to glycogen, but rapid conversion of glycogen to glucose, with glucagon acting more effectively owing to the relative shortage of insulin. This leads to accumulation of glucose in the blood, a condition known as *hyperglycaemia*; there is rapid excretion of glucose in the urine (*glycosuria*), increased and incomplete breakdown of fat for energy release causing the formation of poisonous ketones (*ketosis*), and excretion of these ketones in the urine (*ketonuria*). In extreme conditions, these symptoms are recognized as *diabetes mellitus*, when glucose, acetone, ammonia and a high degree of acidity are found in the urine. The condition can be relieved by regular, carefully calculated doses of insulin, injected into a vein. Excess of insulin causes *hypoglycaemia*, when the blood sugar level falls rapidly with increasing glycogenesis in the liver. The condition is fatal if blood glucose falls below 0.04 g per 100 cm^3, unless there is rapid administration of glucose by ingestion or injection.

Regulation of Muscular Movement

The structure of the three types of muscular tissue, voluntary, cardiac and unstriated, has been described in Vol. I, Chap. 20. The physiology of muscle is discussed on pp. 466–479. Since there are differences in the type of control, the regulation of each type of muscle will be described separately.

Voluntary or skeletal muscles, with the exception of certain of the tongue muscles, are attached to the skeleton by tendons; the bones themselves are joined together by ligaments. Muscle fibres, tendons and ligaments are constantly relaying impulses to the brain from minute proprioceptors. Among and around the individual muscle fibres are spindle organs, whose nerve-endings are sensitive to the degree of contraction of the fibres. In the tendons are small sensory organs which detect the state of tension in the tendons, and further proprioceptors are sensitive to the degree of stretching of the ligaments. When any movement is to be made in response to internal or external stimuli, all three types of sensory organ initiate impulses which are propagated along sensory nerves to the brain. In the brain, information from these three sources is co-ordinated and appropriate impulses are propagated along motor nerves to their endings on the muscle fibres. When an impulse reaches an end-plate (*see* Vol. I, Chap. 20), some acetylcholine is released; this depolarizes the membrane of the muscle fibre by increasing its permeability to all ions. Thus an action potential is initiated and propagated along the muscle fibre; contraction follows. The sequence is therefore: sensory nerve-endings, information to the brain, co-ordination, motor impulses to the muscles, depolarization by acetylcholine, action potential, contraction of muscle fibres.

Cardiac muscle has a built-in system of control and the vertebrate heart will continue beating regularly even when all nervous connections are severed. This type of heart is known as *myogenic* (*see* p. 338). In the following account the mammalian heart is described.

A network of modified cardiac muscle, which has the power of conducting impulses, ramifies through the walls of the auricles from an area in the dorsal wall of the right auricle anterior to the entrance of the superior vena cava. This area, known as the *sinu-auricular node* (*see* p. 338) is stimulated when the incoming blood almost fills the right auricle. When the pressure of the blood reaches a certain value, an action potential is initiated at the node and this spreads very rapidly via the conducting (Purkinje) tissue, and the two auricles contract together. Meanwhile, the conducting tissue, which is the only cardiac muscle to pierce the auriculo-ventricular septum, transmits the action potential via the auriculo-ventricular node into the papillary muscles and into the ventricle walls. When the ventricle is full of blood, the papillary muscles contract first, thus exerting sufficient pull on the chordae tendineae to stop the valves opening into the auricles. Then the two ventricles contract, driving the blood into the aorta and pulmonary artery respectively.

Such a steady state could not persist permanently; the heart must be able to cater for extra demands such as are required during muscular exercise, emotional stress, high environmental temperature and during digestion. In addition, a number of abnormal conditions will cause acceleration, e.g. fever and haemorrhage. Change of rate of beat and hence change of output are influenced by nervous impulses propagated in both sympathetic and parasympathetic systems. Cardiac and vasomotor control centres situated in the medulla oblongata are influenced by impulses from higher centres in the brain as well as from various parts of the body. In particular, sensory nerve-endings in the wall of the aortic arch, and also in the common carotid artery before it branches, are stimulated by a change in blood pressure from the normal. Afferent impulses pass along nerve fibres in the vagus nerve (cardiac depressor branch) and the sinus nerve (branch of the glossopharyngeal) to the control centres in the medulla. The motor outlets have two routes and two opposed effects.

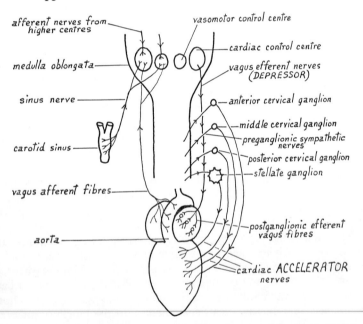

Fig. 6.4. Diagram illustrating nervous control of rate of heart-beat; efferent nerves shown only on right of diagram, afferent nerves on left.

Parasympathetic fibres in the vagus, again cardiac depressor branch ultimately release acetylcholine which causes slowing of the rate o heart-beat. Sympathetic fibres conduct impulses via the spinal cord and the three cervical and stellate ganglia (*see* Fig. 6.4). The post-ganglionic fibres end in the heart muscle, liberate adrenalin, and rate of beat is increased.

In addition to nervous control of the rate of heart-beat, certain substances carried in the blood can affect it. The most obvious substance is adrenalin, secreted by the suprarenal medulla; this has a dramatic effect on rate of beat and output of blood, increasing both considerably. The hormone vasopressin, a neurosecretion from the hypothalamus, is liberated into the blood through the pars nervosa; by its effect on constriction of the blood-vessels, it also has an effect on the rate of heart-beat. Excess of carbon dioxide in the blood affects the cardio-inhibitory centre and also reduces the rate of impulse propagation from the sinu-auricular node and depresses conduction in the Purkinje tissue; the rate of heart-beat is therefore slowed. Reduction of the carbon dioxide pressure in the blood causes the reverse effects. Low oxygen pressure in the blood causes an increase in the rate of heart-beat; under severe lack of oxygen the heart soon fails, being much more sensitive to lack of oxygen than is skeletal muscle.

Finally, cardiac muscle shows a tendency to contract in a definite rhythm, without any stimulation. The resting potential appears to be unstable and there is spontaneous depolarization which sets up action potentials (*see* p. 301). These are conducted easily from one cell to another so that the wave of contraction spreads all over the heart.

Unstriated muscle varies widely in its properties in different parts of the body. In general it is present in thin sheets surrounding cavities; these sheets have the muscle cells arranged in two distinctive lamellae, longitudinal and circular. Like cardiac muscle, there is some power of spontaneous contraction and there is conduction of action potential from one cell to another. However, unlike the other two kinds, unstriated muscle is capable of slow and sustained contraction. All unstriated muscle is innervated by the autonomic nervous system, the parasympathetic and sympathetic components having opposed effects. For example, in the arterioles, sympathetic stimulation causes contraction while parasympathetic stimulation causes slackening and inhibits contraction. This type of muscle is also very senstitive to adrenalin, so much so that some physiologists speak of a *sympatho-adrenal system*. Sympathetic nerve-endings release adrenalin, while parasympathetic endings release acetylcholine. Minute sensory entero-ceptors are present in all unstriated muscle, they are normally sensitive to stretching; afferent nerves pass to the control centres in the medulla, and after co-ordination, the output is controlled in the hypothalamus. A special case is the muscle of the uterus which is stimulated to contract by the hormone oxytocin, a neurosecretion from the hypothalamus. The stimulus in this case is probably the falling output of progesterone which sensitizes the muscle of the uterus to oxytocin.

Regulation in Response to Changes in Light Conditions

There are very few animals which do not adjust their body condition in some way in response to changing light intensity. In nearly all cases, specialized cells either singly or aggregated into compound structures, such as eyes, serve to detect the intensity changes. In some protozoa and in some echinoderms, there is an absence of such discrete photo-receptors, although these animals may still make responses to changing light conditions.

In most animals, the most obvious adjustment is to move the body from one place, where the intensity is too high or too low, to another in which the intensity is more favourable. It is purely a locomotive response brought about by the nervous co-ordinating mechanisms of the body, but can be considered regulatory in the sense that it allows the animal some independence of its environment. In general, it is con-ducive to greater chances of survival, either causing the animal to move from too brightly lit surroundings where it may suffer damage or fall easy prey to others, or taking the animal into more brightly illuminated areas where its food and possible mates are more abundant.

Rapid and intense fluctuations in the amount of light falling on an animal may have other effects besides invoking a locomotive response. First, the photoreceptors themselves may become impaired or permanently put out of action if subjected to too strong a stimulus, and secondly, large doses of radiation of the shorter ultra-violet wavelengths can be extremely harmful to most tissues. Many animals, and particularly those which have to live where they may be exposed to long spells of direct radiation from the sun, generally possess the means whereby they regulate their bodies to protect them from the worst of these effects. Those possessing eyes have the means of regulating the amount of light falling upon the light sensitive cells by muscular control of a diaphragm, the iris, e.g. vertebrates, or by screening with suitably placed pigment cells, e.g. arthropods. Most animals possess the power of developing pigment in chromatophores. These, by concentrating or dispersing the pigment, may regulate to some extent the amount of light which penetrates the body. It should be noted that the same chromatophores may serve other functions such as protection, aggressive coloration, display, and mating functions, these being responses to changing conditions other than light intensity. In the case of sea urchins and some other animals, the movement of pigment in the chromatophores is known to be closely associated with the brilliance of the illumination falling on the body and little if at all with the colour of the background, as is the usual case. Certain crustaceans, inhabiting very brightly lit areas of the Sargasso Sea, possess large quantities of white pigment which is dispersed during the day-time to form a good light-reflecting surface. The physiology of chromatophores is discussed in Chap. 16.

Apart from regulation in response to rapid fluctuations in the amount of light, there is regulation in many animals, particularly terrestrial forms, in response to light changes affecting the day-length or photo-period. Many of these slower and seasonal light changes give rise to rhythmical cycles of activity and inactivity. Some examples are described on pp. 226–228. It has been shown that in many birds, increasing day-length influences the enlargement and maturation of the gonads. In a number of mammals of temperate climates, anoestrus (no visible sex activity) is induced by decreasing day-length, e.g. the European hare; in the yak of northern Asia, anoestrus is induced by increasing day-length. This direct effect of day-length on the oestrous cycle has also been demonstrated in some of the smaller carnivores, e.g. the ferret. In these cases, day-length plays a predominant part, and temperature is of lesser importance; in many other cases, the reverse applies. The process of moulting in mammals is also seasonally controlled partly by the photoperiod and partly by temperature.

This regulation of reproduction and moulting, in response to the photoperiod, is associated with activity of the pineal gland. It is known that *melatonin*, a hormone from the pineal, apart from affecting skin colour in frogs (*see* p. 669), also inhibits maturation of the gonads, an effect opposed to that caused by increasing day-length. In lamprey larvae the change in skin colour from pale at night to dark by day does not take place if the pineal is removed. It is thought that in many vertebrates the pineal plays an important part between perception of light stimuli in the eyes and secretion of hormone releasing factors from the hypothalamus, especially those affecting secretion of gonado-tropins from the pituitary.

In some insects the phenomenon of *diapause* (arrested development) seems to be influenced almost entirely by day-length. According to species, diapause may occur in any of the four stages, egg, larva, pupa or imago. The summer caterpillars of the cabbage-white butterfly pupate, and after a short diapause of about two weeks, the imagines appear. But in the autumn caterpillars, the pupal diapause lasts from October till the following May, a period of about eight months. It has been shown, for some species, that the decreasing day-length leads directly to suppression of certain neurosecretions which normally stimulate the thoracic glands (*see* Vol. I, Chap. 23). These glands then fail to secrete the hormone *ecdysone*, which is essential for further development and for the process of casting the pupal skin. With increasing day-length, in the spring, stimulation of neurosecretion soon leads to further development and the imagines emerge. Experiments demonstrating this effect of day-length on diapause have been carried out for several species under controlled conditions; the temperature is kept constant and the photoperiod is varied.

Osmoregulation

A brief reference was made to this subject in Chap. 3 and it will be realized that some power of osmoregulation is important to an animal if it is to maintain a constant osmotic equilibrium with its surroundings. When the tonicity of these is not the same as that of the animal's body fluids and tissues or when the osmotic condition of the environment is liable to fluctuate, then the animal must make adjustments to its own body fluid tonicity or else run the risk of gaining or losing water to an extent which might be fatal.

Thinking first of aquatic animals, we have marine, brackish and fresh-water forms. There are two ways in which such animals may safeguard themselves against danger from changing osmotic conditions. One is to possess a body fluid which is adjusted in its tonicity to suit that of the medium at all times, the other is to keep a constant body fluid tonicity by some regulatory device, irrespective of the condition of the medium. Animals which use the former method are said to be *poikilosmotic*. They may be described as *osmolabile* since they do not regulate the body fluid tonicity at a constant value but change the fluid so that it becomes equated osmotically with that of the medium. Animals which keep a constant body fluid tonicity are said to be *homoiosmotic*. They may be described as *osmostable* since they regulate themselves so that they keep a constant internal osmotic pressure, whatever the external conditions.

Most marine invertebrates are poikilosmotic. Their blood and body fluids have an ionic constitution which is very similar to that of the sea-water. In most marine situations, the osmotic condition of the water is so constant that the inhabitants are never faced with osmotic problems at all. But if, under experimental conditions, they are removed to media of higher or lower concentration than normal, they lose or gain water according to the usual osmotic laws. Provided the change from normal is not too great, they may survive, but the degree of tolerance shown by different animals is very variable. When placed in a stronger saline solution than that of the sea-water, they will lose water and their volume shrinks. In this way the tonicity of the cells tends to increase towards that of the medium. The cells can only lose a limited amount of water without being impaired in their function, and outside this limit, the cells die. Conversely, when placed in a weak saline solution or pure water, the cells gain water and the volume increases, thus diluting the body liquids to meet the new conditions. Again, this can be tolerated only up to a limit, and in an extreme case, the cells will swell with water up to bursting point. Animals which are known to show these characteristics include the marine protozoa which lack contractile vacuoles, the sipunculid worm, *Phascolosoma*, some species of the polychaetes, *Arenicola* and *Nereis*, and some molluscs such as *Doris*, *Onchidium* and *Mytilus*. In some of these, the body-volume changes are not as great as might be expected from the conditions imposed and there is likelihood that a form of salt exchange occurs between the cells and the solution to assist in reaching an internal and external osmotic balance. Such a mechanism would come within the definition of regulation.

Other marine invertebrates show this ability to a much more marked degree, and in adjusting osmotically to the concentration of the medium, they do not change greatly in volume. They may tend to swell at first, when placed in a dilute solution, but quickly lose body salts and at the same time most of the excess water as their fluids become osmotically balanced with the new medium. Typical examples of such animals are the starfishes and sea-urchins.

The marine fish, although osmotically protected from the sea-water by relatively impervious skins, are not so protected at the delicate gill tissue and it is through this region that a fish runs the risk of losing water to the outside medium. The elasmobranch fish such as dogfish and shark, and the teleost (bony) fish such as herring and cod have distinctly different methods of dealing with the problem. In elasmo-branch fish, it is found that the blood is maintained at a higher osmotic pressure than the sea-water by a concentration of urea, an excretory product. Thus instead of losing water, they tend to gain it. Any gained in excess of requirements is removed at the kidneys, where a urine hypotonic to the blood is produced. It is thought that a special region of the kidney tubule of the elasmobranch fish is concerned with active reabsorption of urea into the blood in sufficient quantity to keep the blood slightly hypertonic to the surrounding water. The teleost fish regulates in quite a different way. Its blood is hypotonic to the sea-water and it must lose water through the gills continuously. To make up this loss, it drinks sea-water in some quantity, but as it does so, it not only takes in water, but considerable quantities of salts which must tend to send up the osmotic pressure of the blood. To prevent this, the teleost fish continuously excretes the excess salts through special cells in the gill epithelium. In such fish, the process of osmoregulation is not carried out by the kidney as it is in most vertebrates and it is a fact that in some fish, e.g. the goosefish, the kidneys are aglomerular, i.e. the glomeruli are vestigial, the work of the kidney being done by tubules only. It is thought that the teleost fish of the sea had their origin in fresh-water ancestors and that the migration to the sea necessitated a water-saving mechanism rather than a water-losing one.

Marine mammals such as the seal and the whale maintain a blood tonicity very similar to that of land mammals. They do not drink sea-water but tend to take in large quantities of salts in their diet. These salts would upset the blood tonicity when absorbed, and the animals are presumed to osmoregulate by producing a concentrated urine in which excess salt is eliminated. In the case of the whale, there are also special buccal glands which may be concerned with the excretion of salt.

In some marine environments, the concentration of the medium is liable to rapid changes. Such are the waters of estuaries and of brackish pools. Here can live only those animals which are able to tolerate the diverse conditions by regulating themselves accordingly. The flatworm, *Gunda ulvae*, crabs of the genera *Cancer*, *Carcinus* and *Eriphia*, and the polychaete, *Nereis diversicolor*, are typical examples. These animals are not wholly homoiosmotic since the osmotic pressure of their body fluids is not kept constant. Instead they tend to vary between iso-tonicity with sea-water and being slightly hypertonic to any dilution of sea-water. In the case of *Carcinus*, in dilute salines it maintains its body fluid hypertonic to the medium by secreting the excess water which it is continuously taking in through the antennary gland and gills. In *Gunda*, which is exposed to alternations of medium concentration twice a day as the tides change, the osmotic regulatory process is really one of water storage. As the medium becomes more dilute, water passes into the cells and the animal swells. At the same time, salts are lost back to the medium. The water which passes in, is eventually secreted into vacuoles in the cells lining the gut, thus eliminating it from other cells so that they do not become diluted. Whilst the water is thus stored, respiratory activity is increased, indicating that energy is being expended in isolating the water. When in full strength sea-water once more, the stored water passes out of the body.

In fresh-water, where the osmotic pressure is always low, animals will always tend to take in excess water and must therefore be equipped with the means of getting rid of it. Protozoa, such as species of *Amoeba* and *Paramecium*, possess contractile vacuoles. In most other fresh-water forms, the organ concerned with osmoregulation is the excretory organ. These have been described in Vol. I for flatworms, annelids and crayfish.

In sponges and coelenterates, e.g. *Hydra*, where there is no special excretory organ, the mechanism which prevents the cells from swelling and bursting is not at all well understood. A hydra, in it natural pond water, does absorb excess. The manner in which this could be eliminated may be as follows. The ectoderm cells actively pass ions into the mesogloea from whence the endoderm pumps them to the enteron cavity, to keep the osmotic pressure there above that of the body wall cells so that excess water now flows into the enteron.

Fresh-water fish do not drink the water but tend to absorb it through the gills. They excrete a fairly large quantity of dilute urine. Some salt is lost in this way and is replaced in the diet and by absorption through the gills. Animals which can live equally well in salt- or fresh-water are described as *euryhaline* (*stenohaline* describes those which can tolerate only very limited osmotic variations). The eel, *Anguilla vulgaris*, is a typical example, and in this case, if it has been living in sea-water and is transferred to fresh, it first gains weight by absorbing water and then loses it by commencing to excrete a hypotonic urine as do other fresh-water fish. If a fresh-water-adapted eel is placed in sea-water, it first loses weight and then regains it by drinking sea-water and excreting the excess salt through the gills in the same way as marine teleosts.

In the case of animals which are adapted to a land environment, the problem is to conserve water rather than to prevent its entry or to get rid of excess. They are in constant danger of drying up. In most of the invertebrates, particularly insects, water retention is achieved by protective external coverings which prevent evaporation and considerable reabsorption of water during excretion, so that the urine is almost solid. The earthworm is not truly land-adapted in this respect and must inhabit comparatively moist places, although some have been recorded as surviving a loss of over 70 per cent of their body water. Any excess water gained, is excreted through the nephridia as urine hypotonic to the blood and coelomic fluid, but containing salts and urea.

Land reptiles are comparable with the insects. They are externally well protected by horny scales and excrete a semi-solid urine, thus conserving water. The birds and mammals are able to excrete a urine hypertonic to the body fluids by reabsorbing in the kidney tubules much of the water which is filtered through the glomeruli.

In the case of birds, the nitrogenous end-product is uric acid which passes down the ureters in saturated solution. In the urodaeum, water is resorbed and the uric acid is precipitated as a pasty white mass which is passed out with the faeces, as a half-white, half-black pellet. Marine birds such as the cormorant, penguin, herring-gull, gannet, albatross, etc., have a diet which is very high in salt content. The problem of the elimination of excess salts has been solved by the evolution of special salt-excreting glands above the eyes (*see* Fig. 6.5). These glands are

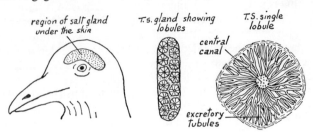

Fig. 6.5. Salt-excreting glands in a herring gull (blood supply not shown).

quite complex in structure, each consisting of a crescentic mass of lobules with each lobule having a radiating arrangement of secretory tubules all opening into a central canal. The central canals lead into a duct which opens into the nasal cavity. The solution excreted is saltier than sea-water, and in fact, it contains ten times as much salt as is excreted by the kidneys. The structure and mode of action of the mammalian kidney are described in Chap. 15; in this connection it should be noted that the kidneys are as much organs of osmoregulation as of excretion.

Amphibians, such as the frog, which may spend the greater part of its life in fresh-water, have been the subjects of much study on osmo-regulation. The findings seem to be that when in the water, the frog gains water osmotically through the skin but not at a very rapid rate, since the skin seems to have a low permeability. Any excess water thus taken in, is got rid of as dilute urine containing some salts. The salt loss is made up partly in the diet and partly by intake through the skin, and the loss is made less than it otherwise would be, by reabsorption from the urine in the distal part of the renal tubules, On land, the frog is subject only to water loss, as are other land animals, and it must have continuous recourse to liquid water to make up any deficit. It is probable that the frog uses its bladder and subcutaneous lymph spaces for the purposes of water storage whilst on the land. Toads are much less reliant on continued access to liquid water and this may be so because of their ability to excrete a much more concentrated urine than frogs, indicating a much greater capacity for water reabsorption in the kidney.

Hormonal Control of Osmoregulation

Many of the examples given in the preceding paragraphs show clearly that osmoregulation involves both water balance and ionic balance. As in the course of evolution various types of organs for excretion and osmoregulation have arisen, so also have mechanisms been evolved for controlling the amounts of water and ions lost to the body from these organs. There are hormones which cause increase and decrease of water loss and ionic loss. These hormones and their actions have been most thoroughly investigated in mammals, but there is little doubt that similar mechanisms exist in all the vertebrates. Among invertebrates, most of the work on hormonal osmoregulation has been done on insects.

In the mammals, at least four hormones have osmoregulatory effects concerned with water balance and ionic balance. Following direct perception of the osmotic pressure of the blood in the brain, nervous impulses to the hypothalamus lead to the neurosecretion of vaso-pressin, which is liberated into the blood via the pars nervosa of the pituitary gland. The hormone has two important effects: it causes a rise in blood pressure and has an antidiuretic effect in controlling re-sorption of water and Na^+ ions from the kidney tubules. The para-thyroid glands appear to be directly sensitive to the concentration of Ca^{++} ions in the blood; the secretion of these glands, *parathormone*, controls the excretion of Ca^{++} and HPO_4^- ions. Renin, a hormone liberated from the juxta-glomerular complexes, (*see* p. 288) affects vasoconstriction and stimulates liberation of mineralocorticoid hor-mones from the adrenal cortex; these tend to diminish loss of Na^+, Cl^- and HCO_3^- ions, and increase loss of K^+ and HPO_4^-. In addition to these, any hormone which affects the concentration of blood glucose, will also affect osmoregulation. Further, any effect on blood pressure must affect kidney filtration.

It is obvious that the combined action of all these different factors makes the process of osmoregulation extremely complex, especially when we do not know the precise mode of action of any one of the hormones. One can only generalize and say that they exert their effects by altering the kidney filtration and resorption mechanisms. Further details of all the hormones mentioned are given in Chap. 8 and kidney filtration is discussed in Chap. 15.

It has been shown that in the frog, the water content of the body is partly controlled by an antidiuretic hormone similar to that of mammals. Removal of the pituitary results in decreased permeability of the skin, and increased loss of water in the urine, whereas the presence of the hormone, from the intact pituitary, increases skin permeability and decreases loss of water in the urine.

The work on insect water-balance control shows that the animals produce an antidiuretic hormone by neurosecretion from some of the ventral ganglia, or the brain. The secretion is released after ner-vous stimulation originating in stretch receptors of the gut, which are themselves stimulated after feeding. The hormone appears to control water loss in two ways; by increased water flow through the Malpighian tubules and by decreasing rectal resorption. With a low concentration of the hormone, both processes are reversed.

Parasites which inhabit the body fluids or tissues of the host must of necessity come to terms osmotically with their medium, but little is known of the mechanisms they employ.

A few animals living in peculiar osmotic environments are worthy of special mention. The brine shrimp, *Artemia*, can live in salt concentration up to 25 per cent. It maintains a body fluid very much hypotonic to that of the salt water, as low as 10 per cent of the medium concentration. It is not certain how this condition is maintained, but low permeability to water, preventing its outward flow, and very rapid and efficient salt excretion may be the explanation. Some desert animals must find great difficulty in keeping a correct water balance of the body. The kangaroo-rat is known to remain healthy indefinitely without drinking. Its water intake is restricted to its food material, chiefly dry seeds. With the small quantity of water in these and from the water released during metabolism of the food, it is satisfied. This is due to the fact that the renal tubules are extremely efficient at water reabsorption and the urine produced is very concentrated by comparison with other mammals. The camel is able to survive long dry periods for a similar reason. It stores water, and when this is exhausted, it can withstand considerable dehydration without harm.

Regulation of the Hydrogen ion Concentration

Animals possessing no internal body fluids which can be maintained at a constant acidity or alkalinity are very much affected by changes in the pH of the medium in which they live. Such are the protozoans, some coelenterates, and flatworms. For example, locomotion in *Amoeba proteus* is affected by changing pH of the water. The animal moves most easily at around pH 7, slowing if the acidity is much removed from this value. Similarly the cilia of some organisms cannot beat in strongly acid media. Such animals have no means of protecting themselves against the worst effects of changing hydrogen ion concentration and quickly die when the changes are too drastic. When an animal possesses a body fluid, the pH of this is regulated at a constant value. Such homeostatic control is achieved partly by the buffering action of substances present in the body fluids (*see* p. 66), but also by the combined effects of the hormones which control water balance and ionic balance (*see* p. 210). For example, lack of mineralocorticoid hormones, due to malfunction or removal of the adrenal glands, results in loss of Na^+, HCO_3^- and Cl^- ions in the urine, and the pH consequently falls; administration of mineralocorticoids rapidly restores the pH to its normal value. The antidiuretic hormone from the pituitary affects pH directly by preventing excess loss of water and Na^+ ions from the kidneys, and indirectly by its stimulating effect on the adrenal cortex. There is evidently a delicate balance between the effects of the hypothalamus (neurosecretion of ADH), the pars nervosa (storage and release of ADH) and the hormones from the adrenal cortex; these factors, combined with the buffering action of substances in the body fluids, achieve a pH which, in a healthy animal, fluctuates about a constant value, and is constantly adjusted so that there is no radical or persistent change.

Regulation of Hormone Production

The production of hormones, which play such a large part in regulating various processes, is itself controlled by a number of mechanisms, the most common of which is the negative feed-back (*see* p. 147). Overall control of most of the endocrine glands emanates from the hypothalamus of the brain, which is the seat of control of autonomic output and also of the activity of the pituitary body. The hypothalamus has nervous connections with all parts of the brain, and may be said to be constantly receiving "information" about the internal condition of the body and about fluctuations in the external environment. To this "information," the hypothalamus acts in three possible ways: by nervous impulses sent out along autonomic nerves to target regions; by secretion of hormones into the pituitary body (*neurosecretion*), or by dispatching *releasing factors* which stimulate secretion of hormones by the pituitary.

In the embryo, the pituitary body is derived from two sources: a downgrowth from the floor of the thalamencephalon which forms the posterior portion called the *neurohypophysis* or *pars nervosa*, and a closed sac formed from a dorsal invagination of the stomodaeum, which forms the anterior portion, the *adenohypophysis*. During further development, the latter becomes subdivided into three parts, the *pars anterior*, the *pars intermedia*, and the *pars tuberalis*. No hormones are produced by the pars tuberalis which finally envelops the neck of the neurohypophysis to form the pituitary stalk through which blood and nerve connections pass from the hypothalamus into the pituitary body (*see* Fig. 6.6).

Neurosecretion by the hypothalamus occurs in several well-defined groups of neurones, whose axons pass through the pituitary stalk into the pars nervosa and there terminate in slight swellings. Droplets of fluid, containing the hormones, pass along the axons and are liberated in the pars nervosa.

During this transfer the hormones are probably allied to proteins which render them temporarily inactive. In the pars nervosa the proteins are detached and the active hormones are absorbed in the blood and tissue fluid. Two important hormones are produced in this manner; they are *vasopressin* and *oxytocin*, each of which has two very distinct

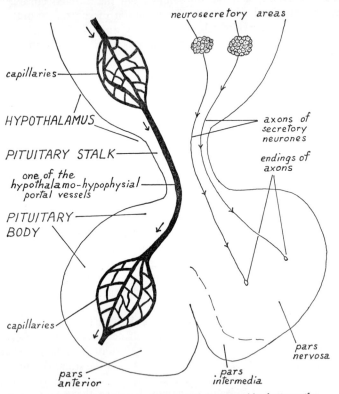

Fig. 6.6. Diagrammatic representation of the relationships between the hypothalamus and the pituitary body.

effects. Vasopressin causes a rise in blood-pressure effected by contraction of the smooth muscle in blood-vessel walls, and also has an *antidiuretic* effect which increases resorption of water and sodium ions from the uriniferous tubules. Secretion of vasopressin is initiated by one or both of two factors; direct perception of osmotic pressure of the blood in the brain and thence impulses to the hypothalamus, and by perception of blood pressure by receptors in the heart and arterial trunks and thence impulses to the brain and eventually the hypothalamus. Cessation of the stimulus in each case is followed by the end of secretion of the vasopressin. This is a direct feedback control; when the blood pressure and osmotic pressure reach a standard value, the hormone production is cut off. Oxytocin causes contraction of the uterus smooth muscle during the birth process, and also causes contraction of the myo-epithelial linings of the mammary alveoli, thus forcing milk out of the pores in the nipples. The initial stimulus in the first case is from stretch receptors in the uterine muscle, and in the second case from receptors activated by the sucking of the infant. In each case nervous impulses finally stimulate the hypothalamus. The secretion of oxytocin ceases when the receptors are no longer stimulated.

The pars anterior produces at least six hormones; the production of each is stimulated by a special releasing factor from the hypothalamus. These factors, also to be regarded as hormones, pass into the pituitary in the blood of the *hypothalamo-hypophysial portal system*. Capillaries in the anterior region of the hypothalamus unite to form veins in which the blood flows down the pituitary stalk; there these veins give rise to a second set of capillaries, thus forming a typical portal system (*see* Vol. I, Chap. 26). From the second set of capillaries, the releasing factors stimulate the secretion of hormones, each hormone emanating from a specialized group of cells.

Growth hormone, *somatotropin*, is probably secreted throughout life, but its output is considerably increased during periods of rapid growth. It also has a diabetogenic effect, acting antagonistically to insulin, by promoting the production of glucagon in the pancreas. Somatotropin may be regarded as a permanent hormone, though its effects on growth are modified after puberty, probably by the action of other hormones such a thyroxin which counteract its activity. *Thyrotropic hormone* stimulates the thyroid to release thyroxin; the presence of thyroxin in the blood in sufficient quantity inhibits production of thyrotropic hormone. This is a straightforward feed-back mechanism. Fall in the basal metabolic rate will start the cycle again. *Adrenocorticotropic* hormone (ACTH) stimulates the production of various "corticoids" in the adrenal cortex. The production of ACTH releasing factor is initiated by perception of salt imbalance, shortage of glucose and certain other factors (*see* p. 213). These corticoid hormones, when present in the blood in sufficient quantity, inhibit production of ACTH. Several *gonadotropic hormones* are produced in the pars anterior; their control is described in Chap. 17.

The production of some hormones is quite independent of the hypothalamus. For example, insulin is secreted in response to a high level

8

of blood glucose with the result that more glucose is absorbed into the liver to form glycogen. As the blood glucose reaches its normal value the secretion of insulin ceases. The presence of acid chyme in the duodenum evokes the production of secretin from the intestinal wall. This circulates in the blood and stimulates the pancreas to secrete pancreatic juice. When the movement of chyme is completed, secretin production is cut off and hence the pancreatic juice ceases to flow. Examples of other alimentary canal hormones are discussed in Chap. 8.

The cases of hormone production which have been given above, are in many cases simplifications. The complete interplay of the various hormones from the endocrine glands is extremely complex and it is rarely possible to isolate the effect of one hormone from that of others. Homeostasis is not a static concept; such a state never exists in living creatures; it is essentially dynamic and is concerned with maintaining not a state, but a multifarious set of activities, all of which are necessary to promote harmonious growth and development. The regulation of hormone production is a vital factor in promoting such a state of constantly controlled change.

ENDOGENOUS RHYTHMS

Not only do plants and animals exist in a background of space that is otherwise occupied by non-living systems, the material nature and physical states of which exercise such marked controlling influences on them, but there is also a background of time into which they fit. To all but human beings this passage of time can be apparent only as a series of cycles of changes in physical and chemical conditions, for example, the endless, unbroken alternation of the light and dark periods of day and night or the longer period sequence or frequency of the seasons or the tidal motions related to the regularly changing phases of the moon. Over the millions of years that living things have existed in rhythmically fluctuating conditions it is not surprising that they have become in many ways conditioned by them or adapted and make responses directly related to them. There are most certainly what can be called rhythms of activity or change under the influence of varying external or exogenous factors and these are sometimes described as *exogenous rhythms*.

At first consideration it might appear that any patterns of activity or behaviour of a plant or animal showing a constant regularity of change is in response to some fluctuating external environmental condition such as temperature or light intensity. If such were the case it would be expected that such rhythms of physiological or chemical activity would disappear immediately the subjects were placed under unchanging, uniform conditions. This is so for a number of events, for example, spore discharge in response to interacting light and temperature and humidity conditions by some fungi. But it is certainly not so for others and there undoubtedly exist what can be referred to as *endogenous rhythms* in which the influence controlling the event originates from within the organism and not from outside it. In such cases, application of an environment of uniform nature does not cause the previous normal activity to change, at least not immediately, and the organism continues to show its rhythmic behaviour for a time. Examples of such phenomena include a wide range. The rhythmic luminescent activity of some algal forms, the dispersion of pigment in some isopods on a regular basis of time, the timing of the emergence of some insects from pupae, the "sleep" movements of leaves and fluctuations in metabolic rates in both plants and animals are among them. All show a periodicity of occurrence not relatable to any known environmental factors. There could, of course, be regular stimulation of an organism by some unknown factor or factors, in which case these rhythms would not be endogenous but it is generally accepted that a rhythm is regarded as such provided it conforms to the following conditions: it continues when the subject is placed in as uniform an environment as it is possible to create; it should be possible to advance or retard a particular momentary state of a rhythm, that is, its *phase*, without changing it in any other way; under anaerobic conditions the phase of the rhythm should be held up or arrested as would be the case for any other internal process dependent upon energy release; it should originate from the application of one stimulus only, such as an abrupt change from continuous light conditions to continuous darkness and there must be no possibility that the rhythm has been impressed by oscillations of the environmental conditions; the frequency of the rhythm should not be exactly twenty-four hours, tending to diminish the likelihood that the rhythm and its frequency is the result of a daily periodic stimulation by an unknown external factor.

Rhythms of activity that meet these criteria are judged to be reflecting periodic variations in cellular processes, that is, are endogenously originated. Such known oscillations in the internal cell workings can continue under uniform environmental conditions for varying periods, from a week or two in plants, up to several months in some animals, the difference in the times of the damping of the oscillations in the two kinds of organisms being due possibly to interference with nutritional processes in plants when they are subjected to continuous uniform conditions, particularly lack of illumination.

Cellular systems that exhibit such endogenous rhythms are commonly referred to as *biological clocks* in the sense that the rhythm is an expression of a system of time measurement. It is the case that at least one of these, namely, that manifesting itself as "sleep" movements in leaves, has been a subject of study off and on since the days of Pliny the Elder (AD 23–79) but it is only during very recent times that the real nature of such a periodic change has been conceived and a new field of biological study opened up, as usual with its own vocabulary. Other manifestations of the ability to measure time include the use of a "sun-compass" for orientation by some insects and photoperiodic responses by plants and animals. Most of what follows refers to endogenous circadian rhythms (*see* below).

In general, a rhythm can be visualized as a wave motion on a time scale (*see* p. 1008) indicating by its outline a rise and fall of activity at regular intervals. Among the terms and definitions now commonly in use, therefore, are some such as *period* and *phase* that describe respectively the time between two successive identical repetitions of a state or condition of the organism that is displaying the rhythm (the time between two successive peaks, for instance, on the wave outline) and the state or condition of the organism at any instant during one cycle of the rhythm (any specified point along the wave track of one oscillation). A *phase change* can be said to occur if a specified state is displaced in time from that normally expected and two similar organisms not exactly coincident for a particular rhythm are said to show a *phase difference*. The period (or wave frequency) of a rhythm having reached a steady state under uniform environmental conditions is referred to as the *natural* or *free-running period* of that rhythm. *Damping* refers to the smoothing out of the contrast between the states of an organism that make the rhythm discernible (the gradual decrease in amplitude of the wave form). It is found in many cases that the natural periods of endogenous rhythms are about one day, not exactly twenty-four hours but most commonly timed between twenty-two and twenty-eight hours. Such rhythms are called *circadian rhythms* (from *circa*—about and *diem*—day). Strictly, they should not be called *diurnal* rhythms because this would indicate a natural periodicity of exactly twenty-four hours but the description is often applied. When a circadian rhythm coincides with or rather is coupled to rhythmic changes in an external environmental factor in the way that most are normally coupled to strictly diurnal changes, then the endogenous rhythm is said to be *entrained* to this period or to have undergone *entrainment*. The forcing oscillation that entrains a biological rhythm is called a *synchronizer*.

From the evidence now at hand the existence of circadian rhythms is by no means universal in plants and animals but examples have been found in nearly all major groups. Notable exceptions are the mosses and liverworts. There is no uniformity, however, even between closely related species as witnessed by the existence of an endogenous rhythm of spore discharge in *Pilobolus sphaerosporus* but not in *Pilobolus crystallinus*.

Where biological clocks are in operation there seems always to be some accruing survival value, for example, the fact that insects such as *Drosophila sp.* emerge from pupae at exactly measured times, just before dawn each day, after the timing mechanism has been set by the previous onset of darkness, gives them two advantages at least. One is that because atmospheric humidity is likely to be highest at this time they are less likely to suffer death by drying out and the other is that, of a batch, all are likely to reach sexual maturity at the same time of day and hence be able to mate successfully. Other advantages that can be gained from synchronization of endogenous rhythms with physical rhythms of the environment are numerous and include the finding of suitable prey by predatory animals that hunt only by day or by night and the pollination of plants by insects, flower-opening coinciding with insect feeding.

When the physiological functioning of the body is forced to break the sequence of its natural circadian rhythms, serious pathological and/or psychological conditions can arise and man is one of the kinds of animals that demonstrate this very clearly. If he moves rapidly from one place to another on the earth's surface so that he undergoes an abrupt change in the timing of environmental conditions such as the day and night sequence to which his functions are entrained, then he invariably suffers disturbances in his metabolic and other physiological processes. There is no doubt that for the well-being of most organisms, biological clocks should be allowed to function without interference.

Current investigation is aimed at elucidating the nature of the basic oscillating system or timing device in the various organisms that possess one but so far this has not been achieved. It is probable, however, that it is not the same in all living systems and this makes the whole phenomenon still more complicated. Some of the characteristics of these timing mechanisms, discovered through studies of one or more of endogenous circadian rhythms, photoperiodic responses or direction-finding by some animals (this dependent on timing via the sun's position relative to the horizon) have been described. It is not possible here to give a comprehensive account of the work so far completed, only to indicate a few of the features of biological clocks.

One characteristic that has become clear is that of all the normally operating environmental factors, light, temperature and oxygen concentration have the most powerful effects on the working of an endogenous oscillating system or timing device. Departures from the normal cycle of variation in either of these can often bring about entrainment of an endogenous rhythm to other than its normal circadian frequency. For instance, the leaves of runner bean, *Phaseolus multiflorus*, previously showing a twenty-four hour cycle can be made to show an eighteen hour cycle of "sleep" movements when subjected to continuous alternation of nine hours light and nine hours darkness. The normal rhythm of "sleeping" is recovered as soon as a uniform environment is supplied. But this is not the case in some other known instances of entrained rhythms. Entrainment is sometimes accompanied by what is known as *frequency demultiplication*, meaning that the period of the entrained rhythm becomes a multiple of the period of the entraining cycle and not equal to it. For example, cycles of six:six, three:three and sometimes two:two hours of light and darkness can produce a rhythm with a period the same as that shown when the organism is exposed to twelve:twelve hour variation. The rhythm of phototactic responsiveness in *Euglena gracilis* can be entrained to sixteen hour periods when subjected to eight:eight hour light-dark cycles.

Some environmental conditions can prevent the timing device from operating at all. Of these, anaerobic conditions, high and low temperatures and very bright light are most effective. Low oxygen tension is known to cause loss of rhythm in the rate of growth of oat coleoptiles and other plant parts. The rhythm of the rate of carbon dioxide metabolism in *Bryophyllum fedtschenkoi* disappears when the plant is placed under high temperature conditions (36°C) but reappears when the temperature is lowered to 25°C. Continuous exposure to very bright light has the same effect. When the dinoflagellate, *Gonyaulax polyedra*, is continuously treated with bright light it is unable to operate its rhythmic luminescent activities but does so again when it is placed in continuous darkness.

The time to complete a full cycle, that is, the period or frequency of a rhythm, and the mechanism that controls this is clearly of paramount importance if the oscillations are to form units of measurement on a time base. Alterations in the period, even slight ones, could have a serious displacement effect in the timing of an event. As previously stated, most organisms show a free-running period of from twenty-two to twenty-eight hours in their normal rhythms and for an individual case deviations from a mean value are usually less than an hour and measured values can generally be replicated experimentally. For animals in general the exactness with which periods are maintained by individuals is quite extraordinary; in a case quoted for a particular rodent its period value for an activity rhythm remained the same within a minute or two for some months. But differences between individuals within species do occur and can be as much as an hour or more. For example periods of the same body rhythms such as urinary excretion, temperature fluctuations and pulse frequency in individual human beings under constant conditions and in complete isolation from the outside world have been shown to vary between 24·7 and 26 hours. It should be remembered that the fact that in normal daily life these endogenous rhythms follow an exact diurnal cycle is, of course, due to their entrainment to a twenty-four hourly cycle by some synchronizer such as the onset of darkness; they are not free-running. Treatment of a subject with continuously unchanging conditions, particularly light, frequently results in changes in the length of a free-running period. But the factor most commonly affecting the periods of circadian rhythms is temperature. This is to be expected if it is accepted that the basic timing system is made up of a number of inter-related chemical reactions each dependent on the temperature of the surroundings for the rate at which it proceeds.

The phase of a rhythm in relation to normal diurnal changes, that is, the setting of the biological clock, seems to be controlled mainly by the changes of light to darkness or the alternations of high and low temperatures. They are the most frequently encountered synchronizers, as it were, for different rhythms but changes of humidity and oxygen concentration can operate in the same way. In the case of light, it is usually a fall in light intensity that is the reference point and in the case of temperature, phase shifts can be caused by rises from lower to higher levels but this is known to depend on the phase of the cycle at which the changes are made. It seems that cycles of high and low temperatures always regulate the animal or plant clock in such a way that the low temperature phase is coincident with the physiological state of the organism that is the normal for night. For organisms subjected experimentally to uniform conditions it seems to be the case that the introduction of changes in phase of rhythms previously showing regular periods can be brought about through similar agencies. How effective any one of these is depends on the point in the cycle at which it is applied. In some cases, it is known that the size of a phase shift when caused by light or high temperature treatment is dependent upon the point in the cycle at which the treatment ends, whereas in the cases where shift is occasioned by treatment with low temperatures and low oxygen supply any retardation of events in the cycle is related to the length of application of the treatment. This latter condition indicates a connection between the operation of the clock and metabolic activity. If this is truly the case then the systems under natural conditions must have built-in temperature change adjusting mechanisms but what these are even if they exist is quite unknown. Other period-modifying treatments are known and include lowering of atmospheric pressure and treatment with various chemicals such as colchicine, theobromine and phenylurethane.

As stated, rhythms that have been entrained to natural environmental fluctuations can continue as free-running cycles under constant uniform conditions for varying lengths of time. One of the best examples of a long sustained rhythm is the daily movement of pigment in the fiddler crab, *Uca*. Another is the daily cycle of spore formation in *Oedogonium*. Such are sometimes described as "self-sustained oscillations." But in all known cases there comes a time when the rhythm "fades," i.e. is not detectable as a change in activity of the subject. Several different factors are known to influence the fade-out time of particular rhythms. Continuous light or far-red illumination as opposed to continuous darkness generally shortens it. By contrast, the fade-out of leaf movement in runner beans is slower in red light than in continuous darkness. In some cases it seems that the fade is due to varying changes in phase of the rhythm in different parts of the organism, these then becoming independent of one another so smoothing out the cycle instead of reinforcing one another to produce a collective peak and trough effect. This is thought to be the case in the rhythmic petal movements of flowers in chicory where, under conditions of continuous light for several weeks, synchrony of movement disappears successively, first between separate plants then between separate capitula on the same plant and then between separate flowers in the same capitulum. The same kind of loss of synchronization of parts in animals can also occur, for example, between separate groups of cells in a kidney. Fade-out of rhythm is also detectable in single-celled organisms. The cyclic luminescent activity and other processes such as cell division and photosynthetic activity of the dinoflagellate, *Gonyaulax*, fades out under continuous light conditions. However, there seems to be no equivalent rhythmical processes in individual cells of higher plants.

When fade-out has occurred a rhythm can be initiated once more by one of a number of possible treatments and a rhythm can be created where none existed, the absence being due to the treatment of the organism with constant uniform environmental conditions from the start of its life. Sometimes a short burst of light in otherwise constant darkness is enough, or the converse of this. Another stimulus can be a change over from continuous light to continuous darkness. Others may be a change in temperature or a change in light intensity.

One of the more interesting and well-studied examples of the application of a built-in timing device by animals is that of direction finding through the use of a "sun-compass." Animals such as ants, bees, birds and spiders are able to take into account in fixing the directions in which they move the changing horizontal bearings of the sun from a fixed point on the earth during the day. It is interesting to note that the sun's changing elevation (vertical angle) from the same point seems to play no part in the steering mechanism. Orientation with reference to the sun's position and the consequent ability to move in a specified direction related to it can be achieved only through use of a timing mechanism that measures, as it were, the apparent movement of the sun across the sky. But not only is this so for the time that the sun is above the horizon. Some birds such as starlings are still able to orient themselves with respect to the real sun even when exposed to an artificial sun during the hours of darkness. They behave as if they were aware of the sun's apparent movement through the full 360 degrees. Starlings are able to use the sun-compass method of direction-finding through the whole arctic summer night. For the accuracy of orientation achieved by many animals such a "clock" needs to be a very good and stable time-keeper. Under experimental conditions the phase and period of the oscillation on which the clock is based can be changed with consequent predictable effect on the orientation capabilities of the animals concerned, indicating that there is no doubt about the fact that direction-finding is related to a time measuring process.

Other rhythmic events of the environment to which animals become adjusted by use of their timing devices include the ebb and flow of tides and the lunar cycle. Numerous inhabitants of the sea show continued behavioural rhythms coinciding with high and low tide periods when these are eliminated under laboratory conditions. For example, the flatworm, *Convoluta*, will burrow into and come to the surface of the sand coinciding with low and high water times of their natural surroundings for some time after removal from them to a constant depth aquarium. Physiological rhythms that are related to the lunar cycle have a period of about 29·5 days or in some instances about half this time, that is coincide with either or both of the spring or neap tides. They cannot be called circadian but there seems little doubt that such rhythms have a relationship with the typical endogenous diurnal case. There appears to be evidence that the rhythms with a lunar cycle period shown by some marine animals are "beat" phenomena reflecting an interaction between a tidal period rhythm and a circadian rhythm. The most frequently quoted cases of such rhythms are the palolo worm of the Pacific and Atlantic coasts that reproduces twice only in the year, exactly during the neap tides of the last quarter of the moon in October and November, and the grunion fish of California that rides the wave crests of only the highest tides to move up the beach to deposit eggs or sperm so that the fertilized eggs can develop in warm moist sand without inundation. At the next spring tide the young have developed sufficiently to hatch and to take to the water as it now reaches them for the first time in two weeks. An example of a plant with a half lunar cycle period is the brown alga, *Dictyota dichotoma*, that in some areas releases its gametes at fourteen to fifteen day intervals with constant regularity.

Various speculations have been made concerning the location of oscillating systems within organisms. Obviously, in unicellular structures, each must have its own clock but in multicellular systems it might be that an oscillatory property is possessed by only certain special components. From some of the evidence available it seems that there are instances where isolated cells, tissues and organs from complex organisms after removal to *in vitro* growth conditions continue to show circadian rhythms in such phenomena as secretory activity, cell division, turgor pressure variation, growth rates and carbon dioxide metabolism. In animals, it is possible to alter the phases and periods of oscillations in different organs, independently of one another. Thus it would appear that separate parts of multicellular structures can each show an independent rhythm. Clearly, though, for any normal complex physiological event to occur in an organism there would have to be a special correlation between the phases of the cycles involved. In which case, this relationship between phases would need to be carefully regulated by a controlling mechanism. This could be simply through a mutual synchronization of the separate cycles or by a control exercised through special structures designed for the purpose.

Search for these in many animals has been made mostly without real positive success although it has been shown that many of the rhythmic activities can be associated with hormone secretion or nervous activity by special body parts. These, however, are not themselves always oscillatory in occurrence and so cannot be responsible for the rhythms shown by the physiological events with which they are connected. One of the apparently more clear-cut indications that a particular structure may have a real control function comes from work with the cockroach, *Periplaneta americana*, in which the rhythm of running activity was studied after the animals had been decapitated. It was found to continue but without the previous daily rhythm until the suboesophageal ganglia of other cockroaches were implanted, when a rhythm, with a phase and period previously shown by the donor animals, reappeared in the headless insects. It was concluded in this case that it was secretions from the implanted organs that at least in part exercised control over the circadian rhythm of the running activity and to reinforce this belief it was shown that if the ganglia only were given a cold treatment at 3°C for a while, this had the effect of bringing about a phase shift in the rhythm roughly equal to the time of application of the cold treatment. But the case was eventually shown to be more complicated than just the involvement of one structure because it was found that a substance secreted by the corpora cardiaca also had an influence on the occurrence and timing of the running activity.

In the higher animals, including humans, it was once considered that a major central controlling influence over the endogenous rhythms might emanate from the brain and central nervous system but there is not a great deal of evidence to support this. Instead there are indications that a number of other organs may also be involved including the adrenal, pineal and pituitary glands, each having some influence on one or more rhythms. Each of these organs shows its own rhythms of activity but it is not wholly certain whether these are acting as pacemakers for rhythms not themselves originating in these organs or whether the glands are really only concerned as part of a mutual synchronization system. Whatever the case for some of the rhythms in higher animals, there is no doubt that others can exist in individual parts in isolation and in the absence of a central control.

In the case of plants, there is no more positive an indication of a single localized controlling influence within the body as a whole than there is for animals. In fact, much experimental evidence seems to suggest the reverse, particularly that derived from isolated tissues in culture. These often show very pronounced circadian rhythms. In view of this, attempts have been made to locate or establish the nature of a rhythm, possibly as a chemical mechanism, inside an individual cell. Work with the convenient giant uninucleate cells of *Acetabularia* has shown that an endogenous rhythm in photosynthetic activity can persist in enucleated cells, so that the oscillatory system is located somewhere in the cytoplasmic structures. But the system must in some way be under control of the nucleus because if the nuclei from two plants showing different phases in the cycle are mutually transplanted then each plant when placed under uniform conditions takes on the phase to which its new nucleus had previously been subjected. The possibility has been suggested, therefore, that nucleic acid metabolism may be involved.

In the case of the dinoflagellate, *Gonyaulax*, it seems that the rhythms of both luminescent and photosynthetic activity are reflections of rhythmic changes in the levels of the enzymes luciferase and ribulose diphosphate carboxylase respectively, suggesting a periodic synthesis of them. This could be explained on the basis of an equivalent periodic synthesis of the messenger ribonucleic acids that encode the enzymes. The fact that RNA and protein synthesis inhibitors are known to have rapid inhibiting effects on the rhythms or can induce phase shifts in them tends to indicate that this could be the case. But not all investigations into the link between various enzyme levels in cells of different plants and animals and the rhythms of their metabolic activities indicate that there is any universally common condition of direct relationship. In some cases, particular enzyme levels do not appear to fluctuate at all as for example, phosphoenolpyruvic carboxylase in leaf tissues of *Bryophyllum* where it might be expected to play some part in carbon dioxide metabolism which itself shows very marked diurnal periodicity. In view of the mass of varied findings about such a complex physico-chemical system an understanding of biological clock mechanisms in terms of biochemical activity is a long way from being achieved.

One other feature of circadian rhythms concerns their inheritance. The question arises as to whether the rhythms exhibited by an individual are represented in the genetic code passed to it by its parents or whether they are imprinted upon it by fluctuating environmental conditions during an early stage in its development. The evidence available, mostly obtained from treating plants and animals with abnormal environmental fluctuations, shows very strongly that the normal rhythms to be expected in a particular species are still present in individuals after many generations, no matter what the previous treatment of their antecedents or the conditions imposed upon them from their earliest stages of development. Crosses between some varieties showing different basic rhythms for the same activity have tended to indicate that the inheritance may even show a simple Mendelian segregation. Similarly, differences in the amplitude of a rhythm can be inherited. There seems, therefore, to be a clear hereditary influence on the circadian rhythms of both plants and animals. It is the case, however, that it often requires a period of exposure to fluctuating environmental conditions, particularly those of light and temperature, before the inherited timing device begins to operate.

Lastly can be mentioned the connection between the biological clock as a day-length measuring device and the control of physiological processes. When these are directly influenced by the measurement of day (or night) length by the organism they are said to be under *photoperiodic control*. By this day-length estimation the organism is able to locate itself in time with respect to the sequence of seasons and thus to adjust or regulate its activities to suit itself best to the forthcoming conditions of its physical environment, in effect, anticipating the favourable and unfavourable periods of the year. This does not mean that all developmental or other changes linked to seasons are necessarily under photoperiodic control; some environmental factors such as temperature in large water expanses are known to be such good indicators of the time of year that their long-cycle fluctuations, that is, their gradual rise and fall during the year, are quite adequate as direct evokers of seasonal responses. But where this is not the case and there is an absence of such reliable indicators of seasonal variation, in order to orient itself seasonally, the organism "measures" the approximate duration of daylight (or darkness), that is, the day or night length. This is done without any regard for light intensity variations that occur naturally within the twenty-four hours; it is based on the incidence or otherwise of light, not its brightness.

Some of the physiological phenomena known to be under photoperiodic control include in plants onset of flowering, germination of seeds, formation of storage organs such as tubers, various phases of vegetative development, the occurrence of succulence, the onset and breaking of dormancy in buds, bulbs and corms, cell division in cambia and the differentiation of tissues and in animals, the enlargement of gonads, the impulse to migrate, the control of diapause in insects, change of fur colour, hair growth and moulting, all in some way related to a change of season. The whole physiological processes by which any of the above events are characterized will not be discussed here but reference can be made to one of them, for example, onset of flowering, which is fully treated as a developmental process in Chaps. 13 and 16. What is important in this present context is the time-measuring property of the organism that becomes manifest through the photoperiodic responses made by it. The timing capacity seems the same for all the organisms that possess it and is related in the same way to all the processes known to be controlled by it. The one factor of significance in all photoperiodically controlled processes that take place under normal environmental day-night cycles, whether the processes are concerned with promoting or inhibiting a change in activity, is the exposure of the organism to a light period of greater duration than a critical value. When this *critical day-length* (cdl) is exceeded, this in some way seems to be registered by the "clock" and this acts to elicit the appropriate response from the organism concerned. Plants and animals that show photoperiodic responses seem able to measure day-length with considerable accuracy even in fluctuating temperature and light intensity conditions except when these have extremely high or low values, that is the timing device is not ordinarily affected by them. Thus the occurrence of the responses is not greatly influenced by variations in conditions other than day-length, unless they are themselves regulated by those conditions. This might be so for growth and metabolic processes that would still proceed but only at rates commensurate with the prevailing temperature and light intensity conditions. Such activities as migration by birds are usually quite independent of the prevailing weather conditions. As might be expected, the critical daylength relating to any particular process, in order to make it a useful measure for locating the seasons, varies with the geographical latitude. An example of a wild rice variety can be quoted for which the cdl is about three and a half minutes longer for every degree northward it is found growing in the northern hemisphere or every degree southward in the southern hemisphere.

The nature of this day-length measuring process has been the subject of a great deal of investigation and there exists many items of information from which a current interpretation, more or less as below, can be arrived at.

Although it is normally referred to as a day-length measuring device, implying a sensitivity to light, the "photo clock" is apparently responsive to the length of a period of darkness. This has been shown to be the case from several sources of evidence one of which is the effect of interrupting the dark period by a light period (*see* p. 545 for an account of the effect of this on the flowering response in plants). An explanation of it could be that the time-measuring process is zeroed from the beginning of either the light period or the dark period and then at a given time from this zero after commencement of the dark period the time-measuring device induces a condition of sensitivity to a further light stimulation. The fluctuations or oscillations in this light sensitivity occasioned by alternating periods of light and darkness constitute the circadian rhythm known as the *photoperiod*. Under natural conditions, the changing lengths of day and night with the time of year can bring into being a rhythm of light sensitivity or photoperiod peculiar to a particular plant or animal to which is coupled one or more physiological processes either to be promoted or inhibited according to the organism and season of the year. In other words, the biological clock set by the light-dark cycles causes a rhythm in the sensitivity or responsiveness of the organism to light with respect to its developmental or other processes, acting as the signal to set them in motion or to bring them to a halt. It should be noted, however, that for organisms in the temperate regions where every proportion of light to dark periods in twenty-four hours (except the longest and shortest days) is repeated twice a year, some other factor(s) must be involved to signal which of the two periods or seasons is the appropriate one for a given response. For plants it has been shown that a vernalization requirement (*see* p. 544), a dose of cold treatment, may also be a factor in determining when a physiological process can be initiated so that this combined with photoperiodic response can fix more precisely which time of the year an event should occur.

It is certain that the whole relationship between the daily light-dark cycles and any actual photoperiodic response is undoubtedly very complicated and cannot profitably be further treated here. The answer to another question, namely through what agency the biological clock controls the variations in light sensitivity, is suggested in Chap. 16, in which the far-red phytochrome pigment system is described as an important photoreceptor in plant photoperiodic responses. In animals, there may be equivalent pigment systems and in some cases these appear to be located behind the eyes although there are many known cases where this is not so. In different species of animals many different organs have been associated with their photoperiodic responses.

From the foregoing outline of some of the features of time-measurement by living systems, it should have become apparent that the possession of such a "clock" is another adaptation to environment enabling plants and animals to adjust themselves to or regulate their behaviour in accordance with daily and other cyclic changes of external conditions. By such means, they are able to perform certain functions or go through certain developmental phases at a time of day or season of the year most suited to that particular activity thus to ensure the greatest chances of individual survival and species perpetuation.

CHAPTER 7

THE WATER RELATIONSHIPS OF LIVING ORGANISMS

WATER is of prime importance to living organisms for two reasons. First, it is an essential part of the structure of protoplasm, and secondly, it is used in many processes both as a metabolite and as a solvent for other substances. Very few living organisms can exist in the continued absence of access to water, but there are many structures produced by living things which can remain alive, often for long periods, without intake of water. Such are some spores, seeds, cysts and eggs; they are formed under particular conditions, for special purposes, usually concerned with reproduction or perennation. In such specialized structures, the protoplasm differs both in composition and activity from the normal state; the water content is low and physiological activity is considerably reduced. Such special cases will not be considered here, but the normal water relationships of fully functional plants and animals will be discussed.

THE WATER RELATIONSHIPS OF PLANTS

The vegetative body of a purely aquatic plant will not be subject to any shortage of water, and thus is not adapted in any special way to withstand even a slight degree of desiccation. Aquatics generally shrivel and die with a few minutes exposure to a drying atmosphere. Neither are such plants specially modified for water absorption. Those which normally have to persist over dry periods must develop specialized structures such as those mentioned above.

In the terrestrial environment, the plant body is subject to drying-out for most of the time, and access to water is restricted usually to a localized region of the environment, namely, the soil. Two requirements are therefore necessary for the success of a plant on the land; it must have an adequate water-absorption mechanism, and it must be able to restrict the loss from its surface. If either fails, wilting and death will ensue under drying conditions. Land plants which are not structurally adapted to meet these two requirements can exist only in one of two ways. Either they must be able to develop and reproduce during short intervals when wet conditions prevail, resting between, or they must be restricted to localities where water is never scarce, so that they cannot be desiccated. The chief members of the former group are terrestrial forms of the algae and fungi, and a few of the higher plants which can

230

successfully perennate underground during dry spells between periods of heavy rainfall, e.g. many geophytes of the near-desert regions. Among those which must have plentiful water at all times are representatives of all the major plant groups. They may be found at the margins of water expanses, and in damp shaded conditions.

Water-absorbing Organs

Plants which are truly aquatic do not possess any parts specialized for water absorption. This is also true of the terrestrial forms of the lower plants, where any cell is capable of absorbing water when it is available; in such plants, continued existence depends on periods of almost completely aquatic conditions. Many possess gelatinous substances which can imbibe and store water. Liverworts and mosses usually develop delicate rhizoids for more intimate contact with the soil water. Both groups lack a waterproof covering on their aerial parts, and thus under rainy conditions, any cell can absorb water. Storage of water may be effected by the development of mucilage cells within the tissues. Pteridophytes and spermatophytes have evolved an elaborate root system which is effective in the absorption of the water and salts required by the whole plant. The highest development of the root system is found in terrestrial angiosperms; their efficient structures for water absorption will now be considered.

Water Absorption by the Roots of Higher Plants

To understand the process as a whole, the conditions at the source of water-supply must be recalled. The soil-water relationships are described in Chap. 6. Two considerations are of greatest importance here. First, the water in a drained soil is held either by surface tension on the minute soil particles and in the interstices between them, or by imbibition in the soil colloids. Secondly, to remove the water from these positions, force greater than that exerted by surface tension or imbibition is required.

If a root system is to be effective in absorbing any great quantity of water, it must be widespread enough to tap a sufficient volume of soil and it must be able to exert absorbing forces greater than those holding the water. To achieve the former end, root systems are invariably much branched and spreading, and in most cases, at the extremities of the roots, epidermal outgrowths form *root-hairs*, which make intimate contact with the soil particles. If we imagine all the roots of a plant laid end to end, the total length would be considerable. A single wheat plant was estimated to possess about 70 km of roots, and a rye plant, 80 km. A single maize plant, unrestricted in growth, had roots occupying over 8 m^3 of soil. These are all comparatively small herbaceous plants; similar figures for large trees run into hundreds of kilometres and thousands of cubic metres.

Not all the higher plants develop root-hairs. Where the outer tissues are not suberized or cutinized, there may be no root-hair formation. Certain conifers and monocotyledons do not develop them, and even many plants which normally produce root-hairs in soil, will fail to develop them in culture solutions. In all these cases, absorption can occur only through the epidermal cells which have not been water-proofed by cutin or suberin. Even where root-hairs are developed, a high proportion of water absorption takes place between the root-hair

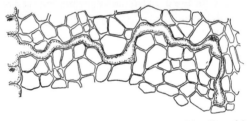

Fig. 7.1. Diagram of root hair in contact with soil particles.

region and the root apex. It is certain, however, that the root-hairs considerably increase the surface available for absorbing water; a tenfold increase is common, and in some cases it may be as much as fifteenfold. As a root ages, the root-hairs become non-functional, due to the formation of a waterproof exodermis under the piliferous layer. Nevertheless, there is no loss of absorbing surface since new root-hairs are constantly developed as the apex pushes on through the soil. The period of maximum function of root-hairs varies considerably; it may be from several days to several months, according to species. In older roots covered by corky periderm, water absorption appears to be impossible, but there is evidence that under very wet conditions, water can enter readily through the lenticels.

It is generally held that the forces which overcome the water-retaining power of the soil are the osmotic forces exerted by the absorbing cells. Under normal soil conditions, the contents of epidermal or root-hair cells can exert a higher osmotic pressure than the soil solution. Thus the slightest fall below full turgidity of such a cell, resulting in a lowering of its water potential, would lead to movement of water into the cell. Such a fall would occur if the outer cells lost water to more deeply situated cells. This simple explanation seems to satisfy the conditions which exist in most circumstances. However, some investigators have reported a difference in hydrostatic pressure between the water in the inner tissues of the root and that in the soil. This may account for some cases of intake of water and its subsequent passage across the root via the cell walls and intercellular spaces. Such a pressure gradient across the root would tend to occur during periods of very active water loss by the leaves. This condition is described as *passive absorption* since the root does not develop the forces concerned, in contrast with *active absorption* when the root does exert the forces. That roots themselves can develop absorbing forces is clearly shown by the fact that some roots will continue to absorb when the aerial parts are either removed or are in an atmosphere where water loss is reduced to nothing. Under such circumstances, water is forced out of cut stems or can be seen exuding from pores on leaf surfaces by guttation (*see* p. 259). The roots are said to be developing a *root pressure*; the process of absorption depends on forces developed in the roots themselves. What those forces are, and how they may be developed, will be considered later.

Loss of Water to the Atmosphere: Transpiration

If a plant loses water by evaporation from its tissues, it is said to *transpire*. The process is one in which the normal evaporation phenomena are evident, and is not therefore a physiological function of the cells. It is purely the effect of a wet surface losing water vapour into an unsaturated atmosphere. Nevertheless, comparisons of the rates of water loss from plant organs and from equivalent free water surfaces, show that there are differences even under identical evaporating conditions. These differences can be explained with reference to the structure of the parts of the plant losing the water.

Sites of Water Loss

Any cell surface not protected by a waterproof covering must be wet, since the aqueous phase of the protoplasm is continuous through the cell membrane to the exterior of the cell. Hence, whenever such an unprotected surface is adjacent to an unsaturated atmosphere, evaporation must occur. In lower plants such as algae, fungi, liverworts and mosses, nearly all the cells are subject to evaporation losses because they are unprotected by waterproofing material. A dry atmosphere will frequently kill such plants since their absorbing mechanisms cannot keep pace with the rate of water loss. This would apply to the higher plants if they were unprotected, but the more fully adapted land plant has developed external layers which reduce such widespread loss. These protective layers consist of cutinized or suberized cells at the surfaces in contact with the atmosphere. Such waterproofing can only be a useful adaptation if carried to a limited degree, since although a complete covering would hinder desiccation, it would at the same time exclude normal activity by preventing adequate aeration of the tissues.

Thus, the successful land plants show a compromise in that the protective layer is perforated at many points; through these perforations they must lose water as long as the external atmosphere is unsaturated. The perforations are the lenticels of stems, and the stomata of leaves and stems. Measurements of water loss through the various parts of plants show that by far the greatest amount is lost through the leaves. The further discussion concerns transpiration from leaves only.

Transpiration from Leaves

The fact that leafy twigs or even separate leaves are losing water to the atmosphere can readily be shown and the quantity of water lost in a given time can also be measured. The simplest method of demonstrating transpiration from the surface of a leaf is to apply a piece of dry cobalt chloride paper to it. It should be covered with glass or mica to protect it from the atmosphere, and the edges should be sealed with vaseline. The change from blue to pink indicates clearly that water vapour is leaving the leaf. Comparisons of rates of water loss from different leaves or from different surfaces of the same leaf can be made by comparing the times taken for the papers to reach a standard pink colour. A suitable apparatus for use with leaves still on the plant in field conditions is shown in Fig. 7.2.

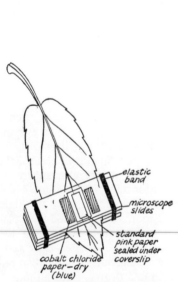

Fig. 7.2. Cobalt-chloride paper apparatus for use in the field.

Fig. 7.3. Continuous weighing method to demonstrate loss of water from a cut twig.

Loss of water from leafy twigs or detached leaves can be measured ›y a weighing method. A simple apparatus for this purpose employs a pring balance which records continuously the weight of the twig and ts water supply (Fig. 7.3).

The potometer is used to measure the rate of absorption by a cut hoot. It is used for transpiration experiments on the assumption that here is a constant ratio between water loss and water absorption. The ate at which water is being taken from the container can be measured ı terms of the time taken for an air bubble or water meniscus in the

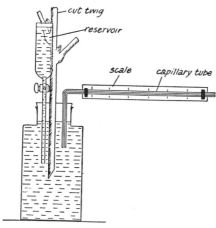

Fig. 7.4. Farmer's potometer.

pillary tube to move a given distance. This quantity is indirectly a ıasure of the rate of transpiration from the twig. Thus the rates ıder different conditions can be compared. If the capillary tube ıameter is known, the volume of water lost by the twig in a given time ı be calculated. Some potometers have accurately calibrated pillary tubes. The potometer shown in Fig. 7.4 is Farmer's pattern. The atmometer is an apparatus designed to measure the evaporating ›wer of the atmosphere. One suitable for operation under field ıditions is shown in Fig. 7.5. It works on the same principle as the tometer. Measurements of transpiration using a potometer and of ıaporation using an atmometer under identical atmospheric conditions ı be used to indicate how closely the transpiration rate of living .ves follows the pattern of uncontrolled evaporation.

All these methods of measuring transpiration from leaves are open to ticism on the grounds that the water is being lost under abnormal conditions. However, they provide useful comparisons if nothing more.

Estimates of the total amounts of water lost by whole plants through their leaves indicate that the process of transpiration is normally very rapid, and the amounts of water involved may be considerable. The loss of water is usually expressed in grams per square decimetre of leaf surface per hour. A common figure for many plants is 1 to 2 g the highest on record is 7·5 g from the leaves of *Eucalyptus*. More striking is the loss of water in litres per day from whole plants;

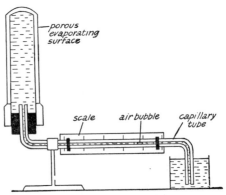

Fig. 7.5. An atmometer.

birch tree with about 250 000 leaves was estimated to lose 360 l per day. A field of maize may account for the equivalent of 25 cm of rainfa per season. Estimates of the quantity of water passing through a plat for every gram gain in dry weight of body are instructive. For instanc peas may transpire upwards of 750 g of water per gram of dry weig gain.

Such estimates indicate that the amounts of water lost to the atmosphere by an area covered by vegetation may far exceed the amounts lost by evaporation from a similar free soil or even a free wat surface. Some knowledge of the magnitude of the quantities involve is important in irrigation problems and in avoidance of the disastro results which follow deforestation. The total leaf surface area of a lar tree is many times that of the area covered by the outline of the tr on the ground; consequently the amount of water lost from this lar area of leaf surface should be greater than that lost from the grour Further, the internal structure of the leaf is such that the surfaces very large numbers of cells are in contact with air-spaces within t leaf, so that there is a tremendous surface area over which evaporati might occur. But the plant is not losing water over this large area as if it were a free water surface; the loss is largely confined to the small pores in an almost waterproof outside covering. This can be demonstrated by comparing the rate of water loss from a leaf surface which is perforated by stomata, with one that is not. Loss of water other than through stomata is known as *cuticular transpiration*; it varies with the thickness of the cuticle. With a thick cuticle such as is found in holly leaves, the amount of water lost through the cuticle is rarely more than

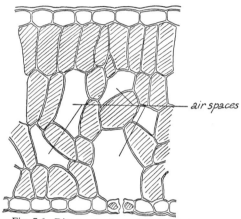

Fig. 7.6. Diagram to show air spaces in leaf tissue.

to 2 per cent of the total lost from the leaf when the stomata are open. On the other hand, leaves of shade plants with thin cuticles such as the wild arum may lose as much as 20 to 30 per cent. There is little doubt that stomatal transpiration accounts for the greater proportion of water lost from a leaf to the outside atmosphere. To appreciate the process more fully, it is necessary to study the nature and behaviour of stomata in more detail.

The Diffusion of Gases through Stomata

The amount of water-vapour lost to a given atmosphere through the stomata depends upon the numbers of stomata and the degree to which they are open. The occurrence of stomata per unit area of a leaf, varies with the species and with the conditions under which the plant grows. Counts of stomata per square millimetre of leaf surface range from 14 in *Tradescantia* to 1200 in the Spanish oak. A common figure for many plants is 300 per mm^2; this is equivalent to 3×10^8 per m^2 of leaf surface area, if the stomata are equally distributed on both surfaces.

However, this is not usually the case; frequently, the stomata are confined to the lower surface only, or are relatively few in number on the upper surface. The area of each stomatal aperture varies with the species but is always exceedingly small. Measurements range from $4 \ \mu m^2$ in Spanish oak to $292 \ \mu m^2$ in *Tradescantia*. Thus the total leaf area through which water vapour can readily diffuse from the internal air spaces to the outside atmosphere is comparatively small. It rarely

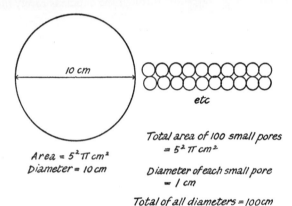

Area $= 5^2 \pi$ cm^2
Diameter $= 10$ cm

Total area of 100 small pores $= 5^2 \pi$ cm^2

Diameter of each small pore $= 1$ cm

Total of all diameters $= 100$ cm

Fig. 7.7. Diagram comparing diffusion from many small pores with total area equal to one large pore when diffusion rate is proportional to pore diameter.

exceeds 0·5 per cent of the total leaf surface even when the stomata are fully open.

The loss from a free water surface is six to ten times as fast as that from an equal area of leaf surface. But if a comparison is made between a free water surface and an equivalent pore area of leaf surface, it is found that water vapour leaves the pores fifteen to twenty times as fast as it leaves the free water surface. The disparity can be explained by the nature of diffusion through small pores. In 1900, Brown and Escombe showed that the rate of diffusion of a vapour through a small aperture is more nearly proportional to its diameter than to its area, whereas diffusion through a large opening or from a large free surface is more nearly proportional to its area. Thus a pore of diameter 10 cm will have an area of 25π cm^2. One hundred small pores covering the same area will each have an area of $0·25\pi$ cm^2, and a diameter of 1 cm each. Therefore, if the rate of diffusion is proportional to the diameter, the 100 small pores of total diameter 100 cm will allow vapour to pass ten times as rapidly as the single large pore of 10 cm diameter.

The reason why diffusion rate becomes proportional to the diameter (or circumference) of the small pore is to be sought in the arrangements of the diffusion shells around apertures. Fig. 7.8 shows the arrangement for different sizes of pores.

They show that whereas the greatest amount of diffusion is over the whole surface of a large pore compared with its edge, for a small pore the greatest amount is over the edge. The smaller the pore, the greater is the circumference to area ratio. Over the edge of the pore, molecules

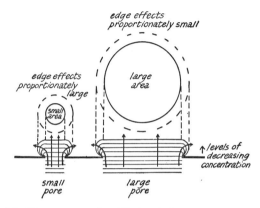

Fig. 7.8. Diagram comparing diffusion shells at small and large pores.

can spread out sideways as well as moving vertically, whereas in the centre the vapour will move vertically. Consequently, the diffusion shells of equal concentration at the edge of the pore will be much closer together than over the centre, giving a sharper diffusion gradient and thus a more rapid flow of vapour. The greater the proportion of edge to area, the greater will be the rate of diffusion. Thus, within limits, the smaller the pore, the more efficient it is in allowing vapour to pass through it in proportion to its area. It is to be noted that these facts are also of considerable significance regarding the intake of carbon dioxide and oxygen by the leaf.

It is true that a stoma should not be considered as a simple perforation in a very thin membrane but as a short tube, and also that if small pores are placed close together they tend to reduce each other's efficiency. Neither of these considerations affects the ultimate result, that the leaf surface does lose water very rapidly through the stomata, and likewise takes in adequate quantities of carbon dioxide and oxygen.

It is obvious that the rate of transpiration from a leaf will be influenced by the degree to which the stomata are open. Careful investigation has shown that the rate is little affected by reduction of aperture from maximum to half, and that even when so near to closure as to have no visible aperture, stomatal transpiration may still be proceeding ten times faster than cuticular transpiration. The mechanism of the opening and closing of stomata is of importance not only

a.

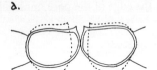

Walls more or less uniformly thickened. Cells elliptical in T.S. when flaccid. Rounded up when turgid and so caused to move apart, e.g. Medeola.

b.

Walls unevenly thickened, with thin "hinges" on faces adjacent to and furthest from pore. Increased turgor causes "opening" of the hinges and rounding up of cells so that pore opens, e.g. Solanum.

c.

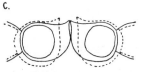

Walls adjacent to pore thicker than elsewhere. Turgor causes thinner walls to distend and thicker walls become bowed away from pore, e.g. Polygonatum.

d.

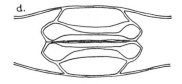

Guard cells dumb-bell shaped. Turgor at thin-walled enlarged ends forces narrow thick-walled regions apart, e.g. grasses.

Fig. 7.9. Stomatal mechanisms. (a), (b), (c) as seen in vertical section through centre of pore, (d) surface view.

in water relationships but also in photosynthetic and respiratory activities.

Stomatal Mechanism. An early study of stomata by von Mohl in 1856 showed that the pores open and close regularly, coinciding roughly with the periods of daylight and darkness. It was also noticed that turgor pressure changes of the guard cells occurred with the same rhythm. It was soon clearly demonstrated that because of the peculiarities of guard cell construction, these turgor changes were effective in bringing about the opening and closing of the pore. All stomata are not constructed in exactly the same way, and thus the precise mechanisms of opening and closing are variable. Some are illustrated in Fig. 7.9. A simplified idea of what occurs in a common case is given below.

The walls of the guard cells bordering the pore are considerably thicker than elsewhere, the thinnest parts being on the sides directly away from the pore. An increase in turgor of a guard cell is accompanied by a distention of the thin wall in a direction away from the pore. This results in an outward pull being exerted on the thicker portion of the wall, which, because of its thickness, cannot distend equally in

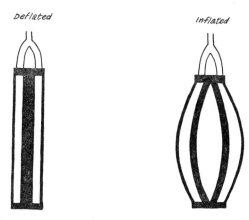

Fig. 7.10. Diagram representing the inflation of eccentrically bored tubing.

the other direction, Thus, any increase in turgor of the guard cells causes a separation of their thickened walls, creating an aperture between them. The process can be understood more easily if one visualizes what would happen at the inflation of two pieces of eccentrically-bored rubber tubing, fixed with the thick walls together and the ends tied, as shown in Fig. 7.10.

It has been firmly established that opening and closing of the pores is due to water movement into and out of the guard cells, and that changes in the osmotic pressure of the cell sap of these guard cells, relative to the surrounding epidermal cells, causes this water movement. Measurements of osmotic pressures of guard cells of widely-open stomata have been shown to be as high as 100 atm, whilst at closure they have dropped to 20 atm; meanwhile, the surrounding epidermal cells have had a fairly constant low value. It is, therefore, easy to suppose that the rhythmical opening and closing is due to the changing osmotic relationships between guard cells and epidermal cells under the influence of light. Many attempts have been made to explain the osmotic changes in guard cells as being due to their possession of chloroplasts whilst the epidermal cells lack them. Thus the rise in osmotic pressure of illuminated guard cells could be explained as the result of the photosynthesis of osmotically active substances. But careful observation of stomatal movement, not only on green parts of leaves but also in the white parts of variegated leaves, in which the guard cells possess no chlorophyll, has shown that all are capable of rhythmical opening and closing; they may even open and close in the continued absence of external carbon dioxide. One of the authors has recorded rhythmical opening and closing of stomata of the spruce,

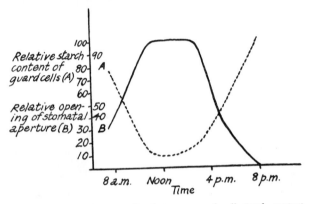

Fig. 7.11. Graphs to show relation between guard cell starch content and stomatal aperture. (*Based on Loftfield.*)

Picea excelsa, up to eight days after removal of the leaf from the tree, and in drying conditions. A simple explanation based on photosynthetic activity of the guard cells cannot therefore account for the known facts, particularly when it is observed that the colourless guard cells of variegated leaves, and even guard cells from unopened bud leaves may contain starch grains. The evidence points strongly to the fact that guard cells in general derive their carbohydrate from sources other than their own photosynthetic activity.

Investigations first carried out by Lloyd, and later by Loftfield, initiated what seemed to be a convincing explanation. It was shown that the starch–sugar equilibrium in guard cells (with or without chlorophyll) is the reverse of what occurs in a normal green cell, when both types are exposed to alternating periods of light and darkness. In the light, the normal green cell slowly accumulates starch as visible grains; in darkness, they disappear. But in a guard cell exposed to light, any starch grains present at commencement of illumination disappear, and reappear during darkness. Fig. 7.11 shows this contrast. The influence of light on the guard cells therefore causes a conversion of their starch to sugar, which is highly osmotically active. Its presence would result in a rapid rise of osmotic pressure and consequent opening of the pore. The conversion of sugar to starch in darkness would have the reverse effect. Loftfield showed that the changing starch-sugar equilibrium was closely followed by the movements of the guard cells in the directions expected. Thus the explanation was to be sought in the conditions affecting the starch-sugar change. Experiments concerning the effect of changing acidity of the cell sap of the guard cells on their opening and closing gave the next lead. It was shown that in an acid atmosphere, the stomata of many plants do not attain maximum opening, or may be inhibited from opening at all. Conversely, it was shown that stomata may be made to open even in darkness in an atmosphere containing ammonia. Thus the pH condition of the cell sap is closely connected with stomatal movement. Values between pH 6 and pH 7·4 in guard cells of open stomata in the light, and values between pH 5 and pH 4 in guard cells of closed stomata in darkness have been recorded. One investigator has stated that the stomata of plants with which he worked were closed and contained starch at all values between pH 2·9 and 6·1, and that they were open and starch-free at all values between pH 6·9 and 7·3; between the values of 6·1 and 6·9 gradual increases in stomatal aperture occurred. It seemed likely therefore that the key to the problem lies in enzyme activity. Diastase hydrolyses starch to sugar most rapidly at the higher pH values, but it cannot also bring about the condensation of sugar to starch under any conditions. The most probable enzyme is starch phosphorylase. At values around pH 7, it brings about phosphorolysis of starch by forming glucose phosphate, whereas at values about pH 5 or lower, starch is reformed from the glucose phosphate. A further known situation also tends to support the hypothesis.

One of the factors known to affect stomatal movement most is carbon dioxide concentration in the vicinity of the leaf tissues and this concentration can be related to the illumination conditions. The effect of changing illumination is to cause a tendency for stomata to close in darkness and open in the light. During light periods the carbon dioxide concentration in the leaf tissues would tend to be low, since photosynthesis would be actively proceeding; during darkness it would tend to be high since photosynthesis would not be proceeding and thus respiratory activity would not be masked. Variation in the carbon dioxide present in the leaf tissues would alter the pH of all liquids. In light, the pH would be neutral or on the alkaline side; in darkness, the pH would be below neutral or on the acid side. The high pH of 7 or more would favour sugar accumulation in the guard cells; thus there would be a rise in their osmotic pressure and this would lead to opening of the pore. The lower pH, below 7, would favour starch accumulation in the guard cells; thus there would be a fall in their osmotic pressure which would lead to closure of the pore.

Unfortunately not all movements of stomata can be explained on this basis and the explanation given above has a number of flaws. It is true that stomata respond very strongly to changes in carbon dioxide concentration around the 0·03 per cent level by staying open in darkness when carbon dioxide is not present and closing in light when it is above normal concentration, but there is no final evidence that movements are directly related to enzyme activity. Further, what can be overlooked is the fact that the starch $\rightleftharpoons$ sugar reaction catalysed by the enzyme phosphorylase cannot immediately lead to a change in the osmotic pressure of the guard cell sap since the total number of particles in solution remains the same (to form glucose-phosphate necessitates the removal of an ion of inorganic phosphate). It would require further reactions involving hydrolysis of glucose phosphate to create the necessary osmotic conditions.

Another condition not properly accounted for on the above basis is the fact that energy must be supplied for the sugar–starch conversion necessary to stomatal closure. Since it has been shown that some guard cells will close in the absence of oxygen and in total darkness it is not easy to correlate closure with an energy supply requirement. Again, the explanation does not cover the opening and closing of the onion guard cells which never contain starch, although soluble polysaccharides could be postulated as playing an equivalent role. Lastly, for a plant such as *Pelargonium zonale*, it has been shown that there are no consistent differences in the starch content of guard cells of open stomata in the light and closed stomata in darkness when any effects due to diurnal rhythms have been eliminated by making comparisons between epidermal strips taken at the same time of day from plants in light and dark conditions respectively.

Other factors besides carbon dioxide concentration are known to affect stomatal movement and to be independent of it. Among these are light quality conditions, with the blue end of the spectrum being most effective in inducing opening. Likewise increase in temperature above about 25°C tends to promote closure under all conditions, independent of carbon dioxide.

There also occurs rhythmic opening and closing of guard cells in many plants even when all external conditions are kept constant, whether light or dark, higher or lower temperatures and so on. Not a great deal is known or understood concerning such autonomous contro of stomatal movement. There is some suggestion that opening and closing is related to rhythmic changes in the starch–sugar equilibrium presumably imposed on the plant by previous alternate light and dark periods but such rhythmic behaviour tends to die away if unchanged conditions are imposed for prolonged periods (*see* p. 222).

When a plant is under water stress, that is, when it is losing more water by transpiration than can be replaced by absorbtion from the soil, stomata tend to close in response to this water deficit irrespective of light or carbon dioxide concentration conditions. Accompanying the closure, sugar in the guard cells is converted to starch (exactly the reverse of what happens in the mesophyll cells under the same conditions) but so far there has been no connection established between this sugar → starch conversion and carbon dioxide concentration changes which would occur if photosynthesis was reduced due to lack of turgidity in the guard cells.

Some of the known responses of stomata to water deficiency in the leaf appear to be the opposite of what is expected. When wilting is very sudden the stomata may open more widely than previously before they eventually close. This can be accounted for by a more rapid drop in tugidity of the surrounding epidermal cells than in the guard cells which consequently swell temporarily and force the pore more widely open. In the other direction, if a water shortage is very rapidly filled, turgor can build up more rapidly in the surrounding epidermal cells causing the guard cells to become more closely applied so tending to close the pore more tightly. The tendency for stomata to remain permanently open during prolonged wilting can be accounted for by a greater fall in turgidity of the epidermal cells than in guard cells.

There is some evidence that the stomata in some plants can be affected by a stimulus transmitted through the tissues and induced to make movements when they would not be expected to do so and the movements seem not connected in any way with variations in carbon dioxide concentration. Postulations as to an electrical or a chemical stimulus have been made but there is little evidence to support either. It seems the case, however, that in *Pelargonium spp.* a heat shock stimulus in one part of a leaf leading to closure of stomata in another part is connected with movement of material in the phloem.

It seems clear that the hypothesis based on starch ⇌ sugar changes in guard cells in response to varying carbon dioxide concentrations will not account for all known stomatal movement phenomena. Others have been proposed, not necessarily more tenable, but some worthy of mention are given here. One concerns the possibility that the changes in turgor of the guard cells are the result of changes in permeability of their membranes in response to a variety of changing conditions so that there may be an active uptake or expulsion of solutes, especially ions, by the vacuolar sap. Another suggests that the changes in turgor are not necessarily related to osmotic effects but that there may be a swelling or shrinking of cell colloids as they imbibe or lose water in accordance with changing pH conditions within the cell. Such uptake or release of water would affect the water potential with accompanying inward or outward flow. Another proposes an "active" pumping of water out of the guard cells under changing conditions to bring about closure, requiring the expenditure of energy. If this were so, lack of oxygen should prevent open stomata from closing. This has never been satisfactorily demonstrated.

There is no doubt that the movements of stomata are under the influence of a number of factors and this makes their behaviour a complex phenomenon to deal with since all the interactions between factors cannot always easily be elucidated. It is not unlikely that stomata can open and close in response to both changing environmental conditions and to some endogenous control mechanisms none of which have as yet been fully explored.

Methods of Measuring Stomatal Apertures

There is no completely satisfactory method of measuring the pores. They may vary considerably on the same leaf and so very many direct measurements must be taken to find an average value. The process is tedious, and even when an average value is found for slit width and length, it is not easy to convert into useful information concerning the capacity of the pores to allow diffusion. Direct measurement is achieved with the microscope, using epidermal strips previously fixed rapidly in alcohol, so that they are as little changed from the living condition as possible.

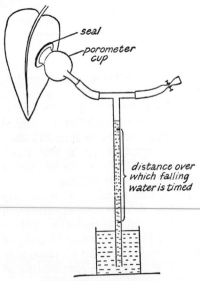

Fig. 7.12. The porometer.

Estimates of the degree of stomatal opening may be made for comparative purposes by using the porometer, as shown in Fig. 7.12. The apparatus really measures the rate at which air can be drawn through a portion of the leaf. The cup is sealed the surface of the leaf and air is pulled through the stomata by creating reduced pressure in the apparatus. The rate at which the air enters the apparatus is taken as a measure of the degree of opening of the stomata. The apparatus can be used to indicate the changes occurring in stomatal opening under different conditions.

Another simple method which gives a rough estimate is to flood the leaf with liquids of different surface tension. Such liquids are xylol, benzene and ethyl alcohol in order of decreasing ability to penetrate the pores. When the stomata are nearly closed, only xylol will enter; when about half open, benzene will enter as well as xylol; when fully open, all three will enter readily.

Factors Affecting Transpiration

Both external conditions and internal factors can affect the rate at which water is lost from a leaf. The chief external considerations are humidity and temperature of the atmosphere, light and wind conditions, water available to the leaves, and atmospheric pressure.

Transpiration has been considered as a purely physical diffusion of water vapour. To comprehend how various factors affect this diffusion by altering the relative water-vapour pressures inside and outside the leaf, it is necessary to digress into the physics of diffusion, evaporation and vapour pressure. The diffusion rate of a gas from one point to another, depends on the difference in pressure of the gas at the two points. This may be called the *diffusion pressure gradient*; it is fixed by the concentration of the gas at the two points. If the concentration changes, the rate of diffusion will change.

If pure water is enclosed in a chamber with a space above its surface, the high velocity of the molecules in the liquid will cause some to shoot out from the surface into the space, where they will continue to move about. They have *evaporated* (*see* p. 968). At first there will not be many molecules in the space but the number will increase continuously even though some will travel back into the water. Gradually, a condition will be reached when the quantity of water molecules in the space is such that equal numbers of water molecules are being evaporated as are re-entering the liquid. At this point, the vapour in the space is in equilibrium with the liquid water, and it exerts a definite pressure known as the *saturation vapour pressure*. If the temperature of the whole system is raised, the amount of water vapour lost to the space before equilibrium is reached is also increased. Thus the saturation vapour pressure increases with temperature.

The inside of the leaf may be regarded as such a system, and at any particular temperature of the leaf, the vapour pressure inside its air-spaces is fixed at saturation value.

9

Humidity of the Atmosphere. Outside the leaf, the water vapour pressure will usually have a value lower than saturation. The ratio of the concentration present to the concentration at saturation vapour pressure is called the *relative humidity*. The lower the relative humidity, the lower will be the vapour pressure at any one temperature. Thus, at 20°C, 100 per cent relative humidity is equal to 23·4 mbar pressure; 60 per cent relative humidity is equal to 14·0 mbar; 20 per cent relative humidity is equal to 4·7 mbar. If the temperature is constant at 20°C inside and outside a leaf, the vapour pressure difference between the two points depends on the relative humidity outside, since the inside atmosphere can be considered constant at 100 per cent relative humidity. At 60 per cent relative humidity outside, this difference is 9·4 mbar, while at 20 per cent outside, the difference will be 18·7 mbar. Thus the rate of diffusion is doubled by lowering the external relative humidity by 40 per cent at this temperature. Any change which occurs in the relative humidity of the atmosphere will affect the rate at which water vapour passes through open stomata, when all the other conditions are constant.

Temperature. Vapour pressure, in addition to depending on water-vapour content, is also influenced by temperature. Thus a rise of temperature of the leaf, or both leaf and air, will tend to increase the rate of diffusion of water vapour through the stomata. We have seen that if the relative humidity inside the leaf is 100 per cent, and outside 60 per cent, then at 20°C the water-vapour pressure difference between leaf spaces and atmosphere is 9·4 mbar. Assuming that the relative humidities remain constant, if the temperature of both leaf and atmosphere is raised to 30°C, 100 per cent relative humidity represents a vapour pressure of 42·4 mbar, while 60 per cent represents a vapour pressure of 25·5 mbar. The pressure difference is now 16·9 mbar at 30°C, which is nearly double the pressure difference at 20°C. Thus with no change in relative humidity inside or outside the leaf, a rise in temperature will cause increased transpiration. In actual fact, when there are no evaporating surfaces near the plant to keep the relative humidity of the air constant, a rise in temperature from 20°C to 30°C would cause a fall from 60 per cent to a value around 30 per cent. Under these circumstances, the vapour pressure inside the leaf would rise from 23·4 mbar to 42·4 mbar, whereas outside the leaf it would not change from 14·0 mbar. The constancy of the outside vapour pressure at a lower relative humidity value is due to the fact that the actual concentration of water molecules in the atmosphere has not changed, and so the vapour pressure will be the same. Under these conditions therefore, the difference in vapour pressure between the leaf spaces and the atmosphere at 30°C would be 28·4 mbar, or treble the difference at 20°C. Thus changing temperature affects vapour pressure difference and hence it affects transpiration.

Light. Light can affect transpiration in two ways. Direct sunlight will almost invariably raise the temperature of a leaf above that of the surrounding atmosphere with consequent effect on the water-vapour pressure difference. A dark green leaf subjected to a long period of insolation may have a temperature as much as 20°C above that of the atmosphere. Thus the vapour pressure inside the leaf would be much greater than that outside, and a rapid rise in transpiration would ensue. Plants habitually in shade are not subject to such large changes in leaf temperature.

Light also affects the movements of stomata. The tendency for the stomata of most plants to open in the light and close in darkness has been previously discussed. The resulting influence of light on transpiration is obvious, and in general, the rate is much higher by day than by night.

Air Currents. The influence of air currents on transpiration is also to be sought in their effect on the vapour pressure difference between the inside and outside of a leaf. In still air, the tendency would be for water vapour to accumulate just outside the stomata. This would reduce the vapour pressure difference and so lessen the rate of diffusion outwards. Moving air would sweep the outgoing vapour away, thus keeping the diffusion gradient high and increasing the rate of transpiration. It is to be noted, however, that in bright sunlight, the tendency to increase transpiration, may be offset by the cooling effect of the wind which would tend to reduce transpiration.

Water Availability. Lack of water supply to the leaves may be due to several causes. The soil water may have reached such a low content that there is none for the roots to absorb. A low soil temperature may cause a rapid fall in the absorption rate of the roots. The root system may be so poorly developed that it cannot keep pace with a sudden heavy loss by the leaves in a drying atmosphere. The soil solution may have increased in concentration so making the absorption of water more difficult. Whatever the cause, when the water supply to the leaves falls, the general tendency is for the transpiration rate to fall. This is due chiefly to the adjustment of the leaf tissues so that the stomata tend to close. However, very frequently this does not happen until severe wilting has occurred, or even sufficient desiccation to cause death. Plants vary considerably in their ability to lower transpiration rate with a falling water supply to the leaves.

As a leaf loses water more rapidly than it can be replaced from below, the osmotic pressure of its cell sap will rise; this will have a slight effect in reducing the vapour pressure inside the leaf, by reducing the rate of evaporation from cell surfaces. This reduction is probably too small to be of great significance.

Atmospheric Pressure. Changes in atmospheric pressure could only affect transpiration rate if they were large. When pressure is greatly reduced, there is a great increase in the rate of evaporation from a wet surface, but the ranges of pressure experienced over the earth's surface at any fixed level would be of little significance. It is probably true that plants at high altitudes suffer increased transpiration when compared with plants at low levels.

Factors of Plant Structure. Plants which show morphological characteristics tending to reduce transpiration, are termed xerophytic. There are numerous examples of such modifications, differing in detail, but in general following a few morphological patterns. The commonest are enumerated below.

1. Reduction in leaf surface area, e.g. *Ulex* (gorse), *Pinus*, *Sarothamnus* (broom). In the case of *Ulex* and *Sarothamnus*, much of the photosynthetic activity of the plant is carried out by stems, which are also accordingly modified. There may be a great reduction in surface area/volume ratio of the plant as a whole, e.g. the cacti and some euphorbias. In these cases an apparent succulence is noticeable and the tissues may store water for long periods. Records show that while leaf reduction may reduce transpiration, small-leaved plants may lose more water per unit area of leaf surface than large-leaved plants when water is plentiful.

Fig. 7.13. *Echinocactus.* Globose stem with spines.

2. Development of a thick cuticle or other waxy covering, e.g. *Ilex* (holly), and many other evergreens.

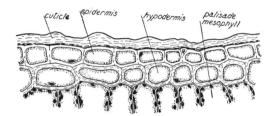

Fig. 7.14. V.S. of cuticle, epidermis and hypodermis of holly.

3. Stomata sunk in deep epidermal pits, e.g. *Nerium* (oleander), *Cycas, Pinus.*

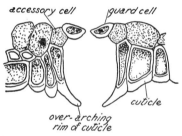

Fig. 7.15. V.S. Sunken stoma of *Cycas.*

4. Hairy blanketing of leaf surface, e.g. *Verbascum* (mullein), *Hippophae* (sea buckthorn).

Fig. 7.16. Hairs of sea buckthorn.

5. Inrolling of leaf surfaces to protect stomata on an inside surface, e.g. *Ammophila arenaria* (marram grass).

Many of these characteristics can be seen in combination in plants favouring drier habitats.

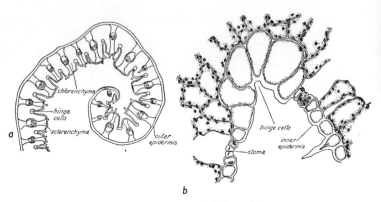

Fig. 7.17. Rolled leaf of *Ammophila*.

Some characteristics of other plants, which are not typical xerophytes, are also of service in preventing drying out. For example, the protection of apices inside bud scales which may be waxy or resinous, prevents not only damage by injury or frost, but also drying-out of delicate parts. Leaf shedding by deciduous trees can also be regarded as a means of protection against excessive water loss during the unfavourable periods of low temperature. The development of mucilaginous substances, which imbibe water copiously, can also help to conserve water. Finally, the production of very high osmotic pressures in leaf tissues would tend to reduce evaporation considerably.

Movement of Water through the Plant

It has been shown that the roots absorb water from the soil and that the aerial surfaces lose it to the atmosphere. Between the site of absorption and the site of transpiration, there must be adequate conduction to keep pace with the loss, and to supply the metabolic needs of all the intervening cells. This conduction has long been recognized as a function of the xylem elements of the vascular tissue. The structure of these elements has been described in Vol. I, Chap. 12 and their efficiency as water-conducting elements need not be further stressed. Here it is intended to discuss only the manner in which water is kept moving through them.

For convenience, we may consider the movement of water through the plant in two stages, namely, a lateral transfer across the roots from root-hairs to vascular tissues, followed by an upward flow through root and stem to the aerial parts, and a distribution of water from the stem system to the leaves where some is lost to the atmosphere by evaporation. It must be remembered, however, that in the plant, each of these phases is part of a single continuous flow.

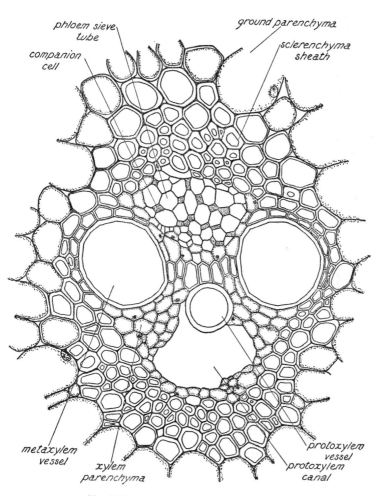

Fig. 7.18. Vascular bundle of stem of maize.

Movement of Water Across the Root

It has been mentioned earlier in this chapter that water absorption might be either active or passive, according to the development of absorbing forces by the root itself or to the formation of a hydrostatic pressure difference between water in the stele and in the soil. In either

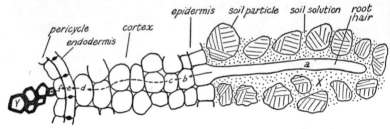

Fig. 7.19. Diagram illustrating passage of water across the root.

case, the forces concerned must be sufficient to keep the water moving across the root to replace any which may be removed through the stele. In some plants the phenomenon of exudation from cut stumps clearly indicates that water is flowing through the root and upwards under pressure. The precise mechanism involved in developing this root pressure is not clearly understood but it is known that only living roots develop it, and that factors affecting metabolism in the root, also affect root pressure. For example, adequate supplies of oxygen, mineral nutrients and carbohydrates are necessary. Chloroforming the root, or lowering its temperature, both suppress root pressure. It is an active phenomenon in that it depends upon living cells.

A possible explanation of this continuous active intake of water by the root may be put forward with reference to Fig. 7.19. The soil solution X is separated from the fluid in the xylem elements Y by layers of living cells of the epidermis, cortex, endodermis and pericycle. If the conditions were such that the solution at Y were sufficiently hypertonic to the soil solution we could visualize a flow of water from X to Y by osmosis where all the intervening cells were acting as a single composite semi-permeable membrane. Movement of water through such a series of cells can be demonstrated by the simple experiment in which a scooped-out, peeled potato is placed in a dish of water. The cavity in the potato contains strong sugar solution (see Fig. 7.20).

From an analysis of the fluid in a wide variety of xylem tissues in roots it seems likely that this condition is achieved because of the presence of dissolved inorganic and organic materials transferred from surrounding cells. The greater the quantity of solute particles in the xylem fluid the lower its water potential (or higher its osmotic pressure, *see* p. 53) by comparison with that of the external soil solution. There would thus be created a water potential difference resulting in a water potential gradient downwards (or DPD gradient upwards) from outside the root to inside the xylem elements so long as the xylem fluid was maintained rich enough in dissolved solutes. This overall downward gradient from outside to in would result in movement of water initially from cell *f* to cell *Y* in response to the lower water potential (higher osmotic pressure) inside the xylem cells. As water is removed from cell *f* so its water potential would be lowered with respect to cell *e* and water would now flow from *e* to *f*. In the same way water would be transferred from *d* to *e* and so on across the root through cells *c* and *b*

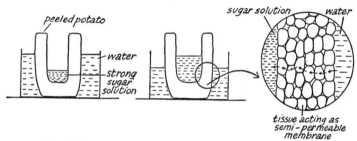

Fig. 7.20. Potato tissue acting as a semi-permeable membrane.

until in cell *a* the water potential fell to a level sufficiently low to cause water to flow into it from the soil solution. There would thus be a continuous movement of water from *X* to *Y* along a downward water potential gradient. The tendency would be for the flow to cease as water movement redressed the imbalance of water potential between the inside and outside of the root, that is as the water potential of the xylem fluid became raised to the same level as that of the soil solution. But in the functional plant there can be two events taking place to prevent this balanced or equilibrium state from being reached so that water continues to be forced from soil solution to xylem tissues. One is the steady "pumping" of solutes from surrounding cells into the xylem elements to maintain a low water potential (high osmotic pressure) there. The other is the negative hydrostatic pressure developed in the xylem by a pull upwards on the water there by evaporational forces acting at surfaces in the shoot. Both these tend to keep the water potential in the xylem below that of the cells outwards to the soil solution. Hence, so long as this difference in water potential is great enough to overcome any resistance to water flow put up by the cortical and other cells, there will be a continuous flow of water from the soil solution to the xylem.

It is the lowering of the water potential of the xylem fluid due to continuous active secretion of solutes into it that can account for root pressure.

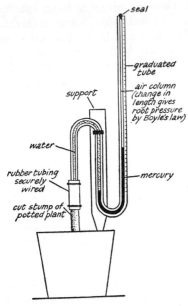

labels: seal; graduated tube; air column (change in length gives root pressure by Boyle's law); support; water; rubber tubing securely wired; cut stump of potted plant; mercury

Fig. 7.21. Apparatus to measure root pressure.

In some plants this upward surge of water can most certainly be demonstrated and the pressure by which it is being raised can be measured. The apparatus is shown in Fig. 7.21. From the length of the air column in the manometer arm at any time, compared with its length at the start, the root pressure can be calculated.

The value of this pressure from below will depend upon the water potential of the fluid in the xylem cells. It will vary according to the rate at which the pericycle cells lose osmotically active substances to the xylem. As has been stated, interference with the metabolism of the root can reduce root pressure to zero. Under experimental conditions root pressure rarely exceeds 2 atm and not all plants seem capable of developing it. In those plants which do, it is generally highest in the spring.

The foregoing has had reference only to a case in which water might be moved across the root and then upward through the xylem as a result of a diffusion pressure deficit created in the xylem by the secretion of salts into it. It does not explain the continuous upward rise of water to the top of a very tall tree nor the continuous flow of water across a root in which no root pressure is ever developed. In the following section the upward movement of water will be more fully discussed, with special reference to evaporational forces at the leaves which can operate to lift water in a purely passive manner.

The Upward Movement of Water: the Transpiration Stream

Many suggestions have been made to account for the upward movement of water. They include root-pressure, capillarity, imbibition pumping mechanisms, and mechanical pull from above. To be completely satisfactory, any explanation must be able to account for a rise of water of up to a hundred metres, the height of some tall trees.

The first two possibilities can be of no significance in such cases. A root pressure of 2 atm could raise water to a height of about 22 m only. Nevertheless root pressure plays some part in the rise of water in certain plants and it can explain guttation in many cases. But it must be noted that guttation may also be explained in terms of active secretion of water by special cells of the leaf, and thus it may occur in the absence of root pressure. A wide variety of plants do not seem to develop root pressure at all; even where it exists, the quantities of water exuding from a cut stump are by no means comparable with the quantities leaving the aerial parts of a similar undamaged plant by transpiration. In many instances a newly cut tree stump will absorb water poured over its surface, indicating that the water it contained before being cut was under strong tension or negative pressure below.

In the smaller xylem elements with a diameter of less than 20 μm, capillarity would cause rise of water 4 m to 6 m. In the large vessels of diameter approaching 500 μm, the rise would be no more than a few centimetres. This capillary rise could be increased by the nature of the xylem cell wall, which contains sub-microscopic interstices. It would possess forces of imbibition more than necessary to raise water to the required height. But the rate of flow through the colloidal substance or through the very minute channels in the wall would be quite inadequate to account for the large quantities of water which move; the resistance to flow is too great.

It has been postulated that pumping mechanisms, due to the secretory activities of the living cells of wood parenchyma and medullary rays, may be responsible for the rise of water. However, it seems that water can quite easily move upwards when all the living cells of the stem have been killed.

The last possibility, that the water is lifted by a mechanical pull from above, was proposed by Dixon and Joly in 1895. The hypothesis is that the water moves upward as a continuous column from roots to leaves, to take the place of water evaporated from the aerial surfaces. The sequence of events is probably the following. The spongy mesophyll cells lose water by evaporation to the leaf spaces and thence to the atmosphere. These cells lose turgor and draw in more water from surrounding cells. Thus there is formed a water potential gradient from leaf spaces to xylem elements in the leaf. Those cells in contact with the xylem elements will gain water from them. This will impose a pull on the water in the vascular system, and thus lift it. The process can be compared with pulling on a rope from above; the xylem cell

walls keeping the water columns intact. Such a lift could be accomplished, provided the following conditions are met.

1. The cells adjacent to the xylem elements in the leaves must
lose water potential sufficient to lift water to the height required,
against the force of gravity, and against the resistance to flow.

2. The cohesive forces between the water molecules in the system
must be great enough to prevent the columns from breaking under the
forces applied to them.

3. The tubes which are conducting the water must be rigid enough
to resist collapse when the water is moving under tension.

4. No air bubbles must enter the system to break the water columns.

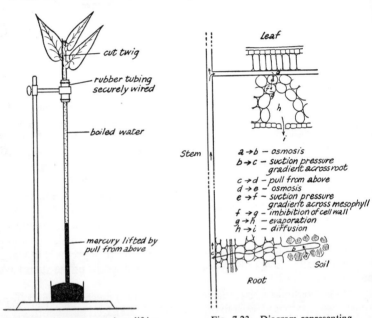

Fig. 7.22. Apparatus to show lifting
power of a transpiring twig.

Fig. 7.23. Diagram representing
movement of water from soil to
atmosphere through the plant.

When the system is examined with respect to these conditions it i
found that there is much in favour of the hypothesis. It is generall
agreed that the water potentials found in the leaves are more tha
adequate. Water potential of the cell sap of many types of leaves ca
be equivalent to 20 to 40 atm but it must be remembered that suc
large forces would only come into operation in a wilted leaf where th
cells had lost their turgor. The tensile strength of a column of wate

seems sufficient to withstand the pull without breaking. Although the equivalent value of 200 atm suggested by Dixon has been shown to be much too high, the more recent estimate of 30 atm is great enough. The structure of the xylem elements is such that they can quite easily withstand the strain without collapsing. Careful measurements have shown that they do tend to narrow in diameter when water is being forced through them under tension. Finally, although air bubbles do enter the system, it seems likely that they can be isolated by the components of the xylem in which they occur. Special mechanisms such as bordered pits may assist in this respect.

One of the more serious criticisms levelled at the hypothesis is related to the formation of air bubbles in the xylem elements under the influence of changing temperature. As the temperature of the water columns is raised, it is obvious that bubbles will appear quite rapidly in many parts of the xylem, as gas is forced out of solution. It is difficult to see why this does not cause almost complete blockage of the tubes. It can only be supposed that gas in solution can escape through the cell walls laterally, before bubbles can be formed in sufficient quantity to cause blockage.

The lifting power of a transpiring twig can be demonstrated by the apparatus shown in Fig. 7.22. Unless the glassware is scrupulously clean, the water boiled, and the whole apparatus fixed so that it receives no heavy vibrations, the mercury column cannot be lifted more than a few centimetres. The highest lift recorded by this method is about 120 cm of mercury, equivalent to about 15 m of water.

This theory is generally accepted as offering the most convincing explanation of the rise of water in root and stem in most cases. The pull from above, transferred through water columns to the xylem in the roots, would create there the equivalent of a water potential low enough to extract water from fully turgid pericycle cells irrespective of whether the root was developing an active root pressure. The total movement of water from soil to atmosphere can be summarized as shown in Fig. 7.23. Since this flow of water is maintained only as a result of the evaporation of water from leaf surfaces, it is often referred to as the *transpiration stream*.

Guttation

This is the exudation of droplets of water from leaf surfaces. It is noticeable in many plants when they are growing under conditions favouring rapid water absorption coupled with very slow transpiration. When the soil is warm and moist and the leaves are cool, in a saturated atmosphere, the exuded water is often mistaken for dew. Such conditions often occur at night in summer.

The water is forced out of the leaf through special structures known as *hydathodes*, of which there are many types. In *Phaseolus* (runner bean), and *Rhinanthus* (yellow rattle), the hydathodes are secretory or glandular hairs on the leaf surface near the veins. Another type, exemplified by *Ranunculus ficaria* (lesser celandine) and *Fuchsia*, consists of colourless, thin-walled, loosely-packed, sub-epidermal cells forming a specialized tissue known as an *epithem*. This is in contact with the terminal tracheids of the minute vascular bundles within the leaf (*see* Fig. 7.24). In *R. ficaria* and *Fuchsia*, the epidermis immediately above the epithem is perforated by one or more special water-pores. These are like large stomatal openings but the guard cells do not move to close the pores. Sometimes the epithem is not clearly defined from the mesophyll, whilst in other cases it may show sharply within a suberized area, or be surrounded by cells possessing Casparian strips

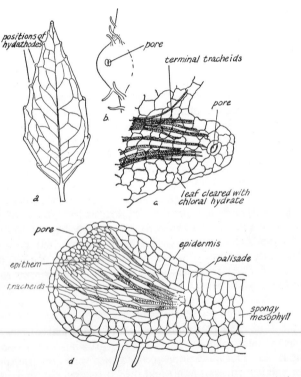

Fig. 7.24. Hydathodes of *Fuchsia*. (*a*) Position on leaf. (*b*) Pore (enlarged). (*c*) Surface view of leaf cleared in chloral hydrate. (*d*) V.S. *Fuchsia* leaf through epithem.

around their walls. The more complicated hydathodes usually occur at the apices of the leaf teeth as in *R. ficaria, Saxifraga* and *Fuchsia*, or at the leaf tips only, as in grasses.

Where guttation is due to glandular hairs, it appears to be a secretory activity. In the more complicated hydathodes, the water leaving the pore does not appear to be actively secreted by the epithem, but merely passes from the terminal tracheids of the vascular bundle through the intercellular spaces of the epithem to the outside, as though it were being forced out of the vascular tissue. This suggests an upward pressure originating in the roots and could very well be the result of root pressure. The exuded water may be rich enough in dissolved substances to leave a residue on the leaf surface after evaporation.

Hydathodes are easily recognizable in young leaves and they are probably more active then.

The Influence of Transpiration on the Plant

Transpiration has several effects on the plant; some of these have been regarded as beneficial. It is often quoted as a function necessary to the well-being of the plant. An important effect often ascribed to transpiration is an increased rate of salt absorption and transport through the plant. If water is being removed more rapidly from the neighbourhood of the roots, then more water would move through the soil to take its place. This would bring more salts into the vicinity of the roots. resulting possibly in a more rapid intake. Then the increased rate of movement of water through the vascular tissues should distribute the salts about the plant more quickly, and the plant should benefit accordingly. However, experimental evidence does not in general bear out these assumptions. There is no certainty that an increased rate of water absorption will cause an increased rate of salt intake; in fact it seems probable that salt absorption is entirely independent of water absorption. Furthermore, although a rapid transpiration stream would undoubtedly carry minerals quickly away from the root, there is evidence that the concentration of minerals passed to the xylem of roots is lowest when transpiration rate is highest. There is also the strong probability that salt transport is not confined to xylem, the phloem being equally responsible.

A definite result of transpiration would be a cooling effect at the site of evaporation. The high specific latent heat of vaporization of water means that much heat is used in evaporating it, with a consequent cooling of the leaves. In still air and full sunlight, leaves do tend to become overheated, and the cooling may be beneficial in such cases. In moving air, the tendency would be much less, and in general, the extent to which the cooling by transpiration saves the lives of many leaves is dubious. In any case, the evaporation of water rarely lowers leaf temperature by more than 3°C to 5°C, and thus at dangerously high temperatures it would be of little significance.

Rapid transpiration has the effect of reducing the water content of a plant; this is responsible for several harmful conditions. In a plant lacking water, the visible signs are general flaccidity, and later, wilting. If prolonged, this may cause death of a part of the plant. The least harmful effects will be to cause a check in growth and to interfere with transport mechanisms. Plants continually exposed to drought have a characteristic stunted appearance and an accentuated woodiness and cuticularization, when compared with members of the same species grown under wetter conditions. This may result in restricted yield of reproductive organs, but it has some advantage in that the plant is physically stronger and probably less prone to attack by parasites.

It is generally agreed that the best view to be taken of this problem is that transpiration is inseparable from the plant structure, which is designed to function efficiently on land, where water is localized and contact with a drying atmosphere must be maintained. Any good or bad effects accruing from the process are purely incidental, and they have balanced themselves out over a long period of adaptation to the land habitat.

THE WATER RELATIONSHIPS OF ANIMALS

Although, in general, animals require water for the same purposes as do plants, the problems of water relationship are quite different for several reasons. In the first place there is considerably less use for water as a metabolite in animals, since there is no photosynthesis, and the large quantities used by plants for this purpose are not required by animals. Neither is there in animals generally, evaporation loss approaching that due to transpiration. Also, animal cells do not normally possess rigid walls, their boundaries merely being the semi-permeable plasma membranes which do not restrict intake of water. Then again, the food of animals, consisting of plant or animal material, contains at least 70 per cent of water. Finally, animals in general are active and have a high rate of metabolism and with the normal respiratory substrate, glucose, can release 60 g of water for every 100 g of sugar oxidized. This, of course, is true of plants, but their metabolic rate is much lower.

Animal cells are either surrounded by an external fluid medium or bathed by internal body fluids, and in view of the permeability of plasma membranes and the lack of any appreciable turgor pressure, the problem for both individual cells and for the whole body is that of maintaining osmotic equilibrium between the cell and the extracellular fluid. Osmoregulatory mechanisms are discussed in Chap. 6.

Intake of Water

All aquatic animals take in water with their food. The food itself will contain appreciable quantities of water. Those without impermeable coverings will take in water over the general body surface if the cell contents are hypertonic to those of the surrounding water. Marine invertebrates are in general isotonic with the sea water; the entry and exit of osmotic water will balance each other. Marine fishes are all migrants from fresh-water; some are still tied to both types of water, e.g. salmon, eel. As described in Chap. 6, the body fluids of marine teleosts are hypotonic to the sea-water and their problem is to conserve water and exclude salts. It has been solved by a combination of several factors. In the first place there is an impermeable covering of bony scales which prevent entry or loss of water over the general body surface. They drink the sea-water and have evolved the method of excreting salts from the gills. Lastly, the glomerular kidney, which was evolved as a mechanism for getting rid of water in early fresh-water vertebrates, excretes dilute urine which is always hypotonic to the body fluids; the water-excreting kidney is almost a liability to marine teleosts. In the elasmobranch fishes the osmotic concentration of body fluids is due to the retention of large quantities of urea; thus a very dilute hypotonic urine is excreted by the kidneys. They appear to drink very little water with the intake of food.

The Amphibia constitute a special case, comparable with lower land plants in the sense that they are never independent of water, or at least of a moist environment. The majority must lay their eggs in water. The soft moist skin is slightly permeable, so in aquatic conditions water will enter osmotically, but at a reduced rate. The kidneys are adapted to excrete a copious dilute urine. Loss of salt is made good by the food and by reabsorption in the distal tubules of the kidney. In many amphibians there is storage of a considerable quantity of water in the allantoic bladder; this is a safeguard against dry periods.

The majority of insects either take in their food in a fluid medium or actively drink water. The chitinous cuticle renders the body surface impermeable. Practically no water is excreted, the nitrogenous excretory substance being mainly uric acid. The greatest loss of water occurs by evaporation; two-thirds of the loss is by way of the spiracles. Hence most insects have developed mechanisms for closing the spiracles; there is also a marked tendency for insects to frequent, and especially to deposit their eggs in, regions of optimum humidity. In general, it is probably true to say that fully terrestrial insects are less dependent on access to liquid water than any other group.

Reptiles, for the most part, rely on the water-content of their food as the sole method of intake. Some few will drink water in periods of extreme drought. Their bony scales protect them from surface evaporation; reabsorption of water in the kidney is so effective that the urine is generally solid or semi-solid. In reproduction, reptiles are either ovoviviparous, or they lay cleidoic eggs protected from evaporation and containing enough water (mainly in the albumin) to satisfy the needs of the developing embryos.

The birds drink water, as well as obtaining the water present in their food. The feathers and the keratinized layers of skin protect them from surface evaporation. The kidneys excrete a hypertonic urine; effective reabsorption of water in the loop of Henlé results in the elimination of almost solid uric acid. Like the reptiles, they lay large cleidoic eggs.

Mammals drink water, often in copious quantities. The hair and the keratinized stratum corneum protect them from surface evaporation, except at the sweat pores. They excrete a hypertonic urine; the loop of Henlé is present, but there is not such effective reabsorption of water as in birds.

Independence of intake of liquid water has been achieved by but few animals. Some insects, such as the clothes moth, meal-worms (larvae of the flour-moth, *Ephertia*), some desert animals such as lizards and the desert rat, rely entirely on the water-content of their food. The camel can store water for long periods, probably for about ten days; it is said that it can exist with no liquid water intake, if provided with good pasture and not worked strenuously.

Output of Water

Except in the case of fresh-water animals the body fluids of which are hypertonic to the medium, there will not be continuous intake of water. For these there will be a constant necessity to eliminate osmotic water and mechanisms ranging from contractile vacuoles to glomerular kidneys have been evolved. The majority of animals pass out their nitrogenous excretory products in solution. In terrestrial animals there is always considerable evaporation from respiratory surfaces and in the mammals the activity of the sudoriparous glands is responsible for the loss of large quantities of water. Most of the mechanism involved will be discussed in Chap. 15.

THE ROLE OF WATER IN THE ORGANISM
AND IN ITS ENVIRONMENT

Living organisms are so constructed that water is essential to their continued existence in two fundamental ways. First, it enters into the composition of protoplasm, which we may very broadly describe as an aqueous system in which substances mostly of extreme complexity are dispersed. It is upon this finely balanced arrangement of matter in a watery phase that all the properties of protoplasm depend. If the water is removed, the molecular architecture crumbles. The protoplast ceases to be alive. No living thing departs from this essential structural pattern, and no other substance can replace water in the living dispersion system. Secondly, within this physico-chemical framework, the functional activities of the protoplasm are such that they can be carried out only in the presence of a substance having the specific physical and chemical properties of water. Some of these properties of water which are recognized as being essential to the normal physiological functioning of cells are listed below.

1. Water is a chemical reagent participating in many of the metabolic activities of protoplasm. Two important examples will suffice to indicate its significance in this respect. It provides the reducing hydrogen in the photosynthetic process of green plants, and it is essential to the multitude of hydrolytic processes which take place in all living organisms.

2. It is the principal solvent of all the substances concerned in protoplasmic activity. All the materials which enter or leave living cells can do so only in aqueous solution. Within the organism, transport takes place only in aqueous solution; the numerous substances which are moved include nutrient materials, excretory products, dissolved gases, inorganic ions, hormones, and many of the protoplasmic constituents themselves.

3. Water readily allows the dissociation of substances dissolved in it, and hence affects their chemical and electrical activities (*see* p. 963).

4. The surface tension of water is high. This property affects activity at the interfaces between a whole protoplast and its surroundings, where a plasma membrane is in contact with an aqueous solution. It also affects the interface relationship between parts of the same protoplast, for example, between the nucleus and the cytoplasm, and between the cytoplasm and some of its inclusions. It is not within the scope of this text to deal with the complicated physical conditions obtaining at liquid interfaces, but it is well recognized that interface phenomena are associated with many physiological processes. Among these are cell membrane permeability, adsorption, imbibition of water by colloids, capillarity and protoplasmic streaming.

5. The low viscosity of water allows of its rapid movement into and

through cells. It is known that any change affecting the viscosity of water, such as lowering its temperature, will decrease its rapidity of movement and hence affect the many physiological processes in which water is concerned.

6. It has a high specific heat capacity and thus will not be subject to extremely rapid fluctuations in temperature which would affect the rate of all chemical changes. Thus it might be said that water offers a fairly stable temperature background.

7. The thermal conductivity of water is comparatively high; this allows of reasonably good heat transference where it is required.

8. Water has its greatest density at 3·98°C. This anomalous property must have saved the lives of countless organisms, for it means that the inhabitants of aquatic environments are not subject to the sudden freezing of the water throughout its bulk. An expanse of water always freezes from the top downwards; this must have been of great importance to living organisms during the cold conditions which have sometimes prevailed in the history of the earth, and which still obtain in many regions, if only seasonally.

CHAPTER 8

CO-ORDINATION

We have stated that an organism can be regarded as an aggregation of organs. But we have also stressed that each part is not an independent unit. All parts of the plant or animal body must develop and function in a manner and at a speed most suited to the common good. There must be harmony in the multiplicity of activities of the parts so that the organism acts as an integrated whole. Before proceeding to the description of the manner of performance of special functions, we must consider two important requirements on the part of the organism by which this necessary high degree of integration can be achieved. The first of these requirements is a system or systems by which all the parts can be co-ordinated with one another and the second is a system for transfer of materials from place to place in the body so that the needs of all parts are met. These topics will be treated in this chapter and the next.

It should be noted here that co-ordinating systems may do more than ensure harmonious development and functioning of all the organs. The organism is subject to changes in both internal and external environments, and parts of the co-ordinating systems may be specially modified for perception of, or response to these changes. For example, specialized receptor parts of nervous systems in animals are capable of detecting changes in conditions and generating impulses which on conveyance to an appropriate part of the organism are able to elicit a response to the changed conditions. We find, therefore, that the co-ordinating systems are always associated with the manifestation of sensitivity. Certain aspects of their role in this connexion are discussed in Chap. 16.

CO-ORDINATING MECHANISMS

There are three known ways in which co-ordination may be brought about—

1. By the passage of electrical impulses from one part of the body to another, by way of specialized cells.

2. By the dispersion of soluble chemical substances throughout the body fluids, these substances having an influence on the activity of special parts in each case.

3. By the purely physical or mechanical effect of one part of the body on another.

In the first case, the specialized cells are *nerve cells* and they form in their entirety a *nervous system* throughout the body. Only animals have evolved such a co-ordinating mechanism.

In the second case, the chemical substances are described as of two kinds, *hormones* and *organizers* (*evocators*). In animals, discrete organs and tissues are responsible for the production of hormones. These organs constitute the *endocrine system*. Organizers are chemical substances which are produced by certain histologically unspecialized tissues to evoke particular developmental responses in others. They are produced in embryonic and regenerating tissues where they influence the mode of differentiation of the developing cells; they are to be regarded as hormones. Substances of regulating and co-ordinating function are produced by plants. They are collectively termed *plant hormones*, but unlike their animal counterparts they are not secreted by specialized glands.

The third case of physical or mechanical effect is peculiar to muscular systems of animals where the pull exerted by one contracting muscle may influence the operation of another.

These systems, where they occur together, do not necessarily work independently of one another; quite often they work in direct association. In animals, nervous excitement may set the endocrine organs working, or hormones may initiate nervous impulses. Neither, where they occur at all, are they of the same compexity throughout the plant and animal kingdoms. In general, the more complicated the body structure of the organism, the more complex are its co-ordinating mechanisms and associated structures.

Within the animal kingdom, the most lowly protozoa show co-ordination of the parts of their non-cellular protoplasmic masses, but with the exception possibly of the ciliates, e.g. *Paramecium*, there are no distinct organelles of co-ordinating function. This means that most of the co-ordinating work is done by undifferentiated protoplasm in an unknown way. Fibres which have been described as co-ordinating in function, called *neuronemes*, have been found in the cells of some ciliates only. These are particularly associated with the locomotive organelles, cilia or flagella, and the two systems together form the so called *neuro-motor apparatus*. There is no evidence as to the existence of hormones or organizers in the protozoa.

In all animals above the sponges, there exist in increasing degrees of complexity, definite cellular systems given over to the transmission of electrical impulses. The simplest of these mechanisms may be called the *nerve net*, and the most complex, the *synaptic nervous system*. The production of hormones and organizers is also a property of a number of groups of invertebrate metazoans. Though in some cases there is little known, a considerable body of knowledge has been accumulated concerning endocrine activity in insects and crustaceans, together with a number of important facts about hormones and their effects in annelids and molluscs (*see* Vol. I, Chaps. 21, 22, 33 and Appendix). On the other hand, the activity of the endocrine gland system of the vertebrates has been the subject of much study and we have considerable knowledge of its co-ordinating function. In most animals, evocators are associated with embryonic development and with regeneration.

In the plant kingdom, with the possible exception of a neuromotor apparatus in some free-swimming algae, there is no mechanism of conduction comparable with the nervous system of animals. It is not a condition essential to their survival. In general, plants have no need for rapid perception of, or response to, changing conditions. The co-ordination effects which can be observed, are those of growth and development; the mechanisms which co-ordinate are hormonal. There has been no apparent increase in complexity of the system with the evolution of higher plants. Substances having a particular effect in lower plants, have similar effects in the higher ones.

PLANT HORMONES

Starling, in 1904, first defined a hormone when he described the gut secretion, *secretin*, as "a substance normally produced in the cells of some part of the body and carried by the blood stream to distant parts which it affects for the good of the body as a whole". Comparable "chemical messengers" called broadly plant growth substances, occur in plants or have been synthesized. They are known to influence a wide range of growth and developmental phenomena including: *growth curvatures in response to stimulation, wound healing, root initiation, onset of flowering, sexuality in flowers, fruit growth, parthenocarpy, abscission of parts, inhibition of and promotion of seed and spore germination, inhibition and promotion of lateral bud development, leaf expansion, cell division, cell enlargement* and *tissue differentiation.* The sorting of these hormones into groups, chemically or by their individual effects, has not yet been wholly accomplished and terminology is very confused. The following is a guide to currently accepted terms.

Plant growth substances includes all organic compounds, natural and synthetic, which, at low concentration, in some way modify growth and development. Their effects are not dependent on their energy content or their nutritive value. *Plant hormones* or *phytohormones* include only naturally occurring growth substances and which truly regulate normal growth. This excludes the synthetic compounds many of which violently disturb the normal growth processes. *Auxins* can be defined as those substances, which in low concentration (< 0.001 mol dm^{-3}) and apparently independently of other growth-promoting substances, are known to stimulate an irreversible increase in size of cells of shoots and to inhibit the same in roots. They may be "natural" or "synthetic" auxins according to origin. They undoubtedly influence other growth activities but it is their effect on cell enlargement only that is commonly used in defining them. *Gibberellins* are substances which have a variety of effects on plant growth and development, sometimes in association with auxin, sometimes not. To be classified as a gibberellin a substance must meet both a chemical and a physiological requirement. It must possess the chemical configuration known as the *gibbane ring* and must be capable of bringing about a reversal of the condition of dwarfism in stems (or alternatively cause the induction of the enzyme α-amylase in the endosperm of barley). Not all substances having gibberellin properties act in the same way. *Cytokinins* are substances active in inducing cell division as distinct from cell enlargement. Other substances now recognized as having some control in plant growth phenomena are *ethylene*, the olefine hydrocarbon (C_2H_4), some *phenols*, the *flavonols* and a substance known as *abscisic acid* or *dormin*. The last three possess properties that make them growth inhibitors. The *accessory growth substances* or *vitamins* could qualify as plant hormones since they occur naturally, are manufactured in one place and effective in another, and are known to be essential to the normal growth of plants.

AUXINS, GIBBERELLINS AND CYTOKININS

One substance only, β-indolyl acetic acid (IAA), has been identified with certainty as a natural auxin by the above definition. This does not mean that no others exist. Other closely related derivatives with similar properties, β-indolyl acetaldehyde, β-indolyl pyruvic acid, β-indolyl ethanol and β-indolyl acetonitrile, have been isolated, but these may not be free active auxins, merely substances which are readily changed into active auxins. There are, however, numerous organic compounds which have auxin effects; many are of great value in agriculture and horiticulture. They include phenyl-acetic phenoxyacetic, 2:4-dichlorophenoxyacetic (2:4-D), α-naphthalene-acetic cinnamic, benzoic and naphthoxyacetic acids. They are used commercially as root initiators, preventers of fruit drop, selective weed killers and producer of seedless fruit.

Auxin activity is remarkable from the point of view of the concentration required for full effect, a few parts per million being sufficient. Different part of a plant do not react in the same way to the same concentration, however Very low concentrations which may elicit maximum response in root cells ar virtually ineffective in shoots, while the higher concentrations which giv maximum effect in shoots and coleoptiles tend to inhibit the response in roots This is now known to be due to the production of ethylene in the root tissue caused by the presence of auxin. Ethylene, therefore, is the direct cause of the inhibition of the root growth, not the auxin. Lateral buds seem to give greatest response to intermediate concentrations. Strong doses of auxin always prove toxic to all cells but no satisfactory explanation of this has yet been found.

In addition to their effects on cell enlargement, by which they are distinguished from other plant growth substances, the auxins are known to have many other growth-regulating influences. The present definition of an auxin is, therefore, probably too narrow and may even be misleading. They are known to act in the inducement of cell division and in some cases as general growth controllers. They may be said to "activate" cambium. Some seeds are said to germinate more speedily and more effectively when given auxin treatment at the correct concentration. Auxins may also act as initiators of new tissues and organs and are used to promote rooting of cuttings, as stimulators of cambial activity in grafts and to bring about the "setting" of greater quantities of fruit. In one plant at least, the pineapple, flowering is undoubtedly promoted by auxins. On the other hand, at concentrations above an optimum for full effect, auxins may act as growth inhibitors. The failure of lateral buds to develop in the presence of an apical bud is said to be due to the movement of auxin from the shoot apex downwards in concentration above the best for lateral bud development. Auxins are known to prevent leaf and fruit abscission in some plants by inhibiting the development of the abscission zone. Auxin sprays are useful in preventing pre-harvest fall of apples and oranges. Another interesting effect of varying auxin levels in plant stems is shown by woody branches growing horizontally. On the underside the xylem develops in the presence of abnormally high auxin level as so-called "compression wood" in which the cells are thick-walled, reddish in colour and heavily lignified. On the upper side, where auxin level is subnormal, "tension wood" with completely contrasting characters is formed. The control of vascular tissue differentiation and regeneration in *Coleus* is also known to be associated with auxin production in the leaves. Auxin limits the regeneration of both tracheids and young sieve tubes near a wounded vascular bundle in the stem. The number of xylem strands that can regenerate in a stem is proportional to the number of young leaves left on the plant. Loss of young leaves can be compensated by application of external IAA to the severed petiole of a leaf. Normal xylem differentiation is also influenced by auxin. Rate of xylem formation in a petiole correlates well with auxin quantity produced in a leaf of *Coleus*.

None of the auxin phenomena can be explained at present in precise physiological terms but a great deal of investigation has been carried out on the effect of IAA on cell enlargement. Nothing for certain has been discovered but it seems most likely that the auxin, probably interacting with a gibberellin (*see* below) influences cell enlargement by having a "plasticizing" effect on the primary cell wall allowing it to be moulded under the influence of turgor pressure during the deposition of new wall substance. Alternatively it may have the action of a co-enzyme concerned directly with wall material metabolism. Another possible mode of action is through the activation of a messenger RNA to promote the synthesis of particular enzymes concerned with the placing of new materials into the wall of a turgid cell.

It is known that not all the IAA present in plant tissues is "free". Some is certainly "bound" to other substances, probably proteins, and these are not extractable in the usual way with ether. This auxin can be liberated through the action of digestive proteinases and has the properties of pure IAA. Another way in which auxin occurs appears to be as a "precursor" substance now recognized as a mixture of compounds containing IAA, inositol and arabinose.

Movement of auxins through plant tissues shows some puzzling characteristics. In young shoots, where auxin is manufactured in the undifferentiated apex, the movement is predominantly in one direction, from apex to base. This polarity has nothing to do with gravity but it is certainly related to the metabolic activity of the cells through which it passes. Movement of auxin in roots does not show the same polarity. In mature stem tissues, movement of auxin seems to occur in the vascular tissues but the movement seems slower than for other materials so transported. For more information concerning auxin transport *see* p. 322.

There are known to be substances in plants which counteract the activity of auxins but which do not destroy them. *Coumarin*, an active germination and root growth inhibitor, is one, and *ascorbic acid* is thought to be another. An auxin-destroying enzyme system has been extracted from young tissues. This enzymatic destruction is by oxidation and the enzyme is a peroxidase functioning as an oxidase. It is more powerfully active in roots than in shoots. It is suggested that such substances could function as auxin concentration regulators within the plant. Auxin is also destroyed by a photo-oxidation process in which very strong light doses are required for effectiveness.

Some compounds, such as 2,4-dichlorophenoxyisobutyric acid, appear to exhibit a kind of auxin activity in that the inhibition of root growth due to auxin is removed in their presence when root elongation is promoted. The nature of this combative effect is not clear since it is considered that the inhibitory effect on root growth by auxin is due to the formation of ethylene in root tissue in response to the presence of the auxin. The anti-auxin effect is therefore probably a secondary one.

The second group of growth regulating substances, the gibberellins, were in fact reported in Japan in 1926, a year before auxin was discovered. This was as a result of an attempt to find out why rice seedlings, infected by the fungal parasite *Gibberella fujikuroi* [*Fusarium heterosporum*], showed excessively long growth of their stems. A complex of substances, called collectively gibberellin, was isolated from a broth in which the fungus had been cultured and shown to contain the active growth-promoting agent. Attempts to separate and purify the components of this mixture were not originally successful but in 1955 a single pure substance having the growth regulating properties previously recorded was isolated and called gibberellic acid. Altogether upwards of twenty related compounds, collectively referred to as the gibberellins (GA 1, GA 2, etc.), have been separated and one or more of them have been found to be produced by higher plants, the ferns, the mosses, algae and bacteria. These gibberellins, previously chemically and physiologically defined on p. 270, have the most profound effects on the growth and development of all classes of plants.

Among the better known of these are the growth effects on stem elongation, on young leaf expansion, on dormancy, on fruit formation and some regulatory and integrating effects on developmental patterns.

One of the most studied of the growth effects of gibberellin is that on stem elongation, most noticeable as the induced "bolting" brought about by gibberellin treatment of dwarf plants. Stem elongation ordinarily can be shown to be due to activity in three zones at or near the apex. There is the cell-producing activity at the apex itself which seems little affected by gibberellin except that it can bring about a break in dormancy of some terminal buds. It is said to counteract the effect of dormin, a growth inhibitor responsible for bud dormancy in some plants. Neither does the effect of gibberellin appear to be in the cell elongation zone at a little distance behind the apex but, in the immediate sub-apical region or intercalary part where ordinarily there is little or no cell division, gibberellin promotes rapid mitotic activity of the cells to produce many more, so that when these extra cells elongate, the length of the stem is greatly increased. It is interesting to note that gibberellins appear not to have any significant effects on the growth and development of roots. This may be correlated with the absence of a sub-apical meristem in the root that cannot be stimulated to produce more cells.

The effect on young leaf expansion is noticeable when gibberellins are supplied to etiolated stems. This causes the leaves to commence growth in size but not as effectively as does full exposure to light. This growth activity of leaves is known to be affected by other factors, including radiation with red light and the presence of cytokinins, so the role of the gibberellins is not clear.

The inhibition of bud growth or bud dormancy which commences in the autumn for deciduous plants is normally only removed or broken by low temperature exposure or by the correct alternating periods of light and darkness. In some plants, noticeably sycamore and birch, gibberellin treatment induces the buds to open without the cold or light/dark exposure. Seed dormancy is also broken by gibberellin treatment in some cases.

Parthenocarpy, or the growth of fruits without fertilization, is also induced by treatment with gibberellins of tomato, apple and pear flowers. In the normal course of events, following fertilization and only when this has taken place, gibberellin is formed in the embryo and surrounding tissues and this stimulates the fruit to grow.

Among the development regulating effects of gibberellin is that concerning the onset of flowering in which a vegetative apex becomes reproductive. This normally happens in response to a special sequence of environmental conditions experienced by the plant such as long days or short days (see p. 544) and subjection to low temperature, i.e. vernalization (see p. 546). Gibberellin treatment is known to be a substitute for the normally required conditions in some cases, particularly long-day plants. For example, if *Hyoscyamus niger* is given gibberellin under short-day conditions it will flower, whereas otherwise it would not. The carrot can be induced to flower without vernalization if similarly treated with gibberellin. It has been thought that the postulated flower-inducing hormone, *florigen*, (*see* p. 545) might be a gibberellin but this appears not to be the case since even if florigen exists, gibberellin does not induce flowering in short-day plants.

Another effect of gibberellins on flower development is in relation to the sexuality exhibited by unisexual flowers, with particular reference to the cucumber. In this plant, gibberellin treatment of the developing flower leads to the promotion of increased maleness, that is the induction of stamens. In a similar sort of way, antheridia can be induced to form in greater numbers than usual on fern prothalli.

Some seed germination processes also appear to be under the influence of the gibberellins. For example, in cereal grains conversion of stored food in the endosperm to a soluble state by the amylolytic and proteolytic enzymes will not normally occur in the absence of the embryo. Gibberellin treatment of grains lacking the embryo is effective in mobilizing the stored insoluble food. There is a diffusible substance obtainable from the embryo, shown to be rich in gibberellins, that will do the same and so there is little doubt that these substances are effective in promoting this food-mobilizing process. The digestive enzyme most concerned is α-amylase and this is caused to appear in the cells almost certainly as the result of a *de novo* synthesis. The effect of the gibberellin in this way is so marked that it can be used as a means of bio-assay for gibberellic activity. Arising from work done on the induction of α-amylase by gibberellins the suggestion has been made that these substances are acting in this case as agents of de-repression of a gene for α-amylase production (*see* p. 802).

There is no doubt that the mode of action of the gibberellins differs from that of the auxins for the two kinds of substance are not interchangeable although they may be required together to produce an effect, as in stem elongation. It has been advanced as an explanation of the apparent dual requirement for such enhanced growth effect that the gibberellins act to stimulate the production of more auxin but whilst this could be so in a number of instances it does not account for those cases where addition of extra auxin fails to elicit the same response as the addition of gibberellin. Clearly, some gibberellin responses are quite independent of auxin, as for instance, the production of α-amylase in cereal grains and the breaking of dormancy. Apart from the possibility of an effect on gene action by gibberellin there is so far no plausible explanation of their mode of action. It is interesting to note that there is some evidence that at least one animal hormone, ecdysone, a promoter of moulting in insects, might act in the same way by de-repressing the genes that lead to the formation of proteins required for the moulting process. One experimenter has shown that a gibberellin possesses a small but measurable activity in an assay based on moulting in locusts. Parallel with this, ecdysone has been shown to stimulate growth in dwarf peas. One further piece of evidence that gibberellins may be acting at gene level is the occurrence of a wide diversity of them allowing for the varied specificity required to activate the genes of a wide variety of species.

Movement of gibberellins in plants is not polar, as in auxin transport, but can be by diffusion. Some transport is by way of the phloem and some the xylem as with other metabolic substances. It is generally considered that they are produced chiefly by the tissues in which their activity is most pronounced, namely where growth and development is occurring.

The third group of plant hormones mentioned previously, known once as kinins but now more commonly as the cytokinins, are of much more recent discovery than the others. In 1954, Skoog and others working with pith of *Nicotiana tabacum* in tissue culture conditions showed that the pith alone grown in the absence of auxin showed little or no change, the pith alone with auxin showed a high degree of cell enlargement but no cell division, but the pith with tissues from the vascular system and with auxin showed rapid cell division and cell enlargement, however. Some substance from the other tissues had clearly stimulated the pith cells to divide and a search for this substance or one that would induce the same effect was successful in that coconut milk, malt extract, yeast extract and autoclaved DNA all showed similar high activity. The first pure substance to be isolated came from herring sperm DNA and this would promote division in tobacco pith cells at a concentration as low as one part in several millions when auxin was also present. It was called *kinetin*. Since then a number of substances have been found with similar properties and are collectively referred to as the kinins or cytokinins.

Kinetin was shown chemically to be dehydrated deoxyadenosine. The cytokinins are defined as hormones that promote *cytokinesis* (cell division) in the cells of plants, irrespective of any other activity they may possess, this being their most characteristic property. Since the isolation of kinetin a number of substances with kinetin activity have been synthesized and some have been found occurring naturally in plant tissues. The juice of coconut tissue and tomato juice, extracts of flowers or young fruits of pears, peaches, plums and apples, horse chestnut, walnut, banana and cambial tissue extracts have all been shown to possess kinetin-like properties. A substance called zeatin has been isolated from young maize cobs. All these cytokinins are referred to as regulators of cell division but have never actually been demonstrated in plants making normal growth; they are demonstrably active in promoting cell division only in tissue culture conditions. Chemically they are purine compounds.

Their effect on cell division is not the only one they can exert on developing tissues. They are known to influence cell enlargement by enhancing it in some cases, and another feature is that they are known to interact with auxins in the control of morphogenesis. For example, their presence may result in very pronounced stimulation of bud growth. This effect is dependent on the ratio of IAA to kinetin in the medium. With a low proportion of the cytokinin, tobacco pith produces no buds in tissue culture but if the cytokinin value is increased proportionately buds may appear in very large numbers. The same kind of effect is seen in some moss protonemata when treated with cytokinins. Roots treated with cytokinins externally show some peculiar responses. A primary root will cease to elongate, may produce numerous laterals and become thickened as cambial activity is vigorously stimulated. Such effects have not been explained.

Other phenomena known to be affected by cytokinins are the dormancy of some seeds and winter resting buds which is broken when the hormones are administered, and the counteraction of the usual dominance of apical buds over laterals. One very striking effect in this latter field for some plants is the removal of dominance from the apical bud accompanied by severe fasciation that is the formation of numerous laterally united stem axes, each with active apical and lateral buds. This condition is known to occur naturally following infection by the bacterium, *Corynebacterium fascians*. Extract of the bacterium, identified as methyl amino purine, has been shown to possess cytokinin activity.

Cytokinins are known to have a retarding effect on the processes of senescence or aging of some leaves, noticeably on the rate at which chlorophyll is broken down and disappears from the leaf. Treated leaves remain green many days longer than those untreated. It has been suggested that this effect is the result of the cytokinin activity to induce the increased flow of material through the vascular tissues to its own location and to prevent the transport of substances away from it. The cytokinin is thought to have a mobilizing effect within the tissues but through what mechanism is not known. There is also some evidence that when senescence is retarded in one part of a plant due to the application of cytokinin, other parts may age more rapidly than normal.

At present, more and more cytokinetic, morphogenetic and other physio-logical effects of the cytokinins on normal plants are being discovered and investigated but so far there is no acceptable comprehensive explanation of them. Although these hormones appear to have their basic effect on cell division and are regarded chiefly as regulators of this process, it has never been shown that normal, healthy, intact plants are actually using them as such. It could be that stimulation to cytokinesis is only a manifestation of the effect of cytokinin on a much more basic plant process. There are indications that this may be connected with nucleic acid metabolism and protein synthesis but precisely in what way is not yet clear.

Mobility studies of cytokinins have produced some conflicting conclusions. From study of the movement of externally applied cytokinins from their site of application it would appear that they possess no high degree of mobility within the tissues. However, what does happen in these cases is that various metabolites tend to accumulate in the area of their presence. By contrast there have been claims made that, in some plants at least, cytokinin is distributed from a site of manufacture and shows a polar movement in the apex base direction but this may be related to auxin transport at the same time. Another finding is that some natural cytokinins are produced in the roots of plants and are carried in the transpiration stream to the sites of their activity.

Ethylene is now classified as a plant hormone. It is well known that ethylene brings about the inhibition of growth of roots and that it is formed in the root tissues as a result of the presence of auxin. It is also active in promoting abscission of plant parts, its activity being counteracted by presence of auxin. Ethylene is also known to affect the movement of auxin by tending to inhibit its basipetal polar and lateral transport but through what mechanism is not understood. It has also been speculated that a feedback control of auxin levels in tissues could be achieved through the ability of auxin to bring about ethylene formation. Increased auxin would increase the ethylene concentration which in turn would prevent auxin from moving away to areas remote from its place of synthesis. The compound, ethylene chlorohydrin, is known to be effective in breaking dormancy in both seeds and winter buds and is reputed to initiate flowering in *Xanthium sp.* when this is

Effect on	IAA	GA	Cytokinin	Ethylene or chloro-hydrin	Abscisic acid
Cell division	promotes		promotes		
Cell enlargement in stems	promotes		promotes		
Cell enlargement in roots	indirectly inhibits			inhibits	
Abscission of parts	inhibits				promotes
Bud dormancy		breaks	breaks	breaks	promotes
Seed dormancy		breaks	breaks	breaks	
Seed germination	promotes	promotes			inhibits
Sexuality in flowers	promotes♀	promotes♂			
Food store mobilization in seeds		promotes			
Root initiation	promotes		promotes		
Parthenocarpy	promotes	promotes			
Onset of flowering		promotes		promotes	
Stem elongation or "bolting"	promotes	promotes			
Apical dominance in buds	promotes		counteracts		
Leaf expansion		promotes			
Bud formation			promotes		
Root elongation			inhibits	inhibits	
Cambial activity	promotes		promotes		
Fasciation			promotes		
Senescence or aging			retards		
Movement of metabolites			accelerates		
Movement of auxin				inhibits	

should be noted that these effects are to be seen in some plants only and may be restricted to certain parts of them.

Fig. 8.1. Comparison of some of the known effects of plant hormones.

growing under wholly non-flowering conditions of temperature and day length. Ethylene is also associated with the induction of the respiratory climacteric, that is the sudden rise in rate of respiration towards the end of the fruit-ripening period, particularly in apples. Other known effects include the induction of epinasty (see p. 620), the stimulation of cells to grow isodiametrically rather than longitudinally, the enhancement of radial growth in roots and stems as a consequence of this, the stimulation of some seeds to germinate, the induction of root hair development, the promotion of flowering in pineapples and the induction of flower fading in pollinated orchids.

There is no doubt that the functioning of the plant is very much under the control of the actions and interactions of the auxins, gibberellins, cytokinins and ethylene and that there are substances such as the phenols and flavonols that modify the control exerted by them, mainly through some inhibitory influence. One such growth inhibitor is the substance *abscisic acid*, otherwise known as *dormin* or *abscisin II*, which is very active at concentrations lower than one part per million. It is found in many woody and herbaceous plants and is regarded as the substance whose presence inhibits the further development of buds. When applied externally to leaves it can have the effect of causing the terminal bud on that stem to rest indefinitely. It interacts with gibberellin in some way in inducing this dormancy. Among its other apparent activities are an abscission-promoting quality and the power to inhibit seed germination. It has been isolated from some fruit walls.

Fig. 8.1. is a table indicating the known effects of the chief plant hormones.

THE ENDOCRINE GLANDS AND ANIMAL HORMONES

Substances of such varied chemical nature have been classed as hormones (see p. 114) that it is not possible to define them by reference to their structure. They are, however, specific substances secreted by localized organs or tissues, and are transported in the body fluids. Each hormone has either a generalized co-ordinating effect in the body or it elicits a particular response from a target organ. For survival, an animal has to adjust all aspects of its metabolism to a fluctuating environment. The nervous system is responsible for the perception of changes and for rapid transmission of impulses, but the hormones actually stimulate metabolic changes so that the individual may regulate itself. For species success, the social group must be co-ordinated especially with regard to reproduction; this is carried out, at least partly, by secretion of pheromones. These are specific substances secreted into the environment; they are perceived by other members of the group and since they have a co-ordinating effect they are now usually described as being in the same general category as hormones (see p. 289).

The embryonic origins of the various endocrine structures and tissues have been described in Vol. I, Chap. 29. Where they are glandular organs, they differ from all other glands in the body in that their secretions are not passed into ducts. All other glands pass their secretions into tubes which conduct these secretions to their sites of action, e.g. salivary ducts, bile and pancreatic ducts, etc. Hence we use the term endocrine, or ductless glands where the secretions are passed into a body fluid for circulation via a vascular system. Most of our knowledge of the effects of hormones has been discovered by carefully controlled experiments which compare the effects of the presence of the hormone with its absence. Thus has been developed a huge and important body of knowledge known as *endocrinology*.

Because of the means of transport of hormones within the body, the development of an endocrine system by an animal could only be effective if its body fluid transport system were well developed. Thus it is not surprising to find that definite organs of endocrine activity have only been described in animals possessing a circulatory system for their body fluids. These include some invertebrates and all the vertebrates. In Vol. I, some account has been given of endocrine activity in annelids, arthropods, molluscs and chordates. Here the mammalian endocrine system is described in some detail; most of the description will apply to all the vertebrate series.

The Vertebrate Endocrine System

This is best described with reference to the individual endocrine structures and their secretions (*see* Fig. 8.2). How the various hormones

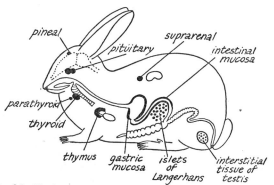

Fig. 8.2. The location of the main endocrine organs of a mammal.

exert their effects is in no case known with certainty. Where there are known points of influence, for example in an enzyme system, it is described on p. 291.

10

The Hypothalamus

This is the centre of control of most of the endocrine glands although it exerts its effects through the medium of the pituitary. By the process of neurosecretion, two hormones are produced in the hypothalamus and secreted along the axons of neurones into the pars nervosa of the pituitary, where the hormones enter the tissue fluids for circulation. In addition to these, the hypothalamus secretes a number of releasing factors which pass into the pars anterior of the pituitary via the hypothalamo-hypophysial portal system (see Fig. 6.4). Each releasing factor stimulates the secretion of a specific hormone. It must be recognized that the importance of the hypothalamus in regulating metabolism cannot be over-emphasized.

The two hormones actually produced in the hypothalamus are *oxytocin* and *vasopressin*. The first of these causes contraction of the unstriated muscle of the uterus during the birth process and also effects the response of the mammary glands to the stimulus of suckling. The myo-epithelial cells of the glands contract, thus expelling the milk. Vasopressin causes a rise in blood pressure by the contraction of the smooth muscle of the blood-vessel walls. Its more important effect, however, is anti-diuretic; in conjunction with other hormones (see p. 211), it plays an important role in osmoregulation, especially concerning the resorption of water and sodium ions from the uriniferous tubules. Deficiency of the hormone, due to disease or injury of the hypothalamus or pars nervosa, results in excessive loss of water from the kidney, leading to intense and constant thirst, a condition known as *diabetes insipidus*. The condition can be relieved by dosage with the antidiuretic hormone.

The Pituitary Body

The origin and development of the pituitary body are described in Vol. I, Chap. 29, and its mature structure is illustrated in Fig. 8.3. The endocrine function of the pars nervosa is mentioned in the preceding paragraphs. The pars anterior produces six hormones, several of which are known as *tropic* hormones because they stimulate both the development and the secretion of other endocrine glands.

The growth hormone, *somatotropin*, is secreted throughout life though there is normally a considerable lessening of its activity after puberty. It exerts its effects by stimulating the rate of protein synthesis. Deficiency before puberty results in *pituitary dwarfism* which fortunately does not affect the mental powers as does *cretinism* (see p. 282). On the other hand, excess of the hormone results in *pituitary gigantism* where there is delayed ossification of the epiphysial cartilages and hence excessive growth of the skeleton. Excess of the hormone after puberty stimulates certain bones to further growth, especially in the jaws, skull, hands and feet. Hence these regions become very strongly marked, a condition known as *acromegaly*. Somatotropin has another quite different effect; it stimulates the production of *glucagon* by the α-cells of the islets of Langerhans, and thus plays an important part in the regulation of blood glucose (*see* p. 199).

Thyrotropic hormone stimulates the development of the thyroid gland in the embryo and promotes the production of *thyroxin*. The *adrenocorticotropic* hormone stimulates activity of the adrenal cortex to produce three types of hormones (*see* p. 283). *Follicle-stimulating* hormone induces the development of Graafian follicles and, in the male, it promotes the maturation of the spermatozoa. *Luteinizing* hormone expedites ovulation in mature Graafian follicles and also induces the formation of the corpora lutea.

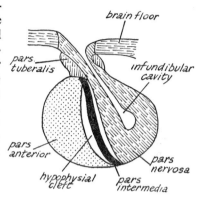

Fig. 8.3. V.S. of pituitary gland of mammal.

It is probably identical with the hormone which, in the male, stimulates the interstitial cells of the testis to produce testosterone (*see* p. 287). *Prolactin* initiates and maintains the secretion of milk in the mammary glands. In conjunction with somatotropin and adrenocorticoids, prolactin also plays a part in the development of the mammary glands. It also stimulates the development of crop glands in those birds which feed their young on crop secretions, e.g. pigeon's milk. It is to be noted that the actual delivery of milk in the mammal in response to suckling is due to the effect of oxytocin which will cause the release of the milk within ten seconds.

The pars intermedia produces two hormones both of which are *melanocyte-stimulating*. They are known briefly as α-MSH and β-MSH. An extract of the pars intermedia known as *intermedin* has been used in experiments. It affects the distribution of melanin granules in the chromatophores of frogs and of certain fishes. Trials with human volunteers have shown that injections of intermedin cause darkening of the skin. When the injections are discontinued, the skin returns to its normal colour. It is evident from this work and from other experiments with mammals that intermedin affects the formation of melanin granules in the chromatophores.

There are no hormones produced in the pars tuberalis.

The Thyroid Gland

The gland produces *thyroxin*, which accumulates in the follicles, but is rendered inactive by attachment to protein to form *thyroglobulin*. On stimulation with the thyrotropic hormone, a special proteolytic enzyme dissociates the protein from the thyroxin which is then secreted into the blood. Thyroxin regulates the basic rate of metabolism (BRM), which is the rate of energy release under resting conditions. Deficiency of thyroxin lowers the BRM (hypothyroidism); excess raises the BRM (hyperthyroidism). Occurrence of the former in young creatures causes retardation of physical and mental development, a condition known as *cretinism*. In adults there may be lesser degrees of hypothyroidism characterized by generally sluggish metabolism, atrophy of the thyroid gland and swelling of the subcutaneous tissues; the disease is known as *myxoedema*. Hyperthyroidism causes almost the reverse condition, high metabolic rate, restless and often feverish activity, rapid heart-beat and general wasting of the body. The eye-balls protrude and the thyroid is usually enlarged; the disease is called *exophthalmic goitre*. Simple goitre, with pronounced swelling of the thyroid, is normally due to lack of iodine in the diet.

Thyroxin has important effects in many growth processes throughout the vertebrates. It is essential for metamorphosis in amphibians and in some fishes. Thyroidectomized tadpoles will not metamorphose; addition of thyroid extract to the water, or feeding with pieces of the gland will induce normal metamorphosis. The hormone is also an important factor in temperature regulation (*see* p. 195). Low temperatures stimulate increased activity of the thyroid by the pathway, hypothalamus—releasing factor—pars anterior—thyrotropic hormone. In this connection, thyroxin seems to affect the basic metabolic rate by uncoupling oxidation from phosphorylation in the respiratory process so that excess foodstuff is oxidized with the dissipation of energy as heat instead of the energy being largely utilized for synthesis of ATP.

The Adrenal (Suprarenal) Glands

Each gland consists of two distinct parts, an inner medulla and an outer cortex. These differ in mode of development, in structure and in function (*see* Vol. I, Fig. 20.52).

The medulla secretes *adrenalin* and *nor-adrenalin* when stimulated through nerve fibres of the sympathetic system. These hormones produce rapid and dramatic changes in the body in response to fear, stress or shock. The whole body is toned up for rapid action and these two substances have justly been called the hormones of "fright, fight and flight." Adrenalin increases heart-beat and heart output. Nor-adrenalin causes constriction of smaller blood-vessels in the skin and viscera, so that more blood circulates to the voluntary muscles. The combined action of both hormones causes a rise in blood-pressure and the heart "thumps." The pupils are dilated, the upper eye lids retracted and the bronchioles are relaxed. In addition, there is increased mobilization of blood glucose from the glycogen in the liver and muscles. The general effect is that the whole body is toned up for rapid and vigorous action in an emergency. The body could not tolerate such a state of stress for more than very short periods and thus it is fortunate that adrenalin is rapidly destroyed in the tissues.

Exposure to low temperature will also cause secretion of adrenalin and this, by liberating more glucose into the blood, will supply the tissues with more respiratory substrate and hence increased heat production.

The adrenal cortex has three histologically different zones, each secreting one particular group of hormones, though all are known as *corticosteroids*. The outer zona glomerulosa secretes *mineralocorticoids*, the middle zona fasciculata, *glucocorticoids*, and the inner zona reticularis produces *androgens* and *oestrogens*.

The chief mineralocorticoids are *aldosterone* and *deoxycorticosterone*. In conjunction with vasopressin from the hypothalamus, and *para-hormone* from the parathyroids, they play a large part in osmoregulation (*see* p. 211). These mineralocorticoids promote retention of sodium, chloride and bicarbonate ions, but increase loss of potassium and phosphate ions. In this connection it is noteworthy that the production of mineralocorticoids, unlike that of the other corticoids, is not stimulated by the adrenocorticotropic hormone from the pituitary, but is associated with the release of *renin* from the kidney (*see* p. 288). Mineralocorticoids also play an important part in wound healing and tissue repair.

The glucocorticoids, *cortisone* and *cortisol*, affect carbohydrate metabolism and particularly the conversion of protein and fat into carbohydrate. This leads to more storage of glycogen in the liver and muscles, and more release of glucose into the blood. They also have an inhibitory effect on tissue repair; hence the rate of healing after damage depends on the balance of mineralocorticoids and gluco-corticoids.

The androgenic and oestrogenic hormones from the zona reticulata induce the male and female secondary sexual characteristics. Over-production in either sex results in precocious sexual development. Again, in one sex, it is the balance of the male and female hormones which is important. Excess of androgens in females leads to varying degrees of masculinity, and excess of oestrogens in males leads to some degree of femininity.

In general, the corticosteroids are of exceeding importance; their combined functions can be summed up as conservation. They enable the animal to meet emergencies and to withstand stress, to repair damage, to mobilize energy and to osmoregulate. It must be emphasized that, as with the action of most hormones, it is the balance that counts. Removal of the adrenal cortex is fatal; underactivity or destruction, for example by tuberculosis, causes *Addison's disease*, which is characterized by decreased ability to withstand infection, injury, fatigue or large temperature changes. The disease is also characterized by pigmentation of the face and hands, now known to be due to melanocyte-stimulating activity by the adrenocorticotropic hormone interacting with certain of the corticoids. Other symptoms are general wasting of the body, muscular weakness and low blood pressure; any crisis leads to a fall in blood sugar level, vomiting, dehydration and finally collapse. In recent years, however, there has been good progress in the treatment of Addison's disease by dosage with various corticosteroid mixtures.

The Parathyroid Glands

These are two pairs of small brown glands lying at the sides of the thyroid, or as in man, embedded in its dorsal surface. The secretion of the gland is controlled by the level of calcium ions in the blood. The hormone, *parathormone*, affects principally calcium metabolism and as such plays a large part in ossification and in later changes that take place in bone. The secretion is stimulated when blood calcium falls below a normal level and four different effects are soon apparent. There is increased absorption of calcium ions from the intestine, withdrawal of calcium salts from bone, retention of calcium ions by the renal tubules and increased excretion of phosphate in the urine to compensate for the high phosphate level which will necessarily follow the withdrawal of calcium phosphate from bone.

Insufficient secretion of parathormone produces the condition of *hypoparathyroidism*, characterized by increased irritability of the nervous system and muscular spasm; the condition is comparatively rare. Complete removal of the parathyroids, which may occur in thyroidectomy, is fatal in a few days. Excess secretion of parathormone or *hyperparathyroidism* sets up the disease *osteitis fibrosa*. The bones are demineralized and become fragile structures which are mainly fibrous and they break very easily. There is copious excretion of calcium and phosphate ions, and often calcium phosphate stones are formed in the kidneys.

It is thought that the parathyroids secrete a second hormone which acts in the opposite manner to parathormone, that is, it lowers a high level of blood calcium. Whether there are one or two hormones secreted, they certainly play a part in osmoregulation in conjunction with vasopressin and the mineralocorticoids.

The Thymus Gland

The gland is large in young animals and it degenerates considerably as the animal matures. In the young, it is undoubtedly a source of lymphocytes which are carried in the vascular system to lymphatic glands where some are retained; they then multiply in these glands. The thymus is thus the source of the immunological reactions to foreign proteins such as are contained in bacteria and viruses. If the gland is extirpated from a young animal, it fails to develop these immunological reactions; dosage with thymus extracts will then restore its resistance. The endocrine function of the gland is to produce the hormone *thymosin* which stimulates the multiplication of lymphocytes in the lymphatic glands and hence there is always an adequate supply in the tissue fluids.

The Pineal Body

In some vertebrates, especially the ammocoete larvae of lampreys and certain reptiles, e.g. *Sphenodon* (*see* Vol. I), the pineal body has an eye-like structure complete with retina. In the lamprey larvae, there is a well-marked change of skin colour, from pale at night to dark by day. This change, however, does not take place if the pineal is removed. It has recently been shown that lightening of a frog's skin is due to a pineal hormone called *melatonin*; there are also indications that this same effect may also occur in mammals. Melatonin has some effect on the maturation of the gonads in mammals; this is an inhibitory effect which is antagonistic to that caused by light exposure. There is considerable evidence that in many mammals, seasonal activity is influenced by day length and light intensity and it is probable that light stimuli affecting the optic retinae stimulate the pineal via the sympathetic nervous system. Secretion of melatonin would then stimulate gonadotrophic releasing factors in the hypothalamus and thus set the whole sexual cycle in action.

The Pancreas

It must be borne in mind that the pancreas is an exocrine gland, secreting the pancreatic juice which plays so important a part in digestion. Its endocrine function is located in isolated groups of cells called islets, which were discovered by Langerhans in 1869 and are named after him. Each islet of Langerhans contains two types of cells, the larger β-cells with alcohol-soluble granules, and the smaller α-cells with alcohol-insoluble granules. The β-cells are sensitive to the blood sugar level and, when it rises above normal, they are stimulated to secrete the hormone *insulin* which promotes the conversion of blood hexoses into glycogen stored in the liver and muscles. Deficiency of β-cells or removal of the pancreas leads to the disease *diabetes mellitus*. Blood sugar rises to such a high level that much of it is excreted in the urine. The liver is depleted of glycogen, and tissue proteins are broken down to release energy, so that there is constant loss of weight. Fat stores are converted into carbohydrate and as a result of excessive fat conversion, ketones are formed and may be detected in the urine. Without treatment, the animal dies in a state of emaciation. However, with injection of regular carefully calculated doses of insulin, a state of normality can be achieved. The α-cells secrete *glucagon* which promotes the breakdown of liver glycogen into glucose. The secretion is stimulated by the growth hormone, somatotropin, from the pars anterior. These two hormones, insulin and glucagon, play a very important part in the regulation of blood sugar (*see* p. 200).

The Ovaries

An account of the oestrous cycle is given on pp. 708–712. Briefly, the cycle may be divided into a follicular and a luteal phase.

The follicular phase is activated by follicle-stimulating hormone from the pars anterior. This induces development of the Graafian follicles and stimulates the secretion of *oestrogens*, probably by the follicular cells. There are several oestrogens, e.g. *oestradiol*, *oestrone* *equiline* and *equilenin* and *oestriol*, the most potent being the first of these. The oestrogens effect changes in the female ducts, especially in the uterus, where there is thickening and increasing vascularity of the lining. It may be said that they activate the reproductive tract in preparation for pregnancy. They also stimulate further development of the mammary glands and the other secondary sexual characteristics. The follicular phase is completed when ovulation takes place. It is to be noted that a simple negative feed-back mechanism causes suppression of the secretion of follicle-stimulating hormone, when the oestrogen level in the blood becomes high enough.

The luteal phase is stimulated by pituitary luteinizing hormone which induces repair of the ruptured follicle and the growth of yellow cells in it to form the corpus luteum. These cells are then stimulated to secrete *progesterone*, probably by prolactin, acting in a luteotropic capacity. Progesterone inhibits further ovulation and hence the next oestrous cycle cannot take place until after birth. The uterine epithelium proliferates further and enters a secretory phase (uterine milk

The mammary glands are stimulated to further development and the uterus is prepared for implantation of the embryo. The uterine muscle is rendered relatively insensitive, thus favouring retention of the embryo. Both the oestrogens and progesterone reach their maximum level at three months (in human pregnancy) and then their production in the ovary declines. They are, however, produced by the placenta during later pregnancy.

Relaxin, probably produced by the corpus luteum, inhibits uterine contraction, but during parturition it stimulates dilation of the cervix and relaxation of the pubic symphysis. (Note that contraction of the uterus is stimulated by oxytocin.)

Contraceptive pills inhibit development of Graafian follicles and suppress ovulation from ripe follicles; they consist mainly of progesterone with small quantities of oestrogens.

The Placenta

At least three hormones are secreted by the placenta. They are *chorionic gonadotropin*, at least one oestrogen, and progesterone. Together, they influence growth of the uterus and mammary glands after ovarian and luteal secretions have waned. It is also thought that they maintain the corpus luteum, which does not become fibrosed until after birth.

The Testes

The endocrine function of the testes is located in the small groups of interstitial cells which lie between the seminiferous tubules. Maturation of spermatozoa is stimulated by follicle-stimulating hormone, while endocrine secretion by the interstitial cells is induced by another hormone from the pars anterior which is probably identical with luteinizing hormone. At least two *androgens*, called *androsterone* and *testosterone* are secreted. They stimulate spermatogenesis and control the activity of the seminal vesicles and the prostate glands, and they also induce the appearance of the secondary sexual characteristics. Both hormones have an anabolic effect, stimulating synthesis of protein, especially in the build-up of muscles. They also play a part in the differentiation of maleness in the embryo.

Removal of testes before puberty prevents development of all the reproductive organs and their accessories. Removal after puberty naturally causes sterility but has no other apparent effect on health. In lower vertebrates, androgens have similar effects to those in mammals. Examples are the development of the breeding pads in male frogs, the combs and spurs of cockerels, crests of newts, the coloration and gonadopodal appendages in some fish.

It is noteworthy that the testes also produce oestrogens, and that the adrenal cortex produces both androgens and oestrogens. The testes of stallions yield the greatest quantity of the oestrogen, oestrone, of any tissue known.

The Kidneys

An account of the mode of action of the uriniferous tubules of the kidneys is given in Chap. 15. At a short distance from the point where an afferent arteriole enters a glomerulus, it is in close contact with a distal tubule. In this region there is a thickened pad formed of cells which have multiplied in that specific area. The pad is known as the *juxta-glomerular complex*. These cells are sensitive to changes in the Na^+ concentration in the blood, and when this concentration falls below a certain critical level, the hormone *renin* is secreted by the cells. This acts on a plasma globulin called *hypertensinogen*, breaking it, by proteolytic action, into smaller polypeptides of *hypertensin*. (Note that the names *angiotensinogen* and *angiotensin* are sometimes used.)

Hypertensin itself acts as a hormone, causing vasoconstriction and also stimulating production of mineralocorticoids from the adrenal cortex. These raise the level of sodium resorption from the distal tubules. This is yet another example of negative feedback control; renin secretion ceases when the sodium level is back to normal and hence the formation of hypertensin and the release of mineralocorticoids also ceases.

The Alimentary Canal

The gastric mucosa produces three hormones; *gastrin* induces continued secretion of gastric juice, the flow of which has been initiated by vagus stimulation following the presence of food in the buccal cavity and later in the stomach; *enterocrinin* initiates secretory activity in the duodenum while *gastric secretin* promotes the secretion of salt solution from the pancreas.

Four hormones are produced by the mucosa of the duodenum. *Secretin* stimulates the continued flow of pancreatic juice rich in salts while *pancreomysin* (*pancreozymin*) induces the flow of juice rich in enzymes. *Cholecystokinin* causes contraction of the gall-bladder which then expels bile into the duodenum. *Enterogastrone* inhibits both acid secretion by the stomach and also the peristaltic contractions of the stomach wall, so that expulsion of chyme into the duodenum is not continuous.

Nerve Endings

Two substances secreted by nerve-endings, which are responsible for exciting adjacent cells, may be considered as hormones. The endings of sympathetic nerve fibres secrete *sympathin*, which is almost identical in its action with that of adrenalin. There are probably two types of sympathin, one of which is inhibitory in its effects, e.g. in causing relaxation in the smooth muscle of the intestine, and the other is excitatory, e.g. stimulating the contraction of muscle of the heart and of muscle in the cutaneous blood vessels. This second type is identical to adrenalin.

The endings of parasympathetic and somatic motor fibres secrete *acetylcholine*, which also plays a part in transmission across synapses from one neurone to another.

Tissue Hormones

Two hormones, *histamine* and *acetylcholine*, have been shown to be present in many tissue extracts. It is probable that both are present, in inactive form, in all the tissues of the body. After liberation, both are rapidly destroyed. When administered by injection, histamine has three powerful effects. Firstly, it stimulates secretion by lachrymal, salivary and gastric glands and also by the pancreas. In this connection it is noteworthy that histamine appears to be identical to gastrin. Secondly, it stimulates contraction of the smooth muscle of the uterus, intestine and bronchioles, and thirdly, it causes a fall in blood pressure, mainly through dilation of the capillaries; at the same time, capillary permeability is increased. The hormone is released from damaged tissue and is one of the main agents causing inflammation. There is acute pain when it is applied to raw surfaces and there is some reason to suppose that it may be the actual mediator in cutaneous pain sensations.

Acetylcholine has very similar effects; it is the most powerful depressor known. There is some reason to believe that these tissue hormones may play an essential introductory part in the healing of damaged tissue, by allowing a plentiful blood supply to the area and by providing a means whereby lymphocytes and antibodies may escape from the capillaries.

Pheromones

It is very probable that all animals secrete substances into the external environment which function like hormones in that they influence the behaviour of other animals, especially those of the same species. Such substances play a large part in many species as a form of social identification; they may also integrate the group and they are certainly concerned in sexual behaviour. In mammals, these *pheromones* are usually produced by specialized sebaceous glands, or they may be present in the urine, as is the case with certain sex steroids which are identical, or almost so, to oestrogens and androgens. Some examples are given here; the reader can doubtless think of many others.

The musk-deer of Tibet and Siberia mark their territory with faeces and also with a pungent oil produced from abdominal glands. The active principle of this oil has been identified and is called *muskone*. In the mating season, it serves as a sexual attractant. Lemurs, for example the indri of Madagascar, mark trees with a secretion from glands in the angles of the jaws. This serves to keep the group together as well as being important to mating.

In Rhesus monkeys, the female secretes a pheromone which excites the male to copulatory behaviour. An immature female, or one with ovaries removed, is treated with indifference. Male dogs respond very strongly to female pheromone which is not only present around the genital aperture but also in the urine of bitches.

Many insects, for example cockroaches and moths, produce pheromones in the females, which attract males for considerable distances. In the case of certain moths, the sex attractant has been shown to be effective over distances greater than 10 km. Wood-lice group themselves in numbers, not because the areas chosen are of sufficient humidity but because they identify a common secretion. Given areas of equal humidity, and separating the animals, they will still return to the group.

Outstanding examples of pheromone control are found in the social insects, bees, ants, wasps and termites. The queen bee secretes a substance, *9-oxodecenoic acid* from mandibular glands, and spreads it over her body. This is licked by the workers and it inhibits oogenesis in their bodies; the same substance attracts the males to the queen. Ants mark their food-trails with secretions from their bodies; this makes it easy for other ants to follow to the food-source.

It is obvious that the concept of pheromones can be extended to include a vast range of behaviour; one could include exudations from ova which attract sperm, or the aggregation into colonies of large numbers of separate amoeboid Myxomycophyta (slime fungi). Here the common attractant is *cyclic adenosine monophosphate* (CAMP) which is secreted by all the individuals. This same substance is exuded by bacteria and attracts these amoebae to their prey. It is believed that CAMP, in some cases in vertebrates, may be the actual intracellular mediator between a hormone and its target tissue.

Pheromones cannot be precisely defined; perhaps they are best regarded as external secretions which induce behaviour conducive to the survival of the species.

The Mode of Action of Hormones

The results of the activity (or inactivity) of hormones are, in most cases, very apparent, and yet it must be stated at once that the complete mode of action, within the cell, is not known for any hormone. The most promising lines of inquiry have been with the use of "labelled" oestrogen and testosterone in affecting the uterus and prostate gland respectively. This has shown, in both cases, that after a steroid molecule enters a cell, it is bound to a large protein molecule which is able to convey the steroid through the nuclear membrane to the vicinity of the DNA chains. There, occurs "switching-on" of certain genes, so that there is RNA synthesis, which in the cytoplasm is translated into enzyme protein manufacture. The enzymes then produce substances which have the observed effects of the hormone. This work has also shown that at effective physiological levels there are several thousand of the steroid molecules in each responsive cell; it has also shown that in unresponsive cells, both the binding and transport mechanisms are absent.

In some cases, the actual effect of a hormone in a cell is apparent, though not the means by which that effect is achieved. Thyroxin has a direct effect on mitochondrial oxidation systems; it particularly stimulates the activity of succinic dehydrogenase (see p. 439). Both glucagon and adrenalin promote the breakdown of glycogen by increasing the actual amount of active phosphorylase present in the cells of liver and muscle. Glucocorticoids increase the rate of protein breakdown and of glycogen synthesis, and here again, in the liver cells there is measurable increase in transaminases and amino-acid oxidases.

In several cases, e.g. glucagon, insulin and adrenalin, there is strong evidence of association of hormone activity with increased production of cyclic adenosine monophosphate. The CAMP may here carry out the function of a binding and transporting agent, ultimately affecting the concentration of the AMP-cyclase enzyme.

In general it may be said that certain facts emerge. Cells which are responsive to particular hormones seem to have the ability to facilitate entry of the hormone molecules, or to have some membrane effect which stimulates a second messenger. The cell responds essentially by "switching-on" genes; this leads to the synthesis of enzymes in the cell and hence production of certain chemicals which give the observed effects. So far, it is not possible to offer any explanation as to why certain cells, albeit with the correct genes, do not respond to certain hormones.

The General Co-ordinating Effects of Hormones

We may summarize the activities of the endocrine system as follows:

1. *Regulation of growth and development.*

2. *Homeostasis; the maintenance of the body in a state of dynamic or balanced equilibrium*, not in a static, unchanging state. A number of examples of this regulatory function are described in Chap. 6.

3. *Regulation of metabolism;* some aspects of digestion; storage of food materials; the normal and rapid utilization of these materials.

4. *Sexual development, reproduction*, and *care of the young.*

5. *The development of skin colouration* and *regulation of changes in this.*

6. *Enabling the body to withstand shock, tension, wounding, etc.* and *to recover from these.*

7. *Together with the nervous system, it provides for effective response to all kinds of stimuli, internal and external.*

ORGANIZERS (EVOCATORS)

In the early development of an embryo, before there is any trace of a nervous system, certain chemicals are produced which seem to direct and control orderly development. The substances are termed organizers and they are of the nature of hormones. Their chemical nature is unknown, though the primary organizer produced by the dorsal lip of the blastopore is thought to be a steroid (*see* Chap. 4).

In some types of eggs, such as those in *Amphioxus*, there are present, even before fertilization, organ-specific areas in the cytoplasm. There is already intracellular differentiation of the maternal cytoplasm. These organ-specific areas are somewhat rearranged at fertilization, and during cleavage, they are distributed to the various groups of cells. Gastrulation further re-arranges these presumptive blastomeres into their definitive regions of the embryo. Other types of eggs, such as those of the frog, do not show any cytoplasmic differentiation into organ-specific areas; but at the blastula stage, presumptive areas are defined.

Wherever this differentiation into organ-specific areas takes place, it is certain that the chemical substances produced by the various areas induce the development of the various parts of the embryo. There must be a large number of these substances produced in orderly succession. Some have very local influence, and some such as the primary organizer have a widespread field of activity. Some of the organizers evoke development of particular structures in undifferentiated tissue; both are sometimes termed evocators. It is noteworthy that both organizers and evocators are specific in their action; a particular organizer will cause the development of specific structures. Examples of such specificity are given in Chap. 13.

Thus, before the development of nervous system or endocrine glands, there is a co-ordinated succession in development which gradually unfolds until all the main organs and tissues of the body are formed. From that point, the nervous system and the endocrine glands take over the roles of co-ordination and control. Few details of this wonderful and comprehensive organizer system are known; the manner in which such different substances are produced in the right succession, and in the right places, is as yet little understood. The possible explanation lies in the distribution of plasmagenes during division of the cells (*see* p. 795). The important fact is, that in every organism, from its inception, there is an inherent system of co-ordination.

NERVOUS CO-ORDINATION

When the functioning of parts of the body is co-ordinated and regulated by the transmission of impulses through a network of specialized cells, we speak of it as nervous co-ordination. It is almost certainly the case that the nerve net of the coelenterates is the most primitive of such systems. Since, however, our understanding of the mode of operation of the synaptic nervous system is much fuller, it is more convenient to consider the functioning of this more complicated system.

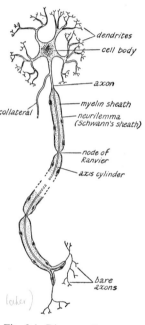

Fig. 8.4. Diagram of a motor neurone.

The Synaptic Nervous System

The unit of such a system is the *neurone*. (*see later*) This has been described in Vol. I, Chap. 20, but a diagram is repeated in Fig. 8.4. These neurone units are arranged to form the complete nervous system of the animal; it reaches its greatest complexity in the vertebrates. The nervous systems of the types dealt with in Vol. I have been described in the appropriate chapters. It will suffice to give a brief résumé of the gross structure of the nervous system of higher animals here. Such a system comprises the following main parts.

1. The central nervous system, consisting of brain and spinal cord.
2. The peripheral nervous system, consisting of cranial and spinal nerves and the autonomic nervous system.
3. The organs of special sense, such as the eye, ear, etc.

Fig. 8.5 shows diagrammatically the interrelationships of these parts. The system as a whole makes the following provisions:

1. That organs, tissues or cells on the peripheral margins of the body, which are capable of detecting changes in external conditions, i.e. the *exteroceptors* such as the eyes and ears, are in communication through

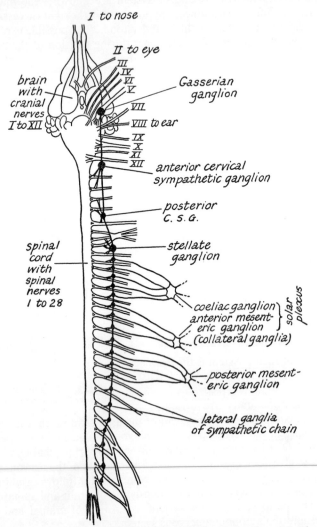

Fig. 8.5. Diagram to show the interrelationships of central, peripheral and autonomic nervous systems of the rabbit.

the central nervous system with other organs capable of making the necessary response, i.e. the *effectors* such as striped muscle and secretory glands. This is achieved through the neurone arrangements in the cranial and spinal nerves and central nervous system (*see* Fig. 8.6).

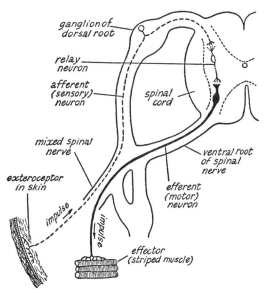

Fig. 8.6. Nervous connexions between an exteroceptor and an effector through the central nervous system.

2. That the sensory nerve-endings in the visceral organs, the *entero-ceptors* of the heart, blood-vessels, alimentary canal, kidneys, gonads, etc., are in communication through the central nervous system with the appropriate *effectors* such as smooth or cardiac muscle and secretory glands. This is achieved through the neurone arrangements in the autonomic nervous system and the central nervous system (*see* Fig. 8.7).

3. That the sensory nerve-endings in the muscles and joints, capable of detecting changing stresses and movements, i.e. the *proprioceptors*, are in communication through the central nervous system with the appropriate effectors such as all the other muscles and joints. This is achieved through the neurone arrangements in the autonomic system and central nervous system, as illustrated previously.

4. That all the parts are in communication with the brain which contains centres capable of exerting *autonomic control* over the activities of many visceral structures, e.g. respiratory movements,

heart beat, etc., and centres capable of exerting directive control over the transmission of many impulses, so that the animal may have a voluntary choice of reaction to many stimuli or may initiate voluntary actions without apparent stimulation. This is achieved through the longitudinal neurone arrangements in the central nervous system and

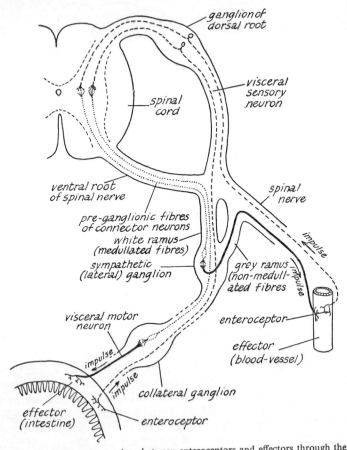

Fig. 8.7. Nervous connexions between enteroceptors and effectors through the central nervous system.

the formation of brain centres associated with specific actions (Fig. 8.8).

The overall result structurally is the extraordinarily complica system of neurone arrangements, involving many millions of un

They have been worked out in the broad patterns outlined above. The result physiologically is the perfect co-ordinating system, in which all parts are intercommunicating with one another in a controlled way. We can liken the nervous system to modern telephone organization, with the central nervous system acting as the exchange. Messages may originate from any subscriber and may be passed to any other subscriber or group of subscribers under supervision of the exchange, or messages may originate from the exchange itself.

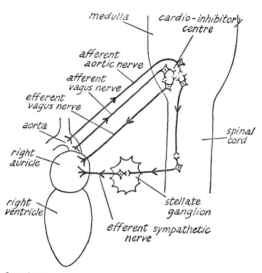

Fig. 8.8. Longitudinal neurone arrangements in the central nervous system, concerned with the heart.

One peculiarity of the system of neurones is that no two are in direct communication with one another by contact. There is every reason to suppose that whilst the very delicate cytoplasmic endings of the individual neurones are very closely adjacent to one another, they do not actually touch. When an impulse is passed from one to another, a bridge of non-nervous tissue must be crossed. Such a gap between neurones is called a *synapse*. The method of bridging this gap will be described later.

The Reflex Arc and its Complications

The simplest possible line of communication between a receptor and an effector would be a single neurone, with its afferent fibre or dendron bringing an impulse from the receptor to the cell body which retransmits the impulse via the efferent fibre or axon directly to the effector as in Fig. 8.9.

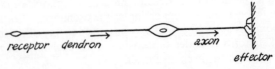

Fig. 8.9. Single-neurone connexion between receptor and effector.

Nowhere in vertebrate animals does such a simple arrangement exist, as far as is known. Neither is there the next possible simple arrangement as in Fig. 8.10.

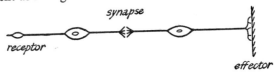

Fig. 8.10. Two-neurone arrangement with one synapse.

The simplest arrangement actually found is the interaction of three neurones as in Fig. 8.11.

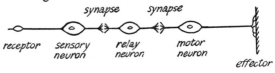

Fig. 8.11. Three-neurone arrangement with two synapses.

Such an arrangement of neurones is to be found in a few cases only as in Fig. 8.12.

If these three neurones are in no way communicating with any others, then the passage of an impulse from receptor to effector is direct and cannot be influenced by any impulses initiated in any other part of the nervous system. Once the receptor has been stimulated, the effector must operate as a result of the stimulation, without any interference, since the impulse never leaves the direct path from receptor to effector, neither does any other neurone from any other part of the body inter-link with this system. A response to stimulation which is known to occur through a neurone linkage such as this, is known as a *reflex action* and the neurone arrangement is called the *simple reflex arc*. Such actions are quite involuntary and unconscious being appropriate responses to particular stimuli calling for no modification by activity of higher centres. Reflex actions by the human body are rare, but the well-known knee-jerks, blinking when an object approaches the eyes, and the contraction of the pupil in bright light, are quoted as such. In these cases, it is very unlikely that the reflex arc is as simple as that described; there are probably more than three neurones affected.

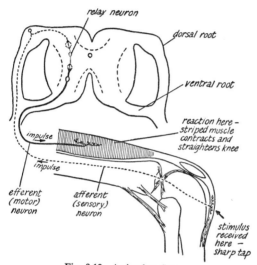

Fig. 8.12. A simple reflex arc.

It is obvious that if a nervous system were constructed on such a series of neurone patterns alone, the activities of the animal as a whole would not be co-ordinated at all. An isolated stimulus could have only an isolated reaction and if all receptors and effectors acted in isolation through simple reflex arcs, the effect would be to divide the body into totally independent parts rather than to weld it into a single functional unit.

Thus, if in higher animals the simple reflex arc exists at all, it is the exception. Actually, when an afferent fibre passes into the brain or spinal cord, it makes numerous connexions with other neurones. Some may be efferent neurones and capable of relaying an impulse to other effectors. Others may be neurones capable of carrying an impulse to other parts of the brain or spinal cord, there to communicate with many other efferent neurones. A single receptor can be connected with many effectors, or one effector with many different receptors. This interconnecting mechanism is of great significance and accounts in large measure for the success of the higher vertebrates. Such animals can make many kinds of response to a single stimulus, and if in possession of the ability to make adjustments to the incoming impulses by directing them along certain paths, there is no limit to the possible combinations of receptors and effectors. These adjustments are made in the central nervous system and are its main function, and the places where it occurs are referred to as *centres*. They constitute the grey matter. In

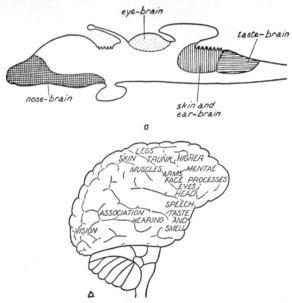

Fig. 8.13. (*a*) Lateral view of a generalized vertebrate brain showing the main centres. (*b*) Human brain, correlation areas.

the brain, there are centres associated with different functional components such as the skin, ear, eye and olfactory organs. Each of these is a centre where impulses are received of a particular kind and where adjustments are made to direct the impulses through many or few circuits as necessary. This function of adjustment and correlation of impulses by the central nervous system is more fully described in Chaps. 26, 27, 28 and 29 of Vol. I.

Conduction of Impulses Through the Nervous System

The transmission of an impulse from receptor to effector through the neurone connecting system must be considered in two steps, first along the nerve fibres and secondly, across the synaptic gaps between neurones and the junctions between fibre endings and effectors.

Transmission Along Fibres

Du Bois-Reymond in 1840, first showed that when a nervous impulse is conveyed along a nerve fibre, it is always accompanied by the passage

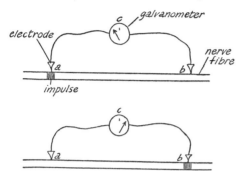

Fig. 8.14. Apparatus for detecting change in potential along a nerve fibre.

along the nerve fibre of a change in its electrical state. The nerve impulse is not an electric current. It is a wave of activity in the nerve which moves rapidly and is accompanied by a change in potential of the nerve. The energy for the transmission arises in the nerve itself and not from the stimulus which initiated it. This wave of activity travels through the sensory fibre into the cell body and then out along the axon. Under experimental conditions, this direction can be reversed, but there is no reason to suppose that this occurs in natural circumstances. When a fibre is stimulated by some means, the impulse is propagated in both directions away from the point of stimulation. As the wave passes, each point it reaches becomes electrically negative to the inactive regions on either side of it. Apparatus for detecting this change in potential and the graphs recording its passage are shown in Figs. 8.14 and 8.15.

As the impulse passes point *a*, the galvanometer registers a deflexion consistent with a negative potential on the outside of the fibre with respect to point *b*. As the impulse passes point *b*, this will in turn become negative with respect to point *a* and current flow is in the reverse direction through the galvanometer. This is in reality a recording of the same wave twice, as it passes two separated points on the fibre. If the points are very close together, the galvanometer readings may be superimposed to some extent. The instrument registers a similar *diphasic wave* as for the passage of a muscular contraction wave (*see* Chap. 12).

The term *action potential* of nerve, is used to describe the record of the potential changes under an electrode when the nerve is conducting an impulse. If an action potential is studied in detail, it is seen to be made up of several components. Following the spike on the tracing, there is a negative after-potential followed by a smaller positive after-potential (*see* Fig. 8.15).

Nervous conduction never occurs without this electrical change, whether it is along a single afferent or efferent fibre or a whole nerve trunk.

One of the basic characteristics of nervous conduction is the "all or none" condition. This means that if an impulse is strong enough to be propagated, the speed of conduction is quite independent of the strength of the stimulus. Response to stimulation varies only with the condition of the nerve. The response of the fibre is affected if it is exhausted by the recent passage of an impulse, or if the fibre is narcotized. It has been shown that after the passage of an impulse, there is an *absolute refractory period* during which no stimulus, however strong,

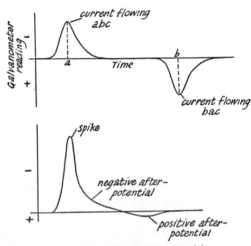

Fig. 8.15. Nerve action potential.

can excite the nerve to conduct another impulse. This period lasts for about one-thousandth of a second. Following this is the *relative refractory period* when the nerve fibre recovers its irritability to some extent, so that an unusually strong stimulus can cause the passage of a further impulse. If this stimulus reaches the requisite strength, the response of the nerve will be "all or none" as before. This relative refractory period lasts for about three-thousandths of a second. If the nerve fibre is allowed to complete its refractory period, it once more becomes capable of responding to normal stimulation. However immediately following the refractory period, there is a brief phase of *supernormal excitability* when a stimulus below threshold value may cause a response by the nerve. Still later, when this supernormal period is over, there follows a *subnormal period* when the fibres are less excitable than usual. The time relations of these changes in excitability in the fibre show that the absolute refractory period coincides with the rise in spike potential and with its fall to the point where the negative after-potential alters the spike shape. The supernormal excitability state coincides approximately with the negative after-potential, and the subnormal period with the positive after-potential.

Since in the undamaged nerve, the magnitude of the conducted impulse is independent of the intensity of the stimulus, some other mechanism must be responsible for distinguishing between weak and strong stimuli. In a nerve trunk, the number of fibres stimulated increases with the size of the stimulus, but in the single fibre, the only gradation with intensity is in the frequency of the recurring impulses in the fibre. A single impulse may cause a muscle-unit twitch, but a stream of impulses following each other along the many fibres of a nerve, can hold the muscle in a state of contraction; there is, however, a limit in frequency beyond which increased rapidity of stimulus will have no further effect. This sustained contraction is known as *tetanus*; the frequency necessary for human gastrocnemius and some other leg muscles is about 30 per second; for human eye muscles, 350 per second; for maintaining posture with the extensors, about 10 per second; for insec twing muscles, the frequency is above 300 per second.

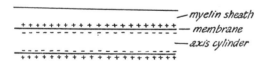

Fig. 8.16. Electrical charges in resting nerve fibre.

Another property of nerve fibres which is independent of the strength of the stimulus is the speed at which the impulse is conducted. An impulse set up by a strong stimulus travels no faster than that set up by a weak one. However, the speed varies between different types of fibre. The so-called *A fibres* (large, myelinated, somatic), *B fibres* (small, thinly-myelinated, visceral) and *C fibres* (small, non-myelinated), conduct at successively slower speeds in the order *A*, *B*, *C*. In mammalian *A* fibres, the speed of impulse conduction is directly proportional to the diameter of the axon. The greatest speed in the largest *A* fibres of man is about 160 metres per second.

In a resting nerve fibre, the axon membrane separates two aqueous solutions which are almost equal in electrical conductivity. Both solutions contain approximately equal numbers of ions but the types of ions and their proportions are somewhat different. Externally, more than 90 per cent of the ions are those of Na^+ and Cl^-, while internally, there is less than 10 per cent of these ions. Instead, within the membrane, the chief positive ion is K^+ while most of the negative ions are a variety of organic particles that are too large to diffuse outwards. At the Donnan equilibrium state (*see* p. 977), the concentrations of the chief ions have the following relative values (inside the membrane is first in each case); K 30:1, Na 1:8, Cl 1:30. This gives rise to a potential difference of 60 to 90 millivolts with the inside being negative with

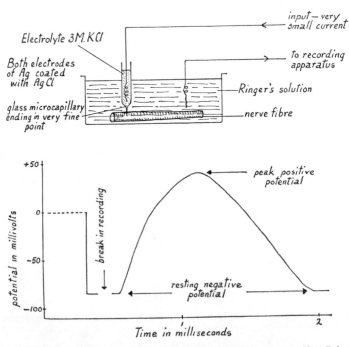

Fig. 8.17. Apparatus for determining the resting potential of a nerve fibre. Below, graph showing changes in potential with a single pulse of current.

respect to the outside. This is called the *resting potential*, and its value can be determined, under any particular conditions, using the apparatus shown in Fig. 8.17, though the experimental details are not easy to arrange. As soon as the fine glass point penetrates the membrane, the recording shows an immediate drop in potential from zero to 70, or so millivolts. There is then a slight break in recording while a pulse of current is passed through the membrane, flowing outwards. The membrane is rapidly depolarized and a potential difference of 30 to 40 mV is set up in the opposite direction. The potential then returns, somewhat more slowly, to its resting value. It is noteworthy that such *bio-electric potentials* have been found to exist for many plant and animal cell membranes.

When a nerve fibre is stimulated at any point, the potential difference across the membrane is lowered and the Na^+ permeability of the membrane increases. The reason for this alteration of ionic insulation of the membrane is not known, but the immediate effect is that Na^+ ions pass through the membrane and reduce the voltage drop across it. The inflow of Na^+ ions builds up until there is a slight internal positive potential. Then the impulse is propagated and this changes the permeability of the axon membrane in front of it. The whole process continues as a wave along the whole length of the axon.

Immediately after the peak has passed a particular point on the axon, the inward movement of Na^+ ions ceases, and there is passage of K^+ ions outward to restore the original negative charge inside the axon. For a few milliseconds after this, it is difficult to set up another impulse. Very quickly, however, the normal permeability is restored and the balance of ions returns to its original state (*see* Fig. 8.17). The actual numbers of ions affected by the whole cycle is so small that the overall composition of the solutions is scarcely affected.

The conditions favouring propagation are, first, that the stimulus has been strong enough to build the action potential to a peak high enough for propagation, and secondly, that the adjacent parts of the

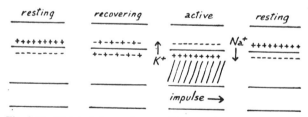

Fig. 8.18. Electrical charges in nerve fibre as action potential passes.

fibre are sufficiently excitable to propagate this action potential. If the first condition is not fulfilled, no impulse is sent out; the nerve fibre obeys the "all or none" law. The second condition varies with the type of nerve fibre and the treatment it has received. Some fibres are less excitable than others, and the action potentials they will propagate will be correspondingly higher. Also, if a nerve fibre is narcotized, the region of narcotization loses all excitability, and no action potential can be propagated past it.

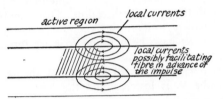

Fig. 8.19. Local currents emanating from a region of depolarization in a nerve fibre.

There is good evidence to show that the local currents set up in the fibre and through its sheath as a result of depolarization are the chief means whereby the fibre in advance of the impulse is made excitable enough to accept the impulse (*see* Fig. 8.19). A nerve fibre, blocked by narcotization at one point, can be shown to have its excitability increased on the distal side of the block, when an impulse reaches the block. Thus, to some extent, the impulse (action potential), is self-propagating in the sense that its presence leads to greater excitability of a fibre in advance of the moving impulse. The phenomenon is called *facilitation*. It is also borne out by the fact that an impulse may be transmitted across the blocked nerve fibre by means of a bridge of electrolyte capable of carrying the local currents set up.

Experiments have also been performed which show that the local currents set up by transmission of an impulse through one fibre can affect the excitability of an adjacent fibre. In artificial conditions with naked fibres, it is possible to stimulate the passage of an action potential through a fibre by passing an impulse along an adjacent fibre. It is doubtful whether such a condition would arise in the living body, but there is little doubt that impulses passing through one set of fibres would facilitate neighbouring fibres.

An alternative to the theory outlined above, involves the activity of *acetylcholine*. Advocates of this theory do not dispute that the impulse is propagated as a result of a flow of current in local circuits, but they put forward the suggestion that the depolarizing effect which accompanies the passage of an impulse is due to the lowered resistance of the membrane. It is suggested that this lowered resistance is due to increased permeability under the influence of the sudden appearance of acetyl choline. This acetylcholine, which is released by the flow of current out of the inactive region, depolarizes the membrane and is then very rapidly inactivated by an enzyme *cholinesterase* which is present in the membrane. The inactivation thus restores the state of polarization

There is some evidence in favour of this view, notably the fact that the enzyme cholinesterase is abundantly present in all nervous tissue. It is also known that a nerve action potential cannot be built up if cholinesterase has been inactivated.

In the case of myelinated fibres with nodes of Ranvier, there is some evidence that each node acts in a special way to transmit the impulse through the internode to the next node. Electron microscope studies show that there is a transverse disc at each node. This might be the effective agent and transmission through such a nerve fibre would therefore be from node to node in a discontinuous way. One node, on being depolarized, would set up currents which would depolarize the next node and thus the impulse would be propagated.

In addition to the electrical effect, there is a very small quantity of heat produced as a wave of activity passes along a nerve.

Transmission Across Synapses and Junctions

An impulse passing from one neurone to another, or from a neurone to an effector, must pass across a gap or synapse. The cell body and its dendrites, but not its axon, are closely invested by axons from other neurones. These axons end in *synaptic knobs*; between them and the cell body there are minute gaps of 10–20 nm. The endings of an effector are similar (*see* Figs. 8.20 and 8.21). It is known that a certain minimum number of these must stimulate the cell body or effector before any action potential is set up. It has already been noted that the conduction of an impulse along a fibre may have as an essential feature the liberation of acetylcholine. Long before this theory was developed, much evidence was accumulated which indicated conclusively that many nerve fibres liberate acetylcholine at their endings when stimulated. This points to acetylcholine as having a predominant role in synapse

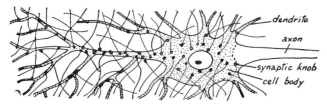

Fig. 8.20. Cell body invested by axons and synaptic knobs from other neurones.

and junction transmission. The acetylcholine theory has been built up largely around the transmission of the impulse across the neuro-muscular junction, where it has been most fully studied. One hypothesis suggests that transmission at the neuromuscular junction is effected by a current flow and that this current flow is dependent for its development on the chemical activity of acetylcholine. Arguments in favour of this, are the far greater concentrations of cholinesterase (the enzyme which inactivates acetylcholine) at the neuromuscular junction than anywhere else in the muscle, and also the fact that such junctions are extremely sensitive to acetylcholine, which depolarizes the end plate region and may give rise to action potentials.

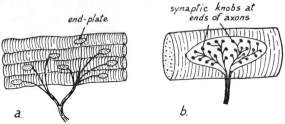

Fig. 8.21. (*a*) End plate nerve endings in voluntary muscle. (*b*) Single end plate enlarged.

Experiments with fibres of the sympathetic nerves of the vertebrates indicate that acetylcholine may not be the only substance associated with junction transmission. It has been demonstrated that a substance with the properties of adrenalin is liberated by such fibres on stimulation and to this substance is ascribed a similar function. Consequently, nerves have been described as *cholinergic*, when known to liberate acetylcholine, and *adrenergic*, when known to liberate the substance similar to adrenalin. The actions of adrenergic nerves are often antagonistic to those of cholinergic nerves and many organs receive a double innervation in which one type of nerve is excitatory and the other inhibitory in its effects.

In addition to the liberation of acetylcholine or adrenalin, in transmission there are marked electrical effects at synapses and junctions when impulses are conducted thus far. Eddy currents are set up and these are held by some physiologists to be strong enough to stimulate the adjacent cells directly. This view is held particularly with regard to the synapses between neurones in the central nervous system rather than nerve-effector junctions. The general concensus of opinion is that impulses are passed as a result of direct electrical stimulation independent of acetylcholine, across the synapses of the neuron system and by contrast the impulse is conveyed to an effector by electric currents dependent for their formation on acetylcholine or some other substance liberated by the nerve endings.

An important property of the synapse, as of the nerve fibre, is that normally transmission occurs in only one direction. The functioning of synapses is profoundly affected by the $K^+:Ca^{++}$ balance in the bathing fluid; K^+ ions excite in low concentration and inhibit in high. At each synapse there is a very slight delay in the transmission, as if the impulse has greater difficulty here, than it has along the fibres. This may be explained as the total effect of several factors: the very fine fibres leading to the synapse will necessarily be slower conductors: the fibres may vary in length, and hence will not be exactly synchronized so that it may require a little time to build up the necessary action potential.

CHAPTER 9

TRANSLOCATION OF SUBSTANCES WITHIN THE ORGANISM

MOST of the physiological processes occurring in plant and animal bodies demand constant transport of materials to and from all parts of the system right down to individual cells, in order that the system can maintain its activity. Within a single cell, it seems certain that diffusion and movement of the cytoplasm are adequate to secure distribution of materials. As the body structure has evolved into more and more elaborate patterns, so the chemical mechanisms and physical apparatus for internal transport of substances have become more complicated. The culmination is seen in the complex vascular systems of the higher animals and plants.

These transport mechanisms must serve to keep all cells of the body linked to the external environment and to one another. By doing so, the system satisfies the needs of the body in the following respects—

1. Each cell is assured of an adequate supply of respiratory substrate and sufficient oxygen to ensure an efficient release of energy for its own purposes.

2. Nutrients can be transported from the site of intake to the regions where they can be synthesized into the requirements of the organism.

3. Each cell can readily rid itself of its own waste products of metabolism.

4. Substances required for the building of new protoplasm can be constantly supplied to the regions of active cell division.

5. Substances to be stored can be diverted to the tissues specialized to store them.

6. The products of secretion can reach the sites of their activity.

7. Reproductive cells can receive the nourishment required for their development.

In this chapter it is intended to indicate how the transport systems of higher animals and plants function to achieve these requirements.

TRANSLOCATION OF SOLUTES IN PLANTS

It is convenient to study the translocation processes in the higher plants under three main headings, namely, transport of nutrients, of elaborated food materials, and of hormones. These are the substances which the plant is actively transporting. Respiratory gases move

310

passively by diffusion; their movements do not need consideration as translocation processes.

Transport of Nutrients

The main classes of nutrients to be considered are water, mineral salts and carbon dioxide. In the case of water and salts, the paths of transport are from roots to leaves, and in the case of carbon dioxide, from atmosphere to leaves. Movement of water through vascular plants has been adequately dealt with in Chap. 7 and needs no further enlargement here. Most moving water is in the xylem tissues passing in the upward longitudinal direction but there will be some lateral escape into adjacent areas as the bulk flow proceeds. A reasonable estimate of the quantity of water that is laterally transferred from the xylem before it evaporates is 5 per cent and that another equal amount is moving downwards in phloem elements, so that at least 90 per cent of all water carried up in the transpiration stream escapes to the atmosphere more or less directly.

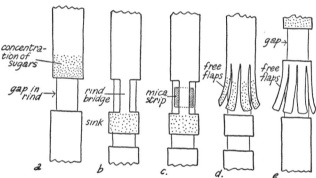

Fig. 9.1. Bark-ringing experiments.

(*a*) Transport interrupted by a complete gap in the rind. (*b*) Transport possible if bridge of rind is left. (*c*) Transport possible even when bridge is separated from xylem. (*d*) Transport into free flaps with no gap above. (*e*) No transport into free flaps if completely ringed above.

Movement of mineral salts has been studied for many years and by numerous investigators; the conclusions they reach are conflicting. Some suggest that the movement of salts is confined to the transpiration stream in the xylem. Others consider that the phloem is the primary channel. Others maintain that mineral salts may travel by the xylem path in some circumstances and by the phloem in others. Most of these conclusions have been reached after examination of exudates from cut stumps or after ringing experiments. In the latter type of experiment, a ring of the phloem-containing tissues is stripped from the stem, or separated from the inner xylem (*see* Fig. 9.1).

The quantities of mineral salts moving through stems thus treated, have been compared with the quantities moving through untreated stems on the same plant. More recent experiments have involved the use of tracer elements; it seems that this technique will eventually yield the most conclusive results.

Examination of liquid from cut stumps, sucked from wounds, spun out by centrifugation from lengths of wood or sucked from the cut end of a woody shoot, shows that it contains minerals in solution. Careful analysis indicates a solid content of from 0·1 to 0·4 per cent with about one third persisting in ash. Apart from differences in proportions with seasons, all the recognized mineral requirements are likely to be present including the macro-nutrients N, S, P, K, Mg, Fe, Ca and the micro-nutrients Cu, Zn, Mo, Mn, and B. Of the organic materials, most are sugars but amino acids and others can be identified. In some experiments involving "ringing," it has been found that minerals such as nitrates and phosphates move upwards just as rapidly through ringed stems as through untreated ones, at least for short periods and when the rings are low on the stem. Such results indicate that the xylem is primarily concerned with mineral transport. If such is the case, it would follow that an increased rate of transpiration would increase the rate of absorption of minerals by the roots by rapidly reducing the concentration of minerals in the root tissues. Experiments to test this tend to show that it is not the case and that increased transpiration has little or no effect on mineral absorption.

Other ringing experiments in which the rings were placed well away from the root region indicated that the movement of some minerals can be slowed down severely when phloem is not present. It is most likely, however, that the removal of a ring of phloem high up on the plant is likely to hamper the movement of food materials to the roots so that salt uptake is diminished. It can also cut down the transpiration rate in leaves above the ring so that even if the minerals were travelling in the xylem sap they would be slowed down by such ringing treatment. Such experiments are therefore not conclusive. It has been reported however that minerals can move upwards through the phloem when being translocated into very young unexpanded leaves or actively growing shoots, where the transpiring leaves have been removed. Not entirely in agreement with this is the result of the investigation into the path of radioactive potassium through stems. In a stem in which the tissues were all intact, the radioactive potassium was found to be more or less equally distributed between xylem and phloem and a similar situation occurred if the phloem was first separated from the xylem and then placed back in contact with it. In a case where the phloem was separated from the xylem by a waterproof layer, so that no we contact occurred, virtually all the radioactive potassium was found in the xylem. This indicates quite clearly that whilst some of the potassium carried by the xylem can readily leak out into the phloem, so long as the two are in contact, no potassium is carried independently by the phloem. In regions where the transpiration stream is very slow, it could well be that some mineral transport is accomplished by the phloem, whereas, in the lower part of the stem near the root, little mineral substance would leak into the phloem, and consequently the phloem transport rate would be very low.

In view of such conflicting observations, it is not wise to be dogmatic. It is possible that the upward path of minerals varies in different plants and in different regions of the same plant and with the type of mineral. Near the roots, the path is likely to be the xylem, near growing tips, the phloem, whilst in between it could be achieved by both. In plants showing root pressure, most minerals may pass through the xylem. Whilst most mineral substances appear to move upward in a free inorganic state, this is not always the case apparently, since some can rarely be detected in this form. Such is the case for the nitrate ion, NO_3^-, which is seldom encountered in plants, nearly all nitrogen in the xylem being in the form of amino-acids and the amides, glutamine and asparagine, with a little in peptides and alkaloids. Likewise some of the sulphur present, although not all, is in the form of glutathione and cysteine and methionine, the sulphur-containing amino acids. Phosphorus is sometimes carried as phosphoryl choline. These findings would point to the fact that some ionic material may be carried as part of more complex organic material, i.e. there may be some carrier mechanism employed in the transport of some inorganic nutrients.

An investigation into the movement of iron into plants from culture media and its subsequent transport through the xylem shows that its mobility may be severely restricted under the influence of certain conditions, so much so that the plants may show chlorosis, a symptom of iron shortage. When the medium is at pH 7 or above, or when it contains a low iron to phosphorus ratio, the iron deficiency becomes apparent. The explanation given is that either or both of the conditions causes the precipitation of iron as ferric phosphate hence rendering it immobile. This seems to fit with the fact that the precipitation of iron is prevented if the substance ethylene diamine tetra-acetic acid (EDTA) is added to the culture solution. This acts as a chelating agent for the iron and removes the restriction on its movement into the roots. A corresponding natural substance may be involved in the movement of iron within the plant tissues.

The movement of carbon dioxide from the atmosphere of the leaf spaces into the cells, must be assumed to be passive diffusion of the carbon dioxide in solution. There appears to be no special mechanism which would indicate active participation on the part of the plant in carbon dioxide transport into the leaf cells.

Transport of Elaborated Food Material

The movement of elaborated food materials through the plant, unlike that of water and mineral salts, must be in various directions, up, down and across. Substances synthesized in the leaves, must find their way into storage and growing regions which may be at a higher level. Such are fruits, seeds and developing buds. On the other hand, such regions may be located at a lower level than the leaves. Such are perennating organs, of which roots, tubers, corms and rhizomes are examples. Food reserves are often stored in pith or in wood parenchyma and must therefore be carried to these tissues. The stored food must later find its way to the regions of active development and must again move in various directions through the plant.

Opinions have been divided as to the paths used in movement of elaborated materials in various directions. It was at one time believed that all movement upward, whether of organic or inorganic material, was by way of the xylem, and that all movement downward, was by way of the phloem. This is certainly not the case.

The evidence from numerous ringing experiments on a wide variety of woody plants, studies of the movement of radioactive substances and analyses of phloem extracts, all indicate that the phloem, largely its sieve tube elements, is mainly responsible for both upward and downward movement of carbohydrate in the form of soluble sugars, chiefly sucrose, and that the food stored in the pith and medullary rays is also supplied via the phloem. Similarly, the evidence leads us to conclude that nitrogenous materials such as amino-acids, together with numerous other soluble organic compounds, are moved about the plant by the activity of the phloem. The xylem seems to be less concerned with transport of food substances. However, exudates not infrequently do contain some organic materials.

That living cells are concerned in food transport is borne out by the fact that rates of transport are very frequently but not always seriously affected by changing conditions of temperature and oxygen supply and by causing damage to the cells. The effect of lowering the temperature of the petiole of a leaf is generally to reduce rapidly the rate at which materials leave it. At temperatures a little above freezing point the translocation process usually ceases, but it rises to a maximum at temperatures between 20°C and 30°C. Above this range, the rate again falls off. A supply of oxygen seems always necessary for continued movement of materials out of a leaf. Experiments have shown that when

he whole of a petiole is deprived of oxygen, the rate of translocation is
'ery much slower than when only part is so treated. Damage to the
:ells by steam treatment will bring translocation to a standstill.

Ringing techniques were employed by Mason and Maskell in 1928 on
he stems of cotton growing in the field. They showed at first that
emoval of a complete ring of phloem early in the day caused a heavy
)uild up of sugars in both phloem and xylem above the ring during the
iext twenty hours, accompanied by an absence of sugars in the tissues
)elow the ring. Thus phloem removal impeded the movement of sugars
lownwards even if they were moving in the xylem, something that could
iot be ruled out at this stage. They went on to show that the daily
ariation of the quantity of sugars in the phloem was much more
losely correlated with the variation of sugars in the leaves than was the
ariation in quantity of sugar in the xylem. Finally, using separation
nd flap ringing techniques they produced convincing evidence that any
ugar in the xylem was due to a lateral movement of it from the phloem
'hich was in reality the true transport tissue.

With the availability of radioactive isotopes much more evidence
1at phloem provides the downward path for elaborated materials was
rovided. Biddulph and Markle were able to show that ^{32}P introduced
1rough cuts into the leaves of the cotton plant was detectable in much
reater quantity in the phloem down the stem than in the xylem so long
5 these tissues were laterally separated. A more sophisticated technique
as employed by Rabideau and Burr in which individual leaves were
:d with $^{13}CO_2$ and allowed to photosynthesize. Movement of the food
1aterials then produced could be traced out of the leaf and into the
em tissues. It was found that if the phloem was not blocked at all by
steam treatment ^{13}C was found both above and below a treated leaf.
the phloem was blocked above the leaf the ^{13}C was found only below
1d if the phloem was blocked below the leaf ^{13}C was found only above

When the phloem was blocked both above and below the treated
af then no ^{13}C could be found in the tissues beyond the blocked areas
ther up or down. To eliminate the possibility that the treatment
ven to the phloem in the blocked areas had interfered with the trans-
irt capabilities of the xylem, it was arranged that whilst $^{13}CO_2$ was
d to the leaves, ^{32}P was fed to the roots. This quite easily moved into
sues beyond any blocked areas in the phloem. The results as a whole
:arly point to phloem transport out of leaves of carbon assimilates.
was later work of Biddulph that showed convincingly that of the
lloem tissues it is the sieve tube elements that are most involved. This
pears to be the case for all higher vascular plants investigated.

An excellent example of a truly ingenious experimental technique is
1t in which the natural parasites of the plant food conducting

channels, the aphids, are used to obtain pure tissue extracts. One of the most useful for this purpose is *Tuberolachnus salignus* on willow. Others are *Acyrthosiphon pisum* on broad bean and *Longistigma caryae* on lime. An aphid when feeding inserts its stylet bundle precisely into a sieve tube or sieve cell, never elsewhere. The method of obtaining phloem extract via the aphid is to allow it to feed for two or three hours, then to anaesthetize it and to sever it from the feeding tube which is still left in the tissues. Sap will exude from the cut end of the stylet stumps and can be taken up by pipette in sufficient quantity for analysis. This shows that there is always a connexion between what the phloem is actually carrying and what may have been fed into it. Another interesting disclosure is derived from the rates at which exudate is delivered. In some cases quantities over five cubic millimetres per hour have been obtained. It is known that stylet tips enter single sieve elements and for this quantity to be exuded from one cell in the time, it would mean that the phloem cell would have to be refilled several times a second.

The Mechanism of Movement of Substances through the Xylem and Phloem

A knowledge of the structure and disposition of these tissues will immediately suggest the probability of differences in modes of transport of materials through them. Xylem vessels and tracheides are elements with thick rigid walls and wide lumina, averaging 20 μm to 200 μm. Phloem sieve-tubes possess thin, elastic, non-rigid walls and small lumina, averaging 10 μm to 50 μm. In any one plant, the total cross sectional area of the xylem elements is much greater than that of the phloem. The vessels of angiosperms are continuous tubes with no cross-walls for much of their distance, and according to species, they range in length from 10 cm to 10 m. In the sieve-tubes, the end-walls are always present, and the length of the tubes ranges between 0·2 mm and 1 mm. Most xylem elements are devoid of living contents and there are frequent inter-connexions between the lumina by means of pits. Normally, they contain a weak aqueous solution without colloid. The sieve-tubes are well filled with enucleate cytoplasm; they have inter-connexions over the restricted sieve-plate and sieve-area regions; they contain much more solid substance including colloids, in solution; they appear to lack the main enzyme systems and their plasma membranes exercise control over entry and exit of dissolved substances.

In the case of xylem, it seems reasonable to suppose that any difference in water potential between the two ends of the system would result in mass flow of liquid from one end to the other. As has been discussed Chap. 7, this change in water potential may be achieved either by increased hydrostatic pressure at the base of a stem due to root pressure, or by a decreased water potential due to rapid evaporation of water at the upper end, or by both. We may visualize transport through xylem therefore mainly as a passive transport actuated by changes in physical conditions.

Similar changes in conditions could not account for movement in the phloem and it has been found that it is impossible to force materials through this tissue in the same way as they may be forced through xylem. Movement in phloem seems to be more allied to properties of living protoplasm. Several mechanisms have been suggested to account for phloem transport, including diffusion, a pressure-flow mechanism, cytoplasmic streaming and others more complicated.

Simple diffusion has never been considered sufficiently rapid to account for the speed of movement or the quantities moved in phloem transport. Calculations on movement of sucrose in the cotton plant, show that in a given time, over 40 000 times as much is moved as would be expected by simple diffusion. Some substances have been reported as moving through phloem over distances varying from 0·1 to 1·5 m in an hour and even faster under special conditions. However, it has been found that movement of materials in the phloem is along a concentration gradient and some kind of diffusion in which the molecules are "activated" has been suggested. The precise nature of this "activation" has never been made clear.

A second hypothesis concerns a mass flow of solution through the phloem under the influence of osmotic phenomena. The principle of such a mechanism is illustrated in Fig. 9.2.

The containers a and b are made of semi-permeable membranes and filled with strong sugar solution and water, respectively. They are

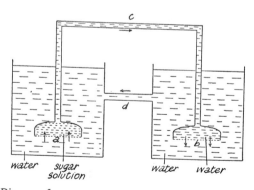

Fig. 9.2. Diagram of apparatus to show flow of sugar solution due to osmotic phenomena.

connected by the glass tubing c, and each container is immersed in a vessel of water; the water-vessels are also joined by a tube d. By osmosis, water will enter the semi-permeable container a, and thus increase the pressure in the system. Water will then be forced out of container b and the solution in a will move along tube c to take the place of the water lost. Thus there will be a gradual flow of sugar solution from a to b along tube c; this flow will cease only when the solutions in a and b are of equal concentration. If we regard the containers a and b as living cells, where a is a permanent source of sugar, and b a site where the sugar is used up or stored in an insoluble form, then the solution could keep moving indefinitely. Such a system of plant tissues involving leaf cells, phloem sieve-tubes, storage cells and xylem vessels, may be visualized, as shown in Fig. 9.3. It could be supposed to function in the following manner.

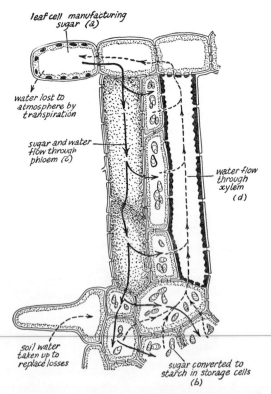

leaf cell manufacturing sugar (a)

water lost to atmosphere by transpiration

sugar and water flow through phloem (c)

water flow through xylem (d)

soil water taken up to replace losses

sugar converted to starch in storage cells (b)

Fig. 9.3. Possible flow of solutes and water through plant tissues on the principle illustrated in Fig. 9.2.

Sugar being manufactured in the leaf cells would give a high turgor there, and the pressure would be sufficient to force solution through the plasmodesmata into sieve-tubes, along these and through the sieve-plates to receiving cells such as may be found in the root cortex. These cells would then use the sugar or convert it into an insoluble product such as starch. The water, still under pressure, would be forced out of the root cells and eventually into xylem vessels, whence it would be transported back to the leaves again. The leaf and root cells in this system are equivalent to containers a and b respectively, that is a "source" and a "sink" whilst the phloem and xylem elements represent tubes c and d. The physical principles of the system are sound and can be used to explain some phloem transport phenomena such as exudation from cut phloem at some distance from cells of high turgor. Investigation into exudation which follows the severance of aphid stylets has shown that the liquid is derived from that flowing directly in the longitudinal path of the stylets and at some distance from them. Exudation ceases quickly if the phloem is severed close above the point of stylet insertion but if the phloem is left intact for upwards of 15 cm from that point, or cuts are made laterally adjacent to the stylets, exudation continues for some while without interruption. Clearly, the liquid passing out of the stylets must have come from some distance away.

A requirement for the mass flow system to work is that there shall be always a downward turgor gradient between the source and the sink. There is some evidence that such a gradient exists in some trees but this is obviously not always the case and there are instances of apparent flow from flaccid to turgid cells as may be seen in a potato sprouting in the absence of water. The old potato tissue, presumably the supplying tissue, is flaccid whilst the young shoot, the receiving tissue, is fully turgid. This movement could still be accounted for by a mass flow system if it is accepted that there need only be a turgor gradient in the phloem cells and this could be set up by the active secretion of substances into these cells from surrounding tissues at the source and equally rapid expulsion of them at the sink end of the system.

Other evidence in favour of mass flow of some kind has been adduced from a microscope inspection of live phloem in sections of *Pelargonium* sp. petioles under controlled turgor conditions. When the ends of the detached petioles were immersed in solutions of varying concentrations the content of the sieve elements could be seen to surge from high to low turgor zones; the direction of the movement and the speed could be controlled by varying the concentrations of the solutions at the ends of the petiole sections.

Evidence derived from the study of the movement through phloem of materials introduced from outside also supports the mass flow idea. It has been found that such substances are conducted out of leaves only under conditions when sugars can be assimilated, that is, in the light or alternatively, only if sugar is added to the leaf cells with the extraneous substance if the leaves are kept in the dark.

But there are objections to the mass flow explanation, one of which has been noted.

Another objection to such a system is the difficulty of finding the forces necessary to push the fluids through the sieve-tubes and fine pores over great distances. Calculations show that it would require forces of several hundred atmospheres to move solutions the length of an average tree. Still further, the mass flow explanation assumes no necessity for active, living function on the part of the sieve element protoplasts, something that does not relate well with the findings concerning the responses they commonly make to varying metabolic conditions such as temperature level and presence or absence of oxygen.

Perhaps the most serious objection to accepting a mass flow theory is the occurrence of the phenomenon of conduction by phloem of different substances at different rates and frequently in different directions at the same time. It is possible, however, to dismiss this objection on the grounds that the different materials are being transported by separate longitudinal series of sieve elements and not by the same cells.

To dispel two of the other main objections, namely, that high turgor forces are necessary to move materials through sieve plates and that the mass flow idea takes no account of phloem transport as an active function, a modification has been suggested in which two kinds of flow are visualized, the mechanical mass flow through the sieve element cavities due to turgor pressure differences as postulated by Munch and an electro-osmotic movement of water, carrying dissolved substances through the pits in the sieve plates as put forward by Spanner. This latter supposes that at a sieve plate there is a polarization effect of electric potential created by the constant circulation of potassium ions from one side to the other, adjacent companion cells being responsible for removing the ions from the cell into which they were being secreted and pumping them back into the cell from which they were being expelled. The electrical gradient so created would produce a uni-directional flow of water and solutes through the sieve apertures by promoting the flow of highly hydrated ions. There are bits of evidence to support the idea, including the fact that sieve tubes do seem to be rich in potassium. There is some evidence, however, that this potassium is being moved through the phloem as are other substances and not being circulated in the areas of the sieve plates only.

Another quite different explanation of phloem transport has reference to a protoplasmic streaming mechanism. The original concept is quite old but stronger support for it has been advanced in recent years. The suggestion is that protoplasm streams through the sieve elements from cell to cell through the sieve plates carrying the whole range of transported materials with it. Whilst cytoplasmic streaming has never been reported in matured sieve elements, it could account for transport through the younger, metabolically active cells.

On the grounds that the movement in both directions simultaneously of phloem solutes can be accounted for only by a protoplasmic streaming mechanism, attempts have been made to substantiate the theory by demonstrating that different substances will move through a single file of sieve tube elements in opposite directions at the same time. The most convincing of these was carried out by Eschrich who showed that honeydew collected from an aphid with its stylets in a single phloem cell contained both fluorescein that had been introduced to the phloem below it and ^{14}C-urea from above it. It should be recognized, however, that the two substances could have arrived in the same file of phloem units by the lateral transfer of one or both of them and that their dual presence does not conclusively support a claim for bidirectional transport by one conducting unit.

Some supporting evidence for the streaming mechanism explanation comes from studies of the transport as an active, living function. It is well known that translocation can be blocked by lowering metabolic rates using respiratory inhibitors, reduced oxygen availability, lowering of temperature and so on. Under the same conditions protoplasmic streaming is also halted and the two events may therefore be connected. However, it has been shown that in isolated sections of willow stems, at temperatures as low as $-1.5°C$, some movement of ^{32}P could be detected using aphid honeydew as the indication. At this temperature, no protoplasmic streaming can be discerned. But whilst there was some longitudinal movement there was virtually no lateral transfer from the sieve elements. It might therefore be that it is this lateral movement that is the more dependent on high metabolic activity.

One piece of evidence, strongly in conflict with a cytoplasmic flow mechanism, is in relation to the relatively slow speed at which this occurs. It has been worked out that the long distance transport of materials in the phloem is up to ten times faster than any measured speed of streaming of the cytoplasm could account for.

An extension to the protoplasmic streaming hypothesis has been put forward by Thaine who reported the visual identification of strands of cytoplasm passing longitudinally through sieve elements of *Primula obconica*. These were continuous from cell to cell through the pores in the sieve plates. He noted unidirectional movement of particles within a single strand but also that not all strands in a single cell showed movement in the same direction. Some of the particles in these trans-cellular strands appeared to be mitochondria and these might be the sites of the release of the energy required to keep the system moving. Others have strongly maintained that such visual evidence is wholly artefact and that what was seen actually occurred in parenchyma cells adjacent to the sieve elements.

Exudates from cut phloem of *Cucurbita pepo*, into which radio-active photosynthetic products were passing from leaves fed with $^{14}CO_2$ at a higher level up the stem, have been reported by Thaine as showing on radioautographic examination the presence of strands of radioactive material nearly 800 μm long (about twice the length of the sieve elements) and about 3·8 μm in diameter. It is suggested that these strands represent long, straight membrane-bound cell components such as the transcellular strands visually recorded.

Other suggestions put forward as possible phloem transport mech-anisms have been "activated diffusion," in which it was postulated that the moving molecules were energized in some way and/or resistance to diffusion was considerably reduced and a rapid movement of materials along protoplasmic interfaces, such as is known to occur along the interface between two immiscible liquids. Neither of these is supported by any real evidence.

At present, there is no generally accepted explanation of the move-ment of materials through phloem tissues that will account for all the observations made. It is probable that the phenomenon is the result of the interaction of a number of mechanisms. Perhaps the most accept-able hypothesis at the moment is that of a mass-flow mechanism driven by electro-osmosis but until it becomes more certain what is the real nature and function of the sieve plate and the protoplasm that appears to pass through its perforations, this can be advanced only tentatively.

Translocation of Plant Growth Substances

Of the three main groups of known plant hormones, the movement or transport of the auxins has been investigated the most fully up to the present, involving work using both isolated sections of tissue and intact plants. Some of the more significant features of auxin movement are as follows. The movement of both the natural and the synthetic auxins under normal conditions is polar, meaning that these substances will move more readily in one morphological direction through the tissues than any other. In the case of auxin in sections of stem internodes, leaf stalks, coleoptiles, hypocotyls and leaves the preponderance of move-ment is from the morphological apex towards the base, that is, basipetal.

In roots the reverse polarity is often exhibited, that is, from base to apex or acropetal. Sometimes auxin will not move basipetally through shoot tissues that have been turned upside down. Auxin certainly moves polarly through parenchyma in all organs and possibly in similar fashion through other tissues as well. For oat coleoptiles, the basipetal polarity of movement disappears when they are treated with narcotics and metabolism inhibitors or are placed in anaerobic conditions but the small amount of acropetal movement is unaffected. Basipetal movement is also speeded up by a rise in temperature. The differences between basipetal and acropetal movements suggests that there are two transport mechanisms concerned in auxin movement. The basipetal movement seems certainly associated with metabolic activity in the cells whilst the acropetal movement is probably a passive diffusion. The fact that the movement of auxin is polar certainly points to the existence of a special mechanism for its transport.

Because of the very small quantities involved the movement of auxin is detectable only by careful assay tests. The Avena curvature test (see p. 615) can be applied or alternatively the movement of auxins labelled with ^{14}C can be traced using radioactivity detection methods. Auxins in oat coleoptiles have been shown to move at rates up to 20 mm per hour and at about 12–15 mm per hour in maize coleoptiles. This is certainly faster than can be accounted for by a passive diffusion mechanism and the suggestion has been made that protoplasmic streaming may be involved. Speeds of movement seem a little on the slow side for this, however. An interesting feature is that when IAA is applied to the cut surface of a decapitated oat coleoptile an electrical impulse passes down the coleoptile and this has been measured at about 14 mm per hour, coinciding roughly with the speed of movement of the auxin. The electrical disturbance cannot be detected in the reverse direction when IAA is applied to the base of a coleoptile but a similar effect is known to occur during the unilateral illumination of coleoptiles when auxin is being displaced laterally. It is not known if auxin is moved as electrically charged molecules under the influence of an electrical potential gradient as these facts might suggest.

Auxin movement is certainly dependent on metabolic activity in the conducting cells and there is the indication that it is "active" in the sense that it can occur against an electrochemical potential gradient (concentration gradient) in the tissues and independently of any passive physical forces such as solvent drag. Movement of IAA is inhibited by other auxins and this is presumed due to competition between them for transport sites within the cells. Movement is also inhibited by lack of carbohydrate and by introduction of respiratory inhibitors such as cyanide, clearly suggesting that metabolic energy is required. Exposure to ethylene for longer than five hours also prevents auxin movement but this is difficult of explanation. It is most likely that the ethylene is not direct in its inhibitory action but has some indirect effect somewhere in the complex system. As stated on p. 277, ethylene is produced by the presence of excess auxin. An interaction between auxin and ethylene may be the means of regulating auxin concentration in the tissues.

There is good evidence that during movement through the tissues there is an irreversible immobilization of IAA and the more auxin there is moving, the greater is the amount immobilized. The auxin appears to become attached by hydrogen bonds to proteins in the cells.

Both non-vascular and vascular tissues seem able to carry the moving auxin polarly but in the latter the movement is comparatively slower than is the movement of other transported materials. Auxins applied externally to a plant can reach the vascular tissues and they are then conducted away from the site of application in a non-polar way at measured speeds up to 50 mm per hour which is more like normal vascular transport activity.

The precise site of transport of auxins within the tissues and cells is not known. There is no real evidence either way as to whether they move from cell to cell through plasmodesmata, that is, through the symplast, nor is there any indication of the parts played by cell membranes or streaming cytoplasm.

Green tissues, darkened for periods, and etiolated stems seem incapable of carrying auxin polarly in the basipetal direction but the property can be restored if they are exposed to light, the effect of which is probably to restore the supply of photosynthesized materials to energize the movement. Darkened sunflower plants unable to carry auxins polarly, have had their transport properties at least partly restored after being sprayed with sucrose solution.

Auxin can move laterally through stem and root tissues (see Phototropism and Geotropism, pp. 606–616) and this is known to happen at the greatest speeds when the organs are placed horizontally. The movement is still distinctly polar, the auxin being able to move about five times faster from the upper to the lower surface than in the opposite direction. The lateral movement is very quickly inhibited by exposure to ethylene, which is not the case for longitudinal movement which requires prolonged exposure to the gas before it is affected. This points to the probability that the lateral and longitudinal movements are under different influences.

With regard to the movement of the gibberellins and the cytokinins, far less factual information has so far been produced. Movement of ^{14}C-labelled gibberellin applied to pea plants has been shown to occur at speeds of up to 50 mm per hour, probably in the phloem. There seems to be no polarity in the movement of gibberellins applied externally to the tissues or of those produced naturally within. Gibberellins produced in the apical 3 or 4 mm of root tissues of several plants, including lupin and sunflower, have been detected in cut stem exudates above steam girdles, indicating transport in the xylem via the transpiration stream.

The fact that it requires direct application of kinetin to laterals in order to break the inhibiting effect of an apical bud, whilst kinetin administered as close as 2 mm away is ineffective, has been taken as indicating that cytokinins do not move far, if at all, from their sites of production. This is also subscribed to by evidence from the movement of externally applied ^{14}C-labelled kinetin which also seems to be restricted. But not all work on cytokinin movement indicates the same lack of mobility. There seems now good evidence that natural cytokinins are manufactured in relatively great quantities in the roots of plants and that they pass into the shoots through xylem channels with the transpiration stream, as seems the case for the gibberellins. Exudates from cut stems show the presence of cytokinins for days after the shoot has been removed. One of the effects of cytokinins in delaying senescence in leaves is very enhanced when the leaf stalks are induced to generate adventitious roots, the inference being that the roots are supplying readily mobile cytokinin to the leaf tissue.

Some experiments are reported to have shown that in detached sections of bean petioles there is a translocation of the cytokinin, benzyl adenine (BA), which is basipetally polar but not so strongly as the movement of IAA. Other work using various parts of a number of plant species has failed to replicate this finding and there has been as yet no explanation of such conflicting observations.

From the results of studies in which the growth substances were mixed with one another and then externally applied it would appear that they can affect the movement of each other. The interaction between IAA and kinetin is to enhance the basipetal polar movement of the former. The same effect occurs if GA is used instead of kinetin. IAA has been reputed to promote the basipetal transport of kinetin and BA.

The conclusions to be drawn from the observations made so far on plant hormone transport in the tissues are by no means clear or precise. Much more needs to be discovered before there can be compiled an acceptable explanation embracing all the various hormone movement phenomena.

Translocation in Lower Plants

In recent years this has become a field of increasing interest to experimenters who have shown that in a number of instances the movements of elaborated materials in algae, fungi and the bryophytes may have some features in common with the equivalent movement of metabolites in higher plants. For example, in the stipe and midrib zones of some of the kelps (large brown seaweeds) there occur cells not unlike sieve elements and it has been shown that photosynthesized substances, labelled with ^{14}C, are moved through these at rates up to 800 mm per hour in a species of *Macrocystis*. Some of these sieve tube-like members, known as "trumpet hyphae" and apparently acting in a lateral transfer capacity, show a cross-wall structure of a nature similar to that of a sieve plate. Through such a wall pass up to thirty thousand cytoplasmic strands through pores lined with callose, *see* Vol. I, p. 245.

Work done with the histologically complex moss, *Polytrichum commune*, has shown that when $^{14}CO_2$ is fed to the leaves, it is incorporated into sugar molecules and there is quite rapid movement of these out of the leaves. Movement of sugar to growing points through the stem is known to be along specialized cell channels known as the *leptoid* tissue, arranged in longitudinal bundles. Leptoid cells, again, can be likened to phloem elements of higher plants, possessing callose on the end walls and denser protoplasts than cells of adjacent tissues.

Translocation in fungi seems very much related to protoplasmic streaming and this can be observed quite readily proceeding through pores in cross-walls, effecting a longitudinal continuity of movement of material from cell to cell along a hypha.

TRANSLOCATION OF SOLUTES IN ANIMALS

The active mode of life of animals has in general necessitated a metabolic rate higher than that of plants. In the simplest animals, there is a condition similar to that of simple plants, but in all the higher groups, a system of endless circulation of a fluid has been evolved. In the vertebrates, the blood conveys food materials, gases, excretory products and hormones to their respective destinations, but in many invertebrates there are separate tubular systems involved in respiration and excretion; they may appear with or without blood vascular systems. Here, some account of translocation mechanisms in the lower animals is given together with a more detailed description of the circulatory mechanism in vertebrates and in the mammal in particular.

Translocation in the Protozoa

Protozoan animals are all of small size and thus diffusion alone is probably sufficient to ensure that all parts of the protoplast are efficiently supplied with necessities, and that the excretory product pass out of the body. Protoplasmic streaming undoubtedly assists in the circulation of materials. Body movements, particularly those where change of shape is effected, such as amoeboid and euglenoid motion, also play some part. In the Ciliophora, the circulation of the food vacuoles distributes nutriment, and possibly oxygen, fairly evenly throughout the protoplasm, while the wide-ranging accessory vacuoles must assist in collecting and expelling excretory substances.

The Use of External Water as a Translocating Agent

In the Porifera and Coelenterata, the living tissue occupies thin layers separated by jelly-like material. No cell is far removed from the water which contains the oxygen, and into which the excretory products diffuse. In the sponges, an inhalant current is set up by the flagella of the choanocytes; the same mechanism speeds the outgoing water through the osculum. Food particles are ingested by the choanocytes and digested intracellularly. Special amoeboid cells wander in the jelly, carrying food materials from the choanocytes to other cells; it is also possible that these wandering cells carry excretory products to the periphery, where they diffuse into the surrounding water.

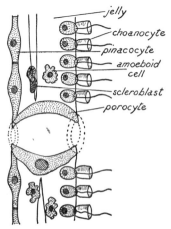

Fig. 9.4. Vertical section through the body wall of a simple sponge.

The hydroid coelenterate forms possess two distinct layers of cells separated by the mesogloea, a thin layer of jelly. Again, all the cells, except the nerve cells in the jelly and the endoderm cells of solid tentacles, are in contact with the external water. Thus for respiratory and excretory functions, each cell is in much the same position as a single protozoan animal; diffusion is adequate. The mesogloea is extremely thin and through it, diffusible food materials pass from the endoderm to the ectoderm. Flagella of endodermal cells ensure mixing of the watery contents of the enteron, while contractions of the body and defaecation through the mouth provide for adequate changing of the internal water. The enteron is sometimes termed a gastrovascular cavity.

The medusoid forms have a definite system of water circulation, the gastro-vascular system (see Fig. 9.5). Owing to the fusion of the upper and lower endoderm layers over a large part of the bell, some form of canal system has become necessary. In its simplest form, it consists of four radial canals leading from the gastric cavity to a circular canal which traverses the margin of the bell. This condition is adequate for very small medusae such as those in *Obelia* (*see* Fig. 9.6). But in a large jelly-fish, this is no longer sufficient and there is a system of single ingoing canals, and a network of outgoing canals, all lined with cilia

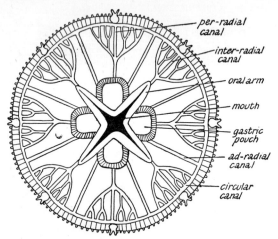

Fig. 9.5. Gastro-vascular system of scyphozoan coelenterate.

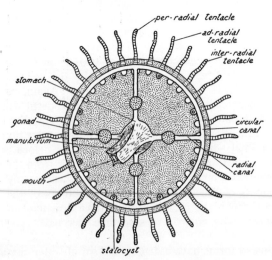

Fig. 9.6. *Obelia* medusa, sub-umbrellar view.

which maintain the flow of fluid. By this method, translocation to and from the endoderm cells is effected. Water enters the gastric cavity, passes into the gastric pouches and traverses the eight ad-radial canals to the circular canal. Then it travels back by branched inter-radial and per-radial canals and leaves the body by the exhalant grooves at the corners of the mouth.

Translocation in Acoelomates

The primary body cavity, the blastocoel, becomes more or less filled with cells budded off from both ectoderm and endoderm. They form the tissue known as mesenchyme, which separates the gut wall from the

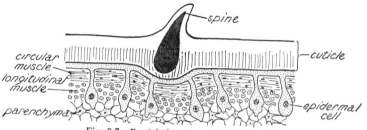

Fig. 9.7. *Fasciola hepatica.* V.S. body wall.

body wall. In some cases, fluid cavities remain among the mesenchyme and form a primitive vascular system. There is no pumping mechanism and there are no cilia, and thus any movement of the fluid is dependent upon locomotive movements of the body. The great majority of acoelomates are either of small size or of flattened shape, hence in the absence of special respiratory organs, diffusion must suffice for entry of oxygen and expulsion of carbon dioxide. The larger acoelomates are sluggish animals, and thus there is no considerable expenditure of oxygen.

The Platyhelminthes are flattened animals, hence the surface area/volume ratio is never low. The profusely-branched mesenchyme cells transmit diffusible food from the gut which is richly branched in most forms. Indeed it may be said, for many of these flatworms, that the caeca of the gut take the food within diffusible distance of all parts of the body. In the cestodes, the gut is absent; the extreme thinness of the body, coupled with the normal habitat in the intestine of the host, ensure that no cells are far from the diffusible food supply. Regulation of water content and probably much nitrogenous excretion is carried out in Platyhelminthes by flame cells. These are at the ends of a much-branched excretory canal system which is ciliated in all the subsidiary canals but not in the two principal lateral canals which lead the fluid out of the body.

In the nematodes, the mesenchyme consists of a few enormous vacuolated cells. The vacuoles are so large that they give the appearance of longitudinal canals (*see* Fig. 9.8).

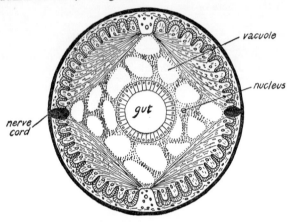

Fig. 9.8. T.S. body of nematode, *Ascaris*.

Undoubtedly they play a part in translocation of diffusible food from the gut to the body wall. Some investigators have described these canals as pseudo-coelomic. Circulation of the fluid can only be effected by body movements. Parasitic nematodes probably move very little and often they live in situations where oxygen is very scarce if not completely lacking; it has frequently been suggested that they are anaerobic in the trophic stage. Flame cells and a canal system provide a means of water regulation and nitrogenous excretion. Free living nematodes are all of small size and restricted to aquatic or damp situations; diffusion is adequate for their respiratory exchanges.

As a group, the rotifers are the smallest metazoan animals; most species are comparable with protozoa in size. Thus, for gaseous exchange and excretion, diffusion would be amply sufficient. Yet they have evolved a surprising degree of complexity for so small a body (*see* Fig. 9.9). There is a highly-differentiated alimentary canal surrounded by a body cavity which is probably primary, i.e. blastocoel. There are two longitudinal excretory canals with side branches ending in flame cells; the two canals open into a cloaca. They are very active animals and their movements must cause circulation of the fluid in the cavity, thus aiding distribution of soluble substances.

The nemertean worms possess a true blood vascular system. Most of the primary body cavity is filled with mesenchyme, but there are several longitudinal blood-vessels connected by transverse vessels.

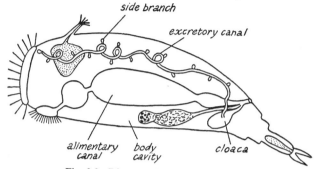

Fig. 9.9. Diagram of the body of a rotifer.

Usually there are three main vessels, one above the gut and two situated laterally; all the vessels have contractile walls and thus peristaltic motion of the blood is achieved, probably assisted by the body movements. Sometimes the blood is coloured red by corpuscles containing the respiratory pigment haemoglobin, but in most cases the blood is colourless. This primitive blood-vascular system seems to be very awkwardly situated in the body. It is too deep to have much value as a carrier of oxygen and too far from the gut to allow of easy diffusion of food materials, though the dorsal vessel may play some part in this

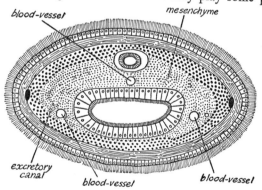

Fig. 9.10. T.S. body of nemertean.

respect. Excretion and water-regulation are again effected by flame cells situated at the ends of lateral branches of the two excretory canals (see Fig. 9.10).

Haemocoelic Translocation

In Arthropoda and Mollusca, the true coelom is considerably reduced, being represented only by the cavities of the gonads and of some types of excretory organs. The primary body cavity (blastocoelic) is filled with blood which directly bathes all the tissues, taking the part of lymph in the vertebrates. There is a contractile heart and certain blood-vessels which do not branch into capillaries but open into the haemocoel. There are special arrangements for refilling the heart.

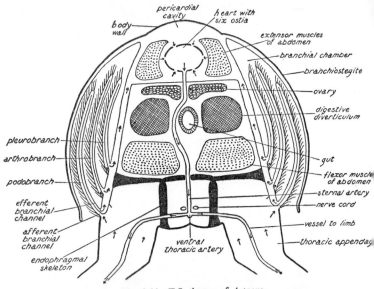

Fig. 9.11. T.S. thorax of *Astacus*.

In both phyla, the respiratory pigment, where present, is haemocyanin and except in the insects there are special arrangements for adequate circulation to and from the respiratory organs. The larger aquatic arthropods are sluggish animals; this is inevitable because of the low pressure and slow movement of the "open" blood system. The great activity of the insects is correlated with the tracheate system of respiration. In Mollusca, the larger members are slow-moving except for the Cephalopoda which possess an extra pair of hearts to give more speed and pressure to the blood flowing from the gills. The most important function of the vascular system in both phyla is the transport of diffusible food materials and of hormones.

Translocation in Echinoderms

Echinoderms have several series of fluid canals, including a so-called blood-vascular system and a water-vascular system. In addition, the coelom is spacious and the gut extends into each arm. In spite of their somewhat large size, there is nowhere a great thickness of tissue. Numerous delicate outgrowths called gills in Asteroidea and Echinoidea, genital bursae in Ophiuroidea, and respiratory trees in Holothuroidea

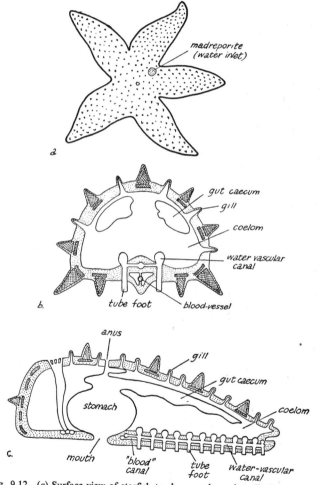

Fig. 9.12. (a) Surface view of starfish to show madreporite. (b) T.S. arm and (c) L.S. arm to show water and blood vascular systems.

expose the coelomic fluid to close contact with the sea-water. The numerous tube-feet probably also subserve a respiratory function. The water-vascular system with a ciliated inlet, circulates the sea-water around the body. Thus adequate respiratory exchange is effected. The diverticula of the gut penetrate deeply into each arm and diffusible food materials have relatively short distances to travel across the coelomic fluid to other tissues. There is no apparent excretory system; probably all the fluid spaces play a part in this function. It is known that in some genera, wandering amoeboid cells accumulate excretory granules and pass through the thin walls of the gills to disintegrate in the sea-water. It appears, on the whole, that none of the fluids has a definite circulatory function. This is correlated with two factors. In the first place, the members of this phylum are all sedentary or very slow-moving, and secondly, there is no great thickness of tissue (see Fig. 9.12).

Translocation in the Blood-vascular System of Vertebrates

All the vertebrates possess a blood-vascular system in which the blood is confined in the vessels. The system carries out translocation of food materials, gases, excretory products and hormones. Pressure is developed by the heart, which may be regarded as a central pumping

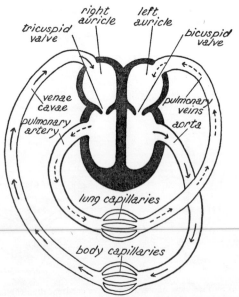

Fig. 9.13. Diagram of mammalian double circulation.

station. The heart pumps the blood into contractile and elastic arteries which lead to arterioles. These break up into networks of capillaries which ramify among the tissues. The return of blood to the heart takes place in relatively non-contractile venules and then veins, well supplied with valves to prevent backward flow. In addition, some of the fluid plus certain leucocytes escapes from the capillaries into the intercellular (blastocoelic) spaces; it is returned to the blood system by lymphatic vessels.

The Heart

The mammalian heart is four-chambered and two-sided, providing the double circulation, pulmonary and systemic. All exits and entrances are guarded by valves, which prevent reflux and ensure that circulation takes place in one constant direction. The thickness of the heart consists of three zones. From outside to inside, they are the epicardium, myocardium and endocardium. The epicardium consists of a squamous epithelium with a little connective tissue rich in elastic fibres. The myocardium constitutes the main substance of the heart; it consists of cardiac muscle fibres bound together by elastic connective tissue.

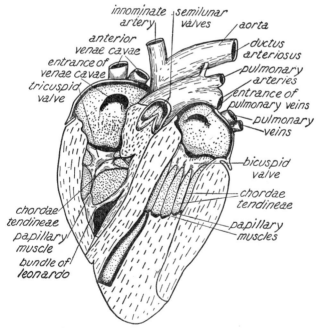

Fig. 9.14. Heart of sheep, ventral dissection.

The endocardium consists of a squamous endothelium beneath which is a small quantity of elastic connective tissue. The heart lies in the pericardial cavity surrounded by coelomic fluid. Contraction of the heart or of any chamber is called *systole*, and relaxation is called *diastole*. The two auricles contract together and then the two ventricles.

The auricles fill with blood until the rising pressure closes the valved entrances of the venae cavae and the pulmonary veins. When the auricles contract, the pressure forces open the tricuspid and bicuspid valves; as the ventricular pressure rises, the tricuspid and bicuspid valves close. Contraction of the ventricles opens the valves of the aorta and pulmonary artery; these vessels are dilated with blood until their pressure rises above that in the ventricles and then the valves close. The successive contractions of auricles and ventricles produce a negative pressure in the pericardium; this is transmitted through the thin walls of the auricles and blood is sucked in from the venae cavae and pulmonary veins. Following the successive contractions of auricles and ventricles there is a pause which is as long as both contractions together. In human beings the average figures are—

Contraction of auricles .	0·1 s
Contraction of ventricles .	0·3 s
Rest period . . .	0·4 s
Total (one beat) . .	0·8 s

Thus there is an endless rhythm of auricular contraction, ventricular contraction and rest. All valves close when the pressure on the outgoing side exceeds the pressure on the incoming side.

In fishes there is a single auricle and ventricle. The ventricular contraction must force the blood through the gill capillaries before it enters the general circulation. Owing to the great resistance set up by these narrow capillaries, the blood pressures in the ventral aorta and dorsal aorta differ considerably. In the dogfish, the pressure in the ventral aorta is about 43 mbar, while the pressure in the dorsal aorta is about 29 mbar. The teleost fishes, with considerably less blood, can maintain a higher blood pressure, but the fall through the capillaries again marked. In the salmon, the ventral aorta shows a pressure of about 100 mbar, while the dorsal aorta shows about 71 mbar.

The three-chambered heart of the amphibian has only partial separation of blood in the ventricle, but the blood returns twice to the heart for each cycle. Hence there is not the sudden fall in pressure found in the fishes. In any of the large arteries of the frog, a pressure

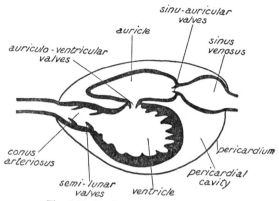

Fig. 9.15. Vertical L.S. of heart of dogfish.

of about 40 mbar is usually found. This pressure is insufficient for a very active terrestrial life, and hence most of the Amphibia are perforce rather torpid animals.

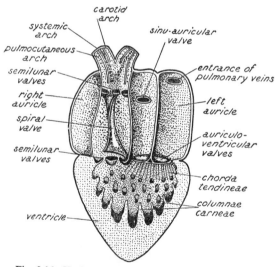

Fig. 9.16. Horizontal L.S. of frog's heart, ventral view.

The four-chambered heart with two distinct channels has enabled a high systemic pressure to be achieved after a low pressure in the pulmonary veins. The condition arises first in certain reptiles, and is perfected in the birds and mammals. In birds, the pressure in the aorta averages 200 mbar; in mammals it is high, but varies considerably. An adult man has a blood pressure in the aorta of about 160 mbar at systole and 107 mbar at diastole. The higher blood pressures and more rapid circulation have endowed the reptiles, birds and mammals with advantages which have made them the most successful of terrestrial animals.

Where blood pressures have been measured in invertebrates, they show considerably lower values. An average for arthropods is less than 13 mbar. Annelids show varying values from 5 mbar to 33 mbar depending largely on the state of activity just before the pressure is taken. Molluscs in general have very low blood pressures, but in the cephalopods pressures up to 80 mbar have been recorded for the octopus and squid. These molluscs have an extra pair of hearts, the branchial hearts; they impart additional pressure to the blood after its passage through the gills.

In all the vertebrates, the contraction is initiated in the heart itself and is independent of nervous control. Such hearts are said to be *myogenic*; they are characteristic of vertebrates and molluscs. They continue to beat when all nervous connexions with the rest of the body are severed, but the beat is inhibited by acetylcholine. Hearts of annelids, arthropods and the lower chordates are *neurogenic*; the beat is initiated in the heart muscle by impulses from nerve cells in or on the heart wall. Neurogenic hearts show an accelerated rate of beat when acetylcholine is administered.

The beat of the mammalian heart is initiated at a point in the right auricle just above the entrance of the superior vena cava; it is called the *sinu-auricular node*. From the node, stimulation spreads through the auricles and ventricles by the specialized *Purkinje tissue*. This is muscular tissue consisting of clear cells often with several nuclei. The cells are placed end to end and striations are apparent only in the peripheral areas. Purkinje tissue forms a network under the endocardium of auricles and ventricles and it is the only muscular tissue to pierce the auriculo-ventricular septum and thus maintain excitation connexion. There are two prominent masses of Purkinje tissue, one at the sinu-auricular node and the other at the *auriculo-ventricular* node.

Although the beat of the heart is initiated by the sinu-auricular node, the rate of beat is controlled by the nervous system. Efferent fibres of the vagus affect both nodes but have no direct action on the muscle of the heart. They have a depressor effect, slowing the rate of beat. The cardiac sympathetic fibres (*see* Fig. 6.6) have a direct action on the ventricular muscle and also stimulate the auriculo-ventricular node; they cause acceleration of the heart-beat. Sensory cells in the wall of the right auricle and in the wall of the aorta, transmit impulses via the vagus afferent fibres. Similar cells in the wall of the carotid sinus send impulses along the sinus nerve. All these afferent fibres are stimulated by increased pressure of blood; their axons terminate in the cardiac control centre. Thus the rate of heart beat is adjusted to suit the needs of the body under various conditions.

This rate is dependent upon many factors among which are the state of the body, whether active or at rest, healthy or diseased, the age and size of the animal, and the temperature both external and internal. The rate is faster for small animals than for larger ones in the same group, for the active state than the resting, for diseased than for healthy

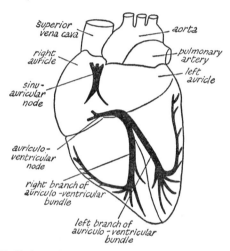

Fig. 9.17. Nodes and main myogenic paths of the mammalian heart.

animals, and for higher temperatures than lower. In the mouse, the maximum rate may reach 500 per minute, whereas in the elephant, it rarely reaches 40. In human beings, the rate fluctuates considerably between different individuals and under different conditions, the average for an adult being 72 beats per minute. In sedentary molluscs, it rarely exceeds 20, while in the octopus it may reach 80. The crayfish averages about 45, while the heart of the water-flea beats regularly at 150. In the domestic fowl there is an average rate of 160, while the tiny humming bird reaches 240 beats per minute.

The Blood-vessels

Arteries have strongly elastic walls composed of the following layers from inside to outside: a squamous endothelium, a narrow layer of connective tissue consisting mainly of elastic fibres, a thick muscular coat permeated by elastic fibres, and a thick layer of areolar tissue again with many elastic fibres (*see* Fig. 9.18). The amounts of muscular and connective tissue progressively diminish from the aorta to the small peripheral arteries. The blood is squeezed along by rhythmical peristaltic waves of contraction originating in the aorta, which received the wave from the left ventricle. The wave can be felt as the pulse where an artery runs near the surface of the body. The blood does not proceed in a series of jerks, because the elastic walls of the arteries exert a steady pressure on it, in between the wave periods. The walls

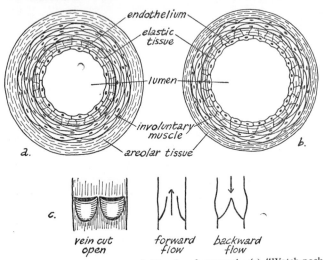

Fig. 9.18. T.S. of (*a*) artery and (*b*) vein of mammal. (*c*) "Watch-pocket" valves of vein.

also absorb shock which would otherwise be transmitted throughout the system and would inevitably burst the smaller vessels.

The final arteries are known as arterioles and are ultimately reduced to an endothelial lining, a single layer of unstriated muscle cells and a thin connective tissue sheath. Control of the amount of blood entering the capillaries, lies mainly in the arterioles. They are supplied with vasodilator and vasoconstrictor nerve fibres which originate in the vasomotor centre in the medulla oblongata. Vasodilator impulses travel through parasympathetic fibres, and vasoconstrictor impulses through sympathetic fibres. Local irritation may sometimes cause local response, but in general, the stimuli are provided by hormones. Adrenalin has a vasoconstrictor action and acetylcholine a vasodilator action; the hormones are liberated by the respective nerve-endings

The hormone adrenalin from the suprarenal glands is responsible for the stimulation of sensory receptors in the common carotid artery, in the aorta, and in many pain endings. These receptors send sensory impulses to the vasomotor centre and appropriate response is transmitted to the arterioles and capillaries by the motor nerve fibres.

Capillaries form a vast network connecting the arterioles with the venules. They are very small vessels ranging from 8 μm to 20 μm in diameter and consist of a single layer of squamous epithelial cells. Their small diameter imposes considerable resistance to the blood flow, and the blood pressure falls considerably. Capillaries are well supplied with nerve fibres, and their cells have some power of contraction. Through the thin capillary walls, diffusible food substances, oxygen in solution and considerable fluid pass out into the intercellular channels, thus directly bathing the tissues. Leucocytes, particularly lymphocytes migrate from the capillaries through the intercellular substance. The total extravascular fluid is called lymph. It is through the lymph that the actual purpose of the blood-vascular system is finally achieved; the fluid which transports everything to and from the cells is brought into close contact with them.

In general structure, venules and veins are similar to arterioles and arteries, but they have less elastic and muscular tissue. Also, along the veins are the so-called "watch-pocket" valves, which prevent reflux (see Fig. 9.18 (c)). It is assisted by the muscular movements of the body; contracting muscles squeeze local veins quite flat, and the valves prevent the blood flowing back. One of the main purposes of exercise is to assist this venous flow especially upwards against gravity. In sedentary occupations where there is little use of the limb muscles, varicose conditions often arise.

The systemic veins finally empty into the venae cavae, and the blood enters the right auricle; blood from the pulmonary veins enters the left auricle. The filling of the auricles is at least partly due to the negative pressure created in the pericardial fluid by the contraction of the heart. This negative pressure acts through the thin auricle walls and thus the pressure in the auricles falls below that in the great veins. The valves are opened and the auricles fill with blood. Thus the whole cycle begins again.

The Lymphatic Circulation

The lymph in the intercellular spaces is driven forward by the pressure of further lymph continuously exuding from the capillaries. These lymph channels merge into small capillaries with walls consisting of a single layer of tesselated epithelial cells (see Fig. 9.19). The lymph capillaries lead into lymphatic vessels with valves like those of veins.

But, unlike veins, the lymphatic vessels do not pass into larger vessels, but all remain of practically the same diameter. Along their course are placed small ovoid lymphatic glands (*see* Fig. 9.20).

Each gland is an open network of lymph spaces separated by con-

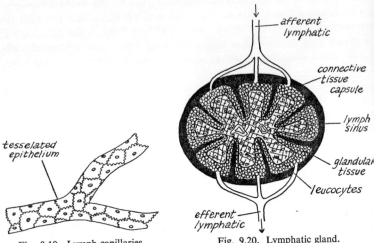

Fig. 9.19. Lymph capillaries.   Fig. 9.20. Lymphatic gland.

nective tissue ingrowths. The central glandular substance and th lymph canals are densely packed with lymphocytes undergoin division; these replenish the supply in the lymph and hence in th

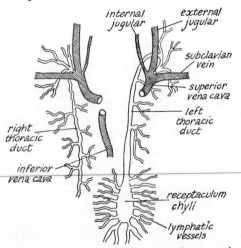

Fig. 9.21. Lymphatics of the thorax of a mammal.

blood. Bacteria in the lymph are retained and destroyed in these glands.

Most of the lymphatic vessels finally open into the thoracic duct which is dilated at its lower end into the receptaculum chyli which receives most of the lymphatics from the gut. The thoracic duct leads forward and finally empties the lymph into the venous system at the junction of jugular and subclavian veins on the left side of the neck. The remaining lymphatics, mainly from the right side of the trunk, head and neck and the right front-limb, lead by a common canal into the jugular-subclavian junction at the right side of the neck. Thus the lymph is returned to the blood.

This complex vascular system is responsible for translocation of all materials round the body. All the cells are nourished and supplied with oxygen, carbon dioxide is returned to the lungs, other excretory products transported to the skin and kidneys, and the vital hormones are rapidly and efficiently circulated, e.g. one circuit measured from neck of the horse, 90 s; from foot of man, 56 s. The fluid carries also the body's defensive mechanism in the various types of leucocyte and there is a mechanism for self-protection in the elaborate system for clotting at wounds. Finally, there are regulatory mechanisms which adjust the system in accordance with the needs of the animal.

CHAPTER 10

NUTRITION

NUTRITION may be defined as the sum of the processes whereby an organism provides itself, or is provided with, the materials necessary for energy release, for growth and repair, for its various secretions, for storage, and for maintenance of its internal osmotic and pH environment.

The materials necessary for these purposes are called nutrients, and there is considerable variation in the range required by various organisms. This can be correlated with the variety of different enzymes possessed by different creatures, for if a certain enzyme or enzyme system is lacking, then the ability to use a particular material for anabolism or catabolism will be lacking also. For example, animals in general cannot synthesize the essential amino-acids, but if supplied with them, they can then make their own special types of protein. In a similar manner, we find that animals cannot synthesize hexose sugars, fatty acids, or glycerol, but if supplied with these materials, they can then manufacture their polysaccharides and fats. In contrast, green plants can synthesize all their complex compounds from simple inorganic materials. Therein lies one essential difference between animals and green plants.

The nutrition of all other organisms depends, directly or indirectly, upon the synthetic powers of pigmented plants. Their ability to elaborate complex compounds with high potential energy, from simple inorganic materials with low potential energy, depends primarily on their possession of a photosynthetic pigment, usually green chlorophyll. The chemical nature of chlorophyll and its location in the plant have been discussed. Because of this essential difference between plants and animals, we tend to think of them as having different nutritive requirements, but they are basically the same. The distinction lies in the fact that the substances initially absorbed are different, and they are treated differently, but in general, the end-products are the same. The elaborate food substances for all living organisms are carbohydrates, proteins, fats and the numerous compounds derived from these, together with water and inorganic ions. Of these, plants and animals have very comparable requirements and tend to use them in the same ways.

THE NUTRITION OF GREEN PLANTS

Chief interest has been directed at the problems of nutrition arising from the cultivation of economically important crop plants. The

344

most of our knowledge today, is largely in terms of the higher plant rather than its more lowly relatives. Much of what follows is based therefore on research into agricultural and horticultural problems. It is interesting to note, however, that most of the information available as to the precise mechanism of photosynthesis, has been obtained from a study of the green alga, *Chlorella*.

Material Requirements: Their Sources and Roles

The raw materials required by green plants are water, carbon dioxide, and a variety of inorganic ions. The so-called *essential elements* or *macronutrients* are ten: nitrogen, phosphorus, sulphur, potassium, magnesium, calcium and iron, together with carbon, hydrogen and oxygen. In addition, many plants require certain other elements, in such small quantities that they are known as *trace elements* or *micronutrients*. They include boron, manganese, zinc, copper, molybdenum and others; a concentration in water of a few parts per million is usually sufficient.

The sources of these substances vary with the environment of the plant concerned. Terrestrial plants must be supplied either from the atmosphere or from the soil; all substances except carbon dioxide enter from the soil solution. Aquatic plants obtain all their raw materials in solution from the water. The following account indicates the states in which the various elements are absorbed, together with references to the principal roles they play in metabolism.

Carbon enters the plant in combination with oxygen as carbon dioxide molecules. It has been estimated that plants synthesize more than two hundred thousand million tonnes of organic matter per annum; all this enormous quantity depends on a supply of carbon dioxide since the chief element used is carbon. It is used either in structural material as carbohydrate, protein and fat, and complexes of these, or as respiratory substrates for the supply of energy.

Hydrogen enters the plant in combination with oxygen as water molecules. Like carbon, it forms part of the structural materials and respiratory substrates; it is associated with biological oxidation-reduction processes, and, as the hydrogen ion, influences the reaction of all cell fluids.

Oxygen enters the plant in two ways; in combination with carbon and hydrogen, as carbon dioxide and water molecules respectively, and as molecular oxygen from the atmosphere. It plays a large part in the structure of organic materials, and in the gaseous form, it is necessary for aerobic respiration.

Although the atmosphere is nearly 80 per cent gaseous *nitrogen*, very few organisms have the ability to utilize it in this form (*see* Nitrogen cycle, p. 350). The higher green plants normally obtain it from the soil solution as the NO_3^- ion, though the NO_2^- and NH_4^+ ions may also be used. It seems that ammonium compounds are nearly as effective as nitrates in the nitrogen metabolism of the higher plants but that nitrites may be toxic in acid solutions. There is also conclusive evidence that plants absorb from the soil soluble nitrogenous materials such as urea, amino-acids and others collectively called *auximones*. Some of these may be necessary for optimum growth, or they may be absorbed solely because they are present in the soil solution as a result of the decay of other organisms. In any case the fraction of nitrogen absorbed as organic material is very small. Many plants obtain nitrogen in organic form from mycorrhizal associations. Very many plants form these partnerships with fungi under natural conditions, and for some at least, the association is essential. Nitrogen is primarily necessary for the formation of amino-acids and hence of proteins, and for the synthesis of purines, pyrimidines and other nitrogenous bases. Nitrogen starvation of a plant is indicated by yellowness of its green parts, stunted growth, retarded flowering and fruiting and general weakness.

Phosphorus is absorbed from the soil solution as the $PO_4^=$ ion and much of it remains in that condition in the plant, where it forms an important buffer in the cell sap. The phosphate grouping is present in a great variety of plant substances; hexose and triose phosphates, nucleotides, nucleic acids and nucleoproteins all contain it. It is associated with the energy-rich bonds in the adenine phosphates; it enters into the structure of phospholipides; some co-enzymes contain it and hence it is important in metabolic cycles. In forms such as phosphoglyceric and phosphopyruvic acids, it is an intermediate of anabolic and catabolic processes. The $PO_4^=$ ions are absorbed in relatively large quantities by actively growing plants; they are removed from the sap rapidly when cell division is taking place, particularly for the formation of nucleotides. Deficiency becomes apparent as very poor root development.

Sulphur cannot be absorbed as the element but is taken up from the soil solution in $SO_4^=$ ions. Sulphur is contained in the amino-acids cystine, cysteine, and methionine, and hence it enters into the composition of some proteins. The enzymes urease and hexokinase require cysteine groups for activation. All the mustard-oil glycosides contain sulphur. The $SO_4^=$ ion is a normal constituent of cell sap. It may be connected with chlorophyll formation, since plants grown in conditions lacking sulphur are often pale green and do not photosynthesise efficiently.

Potassium is absorbed as the K^+ ion and most of it remains as such in the plant, forming the most abundant electropositive ion. It is especially concentrated in meristematic areas and in leaves. Ma functions have been attributed to the K^+ ions; they seem to play a catalytic part in protein synthesis, in chlorophyll formation and in carbon assimilation. There is the suggestion that they are concerned with the electro-osmotic transport mechanism in phloem tissue, *see* p. 300. Culture experiments show that green plants need a constant supply and cannot exist in its total absence. Deficiency results in weak, spindly plants, under-sized seeds and a typical red or purple leaf colouration.

Magnesium is absorbed as the Mg^{++} ion; it is a constituent of the chlorophyll molecule and thus is indispensable. Deficiency leads to chlorosis. It is a catalyst in glycolysis reactions and an activator in phosphatase and carboxylase systems.

Calcium is absorbed as the Ca^{++} ion. It enters into the construction of the plant as calcium pectate in the middle lamella of cell walls. Cells cannot enlarge in the total absence of calcium. It affects permeability of cell membranes and influences the transport of carbohydrate and proteins. It also affects the ability of colloids to absorb water. Deficiency leads to stunting and poor root growth. Crystals of calcium oxalate are of common occurrence in the tissues of many plants.

Iron is absorbed principally as Fe^{+++} ions, though also as Fe^{++}. It is essential for the porphyrin enzymes of respiration, where it acts as an oxidizing-reducing agent, alternately appearing in the ferrous and ferric forms (*see* p. 855). It is essential in chlorophyll formation; lack causes chlorosis. It is often regarded as a micronutrient.

Boron. It has been shown that many green plants absorb the $BO_3^=$ ion. It is necessary for the development of apical meristems, the proper translocation of sugars and is concerned with the uptake and use of calcium. Boron deficiency may show in various ways according to species, e.g. blockage of movement of starch from tomato leaves.

Manganese is probably absorbed only as the divalent ion, Mn^{++}. It is present in all green plants investigated, and is known to be associated with the respiratory enzyme systems. The lack of manganese is indicated by certain deficiency diseases such as grey speck of oats, and blight of sugar cane.

Zinc appears to be universally necessary for green plants and is known to be a constituent of the molecule of the enzyme carbonic anhydrase. Zinc is absorbed as the Zn^{++} ion; deficiency in fruit trees is indicated by mottled-leaf disease.

Copper is very toxic in high concentration, but is a necessary trace element for many plants. It is absorbed as the Cu^{++} ion and may be concerned in respiratory processes. Since copper deficiency can lead to chlorosis, it may be associated with chlorophyll formation.

Molybdenum is essential for normal growth in some plants when the only available nitrogen is in the nitrate form; plants using ammonium compounds do not need it. It is therefore assumed to be concerned with the reduction of nitrates. To nitrogen-fixers it is essential.

Other elements are necessary in trace concentrations. They include *sodium, chlorine, vanadium, cobalt* and *silicon*. Although they are present in many plants there is no evidence that they play essential parts. It is probable that further research will lead to establishing the necessity for other trace substances hitherto regarded as non-essential.

What has been said above concerning the raw material requirements of green plants refers to the plant as a whole. There is mounting evidence from tissue culture work that the requirements for individual tissues are not necessarily the same. For example, the phloem cells of carrot roots can be grown successfully in a solution lacking calcium.

The Intake of Raw Materials

The uptake of water has been fully discussed in the previous chapter. Movement of carbon dioxide into the cells of the leaf from the atmosphere must be considered as a process of diffusion along a gradient, first through the open stomata into the intercellular spaces. Then the gas dissolves in the water on the cell faces and diffuses in solution into the cell. The mechanism responsible for the uptake of inorganic ions by roots and rhizoids has presented much difficulty in its elucidation (*see* Chap. 3).

The Effects of Nutrient Deficiency: Water-culture Technique

It has long been known that plants will grow normally with their roots immersed in solutions of chemicals of the right kinds and in the right proportions. In 1865, Sachs and Knop produced recipes for culture solutions; since that time, they have been little improved. *Hydroponics,* the technique of growing plants in culture solutions, has been used commercially with limited success. The method has been of considerable value in experimental work to discover the effects of the various macro-nutrients, since control of substances and quantities is comparatively easy. It is much more difficult to study the effects of trace element deficiency. The general effects of nutrient deficiency with peas or buckwheat as the plant material can be demonstrated by using the following solutions, based on Sachs' formulae.

1. Normal solution: potassium nitrate 0·70 g; calcium sulphate 0·25 g; calcium phosphate 0·25 g; ferric chloride 0·005 g; magnesium sulphate 0·25 g; sodium chloride 0·08 g; distilled water one litre.

2. Lacking nitrogen: substitute potassium chloride 0·52 g for potassium nitrate.

3. Lacking phosphorus: substitute calcium nitrate 0·16 g for calcium phosphate.

4. Lacking sulphur: substitute calcium chloride 0·16 g for calcium sulphate, and magnesium chloride 0·21 g for magnesium sulphate.

5. Lacking potassium: substitute sodium nitrate 0·59 g for potassium nitrate.

6. Lacking magnesium: substitute potassium sulphate 0·17 g for magnesium sulphate.

7. Lacking calcium: substitute potassium sulphate 0·20 g for calcium sulphate, and sodium phosphate 0·71 g for calcium phosphate.

8. Lacking iron: omit ferric chloride.

9. Distilled water.

All the chemicals used should be as pure as possible. The solutions are placed in labelled jars of capacity about one litre, and covered with black paper to discourage algal growth. Seeds of the plant to be studied, are germinated in dishes of distilled water until the radicles emerge. They are then secured to the under-side of slotted wooden jar covers, as shown in Fig. 10.1, so that the radicles just reach the

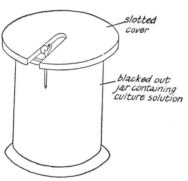

slotted cover

blacked out jar containing culture solution

Fig. 10.1. Water-culture jar.

solutions in the jars when the covers are in position. The jars are left in a well-lit position; the solutions are aerated daily and topped-up with distilled water as required. Trouble from bacterial and fungal infection can be avoided if the jar-covers are flamed prior to use and kept as dry as possible. Over the succeeding few weeks, the effects of lack of various nutrients become evident.

Some Cycles of Raw Materials in Nature

Although there is incessant absorption of these raw materials by green plants, nevertheless, under natural conditions, there is no diminution of the supply available. The distribution varies, but the total supply must remain constant, in accordance with the law of conservation of matter (*see* p. 995). Broadly, chemical substances are utilized by plants; animals utilize plants and some of the raw chemicals; both excrete unwanted materials; both die, and their remains are subjected to the decaying action of micro-organisms; thus, the raw materials are put back into circulation again. There is thus, for all these raw materials, a broad cycle of absorption, utilization, and ultimate return to the raw-material status again. Some specific cycles will now be indicated.

The Nitrogen Cycle

Fig. 10.2 represents the numerous natural changes in the form of nitrogenous matter, as living organisms are known to be associated with it. There are three main sequences of events, as denoted by the

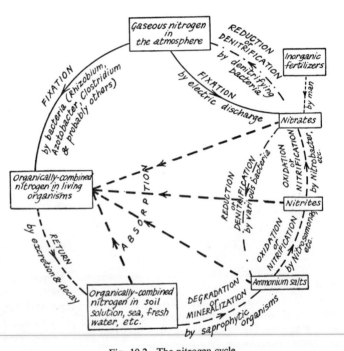

Fig. 10.2. The nitrogen cycle.

three different line forms. They are: fixation or oxidation of gaseous nitrogen into a form usable by higher organisms; the continuous absorption of nitrogenous material from the soil and from water, a its return; the reduction or denitrification of inorganic nitrate by micro-organisms. It must be remembered that these sequences are interlocked and interdependent, but for convenience, each will be treated separately.

Fixation of nitrogen is accomplished in two main ways: as the result of electrical discharges in the atmosphere, and by the metabolic activities of living organisms, chiefly bacteria, which can use gaseous nitrogen in the synthesis of their protoplasm. Certain ionizing phenomena in the atmosphere, such as meteor trails and cosmic radiation, can under some circumstances provide the energy for a reaction to take place between nitrogen and the hydrogen of water or the free atmospheric oxygen. The following sequence of events is believed to occur in electrical-discharge fixation. First, nitrogen and oxygen of the atmosphere combine to form oxides of nitrogen, chiefly nitric oxide. This is immediately oxidized to nitrogen peroxide.

$$\text{(i) } 2NO + O_2 \rightarrow 2NO_2$$

The nitrogen peroxide acts as a mixture of the anhydrides of nitrous and nitric acids.

$$\text{(ii) } 4NO_2 \rightarrow N_2O_3 + N_2O_5$$

With water droplets, nitrous and nitric acids are formed.

$$\text{(iii) } H_2O + N_2O_3 \rightarrow 2HNO_2$$
$$\text{(iv) } H_2O + N_2O_5 \rightarrow 2HNO_3$$

The nitric acid is washed into the soil where, with bases, it forms nitrates. The nitrous acid is unstable; it decomposes to form oxides of nitrogen once more.

$$\text{(v) } 2HNO_2 \rightarrow H_2O + N_2O_3 \rightarrow H_2O + NO + NO_2$$

These oxides of nitrogen recommence the process, until finally all are converted into nitric acid and hence, nitrates in the soil. The quantities of nitrogen fixed in this way vary with the climatic conditions. In Great Britain, an annual total of 4·5 kg per hectare has been recorded at Rothamsted. In other parts of the world, annual quantities ranging from 2 to 22 kg per hectare have been reported.

Organisms which fix nitrogen fall into two main classes. In the first group are symbiotic micro-organisms including bacteria, blue-green algae, and possibly fungi, and in the second are certain free-living bacteria and blue-green algae. By the experimental use of tracer nitrogen, it has been found that certain other organisms can perform fixation also; yeast is reputed to be one.

The symbiotic bacteria most commonly associated with nitrogen fixation belong to the species *Rhizobium leguminosarum* [*Bacillus radicicola*]. They are invariably associated with leguminous plants, and different strains of the bacteria are known to be in alliance with specific legumes, the strains being described according to the host species whose roots they infect. As a result of infection, the familiar root-nodules are produced. The biochemistry of the fixation process is not fully understood since the bacteria are incapable of fixing nitrogen outside the legume nodule. It is assumed that amino-acids are formed by the alliance, the legume supplying the carbohydrate and the bacterium supplying combined nitrogen. The bacteria become parasitic if they are not supplied with carbohydrate. The following outline to the fixation process has been suggested. Apparently, haemoglobin or a closely similar substance may be associated with the process; haemoglobin has been isolated from root-nodules. Through the agency of haemoglobin, nitrogen may be fixed in the form of hydroxylamine, NH_2OH; this could be converted to nitrate by oxidizing enzymes.

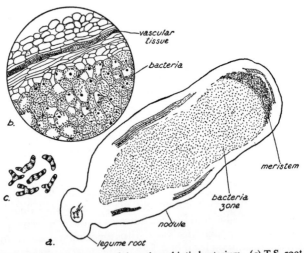

Fig. 10.3. Structure of root nodule and symbiotic bacterium. (*a*) T.S. root and nodule. (*b*) Portion of nodule (H.P. drawing). (*c*) Banded rod form of bacterium.

Thence, provided carbohydrate were available, amino-acids and proteins could be formed. Alternatively, the hydroxylamine could condense with oxalo-acetic acid, a breakdown product of sugars, eventually to form amino-acids such as aspartic acid and serine. Root-nodule bacteria are active fixers of nitrogen under good conditions; estimates state that a nodule can fix up to one and a half times its own nitrogen content per day. Estimates of 350 kg per hectare per year have been made for legumes. The trace element molybdenum is essential to the process. Upon the death of the legume and the bacteria, the nitrogen is returned to the soil in the organic form; it is then converted to nitrate for use by other plants. Other nodule-forming organisms when living symbiotically are known to be capable of nitrogen fixation. They include actinomycete fungi on the roots of *Alnus*, the alder, *Hippophae*, the sea buckthorn, and *Myrica*, the bog myrtle, certain bacteria on the roots of *Casuarina*, the she-oak and possibly phycomycete fungi on the roots of the gymnosperms, *Podocarpus* and *Agathis*. All these organisms cannot fix nitrogen when free-living, but only when in symbiotic union.

The more important free-living bacteria capable of fixing nitrogen in the soil belong to the genera *Azotobacter* and *Clostridium*, the latter usually being more numerous. Another genus, *Beijerinckia*, is now known to include forms similar to *Azotobacter*. When supplied with carbohydrates such as glucose, *Azotobacter chroococcum* can fix nitrogen aerobically or anaerobically. In nature it occurs in close association with the bacterium *Alcaligenes radiobacter*, which seems to supply some of its requirements. Ammonia, nitrites and nitrates are formed as the result of fixation. *Clostridium pasteurianum* can fix nitrogen only anaerobically; again, carbohydrates must be available. There seem to be no reliable estimates of the quantities of nitrogen which these organisms are able to fix but it could be less than one hundredth of that fixed by symbiotic organisms. So far there has not been worked out the full story of nitrogen fixation by these microbes. An "activating" enzyme, nitrogenase, is thought to be necessary to the process through which molecular nitrogen is split into its component atoms of free nitrogen each of which is then reduced to ammonia.

The adoption of tracer techniques in the study of nitrogen fixation has established that the blue-green algae *Anaboena* and *Nostoc* can utilize atmospheric nitrogen in their protein metabolism. Under good conditions for rapid growth it is likely that they can fix appreciable quantities. Both are photosynthetic soil organisms, and are not infrequently found inhabiting cavities in the higher plants. The *Anaboena* species in *Cycas* may have a symbiotic significance. Whether there is appreciable active nitrogen fixation by the blue-green algae so deprived of light is not known. There is no full understanding of how molecular nitrogen is transformed within these algae. It seems possible that the nitrogen is first reduced to ammonia which is then transferred in the form of an amino group to an existing molecule to form an amino-acid. Other photosynthetic micro-organisms are now known to fix nitrogen as a part of their normal activity. They include the genera *Chromatium*, *Rhodospirillum*, and *Chlorobium*; all are pigmented sulphur bacteria. The claims that certain fungi can fix nitrogen are not yet fully substantiated except in the case of a few yeasts.

The second sequence of events to be followed is that of the absorption from, and ultimate return to their surroundings, of nitrogenous substances by higher organisms. Our chief concern here, is with plant activity, since the effects of animals arise indirectly through plants. Large quantities of nitrogenous substances, chiefly nitrates, are removed by plants from the soil or from aqueous surroundings. These substances are ultimately incorporated into the plant as protein. Eventually, the plants (and animals) die, and are decomposed to yield a supply of organic nitrogen to the environment. In this form it is usable only by a limited number of micro-organisms and by higher plants having mycorrhizal associations. However, by the activities of various bacteria and fungi, it is mineralized or degraded into an inorganic form. The sequence of stages is: organic nitrogen → ammonia → nitrite → nitrate.

The first step is accomplished by a wide variety of saprophytic organisms which free ammonia as an excretory product of their own nitrogen metabolism. The last two steps constitute the process of nitrification, in which specialized bacteria are concerned. *Nitrosomonas* can oxidize ammonia to nitrite, according to the reaction—

$$2NH_3 + 3O_2 \rightarrow 2HNO_2 + 2H_2O$$

Nitrobacter further oxidizes nitrite to nitrate

$$2HNO_2 + O_2 \rightarrow 2HNO_3$$

In both these reactions there is a release of energy which becomes available to the bacterium for use in its synthetic activities. *Nitrosomonas* and *Nitrobacter* are thus examples of chemosynthetic organisms. Other ammonia oxidizers are species of *Nitrosococcus*, *Nitrospora* and *Nitrosogloea* and other oxidizers are species of *Nitrocystis*. Some actinomycetes are reputed to be nitrifiers to a small degree.

The third series of events which influences the amount of available nitrogen in the soil is the reduction of nitrate to ammonia and the dentrification of nitrate to gaseous nitrogen, oxides of nitrogen or nitrites, depending upon the type of organism concerned. Ammonia and other gases are easily lost from the soil into the atmosphere. Formation of gaseous nitrogen from nitrate can be achieved by the anaerobic *Thiobacillus denitrificans*, which uses the nitrate ion as a hydrogen acceptor for the oxidation of sulphur.

$$6KNO_3 + 5S + 2CaCO_3 \rightarrow 3K_2SO_4 + 2CaSO_4 + 2CO_2 + 3N_2$$

Amongst others, including species of *Pseudomonas*, *Clostridium* and *Micrococcus*, the bacterium *Escherichia coli* is able to produce nitrites, and later ammonia, using the nitrate ion for reduction while some other substance is oxidized.

$$H.COOH + HNO_3 \rightarrow CO_2 + H_2O + HNO_2$$
$$4H + HNO_2 \rightarrow NH_2OH + H_2O$$
$$2H + NH_2OH \rightarrow NH_3 + H_2O$$

In the last two stages of this sequence the hydrogen is derived from some hydrogen donator within the bacterium. It is probable that the formation of nitrites or ammonia in this way affects soil fertility only slightly, because when oxidizing conditions are again present, the nitrifying bacteria will convert them into nitrates once more.

There is one other way in nature in which nitrogen can be lost from the soil; that is by leaching through the subsoil or into the drainage water carried away by rivers. Such loss can occur in substantial amounts, particularly in areas with sparse vegetation.

Nowadays, man-made factors are entering the nitrogen cycle on an increasing scale. Enormous quantities of nitrogenous fertilizer, at least equal to the total amount fixed in nature, are manufactured by several chemical processes, of which the Haber process is best known. The advance of mechanization in agriculture, has rendered the use of animal power almost obsolete; thus the soil is deprived of a rich source of natural nitrogenous material in faeces and urine. The growth of towns has led to vast sewage systems which usually disperse their effluent into the sea. The large-scale transfer of food from crops has resulted in denuding the soil of fixed nitrogen in many parts of the world. In general, the remedy has always been to increase the production of artificial fertilizers. This solution alone is unsatisfactory; its results are apparent in the "dust-bowls" of the U.S.A. and other regions.

The Carbon Cycle

The total available CO_2 in the atmosphere and in water has been estimated at $5 \cdot 1 \times 10^{13}$ Mg and about one thirty-fifth of this every year is utilized by plants in carbon assimilation. The end-products of this anabolic process, whether used as respiratory substrates or incorporated into the substance of the body, are returned to the atmosphere or to water as a result of respiration or decay. The elaborated products of plants are used by animals as food, and again the CO_2 ultimately returns into circulation. In modern times, the extensive use of fuels, mainly of plant origin, liberates the carbon as CO_2.

The Water Cycle

Water is constantly evaporated from the sea, lakes, rivers and the soil. Water vapour, being lighter than air (0·62:1·00), rises, cools and

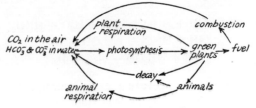

Fig. 10.4. The carbon cycle.

condenses. Coalescence produces droplets and the water descends again as rain, hail or snow. Water is absorbed by all plants and animals. In the case of the higher plants, by far the greater proportion

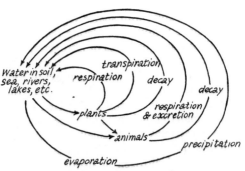

Fig. 10.5. The water cycle.

of the water passes into the atmosphere by transpiration. Indeed, the role of forests as rain-makers is not to be despised. Some water is used in photosynthesis, during which reaction, oxygen is released and the hydrogen retained for reduction of carbon dioxide. The ultimate products of photosynthesis are utilized by plants and animals either as respiratory substrates, when the hydrogen is released as water, or as structural materials, in which case also, ultimate decay produces water.

The Sulphur Cycle

Fig. 10.6 represents in outline the natural processes which affect the distribution and form of sulphur. It is absorbed by green plants principally as the $SO_4^=$ ion. They utilize the sulphur for the synthesis of certain amino-acids and hence proteins, of glycosides, and of certain enzymes. Animals obtain their necessary sulphur compounds by eating plants or other animals. Putrefaction of these sulphur compounds by various micro-organisms in the soil, in water, and in the gut of animals produces hydrogen sulphide and other sulphides. These are oxidized to sulphur, chiefly by pigmented sulphur bacteria such as

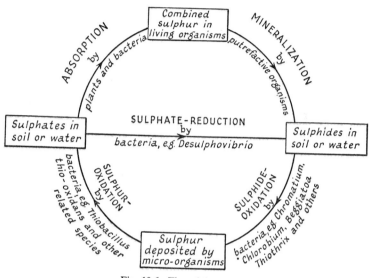

Fig. 10.6. The sulphur cycle.

Chromatium and *Chlorobium*. Sulphur is oxidized to sulphates by bacteria of the genus *Thiobacillus*. Reduction of sulphates to sulphides is carried out by *Desulphovibrio* species (*see* Chap. 36, Vol. I).

Similar cycles can be constructed to show the circulation of other raw materials in nature. Essentially, any such cycle must account for the intake, assimilation into body constituents, and return to the environment.

Carbon Assimilation by Photosynthesis and Factors Affecting It

Carbon assimilation in green plants may be defined as the whole process of synthesis of complex carbon compounds beginning with carbon dioxide and water. Although carbon dioxide may be fixed in organic compounds in darkness by enzymically controlled carboxylation reactions in both plants and animals, in photosynthesis carbon assimilation involves a preliminary photochemical reaction without which the process does not occur. Neither will photosynthesis occur unless certain substances are present and certain factors operative to an appropriate degree. A factor is said to be *limiting* when an increase in its quantity will cause a direct increase in the rate of the process. Some factors influencing the reactions may slow up the rate, not because they are deficient, but because they are present in excess. Such are termed *inhibiting factors*; they may not always be essential to the process at all. We may speak of an *optimum condition* with reference to any factor, when it is present in such a degree that it allows the process to proceed at a maximum rate. In deciding optimum conditions care must be taken to ensure that no other factor but the one under investigation is limiting or inhibiting the process, and the time of application of the condition is having no effect. For example, in photosynthesis, a temperature of 45°C appears optimum for a short time, but if maintained, it slows the process down. Over a much longer period, it is found that temperatures nearer 30°C are optimum. It is to be noted that these considerations are applicable to any physiological process where a number of factors are operative.

Carbon Dioxide Availability

Carbon dioxide is an essential reactant. When no other factors are limiting or inhibiting the process, the rate of photosynthesis is increased with increasing carbon dioxide concentration. The increase is rapid at first, slowing to a steady level at each constant light intensity, and at each temperature level. Under natural conditions of bright sunlight, when the light is not a limiting factor, in a temperature range of 20°C to 26°C and with a good water supply, the small amount of carbon dioxide in the atmosphere is the limiting factor. An increase of this CO_2 content from 0·03 per cent up to as much as 1·0 per cent increases the rate of the process. An increase much above this value does not further increase the photosynthetic rate. This could be due to the effect of the higher concentration of CO_2 on the stomata, tending to close them.

Water Availability

Water is another basic raw material in photosynthesis but it is not possible to show any direct relationship between the rate of supply of water and the rate of photosynthesis. This is due to the fact that water plays so many roles in the plant. Deficiency of water undoubtedly slows the process down, and this may be for several reasons. Flaccid cells do not function normally in any case, and wilting of leaves may induce stomatal closure, so that the CO_2 supply is cut off. Also, when cells are drying out, the rate of penetration of CO_2 through their membranes is reduced. With lack of water, translocation of the synthesized products is severely hampered. On the whole, it does not seem likely that lack of water as a raw material would limit the process until long after the cells are dead.

Mineral Salt Availability

Mineral salts are not essential raw materials for the photosynthetic process, but deficiency of some undoubtedly affects its rate. Prolonged shortage of iron or magnesium will cause chlorosis, and thus the chlorophyll content will be lowered and photosynthesis will slow down or even stop altogether. There is evidence that potassium has some influence; plants deficient in this element show low photosynthetic activity. It is possible that potassium and some of the trace elements may be concerned catalytically in chlorophyll formation, or in one or more of the syntheses subsequent to the initial carbon dioxide fixation.

Light Conditions

There are two light conditions to consider: light intensity (see p. 1013) and light quality or wavelength (see p. 1008). When no other factor is limiting, the rate of photosynthesis under natural conditions is directly proportional to the light intensity. Unless leaves are heavily shaded, on cloudless days, carbon dioxide is likely to become limiting, more so than light intensity. On days of intermittent sunshine, a fall much below full sunlight does tend to limit the process. Too high a light intensity will also slow the rate of photosynthesis; the phenomenon is known as *solarization* and is probably due to destruction of the chlorophyll and inactivation of some other internal substances by photo-oxidation or a photo-inhibitory action. Habitual shade plants are particularly affected by this. It is not likely that there is any definite lower limit of light intensity at which photosynthesis cannot occur, but its measurement at such levels is difficult. The light intensity at which the intake of carbon dioxide for photosynthesis exactly balances its output from respiration, is called the *compensation point*.

Green plants can carry on photosynthesis only when the radiation applied is within the limits of the visible spectrum; they cannot utilize infra-red or ultra-violet rays. The most effective wavelengths are in the red-orange, 600 nm to 700 nm, but the blue-violet end of the spectrum, 400 nm to 500 nm, is also utilized to a considerable extent. The central band is also effective, but less so than the others, despite the fact that spectroscopic examination of leaves or chlorophyll extracts shows very little absorption in this region.

Temperature Conditions

Experiments lead to the conclusion that when no other factor is limiting, temperature affects the rate of photosynthesis by doubling it for every 10°C rise between the range just below 0°C to 35°C (van't Hoff's rule) (*see* p. 983). Below this range, even down to − 6°C, the process continues extremely slowly; above it, up to about 45°C, the rate falls rapidly and death ensues. When the temperature is above 25°C, the time factor becomes operative; although the rate may be doubled by raising the temperature from 25°C to 35°C, the higher rate is not maintained indefinitely, but falls off after some minutes. This effect of temperature is what would be expected in chemical reactions controlled by enzyme catalysts. Under natural conditions, it is certain that temperature is often a limiting factor, particularly during the colder months of the year.

Chlorophyll Concentration

In a leaf, the effect of the quantity of chlorophyll on photosynthetic rate, is difficult to assess. Other than by causing a deficiency of iron or magnesium it is not possible to control the quantity. Comparison of the total amount of chlorophyll in all-green leaves and in variegated leaves of one species, shows that the rate of photosynthesis is proportional to the amount of chlorophyll. In *Chlorella* plants starved of iron, a direct relationship between rate of photosynthesis and quantity of chlorophyll has been shown, photosynthesis increasing proportionately with chlorophyll. When the plants were starved of magnesium, the same proportionality did not exist, though increased chlorophyll tended to increase photosynthesis. This case indicates that magnesium affects the rate of photosynthesis in other ways besides chlorophyll formation.

Other Internal Factors

The influence of enzymes on the rate of photosynthesis is deduced in the first instance from the effects of temperature. High temperature first slows and then halts the process, in the same way as it affects enzyme activity. The presence in suitable states of a wide variety of enzymes is now known to be necessary for the synthesis of carbohydrates to occur. This knowledge has been gained from research on carbohydrate synthesis outside the plant. Cell extracts with adenosine triphosphate (ATP) as a source of energy have proved able to manufacture sugars. However, the living green cell still has the sole prerogative of harnessing light energy for the same ends, and there are thus still unknown factors operating in the natural process.

Demonstration and Measurement of Photosynthesis and the Factors Affecting It

There are two main practical methods of studying photosynthesis. One is to demonstrate or measure the carbon dioxide/oxygen exchange, and the other is to show the building up of organic substance by the tissues. Allowance must always be made for the respiratory activity and possible translocation of material out of the tissues under investigation.

The apparatus shown in Fig. 10.7 can be used to measure carbon dioxide uptake over a given time by a leafy twig.

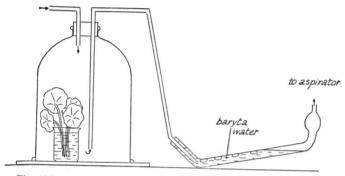

Fig. 10.7. Apparatus to measure carbon dioxide uptake by a leafy twig.

Equal volumes of air are drawn simultaneously over the green leaves exposed to light and then through baryta water, and, as a control, through the same volume of baryta water. At the end of the given period both solutions of baryta water are titrated against standard hydrochloric acid. The difference between the two titrations will give the apparent carbon dioxide uptake by the leaves. The real uptake can only be calculated by adding to this the carbon dioxide evolved by the leaves as a result of respiration in the same time. The whole procedure must be repeated with the leaves in total darkness to find this quantity of carbon dioxide.

Results can be expressed as cubic centimetres of carbon dioxide absorbed in unit time (one hour) per unit area of leaf surface (or fresh or dry weight of leaves). The temperature and light intensity at which the experiment was conducted should be recorded.

Output of oxygen by aquatic leaves can be demonstrated by using the apparatus shown in Fig. 10.8. If the pondweed is well illuminated and a pinch of potassium bicarbonate is added to the water, a test-tube full of gas can be collected in a few days. The gas is not pure oxygen but contains enough to rekindle a glowing splint.

This propensity of cut aquatics to bubble freely may be used to measure the relative rate of photosynthesis under variable external conditions. Under the best conditions, bubbles will appear very readily at the cut surfaces of Canadian pondweed stems and their rate of emission can be counted per unit time. Small bubblers can be fitted over the ends of cut stems to ensure constancy of size of the bubbles, although this is often unnecessary. Variations in intensity of illumination can be achieved by altering the distance between the illumination source and the aquatic shoot. Variations in temperature can be achieved by addition of ice in small quantities to the water from time to time. A suitable arrangement would be as in Fig. 10.9. Relative rates of photosynthesis can be expressed in terms of bubbles per unit time and comparisons of these figures under different conditions clearly indicate the effects of external factors.

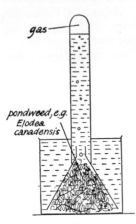

Fig. 10.8. Collection of gas liberated by aquatic plants.

Demonstration of the appearance of starch in a leaf which was previously considered starch-free, is a clear indication of carbon assimilation. A leaf is tested for starch as follows. It is first killed by immersion for a few moments in boiling water. It is then decolorized by repeated immersion in boiling alcohol on a water bath. Washing in cold water then renders the leaf flexible, after its hardening in the

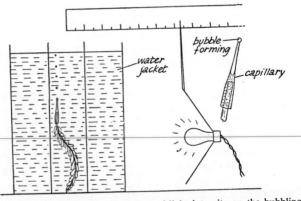

Fig. 10.9. Apparatus to show the effect of light intensity on the bubbling of aquatics.

alcohol. Immersion in an iodine solution indicates the presence or absence of starch by the appearance, or otherwise, of a blue-black colour. A final wash in benzene removes the brown stain of the iodine but leaves the blue-black unchanged. Leaves may be rendered starch-free on the plant by keeping them in total darkness for 36–48 hours.

Using the presence of starch as the criterion by which to judge if photosynthesis has occurred, many experiments can be devised to investigate the factors affecting photosynthesis.

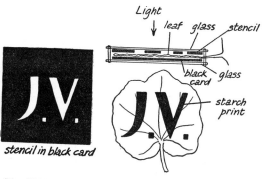

Fig. 10.10. Apparatus to show that light is necessary for photosynthesis.

Fig. 10.11. Apparatus to show that carbon dioxide is necessary for photosynthesis.

To show that light is necessary, use may be made of suitably cut stencils placed over starch-free leaves (*see* Fig. 10.10).

To show that carbon dioxide is necessary, the apparatus in Fig. 10.11 may be employed.

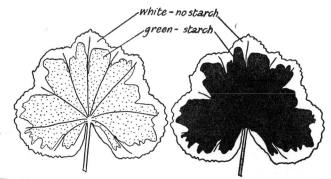

Fig. 10.12. Use of variegated leaf to show that chlorophyll is necessary for photosynthesis.

The use of variegated leaves will enable the experimenter to prove the necessity for chlorophyll (*see* Fig. 10.12).

Perfectly accurate measurements of photosynthetic rate by increase in organic content are difficult to devise. Since, however, over 90 per cent of the dry weight increase in leaves is considered to be due to carbon assimilation, dry weight measurement procedures may be sufficiently accurate for class purposes. The commonest method is that used by Sachs and is known as the "half-leaf" method.

A leaf is selected with corresponding venation on both sides of the midrib. At commencement of illumination a definite area of leaf is cut from the lamina on one side of the midrib. Its dry weight is accurately measured. The remainder of the leaf is illuminated for a given period (some hours) and at the end of that time, an exactly equal area is removed from a corresponding position on the leaf and its dry weight determined. The difference in these dry weights gives the apparent carbon assimilation but does not allow for respiratory activity nor translocation out of the leaf. These can be accounted for by repeating the whole procedure on another similar leaf kept in darkness during the period of the experiment. The gain in dry weight experienced in the first leaf added to the loss in dry weight by the second, gives the total photosynthetic activity which can be expressed in grams of organic material assimilated per hour per unit area of leaf surface (or per unit fresh or dry weight of leaf). Temperature and illumination conditions should be recorded. The experiment is open to many criticisms and a batch of results is required from which to calculate an average result.

To demonstrate the efficiency of various wavelengths of light in promoting photosynthesis, comparisons of rates, measured by any means, can be made, using plants illuminated through coloured filters. Care must be taken to ensure that only wavelength is affected and that intensities are kept constant, otherwise the comparisons have no significance.

The Mechanism of Carbohydrate Synthesis

The equation—

$$6CO_2 + 6H_2O \xrightarrow[\text{chlorophyll}]{\text{light}} C_6H_{12}O_6 + 6O_2$$

has been used for many years to represent the overall photosynthetic reaction. Here it is intended to discuss in more detail the presumed roles of light and chlorophyll in the process and then to outline the possible chemical pathway between the starting and end-points of the equation.

The Chlorophyll Complex

The table below shows the distribution of various pigments known to be associated with photosynthesis in different plant groups. Their chemical nature has been outlined in Chap. 4.

Plant Group	Chlorophylls					Carotenes			Xantho-phylls	Phycobilins	
	a	*b*	*c*	*d*	*e*	α-	β-	ε-		Phyco-erythrin	Phyco-cyanin
Higher green plants	+	+				+	+		+		
Green algae	+	+				+	+		+		
Brown algae	+		+				+		+		
Red algae	+			+		+	+		+		
Blue-green algae	+						+	+	+	+	+
Yellow-green algae	+				+	+			+	+	+
Diatoms	+		+			+	+		+		

The green photoautotrophic bacteria contain chlorophylls somewhat similar to those of higher green plants plus some yellow carotenoids;

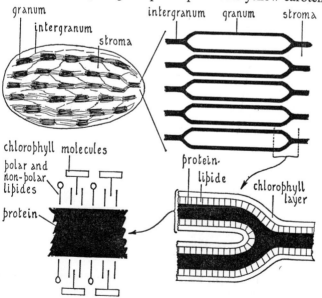

Fig. 10.13. The structure of the higher plant chloroplast.

the purple bacteria contain a different form of chlorophyll known as bacteriochlorophyll together with yellow and red carotenoids.

The pigments are confined to the cytoplasmic plastids, the chloroplasts, except in the blue-green algae and bacteria where they are dispersed

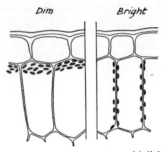

Fig. 10.14. Movement of chloroplasts with light intensity.

evenly through the cytoplasm. The structure of the chloroplast of a higher green plant, as demonstrated by the electron microscope, is shown on p. 5 and Plate 7 and represented diagrammatically in Fig. 10.13 omitting some detail. The pigments constitute about 8 per cent dry

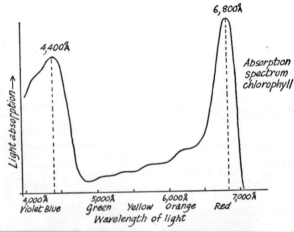

Fig. 10.15. Absorption spectrum of chlorophyll.

weight of the chloroplast, the remainder being proteins, enzymes concerned with photosynthesis and lipides. The chlorophylls are the principal light energy capturing agents in photosynthesis but the carotenoid and sometimes other substances play some part. This they do by transferring what light energy they absorb to the chlorophyll molecules.

In many higher plants the chloroplasts are not fixed positionally within the cells and can be seen to move with the streaming cytoplasm. In some leaves they may take up positions according to the variation in incident light intensity. Fig. 10.14 indicates how chloroplasts in the same cells may be disposed in dim and bright conditions.

Light absorption by chlorophyll can be demonstrated by passing white light through it and examining the spectrum of the emergent light. The experiment may be performed with chlorophyll extracted by acetone, ether or alcohol, though it must be remembered that the chlorophyll may have undergone change as a result of the action of the solvent. Such a transmitted spectrum shows two regions of strong absorption; one is in the red band centred on a wavelength of about 680 nm, and the other in the blue-violet around 440 nm. Few of the intervening wave-bands are absorbed (*see* Fig. 10.15), thus explaining the green colour of chlorophyll. Chlorophylls *a* and *b* show slightly different absorption maxima. Such absorption phenomena can be used to demonstrate the role of chlorophyll as the principal light-absorbing substance.

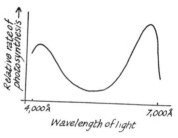

Fig. 10.16. Action spectrum of a green leaf.

If equal amounts of light energy of different wavelengths are allowed to fall on a green leaf, and if the rate of photosynthesis resulting from the light of each colour is measured, the photosynthetic activity of the leaf with each colour may be plotted against wavelength. The resulting curve is called an *action spectrum* (*see* Fig. 10.16). The curve shows that maximum photosynthesis coincides with maximum absorption, which occurs in the red and blue wave-bands, and demonstrates clearly that it is the chlorophyll which absorbs the light energy used in photosynthesis.

The variety of pigments is significant when we consider the habitats of the plants concerned. For terrestrial plants, the red and blue wavelengths are obviously sufficient for photosynthesis. But when the light passes through water of any considerable depth, most of the waves are absorbed, and the wavelengths left are mainly blue and green. Thus it is a necessity for photosynthetic plants living in deeper water to possess pigments which can utilize blue and green rays more efficiently than chlorophylls *a* and *b* can. The brown algae can make good use of green light, and the more deeply-situated red algae can utilize blue. The green algae which possess the same pigments as the terrestrial plants, can live only in comparatively shallow water.

Pigment production is dependent on light, except for the purple bacteria. The fact that green plants lose their chlorophyll in continued darkness, is a matter of common observation. The quantity as well as the quality of the light plays a part in the production of pigments. Thus the total quantity of pigments in aquatic plants tends to increase with the depth of the water. This seems to be true of the land also, since there are shade-loving and sun-loving plants. Also the relative proportion of the various pigments is correlated with the light-penetration in any particular habitat. In the red algae, the deeper the habitat, the greater is the proportion of phycoerythrin. On land, increasing shade increases the proportion of chlorophyll b, which can make greater use of the blue rays which predominate in shady localities.

Light and Chlorophyll in the Mechanism of Photosynthesis

The roles of light and chlorophyll in carbon assimilation have been matters of interest and speculation since Priestley, in 1772, discovered that green plants give out oxygen in daylight, and Ingenhausz, in 1779, proved that light was essential to the process. For a long time photosynthesis was visualized as a mechanism peculiar to green plants, in which light energy was used to split carbon dioxide. The carbon could then be assimilated, in combination with water, as a basic compound, (CH_2O), from which sugars could be formed. The oxygen was given off. The hypothesis assumed that carbon assimilation was the sole prerogative of green plants and could occur by none other than the process of photosynthesis.

Late in the last century, Winogradsky demonstrated that certain bacteria now recognized as the chemosynthesizing forms such as *Nitrosomonas* and *Nitrobacter*, could assimilate carbon dioxide in the dark, without chlorophyll. Engelmann showed that the purple sulphur bacteria could photosynthesize without emitting oxygen. These discoveries, coupled with the consistent failure to isolate a basic (CH_2O) substance, such as formaldehyde, from plants, threw doubt on the validity of the earlier conceptions, but nothing new could be formulated to take their place.

It was not until earlier in this century that the argument was pressed that since some bacteria can photosynthesize without the evolution of oxygen then the assumption that carbon dioxide was split by light energy could not be correct. A new theory was put forward in which it was visualized that the light energy was used to split or photolyse water. It was suggested that the resulting hydrogen was used to reduce carbon dioxide and that the hydroxyl ions, OH^-, produced water and oxygen, which was given off. This was not easily acceptable since the low energy radiation of visible light had never been made to perform such action outside the plant. However, it was to some extent substantiated by the demonstration that all the oxygen evolved during photosynthesis did come from water. This was achieved using the stable heavy isotope of oxygen, ^{18}O (*see* p. 947).

Since that time much more thought has been given to the inference to be drawn from Winogradsky's work, namely, that the reaction $CO_2 + H_2O \rightarrow$ Carbohydrates, is not confined only to cells containing chlorophyll pigments, and that given the energy with which to perform the synthesis, any cell might do it. This has been proved to be the case and all cells have the potentiality of assimilating carbon dioxide to produce carbohydrates, given the right conditions. These are, first, the necessary energy and second, the appropriate enzyme systems. When it is realized that carbohydrate synthesis can be the reversal of the respiratory activity which goes on inside all cells, this should not be difficult to comprehend. All the enzymes necessary are present, only the energy is required.

The essence of the respiratory process (*see* pp. 433 to 442), is that the energy of compounds like starch or sugar is transferred to substances which are the fundamental power units of protoplasm, namely adenosine triphosphate (ATP) and the nicotinamide adenine dinucleotides (NAD and NADP). These are then employed to give their energy and reducing power respectively to the performance of any metabolic process including that of carbon assimilation. The formation of ATP from ADP in respiration is carried out at the cell mitochondria and involves the building of an energy-rich bond when the third phosphate is added (*see* p. 111). The formation of such energy-rich phosphate bonds during respiration is known as *oxidative phosphorylation*. NAD and NADP will readily accept hydrogen (i.e. an electron + a proton) and as $NADH_2$ or $NADPH_2$ become powerful reducing agents (*see* pp. 429 to 431). To produce the ATP and $NADPH_2$ in respiration, carbohydrates are oxidized by removal of hydrogen, ultimately to oxygen, with the formation of carbon dioxide and water. By the reverse process carbon dioxide can be reduced to carbohydrate, provided that the necessary energy and reducing power are available to carry out the synthesis.

A new approach to the mechanism of photosynthesis came from the reasoning that green cells might perform the reverse of respiration, i.e. synthesis, using light to supply the energy and reducing power in the form of ATP and $NADH_2$ or $NADPH_2$. That is, that the function of light in the process is solely the production of these power units and that once they are present all the synthesizing processes can proceed in the dark.

Working on this supposition, Arnon and others have clearly demonstrated that this is indeed the case. There are *photophosphorylating* mechanisms as well as oxidative. Only those cells with the correct pigments can photo-phosphorylate, that is, convert ADP to ATP using light energy. From a number of investigations involving bacteria and isolated chloroplasts from higher plants, it has been shown (1) that isolated chloroplasts can make ATP and $NADPH_2$ in the light and can carry out the complete synthesis of carbon dioxide to carbohydrate: (2) that in the absence of CO_2, but given ADP and NADP, oxygen is evolved and ATP and $NADPH_2$ accumulate, but no carbohydrate is formed: (3) that the enzymes for carbon assimilation when extracted from chloroplasts free from the pigments, can produce carbohydrates from carbon dioxide if the ATP and $NADPH_2$, previously made by other chloroplasts, are added. ATP and $NADPH_2$ from animal sources are also effective in energizing this reaction.

The role of light and chlorophyll in the photosynthesizing cell must therefore be in the production of the power units. A recent interpretation of how this is achieved, so far as it has been elucidated, is as follows.

The process is restricted to those parts of a cell in which the chlorophylls and some accessory pigments occur. In most plants this is within the chloroplasts, the nature of which have been treated on pp. 365–368. In a few cases, where chlorophyll is not confined within membranes, it must occur more or less diffusely through the cytoplasm. In brief, it is a process in which the generation of ATP from ADP and inorganic phosphate and the conversion of NADP to $NADPH_2$ is accomplished, using the energy of visible light to effect the necessary energy-consuming or endergonic reactions. In other words, the overall process is one of electron transport with two main end results, one, a photophosphorylation or the transformation of absorbed light energy by chlorophyll into chemical bond energy of ATP and the other, the generation of the powerful reducing agent $NADPH_2$.

Once these substances are produced they are released to act as intracellular energy source and reducing power respectively, within an enzyme system capable of catalysing the reactions in which carbon dioxide is assimilated into the existing molecular structure. In higher plants, this commonly results in the accumulation of carbohydrate, usually starch in the form of amyloplasts within the chloroplast membranes, *see* plates 5 and 14. When it can be demonstrated that starch is present where none occurred previously, using iodine as the indicator, this is frequently accepted as proof that photosynthesis has taken place, *see* p. 362.

More recently a great deal more has been found out about the light energy absorption processes that set in motion the whole train of events and the ways in which electrons and protons are redistributed during the necessary energy transformations and chemical changes.

Light energy absorption is known to be a property of the chlorophylls, the carotenoids and the phycobilins, but of these only forms of chlorophyll *a* play a direct part in the final energy conversion or photophosphorylating process. The others are excited by exposure to light but must pass their excitation energy to chlorophyll *a* if it is to be used to supply the power for photosynthetic metabolism. They act, as it were, as extra energy collectors. A pigment molecule, when receiving light photons of appropriate wavelength becomes "excited" and in this excitation state an electron somewhere in the molecule is translated from a lower energy level to a higher one, *see* p. 941. How the excitation energy is actually transferred from another pigment molecule to chlorophyll *a* is not certain but a process referred to as "inductive resonance" is cited as a possible mechanism. This is possible only in the direction in which energy passes to chlorophyll *a* because this pigment cannot absorb sufficient energy to excite any of the others by passing it in the other direction. A particular long-wavelength-absorbing form of chlorophyll *a* can be visualized as acting in the role of "energy sink" and collecting excitation energy from all other light absorbing pigment molecules as well as more directly. It is perhaps convenient, therefore, to think of the energy-trapping pigments as forming "photosynthetic units" in which a number of molecules are concerned with light absorbtion, the total energy of which is then directed to a particular chlorophyll a structure which acts as a "reaction centre". If this chlorophyll a cannot transmit its excitation energy in the form of an ejected electron into the photophosphorylating system very quickly (within 10^{-9} s) then the electron drops back into the lower energy level from which it was lifted and the chlorophyll fluoresces, something commonly seen in test-tube solutions of the pigment, see p. 1015. It is significant perhaps that although the other pigments have characteristic fluorescence properties when isolated from chloroplasts and in vitro, only chlorophyll a fluoresces naturally within living cells.

The initial process of "collecting" energy in the form of an excited chlorophyll molecule at a reaction centre having been completed, this energy can now be usefully employed in doing the work of the photophosphorylating processes, provided it can be so channelled within 10^{-9} s of the excitation, otherwise it will be wasted as a fluorescent light emission. The first useful work that requires to be done is the splitting or lysis of water into protons (H^+), electrons (e^-) and free oxygen (O_2). How exactly this is accomplished is as yet unknown but clearly it is an oxidation reaction requiring the activity of a powerful oxidizing agent. It is also a requirement that a reductant shall accept the reducing by-products of the split water molecule, namely electrons and protons, so that these can eventually be used to reduce carbon dioxide through the agency of $NADPH_2$. Obviously, to produce molecular oxygen freely into the atmosphere as is certainly known to be the case, two molecules of water must be involved at each chemical step, thereby making available four electrons and four protons for the eventual reducing reactions.

One possible way of effecting the initial photochemical water-splitting steps is through the agency of electron donor and electron acceptor molecules acting in conjunction with the energy sink molecules in the reaction centre. Each of these latter would need to be closely tied to an acceptor and a donor so that when the long-wave chlorophyll a, having been raised to an excited state by receiving energy from other pigment molecules, releases its excitation energy and returns to its ground state, it takes an electron from the donor molecule, thereby oxidizing it and passes this electron to the acceptor molecule, thereby reducing it. The end result would be the required oxidizing agent that can be employed to split water and the reducing agent that can be used to produce $NADPH_2$. Whether the above very high speed photochemical step really happens is not verified but similar events can occur in systems of organic molecules in vitro. It has to be assumed that something equivalent occurs in chloroplasts. What happens after this, the biochemical changes in which electrons are transferred, ultimately resulting in the formation of $NADPH_2$ and ATP, are much slower and are more readily explained. The explanation, however, is based on the discovery that there are two light reactions concerned in the photosynthetic process, not one as was generally believed. The experimental evidence for this is based on two foundations.

The first of these is that the quantum efficiency of the process, that is, the useful work done per unit of light energy supplied to the system varies according to the wavelength of the light absorbed by the pigments. A curve of quantum efficiency or rate at which the process proceeds against wavelength of light employed shows that although the chlorophyll is absorbing light in the far-red (680 nm and above) there is no photosynthetic work done. The second discovery was that if chloroplasts, irradiated with light at above 680 nm wavelength, were also given light of a shorter wavelength, say 600 nm, at the same time, the photosynthetic output was greater with the two wavelengths together than the sum of the two outputs separately. There is clearly an "enhancement" effect that can only be explained on the basis of two light reactions occurring, one being boosted by the occurrence of the other. That is, both can occur at the shorter wavelength but only one of them occurs at the longer wavelength. Hill and Bendall of Cambridge proposed a scheme of electron transport that would take these findings into account.

To understand their explanation it is necessary first to appreciate the requirements of electron transport systems and this can be achieved with reference to what are known as oxidation-reduction (redox) potentials as referred to on p. 984. Substances with different redox potentials through a range of negative and positive values can be arranged in a sequence to form a "downhill"/"uphill" gradient so that electrons can flow readily downhill from a more negative donor to a less negative (more positive) acceptor without the input of energy, in fact with energy yielded by the system as is the case in oxidative phosphorylation in mitochondria. To flow in the reverse direction, that is, up or against the electrochemical slope, requires that energy shall be supplied and the greater the difference between redox potentials of the donor and acceptor the more energy is required to effect the electron transfer. It is the case that in photosynthesis, electron movement must in the overall be against the gradient since the redox potential of water, the initial electron donor, is $+0.8$ volt whereas that of NADP which acts as the final electron acceptor is -0.3 volt.

The scheme proposed by Hill and others embodied the idea that any "downhill" transport of electrons was between two cytochrome forms that could be extracted from chloroplasts. One of these, cytochrome b, has a redox potential of 0.0 volt and the other, cytochrome c (or a form of this known as cytochrome f), has a redox potential of $+0.35$ volt. The transfer of an electron between these two substances was visualized as the link between two "uphill" transfers energized by two separate light reactions, a first that lifts an electron from water (redox potential $+0.8$ volt) to cytochrome b and second that lifts it from cytochrome c to NADP (redox potential -0.3 volt or an intermediate substance along this path. This proposed electron pathway is shown in outline as the " Z-scheme " on p. 373. There is a good deal of evidence to support this basic concept and most of the detail has been filled in

The two light reactions are regarded as taking place in separate photochemical systems, referred to as PS I and PS II. In each system is a reaction centre to produce an oxidant and a reductant as previously stated. PS I presumed to energize a reaction that gives a strong reducing agent and weak oxidizing agent through the agency of a chlorophyll *a* molecule absorbing at a longer wavelength peak of 700 nm. PS II is presumed to energize the reaction that results in a strong oxidizing agent and a weak reducing agent. The oxidizing agent oxidizes the water. The pigment molecule involved in this is probably another form of chlorophyll *a*, absorbing at a shorter wavelength peak than mentioned above. Some evidence to substantiate the scheme comes from the fact that cytochrome *c* in the chloroplast is reduced under the influence of the PS II shorter wavelength light and oxidized under the influence of the longer wavelength light of PS I. The occurrence of this kind of reversal indicates that cytochrome *c* might lie on the electron transport path between the two systems as conveyed in the "Z-scheme".

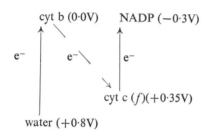

$$\text{cyt b (0·0V)} \qquad \text{NADP (−0·3V)}$$

$$\text{cyt c } (f)(+0·35\text{V})$$

$$\text{water (+0·8V)}$$

Other supporting evidence comes from the fact that dichlorophenyldimethyl urea, DCMU, known to inhibit the flow of electrons from water to NADP, when present in the system allows the oxidation of both the cytochrome types by the PS I action but prevents their reduction by PS II. This would be the expected situation if the DCMU were acting at a point somewhere between PS II and the cytochromes. Further support for this comes from the fact that if an artificial electron donor is supplied to the system to replace the electron source provided by the action of PS II then NADP can be reduced through the action of PS I alone, that is, an electron can be transferred to it.

Still more detail of the electron path from water to NADP can be filled in as follows.

In PS I, most of the light energy (photons) concerned is trapped at shorter wavelengths and passed to long-wave chlorophyll *a* molecules absorbing at about 695–700 nm, known as C_{700}. The energy then is transferred to another special kind of chlorophyll *a* molecule, absorbing at 700 nm, known as P_{700}. This is the final energy trap from whence it performs its photochemical work. If a photon eventually reaches P_{700} this molecule is excited and there is transfer of an electron to an electron acceptor which has yet to be properly identified but is commonly referred to as *ferredoxin reducing substance*, FRS. This means that P_{700} is mildly oxidized to P_{700}^+ and FRS is very strongly reduced to FRS⁻. P_{700}^+ is now in a state where it cannot be excited again until it receives an electron and hence for the time being has lost its power to absorb energy. In fact oxidized P_{700}^+ is colourless showing that it is not absorbing in any part of the visible spectrum. FRS⁻ is a very strong reducing agent and in the photochemical reaction just completed an electron has been moved against the potential gradient, increasing the chemical potential of the system. The efficiency of the energy conversion, light→chemical, is about 50 per cent but some more efficiency is lost when the chemical energy is later used to do the work of reducing carbon dioxide through the agency of NADPH$_2$. In the chloroplast the reduction of NADP is brought about by the action of two other substances to be found in chloroplasts, namely, *ferredoxin* and the enzyme *ferredoxin-NADP reductase*. The former is presumed to be the agent through which FRS$^-$ is oxidized by its electron acceptor. It is a non-haem iron protein. Ferredoxin-NADP reductase is a flavoprotein enzyme which acts as the final agent transferring the electron from ferredoxin to NADP.

Clearly, if P$_{700}$ has lost an electron to FRS, as stated above, it must regain one, that is, it must be reduced before it can absorb more energy and this is achieved through the agency of cytochrome *c* and a copper protein, *plasto cyanin*, both of which are known to be closely linked to P$_{700}$ in the chloroplast. The reduction of P$_{700}^+$ is thought to be effected by an electron transfer along the path from cytochrome *c* to plastocyanin to P$_{700}$ but it could be through plastocyanin→cytochrome *c* →P$_{700}$ or both.

In PS II, oxygen is evolved from water, reducing H$^+$ is made available and electrons are passed into PS I. The oxygen evolution is thought to be the result of the initial photo-reduction of an unidentified substance, possibly *plastoquinone*, via an unknown electron donor which is oxidized very strongly. In some way not yet clear this is able to effect the liberation of oxygen from water. The whole process is known to be dependent on the presence of manganese catalysts. The postulated substance plastoquinone could be associated with another substance, again not fully identified, but probably *plastoquinone A*, which acts in conjunction with it as an electron acceptor. At the end of a PS II event reduced plastoquinone A is formed. It is from this substance that electrons flow via cytochrome *b* into the PS I via cytochrome *c*.

The bridge between PS II and PS I includes the two cytochromes and another unidentified electron donor-acceptor substance designated M. This substance has been shown to exist from work using a mutant strain of a green alga that could not carry out the normal electron transport from PS II to PS I in photosynthesis despite possession of both cytochrome *b* and cytochrome *c*, indicating the necessity of another substance in the electron transport chain between them.

The whole of this electron transport pathway is schematized on p. 375 and the overall result including the reduction of carbon dioxide can be summarized in the equations:

$$2H_2O \rightarrow 4H^+ + 4e^- + O_2$$

$$2\ NADP + 4H^+ + 4e^- \rightarrow 2\ NADPH_2$$

$$2\ NADPH_2 + CO_2 \rightarrow 2\ NADP + H_2O + CH_2O$$

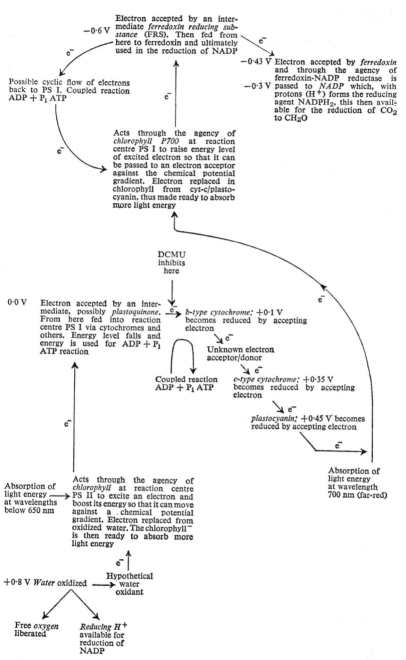

−0.6 V　Electron accepted by an inter-mediate *ferredoxin reducing substance* (FRS). Then fed from here to ferredoxin and ultimately used in the reduction of NADP

−0.43 V Electron accepted by *ferredoxin* and through the agency of ferredoxin-NADP reductase is
−0.3 V passed to *NADP* which, with protons (H⁺) forms the reducing agent NADPH₂, this then available for the reduction of CO₂ to CH₂O

Possible cyclic flow of electrons back to PS I. Coupled reaction ADP + P₁ ATP

Acts through the agency of *chlorophyll P700* at reaction centre PS I to raise energy level of excited electron so that it can be passed to an electron acceptor against the chemical potential gradient. Electron replaced in chlorophyll from cyt-c/plasto-cyanin, thus made ready to absorb more light energy

DCMU inhibits here

0.0 V　Electron accepted by an inter-mediate, possibly *plastoquinone*. From here fed into reaction centre PS I via cytochromes and others. Energy level falls and energy is used for ADP + P₁ ATP reaction

b-type cytochrome: +0.1 V becomes reduced by accepting electron

Unknown electron acceptor/donor

Coupled reaction ADP + P₁ ATP

c-type cytochrome: +0.35 V becomes reduced by accepting electron

plastocyanin: +0.45 V becomes reduced by accepting electron

Absorption of light energy at wavelength 700 nm (far-red)

Absorption of light energy at wavelengths below 650 nm

Acts through the agency of *chlorophyll* at reaction centre PS II to excite an electron and boost its energy so that it can move against a chemical potential gradient. Electron replaced from oxidized water. The chlorophyll⁻ is then ready to absorb more light energy

Hypothetical water oxidant

+0.8 V *Water* oxidized

Free *oxygen* liberated

Reducing H⁺ available for reduction of NADP

13

So far no mention has been made of the formation of ATP resulting from this electron transport system. Theoretically, at least, there are two possible places where ATP formation may be coupled to it, that is, where energy may be used to build a chemical bond between ADP and inorganic phosphate. The first of these is where there is a downhill flow of electrons between the cytochromes and the other involves a possible cyclic flow of electrons between FRS and PS I with the phosphorylation coupled to the downhill transport. The same uncertainty as to the precise nature of the coupling of ATP formation with electron flow in the oxidative phosphorylation executed by mitochondria (*see* p. 440) exists in the case of photophosphorylation in chloroplasts and it cannot profitably be further discussed.

Once the energy and reducing power have become available carbon dioxide can be assimilated in the dark. The "light" reaction is thus one of energy transfer to the power units which then motivate the "dark" reaction.

The Chemical Pathway

The equation $6CO_2 + 6H_2O \rightarrow C_6H_{12}O_6 + 6O_2$ agrees with the observed facts that, in the light, green plants are able to produce sugars from carbon dioxide and water with the evolution of oxygen. Measurements of the molecular quantities involved in the process also indicate the appropriateness of the equation, since for every molecule of carbon dioxide fixed, one molecule of oxygen is given off. A theoretical explanation of the origin of the emitted oxygen, using this equation, shows that one-half should come from the carbon dioxide and one-half from the water. This is not the case, as has been shown, and all the oxygen in the photosynthetic reaction is derived from the water. This can be represented as—

$$CO_2 + 2H_2O \rightarrow (CH_2O) + H_2O + O_2$$

and a simple equation, if it is to be used at all, should be written as—

$$6CO_2 + 12H_2O \rightarrow C_6H_{12}O_6 + 6H_2O + 6O_2.$$

This equation still implies the formation of sugar in one step from carbon dioxide and water. Calvin, Benson and others, using the isotope of carbon, ^{14}C, to trace the "path of carbon" from carbon dioxide to sugar in *Chlorella*, have shown the process to involve many steps. They used paper partition chromatography to isolate inter mediates carrying the tracer carbon which were produced after different periods of illumination of the algae. Their discoveries relating to the chemistry of the synthesis, coupled with those concerning the energy supplied to the system, can be schematized simply as on p. 377—

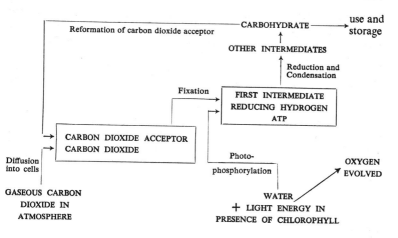

The major steps are—

1. Photophosphorylation producing reducing hydrogen in $NADPH_2$ and energy in the form of ATP. In this process light energy is harnessed to drive the cycle. As a result of this photophosphorylation, oxygen is evolved.

2. Fixation of carbon dioxide by incorporation into a carbon dioxide acceptor molecule and the breakdown of this new compound to a first intermediate product.

3. Conversion, accompanied by reduction, using the hydrogen obtained in (1), of the first intermediate through others to carbohydrate molecules.

4. Regeneration of carbon dioxide acceptor from carbohydrate.

The overall process may be looked upon as cyclic and for each cycle involving six carbon dioxide molecules absorbed from outside, one is acquired for use or storage by the plant. Only the first step is a light reaction. All others are capable of proceeding in darkness. The scheme is based on experiments in which cell extracts were used with adenosine triphosphate (ATP) as a source of energy for driving the cycle. The plant uses light energy to keep the cycle going. The carbon dioxide acceptor molecule in the "test-tube" synthetic cycle, was *ribulose diphosphate*, a five-carbon compound, and this has been extracted from cells.

The first intermediate product is *phosphoglyceric acid*, a three-carbon compound. This agrees with the previous discovery that when ^{14}C was fed to plants for very short periods the first "labelled" compound to appear was phosphoglyceric acid, which contained one radioactive carbon atom in its three. If ^{14}C was fed for longer periods, all the carbon atoms in the phosphoglyceric acid were radioactive. These results indicated two things. First, that the carbon dioxide was being picked up by an acceptor to form labelled phosphoglyceric acid, and secondly, that the first labelled phosphoglyceric acid was itself giving rise to more of the carbon dioxide acceptor, or in other words, the tracer carbon dioxide was being used for the purpose. It has been shown that a five-carbon compound, ribulose diphosphate, can act as the acceptor, and how it can be regenerated from the synthesis products. A scheme representing the photosynthetic process and incorporating these findings, is shown below—

Stage 1. Photophosphorylation in which high energy electrons emitted by chlorophyll under the influence of light are passed through a series of graded steps and their energy used in making energy-rich bonds in ATP and reducing power in $NADPH_2$. The mechanism of the coupling between the formation of ATP from ADP and inorganic phosphate and the electron flow in the chlorophyll system is not understood. There is enough energy available in the "downhill" electron flow from cytochrome *b* to cytochrome *c* to phosphorylate one molecule of ADP to ATP and it is probable that this is one coupling site. Another such coupling may be to a cyclic electron flow around PS I, involving FRS, where ATP may again be generated from ADP and phosphate. Hydrogen ions from the water are involved in the formation of $NADPH_2$.

Stage 2. The carbon dioxide from the atmosphere is united to a five-carbon compound acceptor, ribulose diphosphate, under the influence of a carboxylase enzyme, to make a transitory six-carbon compound which breaks into two three-carbon molecules of phosphoglyceric acid, the first intermediate product.

Stage 3. This involves the formation of carbohydrate from phosphoglyceric acid, presumably by the reverse of glycolysis (*see* p. 434).

Stage 4. The regeneration of the pentose diphosphate involves several steps.

(i) Hexose phosphate with triose phosphate, under the influence of transketolase produces one pentose phosphate (5-carbon) molecule and one tetrose phosphate (4-carbon) molecule. The pentose is ribulose phosphate which is further phosphorylated from ATP to become the diphosphate carbon dioxide acceptor.

(ii) The tetrose phosphate is added to hexose phosphate (6-carbon) and the new substance splits into one heptulose phosphate (7-carbon) molecule and a triose phosphate (3-carbon) molecule. The enzyme concerned is transaldolase.

(iii) The heptulose phosphate with triose phosphate now forms two molecules of pentose phosphate (5-carbon) under the action of transketolase. These are both phosphorylated from ATP and then become two more ribulose diphosphate carbon dioxide acceptors which can re-enter the cycle.

In terms of carbon atom balance the sequence of events may be represented as—

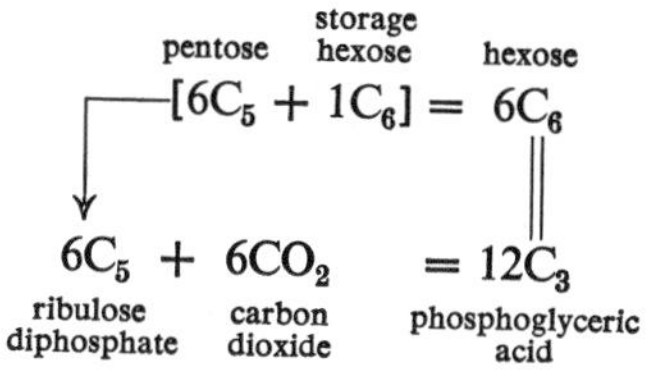

$$\begin{array}{ccc} \text{pentose} & \overset{\text{storage}}{\text{hexose}} & \text{hexose} \\ \left[6C_5 + 1C_6\right] & = & 6C_6 \end{array}$$

$$6C_5 \;+\; 6CO_2 \;=\; 12C_3$$

ribulose   carbon   phosphoglyceric
diphosphate   dioxide   acid

The Synthesis of Other Compounds

The tracer technique employed in following the path of labelled carbon into carbohydrates showed that it may quickly become incorporated with many other types of substances. As stated previously, after very short light exposures the ^{14}C first appears in phosphoglyceric acid. With longer exposures to light, the ^{14}C appears in trioses and in hexose phosphates, appearing in fructose phosphate and fructose before it appears in glucose. The first free carbohydrate in the plants investigated was sucrose. The following are results of such work using the green alga, *Scenedesmus*. With 5 s illumination, 87 per cent of the ^{14}C was present in phosphoglyceric acid, 10 per cent in phospho-pyruvic acid, and 3 per cent in malic acid. Of the ^{14}C present in phospho-glyceric acid, 95 per cent was in —COOH groups. With up to 60 s illumination, 20 per cent to 30 per cent of the ^{14}C present in phospho-glyceric acid was distributed between the other two carbon atoms and not in the —COOH groups. With this longer period of illumination, ^{14}C was also found in triose phosphate, hexose phosphate, sucrose, and the amino-acids glycine, serine, alanine and aspartic acid. With longer periods of illumination, ^{14}C was found in a number of organic acids including succinic, fumaric, citric, malic and glutamic, in glucose and fructose, and in many amino-acids.

Analysis of this and other evidence indicates an important fact. *Carbohydrates are not the only products elaborated as a result of photosynthesis;* there are amino-acids and other organic acids as well, and these in their turn may give rise to proteins, fats and numerous other substances.

A fuller understanding of how carbon from carbon dioxide plays its part in the metabolism of plants can only be gained with reference to a chart showing the anabolic and catabolic changes which are believed to occur. In an abbreviated form this is given below. It incorporates the photosynthetic cycle just outlined and will be described again in more detail in Chap. 11.

The cycle of events in which pyruvic acid is oxidized through citric acid eventually to oxaloacetic acid is known as the Krebs' cycle after its discoverer in animal metabolism.

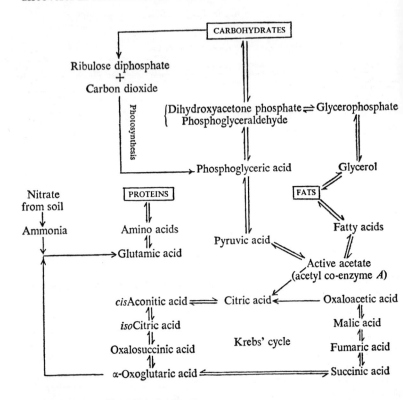

Amino-acid and Protein Synthesis

The principal means of amino-acid synthesis could be by the taking up of nitrogen as ammonia by some of the carboxylic acids of the Krebs' cycle. For example, the formation of glutamic acid is achieved by the addition of ammonia to α-oxoglutaric acid. Alternatively, glutamic acid could be produced from α-oxoglutaric acid and hydroxylamine through the formation of oximino acid.

Now it will be recalled, that for higher plants the chief intake of nitrogen-containing material is in the form of the nitrate ion. The nitrate must therefore be reduced to hydroxylamine or ammonia before amino-acid can be formed. The mechanism of nitrate reduction is most probably first through the agency of an enzyme system known as *nitrate reductase*. This is composed of metalloflavoprotein containing FAD and molybdenum and converts nitrate to nitrite according to the sequence:

$$\text{NADH}_2 \text{ or NADPH}_2 \longrightarrow \text{FAD} \longleftrightarrow \text{Mo}^{5+} + 2e \longrightarrow \text{NO}_3^-$$
$$\text{NAD} \text{ or } \text{NADP} \longleftrightarrow \text{FADH}_2 \longrightarrow \text{Mo}^{6+} \longleftrightarrow \text{NO}_2^- + \text{H}_2\text{O}$$

This step is probably followed by the reduction of nitrite to hypo-nitrite or an equivalent substance in which the nitrogen is in a similar oxidation condition. Possibilities are *hyponitrous acid*, $\text{H}_2\text{N}_2\text{O}_2$, *nitramide*, NO_2NH_2, or *iminonitric acid*, $\text{HN}{=}\text{N(OH)}{=}\text{O}$. The reducing enzyme is again a flavoprotein with an FAD prosthetic group and a requirement for copper or iron ions. The next step is presumably the further reduction of one or other of these substances to *hydroxylamine*, NH_2OH. The reaction would necessitate the presence of a hyponitrite reductase but such enzymes are rarely found in plants so that there is no certainty that it actually occurs. If hydroxylamine is produced this is then presumably reduced to *ammonia* in the presence of NADH_2, hydroxylamine reductase and manganese ions so:

$$\text{NH}_2\text{OH} + \text{NADH}_2 \rightarrow \text{NH}_3 + \text{NAD} + \text{H}_2\text{O}$$

The overall conversion of nitrate to ammonia may be represented by:

$$\text{NO}_3^- \rightarrow \text{NO}_2^- \rightarrow \text{NO or N}_2\text{O}_2 \rightarrow \text{NH}_2\text{OH} \rightarrow \text{NH}_3$$

Ammonia is toxic in quantity and must be combined organically without being stored. It can be incorporated into the system by any of three main reactions, either as an α-amino group, as the amide groups of asparagine or glutamine or as carbamyl phosphate, an intermediate in the synthesis of citrulline and pyrimidines.

To quote the examples given earlier, it is a matter of combining the ammonia or hydroxylamine with a suitable carbon structure, such as an α-keto acid, to form an amino acid. Such an amination is accomplished using FAD as the hydrogen carrier so:

$$\text{R} - \overset{\overset{\displaystyle O}{\|}}{\text{C}} - \text{COOH} + \text{NH}_3 + \text{FADH}_2 \rightleftharpoons$$

$$\text{R} - \overset{\overset{\displaystyle NH_2}{|}}{\underset{\underset{\displaystyle H}{|}}{\text{C}}} - \text{COOH} + \text{FAD} + \text{H}_2\text{O}$$

In the case of ammonia being combined with α-oxoglutaric acid, the amino-acid produced is glutamic acid and its formation may be represented as:

$$
\begin{array}{ccccccc}
\text{COOH} & & \text{COOH} & & & \text{COOH} & \\
| & & | & & & | & \\
\text{C=O} & & \text{C—NH} & & & \text{H—C—NH}_2 & \\
| & & | & \text{glutamic} & & | & \\
\text{CH}_2 & +\ \text{NH}_3 \rightarrow & \text{CH}_2 & \text{dehydrogenase} & \rightarrow & \text{CH}_2 & \\
| & & | & & & | & \\
\text{CH}_2 & & \text{CH}_2 & & & \text{CH}_2 & \\
| & & | & & & | & \\
\text{COOH} & & \text{COOH} & & & \text{COOH} & \\
\text{α-Oxoglutaric} & & \text{α-imino-glutaric} & & & \text{glutamic} & \\
\text{acid} & & \text{acid} & & & \text{acid} &
\end{array}
$$

The formation of glutamic acid involving hydroxylamine and α-oxoglutaric acid may be represented as:

$$
\begin{array}{ccccccc}
\text{COOH} & & \text{COOH} & & & \text{COOH} & \\
| & & | & & & | & \\
\text{C=O} & & \text{CHON} & & & \text{H—C—NH}_2 & \\
| & & | & & & | & \\
\text{CH}_2 & +\ \text{H}_2\text{NOH} \rightarrow & \text{CH}_2 & +\ 2\text{H}_2 \rightarrow & & \text{CH}_2 & \\
| & \text{hydroxyl-} & | & & & | & \\
\text{CH}_2 & \text{amine} & \text{CH}_2 & & & \text{CH}_2 & \\
| & & | & & & | & \\
\text{COOH} & & \text{COOH} & & & \text{COOH} & \\
\text{α-Oxoglutaric} & & \text{oximino acid} & & & \text{glutamic acid} & \\
\text{acid} & & & & &
\end{array}
$$

Other amino-acids can be formed from some of the other acids of the Krebs' cycle. Aspartic acid can be formed from oxaloacetic acid by a process known as transamination in which glutamic acid donates —NH₂ to oxaloacetic acid under the influence of a transaminase. α-Oxoglutaric acid is then reformed and is free to form more glutamic acid. Similarly glutamic acid can transfer its amino group to pyruvic acid to form alanine, another amino-acid. It is not unlikely that other amino-acids can be formed by the same process of transamination. The enzymes concerned are known as transaminases, and one interesting feature of their construction is that they contain a derivative of the vitamin pyridoxine (vitamin B₆). Their prosthetic group is *pyridoxyl phosphate* indicated by PD—CHO. It is with this that the amino group of the amino-acid combines and in this latter there is induced a shift of an electron that frees the acid radical of the amino-acid, but the PD—CHO retains the NH_2 group. The pyridoxyl phosphate is then converted to *pyridoxamine phosphate*, PD—CH_2—NH_2, which can react with a suitable α-keto acid, transfer to it the amino group and thus make it an amino-acid, at the same time regenerating PD—CHO to pick up another amino group and repeat the process. Another example of an amination is the formation of *serine* from *hydroxypyruvic acid*, a derivative of phosphoglyceric acid according to:

$$
\begin{array}{ccc}
\text{COOH} & & \text{COOH} \\
| & & | \\
\text{C} = \text{O} & + NH_2 \rightarrow & \text{H} - \text{C} - NH_2 \\
| & & | \\
H_2\text{C} - \text{OH} & & H_2\text{C} - \text{OH} \\
\text{hydroxypyruvic acid} & & \text{serine}
\end{array}
$$

It would appear that the transaminases occur throughout the whole living world. The amino-acids glutamic acid and aspartic acid are of significance also, in that they can be converted to the amides, glutamine and asparagine, and in this form stored as reservoirs for future amino-acid synthesis.

Assuming that the plant is able to produce the initial donator of amino groups, glutamic acid, it is not difficult to understand that by the transamination process, a variety of amino-acids may be built up, from which its proteins may subsequently be synthesized. Protein synthesis is one of linking amino-acid residues through their α-amino and carboxyl or acid groups by a series of peptide linkages represented by

$$
\begin{array}{c}
\text{—HN—C—} \\
\parallel \\
\text{O}
\end{array}
$$

An outline of protein chemistry is given in Chap. 4. A possible mechanism of protein synthesis at the ribosomes is described in Chap. 19.

The above is a comparatively simple outline of some of the aspects of nitrogen metabolism in plants. There is no doubt that the full detail is far more complex, well beyond the requirements of this level of study.

Fat Synthesis

Fats are esters of fatty acids and glycerol, and these two substances must be formed before any lipoid material can be built up. Glycerol, the alcohol of the fat ester, can be derived from dihydroxyacetone phosphate, a substance on the path of sugar breakdown (*see* p. 436). Fatty acid synthesis is more complicated. The substance "active acetate" or *acetyl co-enzyme A* is the starting point, and this is derived from carbohydrate after its breakdown to pyruvic acid, when with co-enzyme *A* this forms acetyl co-enzyme *A*. This is converted to *malonyl co-enzyme A* by the addition of carbon dioxide, catalysis being through the agency of an enzyme *acetyl Co A carboxylase* activated by the vitamin biotin and manganese ions. Malonyl co-enzyme *A* is highly reactive and is able to react with a further molecule of acetyl co-enzyme *A* to form aceto-acetyl co-enzyme *A* at the same time releasing carbon dioxide and freeing one molecule of co-enzyme *A*. By the removal of water from and the reduction, by the addition of hydrogen, of the substrate attached to the remaining co-enzyme *A*, namely, *aceto-acetic acid*, this is converted to a *fatty acid residue*. Such a residue can now combine with more malonyl co-enzyme *A* in just the same way as could acetyl co-enzyme *A* and by the same hydrogenation and dehydration processes two more carbon atoms can be added to the fatty acid residue chain. This sequence of events can be repeated to form fatty acid residues, bound to co-enzyme *A*, of any length required, the carbon atoms being added two at a time. The fatty acids in combination with *glycerol phosphate* (the substance produced from phospho-glyceraldehyde through dihydroxyacetone phosphate by reduction of the latter using hydrogen accepted from $NADH_2$) produce *phosphatidic acids* and from these, true *glycerides*. The phosphatidic acids can give rise to phospholipides such as lecithin by combination with choline. The true lipides in plants appear as stored oil globules, the oils differing from the fats normally associated with animals in being formed from unsaturated fatty acids, *see* p. 89. The phospholipides are used structurally in the formation of cell membranes.

It will be seen that the reduction process leading to the formation of fatty acids is not the reverse of the β-oxidation process by which they are broken down, *see* p. 443, as once assumed. For more detail of fatty acid metabolism in animals, *see* pp. 417–420.

This method of synthesis accounts for the fact that the natural fatty acids, almost without exception, contain an even number of C atoms. The chief types of fatty acids occurring in plants and animals have been described in Chap. 4, and the fats, waxes and allied compounds with them.

THE NUTRITION OF OTHER PIGMENTED PLANTS

It has been seen that the only pigmented plants not containing one or more of the chlorophylls are the bacteria. The majority of these microbes possess no photosynthetic pigment and thus must exist either chemoautotrophically or as heterotrophes. There are a few however, notably the sulphur bacteria, which contain pigments b which they can harness light energy. The pigments are collectively known as bacteriochlorophyll, but there are several distinct ones such as the green bacterioviridin of the green sulphur bacteria, and the purplish pigment of the purple sulphur bacteria.

The mechanism by which light is used in the energy transformations is one of photophosphorylation (see pp. 369 and 376), and the overall chemistry of the processes by which carbon dioxide can be fixed is well known. If we represent the overall equation of the higher green plant as—

$$CO_2 + 2H_2O \rightarrow (CH_2O) + H_2O + O_2$$

then for the green sulphur bacteria, e.g. *Chlorobium*, the following holds—

$$CO_2 + 2H_2S \rightarrow (CH_2O) + H_2O + 2S$$

Thus the bacterium is using hydrogen sulphide as the source of reducing hydrogen instead of water and is releasing sulphur instead of oxygen.

The purple sulphur bacteria, e.g. *Chromatium*, can perform the same process, but in addition, if deprived of hydrogen sulphide can use elemental sulphur as a reducing agent in the photochemical part of the mechanism thus—

$$2S + 8H_2O + 3CO_2 \rightarrow 2H_2SO_4 + 3(CH_2O) + 3H_2O$$

The overall reaction for the purple bacteria can be represented—

$$2H_2S + 8H_2O + 4CO_2 \rightarrow 2H_2SO_4 + 4(CH_2O) + 4H_2O$$

It can be seen, therefore, that the only essential difference between these bacteria and green plants is a comparatively minor one, in that a different source of reducing hydrogen is used.

THE NUTRITION OF NON-PIGMENTED PLANTS

Plants which lack a photosynthetic pigment include all the fungi and most of the bacteria. The fungi are all heterotrophic, and so are the majority of the bacteria. Some bacteria are chemosynthetic. Special modes of life are treated in Vol. I, Chap. 30.

THE NUTRITION OF ANIMALS

Reference was made in the early part of this chapter to the essential difference between the modes of nutrition of plants and animals. Whereas most plants possess the ability to utilize either light or chemical energy to build their organic food requirements from inorganic materials, animals are unable to do so and must rely entirely on the activities of plants. To this characteristic of animals can be traced most of their structural and physiological differences from plants. It is an interesting study to try to correlate these differences of structure and function with the differing nutritional methods.

Material Requirements, Their Sources and Roles

Also pointed out earlier was the fact that the organic substances used by both plants and animals in their metabolic activities were fundamentally the same. They both require carbohydrate, protein, fats and many other organic compounds but they must be presented to the animal in forms such that its very limited synthesizing powers can further elaborate them. These forms coincide with the organic compounds manufactured from inorganic materials by plants. We can, therefore, summarize the organic nutrient requirements of an animal very briefly as—

1. Carbohydrates.
2. Nitrogen-containing compounds, particularly amino-acids.
3. Fats.
4. Other organic compounds, particularly those described as vitamins.

From these it can derive the materials for growth, repair, respiration, secretion and storage. But the animal must also maintain its ionic, osmotic and pH internal environments within rigid limits; consequently its diet must include inorganic substances which we can summarize as—

5. A variety of mineral salts.
6. Water.

These six groups of substances must be represented in the material intake, i.e. the *diet*, of most animals, in order for them to maintain health and vigour. When they are taken in the correct proportions to meet the animal's needs, we say the animal is feeding on a balanced diet. Over recent years the study of dietetics, particularly in human beings, has led to the eradication of much disease due to malnutrition. The diets of different animal species are as variable as the animals themselves since their requirements are not the same in any two cases. Whilst it is impossible to state the precise rôle of each compound taken in by an individual animal, we may generalize thus—

Carbohydrates act primarily as energy-supply compounds, their oxidation during respiratory activities yielding the energy required for locomotion, heat production and the numerous energy-using chemical reactions of the body. They may be found stored in all parts of the body on a short-term storage principle for this purpose, e.g. glycogen.

They may also enter into the construction of some protoplasmic structures.

Proteins are primarily constituents of protoplasm and consequently are used principally for growth and repair and in the formation of numerous secreted substances, particularly enzymes and hormones. They can enter the respiratory cycles in some cases and can then be used to release energy.

Fats form essential constituents of protoplasm and also are convenient long-term storage substances for later energy release.

Vitamins are difficult to define but may be regarded as accessory food factors, organic in origin, required by an animal for normal healthy function. They do not supply energy nor are they essentially protoplasmic units, but they enter the metabolic activities in some proved cases as catalysts or reactants in chemical reactions. Their absence from a diet leads to the so-called *deficiency diseases*. About fifteen substances can definitely be classed as vitamins, another fifteen may be vitamins, and it is probable that a number of others will be discovered. For convenience, vitamins are divided into those which are fat-soluble, and those which are water-soluble. The former group include A_1, A_2, D_2, D_3, E and K. The latter group include B_1, B_2, B_3, B_4, B_5, B_6, B_{12}, C, P, and the P.P. factor.

Vitamin A_1 is an alcohol, $C_{20}H_{30}O$, derived from the breakdown of plant carotenoids in the liver, thus—

$$C_{40}H_{56} + 2H_2O \rightarrow 2C_{20}H_{30}O$$

β-carotene is twice as productive of vitamin A as α- and γ-carotenes. Vitamin A_2 closely resembles A_1 both in structure and biological effects. It was discovered in the liver-oils of fresh-water fishes. Normally, reference to vitamin A, denotes A_1. The most potent sources of vitamin A are halibut-liver oil and cod-liver oil. It is also obtained from lettuce, spinach, watercress, peas, carrots, butter, cream, cheese, egg-yolk and liver. Deficiency effects of vitamin A are growth failure, and keratinization of epithelial tissues. According to the localization, this latter may result in dry skin, xerophthalmia, pharyngitis, gastroenteritis. Other effects are bone abnormality, nerve degeneration, and night-blindness. Vitamin A is an essential constituent of visual purple, rhodopsin; it is therefore concerned in the synthesis and regeneration of this pigment.

Vitamin B_1, aneurin or thiamine, is the hydrochloride of an organic base, $C_{12}H_{18}ON_4Cl_2S$. Its pyrophosphate is the co-enzyme known as co-carboxylase; it plays an important part in carbohydrate metabolism, especially in the decarboxylation of pyruvic acid to form "active acetate." In its absence, pyruvic acid accumulates, setting up the deficiency diseases, beri-beri, avian polyneuritis, slowing of the heart-beat and certain gastro-intestinal disorders. Thiamine appears to be essential to all forms of life. The chief sources in diet are yeast and whole grains, though adequate quantities can be found in liver, kidney, lean meat, egg-white, spinach, watercress and potato.

Vitamin B_2 is riboflavin, $C_{17}H_{20}N_4O_6$. Riboflavin nucleotides, combined with proteins, form the flavoprotein enzymes. They include several important dehydrogenases, some of which are present in all living cells. Thus B_2 like B_1 provides a co-enzyme necessary for tissue metabolism. There are a number of signs of deficiency, together constituting the condition ariboflavinosis; some of these signs are scaly flaking of the skin about the nose and ears, cracking at the corners of the lips, abnormal redness of the lips and lesions of the eyes.

Vitamins B_3, B_4 and B_5 are growth factors detected in certain animal experiments. Whether they apply to human nutrition is not yet known.

Vitamin B_6, pyridoxin, is sometimes known as the rat anti-dermatitis factor. It is present in yeast, whole grains, liver and milk. An aldehyde of pyridoxin, called pyridoxal when combined with phosphate, is the co-enzyme of certain amino-acid decarboxylases and of the trans-amination enzymes.

Vitamin B_{12}, or cobalamine, is very important in red blood cell formation and is used medicinally in the treatment of pernicious anaemia. The molecule contains cobalt, but the element itself cannot replace the vitamin.

The P.P., or pellagra-preventing factor, is nicotinic acid, $C_5NH_4.COOH$, or nicotinamide, $C_5NH_4CO.NH_2$. The latter is a constituent of the co-enzymes (co-dehydrogenases) 1 and 2, DPN and TPN, which play an important part in carbohydrate oxidation. Pellagra is a serious disease in maize-eating populations; it is still common in south Russia, in the southern U.S.A. and in Italy. Commonly known as "rough skin," its symptoms are inflammation of the tongue and intestine, nervous disorders leading to paralysis, dermatitis, darkening and thickening of the skin. It is probably the most serious of the deficiency diseases. The foodstuffs richest in nicotinic acid are liver, kidneys, yeast, milk, eggs and whole grains.

Vitamin C is ascorbic acid, $C_6H_8O_6$; it is sometimes called the anti-scorbutic vitamin, the deficiency disease being known as scurvy. It was common at one time among sailors and explorers, whose diet lacked fresh foodstuffs for prolonged periods. Nowadays it occurs sometimes in children fed exclusively on sterilized foodstuffs.

Certain dental disorders, and inefficient wound healing are also signs of deficiency. It is suggested that ascorbic acid is an agent concerned in oxido-reduction reactions; it is necessary for wound repair and the production of collagen fibres; and it may play a part in protein metabolism. The richest sources of vitamin C are rose-hips and black currants; it is, however, present in many fruits, in green leaves, in liver and in milk. Mammals, except man and apes, can synthesize it.

Vitamin P, or citrin, is a flavone glucoside first found in lemon and orange juice. It cures a certain type of capillary haemorrhage. It is now known to be present in many fruits.

Vitamin D_2 is calciferol, $C_{28}H_{43}OH$, and D_3 is the natural vitamin D, $C_{27}H_{43}OH$. A number of other substances with similar properties are produced by ultra-violet irradiation of certain sterols. There are at least eight of these substances which have similar effects. The main importance of the D vitamins is their regulation of calcium and phosphate absorption from the intestine. The deficiency diseases are rickets (incomplete calcification of bones) and some forms of dental caries. The liver oils of fishes are very rich in vitamin D, and it is plentiful in many animal fats such as cream, butter and egg-yolk. Among plants it has been found only in the seed fats of the *Cacao* tree (cocoa and chocolate contain small amounts).

Vitamin E includes three closely related compounds, α, β and γ tocopherol with respective formulae $C_{29}H_{48}O_2$, $C_{28}H_{46}O_2$ and $C_{29}H_{50}O_2$. Deficiency causes interference with placental function in females and gametogenesis in males. Lack of E and A leads to sterility. Serious conditions associated with vitamin E are the death and resorption of the embryo, or abortion. Green plant tissues, such as lettuce and watercress contain this vitamin, but the best source of it is the oil extracted from wheat embryos. Whole grains contain moderate amounts.

Vitamins K and K_2 are derivatives of naphthaquinone. Deficiency symptoms are anaemia, increase in haemorrhage, and less ability to clot the blood. The vitamin is associated with the production of prothrombin, a factor in blood coagulation. Good sources of the vitamin are green leaves and tomatoes.

Other substances are known to be accessory food factors and are therefore sometimes called vitamins. Pantothenic acid is known to be a constituent of co-enzyme A and is therefore essential to normal carbohydrate metabolism. It is universally distributed in plants and animals and can be extracted from yeast and liver. *p*-Aminobenzoic acid also has a wide distribution in living things. Yeast is a particularly good source. It affects pigmentation of the hair in rats. Folic acid contains *p*-aminobenzoic acid. It is essential for normal growth in some animals, yeasts and bacteria. Deficiency in animals leads to some forms of anaemia and it is probably concerned in blood cell formation. Biotin or vitamin H is found in yeast and seeds. It is known to be associated with the enzyme system involved in the breakdown of pyruvic acid to oxaloacetic acid. Deficiency shows as dermatitis, lassitude and loss of appetite. Inositol is probably a structural component of subcutaneous tissues and is also a lipotropic factor, i.e. concerned with fat movement. Hexaphosphoinositol or phytic acid is found in seeds, and in quantity reflects directly the amount of phosphate stored. Calcium phytate or phytin on hydrolysis yields phosphate and free inositol. Choline is a lipotropic factor also, and deficiency leads to accumulation of fat in the liver of some animals. It does not appear to be essential to man.

Mineral salts in wide variety constitute another necessary part of the diet. The occurrence and importance of the inorganic salts and ions have been mentioned in Chap. 4 and they are referred to in many instances throughout the text. A brief outline of the utilization of the dietary minerals of higher animals will suffice here. In the first place, it is very difficult to consider them singly since they are all interrelated and balanced in the body, but some are required in greater quantity than others and some are known to be linked individually with specific metabolic reactions. The so-called essential elements are calcium, phosphorus, sodium, potassium, chlorine, magnesium, manganese, iodine, iron, copper, sulphur, zinc, cobalt, molybdenum and fluorine. Aluminium, arsenic, nickel and silicon may be found in many animal tissues but are not considered as having any special function. Of the essential elements, copper, cobalt, manganese, zinc, iodine and molybdenum are required in minute traces only. The elements occurring in greatest quantity are calcium and phosphorus making up between them about 90 per cent of the body ash. They occur mainly as calcium carbonate and calcium phosphate in teeth and bones.

Certain minerals and combinations of them have very definite relationships with the body structure and function. Calcium, phosphorus and fluorine are closely concerned with bone and teeth formation whilst calcium, phosphorus and sodium are largely responsible for maintaining the correct acid-base balance of the body fluids. Calcium and magnesium affect the proper functioning of the soft tissues. Copper catalyses the use of iron, whilst cobalt influences the use of both of these. Sodium, potassium and chlorine occur as ions in the body fluids, sodium in the extracellular fluids and potassium intracellularly. Blood will not clot in the absence of calcium ions and chlorine ions are essential to adequate gastric digestion. Potassium is closely concerned with muscle function and the muscle builds up a concentration against a gradient. Magnesium, calcium and phosphorus all play an important part in the glycolytic activities of cells whilst phosphates occur as phospholipides in nucleic acids and in the very important substance

ADP and ATP. Magnesium occurs in teeth and bones and acts as a cofactor to certain enzymes. Sulphur occurs in the protein of tissues, in the amino-acids, methionine, cystine and cysteine. It is also part of vitamin B_1, and the sulph-hydro enzymes such as glutathione, cannot be built without it. Iron is essential to haemoglobin formation and is part of the cytochrome molecule. The trace elements, although required in only a few parts per million, play equally important roles. Copper deficiency leads to anaemia since iron cannot be used without it. Cobalt is a vitamin B_{12} constituent without which anaemia again results. Manganese is an activator of enzymes, and phosphatase, choline esterase and arginase are all affected by its lack. Zinc is associated with proper growth of hair, the proper functioning of the enzyme carbonic anhydrase, and with the working of insulin in carbohydrate metabolism. Iodine is a constituent of the hormone thyroxine which controls the metabolic rate of an animal. Total deficiency leads to cretinism. Molybdenum in excess is very injurious but is required in trace amounts to act as an activator in enzyme systems. It is known to activate xanthine oxidase, an enzyme which functions in the conversion of an organic nitrogenous base to uric acid during nucleic acid metabolism.

Water is an essential protoplasmic constituent and is involved in most of the chemical reactions of the body.

With the exception of water, and occasionally some of the mineral salts, animals obtain their requirements directly or indirectly from plants. The diet of even the most meat-loving of animals can be traced from a plant source and it is enlightening to follow the food chains of animals back to their initial starting points. All the organic requirements and most of the minerals are obtained through plants, but occasionally mineral salts, particularly sodium chloride are taken directly from the environment. Fish may gulp sea-water and absorb its salts and land animals are often not averse to licking mineral deposits such as rock salt.

Water is almost universally taken by animals independently of other articles of diet, but even so, a good deal must be taken as a constituent of the protoplasmic substances on which they feed. The clothes moth and desert rat are reputed to be able to make perfectly healthy growth on a diet entirely lacking liquid water, obtaining their supplies from the free water and the water in chemical combination with the rest of their food.

It would not be correct, however, to suppose that all animals are totally lacking in the ability to assimilate inorganic substances directly into their bodies by a synthesizing process. When some of the very lowly flagellate organisms are taken from one medium rich in all their organic material requirements and placed in a medium lacking in some of them, growth is at first poor, but eventually it becomes much better. showing that at least some of them have changed their enzyme patterns so that they can manufacture their requirements. An experiment with the protozoan *Colpidium* in a purely mineral medium except for glucose and three amino-acids, indicated that its synthesizing powers were considerable. Mammals normally obtain their nitrogen as protein but if only very small amounts of a few amino-acids are supplied, they can use urea or ammonium salts as a nitrogen source, from which to synthesize other amino-acids and thus protein. Rats have been shown able to do this. It has been found in many cases in higher animals that only a certain number of amino-acids are essential in a diet. For instance, the rat can remain healthy given valine, leucine, phenylalanine, tryptophane, histidine, arginine, lysine, *iso*leucine, methionine and threonine. Man can survive if fed on a diet including only eight of these, not requiring histidine and arginine as essentials.

The ability to utilize carbon dioxide has also been demonstrated in many animals. It has been claimed by some that *Chilomonas*, a protozoan, can grow in darkness with CO_2 as its sole source of carbon. It can certainly live quite healthily in a medium containing glycine or acetate with no other carbon source. There is also evidence that CO_2 can be "fixed" in mammalian liver, according to the following reversible reaction—

$$\underset{\text{pyruvic acid}}{CO_2 + CH_3.CO.COOH} \overset{\text{carboxylation}}{\underset{\text{decarboxylation}}{\rightleftharpoons}} \underset{\text{oxaloacetic acid}}{HOOC.CH_2.CO.COOH}$$

This reaction has been demonstrated using CO_2 with labelled carbon, ^{11}C. The carbon was located later in α-oxoglutaric acid. Thus there is at least a small amount of CO_2 fixation in some animals.

Methods of Obtaining Food Requirements

Among animals there is a wide variety of methods of obtaining food materials. In general, we may classify them as *herbivorous*, *carnivorous* and *omnivorous*. The herbivores feed on vegetation, the carnivores on flesh, and the omnivores on a mixed diet. Every animal shows highly interesting adaptations in connexion with its method of feeding.

Carnivores eat whole animals or portions of them. To capture their prey, they have evolved a wide range of devices, from extensible tentacles to highly-developed powers of locomotion. One or more of the senses is very acute, usually sight or smell. Almost always, powerful weapons of offence have been evolved. As predators, they fall into two groups, hunters and trappers. The hunters pursue their prey and the trappers set snares. Owing to the concentrated nature of the food, most carnivores feed at intervals, which may sometimes extend to several days or even months.

Herbivores form the main prey of carnivores and have evolved many devices facilitating escape. Some rely on great speed, some on the gregarious habit, others burrow and yet others rely on protective coloration. The less concentrated nature of the food implies two things; first, they must eat much larger quantities and secondly, they must spend a great deal of time in feeding.

Omnivores have struck the middle way, and show a mixture of adaptations between carnivores and herbivores. They spend less time feeding than herbivores, but more than carnivores.

The very nature of animal feeding involves a number of essential developments. First of all, there must be an aperture, temporary or permanent, for entry of the food, then there must be a cavity in which the food is suitably treated for absorption. There must also be an aperture for defaecation which may be the same as that for ingestion. Finally, there must be special senses developed for location of the food materials, and special organs for conveying these food materials to the point of entry into the body.

Treatment of the Food Material

When food has been caught or found, it is of no value until it has undergone several stages of treatment by the feeding animal. First it must be transferred to a position where the feeder's body can subsequently deal with it. In all but the most primitive of animals, namely the protozoa and the sponges, the body is constructed to accept the incoming food material into a special part. The lower invertebrates such as the coelenterates may possess only hollow cavities, blindly ending sacs, but all the others and all the vertebrates possess a true *gut* or *alimentary canal* into which the food may be taken, its anterior end being specially modified for this purpose. (The only exceptions are some parasites, e.g. *Taenia*.) The protozoa, being non-cellular, have no such corresponding structures but they are capable of enclosing the food material in vacuoles formed within the cytoplasm. This process of taking in the food is termed *ingestion* and the manner in which it is accomplished constitutes the animal's *feeding mechanism*.

Subsequent to ingestion, the food material must be rendered soluble and hence absorbable. The food is rendered soluble by a process termed *digestion*, in which numerous enzymically controlled reactions, chiefly hydrolyses, are carried out on the food. Once in a soluble state, the degradation products of digestion can be taken into the body tissues of the animal from the vacuole, cavity or gut tube in which they were digested. This is the process of *absorption*. Note that until this

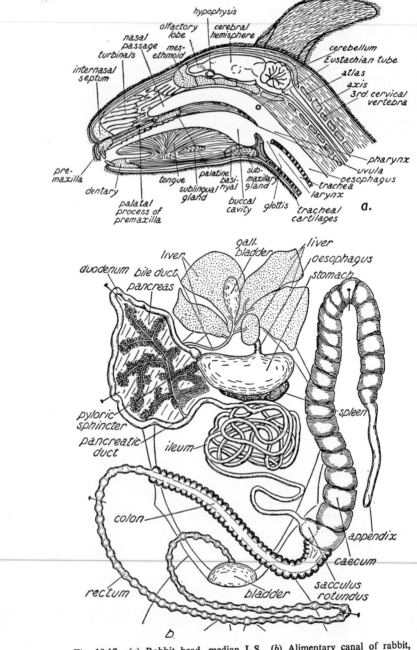

Fig. 10.17. (a) Rabbit head, median L.S. (b) Alimentary canal of rabbit, displayed.

is achieved the food has not become part of the body; and much that is taken in never does, since it cannot be rendered soluble and diffusible. That part which is absorbed contains the nutrient substances of the food and it is that part only which is of value to the feeder. Once in the tissues of the animal it is *assimilated* by the various cells, tissues and organs, and will eventually be used for one of the purposes mentioned previously. Assimilation and use will often lead to waste products which must be eliminated and there are special mechanisms for dealing with these waste products. The unabsorbed portions must also be removed from the site of absorption, for they would otherwise clog the whole system. The animal is therefore able to remove these by a process called *egestion* which will vary according to the construction of the body.

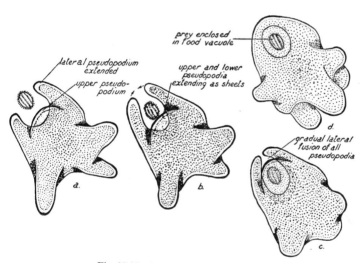

Fig. 10.18. *Amoeba* capturing a ciliate.

his process must not be confused with excretion, since materials can nly be truly excreted if they have once been part of the cellular onstituents. Since these unabsorbed portions have only been lying 1 a hollow or tube within the body, they have never really been part f it.

In order to make use of food material, an animal must, therefore, gest, digest, absorb and assimilate it. The indigestible and unabsorbed ortions must finally be egested. These processes will now be dealt ith in more detail.

Feeding Mechanisms

Within the animal kingdom there have been evolved mechanisms by which the various species are able to deal with every conceivable source of food material. There does not appear to be any correlation between the type of animal and its feeding mechanism. Every major group of animals has representatives which feed in many different ways. Thus the nature of a feeding mechanism seems to depend on the habitat and kind of food and not upon evolutionary relationships with other animals. The alimentary canal as a whole shows similar correlation with the feeding mechanism and type of food. We can, therefore, only describe feeding mechanisms according to the habitats and types of food that animals have to deal with.

According to habitat, there are two distinctly different possibilities. In a fluid medium there are likely to be soluble and diffusible organic substances such as sugars and amino-acids which may be of direct use to an organism without any further treatment. Fluid media are either those of the water habitats of free-living animals in which soluble substances will occur as breakdown products from rotting animal and vegetable matter, or the body fluid habitats provided by the hosts of some parasites, in which the soluble substances will be normal constituents. Such food we may describe as *soluble food*. In the case of free-living animals in natural water habitats, there is only scanty evidence that they are able to make use of this food to any great extent. In laboratory-controlled cultures, it is shown that many of them can, if no other food is available. In the case of the parasites, such as *Taenia*, it is often the only source of food and thus must be used. Therefore *direct absorption* of these soluble substances is the only feeding mechanism which they possess and in such cases, it is usual to find very severe if not complete suppression of alimentary tract development when compared with free-living animals of the same group.

Obviously soluble substances cannot occur other than in a fluid medium, so that in any other habitat, food material must be in solid or particulate form. Particulate food, of course, can occur in water habitats, and it is the case that by far the greater majority of animals whatever their habitat, rely on solid food material for their supply of nutrients other than water. Feeding mechanisms have been evolved chiefly to deal with solid food and Yonge has classified the many methods conveniently as below: the mechanisms are first grouped into three main categories according to the type of food. These are mechanisms designed to deal with:

1. Small particles.
2. Large particles or masses.
3. Fluids or soft tissues.

1. *Mechanisms for Dealing with Small Particles.* These are peculiar to aquatic species only and the multicellular animals adopting this method are either sedentary or very sluggish in their movements. A sticky secretion is often produced in which the minute particles are trapped before ingestion.

(*a*) Pseudopodial. This method is peculiar to protozoans, e.g. *Amoeba*, but it is often used by higher animals as a method of intake of small particles by cells lining the gut; the method is also employed by phagocytic cells in body fluids.

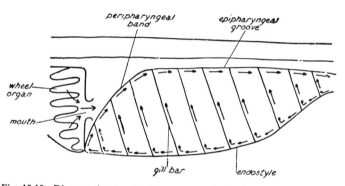

Fig. 10.19. Diagram showing the ciliary currents in the pharynx of *Amphioxus*.

(*b*) Ciliary. Such mechanisms are very widespread in aquatic animals. Beating cilia set up currents which bring large volumes of the medium into contact with the feeding parts and the small solid particles can then be directly ingested as in *Paramoecium*, or carried along ciliated tracts trapped in sticky secretion as in *Amphioxus*. Ciliary feeding mechanisms can be found in the ciliated protozoa, sponges, some annelids, rotifers, polyzoans, phoronids, brachiopods, some gastropods, most lamellibranchs, many tunicates and cephalochordates.

(*c*) Tentacular. Some holothurians possess tentacles around the mouth which catch small falling particles in mucus and from thence they are transferred into the mouth.

(*d*) Mucoid. Some of the sessile gastropods, e.g. *Vermetus*, extend a mucus film from their mouths. The film traps small particles, and at intervals it is retracted into the mouth.

(*e*) Muscular. Muscular contraction may be responsible for setting up currents carrying fine particles through filtering mechanisms. Some jellyfish pass currents by rhythmic muscular movements of the bell through fine pores leading to the enteron. The particles are filtered out in the enteron. Some polychaete worms, e.g. *Chaetopterus*, pass water currents by muscular contraction through secreted mucus bags. Many advanced animals may be considered in this category also. Some whales and fishes employ locomotor or pharyngeal muscles to sweep in currents loaded with plankton organisms, which are then filtered out and swallowed.

(*f*) Setose (hairs). Hair-like appendages or setae may be used to sweep through the water and collect small particles. These are then collected into larger masses by secretions and carried to the mouth. This is characteristic of many aquatic crustaceans, e.g. *Daphnia*, copepods, cirripedes, ostracods and larvae of some insects such as mosquitoes.

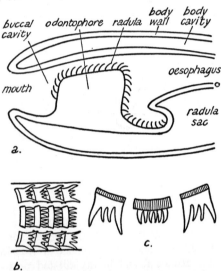

Fig. 10.20. Radula of the whelk, *Buccinum undatum*. (a) Median L.S. buccal region, (b) radula, portion from above, (c) one transverse row of teeth.

2. *Mechanisms for Dealing with Large Particles or Masses.*

(*a*) Intake of static material such as mud and other detritus. This method is usually adopted by burrowing forms of sluggish habits. The mud is swallowed continuously and organic matter utilized whilst the residue is egested in the faeces. Some holothurians use tentacles to force the mud into their digestive tracts. Annelids such as *Arenicola* possess eversible pharynges for taking in the mud. Some echinoderms use the tube feet, and burrowing crustaceans use their mouth parts for the same purpose.

(b) Scraping and boring. These methods necessitate specialized mouth parts, such as the radulae of many gastropod molluscs, e.g. the whelk, limpet, snail and slug, the boring valves of *Teredo*, the ship

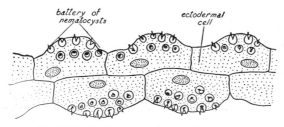

Fig. 10.21. Small portion of a tentacle of *Hydra*.

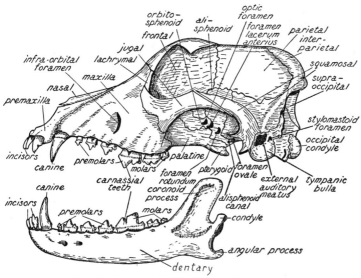

Fig. 10.22. Dog skull, lateral view with lower jaw.

worm, and jaws of the termite, which drill away small portions from a large mass.

(c) Seizing prey.

(i) For seizing only. These would include mechanisms for grasping or impaling only, the prey then being transferred in bulk to the mouth without further treatment. The nematocyst-covered tentacles of *Hydra* afford a good example. Most birds show little or no mastication.

(ii) For seizing and chewing. Here are included mechanisms for taking and breaking up large masses. The toothed jaws of the vertebrates, the mouth parts of many arthropods are excellent examples. In the mammals the teeth are usually specially adapted for different purposes, e.g. incisors, canines and molars.

(iii) For seizing followed by external digestion. Some echinoderms can evert the stomach, kill and digest externally. Similar mechanisms are employed by some carnivorous gastropods, by *Sepia* and by many insects such as *Dytiscus* larvae.

3. *Mechanisms for Taking in Fluids and Soft Tissues*. These usually involve pumping mechanisms and often some piercing or boring

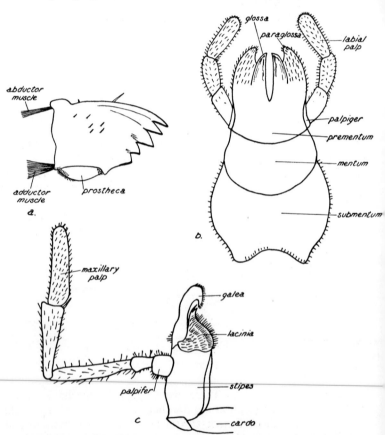

Fig. 10.23. Cockroach mouth parts (*a*) Mandible. (*b*) Labium. (*c*) Maxilla.

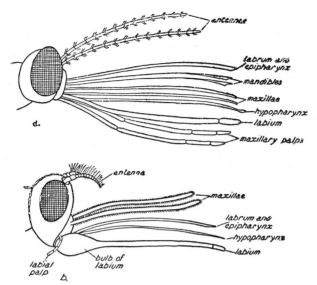

Fig. 10.24. (a) Mouth parts of female *Anopheles*. (b) Mouth parts of *Glossina*.

structure is required. If blood is taken, a substance to prevent coagulation is injected.

(a) For piercing and sucking. The sucking mechanisms of hookworms, leeches, parasitic Diptera, mites, ticks and cyclostomes exemplify this.

(b) For sucking only. The sucking mechanisms of the suctorian protozoa, trematodes, most nematodes, lepidoptera, flies and suckling mammals illustrate this.

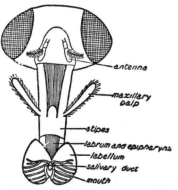

Fig. 10.25. Mouth parts of *Musca domestica*.

Digestion

The structures associated with digestion, as with feeding mechanisms, show a wide variety of adaptation to the nature of the food and the feeding habits of the animal. In general, where digestion is accomplished within a gut tube or tract, this can be divided into regions associated with special functions. There is anteriorly a region which receives the food masses or particles, the mouth and buccal cavity. Usually associated with this are glands which sometimes serve lubricating purposes, e.g. frog, sometimes digestive purposes, e.g. salivary glands of man, sometimes killing or paralysing purposes, e.g. venom glands of reptiles, sometimes anti-coagulant purposes, e.g. blood-sucking insects. Following this is a passage of conduction and sometimes this may be utilized for storage, although in some mammals, face pouches of the buccal cavity can serve this purpose temporarily. The oesophagus and crop of many animals including birds, insects and earthworms perform these functions. Often the crop is only a dilated portion of the oesophagus. Some digestion may occur in these regions in some cases. The next general portion of the gut is concerned with further grinding

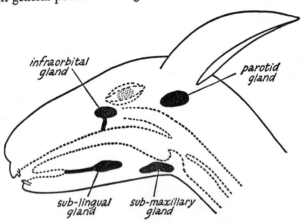

Fig. 10.26. Salivary glands of the rabbit.

or trituration of the food masses followed by digestion. We find such parts as gizzards and gastric mills performing the grinding in many animals and sometimes a filter is associated also to ensure that only the finest of particles can proceed further, e.g. crayfish, earthworm. Subsequent to the grinding, where any is necessary, this region of the gut releases most of the digestive enzymes and it is usual to find multicellular and unicellular glands distributed along it. There may also be diverticula or caeca into which food may pass for digestion before further transfer. They may even assist in the next process of absorption as in the crayfish (digestive gland). In mammals, where no internal

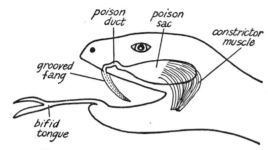

Fig. 10.27. Venom glands of the rattlesnake.

grinding is carried on, the regions of digestive changes are the stomach and duodenum. The stomach in this case also serves as a container and prepares the food into suitable form for further digestion in the intestine. The next region is one of absorption. It often overlaps the preceding one and in some animals is not clearly distinguishable at all. Here the soluble products of digestion are taken into the tissues via the gut lining. There follows then the last region which is conducting and faeces-forming only. It is of considerable importance to most land animals in that it acts to reabsorb water from the gut during faeces formation. In aquatic animals it is frequently very short and of no great importance. It ends at the anus.

Where digestion occurs in such a tract, it takes place outside the lining of cells; the enzymes are secreted into the lumen or cavity. The process is called *extra-cellular digestion*, in contrast with *intra-cellular digestion*, where food materials are taken in by the cells and digested internally. Some animals employ both methods in roughly equal proportions, e.g. *Hydra*, but the majority have developed one or the other almost exclusively.

In general, the very simple animals such as the protozoans, sponges and coelenterates, digest intra-cellularly. There are only a few records of extra-cellular enzymes being produced by these. A hydra can secrete a protease into the enteron, which will cause breakdown of protein there, but all other digestion is intra-cellular. It is possible to consider the engulfing of food particles and the formation of food vacuoles as extra-cellular digestion since the food is separated from the protoplasm by the vacuole membrane and the enzymes are secreted into the vacuoles, absorption of breakdown products taking place through the membrane.

But it is more usual to consider this as intra-cellular digestion. Some more advanced animals also show intra-cellular digestion and this is usually associated with sluggish habits and the taking of small particles of food.

Extra-cellular digestion is associated more with the taking of larger food particles and is a chemical method of breaking up the large particles into smaller ones so that they may be taken in. Where both methods are adopted, extra-cellular digestion usually goes far enough to break down very large molecules into molecules small enough to be absorbed, so that they can then be further broken down inside the cells by internally-secreted enzymes. For example, large protein molecules may be digested externally to proteoses or peptones, and then converted into amino-acids within the cell. From an evolutionary viewpoint, extra-cellular digestion seems to have been derived from intra-cellular digestion, the process of digestion going so far outside the cell, as to make it unnecessary for any further digestion inside.

It is clearly not possible to give a comprehensive account of digestion throughout the animal kingdom, partly because of the numerous types of food substances ingested, and partly because of the wide variety of detail of the processes. Here it will suffice to describe as fully as possible the digestive processes of the mammal and thus to illustrate the fundamental features associated with the process in other animals.

To digest means to reduce or convert to a convenient absorbable form, a more complex substance. (Note that the term may be applied to the breakdown of complex ideas so that they may be absorbed by the mind. Learning is one long process of mental digestion.) Digestion is achieved primarily by the action of hydrolytic enzymes secreted by the gut. Enzymes are specific in their action and can only work efficiently under the proper conditions of temperature and pH. Furthermore they must be discharged into the gut at the right places and times to meet the passing food, so that secretion must be co-ordinated with food intake and transmission through the gut.

The digestive processes are commenced whilst the mammal is masticating the masses of food material in the mouth cavity, for the process is not only one of grinding or tearing, but also one of mixing with the juices of the salivary glands which are stimulated to secrete in quantity when food is being taken. The glands are innervated in part by the chorda tympani nerve (branch of the VIIth cranial crossing the middle ear cavity), and in part by the sympathetic nerves. Artificial stimulation of the chorda tympani only, produces a copious watery secretion, whilst artificial stimulation of the sympathetic only, causes a thick viscid secretion. Stimulation of both produces normal secretion. The stimulation when an animal is feeding is associated with sight taste and smell, but the response in some animals can be made a conditioned reflex (*see* Chap. 16). An analysis of mixed salivary juice from all the salivary glands of man shows it to be largely water containing mainly an enzyme, *salivary amylase* or *ptyalin* and sodium chloride, with smaller amounts of phosphates and carbonates, calcium, potassium and magnesium; sulpho-cyanide is present. The pH of saliva is about 6·7 to 6·8. Its functions are partly lubrication of the inner surface of the buccal cavity and of the bolus of food for ease of swallowing, and partly digestive, since the ptyalin is able to hydrolyse starch and glycogen to maltose and dextrins. It can do this only if starch grains are cooked or broken up to remove the outer covering of amylopectin which cannot be hydrolysed by animal amylases. The optimum pH for ptyalin activity is 6·7. The presence of a separate glycogenase is not considered to be the case.

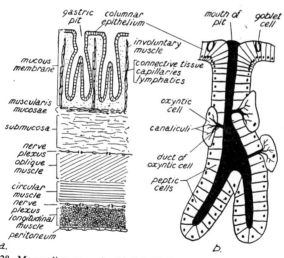

Fig. 10.28. Mammalian stomach. (*a*) V.S. Wall. (*b*) Gastric pit, fundus region.

On completion of mastication the bolus is passed by the act of swallowing into the oesophagus and conveyed by peristaltic action into the stomach. The ptyalin continues its activity for some time, but in the stomach the secretions are strongly acid so that this enzyme is soon inactivated.

Gastric (stomach wall) juices are secreted partly reflexly in response to the stimuli of sight, taste and smell. Pavlov showed in dogs, that the presence of food in the mouth caused copious flow of gastric juice.

Stimulation of the glands appeared to be due to impulses transmitted through the vagus nerve. This first flow is short-lived, but is present when a bolus reaches the stomach. Further secretion is maintained more or less continuously as a response to the stimulation of the glands by a hormone, *gastrin*. This is produced by the epithelial cells in the pyloric region of the stomach when certain food substances are present. It diffuses rapidly into the blood stream and eventually finds its way to the secretory cells of the gastric glands causing them to maintain their secretion continuously (*see* Chap. 8). The gastric mucosa produces two further hormones. *Enterocrinin* initiates secretory activity in the duodenum and *gastric secretin* promotes the secretion of alkaline solution, mainly due to bicarbonate, from the pancreas.

Analysis of gastric juice of man shows the presence of the enzyme precursor *pepsinogen*, and in smaller quantities *pro-rennin* and *lipase*, free *hydrochloric acid*, inorganic *chlorides*, *lactic acid* and *phosphates*. The amount of hydrochloric acid found is variable under test conditions, depending on the test meal and the time of sampling. Juice obtained after fasting contains about 0·1 per cent by weight. The free acid has several functions. First, it activates the pepsinogen to yield the active enzyme *pepsin*, and the pro-rennin to yield *rennin*. It also lowers the pH of the stomach contents to a value of 1·5 to 2·0. Thus a favourable degree of acidity is developed for the action of pepsin. Once formed, pepsin can activate pepsinogen, so that the formation of pepsin becomes autocatalytic. Secondly, it serves to sterilize the stomach contents to some extent, by killing many of the putrefying bacteria ingested with the food. It is due to the action of some of these bacteria that lactic acid is present. It can also bring about the hydrolysis of sucrose to glucose and fructose and split the nucleo-proteins of the food into nucleic acid and protein. The most important digestive process which occurs in the stomach is the hydrolysis of proteins to peptones by pepsin. The optimum pH for the enzyme activity is about 1·8 and in neutral or alkaline medium, the action stops.

Rennin is much more abundant in the gastric secretions of suckling mammals than in adults. It is able to bring about the conversion of the milk protein caseinogen to the insoluble calcium salt of casein in the presence of calcium ions, so forming a curd. Commercial rennet for the making of junket is an extract of the mucous membrane of the stomach of calves. Rennin is also believed to be proteolytic but not a great deal is known of its activities.

The presence of lipase in gastric juice has been variously attributed to secretion by the gastric glands on the one hand and to regurgitation from the duodenum on the other. Its digestive role in the stomach cannot be very great because of the high acidity, since the optimum pH for lipase is about 6·0 to 7·9. It varies for different animals, being 6·3 in the rabbit, 5·5 in man, 8·6 in the horse, and 7·9 in the pig.

Due to the copious gastric secretion and the effects of the enzymes ptyalin, pepsin and rennin, the stomach contents become converted from a rather solid nature to a much more watery condition in which some products of the digestion will already be in true solution. This fluid mixture is termed *chyme* and is ready to meet the next and last series of digestive secretions. Its pH is quickly raised to near neutrality by these when it is passed through the pyloric sphincter to the duodenum into which they are poured.

There are three distinctly different juices entering the cavity of the duodenum. One, called *bile*, is a secretion of the liver; it enters the duodenum via the bile duct leading from the gall-bladder, in which the bile was stored. Another is the *pancreatic juice*, secreted by the pancreas; it enters the duodenum via the pancreatic duct. The third, called the *succus entericus*, is a product of the secretory Brunner's glands and the crypts of Lieberkuhn, both in the wall of the duodenum itself.

Four hormones are produced by the duodenal mucosa. The presence of acid chyme from the stomach induces the production of *secretin*, which stimulates the continued flow of alkaline solution from the pancreas. *Pancreomysin* induces secretion of pancreatic juice rich in enzymes and *cholecystokinin* causes contraction of the gall-bladder (where present), thus expelling bile into the duodenum. Secretion of bile from the liver is initiated by nervous stimulation through the vagus nerve, and the presence of certain foods, and of secretin in the blood, will also stimulate production of bile. *Enterogastrone* inhibits both acid secretion by the gastric mucosa, and the peristaltic contractions of the stomach wall. In the gut we see a progressive series from nervous to hormonal control of secretion. Salivary gland control is entirely nervous; gastric secretion is first nervous, then hormonal; duodenal secretion is almost all hormonal.

The bile juice contains a number of *bile salts* of which the most important are sodium glycocholate and sodium taurocholate. There are also the *bile pigments*, bilirubin and biliverdin, some *mucin*, *cholesterol*, and traces of other substances. The bile salts have the property of considerably reducing the surface tension at fat-water interfaces, and therefore make emulsification of fats a much simpler process. They also tend to keep the fat emulsified once it has become so, thus enabling lipase to work more rapidly. If the fat droplets are small enough, they can be absorbed directly. The bile pigments are excretory products; they play no part in digestion. It is doubtful whether cholesterol has a specific function in the gut; it may be merely another excretory product.

It can be synthesized by a mammal, and is closely allied to vitamin D. The presence of the bile salts retains it in solution; if they fall below the normal level, the typical cholesterol gall-stones are formed in the bile ducts. It is probable that cholesterol is converted to coprosterol in the lower gut by the action of bacteria; it is passed out with the faeces. The pH of the bile is about 8.

Pancreatic juice contains the enzyme precursors *trypsinogen* and *chymotrypsinogen*. In the duodenum, trypsinogen is activated to the enzyme *trypsin* by *enterokinase*, and chymotrypsinogen is activated to the enzyme *chymotrypsin* by trypsin. The juice also contains a carbohydrase, *amylase*, and the fat-splitting *lipase*. The pancreatic juice is distinctly alkaline in man, with a pH of 8·8, due to the presence of *sodium bicarbonate*. Trypsin and chymotrypsin continue the work begun by the pepsin of the stomach and break more peptide linkages in the proteins until only short-chain polypeptides remain. Few amino-acids are produced at this stage, but they are ultimately separated out by the action of peptidases of the succus entericus on these polypeptide chains. The amylase finishes the work begun by ptyalin and thus all starch and glycogen are converted to maltose. The lipase hydrolyses fats to fatty acids and glycerol.

Analysis of succus entericus of man shows it to be a watery secretion containing the enzymes "*erepsin*," *invertase*, *maltase*, *lactase*, *phosphatase*, *nuclease* and *lipase*. It also contains a substance *enterokinase* which serves to activate an enzyme precursor trypsinogen found in pancreatic juice (*see* above). Succus entericus is alkaline, pH 8·3.

"Erepsin" was once thought to be a single enzyme capable of hydrolysing the intermediate breakdown products of protein such as peptones and polypeptides. It was described as a peptidase. It is now known to be a mixture of much more specific single enzymes each of which can hydrolyse different amino-acid combinations of the polypeptide nature. It is, therefore, able to carry on where the protein and polypeptide hydrolysers such as pepsin and trypsin finish. This mixture of peptidases shows a broad pH optimum in the neutral or slightly alkaline zone.

Invertase, maltase and lactase are carbohydrases, that is, they can hydrolyse carbohydrate. Invertase can convert sucrose (cane sugar) to glucose and fructose; maltase can convert maltose to glucose; lactase can convert lactose to glucose and galactose. There is some doubt if the enzymes invertase and maltase are really different, it is probably a case of the same enzyme acting on the two substrates at different rates. These carbohydrates all work best around pH 7.

The nuclease and phosphatase (or more correctly nucleotidase) are concerned with the digestion of nucleo-proteins which are salt-like combinations of nucleic acid and protein. The nucleo-protein is first split in the stomach into nucleic acid and protein. The latter is dealt with as for other proteins but the nucleic acid is further split by the enzymes mentioned. The nuclease liberates the component nucleotides whilst the nucleotidase dephosphorylates these to yield nucleosides, which are glycosides of the nitrogenous bases. It seems probable that a nucleosidase is also present to complete the breakdown of the nucleosides into their base and glycoside components.

It is especially noteworthy that the commonest of all plant polysaccharides, cellulose, cannot be digested by the great majority of animals. The fact is that they are unable to produce the necessary enzyme, *cellulase*. This enzyme has been found in the gut of earthworms, snails and a few wood-boring insects. It is also produced by some of the flagellate protozoa. On the other hand, some bacteria and fungi are very capable of digesting cellulose, so it is not surprising that many of the herbivorous animals make use of their abilities in this respect. The termite is an excellent example. In the hind-gut, it carries bacteria and flagellates which digest the cellulose in the diet. The termites live by digesting the micro-organisms. The termites, after having the protozoa removed, are unable to live on a diet containing only cellulose as a carbohydrate. The ruminant animals such as the cow or sheep also make use of symbiotic bacteria and protozoa, which inhabit the rumen of the gut. In other herbivorous animals such as the horse and rabbit, the caecum contains vast numbers of bacteria which digest cellulose. The cellulose is converted into short-chain fatty acids such as propionic acid. The animal is able to absorb these and from them, it can synthesize glycogen in the liver. Also, the animal benefits by digesting with its residual enzymes the plant cell contents released by the bacterial breakdown of the cellulose walls.

The foregoing account of digestion may mislead the student to the extent of thinking that the process is one of a series of clearly defined stages. This is unavoidable when trying to extricate the activities of a particular enzyme from among all the others, but it must be borne in mind that once the food has passed from the stomach it is subjected to all the remaining juices simultaneously and these in reality form a complete enzyme system which digests food as a continuous process, since the constituents of the system are ordered and regulated to do so.

As the chyme proceeds along the intestine, the enzymes secreted in the juices just mentioned carry on their hydrolysing activities to reduce the food to a still more watery fluid called *chyle*. Once the diffusible derivatives of the food substances are present in the chyle, the next phase of absorption by the gut commences, and all the way along the intestine this will be occurring as digestion proceeds to its end.

SUMMARY OF MAMMALIAN DIGESTION

Food material	Enzymes	Secretion	pH of secretion	Product of digestion
Carbohydrates Polysaccharides Cellulose	—	—	—	None except by micro-organism
Starch	Ptyalin	Saliva	6·7 to 6·8	Maltose
Starch	Amylase	Pancreatic juice	8·8	Maltose
Glycogen	Ptyalin	Saliva	6·7 to 6·8	Maltose
Glycogen	Amlyase	Pancreatic juice	8·8	Maltose
Disaccharides Sucrose	Invertase ⎫			Glucose and fructose
Maltose	Maltase ⎬	Succus entericus	8·3	Glucose
Lactose	Lactase ⎭			Glucose and galactose
Proteins Proteins	Pepsin	Gastric juice	1·1 to 1·8	Peptones and polypeptides
Peptones and Polypeptides	Trypsin ⎫ Chymotrypsin ⎭	Pancreatic juice	8·8 8·8	
	Erepsin mixture	Succus entericus	8·3	Amino-acids
Fats Fats	Lipase	Succus entericus	8·3	Fatty acids and glycerol
	Lipase	Pancreatic juice	8·8	
Fats	Emulsified by bile salts in bile		8	Fat droplets

Absorption

The products of the digestive processes will begin to appear along the alimentary canal from the stomach onwards, but little, if any, of these in the stomach are in an absorbable form. The gut of the mammal is unable, except in certain instances, e.g. absorption of blood plasma, to absorb compounds of greater complexity than hexose sugars, amino-acids, fatty acids and glycerol. Therefore, until the carbohydrates, proteins and fats are reduced to these forms, they cannot pass through the walls of the gut. There will, of course, be compounds of absorbable substances in the food as a general rule and some of these may be absorbed quickly through the stomach wall, e.g. sugars such as glucose. On the whole, however, absorption by the stomach is of no great significance and by far the greater proportion of the absorbed nutrient passes through the wall of the intestine. In most animals,

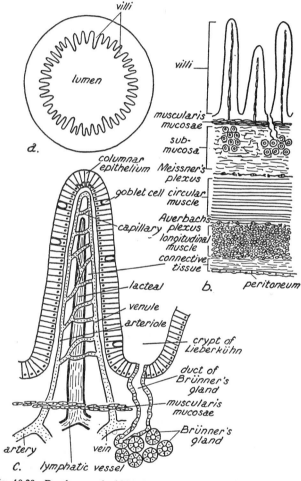

Fig. 10.29. Duodenum of rabbit. (*a*) T.S. (*b*) V.S. wall. (*c*) Single villus.

ome part of the alimentary canal is modified in its form to bring bsorption to as high a degree of efficiency as possible. This entails resenting to the products of digestion a large internal surface area, and there are numerous ways in which this is accomplished. The typhlosole of the earthworm, the digestive gland of the crayfish, the gut caeca of insects, the spiral valve of the dogfish and the folding of

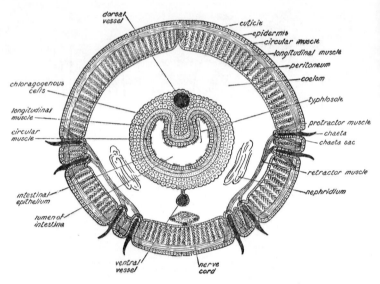

Fig. 10.30. *Lumbricus.* T.S. body in region of intestine.

the intestine wall of the frog are all examples. In mammals the internal surface area of the absorbing region is increased enormously by the occurrence of the villi. In addition, the intestine is highly vascularized so that absorbed materials may be transported rapidly from the site of absorption.

Carbohydrates are absorbed chiefly in the form of monosaccharides the hexoses and pentoses, but the rates at which these are absorbed are quite variable. This is somewhat surprising since their molecular weights are very little different from one another. It must mean that the absorption of sugars from the gut is more complicated than simple diffusion. This is also borne out by the fact that the sugars can be absorbed from very strong concentrations and therefore against a high diffusion gradient. Experiments on the rat have shown that galactose and glucose are absorbed comparatively rapidly, fructose less than half as quickly and the pentoses much less rapidly still. It has been suggested that the rapid selective absorption of glucose and galactose is due to the fact that they are phosphorylated by phosphatases in the cells of the gut lining and thus removed from the end of the gradient. The monosaccharides, once absorbed, are rapidly transferred to the gut capillaries which occur in the villi and are then conducted away.

Proteins are absorbed in the form of their constituent amino-acids into the blood stream. There is no evidence to support the view that

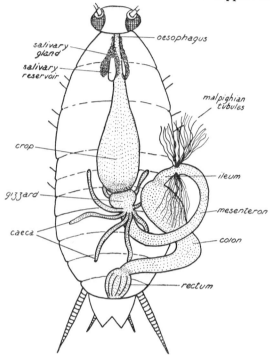

Fig. 10.31. Dorsal view of alimentary canal of cockroach.

rotein breakdown products such as peptones can be absorbed. In ct, experiment supports the contrary.

The problem of fat absorption has not been completely elucidated. could occur with the fat in the form of fatty acids and glycerol or irectly as finely-emulsified fat globules. The action of lipase on the nely-divided fat droplets caused by the emulsifying properties of the le salts is undoubtedly to produce fatty acids and glycerol, some of hich are certainly taken up by the gut cells and transferred to the ood-stream. Analysis of the blood leaving the gut shows this to be e case. But experiment has also shown that fat droplets of less than out 0·5 μm diameter can be directly absorbed. Now, the bile salts alone are incapable of producing an emulsion of fat droplets smaller than about 2·0 μm, but in the presence of fatty acids and glycerides, the bile salts can emulsify fat down to droplets of suitable dimensions for direct absorption. It would seem, therefore, that fat absorption occurs in two ways, (a) as the products of lipase digestion, and (b) as finely divided particles produced as a result of emulsification by the bile salts in the presence of the products of lipase action. This latter absorption component is transferred to the lacteals of the villi and enters the main blood circulatory system via the thoracic duct. Free fatty acids are mainly insoluble in water and their manner of absorption is somewhat problematical. It is probable again that the bile salts assist by their ability to form substances with the free fatty acids which are water-soluble. These substances are combinations of the fatty acids and the glyco- or tauro-cholates, so that when they are absorbed the bile salts are taken back into the blood and can once more be secreted by the liver to serve their purpose over again.

Absorption of other food requirements such as vitamins and mineral salts is in their unchanged condition, since the gut enzymes do not attack the vitamins, and the mineral salts are already in diffusible form. The rate of passage of different mineral salts through the gut wall is variable and there appears to be very definite selective absorption which is probably bound up with the permeability properties of the gut epithelial cells.

Absorption of water is presumed to be largely under the influence of osmotic forces, the contents of the lower gut being generally hypo tonic to the blood in the villi capillaries thus causing water to flow from the gut lumen to the blood.

Assimilation and Fate of Absorbed Substances

Assimilation is the incorporation of the products of digestion of simple nutrients into the complex constituents of the animal. The process is not exactly immediate upon absorption in the mammal since materials absorbed into the blood stream of the gut are directed by the hepatic portal system to the liver and it is in this large organ that the preliminaries to ultimate assimilation and subsequent metabolism are carried out.

Every organism which is unable to synthesize all its own amino-acids requires certain essential amino-acids in its diet, so that its protein requirements can be met. Provided the animal takes in enough food protein to supply its needs of the essential amino-acids, that is sufficient it is then able to convert the essentials into the others that are required It is unlikely, however, that this would be necessary for animals on normal varied diet, since the essential ones are in general the lea common, and so adequate intake of these would be accompanied by adequate intake of the others.

The amino-acids absorbed from the gut are transported through the hepatic portal system to the liver. The primary function of amino-acids is to supply the units from which the proteins needed for growth, repair and secretory activities can be synthesized. The animal's immediate requirements of these are met by the distribution of amino-acids from the liver in the systemic circulation, but almost invariably, the intake of amino-acids is much in excess of these requirements. Proteins and amino-acids are not permanently stored in the animal body so that the excess must be metabolized immediately. This is carried out in the liver, and is, in effect, one of three processes of nitrogen elimination in the form of ammonia, urea or uric acid. The three processes are *deamination, transamination* and *amide formation*.

Deamination is the process of nitrogen removal; it might be brought about by an oxidative, reductive or a hydrolytic reaction. It seems that the process is almost always the first, i.e.

$$R.CH.NH_2.COOH + \tfrac{1}{2}O_2 \rightarrow R.CO.COOH + NH_3$$

This ammonia can be rendered non-toxic by being incorporated into urea, uric acid, or one of a variety of other substances according to species, and eliminated by an excretory process. In transamination, there is, in animals (and plants), an enzyme system for the transference of an amino (NH_2) group from an amino-acid to another molecule. These are the *transaminase* enzymes. This accounts for the fact that only certain amino-acids are essential in a diet, the non-essential ones being synthesized by transamination in the presence of another suitable molecule. Not all the amino groups are eliminated or used in transaminations. It seems that some can be stored in the form of an amide, glutamine, which can serve as a source of amino groups for transaminations when required, or be used as a source of ammonia in the kidney to neutralize excess acids produced during metabolism. The enzyme glutaminase catalyses the glutamine $\rightarrow$ glutamic acid $+ NH_3$ reaction.

After removal or transfer of NH_2 from an amino-acid the residue is a keto acid. Amino-acids fall into three groups according to the kind of keto acid produced. Glucogenic amino-acids yield keto acids which can be converted to glucose and hence enter the normal carbohydrate metabolism of the body. Examples are glycine, alanine, aspartic acid, glutamic acid, arginine and ornithine. In fact about 60 per cent of all amino-acids are glucogenic. Ketogenic amino-acids include leucine, phenylalanine and tyrosine. These form the ketone bodies β-hydroxybutyric acid and acetoacetic acid during their eventual oxidation to carbon dioxide and water. The third group of amino-acids are neither glycogenic nor ketogenic and their fate is not known. They include valine, lysine, tryptophane and histidine.

Carbohydrate metabolism is centred on the build-up and breakdown of glycogen and this occurs mainly in the liver and muscles. Glycogen can be built in the liver directly from the absorbed carbohydrates and from non-carbohydrate sources such as certain amino-acids, glycerol, lactic acid, pyruvic acid, and many other simpler substances. As it is now understood, there are no "essential" carbohydrate food substances in the same sense that there are "essential" amino-acids. All the absorbed hexose sugars, glucose, galactose, fructose and the rarer mannose can be used in the formation of glycogen. Glycogen synthesis occurs when there is a supply of glucose available in an appropriate cell and it was originally considered that because the enzyme phosphorylase could catalyse the build-up of glycogen molecules from glucose-l-phosphate when primed with a short chain of glucose units *in vitro*, it was equally effective in the cell. It is now known that this is not so. In fact, the phosphorylase equilibrium is much too far in the glycogen breakdown direction for it to be possible unless the glucose-l-phosphate concentration is abnormally high. It is now seen that glycogen synthesis is much more complex and involves the activity of a co-enzyme, *uridine triphosphate* (UTP), derived from the pyrimidine *uracil*. UTP is chemically similar to ATP and has "energy-rich" terminal phosphate bonds. This co-enzyme can react with glucose-l-phosphate to form *uridine diphosphate glucose* (UDPG) and inorganic phosphate. This glucose combination with UDP is highly reactive and can undergo a number of changes that the original glucose-l-phosphate cannot. For example, it can be converted into one of the other hexose sugar isomers, such as galactose, very easily. What is important in the glycogen synthesis connexion is the fact that in the UDPG state and under the influence of the enzyme *UDPG-transglycolase*, glucose units can be transferred to short "primer" sugar chains to form the long glucose polymer molecules that constitute glycogen and this at comparatively low sugar concentrations. At each transfer, UDP is set free and this is converted to UTP from ATP molecules. This means that for every glucose unit added to the glycogen chain two ATP bonds are used up since one is used in converting glucose to glucose-l-phosphate and not one only as would be the case with a phosphorylase catalysed reaction.

But the process is still a little more complicated because the product of the synthesis above is a straight chain of glucose units linked only by 1–4 bonds whereas glycogen is a branched structure made by joining one length of 1–4 linked glucose chain to a similar chain via the first and sixth carbon atoms. This last forked combination of straight chain components is built up under the influence of special enzymes of which the detailed function is still obscure but the final result of the trans-glycolase and branching enzyme system is to form chains of glucose units, forking every twelve to eighteen units and composed of a total of up to three hundred hexose structures.

Other sources of glycogen in the liver are the deaminated residues of certain amino-acids. The residues are formed into compounds of pyruvic acid and these the liver is able to phosphorylate, forming phospho-enol-pyruvate which is then converted into carbohydrate by the reversal of the glycolytic processes (see Chap. 11). In fact, any substance which lies on the glycolytic route from carbohydrate to pyruvic acid, or which gives rise to any substance on this route, can be treated in a similar manner. Thus glycerol is a well-known source of glucose through glycerol-phosphate and phosphoglyceraldehyde.

Whilst the liver acts as a central storehouse of glycogen, much is stored in the muscles, and this is built up there at the expense of the central store, which maintains its supplies to the muscles by breaking down its own glycogen to glucose and transferring it to the blood stream, by which it is then transported to the muscles to be re-synthesized into glycogen until used in respiration. The amount of glucose in mammalian blood is carefully regulated to a concentration of about $0 \cdot 1$ g per 100 cm^3. This regulation is a complex process which is discussed in some detail in Chap. 6. The most important factors are the hormones produced in the islets of Langerhans, the β-cells of which secrete *insulin* in response to a rise in blood glucose level, while the smaller α-cells secrete glucagon in response to stimulation by *somato-tropin* from the pituitary. In a healthy animal, the concentration of blood glucose is carefully maintained by the action of these two hor-mones. Serious effects follow excess or deficiency of insulin; they are discussed on p. 286. Two other hormones, *thyroxin* and *adrenalin* can also affect the level of blood glucose (see p. 199).

The capacity of the liver to store glycogen is strictly limited. The liver of the rabbit cannot contain more than about 20 per cent of its weight as glycogen, no matter how rich the diet in carbohydrates. Under conditions of excess carbohydrate intake, the portion which the liver cannot handle as glycogen, is converted to fat and stored in the fat-storage regions of the body.

The fate of absorbed fat has been studied extensively by the use of fats containing "heavy" hydrogen, which forms "heavy" water, and this is easily detectable in tissues and tissue extracts. It has been shown that apparently the immediate fate of absorbed fat is its deposition in the fat-storage regions of the body. These are generally the mesenteries and the connective tissues, particularly under the skin. It is the case that each species generally stores its own particular kind of fat, for example, beef and mutton fats are clearly distinguishable. This is presumably due to the fact that each has its own particular constant diet, for it is possible to change the nature of the fat stored, by controlling the fats in the diet. In those animals where carbohydrates preponderate heavily in the diet, much of the fat is formed from them and this would lead to specific fat formation according to the enzyme systems of the animal.

Carbohydrate not required immediately for respiratory purposes is degraded as far as pyruvic acid which is then decarboxylated to "active acetate" with co-enzyme A. Being in excess of immediate requirements this "active acetate" does not enter Krebs' cycle but is synthesized into fatty acids by a reduction process. With glycerol from dihydroxyacetone phosphate these fatty acids are esterified to fats and stored.

The process by which fatty acids are synthesized is known not to be the reverse of the fatty acid oxidation process mentioned on p. 443. Whereas the oxidative breakdown of fatty acids involves the co-enzyme NAD, the synthetic reductive pathway works best with $NADPH_2$, not $NADH_2$. Also it has been found that when cell preparations are supplied with bicarbonate ion, HCO_3^-, as a source of carbon dioxide they can synthesize fatty acids much more readily than without despite the fact that the carbon dioxide does not enter into the structure of the fatty acids produced as shown by use of labelled tracers. Thus carbon dioxide seems to be acting as a catalyst in the process. Another reason for there being difficulty in reversing the oxidative process is that the equilibrium in the last of this chain of reactions, namely, *thiolysis* under enzyme *β-ketothiolase* activity lies far too near the end point for it to be reversible in the cell with normal substrate concentrations, that is the concentration of acetyl co-enzyme A in the cell is never high enough. It is now thought that most of the fatty acid synthesis that occurs in cells is the result of a preliminary activation of the reactants at the expense of ATP involving another more reactive co-enzyme derived from acetyl co-enzyme A, namely, *malonyl co-enzyme A*. It is in this transformation that the carbon dioxide is used initially, to be released later. The formation of malonyl Co A can be represented as:

$$CH_3CO.S\,CoA + CO_2 + ATP \rightleftharpoons$$
$$CH_2COOH.CO.S\,CoA + ADP + \text{inorganic phosphate}$$

The enzyme concerned is *acetyl co-enzyme A carboxylase* and needs manganese ions and the vitamin biotin for its activity.

The compound malonyl co-enzyme A can now form the β-keto acid *acetoacetic acid*, in combination with acetyl co-enzyme *A* releasing carbon dioxide and co-enzyme *A* according to:

$$CH_3CO.S\ CoA + CH_2COOH.CO.S\ CoA \rightleftharpoons$$
$$CH_3COCH_2CO.S\ CoA + CO_2 + HSCoA$$

The end product so far is the combination of two molecules of acetyl co-enzyme *A* to give one of acetoacetyl co-enzyme *A*. From this, by reduction, using $NADPH_2$, fatty acid can be formed and according to the enzyme complex involved this may be *stearic*, *palmitic*, *oleic* or other acid.

The build-up of fats from fatty acids requires their combination with glycerol to form *glycerides*. There is more than one way in which this is achieved. Some energy from ATP and the co-enzyme *A* are required. By one system glycerol is first activated by phosphorylation to α-*glycerophosphate*:

$$CHOH.2(CH_2OH) + ATP \rightleftharpoons CH_2OH.CHOH.CH_2O.P + ADP$$

This activated glycerol can be used in combination with two fatty acyl co-enzyme *A* molecules to form *phosphatidic acid* which is then converted into *triglyceride*. This calls for the removal of the phosphate hydrolytically by the enzyme *phosphatidic acid phosphatase* (irreversible reaction) to form a diglyceride which can then, with another fatty acyl co-enzyme *A* molecule, produce the triglyceride. All this can be summarized:

$$CH_2OH.CHOH.CH_2O.P + R^1.CO.SCoA \rightarrow$$
$$CH_2OOC.R^1.CHOH.CH_2O.P + HS\ CoA$$
$$CH_2OOC.R^1.CHOH.CH_2O.P + R^2.CO.SCoA \rightarrow$$
$$CH_2OOC.R^1.CHOOC.R^2.CH_2O.P + HS\ CoA$$
$$\underset{\text{acid}}{\underset{\text{phosphatidic}}{}}$$
$$CH_2OOC.R^1.CHOOC.R^2.CH_2O.P + H_2O \rightarrow$$
$$CH_2OOC.R^1.CHOOC.R^2\ CH_2OH + \underset{\text{phosphate}}{\text{inorganic}}$$
$$CH_2OOC.R^1.CHOOC.R^2\ CH_2OH + R^3.CO.SCoA \rightarrow$$
$$CH_2OOC.R^1.CHOOC.R^2.CH_2OOC.R^3 + HS\ CoA$$

(R = fatty acid radical).

In this system one ATP bond is used in making glycerophosphate and three for making co-enzyme *A* derivatives of the fatty acids, making four in all and this reaction path seems to be the one followed in adipose tissue where long-term fat storage is carried out. In other tissue another more direct pathway is followed in which a monoglyceride, *monolein*, can react directly with two successive fatty acyl co-enzyme *A* molecules to give triglyceride without the need of phosphorylation of glycerol.

The enzyme concerned in this pathway is *monoglyceride transacylase*. Three ATP molecules are involved.

Deposited fat is undoubtedly of greater use to the animal as a source of energy-giving compounds than any other substance. Fat is far richer in carbon and hydrogen than carbohydrates and proteins and there is more oxidizable material in one gram of fat, than in the same amount of carbohydrate or protein. Also important is the fact that when oxidized, fat yields about twice as much water as carbohydrate and protein. This is important to animals living under conditions of water shortage, e.g. camel.

Transport of fat to the storage regions is chiefly as neutral fat droplets although a little may be carried as phospholipides. Redistribution of fat from the storage deposits, appears to be totally as phospholipides which are water-soluble. When storage fat is to be metabolized, it is first transported to the liver and there utilized for one of several purposes. Some is incorporated into the protoplasmic structure and becomes a permanent constituent of the cells and tissues. This amount of fat in the body, referred to as the "constant" fat, does not vary from day to day according to food intake, but the remainder, the "variable" fat, is largely used as a fuel in the respiratory activities. There is evidence that small quantities of particular fatty acids are necessary to healthy growth in mammals.

Reference to the fate of water, mineral salts, vitamins, and nitrogenous compounds derived from nucleoproteins has already been made. There is further discussion of these topics in a later chapter (*see* Chap. 15).

Faeces Formation and Egestion

When absorption has been completed by the intestine, there will remain in the lumen undigested residues of food material, particularly cellulose in the cases of many herbivores and omnivores, the secretions of the wall of the gut and the accessory organs, and a good deal of water. As has been mentioned, cellulose may be utilized by micro-organisms of the caecum in some animals with a consequent yield of more absorbable compounds. Beyond the caecum, where it exists, the large intestine or colon, and the rectum of land vertebrates are actively engaged in water reabsorption as a means of water conservation. (Note that in aquatic vertebrates this region is extremely short since there is no need for similar saving of water and the faeces are consequently of a liquid nature.) As a result of this, the food and secretory residues become increasingly solidified as they are passed along. By the time they have reached the end of the rectum, they attain a hardened consistency. At this stage, they are ready for egestion and this is accomplished by the act of *defaecation* by the animal, which is the result of muscular contraction of the rectal walls. The properties and constituents of faecal matter are very variable according to the animal and its diet.

Putrefactive activity by gut bacteria gives the faeces certain characteristics. Indole and skatole, products of tryptophane breakdown, account for the smell. Products of carbohydrate breakdown in the form of CO_2 and CH_4 may accompany the solid part, as may H_2S and NH_3 from amino-acids. Acetic and butyric acids may also result from bacterial activity.

Passage of Food

For the purposes of adequate digestion and absorption, the food material must be actively propelled, and in many instances there must be grinding and mixing of the digestive juices with the food so that the enzymes can work most efficiently. The propulsion of food vacuoles along definite paths around the protoplasm, as in *Paramecium*, is known as *cyclosis*. This propulsive mechanism is not understood; some doubt if it really exists. In most cases, there is no apparent mechanism for the purpose, the food vacuoles streaming with the currents of cytoplasm in no co-ordinated way.

In the metazoa there are two propulsion mechanisms, namely, *ciliary activity* and *muscular activity*. The former, as the sole mechanism, is found only in a few groups of animals such as the bivalve molluscs and the polyzoans. Some kind of muscular activity is found in all others. Occasionally, the muscular activity associated with locomotion is sufficient to keep the food mass on the move through the gut. This is the case in some worms and holothurians, where the gut musculature is very poorly developed. In most cases, though, there is a well-developed muscular wall to the gut (splanchnic mesoderm) and this can operate quite effectively, unaided by other muscles. In such cases, the food is propelled as a result of the movement of *peristaltic* waves of contraction of the muscle along the gut from anterior to posterior. The gut muscles are further capable of performing *pendular movements* which also aid in breaking up and churning the food masses. A brief description of the passage of food through the alimentary canal of a mammal will serve to illustrate these points.

The first stage of mastication in the buccal cavity is one of grinding and mixing of the raw food with the salivary juice. The muscles of the tongue and cheeks keep the food between the teeth, whilst the sideways movement of the lower jaw causes the grinding process. As a preliminary to efficient digestion, this is of extreme importance, particularly to herbivorous animals. The result of mastication is to form within the buccal cavity a *bolus* of food at the back of the tongue, where it is now ready for swallowing. This is pressed between the base of the tongue and the hard palate and with a curling of the tongue from anterior to posterior, the bolus is slipped backwards into the pharynx. This is the act of *deglutition* or swallowing. In the pharynx, it is gripped by the pharyngeal muscles, and passed into the oesophagus. Deglutition really involves three stages. The first is the voluntary act of pushing the food into the pharynx. The second is the conveying of the food through the pharynx. This is a reflex act and is quite rapid. It is accompanied by forward and upward movement of the hyoid bone, which causes several effects. The base of the tongue projects over the larynx, thus protecting it; the glottis is closed, and the vocal cords brought together; the soft palate is lifted to shut off the internal nares and to form a slope under which the bolus may be slipped; the pharynx is rapidly shortened, due to the contraction of the palato-pharyngeal muscles. It should be noted that the epiglottis does not fold over to close the entrance to the larynx completely, but remains erect all the time. The bolus is rapidly passed through the pharynx by the pharyngeal constrictor muscles, and during this part of the process of deglutition, breathing is inhibited by a reflex action. The third stage is that of the passage of the bolus along the oesophagus. The speed at which this occurs depends upon the consistency of the bolus. Whereas liquids are shot along extremely rapidly, more solid masses take up to six seconds

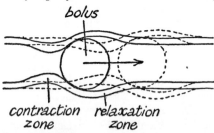

Fig. 10.32. Diagram showing the passage of a bolus through the oesophagus.

in the human being. In the latter case, the mass is moved by peristaltic action of the oesophageal muscles; it proceeds from anterior to posterior as a wave-like motion. The peristaltic movement is controlled at every point by impulses arriving through the vagus nerve and is not a wave propagated by mere muscular conduction. The propagation of the movement is reflex, and once started, it cannot be stopped. The muscular movements cause constriction of the oesophagus behind the bolus and relaxation in front of it, thus squeezing the bolus along (*see* Fig. 10.32).

When the peristaltic wave arrives at the cardiac opening of the stomach, the circular muscle of this region relaxes its normal condition of tonic contraction and allows the food to pass into the stomach. This also appears to be part of the whole reflex, since the opening of the stomach entrance only occurs when the peristaltic wave reaches that region. Liquids which are rapidly shot along the oesophagus are retained there until the slower peristaltic wave catches up.

Only if the food is solid, will it remain in the stomach; liquids are very rapidly passed to the duodenum. Only the pyloric region of the stomach shows any great muscular activity, the fundus serving to hold the main mass and exerting on it a steady pressure sufficient to push it to the pyloric region. The more active pyloric region, by peristaltic contractions, churns and mixes the food with gastric juice as it is received from the fundus, and then at successive intervals, delivers portions to the duodenum through the pyloric sphincter. This does not open at the arrival of every peristaltic wave from the stomach as does the cardiac orifice, but at quite irregular times, presumably coinciding with a sufficient degree of liquefaction of the stomach contents. The movement of the stomach wall seems to be more or less automatic and independent of central nervous control, for even if the vagus and sympathetic nerves which supply it are severed, it still continues its activity.

This appears also to be the case for the small intestine, of which the

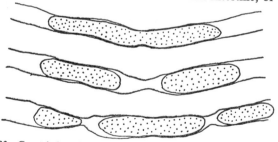

Fig. 10.33. Constriction of the intestine with consequent churning of the contents.

movements are of two kinds, namely peristaltic and pendular. The peristaltic movements are similar to those described for the oesophagus, but much slower. The pendular movements are side to side swinging movements of loops of the gut in a rhythmical manner. Such movement is due to a series of localized constrictions of the intestinal wall, thus causing local variations in the rigidity of the gut and the consequent swaying motions. The result of such movements is to cause breaking or segmentation of a column of food into short blocks. A fresh constriction pattern at new sites quickly follows resulting in the recombination of parts of these short blocks in different ways. In this manner the food is continuously churned (*see* Fig. 10.33). Superimposed on this, is the occasional peristaltic wave which carries the whole series of blocks some distance along the intestine, to be churned again by the pendular movements. The intestine is also innervated by the vagus and sympathetic nerves.

Movements of the large intestine are chiefly peristaltic but even slower than in the small intestine. This allows sufficient time for water absorption in this region.

The act of defaecation is partly voluntary and partly involuntary. The latter is due to strong contractions of the muscular walls of the rectum and relaxation of the sphincter muscles. The former is due to controlled relaxation of the external sphincter muscles and the contraction of the abdominal muscles and the diaphragm. Infant mammals have not the power to control these movements voluntarily, and in them the act of defaecation is purely reflex, voluntary control being learned at a later stage of development.

Vomiting is an act over which there is little or no control, being a complicated reflex action. The cardiac sphincter is inhibited from closing and convulsive contraction of the abdominal muscles and diaphragm causes the stomach contents to be violently ejected. Such a reflex may be set in motion by chemical irritants in the stomach (emetics), physical stimulation of the sensory nerves in the throat, irregular movements as in sea or air travel, and by stimulation from the brain centres as a result of horrifying sights.

Regurgitation of partially digested food for the purposes of feeding young, as practised by some birds (pigeons), is a similar phenomenon but must be due to some other stimulus, possibly the sight of the young with their mouths open.

The time for which food is retained in the alimentary canal depends inversely upon the efficiency of the digestive system. The better the digestion, the shorter is the period of retention. In cold-blooded animals therefore, temperature affects the rate of passage by slowing up the enzyme activity. However, the times of retention are associated with each species and its diet. For example, a meal of blood may remain in a leech for several months, whereas food passes very quickly through caterpillars. The dogfish usually takes 3–4 days to digest and absorb a meal. In warm-blooded vertebrates, the process is faster on the whole. A rabbit passes a meal in about 24 hours, man in 72 hours, a chicken in 20 hours. Food remains in the human stomach for an average of about 3 to 4½ hours, carbohydrates moving out faster than proteins. In cattle, food remains in the stomachs for much longer periods, e.g. in the rumen up to 60 hours, in the omasum 8 hours and in the abomasum 3 hours.

DEMONSTRATION OF THE ACTION OF DIGESTIVE JUICES

The main centres of digestion are the mouth, stomach and duodenum. Whilst the digestive juice of the mouth can be acquired at any time by an experimenter, the juices from the other regions can only be obtained directly by extraction from fresh tissues. There are available, however, commercial products known as liquor pepticus and liquor pancreatini, extracts from stomach and pancreas, which can be used in the following demonstrations. Bile salts powder can also be obtained as a commercial product.

The digestive properties of the secretions are best investigated in a comparative way and one composite experiment will serve this purpose. First obtain the necessary secretions and suitable substrates upon which they can work. These last are a suitable fat such as olive oil, a carbo-hydrate such as a starch sol and a protein such as coagulated albumen suspension or alternatively blood fibrin stained with congo red or carmine. This is achieved by leaving the fibrin in the stain for about a day and then filtering the fibrin off and washing it thoroughly until no more stain is removed. It should then be placed in glycerine if it is to be kept.

Make up three sets of four test-tubes containing respectively fat only, fat plus bile salts, starch sol, protein. Pour some saliva into each of the first four, liquor pepticus into each of the second four and liquor pancreatini into each of the third four. Label each tube according to its contents, plug with cotton wool and incubate at 37°C for several hours. At the end of this time, examine or test each tube for digestive activity. Digestion of fat can be shown by acid reaction of fatty acids and by the presence of glycerol; digestion of starch by the failure to give a colour reaction with iodine and by presence of reducing sugars, and that of the protein by clearing of the cloudy albumen in that case or by release of stain into the solution if stained fibrin is used.

The experiment can be made still more instructive if further sets of tubes are made up in which the solutions are deliberately arranged to have pH values outside the known working range of the enzyme. For example, the effect of liquor pepticus in an alkaline solution and the effect of liquor pancreatini in an acid solution can be tested. A complete set of tubes can be placed in a refrigerator or ice pack to see the effect of low temperature and all the enzyme solutions can be boiled before use in another set to see the effect of excessive heat on the enzymes.

Reference to the table of digestive juices, their content and properties, given on p. 410 should explain the results obtained.

CHAPTER 11

RESPIRATION

IN the previous chapter, it was pointed out that the functioning organism must have a more or less continuous intake of nourishing substances. Among these it finds, or from them it builds, materials of high potential energy which can be used later as an energy source for the work to be done by its body. In this chapter, it is proposed to deal with the processes whereby such substances are made to yield up this energy. We use the term *respiration* to include all those phenomena which are associated with energy release by a living creature.

Originally, respiration was used to describe the exchange of gases between an organism and its environment. It became synonymous with breathing in the case of animals. During the second half of the nineteenth century, it was realized that such external signs were only part of the process, and that they were indicative of much more complicated vital activity involving internal changes in the cells. From that time, the term has had the wider meaning. For convenience of study, we may separate the external signs of respiratory activity from the intracellular processes. We may call the former *external respiration* and the latter *internal* or *tissue respiration*, but it must be remembered that they are both part of the same function, namely, release of energy.

ENERGY CONSIDERATIONS

The living body has often been likened to a machine in that it is capable of converting energy from one form to another and expending it in the performance of work. The analogy is a fair one. The body machine is fuelled: the fuel is used up and work is done on an energy-consumption basis strictly comparable with the performance of a petrol-driven engine. In fact, the body is often a much more efficient transformer of energy into work than any man-made machine. All the functions of the body are dependent upon an incessant energy supply. One of the outstanding characteristics of living organisms is the continuous intake, transformation, and dissipation of energy; when these processes cease, the organism is dead. *See* Appendix for work/energy concepts.

The Necessity for Energy Release

The utilization of energy by a living organism is manifested in many ways. Energy must be used to maintain the organic protoplasm

426

system as an integrated whole within an inorganic system where all its parts would tend to become evenly distributed, as they do after death. Much energy is needed to synthesize the high energy compounds from which the protoplasm is built, thus chemical work has to be performed. Mechanical work in moving the whole or part of a body requires energy for its performance. Heat must be produced to maintain body temperature in some cases. Osmotic work must be done to maintain concentrations of substances inside cells at higher levels than those outside. Sound and light are sometimes emitted by living organisms. Electrical impulse transmission through animal bodies along nerves is another means of dissipating energy, whilst the maintenance of electric potentials across membranes forms yet another.

Energy-rich Compounds

Substances which can be broken down by living things to yield energy are called *respiratory substrates*. Throughout the whole living world, these are undoubtedly many and varied, but there is a remarkable similarity between the majority of plants and animals in the choice of respiratory substrates for normal purposes.

The most widespread respiratory substrate is carbohydrate, followed by fat and protein in diminishing extent. Plants and animals ensure for themselves energy-rich polysaccharides, disaccharides and fats, which may be stored to form a pool from which supplies can be drawn as required. Storage of these substances is a widespread feature of living things and often special structures are formed to contain the stored materials. These storage compounds are not as such immediately involved in energy-releasing reactions, but first are rendered more utilizable by hydrolysis into smaller components. For instance, polysaccharides and disaccharides are first hydrolysed to monosaccharide phosphates, such as glucose and fructose phosphates and these then form the starting points of the energy-releasing processes. In the case of fats, breakdown to fatty acids and glycerol precedes their further degradation to respiratory by-products. Proteins are hydrolysed to amino-acids before yielding their energy. From the metabolic cycle in Chap. 10, it will be seen that all these substrates are at the ends of chains of chemical reactions leading from a common substance, pyruvic acid, and it will be realized later that in reality it is this substance which is the focal point of respiratory activity in all higher forms of life.

The quantity of energy that is released when a complex substance is broken down into simpler substances is equal to the amount of energy that had to be supplied to the simpler substances in order to make them combine to form the compound of higher energy. In green plants, we know that energy from the sun is absorbed to bring about the synthesizing reactions. Sugar can be produced from carbon dioxide and water. If this sugar is broken down by oxidation in the air, it yields once more the carbon dioxide and water and at the same time gives out the energy in the form of heat (*see* p. 1002). This can be accurately measured in a bomb-calorimeter and is thus a measure of the energy released during the oxidation. We can adequately express the power of a substance to release energy by stating the number of joules released on complete oxidation per unit quantity of the substance. The following are the energy values of the major respiratory substrates—

1 g of glucose yields $15 \cdot 7 \times 10^3$ joules
1 g of animal fat yields $39 \cdot 9 \times 10^3$ joules (average)
1 g protein (egg albumin) yields $23 \cdot 9 \times 10^3$ joules.

Another way of expressing this, is in terms of kilojoules per mole (kJ mol^{-1}). The values then are: glucose, $2 \cdot 8 \times 10^3$ kJ mol^{-1}; animal fat (average), $35 \cdot 5 \times 10^3$ kJ mol^{-1}; protein (egg albumin), about $9 \cdot 6 \times 10^5$ kJ mol^{-1}.

Methods of Release of Energy: Aerobic and Anaerobic Conditions

Experiment has shown that in the great majority of cases, efficient release of energy is achieved by living organisms only in the presence of oxygen; the process involves the use of molecular oxygen. On the other hand, absence of oxygen does not immediately halt energy release in those organisms. A few living things can respire only in the total absence of oxygen, whilst a few others can exist in either condition. Organisms which normally require oxygen for their continuance are said to respire aerobically and may be referred to as *aerobes*. These include all the higher forms of plant and animal life. Those which can exist only in the total absence of oxygen are described as respiring anaerobically. They are *obligate anaerobes* and would die in the presence of oxygen. They are chiefly bacterial forms. *Facultative anaerobes* are organisms which, whilst normally requiring oxygen, can exist for long periods anaerobically under certain conditions. Some fungi fall into this class.

These descriptions of plants and animals according to their oxygen requirements for respiratory purposes were derived from the belief that aerobic and anaerobic processes were quite different mechanisms It is now understood that for the higher organisms at least no such separate mechanisms exist; the one is a continuation of the other and both may be going on at the same time. What was regarded as anaerobic respiration is seen to be a long series of chemical reactions in which some energy is released and certain end-products are formed. Aerobic respiration is now understood to be the further breakdown of the anaerobic intermediates by an oxidation process, with release of considerably larger quantities of energy. The aerobic method of energy release is much more efficient than the anaerobic, but it is not necessarily true that an aerobe must immediately die because of the inefficiency of the anaerobic process when it is deprived of oxygen. It is, however, often the case that the accumulated end-products of anaerobiosis prove toxic.

Biological Oxidation and Reduction

The term *oxidation* may denote one of several chemical occurrences (*see* p. 983). These are—

1. A direct addition of oxygen to an element or compound, e.g. $2CO + O_2 = 2CO_2$ (carbon monoxide is oxidized to carbon dioxide).

2. The removal of hydrogen from a hydride or other hydrogenated substance, e.g. $SH_2 + O = H_2O + S$ (sulphuretted hydrogen is oxidized to sulphur).

3. By removal of electrons, e.g. $2Fe^{++} + \frac{1}{2}O_2 + 2H^+ = 2Fe^{+++} + H_2O$ (ferrous iron is oxidized to the ferric form). Alternatively this can be written:

$$Fe^{++} - e^- \rightarrow Fe^{+++}$$

and in reverse for reduction:

$$Fe^{+++} + e^- \rightarrow Fe^{++}$$

The term *reduction* is used to indicate the reverse of any one of these. The most spectacular of oxidations is the burning or combustion of a substance in air, accompanied by the liberation of large quantities of heat. The combustion of a carbohydrate in air is such a reaction and the end-products are carbon dioxide and water. It can be represented by the equation

$$C_6H_{12}O_6 + 6O_2 \rightarrow 6CO_2 + 6H_2O + 2 \cdot 8 \times 10^3 kJ\ mol^{-1}$$

The products of this reaction are identical with those of the aerobic respiratory process where sugar is the substrate. But the two processes, combustion and respiration, are not identical, despite the similarity of the end-products. The breakdown of sugar in respiration is far more complicated, involving a long series of reactions with many intermediate products. The chain of reactions is under the influence of respiratory enzymes produced within the cells where respiration is taking place. Not all the steps involve oxidation processes and even where they do, no direct addition of oxygen to the respiratory substrate or any of the intermediate compounds occurs. Instead the oxidations are of the type in which hydrogen is transferred from the substance to be oxidized to other substances which are thereby reduced. Each such step is under control of one of the *oxidation-reduction enzymes*, of which there are three groups: the *dehydrogenases*, the *oxidases*, and the *peroxidases* (*see* Chap. 5).

Not all such oxidations in cells are concerned with respiration. For example, when tyrosinase, an oxidase, acts on tyrosine in the presence of air, oxygen is absorbed. A second hydroxyl group is introduced into the tyrosine molecule; eventually melanin is produced. Such a reaction is obviously not respiratory, and therefore the enzyme concerned in the oxidation is not acting as a respiratory enzyme. Similarly, polyphenol oxidases which bring about discoloration of plant tissues on injury, by the oxidation of phenols, are not acting as respiratory enzymes.

It is now considered that the oxidation reactions of respiration are under control of specific dehydrogenases acting in conjunction with specific hydrogen acceptors. The hydrogen acceptor is reduced in a dehydrogenase reaction, and then oxidized directly or indirectly by atmospheric oxygen by another oxidizing enzyme or enzyme system.

$$AH_2 \underset{\substack{\text{reduced}\\\text{substrate}}}{} + B \underset{\substack{\text{hydrogen}\\\text{acceptor}}}{} \overset{\text{dehydrogenase}}{\rightleftharpoons} A \underset{\substack{\text{oxidized}\\\text{substrate}}}{} + BH_2 \underset{\substack{\text{reduced}\\\text{hydrogen}\\\text{acceptor}}}{}$$

Dehydrogenase Reaction

$$BH_2 + \tfrac{1}{2}O_2 \overset{\text{oxidase}}{\rightleftharpoons} B + H_2O$$

Oxidase Reaction

Thus the dehydrogenases can act in the absence of oxygen if there i sufficient hydrogen acceptor present. Unless the hydrogen acceptor i oxidized, it can play no further part in the dehydrogenase reaction hence this must cease when all the hydrogen acceptor is used up. Unde normal circumstances, however, the initial or primary hydroge acceptors pass the hydrogen atoms they receive from the oxidize substrates (or the electrons derived from those hydrogen atoms) t other acceptors. There seems to be a variety of pathways along whic the hydrogen atoms or electrons and protons can move before the fin: combination with oxygen is achieved. One of the commonest seems t begin with *nicotinamide adenine dinucleotide* (NAD) and continu through the flavoproteins, *flavine adenine dinucleotide* (FAD) or *flavi: mononucleotide* (FMN), known as flavoproteins because the hydroge acceptor molecules are bound to protein structures, then via *quinor* and finally through the *cytochromes* to *oxygen*, the final step bei: effected by the equivalent of a reaction catalysed by an oxidase enzym in which the hydrogen is combined with oxygen to form water. NAD is otherwise known as *co-enzyme 1* (*see* p. 135) and its derivative NADP as *co-enzyme 2*, the latter being the ester of the adenine portion of the molecule with inorganic phosphate. The hydrogen accepting part of the molecule is the nicotinamide. Flavoproteins are in the chromoprotein class being proteins conjugated with a yellow pigmented structure. The prosthetic group is equivalent to NAD with the difference that riboflavin (*see* p. 102) replaces the nicotinamide and ribose portions of that molecule. Riboflavin is, of course, vitamin B_2 (*see* p. 338) and this explains why the substance is essential in animal diet. Quinone derivatives are the less complex of the hydrogen acceptors and having accepted hydrogen are reduced to hydroquinones.

The last links in the chain of hydrogen transfer, the cytochromes, form a highly complex system of electron transporters. They are described separately below.

Cytochrome Systems

All aerobic tissues, plant and animal, contain three closely allied types of chromoproteins known as *cytochromes*; the main types are distinguishable under the designations a, b, and c. In cytochrome type a there are the recognized forms, cytochromes a and a_3, always it seems in association with one another in mitochondria; in type b, cytochromes b, b_3, b_6 and b_7 have been identified in mitochondria, microsomes, chloroplasts and certain plant cell mitochondria respectively; and in type c the cytochromes referred to as c and c_1 from mitochondria and a form known as cytochrome f from chloroplasts have been isolated. Each of these cytochromes is made up of a protein part, generally a *globin*, and a prosthetic group which is a complex nitrogenous configuration known as a *porphyrin ring* (*see* p. 99) at the centre of which is an iron atom, thus constituting what is called a *haem compound*. The different cytochromes concerned with electron transport in cells are all iron-containing porphyrins, similar chemically to haemoglobin of blood. The blood pigment, haemocyanin, and the chlorophylls are also porphyrin structures but contain, instead of iron, copper and magnesium atoms respectively.

It is the iron of cytochrome, by existing alternately in the ferric and ferrous states, that makes it an effective electron carrier. An iron atom in its ferric oxidation state has lost three electrons and so has a net positive charge of $+++$ ($+3$). In its ferrous state it has lost only two electrons and has a net positive charge of $++$ ($+2$). By the addition of an electron to the ferric ion or the removal of an electron from a ferrous ion, the one state can be converted to the other and the effect is equivalent to reduction and oxidation of the iron atom by the electron changes.

The cytochromes work in conjunction with the hydrogen acceptors and dehydrogenase enzyme systems to complete the transfer of hydrogen from the original oxidizable substrate or primary reductant, such as the breakdown products of carbohydrate, to the terminal oxidant molecular oxygen, but do not actually transfer the whole hydrogen atom, only the electron. The hydrogen ion or proton is liberated into the medium containing the oxidizing system, from whence it is later used in the oxidizing reaction with oxygen to form water.

As electron carriers, the cytochromes form a series of graded energy steps along which the electron moves. Such electron transport factors form *redox couples* of increasing redox potential (*see* p. 984) and an electron is presumed to pass through the cytochrome system in the direction:

$$\text{cytochrome } b \rightarrow c_1 \rightarrow c \rightarrow a \rightarrow a_3 \rightarrow \text{oxygen}$$

that is, in a "downhill" direction requiring no intake of energy to effect it. Cytochrome a_3 at the end of the chain, with reduced ferrous ion is now capable of passing its electron to oxygen and then itself becoming oxidized ready to receive another electron from cytochrome a. The last reaction of all can be represented as:

$$2Fe^{++} + 2H^+ + O \rightarrow 2Fe^{+++} + H_2O$$

in which two electrons, two hydrogen ions or protons and an oxygen atom react together to form water. Because cytochrome a_3 works in this final way to oxidize earlier cytochromes in the series via oxygen it was once regarded as an enzyme and called cytochrome oxidase. Another way of representing the final step in the oxidation process involving cytochromes would be:

$$2 \text{ cyt } a_3 - 2 e^- + O \rightarrow 2 \text{ cyt } a_3 + O^{2e^-}$$
$$2 H^+ + O^{2e^-} \rightarrow H_2O$$

A complete sequence of hydrogen transfer from oxidizable substrate to oxygen can now be seen as:

$$AH_2 \rightarrow NAD \rightarrow FP(FMN \text{ or } FAD) \rightarrow \text{quinone} \rightarrow$$

$$\text{cyt } b \xrightarrow{2e^-} \text{cyt } c_1 \rightarrow ?\text{cyt } c \rightarrow \text{cyt } a \rightarrow \text{cyt } a_3 \xrightarrow{2e^-} \tfrac{1}{2}O_2$$
$$\searrow 2H^+$$

These are not necessarily all the carriers that may be involved and for some oxidation reactions the whole sequence as shown may not be employed, for example, in the oxidation of succinate to fumarate, NAD is certainly by-passed, the first hydrogen acceptor being FAD.

Note that small quantities of HCN inhibit the activity of cytochromes, hence cyanides poison by preventing cellular respiration mechanisms in which the cytochromes are involved. Some narcotic drugs inhibit in the opposite direction and instead of preventing the oxidation of successive cytochromes they slow down the rate at which they can be reduced by accepting electrons from further back along the hydrogen transfer system. This causes a drop in energy levels within the system and has consequent effects on the working of the tissues.

It should now be obvious that the oxidations associated with respiration cannot be compared with combustion in the manner of their execution. In the next section, the complete cycle of events in the conversion of a carbohydrate substrate to carbon dioxide and water will be described. It will then be realized that not only are enzymic oxidations part of the cycle, but that numerous other enzymes such as carboxylases are involved.

RESPIRATION USING A CARBOHYDRATE SUBSTRATE

There is no attempt here to distinguish between plants and animals since it is generally accepted that in principle the respiratory processes are universal. The overall process may be described as a series of successive stages as below.

1. An initial phosphorolysis of a reserve carbohydrate substrate such as starch or glycogen. This process is comparable with hydrolysis, but instead of the elements of water (H:OH) being used to break the bonds of the sugar units, phosphoric acid (H_2PO_3:OH) is used. Phosphorylases are the enzymes concerned in the reactions and the end-products are hexose phosphates such as glucose-1-phosphate.

2. The hexose phosphate is converted finally to fructose-6-phosphate under the influence of phosphohexoisomerase and this is further phosphorylated to fructose-1:6-diphosphate in the presence of phosphatase. Certain metallic ions such as Mg^{++}, Mn^{++} or Co^{++} are necessary in some of these conversions. If the initial substrate is already a hexose sugar, the production of fructose diphosphate still precedes further changes.

3. The 6-carbon (hexose) diphosphate is split into two 3-carbon (triose) phosphate molecules which are then converted by a series of oxido-reduction and transphosphorylating reactions into pyruvic acid.

During these stages, some energy-rich phosphate bonds are built up in the conversion of adenosine diphosphate (ADP) into adenosine triphosphate (ATP). It is the construction of these energy-rich bonds in ATP which constitutes the release of the carbohydrate energy to the organism. It is to be noted that phosphoglyceric acid is one of the intermediate compounds. The whole process is usually referred to as *glycolysis*, but the term has been used by some authors to cover a wider range of reactions. It was once thought to be controlled by a single enzyme, zymase, but this is now known to be a complex system. Altogether, zymase appears to consist of at least fourteen enzymes, and in addition the ions of Mg^{++}, Ca^{++} and K^+ are necessary together with inorganic phosphate, H_3PO_4, to bring about the successful conclusion. All the stages up to pyruvic acid can proceed in the total absence of oxygen.

4. Pyruvic acid enters into further reactions which depend upon the presence or absence of oxygen and in the latter case whether the organism is plant or animal.

(*a*) In the presence of oxygen (aerobiosis), the pyruvic acid enters the tricarboxylic acid cycle (Krebs' cycle) of metabolism as indicated in the previous chapter. In this cycle, a series of biological oxidations under the influence of dehydrogenase enzyme systems, accompanied by decarboxylations controlled by decarboxylases, results in the formation of free carbon dioxide, water and the construction of more energy-rich bonds in ATP. This constitutes oxidative phosphorylation (p. 369).

(*b*) In the absence of oxygen (anaerobiosis), the pyruvic acid is converted through acetaldehyde to ethyl alcohol in the case of plants and to lactic acid by animals. In low concentrations of oxygen, both series of reactions 4 (*a*) and 4 (*b*) may be occurring together. In the continued total absence of oxygen, these end-products prove toxic. No further energy-rich phosphate bonds are constructed.

A schematic representation of the main stages in these respiratory processes is given on p. 435.

DETAILS OF RESPIRATION WITH A CARBOHYDRATE SUBSTRATE

Stages 1 and 2. Phosphorolysis and phosphorylation of substrate to fructose 1:6-diphosphate.

(*a*) With starch or glycogen as substrate.
(i) In the presence of phosphorylase and inorganic phosphate the polysaccharide is converted to glucose-1-phosphate.

$$n(C_6H_{10}O_5) + nH_3PO_4 \overset{\text{phosphorylase}}{\rightleftharpoons} nCHO.PO_3.H_2(CHOH)_3.CH.CH_2OH$$

Scheme representing the main stages in
Aerobic and Anaerobic respiration

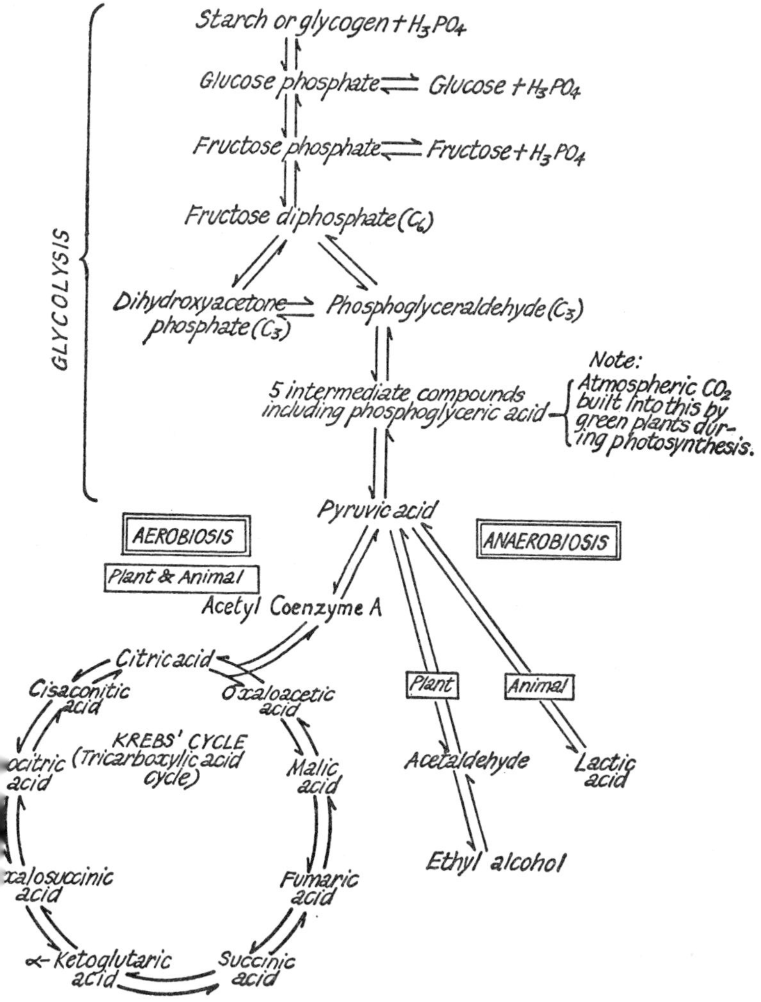

Starch or glycogen + H_3PO_4

Glucose phosphate $\rightleftharpoons$ Glucose + H_3PO_4

Fructose phosphate $\rightleftharpoons$ Fructose + H_3PO_4

Fructose diphosphate (C_6)

Dihydroxyacetone $\rightleftharpoons$ Phosphoglyceraldehyde (C_3)
phosphate (C_3)

Note:
{ Atmospheric CO_2
built into this by
green plants dur-
ing photosynthesis.

5 intermediate compounds
including phosphoglyceric acid —

GLYCOLYSIS

Pyruvic acid

AEROBIOSIS

Plant & Animal

Acetyl Coenzyme A

ANAEROBIOSIS

Citric acid

Cisaconitic
acid

Oxaloacetic
acid

Plant   Animal

ocitric
acid

KREBS' CYCLE
(Tricarboxylic acid
cycle)

Malic
acid

Acetaldehyde   Lactic
acid

xalosuccinic
acid

Fumaric
acid

Ethyl alcohol

α-Ketoglutaric
acid

Succinic
acid

(ii) Glucose-1-phosphate is converted to glucose-6-phosphate in the presence of phosphoglucomutase. Mg^{++}, Mn^{++} or Co^{++} ions are necessary to the reaction.

(iii) Glucose-6-phosphate is converted to fructose-6-phosphate in the presence of phosphohexoisomerase.

$$\overset{\boxed{O}}{CHOH.(CHOH)_3.CH.CH_2O.PO_3H_2} \overset{\text{phosphohexoisomerase}}{\rightleftharpoons}$$

$$\overset{\boxed{O}}{CH_2OH.COH.(CHOH)_2CH.CH_2O.PO_3H_2}$$

(b) With glucose as substrate.

(i) Glucose is phosphorylated to glucose-6-phosphate by the addition of phosphate from ATP in the presence of hexokinase. Note that in this case one energy-rich phosphate bond is lost and ADP is formed. Mg^{++} ions are necessary.

$$\overset{\text{hexokinase}}{C_6H_{12}O_6 + ATP \rightleftharpoons C_6H_{11}O_6.PO_3H_2 + ADP}$$

(ii) Glucose-6-phosphate to fructose-6-phosphate as in (a) (iii).

(c) With fructose as substrate.

(i) Fructose is phosphorylated to fructose-6-phosphate by the use of ATP as in (b) (i). Mg^{++} ions are necessary.

(d) In all the above cases.

(i) Fructose-6-phosphate is further phosphorylated from ATP in the presence of phosphohexokinase to fructose-1:6-diphosphate with the loss of another energy-rich bond. Mg^{++} ions are necessary.

$$\overset{\boxed{O}}{CH_2OH.COH.(CHOH)_2CH.CH_2O.PO_3H_2} + ATP \overset{\text{phosphohexokinase}}{\rightleftharpoons}$$

$$CH_2O.PO_3H_2.COH.(CHOH)_2.CH.CH_2O.PO_3H_2 + ADP$$

Stage 3. Glycolysis of fructose-1:6-diphosphate to pyruvic acid.

(i) In the presence of aldolase (zymohexase), the 6-carbon molecule of fructose diphosphate is split into two 3-carbon molecules, one of dihydroxy acetone phosphate and one of 3-phosphoglyceraldehyde. These two substances are in equilibrium and as one disappears into the next series of reactions, the other is converted to take its place by the action of the enzyme phosphoglyceroisomerase. It is the 3-phosphoglyceraldehyde which is removed with the consequent conversion of dihydroxyacetone phosphate. Thus it is equivalent to converting the fructose diphosphate into two triose phosphate molecules of 3-phosphoglyceraldehyde.

$$\overset{\text{aldolase}}{CH_2O.PO_3H_2.COH.(CHOH)_2CH.CH_2O.PO_3H_2 \rightleftharpoons}$$

$$2CH_2O.PO_3H_2.CO.CH_2OH$$

$$\overset{\text{phosphoglyceroisomerase}}{CH_2O.PO_3H_2.CO.CH_2OH \rightleftharpoons CH_2O.PO_3H_2.CHOH.CHO}$$

(ii) The 3-phosphoglyceraldehyde is converted to 3-phosphoglyceric acid in the presence of phosphoglyceraldehyde dehydrogenase, inorganic phosphate, NAD and ADP. The overall reaction involves a further phosphorylation of the 3-phosphoglyceraldehyde to 1:3-phosphoglyceraldehyde using inorganic phosphate (H_3PO_4) followed by the oxido-reduction of 1:3-phosphoglyceraldehyde using NAD as the hydrogen acceptor. This results in 1:3-diphosphoglyceric acid and this is then dephosphorylated to 3-phosphoglyceric acid by the transference of phosphate to ADP, thus building an energy-rich bond in ATP.

$$CH_2O.PO_3H_2.CHOH.CHO + H_3PO_4 \overset{\text{phosphorylase}}{\rightleftharpoons}$$
$$CH_2O.PO_3H_2.CHOH.CH(OH)(OPO_3H_2)$$

$$CH_2O.PO_3H_2.CHOH.CH.(OH)(OPO_3H_2) + NAD \overset{\substack{\text{phosphoglyceraldehyde}\\\text{dehydrogenase}}}{\rightleftharpoons}$$
$$CH_2O.PO_3H_2.CHOH.COO{\sim}PO_3H_2 + NAD.H_2$$

$$CH_2O.PO_3H_2.CHOH.COO{\sim}PO_3H_2 + ADP \overset{\text{diphosphoglyceric dephosphorylase}}{\rightleftharpoons}$$
$$CH_2O.PO_3H_2.CHOH.COOH + ATP$$

(iii) In the presence of phosphoglyceromutase, 3-phosphoglyceric acid is converted to 2-phosphoglyceric acid.

$$CH_2O.PO_3H_2.CHOH.COOH \overset{\text{phosphoglyceromutase}}{\rightleftharpoons} CH_2OH.CH.OPO_3H_2.COOH$$

(iv) 2-Phosphoglyceric acid is converted to enolphosphopyruvic acid under the influence of enolase. This is a dehydration and during the reaction the energy mobilized is passed to the phosphate bond and causes the free energy value to be raised from 12·6 kJ of ester linkage to 46·2 kJ of enolphosphate linkage. Enolase requires Mg^{++} ions as activator. Sodium fluoride will inhibit the enzyme by trapping the Mg^{++} ions and is thus a respiratory poison.

$$CH_2OH.CHO.PO_3H_2.COOH \overset{\text{enolase}}{\rightleftharpoons} CH_2{=}CO{\sim}PO_3H_2.COOH + H_2O$$

(v) In the presence of phosphopyruvic dephosphorylase and ADP the enol-phosphopyruvic acid is converted to enol-pyruvic acid. Another energy-rich bond in ATP is thus constructed.

$$CH_2{=}CO{\sim}PO_3H_2.COOH + ADP \overset{\text{phosphopyruvic dephosphorylase}}{\rightleftharpoons} CH_2.CHO.COOH + ATP$$

(vi) The enol-pyruvic acid changes to pyruvic acid probably spontaneously.

$$CH_2.CHO.COOH \rightleftharpoons CH_3.CO.COOH$$

Note that the process so far has built three ATP molecules from ADP for each hexose molecule broken down when polysaccharide is the substrate.

This represents no more than 10 per cent of the total energy of the carbohydrate. When hexose sugar is the substrate the net gain at this stage is two ATP molecules.

Stage 4. Further metabolism of pyruvic acid.

(a) In the presence of atmospheric oxygen.

The pyruvic acid is ultimately oxidized to carbon dioxide and water, but only after a long series of reactions forming a cycle has been carried out. This cycle has already been referred to as the tricarboxylic acid cycle or Krebs' cycle. The entry of pyruvic acid involves the use of a "carrier" molecule, oxaloacetic acid, which to put it simply, transports the pyruvic acid around the cycle until it is completely oxidized and the oxaloacetic acid is reformed ready to become a "carrier" once more. The steps in the series of events are catalysed by a complex system of enzymes collectively termed the cyclophorase system. These respiratory enzymes are known to be associated closely with cell mitochondria from which they can be separated individually by suitable treatment. The details of the steps in the cycle follow

(i) The pyruvic acid is first oxidatively decarboxylated and converted to a form of acetic acid known as acetyl co-enzyme A or "active acetate." Co enzyme A is a derivative of pantothenic acid and contains an active SH-group It may be represented by symbols as $HS.CoA$. The acetate will react only in the form $CH_3.COS.CoA$. The oxidation is catalysed by pyruvic dehydrogenase with NAD as the hydrogen acceptor.

$$CH_3.CO.COOH + HS.CoA + NAD \overset{\text{pyruvic dehydrogenase}}{\rightleftharpoons}$$
$$CH_3.COS.CoA + CO_2 + NAD.H_2$$

This is not the whole story of pyruvate decarboxylation because there are two other co-factors and other enzymes involved. One of the co-factors is *thiamine dyrophosphate* (TPP), otherwise vitamin B_1; the other is *lipoic acid* which is alternately oxidized and reduced during the reaction. The co-factors and enzymes are presumed to work in the production of acetyl co-enzyme A and $NADH_2$ as follows. Pyruvic acid first reacts with thiamine pyrophosphate to form acetyl TPP and carbon dioxide. The acetyl group is then transferred to lipoic acid releasing TPP and then is passed from lipoic acid to $HS.CoA$ liberating reduced lipoic acid. Finally, the lipoic acid is oxidized once more by donating H_2 to NAD forming $NADH_2$.

The "active acetate" now condenses through its acetyl group in the presence of water with the carrier molecule of oxaloacetic acid to form citric acid.

$$CH_3.COS.CoA + COOH.CH_2.CO.COOH + H_2O \overset{\text{condensing enzyme}}{\rightleftharpoons}$$
$$COOH.CH_2.CHO.COOH.CH_2.COOH + HS.CoA$$

The co-enzyme A is thus set free to take part in further formation of active acetate and the citric acid forms the first stage in the cycle at the end of which oxaloacetic acid is once more reformed. At the end of this first reaction one molecule of CO_2 has been set free and one molecule of water is formed when the $NAD.H_2$ is eventually oxidized.

(ii) In the presence of the enzyme aconitase which removes one molecule of water, citric acid becomes *cis*aconitic acid.

$$\overset{\text{aconitase}}{COOH.CH_2.CHO.COOH.CH_2.COOH \rightleftharpoons}$$
$$COOH.CH_2.C.COOH{=}CH.COOH + H_2O$$

(iii) *cis*Aconitic acid is converted to *iso*citric acid, again in the presence of aconitase. Here the enzyme adds a molecule of water.

$$\overset{\text{aconitase}}{COOH.CH_2.C.COOH{=}CH.COOH + H_2O \rightleftharpoons}$$
$$COOH.CH_2.CH.COOH.CHOH.COOH$$

(iv) *iso*Citric dehydrogenase oxidizes *iso*citric acid to oxalosuccinic acid using $NAD.P$ as the hydrogen acceptor. A molecule of water is formed when the reduced $NAD.P$ is eventually oxidized.

$$\overset{\text{isocitric}}{\underset{}{\overset{\text{dehydrogenase}}{COOH.CH_2.CH.COOH.CHOH.COOH + NAD.P \rightleftharpoons}}}$$
$$COOH.CH_2.CH.COOH.CO.COOH + NAD.P.H_2$$

(v) Oxalosuccinic acid is decarboxylated to α-oxoglutaric acid, otherwise known as α-ketoglutaric acid, in the presence of oxalosuccinic decarboxylase. One more molecule of carbon dioxide is therefore liberated—

$$\overset{\text{oxalosuccinic}}{\underset{}{\overset{\text{decarboxylase}}{COOH.CH_2.CH.COOH.CO.COOH \rightleftharpoons}}}$$
$$COOH.CH_2.CH_2.CO.COOH + CO_2$$

(vi) α-oxoglutaric acid is like pyruvic acid in possessing a CO (keto) group alongside a COOH (acidic) group and thus can be oxidatively decarboxylated in the same way. Thus in the presence of thiamine pyrophosphate, lipoic acid, Co A and NAD, energy-rich succinyl Co A is formed (*see* reaction (ii) on previous page). This can transfer its energy-rich bond to ADP, releasing Co A and forming succinic acid and ATP.

$$COOH.CH_2.CH_2.CO.COOH + HS.CoA + NAD \rightleftharpoons$$
$$COOH.CH_2.CH_2.CO.S.CoA + NADH_2 + CO_2$$

$$COOH.CH_2.CH_2.CO.S.CoA + ADP + H_2PO_3 \rightleftharpoons$$
$$COOH.CH_2.CH_2.COOH + ATP + HS.CoA$$

The hydrogen acceptor when eventually reoxidized produces another molecule of water. A molecule of CO_2 has been set free.

(vii) Succinic acid is oxidized to fumaric acid by succinic dehydrogenase. The hydrogen acceptor in this case is FAD which becomes reduced to

$FADH_2$. A molecule of water is formed when this is eventually oxidize~~d~~ from air via cytochrome.

$$\text{COOH.CH}_2\text{.CH}_2\text{.COOH} + \text{FAD} \overset{\text{succinic dehydrogenase}}{\rightleftharpoons}$$
$$\text{COOH.CH} = \text{CH.COOH} + \text{FADH}_2$$

(viii) Fumaric acid is hydrated to malic acid by fumarase, thus a molecu~~le~~ of water is used up.

$$\text{COOH.CH} {=} \text{CH.COOH} + \text{H}_2\text{O} \overset{\text{fumarase}}{\rightleftharpoons} \text{COOH.CH}_2\text{.CHOH.COOH}$$

(ix) Malic acid is oxidized to oxaloacetic acid to complete the cycle, b~~y~~ malic dehydrogenase, using NAD as the hydrogen acceptor. When th~~e~~ reduced NAD is oxidized another molecule of water is released.

$$\text{COOH.CH}_2\text{.CHOH.COOH} + \text{NAD} \overset{\text{malic dehydrogenase}}{\rightleftharpoons}$$
$$\text{COOH.CH}_2\text{.CO.COOH} + \text{NADH}_2$$

It is to be noted that to keep the cycle in operation a supply of oxaloacet~~ic~~ acid must be maintained with which the acetyl Co A can combine. This ~~is~~ achieved by reforming the oxaloacetic acid at the end of each comple~~te~~ cyclic event but in some organisms, notably bacteria, there may be anoth~~er~~ source of oxaloacetic acid 'primer''. This is provided by the direct fixation ~~of~~ carbon dioxide to pyruvic acid, catalysed by oxaloacetic decarboxylase.

At every oxidative step in the series of reactions energy-rich phospha~~te~~ bonds of ATP are constructed. The score per molecule of pyruvic ac~~id~~ oxidized to carbon dioxide and water is as follows:

pyruvic acid + NAD → acetyl CoA + NADH₂	= 3 ATP
isocitric acid + NAD → α-oxoglutaric acid + NADH₂	= 3 ATP
α-oxoglutaric acid + NAD → succinyl CoA + NADH₂	= 3 ATP
succinyl CoA → succinic acid	= 1 ATP
succinic acid + FAD → fumaric acid + FADH₂	= 2 ATP
malic acid + NAD → oxaloacetic acid + NADH₂	= 3 A TP

totalling fifteen ATP molecules.

Since each molecule of glucose provides by glycolysis two molecules ~~of~~ pyruvic acid, oxidation of glucose aerobically via Krebs' cycle yields thi~~rty~~ ATP molecules. But to these must be added those formed during the g~~ly~~colytic reactions. During the conversion of glucose to pyruvic acid ther~~e is~~ the formation of NADH₂ when phosphoglyceraldehyde becomes phosp~~ho~~glyceric acid and for each glucose molecule this represents six ATP molecu~~les~~ Add to these the four others produced when ADP is phosphorylated in ~~the~~ steps involving diphosphoglyceric acid and enol-phosphopyruvic acid a~~nd~~ subtract the loss of two ATP molecules when the hexoses glucose and fruct~~ose~~ are originally phosphorylated and the net gain per glucose oxidized is ~~38~~ ATP bonds.

Assuming the energy available to the living system per terminal phosphate bond of ATP is about 33 kJ per mole (lower estimate), the total utilizable energy transferred from glucose to ATP is 38 × 33 or about 1250 kJ per mole of glucose. When it is recalled that by bomb calorimeter measurements the energy released by the complete oxidation of gluose is about 2830 kJ per mole it can be seen that the process is about 1250/2830 × 100 per cent or 45 per cent efficient. This comparatively high efficiency is attained because the oxidation is effected through a large number of stages and the energy is thus liberated in quantities so small as to obviate great local changes in temperature, which would otherwise result in the dissipation of energy (*see* p. 998). Kreb's cycle is illustrated in detail below.

The equation $C_6H_{12}O_6 + 6O_2 = 6CO_2 + 6H_2O$ summarizes the whole sequence of events which occur during the release of energy from a hexose sugar.

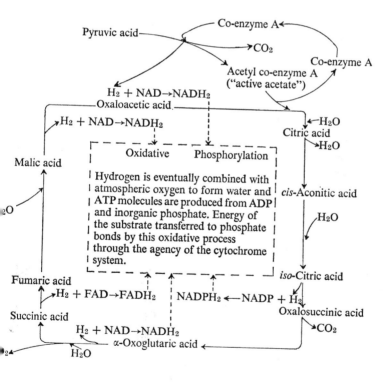

(b) In the absence of oxygen (plants).

(i) Pyruvic acid is decarboxylated by a decarboxylase to form acetaldehyde and carbon dioxide. This is the only source of carbon dioxide in anaerobic respiration.

$$\text{CH}_3.\text{CO}.\text{COOH} \overset{\text{decarboxylase}}{\rightleftharpoons} \text{CH}_3.\text{CHO} + \text{CO}_2$$

(ii) Alcohol dehydrogenase catalyses the reduction of acetaldehyde to ethyl alcohol using $NADH_2$ as the hydrogen donator. Note that reduced NAD is produced during the production of 3-phosphoglyceric acid (Stage 3 (ii)) and that this is once more oxidized in this reaction without the necessity for cytochrome and atmospheric oxygen.

$$\text{CH}_3.\text{CHO} + \text{NADH}_2 \overset{\text{alcohol dehydrogenase}}{\rightleftharpoons} \text{C}_2\text{H}_5\text{OH} + \text{NAD}$$

The overall reaction of converting hexose sugar to ethyl alcohol and carbon dioxide can be represented by—

$$\text{C}_6\text{H}_{12}\text{O}_6 + 2\text{H}_3\text{PO}_4 + 2\text{ADP} \overset{\text{zymase}}{\rightleftharpoons} 2\text{C}_2\text{H}_5\text{OH} + 2\text{CO}_2 + 2\text{H}_2\text{O} + 2\text{ATP}$$

(c) In the absence of oxygen (animals).

(i) Lactic dehydrogenase reduces pyruvic acid to lactic acid using $NADH_2$ as the hydrogen donator. The same condition as in (b) (ii) above obtains with regard to the $NADH_2$.

$$\text{CH}_3.\text{CO}.\text{COOH} + \text{NADH}_2 \overset{\text{lactic dehydrogenase}}{\rightleftharpoons} \text{CH}_3.\text{CHOH}.\text{COOH} + \text{NAD}$$

The overall reaction of converting hexose sugar to lactic acid can be represented—

$$\text{C}_6\text{H}_{12}\text{O}_6 + 2\text{H}_3\text{PO}_4 + 2\text{ADP} \rightleftharpoons 2\text{CH}_3.\text{CHOH}.\text{COOH} + 2\text{H}_2\text{O} + 2\text{ATP}$$

The breakdown of sugar to pyruvic acid by the pathway described above sometimes known as the Embden–Meyerhof–Parnas (EMP) pathway. B this is by no means the only way of respiring carbohydrate. It is consider that most tissues are able to dissimilate carbohydrate by a variety of me bolic pathways, with the probable exception of the tissues of the embryos higher plants which seem to use the EMP pathway only. It is known th in the mature tissues of plants, the initial formation of fructose phospha involves an oxidative glycolysis. This is known as the Pentose Pathway Pentose Shunt and involves the pentose sugars ribulose, ribose and xylulo In some of the lower organisms it seems probable that a dicarboxylic acid cy takes the place of the tricarboxylic acid cycle in the later stages of respirati It must be remembered that the precise course of metabolism in any organi is directly related to the nature of the enzymes it can produce. Consequen it is to be expected that there will be a variety of ways in which respirat processes may be accomplished.

Fermentation

The term fermentation has been used to describe a wide variety of organic reactions. It has been used to describe enzyme activities in general, and enzymes have been called *ferments* in the past. It has also been used in connexion with any activities of micro-organisms whether anabolic or catabolic, whilst others have restricted its use to catabolic activities in which there may or may not be some oxidation. Still others restrict its use to the anaerobic respiratory activities outlined above. This last use seems to be most widely applied, and in this sense it can be used correctly to describe the activity of yeast, for example, in converting hexose sugar to ethyl alcohol.

Respiration Using a Fat Substrate

Most information on this subject comes from a knowledge of fat metabolism in animals. The first stage in the process is the mobilization of deposited fat and its transfer from adipose tissue to the liver. The fat most probably moves in the blood as phospholipides. In the liver it is hydrolysed to glycerol and fatty acids. The glycerol is easily converted by the body to dihydroxyacetone phosphate which can be broken down to pyruvic acid by the usual glycolytic processes, or built up into reserves of glycogen. The fate of the fatty acids is more complicated, depending on whether there is carbohydrate starvation or not. If there is sufficient carbohydrate in the diet, the fatty acids are converted to "active acetate" by what is known as the β-oxidation mechanism. The fatty acid combines with co-enzyme A to form acyl-co-enzyme A, which is converted to β-keto-acyl-co-enzyme A. A reaction between this and another molecule of co-enzyme A results in the formation of acetyl-co-enzyme A or "active acetate," and an acyl-co-enzyme A with two C-atoms less than the original. This series of reactions is repeated over and over again so that the fatty acid is shortened by two C-atoms each time. The "active acetate" formed, condenses with oxaloacetic acid to form citric acid and thence enters the normal tricarboxylic acid cycle of respiration. The net gain of ATP molecules per molecule of six carbon fatty acid derivative respired is calculated at 44 and is thus a better reserve of energy than six carbon carbohydrate molecules. Shortage of carbohydrate would lead to shortage of oxaloacetic acid and the "active acetate" would be unable to enter this cycle. In this case the condensation of "active acetate" occurs to form the ketone body, acetoacetic acid, which may yield other ketone bodies, acetone by decarboxylation or β-hydroxybutyric acid by reduction. Fat metabolism in the absence of carbohydrate leads to the condition of ketosis in which the ketone bodies appear in the blood (acetone) and urine (β-hydroxybutyric acid).

An outline of fat metabolism is given below.

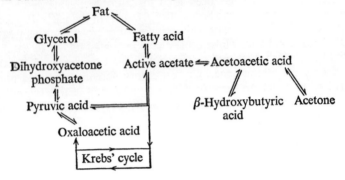

Respiration Using a Protein Substrate

We have seen that protein is not stored as a food reserve in animal tissues. However, it is a storage compound in certain parts of plants such as seeds where it may occur in comparatively large quantities. As far as animals are concerned the respiration of protein as the initial substrate would not normally take place, since any amino-acid intake in excess of replacement requirements would not be built into protein but would be deaminated and disposed of as already described. Under circumstances of extreme starvation in which all carbohydrate and fat reserves have been utilized, proteins of the tissues, particularly muscle, become a final source of energy sufficient to keep the tissues of higher function active for a while. As the muscular tissues waste, the protein substance enter the respiratory activities after hydrolysis to amino acids, which are then deaminated to leave keto-acids, the fate of which has already been explained.

In the case of plants, experiments with detached leaves kept in the dark, that is, unable to maintain a supply of carbohydrate, have shown that during the earlier stages the respiratory substrate is the carbohydrate which was already in the leaf before detachment. Gradually this is used up and the leaves subsequently turn yellow then brown before dying. During this later period, protein material, presumably from the protoplasm, slowly disappears indicating that in the last resort plants also make use of their tissue proteins as respiratory substrates. In normal healthy leaves on the plant there never would be such shortage of carbohydrate and there is no indication that the protein of the tissues then enters the respiratory metabolism. In germinating seeds in which there is a high initial protein content, the protein is rapidly mobilized for use in new tissue construction, and appears as amino-acids and sometimes amides in high concentration. If such seeds are kept permanently in the dark so that there can be no augmenting of their carbohydrate and fat content by photosynthesis, it has been demonstrated that the nitrogen reserves gradually dwindle in the later stages until the etiolated seedling dies. The accumulation of ammonia in some cases indicates a deamination process, the residues of the amino-acid molecules presumably entering the catabolic cycle to be used as respiratory substrate.

THE FORM OF THE RELEASED ENERGY

A comparison of the respiratory cycle of events with that of the photosynthetic cycle indicates at once that they are, in part at least, the reverse of one another. The assimilatory, or anabolic series of events, captures energy and stores it in the form of carbohydrate, fat, protein or any of a wide variety of compounds. Some of these form permanent body structures; the remainder are converted by the catabolic cycle into new forms. Some of the energy appears as phosphate bonds in ATP or other similar substances. It is not clearly understood how these phosphate bonds are later broken and their energy converted to use by the organism, but the dissipation of energy in the forms of muscular work (mechanical), light emission (radiation), and glucose phosphorylation (chemical), are known to be linked with the disappearance of energy-rich bonds of ATP. During respiratory processes, some energy is always lost as heat; this may form a considerable proportion. In the oxidation of glucose it is as much as 35–40 per cent, making the living machine 60–65 per cent efficient from a respiratory or fuel-burning point of view.

Energy Stores

We may regard the living machine as having three energy stores according to their ease of availability. These are—

1. The energy-rich bonds of ATP for immediate use within the cell. ATP is not readily transportable and must be built up by each cell as a result of its own respiratory activities.

2. Soluble, transportable compounds, e.g. glucose, fatty acids, from which ATP molecules can be synthesized during respiration. Only a limited amount of energy can be stored in such forms since the substances are soluble and may give rise to high osmotic pressures within the cells. Fatty acids would affect cell pH also. However, most plant and animal tissues are supplied with these substances.

3. Insoluble, stored compounds, e.g. starch, glycogen, which can yield the soluble, transportable substances as required by the tissues.

GASEOUS EXCHANGES AT RESPIRATORY SURFACES

When the respiratory process is aerobic, its outward signs are the intake of oxygen and evolution of carbon dioxide. Under anaerobic conditions

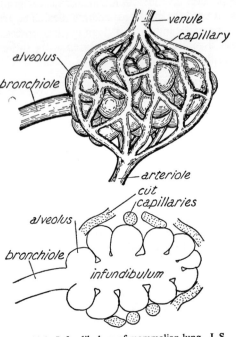

Fig. 11.1. Infundibulum of mammalian lung. L.S

only the latter is taking place. In either case, the passage of the gases must be facilitated. No specialized structures are developed to accomplish this in the case of the non-cellular organisms such as the lower algae and protozoa, and in such forms as the coelenterates, where all the cells are in intimate contact with the outside watery medium. Each cell effects its own gaseous exchanges by diffusion through the cell membrane. But in those cases where most of the body tissues are not in direct contact with the surrounding medium, some specialization of the body is developed, for the purpose of gaseous exchange.

In animals, there is a special *respiratory surface* in contact with all the remaining tissues through a vascular system; the body fluid, as it were, takes the place of the external water. The respiratory surface is highly vascularized and is kept moist, so that oxygen may be transferred rapidly from the medium, through the respiratory area to the body fluid, by which it is then distributed to all tissues. Carbon dioxide is collected from the cells by this body fluid, and transferred to the outside medium via the respiratory surface. Many types of surface have been evolved; they have in common a large area in relation to the volume of the animal. In aquatic animals, we find such variations as the highly vascular epidermal tissue of some annelids, the haemocoelic

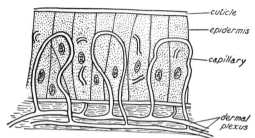

Fig. 11.2. Epidermal capillaries of *Lumbricus*.

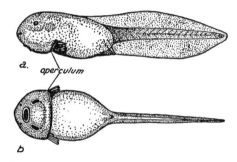

Fig. 11.3. A 9 mm frog tadpole. (*a*) Side. (*b*) Ventral view.

utgrowths of the gills of crustaceans, and the vascularized gills of shes and amphibian larvae.

On land, the skin may suffice as long as it is moist; such a condition found in the earthworm and to some extent in the frog. In the insects, e body is ventilated through a maze of fine tracheae in contact with e atmosphere through spiracles. Pulmonate molluscs have an internal vity, the "lung" of the snail and slug, with access by a single breathing-le. Increase of surface is achieved in the spiders' lung-books and e height of complexity in this direction is achieved by birds and ammals with their highly developed lungs. In all cases, the respiratory rface can only be efficient if it is moist, since gases can only diffuse rough living membranes in solution in an aqueous medium. Various mechanisms have been evolved for keeping the respiratory surface moist and to prevent its drying-out. Some of these have been described in connexion with the terrestrial types in Vol. I.

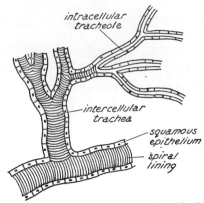

Fig. 11.4. Trachea of cockroach with tracheoles.

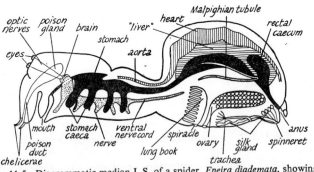

Fig. 11.5. Diagrammatic median L.S. of a spider, *Epeira diademata*, showing a lung-book.

In plants there are the same problems of aeration in water and on land. Plants rely entirely on a ventilating system of channels and air spaces within the tissues. In higher aquatic plants, these are often very highly developed as large cavities in the tissues. In the more lowly aquatics, the surface area presented to the water is amply sufficient for gaseous exchange. For example, the filamentous habit, or the flat blade structure in algae suffices. On land, where more complex and bulkier bodies are developed, the internal ventilating system is in communication with special apertures at the surface such as stomata and lenticels.

Through these, gaseous exchange is readily effected. As with animals on land, such systems inevitably lead to rapid water loss, and drying

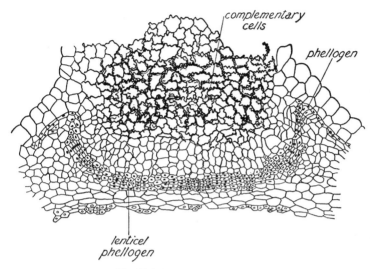

Fig. 11.6. T.S. lenticel of elder.

out of the tissues could easily result were it not for an adequate water supply or special modifications to prevent it.

Respiratory Pigments

The efficiency of the body fluid of an animal in transporting respiratory gases will depend upon the composition of the fluid and the rate at which it flows. Pure water can contain in solution only 1 per cent of its volume of oxygen at saturation; it would be very inefficient, unless present in extremely large quantities. However, the majority of the higher animal groups have evolved substances which combine readily with oxygen at high partial pressures of the gas, and which release it as readily when the pressure falls; such compounds are known as *respiratory pigments*. They include haemoglobin of all the vertebrates, most annelid worms, and some molluscs, haemocyanin of most molluscs, many crustaceans and some arachnids, chlorocruorin of some polychaete annelids and haemoerythrin of the sipunculids. In the vertebrates, the haemoglobin is located in special cells of the blood, the red corpuscles, but in invertebrates, the respiratory pigment is almost always in solution in the blood plasma.

The respiratory pigments make blood a more efficient oxygen carrier for two reasons. First, the pigment enables the blood to carry far greater proportions of oxygen. Vertebrate blood can carry up to 20 per cent of its volume of oxygen at atmospheric pressure and invertebrate blood about half or less than this. The greater efficiency of vertebrate blood is due to the higher concentration of haemoglobin in the red corpuscles than can be achieved when it is in solution in the plasma. An animal without respiratory pigment needs a far greater volume of body fluid than one possessing such a substance. Secondly, all respiratory pigments have the property of becoming almost completely combined with oxygen at pressures well below those present in the atmosphere or saturated water. Consequently a body fluid containing a respiratory pigment is likely to be fully saturated with oxygen when in contact with water or the atmosphere even though these may not contain their maximum quantities. This is in contrast with a liquid taking up oxygen in solution which can do so only in proportion to the oxygen pressure in the medium from which it is being taken up.

Transfer of oxygen to the tissues depends upon the rapidity at which the saturated pigment is dissociated from it when the partial pressure of the oxygen falls at the surfaces of cells where it is being used up. We may picture the overall process as one in which the pigment becomes saturated with oxygen at the respiratory surface, and then proceeds to dissociate from the oxygen as it circulates the body tissues, where the partial pressure of oxygen is maintained below that of the circulating fluid owing to tissue respiration. On dissociation, the pigment is returned to the respiratory surface to repeat the pick-up and subsequent release of oxygen. It is probable that there is never complete dissociation of oxygen from the pigment; it has been stated that the haemoglobin of mammals is as much as 50 per cent saturated in the veins.

The haemoglobin molecule owes its special properties to its capacity to form the compound oxyhaemoglobin by entering into loose chemical combination with oxygen. In the course of the change, the iron in the molecule remains in the ferrous state; it is not oxidized to the ferric condition as might be expected. Other respiratory pigments function in a similar manner.

The transport of carbon dioxide is more complicated than the transport of oxygen. It may be carried in the blood partly in solution, but at least twenty times as much is chemically combined; the majority is in the plasma with only a small amount in the corpuscles. Most of the carbon dioxide is combined with alkalis. In the vertebrate animals this combination and its subsequent dissociation at the respiratory surface, are greatly facilitated by the enzyme carbonic anhydrase found in the red corpuscles.

$$\overset{\text{carbonic}}{\underset{}{\text{anhydrase}}}$$
$$CO_2 + H_2O \rightleftharpoons H_2CO_3$$

The haemoglobin of the red corpuscles also assists in the transfer of carbon dioxide between the tissues and the blood. In the corpuscles, oxyhaemoglobin acts as a weak acid and combines with potassium to form a compound which may be represented thus—

$$K + HbO_2 \rightarrow KHbO_2$$

where Hb indicates haemoglobin. Carbon dioxide from the tissues enters the corpuscles and is hydrated to carbonic acid by the enzyme carbonic anhydrase. Then immediately, the following double decomposition occurs—

$$H_2CO_3 + KHbO_2 \rightarrow KHCO_3 + HHb + O_2$$

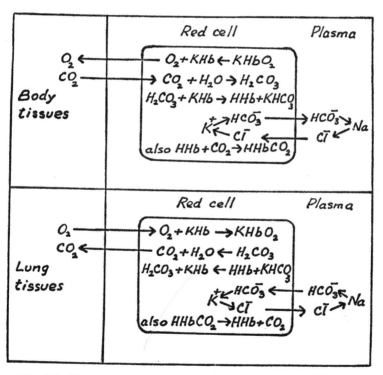

Fig. 11.7. Diagram representing oxygen and carbon dioxide exchange between blood and tissues.

In the plasma, NaCl is present, dissociated into Na^+ and Cl^- ions. The K^+ ions of $KHCO_3$ in the corpuscles cannot pass out, whereas the HCO_3^- ions can. Their place is taken by an equivalent number of Cl^- ions from the plasma. The net result is that KCl accumulates in the corpuscles, and $NaHCO_3$ in the plasma. The whole process is known as the *chloride shift*.

Haemoglobin may also enter into combination with carbon dioxide, by its amino-groups forming carbaminohaemoglobin ($H.Hb.CO_2$). These processes are continuous in the tissue capillaries containing arterial blood, while in the lung capillaries with venous blood, the reverse reactions take place. The $NaHCO_3$ and the KCl are dissociated. The Cl^- ions leave the corpuscles to form NaCl in the plasma. Oxygen enters the red cells forming with HHb, HbO_2 and free H^+ ions which combine with HCO_3^- to form H_2CO_3. This is decomposed by carbonic anhydrase; the free CO_2 escapes into the plasma and is excreted. K^+ ions now enter into combination with HbO_2 to form $KHbO_2$ and the arterial blood circulates ready for the next chloride shift in the tissues. Simultaneously with these changes, the reaction $H.HbCO_2 \rightarrow HHb + CO_2$ takes place. Fig. 11.7 gives a diagrammatic representation of the processes which take place.

Respiratory Quotients

Apart from indicating that tissue respiration is actively proceeding, a measurement of the amount of oxygen absorbed and CO_2 evolved, may suggest whether the process is aerobic or anaerobic. Such a measurement may also indicate the nature of the substrate being respired. By studying the equations for the overall reactions of tissue respiration, we can see how these indications may be obtained.

1. Complete oxidation of carbohydrate (hexose sugar)—

$$C_6H_{12}O_6 + 6O_2 \rightarrow 6CO_2 + 6H_2O$$

This indicates that for every molecule of oxygen used, a molecule of CO_2 is evolved. We can express this as a ratio, CO_2 evolved $/O_2$ absorbed, which in this case is $6CO_2/6O_2 = 1$. This ratio is known as the *respiratory quotient* or R.Q.

2. Complete oxidation of fat (tripalmitin)—

$$2C_{51}H_{98}O_6 + 145O_2 \rightarrow 102CO_2 + 98H_2O$$

In this case, the R.Q. is $102/145 = 0.7$. The R.Q.'s for different fats will of course show slight variation because of differences in molecular composition. A measured R.Q. of about 0·7 thus may indicate that fat is being respired aerobically.

By similar reasoning, the complete oxidation equation for protein substrate, and many other substances, could give us a theoretical value of the R.Q. No concrete value can be calculated for proteins since they vary so much in composition and are difficult to separate in the pure state. Estimates vary between 0·5 and 0·8 for the complete oxidation of proteins.

3. Anaerobic respiration of carbohydrate (hexose)—

$$C_6H_{12}O_6 \rightarrow 2C_2H_5OH + 2CO_2$$

$$R.Q. = \frac{CO_2}{O_2} = \frac{2}{0} = \infty$$

The measurement of an R.Q. can sometimes yield valuable information as to the nature of tissue respiration, but several facts make it necessary to be careful before drawing hasty conclusions. In the first place there are overlapping theoretical values given by fat (0·7) and protein (0·5 to 0·8); also respiration may be taking place with more than one substrate in use at the same time. However, one or two concrete conclusions can be drawn. For example, if the R.Q. is unity, or nearly so, it is fairly safe to assume complete oxidation of a carbohydrate. This is of fairly frequent occurrence in plant and animal tissues. If the R.Q. has an exceptionally high value, it is almost certain that respiration is completely anaerobic. A figure somewhat greater than unity probably indicates a mixture of aerobic and anaerobic respiration. Values of less than one would be far less indicative of precise information; they would have to be analysed carefully in conjunction with knowledge as to what substances are disappearing from the tissues. Very low values might indicate the active assimilation of CO_2 into organic acids. In general, measurements of R.Q. can be very useful when studying the changing respiratory activities under different conditions and at different stages of development, and when comparing the respiratory activity of different tissues and different organisms. The table (see p. 454) gives some R.Q. values for different organisms, organs and tissues, with some possible interpretations.

Demonstration and Measurement of Respiratory Activity

Numerous experiments have been devised to demonstrate various aspects of respiratory activity. Four important experiments are described here.

Subject	R.Q.	Possible interpretation
Germinating starchy seeds ⎫ Leaves rich in carbohydrate ⎭	1·0 1·0	Complete oxidation of a carbo- hydrate substrate.
Wheat seedlings in nitrogen	∞	Anaerobic respiration
Germinating linseed	0·64	Oxidation of a fatty substrate
Germinating peas	3·0 to 4·0	Slow entry of O_2 causing some anaerobic respiration
Germinating peas (testa removed)	1·5 to 2·5	More rapid entry of O_2, but still some anaerobic respiration
Man (average)	0·8 to 0·85	Mixed fat and carbohydrate substrate
Lumbricus terrestris	0·75	Mainly fat substrate
Drosophila (at rest)	1·23	Conversion some carbohydrate to fat: excess CO_2 produced by decarboxylation
Drosophila (flying)	1·0	Complete oxidation carbo- hydrate
Nerve tissue (resting)	0·77	Possibly mainly fat substrate
Nerve tissue (active)	0·97	Almost entirely carbohydrate substrate

Evolution of CO_2 During Aerobic Respiration

Air is forced through the apparatus by a filter pump or an aspirator. The air is freed of CO_2 which is absorbed by soda lime (or KOH) in *A*. This air is tested for CO_2 by the lime water in *B*. At *C* a bell-jar or other suitable container holds the respiring material. Small animals such as earthworms, a frog or a mouse may be used or plant material such as germinating seeds, fruits or potted plants. If green plant material is

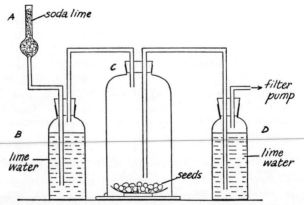

Fig. 11.8. Apparatus to show evolution of carbon dioxide during aerobic respiration.

used, light must be excluded to prevent photosynthesis; a black cloth will effect this. The lime water in D absorbs the evolved carbon dioxide and turns milky. If it is desired to find the rate of evolution of CO_2, D will contain baryta water; after the experiment this can be titrated with standard HCl, and the amount of CO_2 given off in unit time may be calculated.

Evolution of CO_2 During Anaerobic Respiration

The tubes contain mercury, inverted over troughs also containing mercury. Germinating seeds of a suitable size, such as peas, are slipped

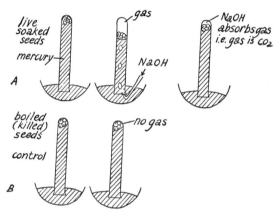

Fig. 11.9. Apparatus to show evolution of carbon dioxide during anaerobic respiration.

under the mouth of tube A. The seeds rise to the top of the mercury. Gas will accumulate in tube A and depress the mercury. When a suitable amount of gas is present, it can be tested with lime water or by absorption with KOH. The rate of evolution of CO_2 may be determined by measuring the volume of CO_2 and calculating the amount given off in unit time. The dead seeds in B act as control.

Measurement of R.Q. for Plant Material

Various forms of *respirometer* have been devised. The pattern shown in Ganong's. A known volume (2 cm^3), of suitable respiring plant material, such as germinating seeds, green leaves, or small ripening fruits, is placed in the bulb A. The manometer contains a saturated solution of sodium chloride. The ground-glass stopper B is turned so that the hole in the neck coincides with the hole in the stopper. Thus the air in the limb C is at atmospheric pressure. The levelling tube D is adjusted so that the saline solution in C is at the 100 cm³ mark. The glass stopper is now turned, to cut off access to the atmosphere. As respiration proceeds, the CO_2 evolved by the respiring

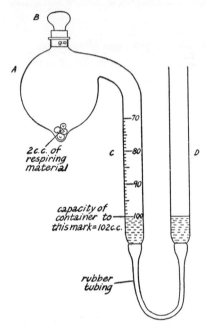

Fig. 11.10. Ganong's respirometer.

material takes the place of the oxygen absorbed. Thus, if the level of the saline solution remains constant, we infer that the volume of oxygen absorbed is equal to the volume of the CO_2 evolved, i.e. the R.Q. is 1. The actual amount of CO_2 evolved, and therefore oxygen absorbed can be measured by the addition of small pieces of caustic potash to the saline solution.

If fats or proteins are being respired, the level of the saline solution will rise. Again, the amount of oxygen used will be 20 cm³ (one-fifth); the actual amount of CO_2 evolved can be found by the addition of caustic potash pellets. R.Q. $= CO_2/O_2 = < 1$.

Measurement of R.Q. for Small Animals

A small frog or a few earthworms are suitable for the experiment. A, B, C, D, E are weighed. The animal is placed in C and its weight found. Then the apparatus is connected up, C is placed in a vessel of cold water to keep its temperature constant, and the pump is run for 10 to 20 min. After the experiment, the weight of CO_2 exhaled is the gain in weight of E. The weight of water given off by the animal is the

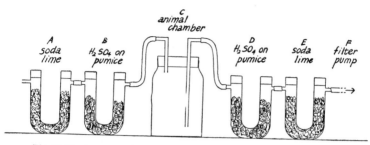

Fig. 11.11. Apparatus for measurement of R.Q. for small animals.

gain in weight of D. The weight of substrate used is the loss in weight of the animal. Hence, the weight of oxygen absorbed can be obtained thus—

$$substrate + oxygen = CO_2 + H_2O$$
$$oxygen\ used = (CO_2 + H_2O) - substrate$$
$$= (gain\ in\ weight\ of\ E + gain\ in\ weight\ of\ D) - loss\ in\ weight\ of\ animal$$

The weights of CO_2 and oxygen thus obtained can be converted into volumes, and thus the R.Q. obtained.

A number of precautions must be taken in practice: the soda-lime and H_2SO_4 containers should be doubled in each case; the animal container C must be carefully dried after the experiment and before weighing; all vessels must not be touched by hand but by strong wooden forceps.

CHAPTER 12

LOCOMOTION

THE two previous chapters have dealt with the intake or synthesis of high-energy compounds and the manner in which these may be treated to release their energy in a form utilizable by the animal or plant. It has been pointed out that this energy is required for the performance of work. In this chapter it is proposed to give some account of the manner in which mechanical work (*see* p. 987) is performed by living things where locomotion is involved.

Some kind of movement is usually readily discernible in most organisms. It may vary between activity within a cell, activity of an organ, and activity of the whole organism. For example, movement may be restricted to some or all of the protoplasm of a cell. The cytoplasm may stream, or some of its inclusions such as chloroplasts or mitochondria may have their positions altered. Chromosomes may move along a spindle. In other cases, whole organs such as an eye or a leg may undergo movement, whilst in others the whole body may be propelled from place to place. It is this change in position of the whole body which is termed locomotion.

Locomotion is much more characteristic of animals than of plants where it is confined to a few of the lower groups. (The growth movements of plants which are not in this category are discussed in Chap. 16. The reason for this is not difficult to find. It is bound up very closely with the differing modes of nutrition. The plant is best able to perform its synthesizing activities when it is in continuous contact with both its light energy source and the solution in the soil which forms its source of inorganic nutrients. A sedentary existence is therefore a necessity for all plants except aquatics, and even here the powers of locomotion are only an advantage when light and nutrient sources are likely to fluctuate rapidly. Animals, on the other hand, have to seek their elaborated nutrient requirement over wide fields in the great majority of cases. Thus locomotion is as much essential to the animal as the opposite is to the plant. There are, of course, other advantages gained by an organism which is freely motile. The seeking of mates, the avoidance of enemies and of over-crowding are such advantages. There is one period in the life history of all organisms exhibiting sexual reproduction (except the most advanced plants) when free locomotion of a part is essential to the continuance of the species. This is the period during which movement of at least one gamete is essential in order to effect fertilization.

The advantages accruing to the animal which possesses powers of locomotion can be considered responsible for the evolution of more complex and efficient organs for movement, and with these, greater complexity and efficiency of other systems associated with them, such as the nervous and vascular systems. It is also possible to regard the requirement for rapid movement as being the underlying cause of the increase in size of multicellular animals, since the larger an animal is, within limits, the greater its powers of rapid progression.

LOCOMOTION IN ANIMALS

In making locomotive movements, animals nearly always make use of specialized body parts. The exception is the movement of a whole cell by the streaming of its protoplasm, in which there appears to be no specialization for the purpose at all. Such movement is known as *amoeboid*. For purposes of description, we shall find it convenient to divide types of movement into two main classes, namely those mainly characteristic of the lower animals, and those peculiar to the more advanced animals. The dividing line between these, seems to be the development of special contractile body tissue, muscle, peculiar to the more advanced forms. It is commonly the case, however, to find a primitive locomotive movement employed by individual cells of the higher animals and by their larvae and gametes so that there is no sharp dividing line at all.

There are two primitive types of movement, namely, *amoeboid*, and *ciliary* or *flagellar*. The movements of higher animals are all fundamentally similar in making use of the contraction and relaxation of muscle fibre. They are thus muscular movements and may be described according to the mode of progression achieved by that means. We may discern the following main types of muscular progression: looping, creeping, swimming, walking (running, jumping), flying.

Amoeboid Movement

This is characteristic of the naked non-cellular masses of protoplasm of many of the protozoans, the slime fungi, some plant and animal gametes and certain wandering cells of higher animal bodies, e.g. leucocytes. The movement is one of protoplasmic streaming without active propulsion by any specialized part of it. This type of progression has been most closely studied in the protozoan, *Amoeba*, from which it derives its name.

The most widely held explanation of this cytoplasmic streaming relates to its property of effecting gel-sol-gel transformations, thus varying the viscosity and fluid pressure at different points within the plasmalemma. Reference to Fig. 12.1 will help to make this clear. At the point where the pseudopodium is to be formed, plasmagel liquefies (solates), thus weakening the cortical wall. The liquid plasmasol is forced into this region, causing a bulge. The force responsible for pushing the plasmasol outwards at the weak point may be due to

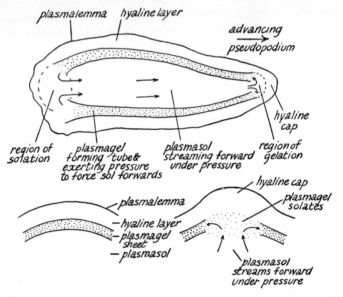

Fig. 12.1. Diagrams of changes occurring in a pseudopodium during amoeboid movement.

contraction of the cortex at other places or may be purely the result of turgor pressure of the cell. As the streaming plasmasol reaches the weakened region of the cortex, it solidifies (gelates) on all sides thus forming gradually a "tube" of solid cortex around the flowing plasmasol. This "tube" may in part be responsible for continued flow of the plasmasol by exerting an inward pressure to force the more liquid contents forward. The fact that the liquid plasmasol is continuously flowing through and forming the tube at the anterior end of the pseudopodium necessitates the liquefaction of the plasmagel at some more posterior region of the animal. Thus we may visualize the cytoplasm flowing through a tube of its own making, gelating in front to form the tube and solating behind to keep up the continuous flow through it. At the extreme tip of an advancing pseudopodium is built up a thickened cap of hyaline cortex which is solid but which may solate enough for particles of plasmasol (endoplasm) to flow into it.

Two other explanations have been suggested regarding the force responsible for moving the plasmasol through the tube of plasmagel. The first of these postulates that contraction of the gelating plasmasol at the anterior region produces a force which pulls the plasmasol towards it. The second suggests that the fixed molecules on the inner face of the plasmagel push the sol molecules along by a type of shearing or ratchet action. This last explanation can be applied to all types of protoplasmic streaming and it also has features in common with muscular contraction (*see* p. 477).

Ciliary and Flagellar Movement

These can be described together since the distinction between a cilium and a flagellum is apparently only one of size and position. The flagellum is relatively larger and is developed singly or in small numbers. Cilia are by comparison smaller and occur characteristically in large numbers upon the cells possessing them. Each type forms a differentiated motor apparatus, and they are capable of performing their work by rapid and usually rhythmical beating. Ciliary and flagellar activity is restricted to an aqueous medium, therefore these structures occur only on submerged or permanently wet surfaces of cells. It is of great interest that cilia and flagella occur in numerous instances in the bodies of higher animals where they perform tasks unconnected with locomotion of the organism as a whole. They may be used in feeding mechanisms where their beating sets up currents carrying minute food particles, e.g. in *Amphioxus*, *Paramecium* and many other examples. In other instances, their activity may maintain circulation of body fluids in coelomic cavities and of food in alimentary tracts, e.g. many jellyfish and molluscs. They may also be used to keep passages or surfaces clear of unwanted debris or secretion, as in the respiratory passages of many animals and the alimentary canals and surfaces of echinoderms. The ciliated tracts in the buccal cavity of the frog have the same function. In the nephridial tubules of the earthworm and in the oviducts of numerous animals, cilia perform the function of keeping materials on the move.

Cilia and flagella are used as locomotive organelles chiefly by small organisms such as protozoa, ciliated larvae, and gametes. In these, acceleration and deceleration can be very rapid and the method of progression is highly efficient. They are sometimes utilized by larger organisms such as small worms, the comb-jellies and even in certain snails, e.g. *Nassa*. In such cases acceleration and deceleration is very slow and the whole movement is never very rapid. As effectors of locomotion, cilia and flagella are invariably co-ordinated in their beating by some mechanism, but this has never been fully elucidated.

In structure and organization, cilia and flagella show remarkable uniformity throughout the animal and plant kingdoms. Each is made up of a sheath containing an axial filament composed of fibrils but special treatment is needed to show this. The sheath may be circular or oval in cross-section and the axial filament may be straight or spiralled within the sheath. It may extend beyond the limits of the

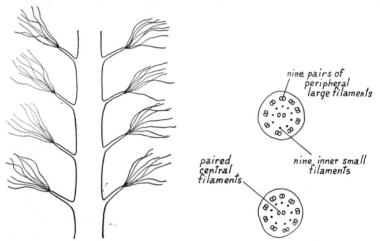

Fig. 12.2. Diagram to show structure of a flagellum based on an electron microscope photograph at magnification × 12,000.

Fig. 12.3. T.S. flagella showing 9 + 2 arrangement of filaments.

sheath as the naked end piece. The flagellum of *Euglena* has attached to the sheath, rodlets or mastigonemes oriented diagonally to give a feathery appearance. The proximal end of an axial filament is always associated with a basal granule which is believed to be derived from the cell centrosome. Many cytologists have described systems of fibrils connecting the basal granules to the nuclear region. In *Euglena*, the flagellum is composed of separate unbranched fibrils spirally twisted on one another and the cilia of *Paramecium* exhibit a similar structure. In all cilia and flagella, the fibrils have a characteristic arrangement with two in the centre and nine arranged in a circle (*see* Fig. 12.3). It is not unusual for cilia to appear as compound structures. Undulating membranes may be regarded as cilia joined laterally, and the undulations as due to metachronal rhythm of the beating cilia.

Movement of a cilium may be one of several kinds. These may be described as—

1. A simple *pendular movement* in which the straight cilium beats backwards and forwards, bending only from its base. The only difference between the effective and recovery strokes is that the former is somewhat faster (*see* Fig. 12.4).

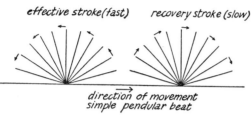

Fig. 12.4. Pendular movement of cilium.

2. A *flexural movement* in which a bending commences first at the tip and passes towards the base. In the recovery stroke the cilium progressively straightens from the base to the tip.

3. A combination of the pendular and flexural motions in which the effective stroke is a pendular movement of the rigid, but slightly bent cilium, and the recovery stroke is a bending at the base of the now limp cilium and the progressive passage of this flexure to the tip (*see* Fig. 12.5).

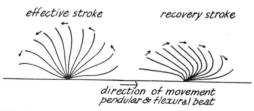

Fig. 12.5. Combination of pendular and flexural movement of cilium.

Flagella also show a variety of movements of which two are most common. These are—

1. An *undulating movement* in which waves pass along the flagellum from base to tip, never the other way (*see* Fig. 12.6).

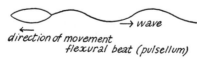

Fig. 12.6. Undulating movement of flagellum.

2. A pendular movement similar in character to that of the shorter cilium (*see* Fig. 12.7). It is not unusual for the same structure to be capable of both types of movement.

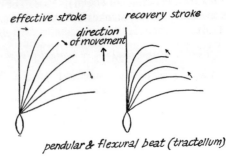

pendular & flexural beat (tractellum)

Fig. 12.7. Pendular movement of flagellum.

These basic movements described above are nearly always complicated by the fact that they do not occur in a single plane. When a flagellum is beating, the tip often describes an ellipse, a figure-of-eight or a pattern even more complicated. An undulating motion of a flagellum normally only exerts a pushing action on the water and consequently the organism, e.g. a vertebrate sperm, progresses with the flagellum trailing. It is then called a *pulsellum*. In the case of *Euglena*,

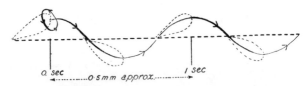

Fig. 12.8. Diagram representing the movement of *Euglena*.

The broken line represents the general direction; the full line (spiral) shows the path traced out by the anterior end in gyration; the small ellipse shows the direction of rotation. Four positions of the organism are shown.

the flagellum acts as a *tractellum*; the organism moves with the flagellum in front. This is because the beat of the flagellum has two components. The wave passes from base to tip around the flagellum as well as along it, thus imparting a rotation to the whole body as well as a thrust. It is this spinning of the body of a euglena around its axis which is considered to provide the chief force in propelling the body forwards instead of backwards. It is as though the whole body were acting as a screw propeller. Part of the traction force may come from the flickering tip of the flagellum, but this is considered small.

The description of ciliary and flagellar movements outlined above is undoubtedly over-simplified since in most cases, cilia and flagella show very great variability and complication of movement. Rarely do they show a simple and uniform activity. Most can execute singly, one or more of a variety of movements or can execute combinations of these under control of some response mechanism within the organism. Efficiency as propelling organelles varies considerably. Simple pendular cilia would appear to be the least efficient, but they serve well in larger animals for sweeping liquid past a motionless surface. In undulatory movements of flagella, the thrust exerted on the water is proportional

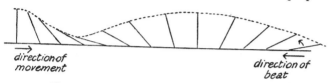

direction of movement

direction of beat

Fig. 12.9. Diagram representing metachronal rhythm.

to the speed of transmission of the wave along the flagellum. The wave must progress from base to tip of the flagellum and be thrown off the end, as it were, in order to produce any thrust at all. A stationary wave, such as would occur if both ends of the flagellum were fixed, could not exert any pull or thrust, since they would be equalized along the whole length.

Very little is known of the co-ordination and control mechanisms which govern the activities of cilia and flagella. In many cases it seems purely automatic. Small strips removed from protozoan surfaces will show continued ciliary beating, and even sperm without heads can continue locomotion. The control therefore is likely to be in the cilium itself and the surrounding cell parts since connexion to the basal granule is essential.

The beating of associated cilia in a common direction in regular sequence is known as *metachronism*, and the cilia are said to exhibit *metachronal rhythm*. The effect is due to each cilium beating slightly out of phase with those on either side (*see* Fig. 12.9). The visual effect is reminiscent of a field of corn being disturbed by the wind. Waves which pass over the beating cilia have the crests formed when the cilia are at the peak of the effective stroke and the troughs formed when the stroke is completed. On the whole, the direction in which the waves pass, is constant; it seems to be an inherent property of the cilia themselves. In many protozoa, some molluscs, and sea-anemones, there is, however, very definite reversal. In chordates, the only case of reversal known, is found in amphibian larvae. The direction of beat cannot be changed in the cilia of the frog's buccal cavity even when a strip of the ciliated epithelium is removed, rotated through 180°, and grafted back again.

The reasons why cilia and flagella beat are also difficult to find. There are two possibilities. Either the moving force comes from within the parent cell and the cilium is purely passive, or the cilium or flagellum possesses its own independent contractile powers. The latter seems more probable. Detailed structure, revealed by X-ray diffraction and electron micrography, indicates an organization of molecules comparable with that of muscle fibrils. Furthermore, a protein similar to actomyosin, the contractile substance of muscle, has been extracted from flagella. It is suggested that bending at any point may be due to the contraction of five filaments on one side while the other four are relaxed. The centre filaments may subserve the function of transmission along the length. Any type of ciliary or flagellar movement could be due to similar localized contractions. Continued beating seems to be related to metabolic activity of the cell, since it can only occur for long periods in the presence of oxygen.

Muscular Movement

The development of specialized muscle tissue is a feature common to invertebrates and vertebrates and it is possible to trace an increasing degree of differentiation of this tissue upwards through the animal kingdom. Nearly all cells show a contractile property, without being specialized, since the whole of the protoplasm is able to contract, but in some cells of the protozoans there are to be encountered special parts of the cytoplasm which are differentiated as contractile threads called myonemes. Their function is not one of locomotive activity except in a few instances, e.g. some parasites. By contraction and relaxation, they may alter the shape of a cell and are thus, as for example in *Monocystis*, responsible for what is called *gregarine movement*, its only means of locomotion. Myonemes have been likened to the muscle fibres of multicellular animals and in *Stentor* have been reported as being striated in a similar manner to the muscle fibres of many higher animals. They tend to lie just below the surface of the cell and parallel to it and may be longitudinal and transverse so that cell shape may be altered in any direction. Occasionally, as in *Vorticella*, they are in a specially differentiated part of the cell, the stalk. They resemble muscle in their relatively rapid contractions as opposed to the much slower contraction of undifferentiated protoplasm.

The nearest approach to a contractile tissue, but having also other functions, is to be found in the musculo-epithelial cells of the coelenterates. Here most of the ectoderm and endoderm cells possess

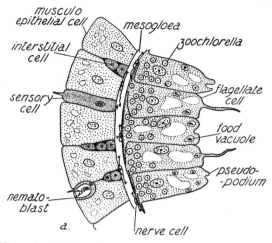

Fig. 12.10. Portion of body wall of *Hydra* seen in T.S. The endoderm cells show striated muscle tails.

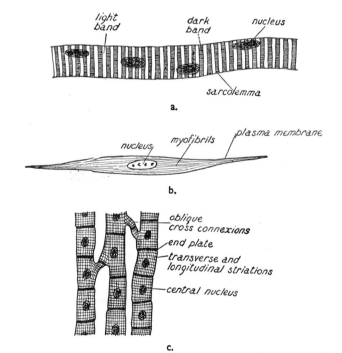

Fig. 12.11. (*a*) Single fibre of voluntary (striated) muscle. (*b*) Single fibre of unstriated muscle. (*c*) Small piece of section of cardiac muscle.

"muscle-tails," which are contractile fibres overlying the mesogloea. The muscle-tails are specialized parts of the cytoplasm, as are myonemes, but their co-ordinated contractions and relaxations are responsible for the locomotion of many of the hydroids and medusae.

Animals above the coelenterate grade of organization have developed true muscular tissue with the special property of contraction. There are three kinds: skeletal and cardiac, both striated, and unstriated or smooth. Intermediates between skeletal and smooth muscle occur in some invertebrates, where smooth muscle cells may possess a few striations in their central parts. In most of the simpler invertebrates, the muscle cells are of the unstriated variety, but skeletal muscle occurs in annelids, some molluscs, e.g. cephalopods, and nearly all arthropods, particularly insects. In the invertebrate chordates, the myotomes contain only striated muscle fibres, e.g. *Amphioxus*. The graduation from simpler to more complex invertebrates shows an increase in the proportion of skeletal muscle and this increase is paralleled by an increase in activity. It should be noted that much of the skeletal muscle of invertebrates is spirally striped, not transversely striped as it is in the vertebrates. However, insect muscle striations are transverse. In the vertebrates, all locomotive movements are effected by skeletal muscle, while unstriated muscle is associated mainly with digestion, excretion, reproduction, and the control of blood-vessels. This smooth muscle usually occurs in sheets, and it is generally characterized by slow, rhythmic contraction. The cells are separated but nevertheless co-ordinated. Skeletal muscle occurs in large blocks, often spindle-shaped, and is capable of rapid, almost explosive contraction. Unlike smooth muscle, cardiac muscle and skeletal muscle are not cellular but syncytial.

Muscles can be divided into two groups according to their function. There are the so-called *phasic muscles* or muscles of movement, and the so-called *holding muscles*. Phasic muscles usually form part of a lever system and have their origins and insertions on endoskeletal or exoskeletal structures or on skin. Typical examples are the muscles responsible for moving the body appendages such as legs, wings and fins. Such muscles occur in antagonistic couples, the contraction of one being accompanied by non-contraction of the other. Movement of the system can thus be effected in opposite directions.

Holding muscles are arranged at the periphery of hollow structures and have no strictly limited origins and insertions. One portion of the muscle is inserted into and hence pulls on another part of itself. They occur in the vertebrate animals in such places as the bladder, ureter and stomach walls. They occur in invertebrates sometimes as part of the body wall, e.g. annelid worms. They also, are paired structures, usually being alternate sheets of fibres with axes around the hollow

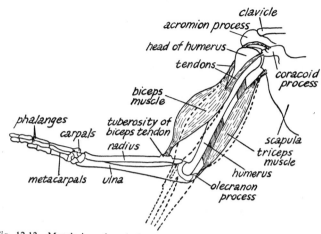

Fig. 12.12. Muscle insertions in human arm. Broken lines show the effect of straightening the arm.

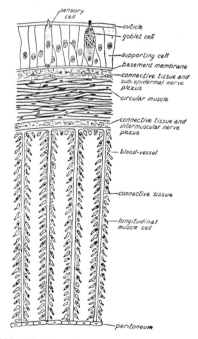

Fig. 12.13. Histology of the body wall of *Lumbricus*. T.S.

structures, i.e. circular muscle, and along the hollow structure, i.e. longitudinal muscle. Reciprocal movements of the wall can thus be achieved by alternate contraction and relaxation of the sheets.

Both these kinds of muscle are under nervous control. Within a muscle, motor nerve fibres branch among the muscle fibres. In mammals, one such fibre may serve several hundred muscle fibres, but in some invertebrates it is not unusual to find a single nerve fibre supplying a whole muscle. Many muscles receive several different types of nerve fibres. There may be separate innervation for fast contractions, slow contractions and inhibition of contraction. Neuromuscular junctions are discussed more fully in Vol. I, Chap. 20.

There has been a great deal of modern experimental work carried out on the ultrastructure of muscle, its contractile properties and its biochemistry. Some of the main conclusions are described below.

Ultrastructure

Apart from the usual components of protoplasm (*see* Chaps. 1 and 2) muscle is distinctive because of the presence of its contractile proteins *myosin* and *actin*. Both have been successfully extracted in reasonabl pure form, enabling further details of their properties to be elucidated Myosin is extracted from fresh minced muscle by a solution containin 0·01 mol sodium pyrophosphate, 0·1 mol potassium phosphate (buffere pH 6·5), and 0·47 mol potassium chloride per litre. The molecule are about 0·15 μm in length, each having a very distinctive "head" an "tail" region (*see* Fig. 12.14). The head region consists of *heav meromysin* (relative molecular mass about 350 000), and the tail regio of *light meromysin* (relative molecular mass about 150 000). Th former appears to be of globular protein nature consisting not only

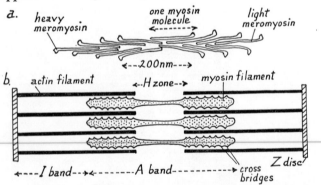

Fig. 12.14. (*a*) Arrangement of myosin molecules in a dilute solution. (*b*) Portion of one sarcomere in optical section, diagrammatic.

the binding cross-bridges (*see* p. 477), but also of the enzyme for splitting ATP (adenosine triphosphatase).

Light meromysin consists of a double helix of globular protein units and is probably mainly structural, serving to maintain the filamentous shape, but also providing the binding sites for the cross-bridges. When a solution of myosin molecules is diluted, the molecules aggregate to form rods, visible in an ordinary light microscope. The actual arrangement of the units is shown in Fig. 12.14 (*a*). Electron micrographs show that the lateral projecting portions (cross-bridges) on myosin filaments are about 45 nm apart; the portions which do not interdigitate with the actin filaments appear to have no cross-bridges and form the "H" zone (*see* Fig. 12.14 (*b*)).

Actin is extracted by 0·6 mol per litre potassium iodide solution, followed by dialysis (*see* p. 973). It is then found to consist of spherical molecules (relative molecular mass 60 000), with one ATP molecule firmly bound to each. It should be noted that while the ATP units lie on the actin filaments, the ATP-ase units lie on the myosin filaments. These roughly spherical molecules are arranged in chains to form the actin filaments.

The sarcoplasm between the myofibrils in a muscle fibre contains a meshwork of tubules, *the endoplasmic reticulum*. Transverse tubules have openings on the surfaces of the fibrils; the tubules then ramify inwards in the "Z" discs. Each transverse tubule passes between paired vesicles which branch profusely in the longitudinal direction (*see* Fig. 12.15). The system of tubules and vesicles has the function of controlling the level of Ca^{++} ions in the sarcoplasm between the filaments; a certain critical level of Ca^{++} ions being absolutely essential for the splitting of ATP and hence for contraction to take place (*see* p. 477).

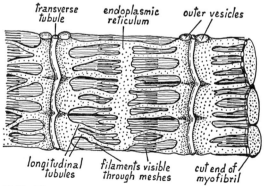

Fig. 12.15. Diagram showing the interfibrillar endoplasmic reticulum.

Apart from ATP, the sarcoplasm contains two other utilizable sources of energy, phosphocreatine and glycogen. Skeletal muscle is syncytial with peripheral scattered nuclei. Cardiac muscle has an essentially similar structure, with minor differences. It is syncytial, but the nuclei are more frequent than in skeletal muscle; the "Z" discs are not at the same level throughout all the fibrils and the filaments seem to be continuous through the "Z" discs. In unstriated muscle cells, the fibrillation is barely perceptible; though both actin and myosin are present, they do not appear to be arranged in permanent filamentous form. The following description of various phenomena, which take place in muscle, applies specifically to skeletal muscle, unless otherwise stated.

Stimulation and Response of Muscle

Skeletal muscle will contract only in response to certain specific stimuli. In life, the stimulus is provided by a nervous impulse, which causes release of acetylcholine at the motor end-plates. In the case of unstriated muscle, hormonal stimuli may act directly, e.g. adrenalin. Both of these may apply to cardiac muscle at particular times, though there is an innate system of stimulation independent of both the nervous system or of hormones (see p. 338). Under experimental conditions electrical, thermal, chemical or mechanical stimuli may be utilized. Electrical stimulation has proved to be the easiest to investigate, though the results obtained are generally applicable to the other forms of stimulation.

An electrical stimulus can be applied to a muscle through its nerve supply, or by inserting electrodes directly into the muscle. In the latter case, any possibility of nervous stimulation through fine fibrils can be obviated by blocking the neuromuscular junctions with curare. Many forms of apparatus have been devised for stimulation and recording of results; one of the simpler forms is shown in Fig. 12.16. It must be noted that in any extended work on fresh muscle, the tissue must be bathed in an isotonic solution, e.g. Ringer's, and must also be amply oxygenated.

When a skeletal muscle, or a portion of one, is given a single adequate stimulus, the response is called a *twitch*. The kymograph recording will appear as in Fig. 12.16. There is a short time between stimulation and contraction called the *latent period*; the upward part of the curve represents the phase of *contraction* and the downward part corresponds to the period of *relaxation*. The graph shown in Fig. 12.16 illustrates the results obtained with fresh calf muscle (gastrocnemius) of a frog and the total time of the cycle, 0·5 s is divisible into a latent period 0·05 s, contraction 0·2 s, and relaxation 0·25 s. Under the same conditions, a series of stimuli separated by 0·5 s. intervals will give the curves shown on the kymograph recording. By suitable experiments it can be shown that the stimulus must reach a certain minimum strength for any contraction to occur. Once this threshold is passed, a single muscle fibre will obey the "all or nothing" law, i.e. no increase in stimulus will produce increase in tension. But, for a whole muscle, increase in the strength of the stimulus will cause more and more tension as more fibres are brought into use. It is important, therefore, in experiments on whole muscles that the stimulus should reach supramaximal value. If further experiments are performed with the same material bathed in solutions at other temperatures, it can easily be shown that the time for the twitch cycle is reduced by a factor of 2 or 3 for every 10°C rise in temperature.

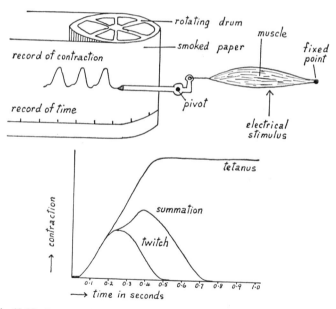

Fig. 12.16. Apparatus for measuring and recording isotonic contraction. Below, graph showing twitch, summation and tetanus.

If a second stimulus is applied before the response to the first is finished, the phenomenon known as *summation* occurs and the graph will appear as shown in Fig. 12.16. If the stimuli are repeated with regular high frequency, the highest tension is maintained, as shown in Fig. 12.16. This is known as *tetanus* and it can be maintained in muscle until fatigue supervenes.

The experiments described (p. 473), refer to *isotonic* contraction, when the tension of the muscle is held constant, or nearly so, and the length is recorded. With suitable apparatus, *isometric* contraction can be registered; in this case the length of the muscle is kept constant while the tension of the muscle is recorded. The graphs in either case will be of similar shape and significance.

The stimulus rate at which tetanus occurs varies widely, both between species and between various parts of the body; for example, in the gastrocnemius and other muscles of the human leg, it will occur at a rate of 30 stimuli per second, whereas in the human eye muscles a stimulus rate of 350 per second is necessary. For insect wing muscles the rate is often above 300 per second.

Mechanical and Thermal Properties

Muscular tissue is both extensible and elastic. If a resting muscle is suspended and subjected to hanging weights of increasing magnitude, the successive increases in length become less and less. Hooke's law is therefore not obeyed. If the stretching forces are now removed, the muscle regains its previous resting length, provided the weights were not of such magnitude as to cause tissue damage. This shows that the muscle has a high degree of elasticity and also of extensibility. The work done by a muscle can easily be measured by causing it to lift weights in the suspended position. Work done = force × distance lifted (*see* p. 987). Muscles normally work in this way by pulling on the lever systems provided by the skeleton. The energy expended is the sum of the mechanical work plus the heat dissipated (both are measured in joules). But a muscle can be activated without doing any mechanical work, as, for example, when a suspended muscle is attempting to lift a weight which is much too great for it. The tension in the muscle increases but no mechanical work is done; consequently, all the energy is dissipated as heat. This is isometric contraction. It is used, for example, in those muscles which maintain the body's posture. The extensors at the front of the thigh, for instance, exert a steady pull on the leg and hold it in the extended position. In such cases, all the fibres of a muscle are not tensed at the same time, the contraction of some fibres being replaced by that of other fibres, so that none become fatigued. On the other hand, isotonic contraction occurs, for example, when the forearm is lifted at the fulcrum of the elbow joint, mainly by contraction of the biceps muscle. Since muscles almost invariably work in groups, it is obvious that at any moment of movement, some are exhibiting isometric contraction and some isotonic.

It is a matter of simple observation that muscles liberate heat when they contract isometrically or isotonically. Exercise is a very effective means of raising body temperature and the effort exerted in lifting an immovable weight will soon cause profuse perspiration. This production of heat is an important factor in the maintenance of body temperature (*see* Chap. 6). The heat produced by a single muscle can be calculated from rise in temperature measured by sensitive thermopiles. For a single twitch, the rise is very small, of the order of 0·002°C, but in life, when movements involve numbers of muscles, the quantity of heat released is very great. For example, a man weighing 80 kg can raise his body temperature by 2°C in a few minutes of violent exercise. The total heat produced is about 0·7 megajoule, equivalent to power output of 200 W; most of this heat is produced by the muscles. Heat production by a muscle contracting can be divided into three stages; *initial heat* detectable immediately after stimulation, *delayed anaerobic* heat during continued contraction and relaxation, and *delayed aerobic heat* or *recovery heat* produced only in the presence of oxygen during recovery. These three phases appear to coincide with the phases of chemical activity (*see* p. 478): the breakdown of phosphocreatine (phosphagen), glycolysis of glycogen to lactic acid (anaerobic), and thirdly a combination of oxidation of lactic acid to carbon dioxide and water, together with synthesis of phosphocreatine from creatine and phosphoric acid.

The efficiency of muscles as machines for converting chemical energy into mechanical work has been variously calculated at 20 to 25 per cent, for a number of vertebrates. This compares unfavourably with the efficiency of a steam turbine, 35 per cent, a diesel engine, about 40 per cent and fuel cells, 70 per cent. It must be remembered, however, that the heat dissipated in the case of muscles is often of great advantage to the body.

Electrical Properties

Electrical changes in muscle fibres can be demonstrated in the same manner as that used for nerve fibres (*see* p. 301). A very small current is passed through the muscle coinciding with the wave of contraction and moving in the same direction (*see* Fig. 12.17). As the contraction wave passes *A*, a current momentarily flows in the circuit in the direction *ABG*, showing that *A* is negative to *B*. As the contraction wave passes *B*, a current momentarily flows in the circuit in the direction *BAG*, showing that *B* is negative to *A*. If electric potential is recorded against time a *diphasic curve* is obtained, similar to that obtained with nerve fibres (*see* Fig. 2.18).

The determination of membrane potential is achieved in exactly the same manner as that described for nerve membrane potentials in Chap. 8. Its value for muscle varies between 50 and 100 mV, with the inside of the fibre negative to the outside. Using the apparatus shown on p. 304, as soon as one electrode penetrates the membrane, an immediate drop in potential from 0 to −70 mV or so is registered. That is the *resting potential* of that particular muscle fibre under the conditions of the experiment. There is then a slight break in recording, while a pulse of current is passed, then the membrane is rapidly depolarized and a potential difference of 30 to 40 mV is set up in the opposite direction. The potential then returns, somewhat more slowly, to its resting value. Such a sequence is called an *action potential*.

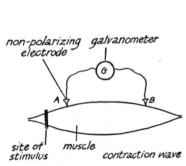

Fig. 12.17. Apparatus for detecting electrical changes in contracting muscle.

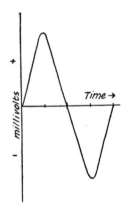

Fig. 12.18. Diphasic curve.

The resting potential, as with nerve fibres, depends on the perme ability of the membrane to various ions in solution in the fluids outside and inside the fibre. In this case, the Donnan equilibrium (*see* p. 977 is reached when the concentrations of the main ions have the following relative values (inside the membrane is first in each case); K^+ 30:1 Na^+ 1:8, Cl^- 1:30. The membrane evidently has selective permeabilit which can be destroyed by various stimuli such as electrical impulse and acetylcholine. Local currents set up by one action potential stimu late the depolarization of neighbouring regions, and thus an *actio potential wave* travels along the fibre. When the inside of a muscl fibre becomes less negative, the filament mechanism undergoes con traction. In life, the action potential in a muscle fibre is not initiate at a single point, but is amplified by the specially-shaped end-plates, s that there is a release of acetylcholine over a larger area. It is the acety choline released at the end-plates which causes depolarization, des troying the selective permeability and thus setting up an area of actio potential. Within the muscle fibre, there is a conducting system which translates depolarization into contraction.

The Contractile Mechanism

The energy for contraction is supplied by the splitting of ATP to ADP and phosphate. This splitting of ATP will not occur if the concentration of Ca^{++} ions in the sarcoplasm is less than a certain critical value. Most of the calcium in resting muscle is present in the outer vesicles of the endoplasmic reticulum. When an action potential is set up by depolarization of the membrane, one immediate effect is the release of Ca^{++} ions into the sarcoplasm. It is thought that electrical signals through the transverse tubules stimulate this process. When the action potential wave has passed a particular region, the Ca^{++} ions are actively pumped back into the outer vesicles. The process is certainly very rapid but the mechanism which achieves it is not yet known.

The presence of sufficient Ca^{++} ions activates the ATP-ase locations in the heavy meromyosin; ATP is split and the energy thus provided at the binding sites promotes the formation of cross-bridges from myosin to actin filaments. There is clear evidence to show that none of the filaments actually shorten during the contractile process, but that the myosin filaments slide further and further between the actin filaments. Consequently the distance between two consecutive "Z" discs becomes less, but there is no change in the volume of the sarcomeres; the area of cross-section becomes greater. There is no convincing evidence as to how the sliding occurs; the cross-bridges are obviously important in this connexion. One attractive theory (Davies, 1963) proposes that, at rest, each bridge is held out in an extended form by repulsion between an ionized negative group at its tip (outer end of the head of a heavy meromyosin molecule) and a fixed negative charge at its base, where it is attached to the myosin filament. It is further postulated that Ca^{++} ions form links by electrostatic attraction between the outer ionized negative groups on the myosin filaments and negatively charged sites on the actin filaments. The myosin and actin filaments are thus joined together by a series of calcium links. At the same time, the repulsive force causing the extended condition is neutralized, so that the bridges shorten and the actin filaments are pulled a short distance along the myosin filaments. It is supposed that this pulling effect is repeated several times during the course of one contraction. The stumbling block to this theory is that it hardly accommodates a contraction in length of a muscle fibre by 30 per cent or more, which is common.

Another theory proposes attachment of the cross-bridges to the actin filaments (by an unknown mechanism), contraction of the bridges, then detachment followed by re-attachment further along the actin filament. It is comparable with the process of rowing an "eight," where the oars contact the water, the pull moves the boat along, the oars break contact with the water and are inserted further along the stream. To achieve maximum contraction of a sarcomere, the process would have to be repeated six or seven times. This is an attractive theory but is unfortunately not yet supported by evidence.

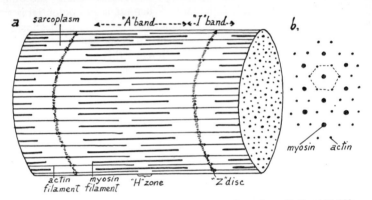

Fig. 12.19. (*a*) Detailed structure of a small portion of myofibril, × 20 000. (*b*) Arrangement of the filaments in L.S.

To summarize, a contraction of a muscle fibre depends on the following sequence: the existence of a resting potential due to selective permeability of the membrane; depolarization by destruction of selective permeability, setting up action potentials; some form of signal which increases the concentration of Ca^{++} ions in the vicinity of the cross-bridges and, finally, an adequate supply of energy.

The Biochemistry of Muscle

The immediate energy necessary for muscular contraction is supplied by hydrolysis of ATP; the amount of this present in muscle is very small, sufficient only for seven or eight twitches, and the supply is therefore rapidly exhausted. The other chemical reactions which occur ensure a constant rebuilding of ATP molecules. The second energy reserve is phosphocreatine (phosphagen), together with the enzyme necessary to split it, *creatine phosphotransferase*. By the splitting of phosphocreatine molecules, phosphate groups are liberated which combine with ADP to reform ATP.

$$ADP + phosphocreatine \rightleftharpoons ATP + creatine$$

(It is noteworthy that in non-chordates, the comparable second energy source is arginine phosphate; the echinoderms are, however, in this matter as well as in others, closely linked with the chordates, since they possess phosphocreatine.) There is normally enough phosphocreatine present to secure the rebuilding of ATP molecules during actual contraction. The phosphocreatine must, however, be regenerated eventually, and this proceeds during the recovery period which continues after contraction has ceased. The energy in this case, the third source, is derived from glycogen, which is present in ample quantities in muscle. The essential part of the process is the rephosphorylation of ADP to ATP. When the ATP concentration increases slightly, the phosphocreatine ⇌ creatine reaction proceeds in the opposite direction, thus restoring the reserve of phosphocreatine. In the presence of ample oxygen, aerobic respiration proceeds in the muscle with glycogen as the respiratory substrate. The total results of glycolysis plus the Krebs' cycle is the production of 38 energy-rich phosphate bonds per molecule of hexose. However, in life, the supply of oxygen to contracting muscles soon becomes insufficient and anaerobic respiration occurs. Glycolysis proceeds as far as pyruvic acid which reacts with reduced NAD to form lactic acid (see p. 442).

$$CH_3CO.COOH + NADH + H^+ \rightarrow CH_3CHOHCOOH + NAD^+$$

By this anaerobic respiratory process, only 2 energy-rich phosphate bonds are produced per molecule of hexose. Most of the lactic acid is removed from the muscle in the blood circulation and oxidized via the Krebs' cycle after the reverse of the above equation, i.e. the formation of pyruvic acid from lactic acid. This process does not take place in skeletal muscle but in other organs, the kidneys, liver, and notably, in the cardiac muscle of the heart.

According to the amount of lactic acid which has accumulated in the body, additional oxygen is needed to oxidize it. The amount of oxygen needed constitutes the *oxygen debt*, and after hard, sustained exercise, this may be as much as 20 litres (dm³). Muscular *fatigue* is due to the accumulation of lactic acid. Normally, the amount present in blood varies from 5 to 20 mg per 100 cm³. After severe exercise, this may rise to 200 mg, and at about 250 mg per 100 cm³, the muscles will fail to respond to stimuli. It can be shown that fatigue affects the end-plate long before actual muscular collapse is caused. When a muscle is stimulated through its nerve supply until it fails to respond, direct stimulation of the muscle, by-passing its nerve supply, will still produce contraction. The myoneural junction is more susceptible to fatigue than the muscle itself.

Modes of Progression by Muscular Contraction

Animals have evolved a great variety of methods of locomotion in adaptation to their particular environments. As has been stated, apart from amoeboid, ciliary and flagellar movement, all other types are essentially the same: they are based on muscular contraction. The main types are looping, creeping, swimming, walking (running and jumping), flying and gliding. It must be pointed out that a great many animals can employ more than one of these modes of progression.

Looping

This type of locomotion is effected by organisms capable of holding on to a substrate by both anterior and posterior ends, either separately

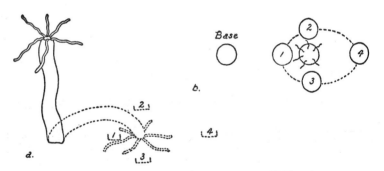

Fig. 12.20. Movement of *Hydra*. (*a*) Side view. (*b*) Plan view.
The numbers show some of the positions in which the base may be placed.

or simultaneously. The substrate is gripped by the anterior end while the posterior end is freed; by contraction of longitudinal fibres and relaxation of circular fibres, the whole body is shortened and looped upwards. This has the effect of pulling the posterior end forward, nearer the anterior end. The anterior portion is now loosened, and a grip is taken by the posterior part; the circular muscle fibres contract and the longitudinal fibres relax, and thus the anterior end can be extended forward. Continued repetition gives the effect of progression in a series of loops. Sometimes, the posterior end may travel through 180° and be placed in front of the anterior end; thus a somersault may be achieved. Looping movement is only normally carried out in invertebrates: it can be observed in *Hydra*, leeches, certain caterpillars and starfishes.

Creeping

Creeping movements are executed over a solid substratum and the whole of the lower surface is in contact with the substratum all the time. It is effected by peristaltic waves of contraction and elongation of the circular and longitudinal muscles of the body wall. Successive waves pass from anterior to posterior along the whole length of the body; reversal can usually be effected. It is best exemplified in the

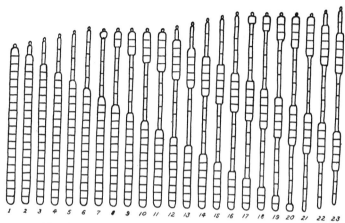

Fig. 12.21. Locomotive movements (from rest) by *Lumbricus*.

earthworm; a detailed account of this is given in Vol. I, Chap. 21. It can also be observed in the snail and slug.

Swimming

Swimming as a mode of progression has been developed by most of the animals living in an aquatic medium. It is achieved in higher animals by one of two methods, sculling and rowing, comparable with the methods of moving a small boat. In sculling a boat, a man stands in the stern and displaces masses of water alternately to left and right with an oar. This is almost exactly paralleled by the method of swimming adopted by most aquatic vertebrates, such as fishes, whales and seals. In fishes, the swimming organ is the post-anal tail, which consists of strong W-shaped segmental blocks of muscle dovetailed into one another, and centred on the flexible backbone. Forward movement occurs as a result of the thrust obtained by the hinder part of the body on the water, due to waves of muscular contraction and relaxation passing back from the anterior region. The same principle is adopted by some swimming invertebrates, e.g. annelid worms. In whales and

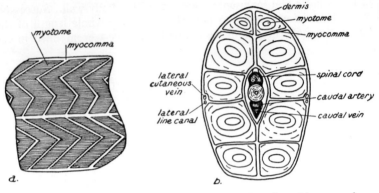

Fig. 12.22. *Scyliorhinus.* (*a*) Lateral view of trunk region, skin removed. (*b*) T.S. tail region showing muscle cones.

other aquatic animals with horizontal tail flukes, the stroke is up and down, instead of side to side.

Rowing is achieved by the simultaneous exertion of pressure on both sides of the body by suitably developed appendages acting like oars. The duckbill platypus rows with webbed fore-feet, the turtle with

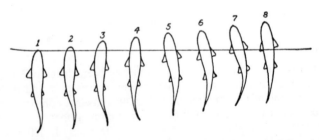

Fig. 12.23. Successive positions in the movement of a dogfish.

broad paddles, surface birds with webbed hind-feet, the frog with both front and hind-limbs. Some invertebrates adopt a similar method; the water boatman, *Notonecta*, swims using its long third pair of legs as oars

There are various other methods of obtaining thrust on the water. The crayfish's abdomen is greatly flattened at its extremity into a type of paddle. It can be pulled forward rapidly under the body, thus causing a backward movement of the animal by displacing a mass of water towards the head. In cuttlefishes, the principle of jet propulsion

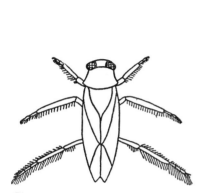

Fig. 12.24. A water boatman, *Notonecta*.

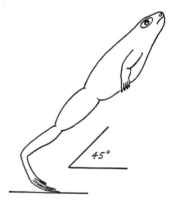

Fig. 12.25. Frog leaping; about to leave the ground.

is adopted. They expand the mantle cavity and allow it to fill with water. It is then closed, and the water is expelled forcibly through a narrow funnel by muscular contraction. As the jet of water comes out, the body is driven rapidly through the water with the head and tentacles behind. The dragonfly larva uses a similar method.

Walking, Running and Leaping

These modes of locomotion are all variants of the same method. Fundamentally, it is achieved by making use of special appendages which can act as levers against a firm substratum. A cockroach scuttling into a dark corner, a frog jumping through the grass, a crayfish walking on a sandy river bottom, and a greyhound coursing at full speed, are all using levers which propel the body forward by pressing against a resistant surface. The details vary with anatomical structure of the appendages, many of which have been described in Vol. I.

Flying and Gliding

These are achieved in air. Flying is, in principle, similar to the rowing motion in water, but the body is equipped with highly specialized appendages called wings, and is of especial lightness. Birds, bats and insects have achieved the greatest perfection in flying. Details of flight are described in Vol. I, Chaps. 23 and 28. Gliding has become effective in a few vertebrates such as the flying squirrel, flying fish, the gliding opossum, the flying frogs of Borneo, the flying lizard and the flying

Fig. 12.26. Cockchafer in flight.

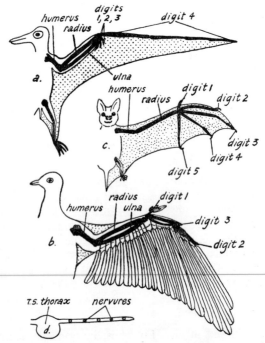

Fig. 12.27. Wings of (*a*) pterodactyl, (*b*) bird, (*c*) bat and (*d*) insect.

snake. All produce some form of expanded horizontal membrane which confers upthrust on the body from the air and thus enables them to glide.

TABLE OF COMPARATIVE VELOCITIES
(In km per hour)

On land		In water		Bird flight		Insect flight	
Caterpillar (ruby tiger)	0·08	*Euglena*	0·0045	Jay	32	Mayfly	1·8
Snail (*Helix pomatia*)	0·11	*Paramecium*	0·009	Blue tit	34	Alder beetle	3·2
Housefly (crawling)	0·19	Shrimp	0·4	Cuckoo	39	Mosquito	5·0
Snail (*Haliotis*)	0·32	Octopus	6·4	Goldfinch	40	Cranefly	7·0
Spider (*Tegenaria*)	1·84	Man	6·4	Blackbird	48	Housefly	8·0
Seal (on land)	16	Eel	12	Sparrow	56	Cabbage white butterfly	8·8
Snake (mamba)	32	Perch	16	Little owl	64	Cockchafer	10·6
Man	36·5	Seal	29	Gannet	77	Red admiral butterfly	14·1
Elephant	40	Sperm whale	32	Pheasant	96	Locust	16
Rabbit	56	Salmon	40	Teal	109	Wasp	19
Greyhound	64	Dogfish	48	Pigeon	152	Honey bee worker	24
Racehorse	77	Whale (Rorqual)	48	Golden eagle	192	Honey bee Queen	32
Lion	80	Dolphin	59	Vulture	208	Honey bee drone	40
Antelope	96	Swordfish	96	Peregrine falcon (diving)	320	Horse fly	48
Cheetah	112	Sailfish	109	Spine-tailed swift (diving)	352	Dragonfly	80

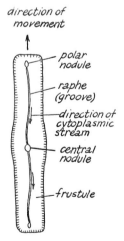

Fig. 12.28. Diagram representing movement of the diatom *Pinnularia*.

LOCOMOTION IN PLANTS

Free motility of whole plants is found only in some algae, slime fungi and bacteria. Motile reproductive structures such as zoospores are common in lower groups and motile gametes are encountered in all forms up to and including the gymnosperms (*Cycas*). In the majority of cases, the structure can move only in water or water-films and does so by means of cilia or flagella comparable in every way with the corresponding animal structures. Slime fungi creep over wet surfaces in a manner which may be described as amoeboid movement, although it does not agree in every respect with this mode of progression. Many of the boat-shaped diatoms such as *Navicula* and *Pinnularia* can move over the solid substratum of their aquatic environment by streaming the cytoplasm along a series of slits in the outer siliceous covering or frustule. Where the moving cytoplasm comes into contact with the water, it gains thrust and the diatom is propelled forwards.

The retention of motile characteristics by the reproductive parts of terrestrial plants may be regarded as a link with aquatic ancestors. Only in the most highly evolved land plants, the angiosperms, are all signs of locomotion lost.

CHAPTER 13

GROWTH AND DEVELOPMENT

LIVING organisms grow in size, undergo changes in form, and their numbers vary from time to time. Of all their characteristics, growth and development are least easily explicable, the nature of these being extremely complicated and affected by a multitude of factors. The combination of growth and changing form is referred to as *morphogenesis*. A typical morphogenetic event is the change that occurs at a plant stem apex when it ceases vegetative activity and commences to form reproductive parts. The study of morphogenesis has two main aspects; one relates to the nature of the transformation or developmental processes themselves and the other is concerned with the nature and action, including the timing, of the conditions that trigger off the morphogenetic events.

THE MEANING OF GROWTH

The changes in a population can be appreciated by merely counting the numbers. Hence it can be determined whether the numbers are increasing, remaining stationary, or decreasing. Then, with a knowledge of the environment, and of the individual organism, we may be able to explain these changes. But for a single organism, the meaning of growth is not so clear-cut. Its bulk tends to increase and at the same time there are morphological and physiological changes. To the quantitative increase in bulk, we may apply the term *growth*, and to the qualitative changes, we apply the term *development*. Both are very closely interrelated.

It is difficult to determine what exactly constitutes true growth, when defined as increase in bulk. All such changes certainly do not constitute growth for which there must be an increase in the protoplasm and its normal secretions. Fluctuations such as are found in water content, and in stored food materials, have no permanency; they are short-term variations, and do not represent true growth. The cellular constitution of the body must be considered separately from the products of its cells' activities.

Growth in an organism is generally in the positive direction, that is, there is addition to the existing body material and anabolism outweighs catabolism. Occasionally growth may be negative, in which case there

487

is loss of body material and catabolic processes are faster than anabolic processes. However, an increase in size alone does not necessarily indicate true positive growth, since the volume of an organism may be considerably affected by variations in the impermanent body substances. Thus a dry seed imbibes water and invariably increases its size, but it has not thereby achieved any true growth.

It is impossible to study the changes in cellular make-up alone because we cannot separate them easily from fluctuations in body substance due to cellular activity. Hence, for practical application, it is difficult to make a hard and fast definition of growth. It is customary to investigate two criteria; they are dry weight and size. It may be said that an organism has achieved growth when there is change in its dry weight, or change in its size. In general, the changes will be increase and will always be accompanied by development.

In multicellular creatures, most of the changes in bodily form which occur in a life history are due to two factors. First, various organs commence growth at different times in life. Secondly, individual organs may grow at different rates from one another and from the growth of the body as a whole. An organ which grows at the same mean rate as the whole body is said to exhibit *isometric growth*; if its rate of growth is different from that of the whole body, it shows *allometric growth*. Often the last organs to be differentiated in the body are those concerned with sexual reproduction. At that time the growth rate of the rest of the body will be diminishing, hence the reproductive organs show allometric growth. The result of such differential rates of growth of various organs is that the body shows continual changes of form.

GROWTH OF A POPULATION

Studies of the growth of wild populations are of interest and importance since by making them, we may bring to light hitherto unknown facts which control the reproduction and distribution of a species. In human affairs, the census is of great value in predicting future needs, and then making plans for them. It also helps us to appreciate the mathematics of growth.

A convenient subject for theoretical study is the growth of a yeast population in a suitably prepared growth medium. A single yeast cell has the power of budding or division into two cells at regular intervals. Thus after the first interval, there will be two cells, and after the second there will be four cells. The culture continues to double its numbers as long as none of the cells die or lose their power of division. The population increases according to the *exponential law*, and it may epresented graphically as in Fig. 13.1 (*A*). The mathematical expression vhich fits the curve is—

$$N_{t_2} = N_{t_1} e^{kT}$$

vhere N_{t_2} = the total number present at any time t_2, the end of the time interval over which the calculation is made;

N_{t_1} = the total number present at the commencement of the time interval over which the calculation is made;

e = log base (2·718);

k = a constant representing the efficiency of the population to increase its numbers, i.e. rate of division of the yeast in this case;

T = the time interval between t_1 and t_2, i.e. the time over which the calculation is made.

Both the mathematical expression and the graph show that the owth rate increases more and more rapidly with time, and that this

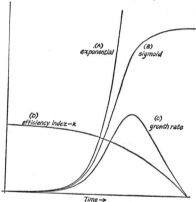

Fig. 13.1. Growth curves. *A*. Exponential. *B*. Sigmoid. *C*. Growth rate against time. *D*. Efficiency index against time.

e at any time is proportional to the numbers already present. The dition is paralleled by the way in which a sum of money would ease by compound interest if the amount to which the principal ws is calculated from instant to instant, instead of in yearly intervals. he equation, N_{t_2} represents the amount after a time T; N_{t_1} represents principal and k represents the rate of interest.

the yeast population continued to increase according to this law, ould become an infinitely large number. But in any culture, long re this possibility is reached, factors begin to operate in slowing growth rate. An actual curve, obtained from experimental data several days, would appear as in Fig. 13.1 (*B*). In the early stages, the growth of the population follows closely the exponential law, but gradually the growth rate, instead of regularly increasing diminishes and eventually becomes zero. Such an S-shaped curve is said to be of *sigmoid* pattern and is characteristic of any growing population under natural conditions. There is no single mathematical expression by which it can be accurately described over its whole length. Initially it obeys the equation $N_t = N_0 e^{kT}$ where k has a constant value, but later falls away to zero indicating that the yeast population is no longer efficient at increasing its numbers, which remain at a steady figure. A graph showing the rate of growth with time will appear as in Fig. 13.1 (C); the value of k plotted against time will give a curve as shown in Fig. 13.1 (D).

The difference between the theoretical case and the actual case is due to the conditions under which the yeast culture is developing. At first every cell has equal and optimum conditions; all material requirements are adequate, and the toxic products of fermentation diffuse away rapidly enough to cause no interference with metabolism. Later material requirements become inadequate and the rate of metabolism falls as substances like ethyl alcohol accumulate. Hence the rate of reproduction becomes less and less; some of the cells die, and slowly the growth rate decreases to zero. The population has reached the highest possible value under the existing conditions.

GROWTH OF MULTICELLULAR ORGANISMS

Here, the number of cells cannot be counted, neither can the amount of basic cellular substance be measured. Hence, we must measure some quantity which is, as nearly as possible, directly proportional to true growth. Since increase in size usually accompanies increase in the body substance, we might measure some linear dimension, area or volume of a whole body or one of its parts. This is convenient, since we can use the same organism or one of its organs without damaging them throughout the course of the experiment.

Growth is also accompanied by increase in body weight and this can be measured at successive intervals on the same organism. Since, however, much of the body weight is made up of water, and since the quantity of water is subject to considerable fluctuation, these measurements of total weight are of little use. Thus we must measure dry weight, and because this can be done only once on any organism or organ, a further complication is added. Hence it is necessary to weigh groups of individuals selected from a large population at regular intervals, so that the average dry weight of one can be calculated.

Thus, in practice, we have two methods. They are measurement of a linear dimension on a live organism or on one of its organs, measurement of dry weight of representative samples of a population. Some examples of experiments in which such methods are used for studying growth rates are described below.

Measurement of Growth of a Root

A germinating broad bean seed with a straight radicle about 1 in. long is selected. It is marked with fine indian ink lines at equidistant short intervals (2 mm is suitable), starting from the apex. The length

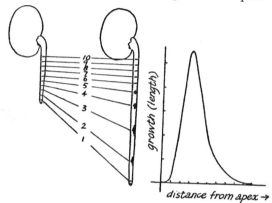

Fig. 13.2. Growth of a bean radicle.

of each marked zone is measured with a moving microscope at equal time intervals (12 or 24 hours). During the experiment, the root should be kept in an undisturbed vertical position to avoid curvatures. From the measurements the following curves may be plotted: total length against time; individual zone lengths against time; total increase in length of each zone against zone sequence. From the curves the following information may be extracted: average growth rate of the whole root; average growth rate of each zone; distribution of growth over the whole root.

Measurement of Growth of a Stem from its Internode Lengths

The stem of a young elder or similar shoot, on the plant, is marked with indian ink at each node from the tip downwards. The lengths of the internodes are measured at equal intervals of time, a metre rule giving sufficient accuracy. The measurements are recorded, the same curves constructed and interpreted as before.

Measurement of the Growth in Length of a Whole Stem

A convenient apparatus is the *auxanometer* (see Fig. 13.3). It consists essentially of a carefully pivoted lever with unequal arms. The shorter arm is connected by cotton to the shoot apex, the cotton being kept taut by weights suitably distributed on the longer arm. The slight upward movements of the shorter arm during growth are magnified by the downward movements of the longer arm, in proportion to the arm lengths. These movements are recorded by the marks made by the pointed end of the longer arm on smoked paper carried on a revolving drum. The drum is controlled by a mechanism which rotates it for a

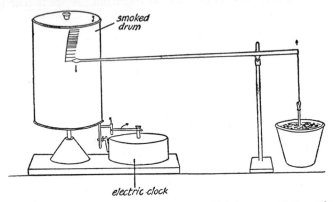

Fig. 13.3. Diagram to show the principle upon which the auxanometer works.

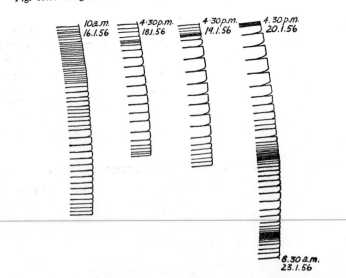

Fig. 13.4. Tracings of the drum record of growth in length of a petiole of *Pelargonium*.

short distance and then causes it to recover its original position. Thus the continuous line which the pointer would make on the smoked paper, is interrupted by a series of lines at right angles to it, and a record as shown in Fig. 13.4 is obtained. The distances between the lines

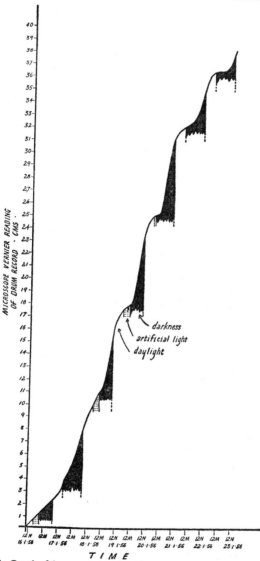

Fig. 13.5. Graph of drum record against time for the tracings shown in Fig. 13.4.

represent the growth in length made by the shoot in equal intervals of time. They can be measured accurately with a travelling microscope and permanent records can be made by coating the smoked paper with melted paraffin wax. From the measurements obtained, a growth curve of length against time can be plotted, and thus the average rate of growth may be calculated. The apparatus is very convenient since it can be left for a period of days, and can also be used to study the effect of different conditions of light, temperature and other factors on the rate of growth.

Measurement of the Growth of a Leaf from its Area

The area of a leaf, while still on the plant, can be measured by taking a photographic print of it on daylight paper in a suitable printing frame. The process is repeated at regular intervals. The leaf shapes are outlined in ink and their areas measured with a planimeter or by reproduction on squared paper. A curve of leaf area against time will give information about its growth rate.

Measurement of Growth in Length of Animals

Similar measurements may be made on the lengths of whole animals or on parts of animals. The height of an animal, the length of a caterpillar or of a fish, may be recorded at regular intervals. Similarly for parts of the body, length of a part, circumference of a head, or area of the foot may be used. In the case of some animals such as a fish, earthworm or frog, a volume measurement does not present much difficulty. From the information thus obtained, curves may be plotted to give the growth rate.

Measurement of Growth of a Plant by Dry-weight Method

A large number of seeds, such as peas, may be grown on wire trays over flat dishes containing culture solution. The seeds must be covered with damp muslin until their radicles reach the liquid. At regular intervals, a convenient number are removed (fifty would be suitable) and their dry weight found. The average dry weight can be calculated. From the results, a curve of dry weight against time may be plotted and from this the growth rate can be discerned.

In a similar manner, the growth of cereal seedlings in sand or sawdust may be determined. The technique has also been used for whole plants under field conditions.

Measurement of Growth Rate of an Animal by Dry-weight Method

The choice of material is somewhat restricted, but it may be performed with insect larvae, tadpoles or fish fry. Representative samples of the population are taken at regular intervals and their dry weight determined. From the results, the curve of growth against time for an average individual may be drawn.

The Interpretation of Growth Measurements

From any of these experimental data, information may be obtained about three features. They are: the *growth pattern* as a whole, the

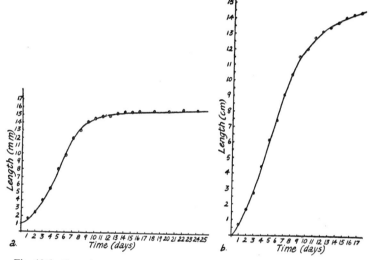

Fig. 13.6. Growth curves. (*a*) Length of *Tradescantia* internode against time. (*b*) Length of bean radicle against time.

rate of growth at any time during the experiment and the *relative growth rate*. In the first place, it is necessary to plot the curve for the value of the unit measured (length, area or dry weight) against time. This will indicate the growth pattern. Examples of such curves are given in Fig. 13.6.

The slope of such a curve at any point will give the growth rate at that point. If growth rates at regular intervals are plotted against time, the changes in growth rate over the whole period can be visualized.

Relative growth rate at any time is found by dividing the growth rate at that time by the amount of growth already made. It is a measure of the efficiency of the organism in adding more material to its body, and it represents the quantity *k* in the compound interest equation. Thus a curve of relative growth rate against time can be plotted.

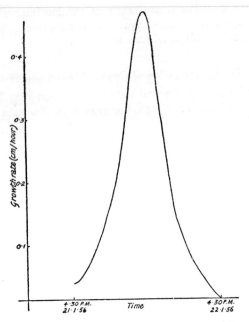

Fig. 13.7. Growth rate of *Pelargonium* petiole. From data of drum recording in Fig. 13.4.

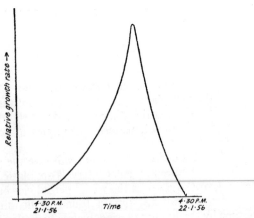

Fig. 13.8. Relative growth rate of *Pelargonium* petiole. From data of drum recording in Fig. 13.4.

The Growth Pattern in Plants

Whenever such methods are applied to the study of growth in higher plants or their organs, a remarkable constancy of the total growth pattern is exhibited. The curves of total growth against time nearly always show the sigmoid shape as in Fig. 13.1 (*B*), or variations of it.

If we follow the growth of an intact annual plant throughout its life cycle by means of dry weight measurements, we shall find the following sequence. The dry weight of the very young plant tends to decrease slightly, as the food reserves in the seed are used up. This loss

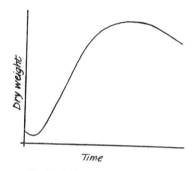

Fig. 13.9. Growth of an annual plant.

is made up when the first leaves develop and function photosynthetically. Gradually the rate of increase in dry weight becomes more and more rapid until it becomes fairly constant at a high value. This high rate is maintained until maturity is approached and then the rate declines slowly and approaches nothing. After a period of ageing, the dry weight may even decrease slightly.

The growth in area of a broad-bladed dicotyledonous leaf, for example, exhibits much the same pattern, except that there is no initial loss. There is a gradual increase in the growth rate, followed by a rapid approach to the maximum. Then there is a slow fall in growth rate as maturity is reached, and finally the growth rate is zero and the leaf enlarges no more.

Examples such as these illustrate the pattern of *limited* or *definite growth*. Annual plants, dicotyledonous leaves, and stem internodes in general reach a maximum size beyond which there is no increase, but this is not the case for all plants or plant organs. Perennial woody plants presumably never reach a maximum size, but each year they add to the previous year's growth. If the growth of the whole plant is measured over a succession of years, the total curve corresponds to a series of sigmoid curves added together, as in Fig. 13.10. Similarly, the narrow-bladed leaves of the monocotyledons never quite cease growth, so that the top of the sigmoid curve is never quite flat. Such growth may be described as *unlimited* or *indefinite growth*. It appears to be the case also for many of the lower plants such as the algae and fungi; probably their growth never quite ceases as long as they remain alive.

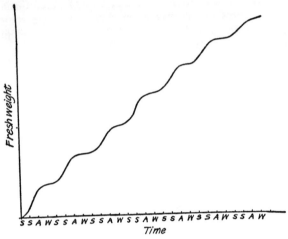

Fig. 13.10. Growth of a perennial woody plant.

The Growth Pattern in Animals

Measurements of the total growth of multicellular animals and their organs, show that the greater majority of them never quite cease growing. This is particularly the case with invertebrates, and their growth pattern is represented by part of a parabola rather than an S-shaped curve. Such a pattern, obtained by measuring the length of a lobster at regular intervals is shown in Fig. 13.11. Many of these lower animals exhibit unlimited growth, though the growth rate diminishes with age. It must not be thought, however, that such animals can or do live for ever.

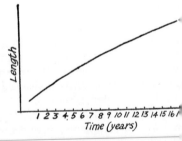

Fig. 13.11. Growth of a crustacean (lobster).

As the growth rate slows towards zero, the metabolic rate slows with it, and in a natural population, the law of survival of the fittest eliminates the aged.

In the case of higher land animals such as the birds and mammals, it is true to say that growth is limited. A maximum size is reached beyond which no growth is made. The same is true of the insects. The growth pattern in all these groups is similar to that of annual plants. Such a growth pattern is probably related to the land habit,

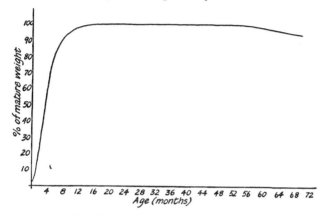

Fig. 13.12. Growth of a rat over life span.

where unlimited growth would present mechanical problems. Nevertheless, in some groups, particularly the reptiles, there appear to be cases of unlimited growth. A good example is seen in the giant tortoises.

The exact shape of the growth curve of a mammal depends upon the speed at which sexual maturity is reached. In the case of the rat, the curve is a single continuous sigmoid shape and there is little diminution

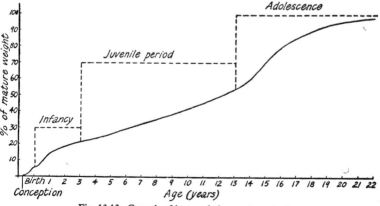

Fig. 13.13. Growth of human being up to maturity.

of the growth rate before sexual maturity is reached. On the other hand, the growth curve of a human being after birth shows two distinct phases (*see* Figs. 13.12, 13.13).

Interpretation of the Growth Pattern

We shall confine ourselves here to the most commonly demonstrated total growth curve, namely the S-shaped curve of limited growth, shown over a complete life cycle, as in Fig. 13.14. Such a curve can be divided

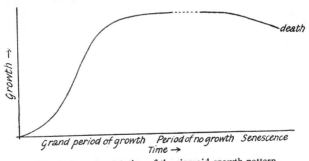

Fig. 13.14. Interpretation of the sigmoid growth pattern.

into three parts. First, there is an increasing rate of growth up to a maximum, which is maintained for a while and then falls away to nothing. This period is known as the *grand period of growth*. Secondly, there follows a phase over which the growth already achieved remains practically constant, and the growth rate is zero. Thirdly, this steady state merges into a condition when the body loses some of its substance as it continues to age. We may say there is negative growth. This is a period of *senescence*, which ends in death.

This broad description of growth pattern will fit many organisms, though no two species will show identical curves. The periods of grand growth, constancy and senescence will be very variable in length. Finally, it must be borne in mind, that some organisms may never cease growing, and they will not therefore have the type of growth pattern described above.

Mathematical Interpretation of the Sigmoid Curve of Growth

Such a curve, as shown in Fig. 13.14, does not obey any single mathematical law over its whole length. However, the phase of greatest interest is the grand period of growth, and here, the curve is essentially similar to that for the growth of a population (Fig. 13.1 (*B*)). At first it follows the compound interest law and then the growth rate slowly diminishes as the efficiency of the organism in adding more substance to itself, falls away to nothing.

The growth rate is the amount of growth made in unit time. It is proportional to the size or dry weight already attained, and also to the difference between its size at any moment and the final size it will achieve. This is comparable with the interpretation of a chemical reaction, the rate of which is dependent upon the amount of product already made and also upon the amount of unused reactants. If the growth rate is proportional to the amount of substance already made, then it should increase as the organism gets larger. If it is proportional to the difference between its present size and final size, then the rate should decrease as the organism gets larger. This is, in fact, what appears to happen over the grand period of growth. It can be expressed mathematically by the equation—

Growth rate = constant × present amount of growth (final amount of growth − present amount of growth)

i.e.
$$\frac{dx}{dt} = C \times x_{present}(x_{final} - x_{present})$$

Such an equation gives a curve of sigmoid shape resembling some growth curves.

Morphological Interpretation of the Sigmoid Curve of Growth

We may attempt to construe the growth pattern of an organism in terms of the changes occurring in the cells during the building of its body. In annual angiosperms at least, the major events are the following. At first, there is a period of very rapid cell division during which the body substance is increasing at a rate proportional to the substance already present. All the cells are capable of division and they divide repeatedly, rather like unicellular organisms in good environmental conditions. Gradually, as the body enlarges, changes occur in some of the cells. They lose their power of division, and become enlarged and differentiated into a wide variety of forms. The regions of new cell formation become localized and thus the addition of new cells is no longer directly proportional to the number already present. When they are differentiated, the great majority of the cells cease to contribute to the total growth. Then, as the reproductive phase is passed, the localized meristems slow up in their activities and growth either ceases completely or dwindles to a very small amount. The grand period of growth may thus be visualized as portraying this sequence of changes in cell activity. The early part, where growth appears to be exponential, represents the period of rapid cell division, so that the growth rate becomes faster and faster. A high rate of growth is reached and maintained as large numbers of cells enlarge and differentiate. This phase is represented by the long, more or less straight climb of the growth curve. The gradual flattening of the curve coincides with the period when meristems are becoming less and less active and nearly all the cells have completed their differentiation (*see* Fig. 13.15).

Animal growth curves, in general, follow the sigmoid pattern, but caution has to be used in interpretation. In the higher animals, all

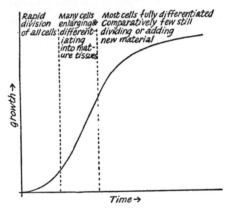

Fig. 13.15. Morphological interpretation of the sigmoid growth curve.

cells, except possibly nerve cells, retain the power of division. There is continual disintegration and replacement of cells. A generalized curve for mammalian growth is shown in Fig. 13.12. There is a phase of rapid cell multiplication followed by differentiation, but there is comparatively little cell enlargement. As various organs are differentiated there is rapid cell multiplication in these organs. In the majority of terrestrial animals, at least, growth slows down as differentiation is fully accomplished, and gradually the curve flattens out. For a varying period, there is no growth; here, replacement and disintegration exactly balance each other. The final downward slope indicates that replacement lags behind disintegration and there is negative growth. This phase of senescence continues until death.

The growth curve for many coelenterates which maintain the hydroid form would appear as in Fig. 13.16. At first there is rapid cell division, then differentiation, but after this the graph will be a straight line until the animal dies. There is no further growth, but efficient replacement by the interstitial cells. Sea-anemones of the genus *Sagartia* were kept at Edinburgh University for eighty years; throughout the whole period there was no change in appearance or size, apart from temporary fluctuations due to budding.

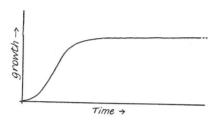

Fig. 13.16. Growth curve of coelenterate.

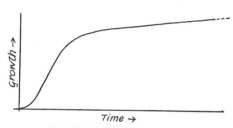

Fig. 13.17. Growth curve of fish.

In a considerable number of aquatic animals such as certain fishes and some marine crustaceans, there appears to be no cessation of positive growth, though the rate becomes extremely slow (*see* Fig. 13.17).

Physiological Interpretation of the Sigmoid Curve of Growth

Neither the mathematical nor the morphological interpretation of growth curves assists us greatly in understanding the physiological problems of growth. Total growth and rate of growth of an organism are influenced by many factors. In many cases, the environmental factors have been studied in great detail and their effects analysed. Changes in external conditions only, do not affect a growth pattern for more than a short time. Certainly, abrupt changes in conditions such as nutrient materials, light and temperature, do seriously interfere with the normal growth pattern. But, as long as conditions are favourable, some factors, inherent in the organism, cause it to complete its growth in a manner peculiar to itself. Its cells divide and differentiate under influences of which we know very little. Until we have considerably greater knowledge of the reasons for the behaviour of cells, why and how they act in particular ways, we cannot pretend to understand the physiology of growth.

Distribution of Growing Tissues

It is often regarded as one of the major distinctions between plants and animals that the tissues capable of continuous addition of new cells are quite differently distributed in the two kingdoms. In all but the lowliest of plants, new cells are added from special regions known as meristems. As the new cells differentiate and mature, they lose the power of division in most cases. In all but a few groups of animals, such localization is absent; each tissue, as it is differentiated, acts as its own source of cells, so that growth may occur in all regions of the body simultaneously and continuously.

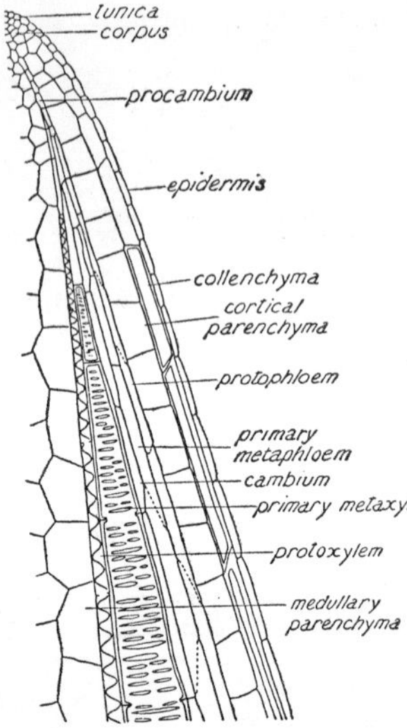

The exceptions in plants are to be found in the Thallophyta, where differentiation into tissues is often absent, so that the body may be composed of a filament or other structure in which all the cells are identical, and all have the power of dividing at any time to add new substance to the body. Such *intercalary growth* is generally the condition in the algae. In plants of higher grades of organization, cell division, except under special circumstances, is restricted to a single apical cell or group of cells,

Fig. 13.18. Diagram to show sequence of differentiation of stem tissues from apical meristem.

and growth is then described as *apical*. The exceptions in animals include the colonial coelenterates and the polyzoans, where additional members of the colony are added from fixed apical regions only. These may be said to exhibit apical growth as in plants; it is quite distinct from the interstitial growth characteristic of most animals.

The meristems of higher plants are present in the developing embryo at the apices of shoot and root. They are responsible for the production of cells which give rise to all the organs of the primary body. Where secondary growth is a feature of development of the vegetative body, new lateral meristems arise. They are known as cambia, one arising in the stele and one in the cortex to form periderm. The apical meristems are responsible for increase in length, and for the production of branches, leaves and flowers. The lateral meristems, by dividing parallel to the longitudinal axis of the stem or root, give rise to increases in girth.

In the maturing animal, all the organs and systems are growing an developing within themselves. Each is capable of producing its own new tissue, so that the production of new cells for the body as a whole, is nowhere localized. There are a few special regions which differ from the rest of the body in their capacity to produce new cells. These are to be found in the skin and certain other epithelial regions, where germinative layers must continually replace cells which are sloughed

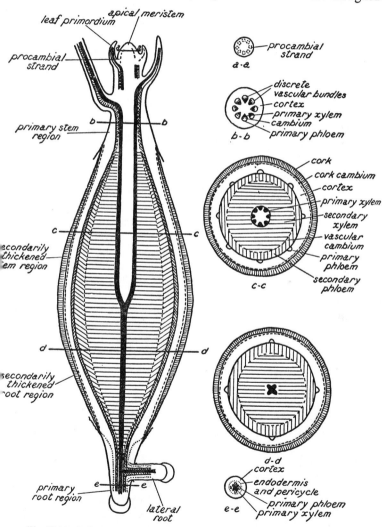

Fig. 13.19. Apical and lateral meristems in dicotyledonous angiosperm.

from the surface, in the gonads, where gametes have to be produced over prolonged periods, and in tissue such as bone-marrow, where blood corpuscles are continually produced.

In the majority of lower animals, and even in many fishes, amphibians, and reptiles, it seems that interstitial growth continues throughout the whole period of life. The rate becomes slowly reduced during the later periods of the life, but positive growth never ceases. In the higher vertebrates, the birds and mammals, a maximum size is reached; at that point, replacement and destruction of cells balance one another, and no positive growth is made. This is probably bound up with the fact that the endoskeleton ceases growth when the epiphysial cartilages are ossified, so rendering the bones incapable of any further elongation

Factors Affecting Growth

These factors may conveniently be divided into two categories; those of the external environment and those which operate within the organism.

External Conditions

Factors of the external environment known to influence growth and growth rates are, nutrient availability, accumulation of the products of metabolism, temperature, light intensity and duration, and pH.

Nutrient Availability. Body substance can be synthesized only when the necessary materials and energy required for the building processes are available. Every organism has its special nutritional requirements Green plants can manufacture most if not all their body substance from relatively few inorganic materials, whilst plants without photosynthetic pigments, and animals, must be supplied with all the compounds they are unable to synthesize. Deficiency of any one essential requirement results almost always in the cessation of growth, if not in death, even though all the other necessary substances are abundantly present Occasionally, an animal may overcome such a shortage by synthesizing an essential substance from other food components. For example, the rat can synthesize arginine, which is essential to growth, from other substances. It can do this only slowly, so that a diet deficient in arginine results in retardation but not complete cessation of growth. There are no known comparable cases in green plants; one essential element cannot be replaced by another, though practically all plant organs will continue to make growth if supplied with suitable organic nutrients in place of the essential elements.

Accumulation of By-products of Metabolism. Growth may be inhibited by the presence of toxic substances. Where these are formed as by-products of metabolism, their increasing concentration, if allowed to continue, will soon retard growth. The majority of plants and animals do not normally suffer in this way. Free-living animals are able to convert such toxic products into innocuous waste substances which they can excrete, so that their growth is not affected. Such is the case with ammonia, which is converted into urea, uric acid, or urates. Free-living plants do not form comparable toxic by-products. In the case of many parasitic and saprophytic organisms, the position is somewhat different. The fungi and bacteria especially are affected by the increasing concentration of the products of their putrefying activities. Yeast, for example, ceases to be active as the concentration of alcohol approaches 15 per cent, even though all nutrients are present in adequate supply.

Although carbon dioxide is not toxic, its presence above a low concentration is detrimental to metabolism. Thus, if an organism builds up a high concentration in its own vicinity, its growth rate will be retarded, and death will ensue. Only rarely would motile organisms be in danger of such consequences, but fixed organisms might easily be affected; thus we find that root growth in higher plants is rapid only in well-aerated soils.

Temperature. A very important external factor controlling growth rates is temperature. All biological processes are accelerated as the temperature rises from a minimum value at which no growth occurs, to a certain point beyond which retardation of growth occurs. These effects are due, at least in part, to the effect of temperature on the enzymically controlled reactions which occur in metabolism. When these are proceeding rapidly, more body substance can be built in a given time. The process is not quite as simple as this in some cases; there are very definite seasonal rhythms in growth which cannot be related to temperature. For instance, children in the Northern hemisphere are reported to grow most rapidly in autumn and least in spring. On the other hand, in perennial plants, the most rapid rate of growth usually occurs in spring. Sudden changes of temperature may have an effect on growth, irrespective of the general temperature level. Many plant organs, when suddenly subjected to a rapid rise in temperature, show at first an accelerated growth rate, followed later by a considerable slowing down. This has sometimes been considered a growth-rate response to a changing temperature stimulus; it has been termed a *thermo-growth reaction.*

Light. While most animals and many fungi and bacteria can live through their whole life cycle in darkness, it is obvious that visible light is essential for the growth of photosynthetic plants, since it supplies the energy by which all new tissue is synthesized. Apart from this, variations in light intensity and duration are known to have marked effects on the growth rates of many organisms. In green plants, where the effect of light on growth has been studied in greatest detail, it has become obvious that total absence of light has two effects. First, the plants become etiolated; they show considerable and abnormal internode elongation, and organs such as leaves, never expand beyond a rudimentary size. Secondly, the plants are unable to produce chlorophyll. Short periods of light, far below the requirements for normal photosynthesis, will subdue both effects. Different species have different light requirements for normal growth. It has been shown that black-currant plants make normal growth with a constant day-length of $13\frac{1}{2}$ hours. With 17 hours day-length, they make rather more growth. But with only 8 hours day-length, growth is entirely suppressed; the plant passes into a state of quiescence, though growth will be resumed immediately on return to long day-lengths. It seems that each plant species will grow best within a limited range of light intensity and duration. In general, plants in the equatorial regions grow best in the lower ranges of intensity, while as the poles are approached, much higher light intensities are necessary for maximum growth. Extremely high light intensities may kill the plant by destroying the chlorophyll.

The quality, or wavelength, of the light is also a factor affecting growth in green plants. Photosynthesis may be satisfactorily carried on with red light or blue light alone, but the growth of the plant as a whole is not normal in either of these. Plants grown in the longer red wavelengths tend to be weak and spindly, whilst in the shorter blue and violet wavelengths, they tend to be very squat in structure. There must be some physiological processes which are, at least partly, controlled by light.

Other radiations outside the visible spectrum (see p. 992) influence growth, particularly the shorter wavelengths. Ultra-violet radiation of wavelength 185 nm to 300 nm causes cessation of growth, and if continued, it is lethal. X-rays have the same effect. Experiments on developing insect eggs, on insect larvae and on sea-urchin eggs, show that ultra-violet and X-rays retard growth by affecting cell division. With sufficient intensity, cell division is completely inhibited, and it seems that the rays act principally in preventing formation of the spindle.

Hydrogen ion Concentration. The pH of the fluid in contact with a cell has a profound effect on all its activities. The higher grades of animals produce their own body fluids, and so control from within, the pH in which their tissues grow. But the lower animals and all plants must grow in the pH of the environmental fluid, and each species grows best in a given pH range. This range is narrow and any marked change in pH results not only in the cessation of growth but of all other activities, and death ensues. In some cases, especially in small bodies of fresh water, there are marked seasonal pH changes. Organisms counter these changes either by producing resistant eggs before they die, or by forming resistant resting-stages; in either case, there is no growth until the pH becomes satisfactory again.

In the case of organisms living in soil, the pH of the soil solution has a considerable effect on growth. Each species has a tolerance, but cannot make growth outside its limits. Very high or low pH values inhibit the growth and functioning of nearly all roots; hence the rest of the plant is affected. The effect is due to the influence of pH on enzyme activity. The pH also affects the availability of some mineral nutrients. For example, at pH 6, inorganic iron precipitates from solution as $Fe(OH)_3$ and is deposited as Fe_2O_3, in which form it is non-available. Other elements, such as manganese are precipitated at high pH values. Further, for all soil organisms, pH changes may seriously affect the permeability properties of cell membranes; thus the rate of entry of substances is altered, and hence the growth rate.

Internal Factors

Even when external conditions are known to be at their best for growth, an organism still follows a growth pattern peculiar to itself. This indicates that the growth rate is affected by factors operating within the organism. The rate varies with the stage of development along pre-determined lines. Not much is known about the growth-controlling influences, but what information has been gained, points to some form of chemical control by specific substances. These may be placed in two categories, namely, those which may be called *tissue substances*, and the more clearly defined *hormones*.

Tissue culture experiments tend to show that no tissue can grow except in the presence of extracts of the same tissue. This indicates the presence of one or more substances in the tissues themselves which govern the addition of new cells. As to their nature, there is a little evidence which points to their being nucleic acids or allied substances.

Both plant and animal hormones are known to affect growth in one or more ways. The auxins in plants influence cell division, cell elongation, regeneration of tissue at cut surfaces, growth of ovaries into fruits, the development of buds and roots and the growth of abscission layers in leaves. In most of these cases, the manner in which the auxins operate, is not understood. In vertebrate animals, the hormonal secretions of some glands have comparable influences on growth. The thyroid gland produces thyroxin in a concentration which promotes normal growth; in its absence, no growth at all is possible; excess of it curtails growth considerably. The mechanism of this control has not been elucidated, neither has that of the growth substance which has been isolated from the pituitary gland. The latter is known to promote general symmetrical increase in growth in young vertebrate animals; it may be due to ultimate cessation of secretion of this substance that growth finally ceases in these animals. The secreted substance is known to be bound up with protein synthesis. The sex hormones also have some effect on growth. Castration of cattle causes at first a retardation of the growth rate, and this is followed by a rapid acceleration. Removal of the ovaries has the retardation effect but it is never followed by the acceleration.

It will be necessary to amass much more information about all the physiological processes before there can be any clarification of the ways in which growth is influenced from within the organism.

REGENERATION, WOUND HEALING AND TISSUE REPLACEMENT IN ANIMALS

Each of these entails the formation of new cells to replace those which have been removed or damaged. Thus they may be classed as growth phenomena.

Regeneration is the process by which an organism regains its normal form when this has been altered by the loss of a part. For example, if a tadpole's tail is removed, a new tail will grow, in every way similar to the one removed. Regeneration of the tail occurs even if the animal is not supplied with food, in which case the new tail grows at the expense of materials in other parts of the body. The healing of wounds and recovery from starvation are other examples of the same phenomenon. In vertebrates, some parts of the body such as the outer layers of skin, and blood cells, die continually; they are replaced by regenerative processes. It may be the case in mammals that all protoplasm is continuously regenerating, since it has been shown with radioactive isotopes that the protoplasm is forever changing its components such as amino-acids, using materials from its food.

All animals possess the power to regenerate lost parts, but the extent to which this can occur becomes less as complexity increases. In the simpler multicellular animals, the whole body can be regenerated from a very small part. In the higher vertebrates, only very small parts of the body can be regenerated. As a general rule, the more primitive the structure of an animal, the greater its power of regeneration.

The process varies considerably with the extent of the damage done to the body, but may be said to consist of three interrelated parts. Immediately after damage, there is the formation of a mass of undifferentiated tissue in the injured region. This mass is formed both from local cells which de-differentiate or revert to greater simplicity, and from cells which migrate from other parts. The pad of undifferentiated cells is called a *blastema*. Differentiation of the blastema into various tissues will not take place unless there is a sufficient concentration of the growth hormone present, and it is known, in many cases, that other hormones may be implicated, some having an inhibitory effect. At first, this process is often undetermined. In vertebrates, if a regenerating pad of tissue is removed from its site of origin and grafted elsewhere, it forms organs which would normally be present in the new region.

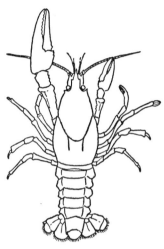

Fig. 13.20. Crayfish regenerating cheliped.

Thus a tail bud, in early development, grafted on the stump of an amputated limb, produces a limb. But, as regeneration proceeds, the types of tissue become determined, so that a tail bud grafted to a limb stump after a short period of re-differentiation, will produce only a tail.

Normally, the tissues which form any organ in regeneration, are derived from similar tissues in the parent body; ectoderm produces ectoderm, and endoderm produces endoderm. However, this is not always the case. Reorganization of the tissues is effected by the differential growth of the regenerating parts. There is just the same controlled growth of the regenerating organ as of the whole body, but it is more rapid. The growth rate of the new part seems to depend on the total damage. For instance, two inches cut from the posterior end of an earthworm will not take twice as long to be regenerated as would one inch. Then again, the speed of regeneration varies with the nature of the organism. Thus, in the crayfish, a limb will not be fully re-formed until the animal has moulted at least once. In crustaceans generally, the regeneration process is also partly dependent on two hormones. One, from the sinus glands, has an inhibitory effect on regeneration, and the other, from the Y organs, has an accelerating effect. The actual rate of regeneration is, therefore, at least partly determined by the balance of these two hormones.

Abnormal regeneration may sometimes occur, a striking example being afforded by the prawn. If a whole eye, eyestalk, and optic ganglion are removed, the animal will regenerate an organ like an antenna.

In many coelenterates and sponges, the whole body may be regenerated from a small part of it. If a hydra is cut into small pieces, or a sponge forced through a fine sieve, each small piece will regenerate a complete animal. It seems that bud formation which occurs as a process of asexual reproduction in many lower animals, is in a sense a form of regeneration. It involves the same processes but is not initiated by the same means.

Regeneration of parts of the body in certain invertebrates has been used to elucidate the general differentiation and organization phenomena encountered there. The planarians (free-living flatworms), have been extensively used as material, and as a result of the investigations, certain important principles have emerged. First, there is a very distinct polarity along the anterior-posterior axis of the body. If a flatworm is cut in half transversely, the front half grows a tail backwards, and the hind half grows a head forwards. If a piece is taken out of the middle by two transverse cuts, the front end of it always produces a head and the hind end always produces a tail. Secondly, it is found that in regeneration in a separated piece, more posterior organs are regenerated only if more anterior organs are already present, either by being in the piece at the time of its separation, or by having regenerated there. In a piece taken from the middle, the head develops first and then the tail, never the tail first. This is explained by saying that the head is "dominant" to all the rest of the body, and extends its influence to all other parts; thus there is an anterior-posterior axial gradient. Thirdly, the

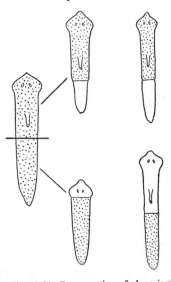

Fig. 13.21. Regeneration of planarian

regenerated body is organized so that it uses up the tissue available in the piece, and each organ is re-formed of a size appropriate to the size of the piece.

Experiments on other animals have disclosed features of interest. They indicate that the precise control of organization of a new part varies in different animal groups. There is evidence that in some animals, special tissues may control the regeneration of large parts of the body. If the anterior part of an earthworm is removed and then the nerve cord cut away from several segments in the anterior region of the remaining part, the new head is regenerated from the first segment in which the nerve cord is intact. The anterior segments, which lack the nerve cord, soon degenerate. Marine annelids of the genus *Sabella* show a peculiar feature of regeneration in that a central piece of the body does not produce a head first. They regenerate the head and intervening regions at the same time, and occasionally, the head may not be regenerated at all. Here there is no such definite axial gradient as in the planarians. This is further exemplified by annelids of the genus *Perionyx*, in which an anterior piece of the body will produce a head at both ends. It has been shown in various nereids that for successful regeneration both the blastema and a sufficient concentration of growth hormone must be present (*see* Vol. I, Chap. 21).

In the higher vertebrates, regeneration is much more limited. Whole organs are not regenerated. It is true that if a portion of the liver is removed, the surrounding hepatic cells will divide and produce tissue which resembles liver tissue. The same is partly true of the kidney. Pregnancy leads to regeneration of uterine muscular and mammary secretory tissue, but if either of these is removed experimentally, they are not regenerated but only repaired with scar tissue. Strictly speaking, regeneration in higher vertebrates is confined to epithelial and endothelial layers, connective tissue systems and blood-vessels. In all other tissues, a defect is repaired, but the damaged or missing types of cells are not regenerated. In skin, active division of deep-seated germinative cells regenerates those which are lost at the free surface. But the specialized elements of skin such as hair-papillae, touch corpuscles, other sensory organs, sweat glands, sebaceous glands and hair follicles are not regenerated. Glandular endothelia can be regenerated as long as there are corresponding cells still present. Endothelia of blood-vessels can be regenerated and will form new capillaries, arterioles and venules; they form the "granulations" in a wound prior to healing. Connective tissue regeneration accounts for the formation of scar tissue in repair of a defect, for the production of adhesions in serous cavities, and for the repair of bones after fracture. Nerve cells cannot be regenerated; when they are destroyed, they cannot be re-formed. However, nerve fibres may develop outwards from the undamaged nerve cell and thus return the cell to its normal function. In the vertebrates generally, for successful regeneration a delicate balance of a number of hormones is essential. The growth hormone, somatotropin, is obviously involved and, in addition, the mineralocorticoids which have an accelerating effect. On the other hand, the glucocorticoid hormones have an inhibitory effect on regeneration, except in the early stages. There is also some evidence that androgens (positive) and thyroxin (positive) are both involved. It is likely, that as in other regulatory processes, a number of hormones are implicated at various stages, the effects of the hormones varying according to the age at which regeneration is taking place.

A wound may be a contusion (bruising), an incision (cutting), or a laceration (tearing). In bruising, the superficial layers are not laid open but the cells and blood-vessels beneath the surface are damaged by crushing. The area contains loose dead cells and released blood corpuscles which give the bruise its colour. The damage is repaired by the gradual absorption into the system of the substances resulting from the self-digestion (autolysis) of the dead cells, and by the re-formation of damaged blood-vessels. Reaction to an incision or laceration depends on the extent of the injury and on the organs concerned. During healing, there are several interrelated processes involved. First of all, blood released from cut vessels clots over the opening. Inflammation quickly ensues; it involves dilatation of blood-vessels, increased flow of blood to the area and alterations in the walls of capillaries which enable plasma and leucocytes to pass into the tissue spaces. The area becomes hotter than its surroundings, and is red and swollen as a result. Dead cells undergo autolysis and are replaced by cells derived from adjacent tissues which are capable of regeneration. Such are the epithelial layers of skin and the connective tissues of the dermis. The whole process seems to be initiated either by substances released by damaged cells, or by substances produced during the autolysis of cells. Thus it is self-regulated.

Cell and tissue replacement is a continuous process in parts of all animals. For example, skin cells are sloughed and blood corpuscles become effete; they must be replaced. There are specialized tissue regions in the body provided for this purpose. Such are the germinative layer in the skin, the bone marrow and lymph nodules.

REGENERATION, WOUND HEALING AND TISSUE REPLACEMENT IN PLANTS

As with animals, regenerative powers in plants are associated to some degree with primitiveness. In the lower groups such as the algae, fungi, liverworts and mosses, it is frequently the case that whole bodies may be re-formed from very small pieces of any part of the plant, even sometimes from single cells. This is comparable with the "reconstitution" of animal bodies from small pieces, as is found in the coelenterates and sponges. This complete reconstitution from any cell is, in general, confined to those plants in which tissue differentiation has not reached a high level (but *see* pp. 536–542 for reference to further development of single cells, tissues and organs in sterile culture). Thus a single cell of a filamentous alga such as *Spirogyra* can produce more of its kind and so regenerate a whole filament. Small isolated pieces from any part of a *Fucus* thallus can regenerate the whole structure, given the right conditions of nourishment. Chopped-up pieces of *Pellia* and *Funaria* plants can do likewise. In the higher vascular plants where tissue differentiation is exhibited to a much higher degree, the power of reconstitution and regeneration varies considerably with the type

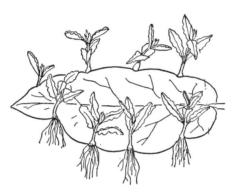

Fig. 13.22. New plants developing on *Bryophyllum* leaf.

of tissue isolated and from plant to plant. A piece of *Begonia* or *Bryophyllum* leaf may regenerate a whole plant, while from many higher plants it is not even possible to strike cuttings.

If a higher plant be dissected into pieces, the power of the separate parts to regenerate the missing pieces varies according to species, according to the type of piece, root, stem or leaf, and sometimes according to its size. In any case, it is obvious that only parts containing living cells can regenerate anything; pieces of bark or xylem vessels cannot form new cells.

If a root-tip is cut off, the remaining cut stump will sometimes re-form a meristematic region and continue growth normally. The root-tip itself, if carrying some differentiated tissue, can be grown in a culture medium, where it will proceed to develop a whole root system. Such a separate root-tip, however, lacks the power of regenerating any tissue at its cut surface. Experiments illustrating these phenomena have been carried out on excised tips of pea roots. If the whole or part of a root is cut off, it will often initiate bud primordia in its outer tissues. These buds will later form a complete aerial shoot system. An apical bud removed from a stem tip and placed in a culture solution can regenerate a whole plant; such a process is found in *Tropaeolum* and lupin. The remaining stump does not re-form a new meristem at the apex, but one of the lateral buds assumes the apical position and proceeds to develop, whereas it was previously inhibited from doing so. If a plant of *Solanum nigrum* has its apical and lateral buds removed, it will develop new bud primordia at the cut apex in the callus tissue which results from the wounding. The same will occur with *Populus nigra* and *Nicotiana* species. The addition of the purine adenine to a culture medium in which wound callus is growing, will in some cases promote bud formation.

A portion of the stem alone will very frequently regenerate a whole plant. This has important practical applications in growing cuttings of numerous plants. In such cases, there is a remarkable parallel with animal regeneration, in that such pieces of stem exhibit distinct polarity. The original upper part always tends to develop bud initials and the original lower portion tends to produce root initials.

In general, leaves are less able to regenerate than other parts, but, as in the case of *Begonia* and *Bryophyllum*, some are capable of producing a complete plant. There is little doubt that natural methods of vegetative propagation such as the formation of stolons, bulbils and gemmae, are special adaptations of this power of regeneration.

Wherever regeneration of plant parts has been studied in detail, it has been found that auxins invariably have a role in the process. Root formation, bud initiation, meristematic activity and subsequent cell growth and differentiation, are all known to be affected by the presence or absence of auxins.

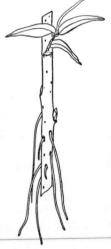

The healing of wounds in plants is accomplished by the formation of new tissue. This is initiated by chemical substances liberated from the injured cells. The cells which respond to the wounding by meristematic activity are not those of the wounded layer itself, but a little below it. Thus the liberated chemicals have to move through the tissues and accordingly are considered to be hormones. Haberlandt first demonstrated their presence and called them *wound hormones*. When a potato tuber is cut and left undisturbed, the cut surface will become covered with corky cells, thus healing the wound. But if the cut surface is washed immediately after cutting, no corky cells are formed. If now a washed cut surface is smeared with juice from another cut surface,

Fig. 13.23. Roots developing on willow stem cutting.

normal healing results. Experiments of this type show that the hormone concerned is liberated by the cut cells. The only wound hormone definitely

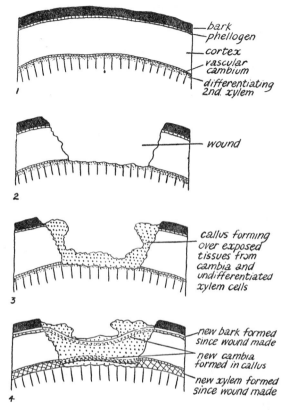

Fig. 13.24. Diagrams representing formation of callus at a wound in a woody stem.

recognized is *traumatic acid*, which can be extracted from injured bean pods. It seems to be specific to the bean since it will not bring about healing in other plants.

In some cases, the only result of meristematic activity in the wounded region is to produce a layer of corky cells sealing the wound. In other cases, there is a considerable development of new tissue known as *callus*. It is formed of soft parenchymatous tissue which, in many instances, merely becomes suberized on the outside to form a periderm which seals the wound. Callus may also differentiate into new lateral meristems, cambium and phellogen, continuous with the original ones. These new meristems proceed to form vascular tissue and bark at the same pace as the adjacent uninjured meristems. Thus not only is the wound covered, but the injury is rendered completely unnoticeable. When callus tissue is grown in culture solution, it will in some cases produce complete new organs such as buds and roots.

Tissue replacement is a necessity in those plants which continuously shed their outer layers. Thus bark is replaced in a manner quite comparable with replacement of skin cells in animals. In plants, the phellogen of the periderm continuously forms new cells which become suberized to replace those lost.

THE CYCLE OF DEVELOPMENT

A new organism commences development as the result of some reproductive process, which may be sexual, asexual, or vegetative, according to the type of organism. Some may reproduce by all these methods. Thus the starting-point for a new individual may vary, causing the exact course of development to vary also. Sometimes, in a single species, the sequence of events may be widely different according to the mode of origin, but invariably the mature organism is true to pattern. In this section only special cases will be mentioned, as it is not possible to generalize over the wide variety of types of development.

The precise course of development for any organism is controlled by two sets of factors. There are internal influences which are inherited, and there are the conditions of the external environment, which may modify the action of the inherited factors. If we assume that development is determined by inheritance, we must associate it with genes controlling the production of protoplasmic materials and the way in which they are distributed throughout the body during its development. There is good evidence that this is the case from genetical experiments on the inheritance of ability to synthesize specific proteins and enzymes. For example, it has been found that a single gene controls the synthesis of the enzyme tyrosinase in the fungus *Neurospora*. There is also evidence that such controlling genes are not always confined to the nucleus. It has been shown, in a few cases, quite conclusively, that the cytoplasm of the cell can be just as efficient as the nucleus in transmitting inheritable characters. Leaf-colour inheritance in *Oenothera*, and the form of the capsule in *Funaria* are known to be cases. This has led to a conception of development as being due to the distribution of nuclear genes, and the varying distribution of plasmagenes which occurs when cells divide (*see* Chap. 19). The determining mechanism begins to operate with the first cleavage of the zygote. The nuclear genes are constant in every cell of any particular organism, but the plasmagenes may vary. The different combinations of these may determine the manner in which the cell will develop.

Development of the Flowering Plant

Confining ourselves to the development cycle initiated by the formation of a zygote, it is possible to distinguish in the flowering plants a fundamental series of events which is common to all. Variations in the sequence occur chiefly with regard to the time taken to complete the life cycle. The simplest cycle of events, exhibited by most annual flowering plants can be divided into five stages.

1. *Development of the Zygote into the Embryo, and Seed Formation.* This involves a period of rapid cell division and the differentiation of the embryo and other parts of the seed. Finally, the seeds are shed and dispersed.

2. *Period of Inactivity or Dormancy.* During this time, when it applies, growth and development are apparently at a standstill, though physiological changes may be occurring in the seed. They are sometimes necessary preludes to the next period of activity.

3. *Germination of the Seed.* This is a period of extremely rapid metabolism as a result of which the embryo enlarges considerably at the expense of stored food material. When germination is successfully completed, the plant is self-supporting.

4. *Period of Vegetative Development.* From the young seedling, the primary body is developed; secondary growth may be made according to species. During this period, the root and shoot systems expand vigorously. Special structures may be formed to effect vegetative reproduction.

5. *Phase of Reproduction.* This may overlap the previous phase to a greater or lesser extent. It involves the formation of the flower which produces the asexual spores called pollen grains and embryo sacs. When fertilization is effected, the cycle commences again. The parent plant dies.

Biennial flowering plants complete their cycle in two seasons. In the first there is the building of the vegetative body, and in the second season there is further vegetative development followed by reproduction, after which the plant dies. Perennial plants may take several years to reach reproductive maturity and thereafter, interrupted by rest periods, continue to make vegetative growth and reproduce yearly.

Development of the Zygote into the Embryo: Seed Formation

Division of the zygote and differentiation into an embryo is known to follow a distinct pattern for each species. It results in a structure with definite parts and distinct polarity. From a knowledge of cell position, it is possible to predict what parts of the dividing cell mass will form the finished structures of the embryo, but so far, it has been found impossible to explain them except vaguely with reference to the distribution of plasmagenes. Tissue culture experiments have shown that a period of contact with the maternal tissues is essential for development. This may point to a chemical influence being exerted by the parent, but no details are known. Supply of nutrients of the correct kind and the correct conditions of temperature, osmotic pressure and pH are all known to be essential for normal development. Angiosperm seed formation is described in Vol. I, Chap. 16.

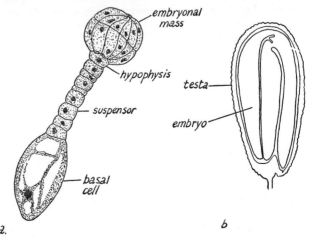

Fig. 13.25. (a) *Capsella* embryo expressed from young seed. H.P. drawing.
(b) L.S. *Capsella*, mature seed. L.P. plan.

Germination and Dormancy of Seeds

No seed will germinate unless certain conditions are prevailing. There must be a supply of water, a supply of oxygen, and the temperature must be within certain limits of range. Even when these three needs are met, some seeds will not become active unless they have been subjected to certain light conditions, and others fail to spring into growth until they are internally prepared to do so. The breaking of dormancy by a seed is thus in some cases only possible when both external and internal conditions are favourable.

In the "dry" state, all seeds contain some water, but the proportion by weight is usually in the neighbourhood of 10 per cent only. The need to increase this water content is explained when it is realized that during the germination period, the protoplasm is probably more actively metabolizing than at any other time. Water is required convert the protoplasm to a comparatively highly aqueous condition, the only condition in which its enzymes are fully functional; the water is required for the hydrolysis of complex stored food reserves into simpler mobile substances, and to act as the solvent through which these can be diffused to the developing parts.

Water is taken in by the seed, in the first instance, largely by imbibition through the walls of the testa cells, but there are many instances where a caruncle or aril may expedite this water passage into the living tissue of the embryo and endosperm (if present). The living cells will undoubtedly take in water by osmosis, and as more and more substances with high osmotic activity are produced in those cells, the rate of intake of water will progressively increase until all are fully turgid. The swelling process causes rupture of the water-softened testa. The rate at which water is absorbed varies with the species, and there are notable examples of leguminous seeds in which the testa is to all intents and purposes impermeable for weeks, months and even years. A scratch with a file on the testa of such examples as lupin, sweet pea, gorse and clover seeds, will allow water to penetrate quickly and so hasten the germination. The rate of entry of water also varies with the temperature, being much quicker at higher temperatures. Some seeds such as barley can be damaged by too long an exposure to wet conditions and are said to suffer "soaking injury."

The necessity for continued intake of oxygen by a germinating seed is readily understood from a knowledge of the simple facts of respiration. From the stored food reserves of the seed, the young plant must obtain all its supply of energy by which to perform both chemical and mechanical work. It must constantly synthesize new protoplasmic materials and physically drive its root downwards and its shoot upwards. The energy required for these purposes is released as a result of oxidation processes which cannot commence when oxygen is not available (see Chap. 11). Seeds can respire for a short while, as indeed can most tissues, in the absence of oxygen, but the rate of energy release is very low and the end products of the anaerobic process are toxic. Carbon dioxide in high concentration is an inhibitor of germination for many seeds most of which will not, therefore, germinate under anaerobic conditions such as full immersion in water. There are some, however, such as rice that are unaffected by such conditions and germinate quite readily in the inundated paddy fields. Entry of oxygen is presumably by diffusion from the outside atmosphere, into which is released the unwanted carbon dioxide. The very rapid rate of respiration by germinating seeds is indicated by their rise in temperature when energy is lost as heat. The well-known experiment in which surface-sterile, soaked, live peas are placed in a vacuum flask, and their temperature after two or three days compared with that of similarly treated but killed (by boiling) peas, is sufficiently convincing of this.

The limits of temperature range within which seeds will successfully germinate, vary with species. The lower limit is freezing point in many cases, but in some such as cucumber, marrow and others of the Cucurbitaceae, the lowest temperature at which germination will occur may be as high as 18°C. This discounts the explanation of the lower limit as being fixed by the freezing of water within the seed. As temperature is increased above the lower limit, so the rate of germination is quicker as long as no other factors limit the process. This is to be expected if we consider that the rate of germination will be dependent, partly at least, on the rate of enzyme action. That there is an upper limit of temperature beyond which no increase in rate of germination occurs, can be readily appreciated when the denaturing effect of high temperature on proteins is recalled. Indeed, if the temperature of the seeds is raised to too high a value, the rate of germination gradually falls away to zero and the seeds are killed. This upper limit of temperature also varies with the species, being in the region of 40°C for the majority. Optimum temperature for germination is about 5°C below this; temperatures just above or below the optimum tend to slow the process without bringing it to a halt. Some seeds germinate most rapidly in constant temperature conditions, others do better in exposures to alternating higher and lower temperatures.

Seeds vary in their reactions to the presence or absence of light. Most are quite unaffected by light or darkness, that is, they are *light-indifferent* or are said not to be *photoblastic*. However, in some species, the seeds will not germinate unless exposed for some period to illumination. They are said to be *light-sensitive* or *positively photoblastic* and include among others, species of the genera *Veronica* (speedwells), *Rhododendron*, *Lythrum* (loosestrife), *Nicotiana* (tobacco), *Betula* (birch), *Digitalis* (foxglove), *Lactuca* (lettuce), *Epilobium* (willow herbs), and *Viscum* (mistletoe). The quantity of light necessary to cause germination when all other conditions are fulfilled, is sometimes very small. A few hours in diffuse light is often sufficient, and in the extreme case of *Lythrum salicaria*, an exposure to a light intensity of 730 lux for one-tenth of a second is all that is required. The response seems to be dependent on quantity of light energy received, i.e. intensity × time, rather than just intensity.

Light-hard seeds or those *negatively photoblastic* are seeds in which germination is slowed or prevented by the presence of light. Species of the genera *Phlox*, *Nigella*, *Lamium*, *Helleborus* and *Allium* (onion) are examples.

The effects of different wavelengths on photoblastic seeds have been studied and in fact led to the discovery of the photoreceptor substance, *phytochrome, see* p. 542. For positively photoblastic seeds of lettuce and others, exposure to red light at 650 nm promotes germination, whilst a wavelength in the infra-red (or far-red) at 730 nm inhibits it. The two conditions seem to counteract the effects of one another. To fit such a state of affairs it was postulated that there exists a substance occurring in two states, according to the light conditions to which it had been previously exposed. In one state (P650) it absorbs light in the red region and in the other (P730) it absorbs it in the infra-red. At each exposure it absorbs radiation and becomes converted to the other form. Thus, when lettuce seeds are exposed to red light this converts any P650 to P730 and sets up a reaction chain leading to germination. But when P730 is exposed to infra-red radiation this causes the re-formation of P650 which halts the pre-germination processes. The difference between positively and negatively photoblastic seeds seems to lie in their different reactions to red and infra-red radiation. In the positive cases the germination-promotion effects of red wavelengths tend to predominate whilst in the negative cases it is the infra-red that overrides the other. The effects of the shorter blue wavelengths have been studied, but tend to indicate a very complex relationship between seed germination and this kind of radiation. The responses of some photoblastic seeds appear to have some dependence on temperature conditions. For example, lettuce seeds over the range 10°C–20°C are light insensitive but become light-requiring at 20°C–30°C, whilst at 5°C germination is inhibited in both light and darkness. In some species of dock, an alternation of daily periods at 15°C and 25°C removes the light requirement which is present when the seeds are kept at 25°C all the time. In some cases there appears to be what can be called a form of *photoperiodism* in the germination responses. Some seeds, given short daily exposures to light show a higher germination rate than if they are exposed to long periods of light every day. Some show the opposite response, that is, there appear to be some "short-day" and some "long-day" seeds, *see* p. 544.

Another factor that affects the light sensitiveness of some seeds is their age. Freshly gathered seeds of lettuce, for example, show all the special features of the positively photoblastic condition, but after being kept in storage for some months are no longer light-requiring at all. There is a similar loss of sensitivity to light exposure by the negatively photoblastic kinds.

That the germination response made by some seeds to the presence or absence of light is a complex phenomenon is further accentuated by the fact that the presence of certain chemical substances can completely replace any light requirements. Substances as widely varying in nature as gibberellic acid, potassium nitrate and thiourea are known to be effective in different cases. Kinetin will make lettuce seeds respond much more readily to very brief exposure to light whilst coumarin inhibits their germination. It is also the case in a number of species, e.g. *Betula pubescens*, that isolated embryos are quite without any special light requirements and will develop equally well in the light or in darkness whilst intact seeds are positively photoblastic.

If one or more of the conditions required for seed germination is not provided in the environment then a seed will have imposed upon it a period of rest or quiescence after its dispersal from the parent. But even when all known requirements of the surroundings are met favourably, for many seeds there is still an innate dormancy or period of growth inactivity which must be broken before germination can occur.

The period of seed dormancy varies considerably with different species. Seeds of some plants are capable of germination as soon as they are shed from the parent; by contrast, some seeds cannot be made to germinate under any conditions until the "rest" has been completed. The dormancy period may be weeks, months or even years. In a few species, e.g. *Taxus*, peas, beans, maize, if the seeds are given favourable conditions of water and oxygen supply and are within the correct temperature range, they will develop as soon as they are shed, but if such conditions are not encountered soon after dispersal, the seed enter a period of dormancy which cannot then be broken for some time. This is a *secondary dormancy* as distinct from a *primary dormancy* in which the seed will not germinate immediately on dispersal whatever the circumstances.

Investigations have shown that a primary dormancy may be due to one cause or to a combination of several causes. In a few cases, when the seed is shed, the embryo is not mature. No further activity of the seed can take place until embryo development is completed. Completion of development occurs only when the seeds are given favourable germination conditions, and hence germination is delayed as compared with other seeds in which the embryo is mature when the seeds are shed.

In some cases, the testa may be very impermeable to water and/or oxygen; in either case, metabolic activity of the seed cannot begin until adequate supplies of both materials can enter readily. Generally, the testa becomes more and more permeable as time passes and the seed slowly responds by breaking dormancy. It can be hastened artificially if the testa is removed or in some way mutilated.

Even when water does enter readily, the testa may be tough enough to resist rupture, and consequently it may prevent entry of sufficient water. Alternatively, it may cause the retention of the radicle and plumule

In time, the seed coat becomes less resistant and the period of dormancy is ended. But there are some cases where the seed coat plays more than a mechanically resistant or passive role in dormancy. In some of the cases where it is not the embryo that is dormant, as witnessed by the fact that it will develop at any time without special treatment if removed from its covering layers, then the testa is certainly responsible for the failure to germinate through some special requirement of its own. This may be through a light requirement as in *Betula pubescens* mentioned previously.

In all the instances of dormancy mentioned above, the condition is more or less forced on the seed by reason of its immaturity or its structure and may be referred to as *relative dormancy*. But cases are known in which the dormancy is of a different nature. Seeds of *Crataegus* (hawthorn) and the apple, when shed, have fully mature embryos, but even when the outer coverings are broken or removed, no activity can be induced despite the prevalence of the most favourable of germination conditions. Such seeds are said to exhibit *true dormancy* and require a period of *after-ripening* of the embryo. It is assumed that this is some chemical or physical change which is under the influence of internal factors. In the case of *Crataegus*, if the endocarp and testa are removed, the after-ripening period can be shortened from approximately one year to one month, and the process is also favoured by a temperature around 5°C. A number of seeds can be said to exhibit a similar chilling requirement in that a period at a low temperature is effective in shortening the dormancy period. Temperatures between 0°C and 5°C are most effective so long as the seeds are fully soaked and supplied with oxygen. In some seeds, e.g. water plantain and other aquatics, it appears that only the testa of the whole seed requires the chilling since the embryo will germinate without it when removed from the seed. In the case of the peach it is the embryo that requires to be chilled. In the hazel and apple both seed parts require the low temperature treatment, but if the embryo is removed from the testa, its dormancy can be broken by a shorter period of chilling than is required for the whole seed. In some cases, such as acorns, the chilling is related to the development of the plumule only. In these, said to exhibit *epicotyl dormancy*, the radicle emerges without the need for low temperature treatment but for the plumule to show a chilling period must be experienced. In the genera *Convallaria* and *Polygonatum* some species show a double chilling requirement related first to the emergence of the radicle followed by a second chilling requirement for the emergence of the plumule. The seeds are thus said to be *"two-year"* seeds, needing two winters to break dormancy under natural conditions. The genus *Crataegus*, mentioned above, is also a "two-year" form but for a different reason. In this case, the chilling given by the second winter is the effective one after the seed has been properly wetted during the first year through its stout endocarp, the chilling of the first winter being ineffective on the dry embryo. Treatment with some chemicals such as ethylene chlorohydrin and nitrates can replace the chilling requirement for some seeds, it is reported. Treatment with acids also tends to speed the after-ripening process, apparently causing a lowering of the pH of the hypocotyl tissues of the embryo. Treatment with gibberellins and kinetin is known to break dormancy in some species, e.g. lettuce, but through what agency is not known.

Secondary dormancy has been the subject of much investigation, but no clear-cut reasons for its occurrence have ever been put forward. It can be forced on some seeds by artificial treatment with high concentration of carbon dioxide, and on others by subjection to temperatures well below the lower limit, with the other germination factors favourable. A suggestion to explain secondary dormancy, which may have some foundation, is that if a seed does not immediately find favourable conditions on dispersal, then its seed coat undergoes a change of permeability to water and gases which delays the germination when good conditions are later encountered.

A problem which must come to mind when considering germination of seeds is the length of time a seed can remain in its dry, dormant state before losing its capacity to germinate at all. How long can it remain viable, and how does it retain viability in the dormant state? Investigations into the first part of the problem have shown that there is very great variation between different plants. In some species of *Oxalis* the seeds are not "dry" when shed, and will remain viable for but a short time after dispersal. They are quickly killed if subjected to drying conditions. From this case of very short retention of viability, examples can be found ranging from a few days to many years. The seeds of most British plants can live from the autumn in which they are shed till the following spring at least, whilst many can persist over several years. Examples of proven cases are given below—

Some species of willow, a few days; poplar, a few weeks; elm, beech and birch, a few months; oak, alder, hornbeam, larch, maple, lime, one to three years; ash, Scots pine, four years; spruce, five years; wheat and barley, ten years.

There are well authenticated cases of seeds remaining viable for much longer periods than these. Perhaps the most remarkable proven cases are those of seeds of *Cassia bicapsularis* and *Cassia multijuga*, stored dry in a museum. The seeds germinated readily when sown after 115 years and 158 years respectively. Other cases of extreme retention viability have been recorded, including one of at least 160 years for the Indian lotus. The most extravagant claims of all, concerning the germination of wheat grains taken from Egyptian tombs, and at least several thousand years old, are entirely without foundation.

The morphological and physiological differences between short and long-lived seeds have never been explained in any detail. It is true that the most long-lived possess very hard seed coats which are highly impermeable to water and gases, but there are well authenticated cases of seeds with soft testas lying dormant deep in a soil in wet conditions for as long as 25 years.

The morphological changes which accompany the development of an embryo into a seedling have been described in Vol. I, Chap. 17, and will not be treated here, but accompanying the rapid change in form are biochemical transformations which should receive notice. The absorption of water by the seed has already been mentioned. Once this has been accomplished, the first chemical process is the conversion of insoluble storage compounds into a soluble diffusible form. The process is hydrolytic and is under enzyme control. Generally, food reserves are of several kinds in a seed, but differ in proportions with species. Most seeds store fat as the chief food reserve, a few, no more than about 10 per cent, utilize starch in great quantity, whilst very few store high proportions of protein. In occasional instances, the bulk of the reserve may be hemicellulose. Whatever the reserve may be, its energy content and substance are needed for the rapid building of new cells at the apical meristems of the embryo axis. The first essential step in the utilization of the reserves must be the development and activation of enzymes capable of catalysing the necessary chemical reactions. In fatty seeds, although some fat may be translocated in a finely emulsified form, most is split by lipase into fatty acids and glycerol, and in starchy seeds diastatic enzymes produce sugar. In the former case, it is likely that most of the fatty acids and all the glycerol are quickly converted into sugar, since glycerol cannot be demonstrated in germinating seeds. The reserve materials gradually disappear from the storage tissues (cotyledons, endosperm or perisperm) and if sections of cotyledons of germinating bean seeds are examined, this can be verified from the disintegrated appearance of the starch grains. Some of the sugar so produced is respired in all parts of the seed and there is always a rapid increase in respiration rate and a loss in dry weight of the seed as carbon dioxide and water are formed. The remainder, after translocation to the growing cells, is utilized with other substances in the formation of new cell material. Little, if any of the protein content of a dry seed is lost during the germination stages and thus is rarely used as a respiratory substrate. Wherever there is any stored protein it is hydrolyzed by proteinases into its constituent polypeptides and amino-acids and transferred to the sites of meristematic activity to be incorporated into new protoplasm. Seeds rich in stored protein, e.g. legumes, show a tendency to form amides such as asparagine during protein metabolism. This can be demonstrated in lupin hypocotyls in a crystalline form. It is probable that the $-NH_2$ groupings are used later in development for protein synthesis when more carbohydrate is being manufactured by the green seedlings. It has been suggested that the temporary storage of $-NH_2$ groups in this way will allow for the manufacture of amino-acids not represented in the seed, but needed for the building of specific proteins in the meristems.

Until the young radicle has penetrated the soil, any mineral requirement must come from within the seed, and all seeds examined show a

100 g HEMP SEED (analysed by Detmer)

	Seeds	10-day seedlings	Gain or loss
Total dry weight (g) . . .	100	94·03	− 5·97
Fat	32·65	15·20	− 17·45
Starch	0·00	4·59	+ 4·59
Protein.	25·06	24·50	− 0·56
Cellulose	16·51	18·29	+ 1·78
Undetermined . . .	21·28	26·95	+ 5·67
Ash	4·50	4·50	0·00

1,000 SEEDS OF LUPIN (analysed by Beyer)

	Seeds	7·5 cm seedlings	Gain or loss
Total dry weight of substances analysed (g)	80·200	77·732	− 2·468
Fat	4·832	3·439	− 0·393
Starch, cellulose and pectin .	8·869	9·253	+ 0·384
Protein.	49·075	43·097	− 5·978
Asparagine	0·000	2·612	+ 2·612
Sugar, gum, alkaloids . .	14·040	15·698	+ 1·658
Ash	3·384	3·633	+ 0·249

proportion of mineral substances free or combined with organ[ic] materials. Organic phosphates are universally present, and durin[g] the early stages of development, these are broken down to release th[e] soluble inorganic phosphates which are then transferred to the growing regions to be incorporated into nucleo-proteins, amongst other things. All the other essential and trace elements can be demonstrated in seeds and in a few species there may be found quite high concentrations of mineral substances which have no obvious value.

Many analyses of seeds and seedlings have been made in attempts to decide the fate of the stored materials. On p. 528 are given in an abbreviated form the results of two of these analyses. A great deal of work has been carried out on the respiratory activity of germinating seeds and it is clear that there are varying patterns of oxygen intake, carbon dioxide output and hence R.Q. values for different species. There seems frequently to be an anaerobic glycolysis or glycolytic phosphorylation (*see* p. 434) resulting in the formation of ethyl alcohol. Inorganic phosphate is used up. The enzymes required for this breakdown process have been isolated from many sources. When adequate aeration is provided the pyruvic acid produced by this glycolysis is converted via the tricarboxylic acid cycle to carbon dioxide and water. The enzymes required in this sequence of events are located in the mitochondria of the cells, but they seem active only in adequately hydrated protoplasm. Thus ATP is absent from dry seeds but slowly builds up during normal germination processes. One of the oxidation-reduction series of enzymes, cytochrome oxidase (more properly cytochrome a_3), is commonly present suggesting the common flavoprotein-cytochrome chain of hydrogen or electron carriers necessary to the oxidation process (*see* p. 430). Other electron transport systems involving pyridine nucleotide, glutathione, ascorbic acid and ascorbic acid oxidase have been demonstrated and there may be others. In some seeds pentose phosphates are used in place of hexoses as respiratory substrates.

The enzymes required for the rapid speed of metabolism during early germination are clearly not all present in dry seeds and some are known to be synthesized at this time. One of the more clearly established instances of this is the formation of α-amylase in the aleurone layer of barley endosperm under the influence of gibberellic acid, a clear case of enzyme synthesis under hormonal control (*see* p. 805).

Among other metabolic activities there seem to be processes leading to an increase in RNA content in seeds during germination and coupled with this the ribosomal system develops rapidly in endosperm cells.

Germination, then, is the period of greatest activity in the life cycle of a higher plant. Structural changes and metabolism are proceeding at the greatest rate. But, as is shown by the growth curve given earlier, no true growth of the whole seed as expressed by dry weight measurements is made. The stored materials of the seed are merely converted into new living substance with some loss of the former as respiratory by-products. True growth of an embryo is made, however, when all the food is stored in endosperm tissue, and if the curve represented dry weight of such an embryo only, no initial loss would be recorded.

Period of Vegetative Development

Once the seedling has established itself, growth of vegetative parts is rapid if conditions are favourable. Root development may occur in several places. In the first instance development takes place from the primary meristem, and later from the pericyclic regions. This may be followed by development of roots from stem, petiole and even from leaf tissues in some cases. It is always associated with the auxin β-indolyl acetic acid, which is also involved in many other phenomena. The part played by auxin in initiating root formation seems to vary according to the nature of the parent organ. The formation of lateral roots on a primary root seems less directly under auxin control than is their appearance on stem cuttings. In the former case, treatment with auxin causes a first burst of lateral root initiation but subsequent further treatment has no added effect. This suggests a requirement in addition to auxin which is used up in the formation of the first sprouting of laterals. This second substance, sometimes referred to as an un-identified *rhizocaline*, could be produced at the primary root tip. This is suggested by the fact that a primary root, so long as it keeps on growing after an auxin treatment, can continue to produce laterals at its usual slower rate in the newly developed untreated zone. But the factors controlling lateral root formation are by no means clarified yet because it has also been found that some of the accessory growth substances are promoters of lateral root development. Generally, it may be said that in the formation and subsequent development of lateral roots, auxin is one only of several interacting factors.

In the cases of stem and leaf tissues, however, auxin seems to play a much more individual role in initiating root formation and although the response to auxin treatment is very much dependent on such factors as species and age of the tissues, the temperature at which the tissues are kept and even the season of the year, no specific internal chemical factors appear to interact with auxin in the initial processes of adventitious root formation.

More recently it has been suggested that the cytokinins may also be involved in root initiation. There seems little doubt that these substances are produced by roots and there is evidence that at certain concentrations they inhibit elongation of the primary root whilst promoting the formation of laterals. The induction of large numbers of root initials in the leaf tissue of one of the water ferns, *Marsilea drum-*

mondi, when treated with kinetin in sterile culture has also been recorded. Auxin treatment failed completely to produce a similar effect.

The primary meristem at the apex of the shoot is also embryonic in origin. It gives rise to the rest of the plant axis and, in contrast to most root apices that produce from themselves more root tissue only, it is more versatile in being able to produce the whole range of tissues and organs. Many kinds of excised shoot tips can regenerate whole plants, including stems, roots, leaves and flowers. The development of subsidiary branches, initially as buds, whether vegetative or floral, usually takes place just behind the growing apex but not uncommonly buds may appear almost anywhere on the parent plant. Bud development is thought to be influenced by a very delicately balanced interaction between auxin and cytokinin, auxin tending to inhibit, cytokinin tending to promote, the final result depending on the ratio of their concentrations in the tissues. It has been shown, for instance, in the case of tobacco pith tissue in sterile culture, that with an auxin:cytokinin ratio of 100:1 ($2 \text{ mg dm}^{-3} : 0.02 \text{ mg dm}^{-3}$) there is produced only an amorphous, undifferentiated mass of cells, called callus, but when the ratio auxin:cytokinin is lowered, either by reducing the auxin or increasing the cytokinin, buds are formed that can eventually grow into whole new plants. It is interesting that if the ratio is increased, roots are developed by the callus.

On whole, in naturally growing plants, once buds are initiated behind the apex, they commonly remain dormant as laterals unless some change in the shoot apex occurs. This growth inhibition due to the presence of the apex, or *apical bud dominance*, is an example of what can be referred to as a *growth correlation*, this being defined as a growth interrelationship or correlation between two parts of a plant in which the growth and development of one is influenced by another on a reciprocal basis. The dominating influence of an apical bud over subsidiaries produced by it shows up as this growth-inhibiting influence but there are other forms of apical dominance, for example, the influence of the continuous development of the main shoot on the development and subsequent orientation of laterally produced structures such as leaves and branches, including rhizomes and runners. Apical dominance over subsidiaries is exercised by buds has been the most extensively studied of these growth correlations and the growth retardation in laterals has long been considered to be due to the direct growth-inhibiting effect of auxin produced by the apical bud on the laterals. But there are some indications that auxin produced apically may not be acting quite so directly as supposed. There exists the possibility of an interaction between auxin and a more specific inhibiting substance but so far this has not been fully substantiated. It is the case, however, that cytokinins are in some way involved as is evidenced by the fact that the inhibition of growth of axillary buds can be removed for at least short periods by treatment with kinetin. It seems that the axillary buds will show continuous growth only when both auxin and cytokinin are present in adequate quantities and in the right proportions. When whole plants are treated with a gibberellin this can result in a stronger than normal apical dominance effect, but whether this is due to a direct participation of the hormone in the process or not is not understood. Apical dominance is also known to be influenced by external environmental conditions, including factors such as nitrate availability in the soil, light intensity and the photoperiod, temperature and even gravitational forces. It is likely, however, that these factors are operative only as affecting hormone concentrations and their rates of movement through the tissues.

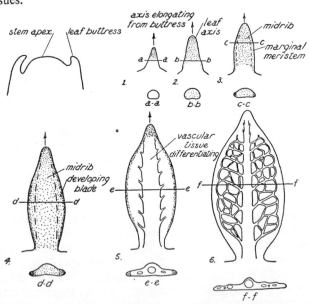

Fig. 13.26. Diagrams to show development of a simple dicotyledonous leaf from a leaf buttress.

Leaves, in whose axils the buds normally develop, also commence their formation as initials at the stem apex. In many cases, the leaf meristems cease activity after a time and the leaf may remain in miniature, protected inside the bud-scales until bud-burst. The mature leaf in dicotyledons at least, is an organ of fixed growth. Its tissues arise, enlarge and differentiate from those formed in the early stages of it development. When expansion is completed there is no further growth. Control of leaf expansion is another complex physiological phenomenon not yet fully understood. Under experimental conditions, a number of factors have been indicated as having some possible promoting influence on the expansion of etiolated bean leaves. The hormones, gibberellin and cytokinin, might in some way be involved as well as light in the red waveband, but so far there is no real knowledge of how any such factors operate.

Three areas of study within the phenomenon of vegetative development are worth enlarging upon, namely dormancy of buds and other organs, the nature of the control of differentiation of parts as disclosed by tissue culture investigations and the apparent control exercised by light other than through photosynthesis, on growth, development and differentiation of plant parts, through photomorphogenetic effects in which the substance phytochrome is involved.

Dormancy of Buds and Other Organs

As in the case of seeds, plant parts may be said to be dormant when their growth activity is temporarily brought to a halt either through adverse external conditions such as low temperature or through some internal factors operating to inhibit the growth processes. In the former case the suspension of activity is imposed from outside and lasts only as long as the conditions are unfavourable. In the latter case, however, the dormancy is an innate condition, part of the natural sequence of internal growth control events. Even if external conditions are at their best or most favourable, the dormant period continues until internal conditions are likewise.

The fully and truly dormant condition in plant organs such as buds not attained suddenly in one immediate step from actively growing to a state of unbreakable quiescence; rather, the condition is slowly arrived at during a period of pre-dormancy when the tendency to lower growth can be reversed by the application of various treatments such as raising the temperature above normal. But once the true dormant state is reached at its deepest then the possibility of breaking it by any externally applied conditions is very much reduced or even altogether removed. In a similar way, full growth activity from full dormancy is reached gradually during a post-dormancy period. In some cases, where the dormancy can be described as relative, meaning that it does not reach the virtually unbreakable state of true dormancy, then the application of certain unnatural conditions such as large changes in the normal photoperiod may induce activity.

Sometimes it is possible to reverse the changes leading to the breaking of dormancy after they have started and to establish the deep resting condition for a second time. The application of abnormally high temperature may do this and the state is said to be one of *induced dormancy*.

Dormancy in vegetative buds occurs very commonly and not only in plants of temperate and colder climates. Nearly all woody plants, including those of the tropics, develop buds in which the young leaves reach primordial condition and then, either under the protection of special scales or otherwise, cease further growth for a period. The formation of the typical apical or axillary winter resting bud results from the cessation of the axis to elongate whilst the apical meristem is still producing leaf primordia, the lowermost and thus the outer of which are most often modified as protective scales. The bud fills up internally for a period with leaf primordia at closely set nodes and is said to "swell." But eventually this formative activity ceases and the bud "rests." In a few cases of terminal buds, for some unknown reason, the whole bud apex then dies and is shed, leaving further extension of the main axis to a subsidiary bud. It is this occurrence that produces the characteristic sympodial growth of the elm, for instance (*see* Vol. I, p. 383).

In addition to the morphological state, there are certain well-known physiological conditions of bud dormancy for some of which at least partial explanations can be given. For example, there is no doubt, as mentioned before, that after an axillary bud has been formed it is inhibited initially from further development by auxin production at the terminal bud and in fact will remain quiescent so long as the terminal bud is present or auxin is supplied at the apex. However, after an initial period of inhibition by the apical bud, an axillary bud enters a phase of true dormancy and will not develop whether its shoot apex is intact or not. Thus a lateral bud is not from the first truly dormant, but later becomes so. Exactly what its internal state is when in this condition is not really known, but there are definite connexions between external conditions and the onset of the true dormancy. There are differences between species in the conditions known to induce dormancy but for many woody plants day length is a very significant factor. An artificial change from long days to short days simulating the arrival of the winter season certainly promotes the dormant state in birch but this does not seem to be the case for ash or apple. In the case of birch, if the long dark periods are broken by only short light periods then the short-day effects are nullified and this points to the possibility that the photoreceptor substance phytochrome (*see* p. 542) may be involved in some way. Under natural conditions, in species where growth in length of an axis may continue well into autumn, for example poplar and larch, there is the strong likelihood that day length may have a real effect in promoting dormancy, but in others where elongation the axis ceases before the end of summer, this is hardly likely to be so. What induces bud dormancy in these species is not known.

Not only can bud dormancy be induced in some plants by the application of certain conditions but it can also be broken. One of the most effective dormancy breakers is the chilling effect brought about by lowering the temperature to below 10°C for periods up to as long as a thousand hours. In nature, adequate doses of chilling are supplied through freezing winter conditions, but if these are not experienced the normal leaf development may be irregular and delayed. Another known dormancy-breaking treatment is the application of long-day periods after a phase of short-day treatment. More artificial treatments are also effective including exposure of buds to chemicals such as ethylene chlorohydrin, dinitrophenol and thiourea. Application of gibberellic acid is also known to overcome dormancy in the buds of some woody plants in which there is a reputed rise in gibberellic acid content as a result of chilling treatment. It is also reported that buds kept at about 40°C in a water bath for some hours will shoot if external conditions are then made favourable.

The true dormant state is achieved by organs other than buds and in general the breaking conditions are the same. Corms, rhizomes and tubers tend to regain activity in response to chilling treatment but potatoes are said to react more strongly to a temperature of about 2°C. Onion bulbs have a dormancy promoted by long-day conditions, that is the bulbs ripen in summer, but this quiescence is short-lived if the bulbs are stored at cool temperatures. The resting tubers or turions of some aquatics such as *Hydrocharis sp.* and *Stratiotes sp.*, in which resting appears to be induced by short-day and warm temperature conditions, have their dormancy period likewise shortened by a dose of chilling at 10°C or below. The same applies to the resting buds of the insectivorous butterwort, *Pinguicula vulgaris*, the dormancy of which can be induced by a short-day treatment. A summer dormancy period has been reported for the liverwort *Lunularia cruciata*, a native of the Israeli desert. It is said to be promoted by long-day conditions and broken by short-day treatment. This would be a survival advantage in the hot, dry conditions of the Israeli summer. Rhizomes of lily-of-the-valley, *Convallaria majalis*, become dormant during long warm days of summer and require a period of chilling the length of which varies with the actual temperature of the chilling period (one week at -5°C–2·0°C up to three weeks at 5·0°C) before the dormancy is broken. Gladiolus corms will break dormancy with a chilling dose of as little as twenty-four hours at 0°C–5°C. In buds of the water soldier, *Stratiotes aloides*, the dormancy condition can be broken by a chilling period after which they will develop if kept within the natural temperature range. If, however, they are quickly raised to a temperature much above this and kept there they will re-enter the dormant state and then require a second chilling treatment before growth will commence. This inducement of a second period of dormancy is known to occur in the buds of some woody plants also.

Because there are some similarities and parallels between the dormancy phenomena exhibited by seeds and buds it is possible that the two have some common basis. What this is, however, can only be conjectured at present. Two possibilities have been put forward, either there is some interference with the gaseous exchanges of buds and seeds occasioned by the formation of their external protective layers, or their dormancy is under the control of endogenous hormonal substances. Whereas there is little direct evidence to substantiate the former, there are some indications that the latter may have some real significance. In terms of hormones, it could be reasoned that dormancy is induced as a result of the removal or deficiency of a growth-promoting substance or alternatively as a result of the presence of a special growth-inhibiting substance, or a combination of these two conditions. Substances in the former category might be auxins, gibberellins or cytokinins. Of these three, auxins appear to play no part in dormancy changes but the others may since they have both been shown to appear and increase in quantity in buds as dormancy breaks. Claims have also been made for the activity of a growth-inhibiting substance known as *abscisic acid* (*dormin* or *abscisin II*). This substance, active at concentrations of less than one part in one million, when applied to the leaves, is known to bring on dormancy in young woody plants that are actively growing under summer long-day conditions. The substance has been extracted from sycamore buds but has been shown to be present in herbaceous plants also and may therefore have other functions besides being a growth inhibitor. There is, of course, the possibility that bud dormancy is controlled by a balance between the action of growth promoters such as gibberellic acids and a growth inhibitor such as abscisic acid, acting as an anti-gibberellin. So far it has not been clarified as to whether this is the case or whether the substances are quite independent of one another. It is perhaps of significance that seed dormancy in some cases seems linked to the presence of an inhibitor such as abscisic acid and its interaction with the growth promoting gibberellic acid.

Control of Differentiation as Elucidated by Sterile Culture Techniques

Another extremely complex physiological event is the differentiation of mature structures from cells produced at meristems by dividing initials. The technique of growing organs, tissues and cells in isolation from other parts under rigorously controlled and sterile conditions has provided a practical means of attacking the problems with some interesting discoveries. The first successes came from studies of the development of whole organs, particularly excised roots, grown under controlled culture conditions. This was followed by the successful culturing of blocks of storage tissue that formed what are referred to as *callus cultures*, where callus describes a proliferation of more or less similar cells of a parenchymatous nature. More recently it has been found possible to culture separate cells that break free from a callus surface, suspended in a sterile liquid culture medium. In all such experiments involving the use of artificially prepared culture media the effects of nutritional, hormonal and other factors on the growth and development of isolated parts from a wide variety of sources have been studied, providing an increasing body of observations from which the answers to all the complex questions relating to development may eventually be obtained.

Organs grown successfully in culture include root and shoot apices, leaves, flower parts, ovules and embryos and even fruits. Nutritional requirements, although not identical for all cases, have certain features in common. For example, any organs lacking chlorophyll demand a carbon source such as sugar (glucose or sucrose) and all organs require supplies of all the macro-nutrients and trace elements together with accessory growth substances such as thiamin, pyridoxin and nicotinic acid. Roots have the strongest need of the vitamins since under normal growth conditions these are supplied from the aerial parts of the plant. Some species also show an auxin requirement for continued healthy development and must be regularly subcultured for long survival. Whereas most kinds of excised roots continue to produce more root tissue only, there are some such as dandelion and dock that will very readily regenerate shoot buds.

Isolated shoot apices with their leaf primordia can frequently be made to regenerate whole new plants with fully formed adventitious root systems. The exact culture medium requirements varies with the species of apex. For example, small sections of angiosperm shoots often require the basic medium to be augmented with a general organic nitrogen source as well as specific amino-acids and accessory growth substances, whilst some fern apices can thrive on the basic medium of carbohydrate and minerals. Young leaves from several sources, such as tobacco and sunflower, will grow into smaller versions of their normal form if provided with sucrose and a range of mineral salts.

Work with whole ovules and isolated embryos has shown considerable differences in the requirements of the embryos for full development. Generally speaking, a whole ovule in contact with a nutrient medium and carrying some placental tissue will support the development of a whole embryo from the zygote. It will do less well without the placental tissue. Fertilized ovules from different species will all do equally well when grafted to the same placental material indicating that their physiological requirements are by no means peculiar to species.

In the case of isolated embryos, however, these can be much more exacting in their nutrient and hormonal requirements and unless they are approaching maturity will make no progress on a basic sugar and mineral medium. It appears that embryo development is much more under control of hormonal factors than nutritional ones. It has been shown that many kinds of immature embryos can develop in the presence of the liquid endosperm of coconut (coconut milk) but otherwise will not. This liquid is known to contain the hormonal substances cytokinins, probably also auxins and gibberellins, some sugar alcohols such as myoinositol and some leuco-anthocyanins and all these substances could be essential requirements. It is only as the embryos approach a fuller development that they require a less varied supply of nutritional and hormonal substances. Some very young embryos have been brought to full form in a basic medium of sugar, minerals and accessory growth substances augmented with kinetin, indolyl acetic acid and adenine sulphate.

Tissues composed of comparatively simple, homogeneous systems of cells exist in a wide variety of forms from endless sources. With few exceptions they can be grown under sterile culture conditions. Success with each form, varying from storage parenchyma from tap root through medullary parenchyma to palisade cells of leaves, yields a little more information concerning the processes of cell differentiation since not only can the parent mature tissue be kept alive in a suitable medium but it can be induced to revert to a dividing state or dedifferentiate and produce many new cells in the form of a callus of relatively undifferentiated components. Medullary ray cells from a fifty-year-old annual growth zone of lime have produced such a callus, so not even age seems to destroy this regeneration potential in some cases. But not all tissues do equally well under the same conditions of nourishment; there is a wide diversity of demands according to species and the type of tissue from within one species. In the general case of parenchyma from the medulla or phloem zone of a stem the nutritional requirements include the usual carbon source and mineral salts together with a source of organic nitrogen in the form of amide (glutamine) or amino-acid, the accessory growth substances thiamin, pyridoxine and nicotinic acid an auxin (2:4-D) and the sugar alcohol, myoinositol. Cytokinin may also be an essential ingredient. The hormone requirement may be due to the fact that callus tissue is not a centre for hormone metabolism is is meristematic apex. The accessory growth substance requirement s not a feature of green callus derived from chlorophyllous leaf palisade, out is very much so for other colourless tissues. An auxin requirement oy some callus tissues disappears with successive sub-culturing and the issue is said to become autotrophic for auxin or habituated, that is, an synthesize its own. When habituated in this way the callus appears nore like that found in zones of plants infected with the bacterium of rown gall, *Agrobacterium tumefaciens*, and it is the case that culture-rown callus derived from gall tissue, in which the bacteria have been illed, is itself self-supporting for auxin and continues to be so through ll succeeding derivatives.

The most recent successes in the field of sterile cultures has been with eparate individual cells, but these are much more difficult to handle han whole organs or tissue masses. One of the difficulties is that free ells tend to lose some of their material content to the liquid medium in hich they grow and their progress of growth and development is thus etarded. Another peculiarity is that when a free cell undergoes succes-ive divisions, the new cells do not always separate but tend to adhere s a mass equivalent to a tissue. Such clusters of cells must be regularly ltered off to maintain the free single cell condition in the culture. Once ee, though, plant cells rarely adhere to other free cells as do animal ells in culture.

When grown in isolation, cells seldom retain their normal forms and nd all to be alike from whatever source they may have come. Their netabolic activities are likewise not usually comparable with those of e parent cell source. Nutrient requirements of free cells are often very omplex and the liquid endosperm of coconut is commonly an essential gredient, indicating a condition comparable with that of very young nbryos for which some requirements have not properly been defined. can only be presumed that in normal conditions in the plant these quirements of free cells in culture are supplied by adjacent cells and sues.

Success in keeping organs, tissues and cells alive, growing and veloping in culture conditions, has enabled many detailed studies of e factors involved in differentiation of parts to be made. The follow-g summarizes some of the more significant findings.

Many living cells, possibly all, exhibit *totipotency*, meaning that each, en the correct nutritional and hormonal conditions, can give rise to the forms of tissues found in the mature plant and can do this even ough they have once been fully differentiated themselves. This licates that the complete genetic potential of the species is retained by ry one of its cells despite the fact that they may already have matured particular forms and with particular manners of functioning. The evidence for this comes from work with the carrot where it has been shown by Steward that whole carrot plants can be grown from *adventive embryos* (*see* p. 541), themselves derived from individual cells of root, stem, hypocotyl, leaf stalk or embryo tissues. Only the mature, green cells of the leaf lamina show some inability to regenerate whole plants in this way. There seems to be some requirement for total isolation of the parent initiating cell for it to show a totipotentiality because the cells of a callus mass are much restricted in their pattern of development whilst surrounded by others. What constitutes this restriction and prevents all cells of a callus growth from regenerating whole plants is not known.

It has been shown that the differentiation of organized root and shoot meristems from callus groups is at least partly influenced by the interaction of the hormones auxin and cytokinin. These substances are known to be involved in the initiation of cell division but it is now also clear that the proportions in which they occur has some considerable bearing on how the newly formed cells will eventually differentiate to create an organized growing point. It is found that when the ratio auxin:cytokinin is high a callus tends to produce root primordia but when this is reversed then stem apical meristems are formed. The effects of the interaction of the hormones are subject to modification by other influences, particularly nutritional factors, and the whole process of root-shoot initiation is probably under the influence of other factors besides auxin:cytokinin proportions. There is some evidence that the phytochrome system (*see* p. 542) may be involved since adventitious root formation on pea stems is possible only in the absence of red light.

The effects of differences in the hormone concentrations on meristem initiation is also illustrated by what occurs when stems are subjected to varying treatments. When a stem is supplied with a solution of cytokinin only it commonly reacts by the formation of numerous buds at its morphological apex but with little root growth at its basal end. If auxin is used instead, the reverse occurs; roots are developed in large numbers at the base whilst there are few if any buds formed at the apex. This does not necessarily mean that the separate hormones have individual, separate effects, polarized on stem or root sections. It might well be that the polarization of bud and root formation on portions of the main axis is a reflection of the movement of the two hormones in opposite directions through the tissues. In chicory root, for example, it has been shown that just prior to the formation of bud and root primordia at opposite ends of a cutting the distribution of auxin and cytokinin is altered; auxin concentrates at the basal end whilst cytokinin is higher at the apical zone.

Why some cuttings initiate primordia so much more readily than others is not at all understood but in the cases of stems that will regenerate roots it is well known that they can do so readily only if a leaf and bud are left intact on the stem cutting. This again suggests that root formation on stems is influenced by hormone produced in the green tissues.

When root tissue is kept under culture conditions it is rarely capable of initiating any other than more root tissue and again auxin appears to be a factor in promoting this condition. Primary roots when treated with auxin tend to grow more slowly than normal but produce many laterals which themselves remain stunted under continued auxin treatment. This again points strongly to a connexion between auxin and root primordia initiation and their subsequent growth and development.

Among the more fascinating studies of plant development have been those with embryos formed from vegetative cells, otherwise known as *adventive embryos*. These were first produced in culture from phloem parenchyma of carrot root by Reinert in 1959. When certain nutritional and hormonal conditions of the medium were operating it was found that individual cells acted as though they were zygotic cells and proceeded to develop into normal looking embryos. These, when transferred to more suitable media, were able to grow into whole plants of quite normal form and function passing through all the usual developmental stages. Since then adventive embryos have been produced from other species. The best tissue source has been the scrapings of immature ovular embryos from which single cells are quite capable of becoming embryos, indicating totipotency. But other tissues can also be induced to form embryos under suitable conditions provided that they are not fully matured.

The differentiation of vascular tissues in isolated plant parts in culture and in callus culture has been another subject of study. When a bud of chicory is grafted to a mass of callus this induces the differentiation of xylem and phloem cells from callus, continuous with those forming in the bud axis itself. Tests with hormones have indicated that the stimulus arising in the shoot primordium is probably of auxin nature, both indolyl acetic acid and naphthalene acetic acid being effective in promoting vascular tissue formation in callus in the absence of a bud graft. This agrees with the finding that when a vascular bundle in the stem of a plant such as *Coleus sp.* is severed its cut ends are joined up with new xylem and phloem tissues as a result of cambial activity, but only if indolyl acetic acid is supplied from leaves above the severed zone.

The above is but a very brief introduction to a rapidly expanding field of study but how relevant are the findings of tissue culture work to the growth and development of normal intact plants is not sure. It is tempting to think, however, that there may be some fundamental similarities and that the controlled culture material will eventually yield the natures of these basic processes at least.

Photomorphogenesis

The phenomenon of *photomorphogenesis*, that is, the initiation and development of plant parts under the influence of light, is now well recognized and is known to be bound up with the occurrence of the photoreceptor pigment, *phytochrome*. This is a substance existing in two forms interconvertible through the action of light so:

$$\text{red light at about 660 nm}$$
$$(\text{r-phytochrome}) \; P_r \text{ or } P_{660} \rightleftharpoons P_{fr} \text{ or } P_{730} \; (\text{fr- phytochrome})$$
$$\text{far-red light at about 730 nm}$$

P_r has an action spectrum maximum near light wavelength 660–665 nm, in the red, whilst P_{fr} has an action spectrum maximum near light wavelength 725–730 nm, in the far-red. The equation above indicates that when light of the appropriate wavelength is absorbed by either of these forms it is converted into the other. The pigment has been isolated from very low concentration in plant tissues of wide variety, from algae to higher plants, from green and colourless parts and of primary and secondary growth. The precise location of the substance inside cells is not known, but it is a protein with a pigment group and purified phytochrome after irradiation with far-red wavelengths is blue-green in colour, whilst if this is treated with red light it changes to a light green hue. The molecular mass of the protein has been given as about 60 000 and the pigment component of the molecule is very similar to that of the substance c-phycocyanin of blue-green algae and chemically is a "bile" pigment of the bilitriene class. When phytochrome is produced in seedlings developed in total darkness it is always of the P_r or P_{66} form which when treated with red light, through some photochemical step, becomes P_{fr} or P_{730}. Whilst P_{660} is stable in darkness, P_{730} is not and if tissues containing it are kept for long in total darkness the P_{730} gradually disappears either by being used up, destroyed irreversibl or by being converted to P_{660}. Of the total phytochrome present i tissue not more than 80 per cent ever seems to be present as P_{730} no matter what the conditions and even when the tissue is treated with 660 nm radiation, only 80 per cent of the P_{660} becomes converted t P_{730}. If this is then treated with far-red radiation at 730 nm about on per cent only remains unconverted to P_{660}.

P_{730} is the physiologically active form of the photomorphogeneti substance and there are a number of plant growth and development processes that are known to be under its influence. The onset of flowering as a photoperiodic response is one of the more well known (*see* p. 544). Etiolation, the effect on plants brought about by total darkness treatment, is also known to be linked with phytochrome. Etiolated seedlings show among other conditions excessively lengthened internodes and failure of leaf primordia to expand. When treated with red light wavelengths at which P_{660} becomes the active P_{730}, any newly formed internodes tend to be shorter than previous ones and the leaves to expand showing a growth-inhibiting effect in the one case and a growth-promoting effect in the other. The control of internode length in plants grown in light has been shown to be associated with phytochrome action and the location of the photoreceptor in this case is in the internode concerned.

Mohr and others have contributed to a list of all the photoresponses known to be made by light-grown mustard seedlings, *Sinapis alba*, due to the formation of P_{730} in the cells and not shown by dark-grown seedlings in which P_{730} is absent. The list is quite lengthy and includes: inhibition of growth in length of hypocotyl; enlargement and unfolding of cotyledons; hair formation on hypocotyl; opening of the plumule hook; formation of leaf primordia; differentiation of leaves; formation of xylem and phloem elements; synthesis of anthocyanin; formation of plastids in cotyledonary mesophyll; differentiation of guard cells in epidermis of cotyledons; increase in rate of synthesis of RNA and protein in cotyledons; changes in rate of disappearance of storage compounds in embryo.

Certain biochemical activities in plants are also thought to be linked to phytochrome occurrence, for example, synthesis of pigments in tomato and apple surface layers, synthesis of flavonoid substances, chloroplast protein and NADP triose phosphate dehydrogenase, any of which may have an effect on the development of the plants concerned.

Phytochrome has also been suggested as an active agent in some tropic responses, is said to be associated with the "sleep" movements of the leaves of the sensitive plant, *Mimosa pudica*, and to be concerned with the movements of chloroplasts in the cells of the algae *Mougeotia* and *Mesotaenium spp.* in response to light conditions.

Because of the wide variety of responses made to the presence of the active P_{730} there is no generally acceptable explanation of its physiological role. It seems that the P_{730} has some sort of "trigger" action in that it can set other processes in motion. These other processes are, as it were, predetermined in the plant cells and tissues concerned according to their particular natures and states of development but they can proceed only when the active P_{730}, a product of red light treatment of dark produced P_{660}, is present in the cells. In other words, unless the plant is exposed to light of the appropriate waveband, thus converting the photoreceptor substance to its active state in which it can promote the necessary physiological activities, a plant is unable to grow and develop in its normal way. Hence we may speak of photomorphogenesis as the control exercised by light over the growth, development and differentiation of plants other than through photosynthesis.

Phase of Reproduction

The reproductive phase commences with the formation of flower primordia. It continues with their development to maturity and the production of pollen or ovules or both. Then the phase is completed with the ripening of the fruits.

As flowers develop, the shoot apices at which they are formed undergo morphological changes. Such differentiation of parts, or morphogenesis, from meristems, is one of the most fundamental of all biological phenomena but the physiological processes involved are far from clear. Morphologically, the apex becomes converted into a bud, enclosing the young floral parts, which on later expansion completes the flower formation. The factors which influence the onset of

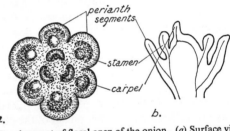

Fig. 13.27. Development of floral apex of the onion. (*a*) Surface view. (*b*) L.S.

flowering, vary with species. With the exception of one or two rarities such as *Arachis hypogea*, which will produce flower primordia in the axils of its cotyledons in the seed, no plant will flower until a certain minimum of vegetative growth has been made. It is well known that for some plants the onset of flowering is influenced by external conditions such as relative lengths of day and night and chilling experiences. The response by flowering to relative lengths of alternating light and dark periods illustrates *photoperiodism*, an ability to "measure" the duration of photoperiods. Plants which require certain low temperature experiences before they will flower are said to require *vernalization* or conversion from winter to spring (verna) condition.

With regard to photoperiodism and the flowering process, plants can be divided into three classes. *Short-day plants* are those which will flower only if the daily illumination is shorter than a certain period, e.g. cockle-bur (*Xanthium*), see p. 714. *Long-day plants* will flower only when the daily illumination exceeds a certain period, e.g. many summer-flowering annuals. The third class, *day-neutral plants*, are not influenced by the length of daily illumination, e.g. tomato. The photoperiodic flowering response is one of the most complicated of physiological phenomena and has proved very difficult to elucidate, but certain generalizations seem to have emerged from a vast amount of experimental work involving the artificial control of light and dark periods.

Short-day plants will flower if the light period is less than a critical amount during a 24-hour cycle of light and darkness. They are really responding to a dark period of greater than a minimum duration and could be called long-night plants. Long-day plants appear to flower in response to light periods greater than a certain minimum value, but again, the response is to dark periods of less than a maximum value. They could be called short-night plants. In both cases, interaction of the light and dark periods very frequently affects the response. For example, if the dark period is interrupted by only a short light period, short-day plants are inhibited from flowering, whilst in long-day plants, a similar interruption promotes flowering. Light of orange-red wavelengths, about 660 nm, is most effective in this and light of longer, far-red, wavelengths about 730 nm, reverses the effects. This means that these light interruption responses can be placed in the same category as other light phenomena in plants which are known to be controlled by a so-called photomorphogenetic pigment system. The pigment system, called *phytochrome* (see p. 542), has been thoroughly investigated and this is assumed to be the photoreceptor substance of photoperiodic responses. There is some evidence, but not wholly convincing, that a flowering hormone, or floral stimulus, referred to as *florigen*, is manufactured in the leaves and moves to the apical meristems, there to initiate in some way the morphogenesis. The biochemical connexion between phytochrome and florigen has not been established; indeed, no substance with the properties of the postulated florigen has as yet been isolated and some authorities have denied its existence, postulating instead that its so-called flower-promoting effects are really the result of the removal of the inhibiting effects of other substances. Flowering in most of the photoperiodically sensitive plants is an *inductive process*; once treatment which promotes flowering has been given, the plant continues to flower even though conditions which inhibit flowering are introduced. However, usually more than one, sometimes many, photoperiodic cycles are required to bring about this condition. The cockle-bur is an example of an extremely sensitive short-day plant and one cycle of short-day/long-night is sufficient to bring it to flowering. But the flowering response is not often a simple one such as this; a combination of responses is often shown by a single plant. Many plants are known in which the response is day-neutral at one temperature and sensitive to day length at another temperature. Some species require a combination of day lengths in a particular order.

A requirement for vernalization is exhibited by a range of plants including annuals, cereal (wheat) and non-cereal (pea), biennials (carrot) and perennials (chrysanthemum) meaning that they must spend a winter in the soil in the vegetative condition, before they will flower in the following season. The so-called "winter" wheats are typical of plants requiring vernalization. They are planted in autumn and allowed to winter in the soil. If winter wheat is planted in the spring, it develops vegetatively but produces no flowers. If germinated artificially and subjected to a temperature of about 5°C for a few days before planting out in the spring, the wheat will develop

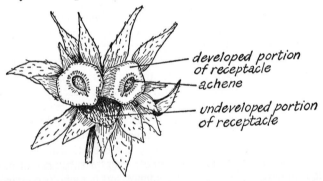

developed portion
of receptacle
achene
undeveloped portion
of receptacle

Fig. 13.28. Strawberry showing development of succulent receptacle only in region of fertilized ovules.

and flower quite normally. The "spring" wheats, on the other hand, are planted in spring to flower in the summer; they require no cold treatment, acting as annuals, while the winter wheats act as biennials. It has been shown that this difference is inherited in Mendelian fashion. It seems that cold treatment of the winter wheat brings about some physiological change in the shoot apex, which induces a capacity for flowering by "preparing" the apex. The chilling process does not itself initiate flower primordia and the evidence for the production of a flowering hormone as a direct result of the chilling is inconclusive. The phenomena of photoperiodism and vernalization are referred to again in Chapter 17.

In general, after the floral parts have been differentiated, they continue development irrespective of conditions, but this is not true for all investigated cases. Complete development of fruit from the ovary is in most cases entirely dependent upon pollination. It has been shown that ovary development virtually ceases after ovule production unless ripe pollen is allowed to germinate on the style. Abscission of the ovary may occur in a plant which is not pollinated, and it has been shown that auxin is once more involved. An unpollinated ovary may often be induced to develop by application of auxin. The fruit will not produce fertile seeds but will be completely formed in other respects. Such parthenocarpic development is not uncommon naturally and is frequently employed in the artificial formation of seedless fruits such as grapes, bananas and oranges. Pollen grains do contain auxin but this is generally insufficient for full development of the ovary. Apparently the auxin comes from the developing seeds after the ovules have been fertilized. Little is known about the manner in which this auxin is produced. From tissue culture experiments, it seems plain that other substances besides auxin are involved and that these have their origin in the maternal tissues. The stimulus for seed development is undoubtedly bound up with fertilization; the way in which it exerts its effect, is another enigma of physiology. Parthenogenesis does occur in flowering plants, but so far this has shed no light on normal processes.

When growth of the fruit is completed, it ripens. This is generally recognized by changes in colour, softening of the tissues and increases in sugar content, among other things. Provided that a certain stage of development has been reached, a fruit may ripen off the parent plant. Such is found in the case of the tomato, but most fruits will not ripen unless given certain treatment. In many fruits, ethylene will induce ripening after separation from the parent plant. It is frequently used for apples, pears, citrus fruits and bananas, and in this connexion it is interesting to find that many fruits produce ethylene during the early stages of ripening. Ethylene may be the substance which initiates the physiological processes which bring about fruit-ripening.

It has become clear from investigations of the major developmental phenomena that hormonal control is involved in most of them. But, though we may recognize the presence of an auxin and note its effects, plant physiology has not reached the stage where these are understood.

Development of the Vertebrate Animal

The course of development of the vertebrate animal over its full life-span may be summarized as a series of successive and continuous stages.

1. *Development of the zygote into the embryo,* known as *gestation* in the mammals. This is a period of rapid cell division (cleavage), followed by cell re-arrangement (gastrulation), and then organogenesis (laying down the systems and organs of the adult structure).

2. *Hatching, or the release of the embryo from its membranes.* In the mammals, this is followed by birth, when the attachment of the embryo to the mother is severed.

3. *Growth of the embryo into the adult, and the attainment of sexual maturity.* This may include a period of metamorphosis.

4. *The reproductive phase* which may be accompanied by further growth. This may occupy a relatively long period in the life-span.

5. *Senescence,* terminating in death.

Development of the Zygote, Gastrulation and Organogenesis

The zygote is a relatively enormous cell, and the rapid cell division which begins its development does not result in any growth. Up to the formation of the blastula, any increase is mainly due to the watery fluid which fills the blastocoel. The stage at which presumptive areas are delimited varies in different groups. In *Amphioxus*, the cytoplasm of the zygote is already mapped out into areas from which specific parts of the animal will be developed. In Amphibia, this determination does not begin until the blastula is formed. In any case, this early differentiation is due to distribution of chemical substances, none of which has been identified.

During gastrulation, there is further cell multiplication and considerable re-arrangement of the cells, resulting in the delimiting of the three types of cell layers, ectoderm, endoderm and mesoderm. Masses of cells move; the rates at various points and various times differ; the essential movement is amoeboid. There are cells which attract each other and cells which repel each other; it is suggested that this is due to surface changes in the cells. The physiology of these processes is little understood at present; such complex yet orderly mass movements imply intricate organization, and the pattern differs for every species.

From considerable work on amphibian embryos, we know that from the beginning of gastrulation, certain chemicals are produced which have determining influences on the future course of development. The primary hormone is developed from cells in the dorsal lip region. The chemical substance produced diffuses into surrounding cells and hence there is set up a "field of influence," which is strongest in the centre of the mass and gradually fades with distance from the centre. This primary organizing hormone is said to be a steroid, and the dorsal lip region which produces it, is termed the *primary organizer*. The tissue invaginated from the dorsal lip is destined to form the notochord and later, other axial structures. If a small portion of this primary organizer tissue is grafted on another embryo amphibian, it will induce the formation of axial structures, wherever the graft is made.

Later, *secondary organizers* induce the initiation of various organs,

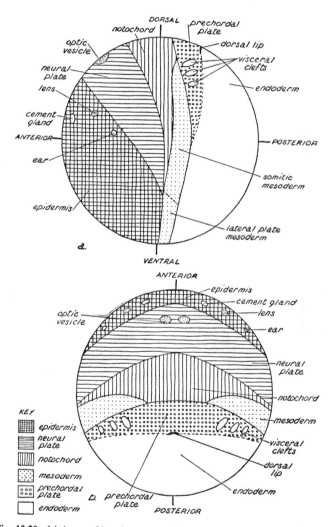

Fig. 13.29. (*a*) Anuran blastula, presumptive areas, side view. (*b*) Same, dorsal view.

then *tertiary organizers* induce the formation of the various parts of an organ. Grafts and treatment of undifferentiated tissue by cell extracts, have shown that there is almost an endless succession of these organizers produced. Some, which also induce the formation of a structure from another type of tissue, are called *evocators*. Thus the optic cup will

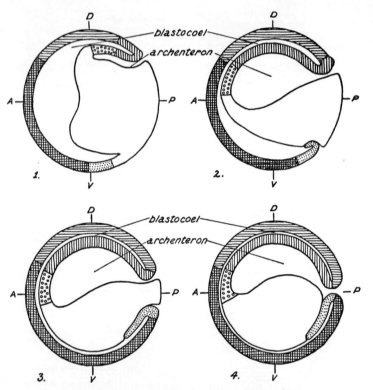

Fig. 13.30. Four stages in gastrulation of frog embryo.

always cause the ectoderm above it to form a lens; if the first lens is removed, another will develop; if the optic cup is transplanted to another position beneath the ectoderm, a lens will develop there, but not in the original position. The optic cup is thus the evocator of the lens. Similarly the nasal placodes are evocated by the front of the fore-brain, the hypophysis of the pituitary body by the front end of the notochord; a host of such examples are known. It is thought that these chemicals responsible for physiological differences in cells are substance which vary according to the plasmagenes and hence to the enzymes produced by different types of cells. Morphologically, it has been observed that cells destined to come together in any organ or tissue, tend to stick to one another and not to other cells; the differences are due to surface changes.

In the mammals, during embryonic development, we have additional effects due to hormones which diffuse from the maternal circulation into the embryo, via the placenta.

Hatching: Birth

From a physiological point of view there is little to say about these processes. Essentially the embryo grows to such a size that it can burst its bonds. Where there are hard-shelled eggs, as in the birds, a special device enables the embryo to break the shell. In placental mammals, the embryonic membranes are burst, partly by the pressure of the growing embryo, partly by its changes of position and partly by muscular pressure from the uterine wall. The birth cannot take place without considerable assistance from the mother; this takes the form of a widening of the pelvic aperture and purposive contractions of the uterine wall. The umbilical cord which provided the physical connexion between mother and offspring has to be severed. Normally it is bitten off by the mother.

Growth to the Adult Stage: Attainment of Sexual Maturity

Further development of the young animal is now dependent upon a combination of external and internal factors. Even in an optimum external environment, no growth will take place without the internal regulation afforded by the hormones. A fuller discussion of these is given elsewhere (*see* Chap. 8); here it will suffice to outline the essential

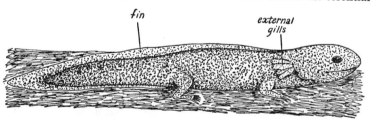

Fig. 13.31. The axolotl.

activities of the endocrine glands in promoting growth and metamorphosis, where the latter occurs.

The term metamorphosis means change of shape or form, and in this general sense, every organism undergoes metamorphosis during its development from the zygote. In zoology, the term is restricted to changes which occur after the embryonic period has ended. Thus the butterfly and the frog metamorphose, but the rabbit does not. Not only does the body form differ in young and adult but the habits and mode of life are different also. Among vertebrates, metamorphosis occurs in the cyclostomes, e.g. the lamprey, in a few groups of fishes, e.g. the flatfish, and in the amphibians. It is universal in hemichordates, urochordates, and cephalochordates, and is common in all the great invertebrate phyla.

There has been considerable study of the physiological changes which occur in various animals during the critical periods of metamorphosis. It was soon found that frog tadpoles will metamorphose precociously if fed on the thyroid gland or its extract, derived from any vertebrate animal. If the thyroid is removed from a tadpole, it continues to grow, but does not change into a frog. The effect of feeding thyroid to such tadpoles has been shown to be quantitative up to a point. If the doses of the extract are gradually increased, there are more rapid and abrupt changes, but there is a maximum beyond which increased dosage has no effect. External factors may also affect the change. High temperatures accelerate it, while low temperatures may completely prevent it. Diet also plays some part; excess of fat retards metamorphosis, while protein-rich food may accelerate it. Since iodine is an integral part of the thyroxin molecule, then the supply of this element has profound effects on tadpole metamorphosis. Thus amphibian larvae do not metamorphose in waters which have insufficient iodine.

The axolotl is an excellent natural example of retarded metamorphosis. Normally, it spends its whole life in the water and reproduces while in the larval form, an example of *neoteny* (*see* Chap. 17). A very small quantity of thyroxin from any vertebrate will cause it to change into a land salamander in about two weeks. It loses its external gills, the fin on its tail disappears, and it darkens in colour. The absence of metamorphosis under natural conditions is not due to lack of a thyroid gland. It is present, and if grafted into a thyroidectomized tadpole, it will cause metamorphosis. The explanation appears to lie in the difference between urodele (tailed) and anuran (tail-less) amphibians. Whereas in frog tadpoles, a hormone is continuously secreted by the thyroid into the blood, to build up its effect towards metamorphosis, in the salamanders, all the secretion is stored in the gland, only to be released by "trigger" action of a pituitary hormone. Under natural conditions, this pituitary hormone is never released.

Other aspects of metamorphosis have been investigated in frog tadpoles and they show interesting results. Different species exhibit different rates of change and this is partly due to the fact that there must be a definite concentration of thyroxin in the blood before the change can begin. If the thyroid gland grows slowly by comparison with the rest of the body, metamorphosis is a slow process. If the tissues of the body are not sufficiently sensitized to the hormone by a correct degree of development, the change will not take place, however great the concentration of thyroxin. Experiments on the tail have shown that the age of the tissue in the tadpole is important. Normally, the tail shrinks towards the end of metamorphosis and is finally absorbed. If a large part of the tail is cut off just before the stage of shrinking, the missing part is regenerated and it will do the same again if a further portion is removed. When a tadpole so treated is given thyroid extract, the original base of the tail shrinks first, followed by the regenerated parts in the order of their regeneration. Thus, two internal factors may control the rate of change; the concentration of the hormone and the degree of development of the tissues.

Thyroid activity is governed by a special thyrotropic hormone from the anterior lobe of the pituitary. The secretion of this hormone is itself induced by a releasing factor from the hypothalamus; the factor reaches the pituitary in the blood of the hypothalamo-hypophysial portal system (see p. 280). The hypothalamus is therefore the time-keeper of metamorphosis; after summation of a variety of external and internal stimuli, if the time is ripe, the releasing factor is produced. This activates secretion of thyrotropic hormone from the pituitary, and this hormone stimulates the thyroid to secrete thyroxin. In the absence of either the pituitary or thyroid, metamorphosis will not take place except, in rare cases, in the presence of large quantities of an acceptable iodine compound. This indicates that some other tissue may be capable of utilizing iodine in the absence of the thyroid.

It is also clear that thyroxin does not influence development during the early stages of an animal's life, since the egg will segment and gastrulate in both the presence and absence of thyroxin. These early phases are under the control of organizers. But there comes a stage when further development cannot occur except in the presence of thyroxin or some equally acceptable iodine compound.

Although the body tissues are subject to the influence of hormone during metamorphosis, the gonads are not. They develop as long as body size is commensurate, and thyroxin does not affect them. Thyroidectomized male tadpoles of some species have been known to produce ripe sperm, and the females, eggs, at a stage far beyond that normally reached before metamorphosis. This is an exact parallel to the natural condition of the axolotl, which can reproduce without ever reaching the metamorphosed condition.

In insects, metamorphosis may be accompanied by moulting (Holo-metabola), or there may be periodic moulting without metamorphosis (Ametabola and Hemimetabola). It will be remembered (*see* p. 205) that in many insects, the phenomenon of diapause or arrested development may occur in any one of the four stages, egg, larva, pupa, imago. The stage at which diapause occurs, is in many cases directly linked with day length. During diapause, especially in pupal stages, there is suppression of a neurosecretion from the brain, which is released into the blood from the corpora cardiaca (*see* Fig. 13.32). This neuro

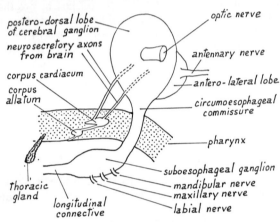

Fig. 13.32. Lateral view of anterior part of nervous system of cockroach and endocrine structures (*partly after Buch and Keister*).

secretion is a trophic hormone which stimulates the thoracic glands secrete ecdysone, a hormone which is essential for further development within the pupal skin and for casting this skin. Also an antagonist hormone, *neotenin*, from the corpora allata, prevents metamorphosis in the immature stages. Thus there is progressive differentiation and development during each instar, then during the last one or two moults production of neotenin gradually ceases, and ecdysone then stimulates metamorphosis.

There seems to have been little research into physiological aspects metamorphosis in groups other than insects and amphibians.

The pars anterior of the pituitary produces the hormone somato-tropin, which is essential to normal growth. Its secretion is dependent upon a releasing factor carried in the portal system from the hypothalamus. Somatotropin is of lesser importance in the early stages an animal's free life, but in the later stages, it becomes of great significance, acting as a direct promoter of protein synthesis. Hence, remo or disease of the pars anterior results in a rapid cessation of growth. Apart from somatotropin, the pars anterior produces tropic hormones which stimulate other endocrine glands to secrete their particular hormones, some of which affect growth. The most important of these, in this connexion, are thyroxin, insulin, thymosin, parathormone, mineral-ocorticoids, glucocorticoids, androgens and oestrogens (see Chap. 6).

The thyroid gland is essential to growth. In the earliest stages, it appears to exert its influence independently of the pituitary hormones. Thyroidectomy in mammals results in almost entire cessation of growth. Administration of thyroxin will restart growth, but only if the pituitary is functional. If pituitary and thyroid are removed, thyroxin admini-tration alone will not promote growth.

The primary feature of the attainment of sexual maturity is the growth and differentiation of the gonads until they are able to produce gametes. These processes are always accompanied by development of the secondary sexual characteristics which give the different appearance and behaviour to the male and female. These phases in the life-history are regulated by three gonadotropins produced in the pars anterior in response to releasing factors from the hypothalamus. The production of two gonadotropins, follicle-stimulating hormone and luteinizing hormone, is stimulated by the two respective releasing factors, but the secretion of the third, prolactin, is inhibited by the releasing factor (see Chap. 17). The gonadotropins are not sex-specific or even species-pecific. Removal of the anterior pituitary will arrest the development of the ovaries in the female, or the testes in the male. Administration of pituitary extract from either source will stimulate further development in either sex. Gonads of reptiles are stimulated by hormones from mammals; sex activity in amphibians is stimulated by pituitary extracts from many sources, including fish, mammals and even human pregnancy-urine. Thus, provided that there are favourable conditions in the external environment, and this involves many factors, the development of an animal is regulated by its hormones. These are themselves regulated by the hypothalamus; in the control of metabolic processes, the great importance of the hypothalamus cannot be over emphasized.

The Reproductive Phase

Growth continues in the vertebrates after sexual maturity is achieved, often for long periods, though there is a gradual waning of the rate of growth due, at least in part, to the lessening secretion of somatotropin. Reproduction is regulated at three levels: (1) hypothalamus; (2) pars anterior; (3) target organs. The latter are the glands stimulated by gonadotropins from the pars anterior. The hormones concerned are androgens, oestrogens and progestins; both sexes produce all three types.

In response to interstitial-cell-stimulating hormone (which is probably identical with follicle-stimulating hormone in females), sperm maturation is stimulated. The testes secrete androgens, the principal one being testosterone in response to luteinizing hormone; smaller quantities of androgens are also produced by the cortex of the adrenal glands and by the ovaries. These hormones induce spermatogenesis and they control the activity of the seminal vesicles and the prostate glands. They also stimulate the development of the male secondary sexual characteristics; well-known examples are the comb of the cockerel, the antlers of the stag, the low-pitched human male voice and the swollen first digit of the male frog.

In females, follicle-stimulating hormone stimulates the ovaries to produce oestrogens from follicle cells; lesser amounts are produced by the adrenal cortex and by the testes in males. The principal oestrogen is oestradiol. Follicle-stimulating hormone promotes the development of Graafian follicles; the oestrogens activate the reproductive tract in preparation for pregnancy, stimulate development of the mammary glands and the secondary sexual characteristics; well-known examples are the smaller comb of the hen, the lack of antlers in the hind, the higher pitch of the human female voice and the brown plumage of the female blackbird.

After rupture of the ripe Graafian follicles, repair and growth of the corpora lutea is activated by luteinizing hormone from the pars anterior. The corpora lutea, stimulated in many mammals by prolactin from the pituitary, secrete progestins, the chief one being progesterone. This inhibits further ovulation, is essential for further development of the uterus and for implantation, and induces further growth of the mammary glands.

Production of pituitary gonadotropins and of oestrogens by the ovaries, wanes during pregnancy. The placenta then develops endocrine function and secretes a gonadotropin, considerable quantities of oestrogens and some progesterone. These promote continued growth of the uterus and mammary glands and assist in maintenance of the corpora lutea until birth.

During birth, relaxin, from the corpora lutea, induces dilation of the cervix and relaxation of the pubic symphysis, thus facilitating widening of the pelvic girdle. Oxytocin, a hypothalamic neurosecretion liberated via the pars nervosa, induces muscular contractions of the uterus.

The development of the mammary glands is first promoted by oestrogens, which induce nipple and duct growth; later, oestrogens and progesterone promote growth of the lobule-alveolar system; the latter is further stimulated by prolactin, which is finally instrumental in bringing the glands into a secretory state. The actual "let-down" of milk is induced by oxytocin within ten to fifteen seconds of the stimulus of suckling.

The control of the reproductive cycle thus depends on three levels of hormone production, the secretions of the hypothalamus providing the initial stimuli. During growth there is gradually increasing production of gonadotropins up to sexual maturity. At this stage, further development of the cycle is controlled by androgens, oestrogens, progestins; the latter stages, to complete the cycle, being also influenced by prolactin, relaxin and oxytocin. In most vertebrates, reproductive activity is a cyclical process which usually coincides with the season of the year most favourable for the offspring (see p. 704). The cycle is initiated in some animals by the photoperiod, in others by temperature, in most by a combination of both. In some mammals, such as the rat, mouse and man, the cycle is rhythmical and independent of solar seasons.

Throughout the long period of reproductive activity, general metabolic processes are controlled by the balance of hormones. Gradually, anabolism ceases to exceed catabolism, and for a long period a balance is preserved where construction is equal to destruction of cells. The only major changes which alter this steady state are variations in the amount of fat deposited in the tissues.

Senescence and Death

Eventually catabolism outweighs anabolism; cells are not replaced as fast as they are destroyed; the growth rate is negative. This may be concealed when weight analyses alone are considered, by excessive deposition of fat. The immediate effect of the degenerative process appears as lack of ability of the cells to divide; thus cells which are destroyed, are not replaced. The balance of hormones changes, sexual activity ceases. Any attempt to explain the downward slope of the growth curve must, as yet, be speculative.

In natural conditions, the onset of senescence invariably means early death; any loss of vigour makes the animal easy prey; it cannot compete effectively in the struggle for existence. In unnatural conditions, life may be prolonged. Frogs can be kept for twelve or thirteen years under controlled conditions; in the wild state, they rarely exceed five years. The human expectation of life has gradually been extended, but it must be remembered that the conditions are highly artificial. Yet, in spite of every protection from natural hazards, man grows old. The examination of the causes of senescence is the basis of the study known as *gerontology;* as yet, biologists have explored only the fringes of the subject.

CHAPTER 14

SECRETION AND STORAGE

WHEN a cell actively separates substances from its own protoplasmic fluids, it is said to secrete, and the substances it so separates are its secretions. This is the widest meaning that can be given to the term.

In most instances of secretion, the cell has previously elaborated its own product from substances delivered to it and is one of many types of secretory cells specially modified for the production of substances of great physiological significance. These include digestive juices and hormones. Some would restrict the use of the term secretion to such instances as these, but others have more commonly used it to include any active discharge of material whether formed within the cell or not. Thus cells of kidney tubules and sweat glands are described as secretory cells when they actively discharge metabolic by-products on behalf of the rest of the organism.

Cells may also be said to secrete when they discharge substances into vacuoles within their own cytoplasm, i.e. they secrete intra-cellularly. This is of frequent occurrence in both plants and animals. In plants, it may be followed by the subsequent reabsorption of the secretory protoplasm into neighbouring cells leaving the secretion isolated as an intracellular substance.

This chapter will be concerned chiefly with cases in which the secretory cells are actively producing and discharging substances of physiological significance. In such cases, the secreting cells usually form special structures known as *glands*, though there are many instances of single scattered cells performing a secretory function. In animals, if not in plants, the secretory activities of both glands and of individual secreting cells are generally co-ordinated with the activity of the rest of the body. Their period and rate of secretion are regulated by the body according to circumstances. Knowledge of secretion of this kind is much greater from the aspect of the chemical nature and function of the discharged substances than from the mechanism of the secretory processes adopted by the cells. It is thus intended to survey the secretory activities of living things, first from the functional significance of the secretions and second, from the nature of the processes involved, treating animals and plants separately.

558

SECRETION BY ANIMALS

In animal bodies, we may find secretions performing functions associated with all the main physiological activities and also being used to protect and give mechanical support to other parts. These functional properties of secretions are set out briefly below. Many of them are referred to in much greater detail in Vol. I.

The Functions and Nature of the Secretory Substances

In feeding, many animals rely on secretions of mucus for the trapping of fine particles of food which can then be carried by ciliary or flagellar movement into the alimentary canal. It is the most common of the filter-feeding mechanisms among the invertebrates and is exhibited also in *Amphioxus* and the tunicates among the chordate animals. Most animals also rely entirely on secretions of the alimentary canal for the digestion of food. Enzymes are secreted by cells of the gut wall either from specialized areas or from distinct secretory organs, the digestive glands, so that digestion is carried out extracellularly in the lumen of the gut. In addition to aiding in ingestion and digestion, other secretions, chiefly water and mucus, assist in keeping the gut moist and lubricated so that the food may more easily be passed along the canal by peristalsis.

To make possible gaseous exchange with the atmosphere at respiratory surfaces, cells of the surface or of special glands below it secrete mucus or some other moisture-holding substance. It is only in solution in this moist covering that oxygen and carbon dioxide can be fully exchanged. All terrestrial animals exhibit this secretory activity and where the respiratory surface is the skin, the phenomenon is particularly noticeable, as for example in the earthworm and frog. Respiratory passages are likewise moistened by similar secretions which aid in cleaning inhaled air by trapping small particles which can then be removed to the outside by cilia beating in the mucus.

Where excretion is associated with the draining of the coelomic cavity by special ducts such as the nephridia of earthworms and *Amphioxus*, the peritoneal cells must continuously produce a fluid in which the excretory products can be passed to the exterior in solution. Similar activities must occur in the flame cells of lower invertebrates. Vertebrate excretion is effected by two distinct mechanisms, namely, filtration and secretion. In renal tubules, secretion of some unwanted substances into the urine is carried out; the urine is filtered off by the Bowman's capsules. In sweat glands, secretory cells discharge similar substances at the skin surface. Controlled secretion by sweat glands also helps to regulate loss of body heat. A great deal of the co-ordinating and regulating of the body of an animal is effected by the secretion of hormones from specialized glandular tissues directly into the blood stream (*see* Chap. 6).

In the reproductive processes, many uses are made of secretions. Where gametes have to traverse ducts, their passage is assisted by lubricating fluids secreted by cells of the ducts. Examples are seen in Cowper's and other glands of the mammalian genital tracts. Ova are frequently surrounded by layers of nutritive substance deposited by special cells of the oviducts (e.g. albumin of frog and chick), and may receive outer protective membranes or shells also secreted by the cells of the oviducts or by glands associated with them. Where the ova develop in a uterus, as in mammals, their initial nourishment is derived from uterine secretions. When the young are born they are further suckled at mammary glands which secrete milk. Some external secretions called *pheromones* play a vital part in certain activities of animals, particularly in mating.

As strong protective and supporting substances, secretions are very numerous. They include calcareous or siliceous spicules, cartilage (chondrin), bone salts, chitin, dentine, enamel, conchiolin, pearl, collagen, elastin and keratin. Lubricating secretions other than mucus are common and include synovial fluids, pericardial fluids and serous fluids.

Secretions may be put to a wide variety of uses besides those mentioned. As a means of passive defence, acrid-tasting secretions of the skin and other parts may serve to deter predators. Pigment secretion may serve to camouflage. As a means of active defence, the "ink" secreted by cephalopods is a notable example, as are the vile-smelling fluids of the skunk and devil's coach-horse beetle. Venoms secreted by many animals, particularly reptiles, predatory insects and spiders, are used for active offensive purposes. Silk secreted by spiders is used to catch prey; it may be secreted by insect larvae to form protective cocoons. Silk may even be used as a means to further passive progression from place to place, as for example, gossamer threads. Other less widely occurring secretory activities include the production of luminescent substances, wax secretion by honey bees, and byssus threads by mussels.

Most of these are interesting adaptations of cellular activity to meet special needs of the animals producing them. The student will un-doubtedly encounter many such instances.

Secretory Processes by Animal Cells

Many of the precise details of the internal chemistry of cells are not fully understood, and there is little definite knowledge about the active passage of materials through cell membranes against diffusion gradients. Because of these difficulties, little detailed work has been done in the study of secretory processes by animal cells. We may draw attention to some generalizations which have been made and state what is known of secretion by cells in one or two specific instances.

The whole process is invariably very complicated but ordinarily may be divided up into three main stages. The first involves the reception of material from which the secretion is to be elaborated, together with water and salts. In higher animals, there are frequently substances of hormonal nature, which may stimulate or regulate the cell in its function. This preliminary reception is presumably comparable with the intake of materials by any cell and need not be further discussed here.

Then follow the chemical reactions which lead up to the formation of the secreted material. The chemistry involved in the formation of secretions is undoubtedly complicated, and in no case is it completely understood. The Golgi apparatus is enlarged in actively-secreting cells, and mitochondria are concentrated around both the Golgi body and adjacent portions of endoplasmic reticulum; this concentration of mitochondria is always found in regions of high energy expenditure. In the salivary glands of vertebrates, there is evidence of increased oxygen consumption and a rise in temperature during secretion, indicating increased metabolic acitivity.

Perhaps the most thoroughly investigated and best understood example of secretion in animal cells is that of the production and extrusion of α-chymotrypsinogen from pancreatic cells. The enzyme in question is formed in, or on, the ribosomes of rough endoplasmic reticulum, presumably after the usual sequence of DNA → messenger RNA and transfer RNA. From the endoplasmic reticulum, the material is transferred to the Golgi apparatus in small vesicles. Coalescence of the vesicles now occurs and the enzyme material becomes enclosed in larger vesicles called *primary lysosomes* which are budded off from the Golgi body. The bounding membrane of such a vesicle is thicker than that of most cell organelles, being about 9 nm across.

The third phase of secretion, extrusion, then follows. In the case of the above pancreatic secretion, the vesicles are transferred, by an unknown mechanism, to the plasma membrane at the free border of the cell. There, the vesicle membrane fuses with the plasma membrane, and the secretion in granular form (*zymogen granules*) is released into a pancreatic saccule. This process, with the necessary chemical differences, appears to be the method by which all enzymic secretions are elaborated and liberated. The isolation of such secretions in membrane-bound vesicles is often essential to prevent cytolysis.

There are other ways in which extrusion of secretions occurs. In the case of the well-known goblet cells found in the respiratory tract of mammals and in the digestive glands of snails, etc., the cell appears to burst open and a large proportion of the cytoplasm may be lost. In some cases, for example, the secretions of sebaceous and mammary glands, whole cells are lost; each one, therefore, functions only once and must be replaced by rapid cell division. There are doubtless other methods of extrusion of secretions from cells; the three examples given above are well authenticated. The vesicular method is probably the most widely spread.

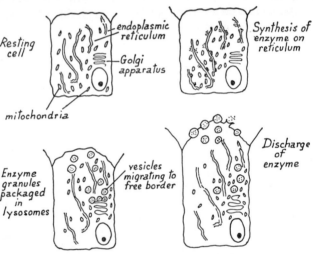

Fig. 14.1. Diagrams representing secretion by cells of the pancreas.

Fig. 14.1 represents diagrammatically the main stages in secretory activity of pancreas cells.

The Regulation of Secretory Activity

Secretory activity by some glands is known to be regulated in one or both of two ways. Stimulation or inhibition of the process may be caused by nervous action or by the action of hormones. Excellent examples of both these have been given in the section on digestion in Chap. 10. Other secretory surfaces appear to function more or less permanently and at a more or less constant rate, when conditions are normal. There are many instances where abnormality may lead to considerable increase in activity, as witnessed by a cold in the head.

In some glandular structures such as the salivary glands, it is not unusual to find muscular tissue forming part of the gland. In these cases nervous control is extended to these muscles, so that they may assist by their contractions in squeezing the secretion free of the gland.

SECRETION BY PLANTS

Plant secretions have uses corresponding to those of animals in many cases. The more significant functions are described below.

The Functions and Nature of the Secretory Substances

In one sense of the term, we may say that practically every plant cell shows secreted products in that each possesses an externally secreted wall and a watery vacuole produced by the protoplasm during its differentiation. Active purposive secretion by mature plant cells and tissues, however, is often a less clear-cut process than in animals. There are specialized glands in many plants concerned with external secretion, and it is also a common occurrence to find that many living plant cells secrete substances internally. It frequently occurs just prior to their death and eventual breakdown. Thus, whilst it is possible in some cases to find a physiological or structural significance for plant secretory activities, there are equally as many instances where secreted materials have no clear purpose. The generally recognized functional purposes of secretion by plants are here briefly outlined.

In the nutritional processes of plants, one form of secretion is common to all, in that both green and non-green plants secrete the elaborated food materials in the form of solid grains or crystals and sometimes as soluble products in cell sap. Non-green plants, however, are much more active than this. They must of necessity secrete externally the enzymes by which dead organic remains and host protoplasm can be digested. These enzymes vary with the fungal species, and the ability of a fungus to live on a given substrate is dependent upon whether or not it can secrete the necessary enzymes to break the substrate down. Penetration of host tissue through cuticular coverings by parasitic fungi, was once thought to be the result of enzyme or acid secretion, but is now considered to be more a mechanical penetration than a chemical one. The killing of host cells by parasites, where this is a feature of the parasitic attack, is certainly achieved by the secretion of enzymes or acidic chemical substances by the parasite. Such enzymes include pectinases which destroy the middle lamellae and so commence cell destruction. Not all parasites kill the host cells, however. Some produce haustoria in the cells and the host remains alive. Toxic substances are also frequently secreted by parasites; they aid in killing host cells. The mechanism of such secretory processes is not understood, since often it involves the continued passage of the secretion through the cell membrane and wall of the secreting fungus against a concentration gradient. The ability to continue secretion under such conditions must be bound up with cell permeability. Externally secreted sticky materials and enzymes are produced also by insectivorous plants which capture and digest animals outside their cells. In such cases as *Drosera* and *Pinguicula*, the secretions are products of specialized glands in the form of hairs with different structure according to species.

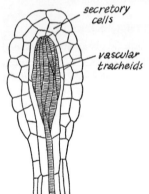

Fig. 14.2. Secretory gland on "tentacle" of *Drosera* leaf as seen in transparency. H.P. drawing.

The active exudation of water through cell walls and membranes by plants may be considered a secretory activity. It is not uncommon in both higher and lower plants, and some significance may be attributed to it. In flowering plants, guttation is of frequent occurrence; it is associated with special glandular structures called hydathodes (*see* Chap. 7). In the fungi, many mycelia delay or prevent drying-out by secreting water droplets along the hyphae. Many forms of plants secrete water internally as a storage material to resist drought conditions, as for example in the succulents such as cacti. It has been suggested that the essential oils secreted by certain leaves, may slow stomatal transpiration by lessening the rate of water vapour diffusion.

One of the chief methods by which plants are able to eliminate unwanted metabolites is by a process of intra- or extracellular secretion. Many of the isolated granules and crystals found in cells are almost certainly of this nature, but it is extremely difficult to be sure in most cases, since their permanency is often variable. Such secretion may be only a short-term excretory mechanism and thus is more in the nature of storage.

Differentiation of cells and growth of parts and the regulation of reactions to stimuli are undoubtedly bound up with secretions of hormones or auxins in plants.

In the reproductive processes of many plants, secretions also clearly play a part. Where fertilization is effected by swimming gametes, it is known in many cases that female organs are attractive to the male gametes because of their secretions. This is well illustrated in mosses and ferns. In the flowering plants specialized glandular structures, the nectaries, exude sugary substances, thus attracting insects for pollination mechanisms. Pigments and scents may also serve a simila attracting function. Stigmas secrete sugary substances for the better collection of pollen grains and as a nutrient for the initial germination of the latter.

Structural and protective secretions are common to practically all plant cells (very few are naked protoplasts). Cellulose of the cell walls is the commonest form, but this may become impregnated with other substances such as lignin, cutin or suberin, which strengthen and protect. Cuticularized or suberized outer coverings serve to withstand adverse external conditions as well as infection by parasites. Sometimes the protective substances may be of a resinous or gummy nature, as on the scales of many dormant buds. Another form of protection may be provided by the secretion of acrid-tasting substances or toxic compounds such as the glycosides, alkaloids and tannins, which may deter animals from consuming the plants. Wound gums are often formed at damaged plant surfaces.

Mechanism and Regulation of Secretory Processes

Whereas the formation of the secretion and method of discharge by glandular cells of animals can be said to conform to a general pattern, no such generalizations can be applied to plants. There is very little known about either the cytology or the chemistry of the elaboration of the secretions or of the mechanisms of discharge. We may illustrate the processes by quoting one or two better-known examples of plant secretion.

In the case of the discharge of oils into cavities in the rind of citrus fruits, the process is one of cell breakdown or lysis and the cavity so formed is called a *lysigenous cavity*. The cavity is first commenced by the lysis of a cell or group of cells within the rind tissues, with the consequent release of oil intercellularly. Other cells surrounding the cavity do likewise and eventually the secreted oil becomes a large intercellular droplet.

The discharge of resin into the ducts of the conifers is effected by a different method. The ducts are *schizogenous*, formed by rearrangement of young cells to form a large intercellular cavity or duct, and the secretory cells line this duct into which they continuously pour their secretions.

Laticiferous ducts are of two kinds. One is known as non-articulate, as in the Euphorbiaceae, in which the duct is composed of individual soft-walled, coenocytic cells ramifying for long distances through the plant. The other is referred to as articulate, as in the Compositae, *Hevea* (rubber plant) and in the Papaveraceae, in which the ducts are formed as continuous passages throughout the plant as a result of the breakdown of intervening cross-walls between adjacent cells. These ducts are lined with coenocytic protoplasts. In both kinds of latex ducts, the protoplast lining the cavity secretes the fluid into it as into a vacuole.

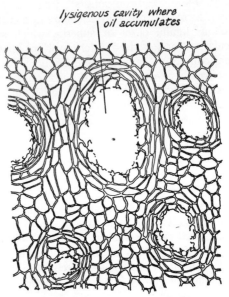

Fig. 14.3. Lysigenous cavity in rind of orange.

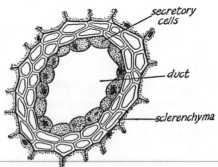

Fig. 14.4. Resin duct of *Pinus*.

STORAGE

It has been stated that its elaborated food materials can be used by a organism as energy sources, as materials for the addition of new bod substance and for the manufacture of essential secretions. Plants and animals, however, are so organized that the whole of the nutrient intake is not immediately put to one or other of these uses. Under good

duct

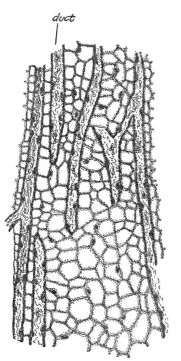

Fig. 14.5. Laticiferous ducts of
Euphorbia.

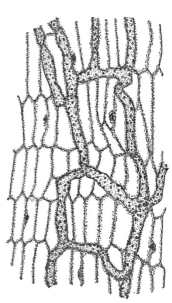

Fig. 14.6. Laticiferous ducts of
Taraxacum.

nutritional conditions, organisms are able to reserve quantities of nourishing substances for use at subsequent periods. Such a storage mechanism has the advantage that continuous activity is not dependent upon constantly favourable nutritional conditions. Also, the organism may be in such a condition that it cannot avail itself of external nutrient; such are dormant seeds, spores, and developing embryos in egg membranes.

Purposes of Storage

There are two main purposes for the storage of food reserves by living things. These are, first, to tide the plant or animal over periods of poor

19

nutritional conditions, and secondly, to start a young organism in life with sufficient material for its initial requirements.

The first of these may be considered from both the cause and duration of the period of poor conditions. In temperate and colder climates, the seasonal changes make it such that there are alternating periods of good and bad nutritional conditions. During the warmer seasons, when the environment is conducive to rapid metabolism and supplies of nutrients are much more readily available, an organism is in little danger of extinction by starvation. During the colder periods, the reverse is the case. Thus, many organisms whose life span extends over more than one favourable period would be unable to survive if they had to rely on continuous nutrient intake. The capacity to store is thus an advantage, in that excess food gathered during the better conditions can be put to use when none is externally available. Such adaptation has led to the perennating habit in plants and the hibernating habit in some animals which are unable to migrate to better environments. Perennation and hibernation are similar phenomena in that they serve the same end by reducing the normal activities of plants and animals to a minimum, without death occurring. Perennating plants often reduce the body material by such processes as leaf shedding or by the die-back of aerial parts, and they develop specialized organs wherein the material for future use is stored. Hibernating animals do not shed parts nor wither, but merely reduce their metabolic activity to a very low level. In both cases, before renewed feeding can commence, much material and energy must be expended. Plants must produce new synthesizing organs and animals must move to find new food sources. The stored food from the previous good season, therefore, maintains life during the resting condition and gives the start for new activity.

Another source of interference with continuous nutritional activity in addition to climate, lies in the fact that nutrients may not always be readily available even under the best of climatic conditions. This is far less applicable to plants than to animals, since their inorganic requirements are more or less evenly distributed through the environment and can be continuously absorbed. It is true that conditions of synthesis change fairly abruptly in every twenty-four-hour period and thus the plant tends to accumulate elaborated food materials in its synthesizing organs during the daylight hours, but the certainty with which day follows night is infinitely greater than the certainty of an animal's next meal. The majority of animals spend most of their time in endeavours to find food. Intervals between successes may be prolonged, so that storage may serve a useful short-term as well as long-term purpose in such cases. Such storage is accomplished by the stocking up of high-energy compounds in the tissues as quickly as possible after each meal. Thus, for a period, full activity can be maintained without the necessity of calling on more permanently secreted reserves which could only exist if current needs had been continually met.

These two levels of storage, namely, the more permanent and the less permanent have been discussed previously under energy stores in Chap. 11.

The second purpose of storing food, namely, to initiate activity in young organisms, has obvious advantages. Chance of survival in most cases is entirely dependent upon it. Few reproductive bodies, whether sexually or asexually produced, could immediately lead a fully independent existence. They are therefore provided with the materials from which they can derive energy and make growth until such time as they are developed sufficiently to be self-supporting.

Storage Substances

A substance cannot be regarded as a food reserve until it has been shown to accumulate in a tissue and then be maintained there at a more or less constant high level of concentration, later to disappear as it is used by metabolic processes. Thus we may disregard substances which accumulate as the products of irreversible synthetic activities and which may be classified more aptly as waste products, and also the substances which may play a part as accumulated secretions serving some useful purpose other than as nutritive materials. We may therefore eliminate many of the compounds occurring in plants such as the tannins and alkaloids, so frequently found in fruit coats, seed coverings, falling leaves and other tissues, and also the outer protective coverings of many animals and their eggs. The difficulty of deciding whether an accumulated substance belongs to one or other of these categories has already been mentioned.

It is very noticeable that whereas a substance actively engaged in metabolism must be readily soluble and diffusible, and thus mobile, a storage substance is either insoluble in water or else composed of such large molecules that it exhibits colloidal properties, and is therefore immobile. It cannot diffuse out of the cells of storage tissue. In both animals and plants, the substances which constitute true food reserves are varied and numerous, but nearly all belong to the carbohydrates, fats or proteins.

The properties of all the main food reserves have been described in Chap. 4 and will not be further elaborated here. The following table summarizes the common reserve substances, their form and their occurrence in plant and animal tissues.

Reserve substance	Form	Occurrence in plants	Occurrence in animals
CARBOHYDRATES Chiefly polysaccharides but occasionally as di- and monosaccharides.		Most plants, in a wide variety of tissues and often in several forms	Most animals, but usually in one form only and most frequently in muscle and liver
Starch	Grains	Almost universally distributed	
Glycogen	Granules or in colloidal solution	A few fungi, e.g. yeast	Universally distributed in vertebrates and common in invertebrates
Inulin	Colloidal solution or as sphaero-crystals	Storage organs of Compositae, e.g. dahlia, artichoke, chicory	
Hemi-cellulose	Depositions in cell walls	Endosperm or cotyledons of some seeds, e.g. date, ash, vegetable ivory, lupin	
Sucrose	True solution	Some swollen roots, e.g. beet, carrot, parsnip	
Glucose	True solution	Some bulbs, e.g. onion	
Fructose	True solution	Some bulbs, e.g. onion	
FATS Too many to be listed in any detail. Each tissue of each species maintains its fat at fairly constant composition and each differs slightly from the others	Frequently as free neutral triglycerides as fat or oil globules, may also be combined in various ways	Chiefly restricted to reproductive bodies, e.g. spores and seeds, where they may be the chief food reserve, some algal and fungal cells in the absence of carbohydrate	Adipose tissue of all vertebrates except amphibians where it may be restricted to fat bodies, e.g. fish liver, all animal eggs, nearly all invertebrates in a wide variety of tissues
PROTEINS Too many to be listed in any detail	Colloidal solution or in solid form in the aleurone grains of plant seeds	Chiefly restricted to reproductive bodies, e.g. spores and seeds	Chiefly restricted to eggs; liver protein may act as reserve in some cases

TABLE OF RESERVE MATERIALS FOUND IN PLANTS AND ANIMALS—(contd.)

Reserve substance	Form	Occurrence in plants	Occurrence in animals
MISCELLANEOUS COMPOUNDS			
Latex	Milky fluid with a variety of granular inclusions	Laticiferous vessels of many Compositae, Euphorbiaceae and Sapotaceae	
Asparagine Glutamine } amides	Crystals	Cells of many tissues	
Creatine or arginine phosphate	Solution		Short term phosphate stores; vertebrate and invertebrate muscle respectively
Lecithin } phospho- Cephalin } lipides			Egg-yolk, nervous tissue
Vitamins	Solution	Many plant seeds	Occasionally in solution in fatty deposits and in eggs
Mineral salts	Solution	Plant reproductive structures	Eggs

TO DEMONSTRATE SECRETORY AND STORAGE SUBSTANCES IN PLANT AND ANIMAL ORGANS AND TISSUES

In Chaps. 4 and 10 will be found details of chemical tests which can be applied for the identification of the more important plant and animal substances. Most of these tests were described for use with comparatively pure reagents but many of them can be applied to the tissues directly or to their extracts. Since, however, the tissues will invariably contain mixtures of complex substances it is better first to have seen the characteristic reactions separately, so that these can be recognized more easily during this work.

When the substances under examination are externally secreted and can be collected in quantity from the organism, e.g. urine and milk, the tests can be applied to samples in test-tubes. Where the substances are secreted or stored internally, tissues or their extracts must be examined in various ways. The main methods of treatment which can be applied directly to tissues are—

1. Microscope examination of thin, untreated sections for solid secretions or storage substances which are clearly recognizable by their form. Starch grains, aleurone grains and crystals of various kinds can be recognized in this way.

2. Chemical, staining, or other tests with thin sections on the slide or in a test-tube with subsequent microscope examination. By selection of the right tests, substances observed in (1) can be verified and others demonstrated.

3. Chemical or staining tests on juices expressed from the tissues by squeezing. If certain substances are to be extracted in a relatively concentrated or pure form special extraction methods may have to be employed.

Simple Examination of Urine

Urea. To some urine in a test-tube add some alkaline hypobromite. This should be freshly made by adding slowly 10 cm³ of bromine to 100 cm³ of 40 per cent sodium hydroxide, keeping the mixture cool. Nitrogen is evolved when the urea is oxidized to nitrogen, carbon dioxide and water. The CO_2 forms sodium carbonate.

Sugar. Apply Benedict's test (*see* p. 120). Normal urine should show no reducing sugar.

Protein. Boil a small quantity of urine in a test-tube and add one drop of fairly strong acetic acid. Turbidity indicates the presence of protein which should not be present in normal urine.

Urinary Deposits. Centrifuge about 10 cm³ of urine for some minutes. Collect some deposit and examine under a microscope. Uric acid shows as yellowish or reddish-brown crystals of diverse shape, often aggregated into rosettes. Colour is due to urinary pigments since uric acid is colourless. In a crucible, add a little dilute nitric acid to the deposit. Gently evaporate to dryness. When cold, add one drop of dilute ammonia solution. A purple colour indicates the presence of uric acid or urate (purine derivatives). The colour is due to the formation of murexide and is deepened by the addition of one drop of dilute sodium hydroxide. Uric acid occurs only in acid urine.

Among other deposits in acid urine may be seen crystals of calcium oxalate. In alkaline urine may be seen crystals of ammonio-magnesium phosphate (triple phosphate) which resemble coffin-lids, and calcium-hydrogen phosphate.

Sodium Chloride. To 10 drops of urine in a test-tube add one drop of 20 per cent potassium chromate solution. Slowly add a 2·9 per cent silver nitrate solution drop by drop until a brick-red precipitate appears. The number of drops of silver nitrate required is equal to the grams of sodium chloride per litre of urine.

Simple Examination of Milk

To some milk add an equal quantity of water and a few drops of dilute acetic acid. Warm the mixture and filter. The residue is the protein casein which can be demonstrated by a protein test (*see* p. 123). If the filtrate is heated, lactalbumen, another protein, is precipitated. Filter this off and test the filtrate for reducing sugar by Fehling's test (*see* p. 120). The sugar present is lactose.

Skim the top from some standing milk and test for fats with osmic acid.

Simple Analysis of Mucus

Test the slime of a fish for protein and then apply the α-naphthol test to show that it contains a glyco-protein, mucin.

Demonstration of Calcareous Substances

Rapid evolution of carbon dioxide, when hydrochloric acid is in contact with the substance, usually indicates its calcareous nature. Suitable material for testing include egg shell, shell of snail, coral, exoskeleton of crayfish, calciferous glands of earthworm, calcareous patches at insertions of spinal nerves of frog, bone, cystoliths, otoliths.

Demonstration of Enzymic Secretions

These can be demonstrated conveniently only by their action on suitable substrates and the methods have already been described in Chaps. 5 and 10. Extraction of enzymes from fresh tissue can be carried out as below.

"Gastric juice" can be obtained from the stomach of a freshly-killed pig. Open it and rinse thoroughly with water. Scrape the inner surface of the cardiac end with a knife and grind the scrapings with sand and about 20 cm³ glycerine. Allow to stand overnight then pour off liquid and add 2 cm³ 0·5 per cent hydrochloric acid.

"Pancreatic juice" can be obtained in a similar way from fresh, minced pig's pancreas (sweetbread). Do not add hydrochloric acid in this case.

Zymin, containing some of the zymase enzymes can be prepared from fresh yeast. Wash the yeast thoroughly in distilled water and filter it off. Stir about 5 g of the washed yeast into 60 cm³ absolute alcohol and 20 cm³ ether. Leave for a few minutes and then filter the yeast off again. Wash the residue through with more ether and then spread it out and allow it to dry. The dried powder is zymin.

Lipase can be prepared from castor oil seeds which have just commenced to germinate. Cut up the endosperm into small pieces and wash out the oil with ether. The residue is ground to a pulp in 0·5 per cent acetic acid. This frees the enzyme. The mash is filtered and washed until the filtrate gives no acid reaction. The residue is then shaken up with a small quantity of water. This suspension contains the lipase.

To Demonstrate Plant Structural Materials: Cellulose, Lignin, Cutin, Suberin

In all cases except the last, thin sections must be cut with a razor and stained on a microscope slide or in a suitable dish and examined under the microscope. All the sections cut by the student in his plant anatomy studies should be subjected to the following simple procedures.

Staining with chlor-zinc-iodide (Schulze's solution) often shows cellulose directly by the blue-violet colour reaction, but the sections may need a previous few minutes treatment with 5 per cent potassium hydroxide solution to remove fatty or other materials which will hinder the cellulose reaction. Lignified cellulose stains yellow in Schulze's solution.

Sections stained in phloroglucin acidified with hydrochloric acid show all lignified cell walls a bright red. In aniline sulphate or aniline chloride they stain yellow.

Cutinized and suberized cell walls can be demonstrated by staining in Sudan III and washing in 70 per cent alcohol. Cutinized parts will retain the dye, as will any fatty or waxy substance. Leaves of holly, ivy *Aucuba*, *Pinus* and laurel all have thick cuticles which show well when stained by this method. Suberized cells can be found in almost any bark tissue or in bottle cork.

Demonstration of Other Secretions by Plants and Animals

Ethereal Oils

Cut sections of orange rind (not in alcohol) and to them on a microscope slide, add a drop of 1 per cent osmic acid. Add a drop of water after a minute or so. Cover and examine. The essential oil droplets are stained black. Note the lysigenous cavities into which the oil is secreted. Alternative material includes fruits of members of the Umbelliferae, leaves of thyme, marjoram, lavender. Note that this demonstration does not distinguish the ethereal oils from the lipides. These can be demonstrated by similar means in sections of plant and animal tissues as described on p. 123. To distinguish ethereal oil from lipide, wash the sections thoroughly in alcohol before applying the test. The ethereal oils and castor oil (a lipide) are readily soluble in alcohol and will not be demonstrated whilst the remaining lipides, not soluble in alcohol, will be unaffected.

Resin

Cut sections of young *Pinus* stem and immerse them in a saturated solution of copper acetate for about one week. Remove and wash in water. Examine. The resin deposits are an emerald green colour. The stem of ivy is also suitable material for this demonstration.

Pectic Compounds, Mucilage

Ruthenium red stain can be used to demonstrate the middle lamella in many plant tissues. The demonstration of mucilage in the walls of linseed testa has been described on p. 122.

Tannins

Cut sections of wood, bark or galls of the oak and fix in copper acetate (7 per cent in alcohol) for about a week. Place in 0·5 per cent aqueous ferrous sulphate solution for three minutes; wash in water and mount. Cells containing tannins will have their contents stained a blue-black colour.

Latex

In fluid latex there may be many substances in solution or suspension including mineral salts, sugars, caoutchouc (rubber), starch grains, alkaloids and various nitrogenous substances. Fresh latex from spurge, dandelion, and poppy, expressed on a slide, should be examined for suspended particles. Addition of iodine will show up any starch grains, and protein particles are stained brown. Due to the presence of caoutchouc granules, a fresh mixture of saturated aqueous sucrose solution and concentrated sulphuric acid in equal parts, will impart a purplish colour to the latex. Treated with fresh alcoholic solution of alkannin, the rubber particles stain red. In sections of the above plants, treated with either of these stains, note the network of laticiferous vessels in the tissues.

Glycosides, Saccharide Derivatives

Tests for these substances have been described in Chap. 4.

To Demonstrate Food Storage Compounds

Starch

Examine untreated sections for starch grains. Irrigate with iodine. The grains become blue-black. Good material includes potato tissue, pea and bean cotyledons, cereal grains, cortex and medulla of many plant stems, rhizomes and tubers.

Glycogen

Using liver of a freshly-killed rabbit, drop a small piece into boiling water to prevent enzyme action on the glycogen (autolysis).

Smear cells from the liver on a slide and stain with dilute iodine. Glycogen granules show darkly stained under the microscope.

Inulin

This occurs in solution in cell sap of tubers of dahlia, Jerusalem artichoke, root of dandelion and some other members of the Compositae. It will not be observed in fresh sections mounted in water, but if the water is removed and replaced by alcohol the inulin is precipitated in a granular form. If chunks of the tissues mentioned above are stored in alcohol for about a week, the inulin crystallizes out as very characteristic sphere-crystals located on the cell walls. Sections of tissues so treated should be mounted in glycerine for examination. The inulin can be demonstrated by staining the sections in alcoholic orcin followed by heating in dilute hydrochloric acid, when it will appear orange-red in colour.

Hemicelluloses

Whilst there are no specific chemical tests for such materials, a section of date stone mounted in dilute iodine shows the very thick cell walls, in which the hemicellulose is deposited, coloured purplish by the iodine.

Sugars

These occur in solution in the cell sap of many plant tissues and cannot be observed in sectioned material. They can be tested for in expressed juices by the chemical tests outlined in Chap. 4. Suitable material includes most ripe fruits (grapes are very good), carrot root, onion and other bulbs for glucose, and beetroot for sucrose. The sucrose in beetroot can be demonstrated by first hydrolysing it by boiling the juice with hydrochloric acid and then applying Fehling's test.

Lipides

Fats in tissues can be demonstrated quickly by squeezing and smearing the tissue across clean filter paper. A grease-spot is formed which turns brown or black when 1 per cent osmic acid is added to it. Fat globules in plant cells are readily stained with Sudan III or osmic acid. Sectioned almonds, brazil nuts, sunflower seed, linseed and castor oil seed will all show plenty of oil stored in the cells. Adipose tissue from rabbit, fat bodies of the frog, fatty material from the cockroach's abdomen, oil from the dogfish's liver also show the characteristic fat reactions (*see* p. 123).

Proteins

Proteins form the major part of the non-aqueous portion of proto-plasm and can therefore be demonstrated in any tissue (except where only the cell walls survive, as in wood and fibre). Pulping the tissue and washing through with water gives a suspension which on testing with the biuret, xanthoproteic or Millon's test (*see* p. 124), indicates the presence of protein. In some plants, protein may be stored in the form of aleurone grains. These can be demonstrated very easily in castor oil endosperm and brazil nut, where they are plentiful and quite large. Cut sections in alcohol and dissolve out any remaining oil with ether. Wash again in alcohol and mount in thick glycerine. A per-manent preparation can be made by staining the sections with eosin *Y* after they have been fixed in a concentrated alcoholic solution of picric acid for several hours. After fixing, wash in alcohol and apply the stain in 90 per cent alcoholic solution for two or three minutes. Wash

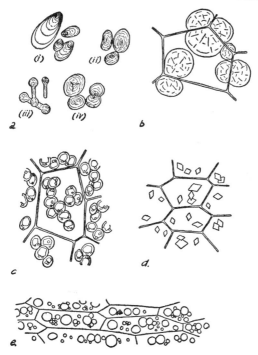

Fig. 14.7. Storage compounds. (*a*) Starch grains of (i) potato, (ii) wheat, (iii) *Euphorbia* (from latex), (iv) pea. (*b*) Inulin of *Dahlia*. (*c*) Aleurone grains of castor oil. (*d*) Asparagine crystals of lupin hypocotyl. (*e*) Oil globules from sunflower cotyledons.

in absolute alcohol (twice), clear in clove oil and mount in Canada balsam. The grain matrix is dark red, the crystalloid yellowish and the globoid uncoloured.

Aleurone grains also occur in cereal grains (in the outermost layer of the endosperm), pea, bean, sunflower, lupin cotyledons and many other seeds, but they are characteristically small, indeterminate structures which appear somewhat refractive in untreated sections. They frequently stain brown with iodine.

The proteins in wheat flour can be extracted simply by enclosing the flour in fine muslin and washing and kneading it under a running tap until all starch has been washed away, i.e. until the tap water ceases to be clouded. When the muslin is opened, the content is a yellowish, sticky substance called gluten. The protein gliadin can be extracted from this by boiling with alcohol in which it is soluble. If the alcohol is decanted and evaporated, the gliadin remains as a residue. The undissolved portion of the gluten contains another protein, glutelin, which is soluble in dilute acids and alkalis only.

Animals do not store protein in their tissues generally but the eggs of many animals contain it and may be covered by albuminous secretions which serve to nourish the developing embryo. The most easily available of these is the hen's egg and the egg white will give positive reactions to all the protein tests.

Amides: Asparagine, Glutamine

Grow some lupin seedlings in darkness and examine sections for the rhombic plate-like crystals of asparagine. This substance is the amide of the amino-acid aspartic acid. The corresponding amide of glutamic acid, glutamine, accumulates in some plants such as *Cucurbita* and can be seen in sections of stems.

CHAPTER 15

EXCRETION

FROM the foregoing chapters on nutrition, respiration, growth, secretion and storage, which are concerned with the metabolic activities of plants and animals, it will be realized that there are processes involved which must lead to the intake and formation of numerous substances which cannot serve useful purposes. Organisms cannot always exercise such strict control over their intake of nutrients that their requirements are met exactly. Neither can their methods of dealing with the absorbed materials utilize every atom and molecule without wastage. Thus, for two main reasons, a plant or animal will tend to accumulate within its body, substances which are of no use to it. In the first place, excess or even unwanted materials may be absorbed. Secondly, the chemical processes involved in converting absorbed materials to useful purposes, often leads to unusable by-products. In most cases, such accumulations would tend to clog, if not to destroy completely, the protoplasm which produced them. They must, therefore, be removed from contact with the protoplasm. Any physical or chemical mechanism which serves to eliminate unwanted substances from further participation in the activities of the organism, is termed excretion. Efficient excretory mechanisms are of prime importance to all living creatures, since many of the excretory substances are highly toxic, and if not eliminated, they would destroy the organism in which they originated. It must be noted that the term excretion can be applied only to a process which eliminates substances which have been taken up by cells or have been formed as a result of their activities. Thus, although animal faeces contain some truly excreted substances as a result of the excretory activities of the gut, they are largely composed of material which has never been part of the cellular constituents of the body.

Excretory products are of numerous different kinds and they vary from species to species, and even from time to time in the same organism according to the food supply and the condition, healthy or otherwise, of the organism. They may be taken in as foods, or with foods but not required as such. They may be products of cellular decomposition, or substances formed as a result of normal metabolic processes, or the result of some disease condition.

In plants, there are not, generally speaking, any special body systems or elimination of waste materials; they are usually taken out of

579

contact with the protoplasm by storage in insoluble form, either within cells or in the cell walls or intercellular spaces as secretions. In animals, on the other hand, specialized organs and organ systems are formed for the purpose of efficient excretion.

EXCRETORY SUBSTANCES AND THEIR ORIGINS

There are five major processes from which unwanted substances may be derived. They are—

1. The absorption of substances with nutrients but not required as such.

2. The absorption of nutrient substances in excess of requirements.

3. Numerous chemical reactions which constitute the normal metabolism of the cells.

4. Osmoregulatory processes.

5. The breakdown or lysis of protoplasmic constituents.

Materials Absorbed Unnecessarily

It is said that there are few, if any, of the natural elements that have not been found in some plant or animal at some time. From a knowledge of the nutrient requirements of a very wide variety of plants and animals, it must be concluded that many of these substances are serving no useful purpose. They have entered the bodies of organisms by means similar to those by which the normally necessary substances have entered. In plants, such substances often accumulate in solution in cell sap, in concentrations far above those of the external environment. Sometimes they occur as insoluble inter- or intracellular secretions. It is not always possible to judge whether such secretion is essential to normal cell activity or whether it is a process of excretion.

Some remarkable instances of accumulation of seemingly unnecessary materials have been discovered in plants. Brazil nuts sometimes contain large quantities of barium, and the walls of some fungi show the presence of barium sulphate. Aluminium oxide may make up as much as 30 to 80 per cent of the wood of the Australian oak (*Orite excelsa*). Silica is found in many grasses, and in the bamboo it may exist as large loose granules. The extracted product is known as tabasheer and is used medicinally in the East.

In animals, the constituents of the food will be absorbed if they can be rendered diffusible. Consequently animals must absorb at least small quantities of unnecessary elements and compounds. Thus the excreted substances may vary from day to day in the same animal according to its food. Much of our knowledge of the excretory mechanisms of animals has been derived from experiments in which they have been fed with innocuous but unnecessary chemicals.

Nutrients in Excess of Requirements

In the case of autotrophes, all organic requirements are synthesized internally from inorganic materials. Thus there will never be more than can be used or stored. The whole synthesizing and utilizing mechanism can be regulated from within. But it is generally the case that plants tend to absorb excessive amounts of the essential inorganic ions. This can lead to alterations in the reaction of cell fluids and may prove harmful. Conditions can arise where there may be excesses of cations or anions accumulating within the cells. Such would happen if nitrate, absorbed with equivalent quantities of cations, were rapidly metabolized to leave the cations unbalanced. Similar effects could occur if ammonium ions, absorbed with anions such as chloride or sulphate, were rapidly used up. The balance is usually maintained by the presence of organic acids. Calcium salts of organic acids are barely ionized and thus formation of such a salt not only assists in restoring ion balance but also in reducing the total amount of ionization. Oxalic acid is a common example of such organic acids, and calcium oxalate crystals in cells may be regarded as serving a useful excretory purpose by reducing cation excess, and taking the acid and calcium out of solution. Both calcium and magnesium may be similarly taken out of solution by combination with pectic acid.

Leaf abscission can regularly aid removal of metallic ions from plants. It has been shown that there is continuous increase in quantity of calcium, iron, manganese and silica in beech leaves up to leaf-fall. None of these pass back into the plant, unlike potassium, nitrogen and phosphorus which are translocated out of the aging leaf.

In the case of animals, the same conditions do not obtain. The heterotrophic animal must take its mixtures of food substances as it finds them. Carbohydrates and fats present no difficulty, provided they can be digested. They can be stored conveniently. Proteins, mineral salts, vitamins and water, cannot be held in a similar way, since in excess they have detrimental effects on other physiological processes. Mineral salt and water content of the body fluids must be maintained within small variation from fixed values, not to interfere with the osmotic relationships of the cells. Excess minerals cannot usually be organically combined and thrown out of solution in the form of grains or crystals as in the plant, because of the nature of animal tissues and the effect of such accumulations upon them. The "stones" in kidney, gall-bladder, etc., are the harmful result of such accumulation. Consequently, the animal is continuously eliminating mineral substances in solution.

Excess protein and other nitrogen-containing compounds such as the purine and pyrimidine bases of nucleic acids are also potential sources of danger. They cannot be oxidized to release energy in their existing form, neither can they be stored conveniently. The processes involved in making part of the molecule of use in other metabolic events such as oxidations by conversion to carbohydrate, lead to the production of toxic nitrogenous compounds which must be eliminated. Throughout the animal kingdom, there is a wide variety of such excretory products and they vary with the source of the excreted compounds. Protein breakdown products include amino-acids, ammonia, urea, uric acid, trimethylamine oxide and guanine.

Amino-acids, greatly in excess of requirement, may be taken in directly from the gut as the products of protein digestion. They may also result from the internal degradation of proteins. In a few cases, amino-acids may be excreted without change, but in the majority of cases, they undergo deamination, resulting in the formation of ammonia (*see* Chap. 10). This is apparently always an oxidative deamination in animals and may be summarized by the equation—

$$R.CH.NH_2.COOH + \tfrac{1}{2}O_2 \rightarrow R.CO.COOH + NH_3$$

The residual part of the amino-acid molecule is easily oxidizable or can be built up into carbohydrate reserve. The equation below, indicates what happens in the case of a specific amino-acid, alanine.

$$\underset{\text{alanine}}{CH_3.CH.NH_2.COOH} + \tfrac{1}{2}O_2 \rightarrow \underset{\text{pyruvic acid}}{CH_3.CO.COOH} + NH_3$$

It will be recalled that pyruvic acid is an intermediate product in glycolysis. The ammonia formed by the deamination of amino-acids is very toxic and must be eliminated. If the organism is small and bathed continuously in a fluid medium, the ammonia, being highly soluble, may diffuse into the surrounding fluid in an unchanged state. This happens in many invertebrates which are therefore described as *ammonotelic*.

Where the organism is large, and free diffusion from its surface is impossible, the ammonia must be converted into an innocuous form. The two chief excretory products from this source are urea and uric acid. Hence some animals are described as *ureotelic* and some as *uricotelic*. Ureotelic animals are able to form urea from ammonia and carbon dioxide as outlined here.

The liver of many animals contains an enzyme arginase which

catalyses the first stage in the formation of urea from the amino-acid arginine, thus

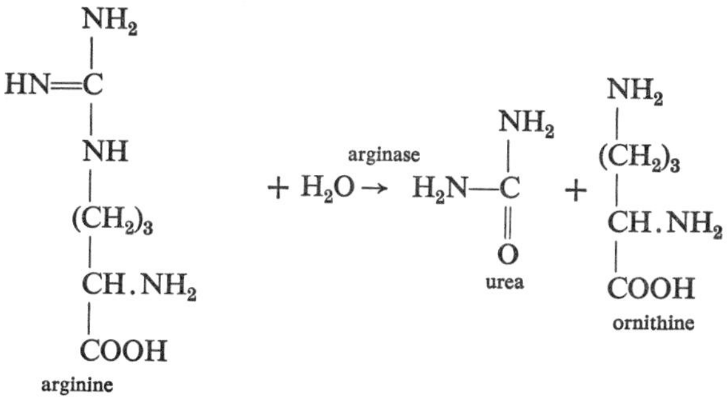

$$
\begin{array}{c}
\mathrm{NH_2} \\
| \\
\mathrm{HN}{=}\mathrm{C} \\
| \\
\mathrm{NH} \\
| \\
\mathrm{(CH_2)_3} \\
| \\
\mathrm{CH.NH_2} \\
| \\
\mathrm{COOH} \\
\text{arginine}
\end{array}
\quad + \mathrm{H_2O} \xrightarrow{\text{arginase}}
\begin{array}{c}
\mathrm{NH_2} \\
| \\
\mathrm{H_2N}{-}\mathrm{C} \\
\| \\
\mathrm{O} \\
\text{urea}
\end{array}
\quad + \quad
\begin{array}{c}
\mathrm{NH_2} \\
| \\
\mathrm{(CH_2)_3} \\
| \\
\mathrm{CH.NH_2} \\
| \\
\mathrm{COOH} \\
\text{ornithine}
\end{array}
$$

The ornithine thus formed is then converted to citrulline by the addition of ammonia and carbon dioxide, with the loss of water. The enzyme concerned has not been isolated.

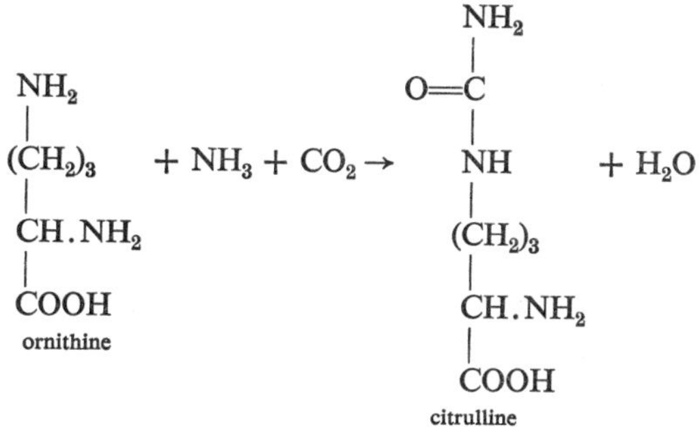

$$
\begin{array}{c}
\mathrm{NH_2} \\
| \\
\mathrm{(CH_2)_3} \\
| \\
\mathrm{CH.NH_2} \\
| \\
\mathrm{COOH} \\
\text{ornithine}
\end{array}
\quad + \mathrm{NH_3} + \mathrm{CO_2} \rightarrow
\begin{array}{c}
\mathrm{NH_2} \\
| \\
\mathrm{O}{=}\mathrm{C} \\
| \\
\mathrm{NH} \\
| \\
\mathrm{(CH_2)_3} \\
| \\
\mathrm{CH.NH_2} \\
| \\
\mathrm{COOH} \\
\text{citrulline}
\end{array}
\quad + \mathrm{H_2O}
$$

By the further addition of ammonia to the citrulline, arginine is formed to commence the cycle again.

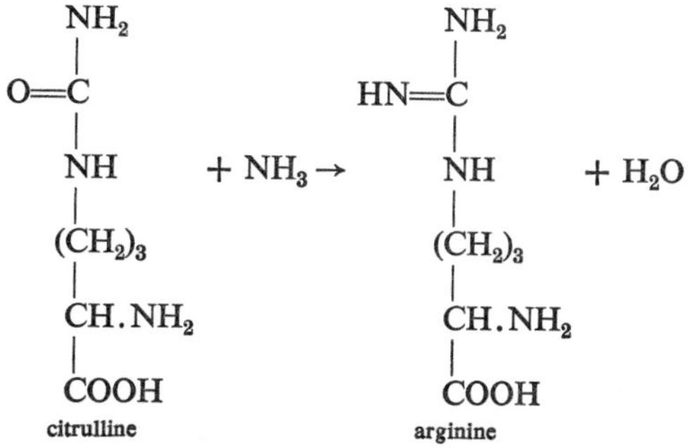

$$
\begin{array}{c}
\mathrm{NH_2} \\
| \\
\mathrm{O}{=}\mathrm{C} \\
| \\
\mathrm{NH} \\
| \\
\mathrm{(CH_2)_3} \\
| \\
\mathrm{CH.NH_2} \\
| \\
\mathrm{COOH} \\
\text{citrulline}
\end{array}
\quad + \mathrm{NH_3} \rightarrow
\begin{array}{c}
\mathrm{NH_2} \\
| \\
\mathrm{HN}{=}\mathrm{C} \\
| \\
\mathrm{NH} \\
| \\
\mathrm{(CH_2)_3} \\
| \\
\mathrm{CH.NH_2} \\
| \\
\mathrm{COOH} \\
\text{arginine}
\end{array}
\quad + \mathrm{H_2O}
$$

The arginine may then be split into urea and ornithine, to repeat the cycle indefinitely as long as ammonia and carbon dioxide are available. The whole sequence is known as the *ornithine cycle of urea synthesis*. It may be represented simply as shown below.

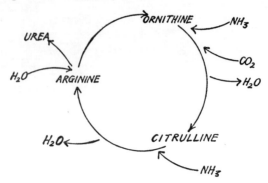

The Ornithine Cycle

The overall reaction may be written thus—

$$2NH_3 + CO_2 \rightarrow NH_2.CO.NH_2 + H_2O$$

It is almost certainly the case that the synthesis of urea is not quite so straightforward as the preceding would suppose. Since the ornithine cycle was first proposed closer investigation has shown that the full series of events in urea synthesis is most probably as given below.

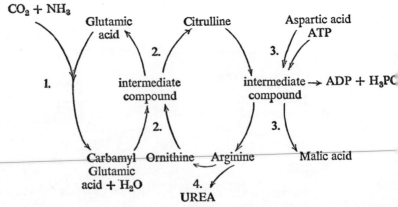

1. glutamic acid + CO_2 + NH_3 → carbamyl glutamic acid + H_2O
(with ATP, Mg ions and in aerobic conditions only)

2. carbamyl glutamic acid + ornithine → $\overset{\text{transcarbamylase}}{\underset{\text{compound}}{\text{intermediate}}}$ → glutamic acid + citrulline

3. citrulline + aspartic acid + ATP → $\underset{\text{compound}}{\text{intermediate}}$ + ADP + H_3PO_4 → arginine + malic acid

4. arginine $\overset{\text{arginase}}{\longrightarrow}$ ornithine + urea.

In some ureotelic animals this cycle does not necessarily operate; there are probably methods of urea synthesis which are not yet known. The urea, once formed, is excreted in solution in the urine, a process not economical of water.

In uricotelic animals, notably birds, terrestrial reptiles and some insects, the ammonia is converted to uric acid. This is not toxic, but is highly insoluble. It can be stored to a small extent but is usually excreted in crystalline form. This mechanism has the advantage that there is far less water lost with the excretory product. The mechanism of uric acid formation from ammonia does not appear to be known.

Trimethylamine oxide is excreted by some marine fishes. Its synthesis from ammonia may be by methylation forming trimethylamine; this may then be oxidized to form trimethylamine oxide. Guanine is excreted by spiders, earthworms and some other animals. It is even less soluble than uric acid. The method of synthesis is not known.

The purine and pyrimidine bases of nucleic acids may also lead to nitrogenous excretion, and it is thought that about 5 per cent of excreted nitrogen comes from this source. The excreted products include purines such as adenine and guanine in some animals, whilst in some others, adenine may be excreted in an oxidized form. Guanine may be excreted as such, but in most mammals it is deaminated before excretion. Uric acid may be formed as the excretory product by an oxidative deamination of adenine or guanine. Allantoin is formed from uric acid by the enzyme uricase which opens the pyrimidine ring, and thus may feature as an excretory compound, particularly in vertebrate embryos, or it may be further oxidized to allantoic acid by allantoinase. The molecule of allantoic acid can be further broken down into urea and glyoxylic acid before excretion. Finally this urea may be converted to ammonia and carbon dioxide by the enzyme urease, which occurs in a few animals (note that urease is a very common enzyme in plants). The products of excretion of nucleic acid metabolism will vary between the compounds mentioned above, according to the enzyme systems possessed by different animals.

An examination of excreted matter shows the presence of other nitrogenous compounds in many instances. These include hippuric acid and ornithuric acid which are formed from benzoic acid in the diet. Creatine and creatinine are also sometimes found. Creatine is important in muscle action (*see* Chap. 12), and may be excreted occasionally, but is more often converted to creatinine. Certain nitrogenous pigments resulting from breakdown of the blood pigment haemoglobin are excreted, particularly in mammals. The orange-red bilirubin and the green biliverdin are the most common. The pterins in butterflies' wings are sometimes regarded as excretory products. Other secretions containing nitrogen, such as chitin and keratin, could be regarded as excretory products. However they serve such useful functions that it is better to omit them from the list of excretory substances.

By-products of Chemical Reactions

When fats, carbohydrates and proteins are catabolized the end products are carbon dioxide, water and ammonia. They are truly by-products, but are not always necessarily excreted. Carbon dioxide released in respiration may be "fixed" all over again by plants and animals. The water derived from oxidation processes can be of use in conserving the water supplies of many land organisms. For example, the clothes moth and the kangaroo rat are reputed to exist without recourse to liquid water. Ammonia, formed by the deamination processes outlined previously, need not all be discarded. It can be used in transamination processes to form new amino-acids. In plants it need not be eliminated at all, but can be stored as the amides, glutamine and asparagine. Thus whilst not all metabolic by-products are necessarily waste, in many instances there will be a surplus of them and they will then be expelled to maintain balanced conditions within the organism.

Carbon dioxide is normally excreted by diffusion through the surfaces of cells into the surrounding water or atmosphere. In lower aquatic animals and plants this is a direct process and may go on all over the surface of the organism. Higher animals have evolved respiratory surfaces which are therefore partly excretory in function. Higher plants possess systems of intercellular spaces communicating with the outside world. Sometimes such excreted carbon dioxide can be usefully employed. It can be combined with calcium to form chalk coverings as in corals, the shells of some invertebrates and the deposit on some algae such as *Corallina*, and *Chara*. Internal secretion of calcium carbonate may also be regarded as a method of excreting carbon dioxide but can also serve the purpose of removing calcium ions from solution.

There are numerous instances of compounds accumulating in organisms as a result of irreversible synthetic activities. It is frequently necessary for them to be eliminated. Sterols are continuously formed in small amounts in vertebrate animals. The animal cannot degrade them for further use and they are excreted as by-products from the liver. In plants, similar accumulations of synthesized material can be got rid of during leaf shedding or by secreting them in cells where they cannot interfere physiologically. Unwanted active compounds can be rendered physiologically inactive in the plant by combining them with other compounds, thus altering their molecular structure. Conjugation with sugars to form glucosides and other compounds can eliminate substances such as ethylene chlorophenol, chloral hydrate and ethylene chlorohydrin. Animals may occasionally secrete similarly but this is not the general rule, since they are equipped with efficient excretory systems.

Although many of these apparently unwanted synthetic by-products appear in quantities far beyond any chemically functional value, they are not necessarily always without some use. The alkaloids of plants are typical examples. Nicotine in particular is considered to be important in the tobacco plant root, where it is associated with nitrate absorption. It may be that many plants employ nicotine in the same way, but whereas the majority possess the enzymes for degrading it and hence disposing of it for other purposes, the tobacco plant lacks these enzymes and must accordingly accumulate it.

Tracey in his *Principles of Biochemistry* has adequately summed up this difficult problem. The following are his words—

The essential metabolic activities of an organism may be regarded as occurring in a great cycle, itself composed of cycles of lesser cyclic changes. Fed into this metabolic cycle are energy and raw materials for synthesis. Leaving the cycle are excess raw materials, products of synthesis used in growth, products of irreversible synthesis, substances lost by leaking, products needed to make good losses from the organism and products from which energy can no longer be derived that under some circumstances re-enter the cycle. All these may be regarded as by-products of metabolic activity and vary in their nature according to the needs of the organism. Whether particular by-products are regarded as excretory products or not is to some extent a matter of choice in our present state of knowledge.

Products of Osmoregulation

The maintenance of exact osmotic equilibrium between cells and their surroundings is essential. The osmotic condition of cells, whether hyper-, iso-, or hypotonic to the surrounding fluids, depends upon the nature of the fluid and the osmotically active cell contents. Any process designed to maintain this constancy is an osmoregulatory process.

In plants, by virtue of their construction, the problem is automatically solved. Although the internal concentration of the cells is greater than that of the outside watery solution, they are able to prevent the entry of excess water by reaching turgidity. In such a condition, resistance to water entry is applied by the inability of the cell wall to stretch farther.

In animals, where cell walls are lacking, the problem is solved by other means. In those animals in which the cells are all in direct contact with the outside watery medium and whose cell fluids are hypertonic to it, such as freshwater protozoa, sponges and coelenterates, there will be a tendency for water to enter in excess of requirements. This must be constantly removed.

In those animals in which the cells are not in direct contact with the external medium but with body fluids such as in most metazoans, both the body and cell fluids must be osmoregulated relative to one another. If this necessitates keeping the body fluids hypertonic to an outside watery medium, as it always does in the case of fresh-water animals then some mechanism must rid the body of water which will be constantly entering it by osmosis. In the case of the many marine animals where the body fluids are roughly isotonic with sea-water, the problem is to maintain this condition by eliminating the excess mineral substances which inevitably enter the body. In land animals, excess mineral salt and water may be taken with the food and both must be kept at the correct level. We find, therefore, that all animals have the need to excrete water and mineral salts.

Osmoregulatory systems have been mentioned in Chap. 6 and in connexion with specific animal types in Vol. I. They are usually the same organ systems responsible for separating the other excretory products from the body. They will be summarized later in this chapter

Products of Protoplasmic Breakdown (Lysis)

Since protoplasm is composed largely of water, protein, carbo hydrate, fat and mineral salts, its breakdown will result in product similar in nature to those already mentioned. Most of these ar undoubtedly used in other parts of the organism and are therefore no excreted. There are however certain particular protoplasmic con stituents of some organisms which are continuously being lysed t yield substances of no further value. One is haemoglobin of vertebrate The death of a red blood cell is accompanied by the degradation o

SUMMARY OF THE MORE COMMON ANIMAL EXCRETORY PRODUCTS
AND THEIR DERIVATION

Excreted Substance	Derivation	Type of Animal
NITROGENOUS COMPOUNDS	From excess amino-acid, protein breakdown, purine and pyrimidine bases of nucleic acids	All
Ammonia (highly water soluble but very toxic)	(a) Deamination of amino-acids	Many aquatic animals
	(b) Deamination of purine derivatives such as guanine	Crustaceans and a few others
	(c) Breakdown of urea (derived from allantoic acid) by urease	Sipunculids, a few crustaceans, and some gastropod molluscs
Urea (highly water soluble, non-toxic)	(a) Deamination of amino-acids to form NH_3 which is then converted to urea in the ornithine cycle (some cases only)	Many aquatic animals where urea is removed in solution, e.g. elasmobranch and teleost fish, amphibians; land animals not adapted for water conservation, e.g. mammals
	(b) From breakdown of allantoic acid	Known to occur in the mussel, elasmobranch and teleost fish
Uric acid (highly insoluble, non-toxic)	(a) Presumably synthesized from NH_3 from deamination of amino-acids (process unknown)	Many terrestrial animals adapted to water conservation, e.g. insects, birds, reptiles, some mammals
	(b) Oxidative deamination of guanine or adenine	
Trimethylamine oxide	Methylation of NH_3 from deamination of amino-acids, followed by oxidation	Some marine fish
Guanine	(a) Synthesized from NH_3 but process unknown	Spiders, earthworms
	(b) Directly from breakdown of purine and pyrimidine bases of nucleic acids	Some mammals
Adenine	(a) Directly from breakdown of purine and pyrimidine bases of nucleic acids (not oxidized)	Some mammals
	(b) Oxidized before excretion	Some mammals

SUMMARY OF THE MORE COMMON ANIMAL EXCRETORY PRODUCTS
AND THEIR DERIVATION—(contd.)

Excreted Substance	Derivation	Type of Animal
Allantoin	From uric acid	Most mammals; Diptera
Allantoic acid	Oxidation of allantoin	Some teleost fish
Hippuric acid ⎫ Ornithuric acid⎭	From benzoic acid in the diet	Some mammals
Amino-acids	Protein breakdown	A few animals occasionally excrete amino-acids in unchanged form in the urine
Creatine ⎫ Creatinine⎭	From phosphagen in muscle	Some mammals
Bilirubin ⎫ Biliverdin⎭	Lysis of haemoglobin	Mammals
CARBON DIOXIDE	(a) Respiratory metabolism	All animals
	(b) Breakdown of urea	Animals containing the enzyme urease; sipunculids, a few crustaceans and some snails
WATER	(a) Osmoregulatory processes	Most animals
	(b) Hydrolytic metabolism	Most animals
MINERAL SALTS	Osmoregulatory processes	Most animals
MISCELLANEOUS ORGANIC COMPOUNDS Cholesterol	Excess intake or synthesized in excess	Mammalian bile; occasionally small amounts in mammalian urine
Animal hormones	Excess production	Rarely; female urine during pregnancy
Plant hormones	Taken in with vegetable diet	Some mammalian urine

its contained pigment molecules. The haematin part of the haemo
globin molecule has its iron removed and is thus converted to haemo
bilirubin. This change occurs mostly in the cells of the spleen and in
the Küpfer cells of the liver. The haemobilirubin is circulated in the
blood plasma and finds its way to the liver cells where it is converted
to bilirubin and its oxidized form, biliverdin. When bile is released

into the intestine, its pigments, these two substances, are reduced by bacteria to hydrobilirubinogen (stercobilinogen) some of which may be oxidized to hydrobilirubin (stercobilin). Much of this mixture is passed out with the faeces and is thus truly excreted. The faeces owe their dark colour to their presence. Aggregation in the blood due to failure to eliminate in the bile, leads to jaundice. Some of the stercobilinogen is reabsorbed from the intestine and so returns to the liver, where it may be either oxidized to bilirubin and pass out once more with the bile or be circulated in the blood to be excreted eventually by the kidney as urobilinogen in the urine.

EXCRETORY METHODS IN ANIMALS

Primarily, all the systems of animals concerned with excretion of nitrogenous substances employ the same principle. They involve surfaces through which unwanted materials can be passed directly or indirectly to the outside. Complexity of the system increases with complexity of the body in other respects. In the most lowly of animals, where all the cells are in contact with the outside medium, excretion takes place by diffusion from the cell surfaces. This applies to the protozoa, sponges and coelenterates, but in some cases the contractile vacuoles possessed, may excrete soluble substances dissolved in the water which they pass to the exterior. Analysis of contractile vacuole contents extracted by micro-pipette, has been recorded as showing the presence of ammonia, urea and carbon dioxide in small quantities. Pigments taken up by protozoan protoplasm have also been seen to colour vacuole contents, thus indicating an excretory function. In some sponges such as *Grantia*, certain cells are formed and function exclusively in an excretory capacity. They gather within themselves granular excretory products, move to the surface and disintegrate to discharge their contents. Such a method is known as intracellular excretion and it is probable that most animals employ the method to a greater or lesser degree.

Among the higher invertebrates, only one group relies on direct transmission of excretory substances to the exterior without specialized organs. These are the echinoderms and their excretory matter is diffused away from the body at the epithelium of the respiratory and digestive structures.

Many of the invertebrates also employ what may be called storage excretion, since within their cells and tissues are frequently found isolated particles and granules of no apparent functional significance. However, most of them rely on special excretory organs.

The simplest form seems to be the flame-cell system characteristic of the flatworms. A modification of this is seen in the collections of solenocyte cells in *Amphioxus*, which, acting like flame cells, pass their excreted products collected from the coelomic fluid, into a common duct which communicates with the outside via the atrial chamber.

In coelomate worms, the active excretory structure is the nephridium. This is fundamentally a duct, opening internally into the coelomic cavity where it often forms a wide-mouthed ciliated funnel, and opening externally at an excretory pore. Parts of the duct may be ciliated as well as the funnel, and the whole duct may be differentiated into distinct regions of slightly differing function. Some parts of the duct may be glandular and excrete directly from the vascular system, but the main purpose is to remove coelomic fluid to the exterior, carrying with it the excreted products which have been deposited by the break-up of special excretophores. These are cells formed by the coelomic epithelium and in them excreted substances accumulate as they are extracted from the blood. As these cells mature, they break off into the coelomic cavity and disintegrate, with the result that their accumulated products are released into the coelomic fluid. Amoeboid phagocytes in the coelomic fluid play an active part in their destruction. The excretophores of the earthworm are known as the chloragogenous cells. Nephridia, modified in various ways, are common excretory structures in the invertebrates. They occur in the molluscs and crustaceans. In *Astacus*, the excretory green gland is a modified nephridium. In insects and spiders, the condition is somewhat different. The excretory organs are the Malpighian tubules which are hollow tubes formed as diverticula of the hind-gut. Unwanted matter, chiefly uric acid and urates, are removed from the haemocoelic cavity into which they project. The materials are passed into the gut to be removed with the faeces. Work on water-balance control in insects has shown that an antidiuretic hormone, produced by neurosecretion from the brain or from some of the ventral ganglia, controls water loss in two ways. If there is sufficient secretion of the hormone, there is increased water flow through the Malpighian tubules and decreased rectal resorption. A low concentration of the hormone causes a reversal of both processes.

In the vertebrates, the common excretory organ is the kidney although this varies in construction and function within the classes. The kidney, primitively, is a collection of coelomoducts (renal tubules) opening internally to the coelomic cavity by a ciliated funnel and emptying to the exterior by a duct common to all. The walls of the tubules are glandular and serve to secrete unwanted matter into the ducts from associated blood-vessels. Such a structure, except for its compactness, is essentially similar to the nephridium of some of the invertebrates. The condition is seen in some fish, e.g. the goose fish, and in the developing embryos of most vertebrates as the pronephros. It is known as the aglomerular kidney.

In more advanced vertebrates, a condition arises in which Malpighian bodies are formed along the renal tubules. These are constructed of an expanded flask-like chamber invaginated to form an inner space surrounded by two thin walls. This constitutes the Bowman's capsule. Within it, is a knot of blood capillaries forming the glomerulus. When Malpighian bodies are formed, the open coelomoduct may be sealed off and all excretion be effected by transfer of material from the blood in the glomerulus, through the Bowman's capsule and into the duct. Such kidney structure may be seen in the mesonephros of dogfish and frog. A kidney composed of such tubules is known as a glomerular kidney.

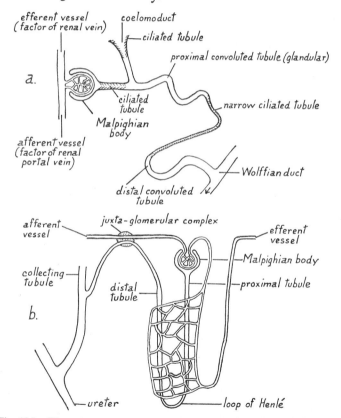

Fig. 15.1. Diagrams to compare, (a) amphibian and (b) mammalian renal tubules.

In mammals, birds and reptiles, the kidney is composed of renal (uriniferous) tubules in which the open coelomoduct is always closed. It really does not exist. Such a kidney is referred to as a metanephros and every tubule ends in a Malpighian body composed of Bowman's capsule and glomerulus. The tubule is differentiated along its length and to each part are ascribed different functions. The mode of action of fish and amphibian kidneys is different from that of the higher vertebrates and this may be correlated with differences in structure of the renal tubules. These structural differences are represented diagrammatically in Fig. 15.1, and the series in Fig. 15.2 represents the structure of the various nitrogenous excretory organs mentioned above.

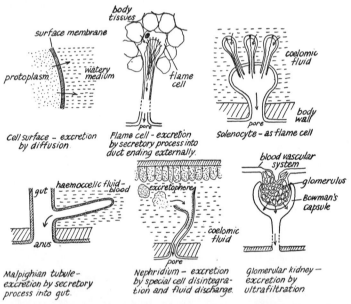

Fig. 15.2. Diagrams representing a variety of organs of nitrogenous excretion.

Excretion of carbon dioxide is associated with respiratory surfaces and these have been discussed in Chap. 11.

Mammalian Kidney Function

The mammalian kidney is undoubtedly one of the most active organs of the body. It has been estimated that of the total blood distributed by each heart beat, between one-third and one-quarter passes through the kidneys. Roughly similar quantities pass through liver and brain whilst the small remainder is distributed through the rest of the body.

Functioning of the renal tubule will be best understood with reference to Fig. 15.3. This represents a tubule straightened out and indicates the different processes occurring along it.

It may be appreciated that great technical difficulty confronts the physiologist attempting to analyse the contents of a renal tubule at different points. Liquid is first extracted by micro-pipette from any point that the pipette may enter. The position of this point is then established later by careful dissection and microscope study.

Blood enters each glomerulus by way of a factor of the renal artery and carries with it in the plasma, blood colloids (proteins), glucose, amino-acids, urea and other excretory compounds and a wide variety of mineral salts. Between the glomerulus and the Bowman's capsule is carried out a process of *ultra-filtration*. This is filtration under hydro-static pressure and is comparable to what happens when water containing colloids and crystalloids is filtered under pressure through a

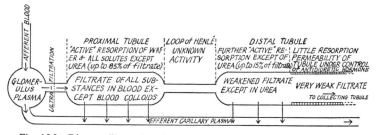

Fig. 15.3. Diagram illustrating the processes occurring along the mammalian renal tubule.

membrane containing minute pores. The crystalloids in solution pass through the membrane and the colloids are retained. The hydrostatic pressure is built up in the glomerulus above the normal blood pressure by reason of the fact that the efferent glomerular capillaries are narrower in diameter than the afferent, so constituting a "bottle-neck." The walls of the glomerular capillaries and the double wall of the Bowman's capsule constitute the membranes and here it is found that the filtrate passes intercellularly by minute spaces between the cells and not through them, so that there is no selective filtration except by the sizes of the pores, which are small enough to prevent the passage of colloids. It is generally held that all crystalloid components of the plasma freely enter the capsule. Altogether about 20 per cent of the afferent blood plasma is passed into the proximal renal tubule in this way.

As the filtrate moves through this part of the tubule, still under pressure from the capsule end, a rapid process of *resorption of filtrate* into the efferent capillaries enveloping the tubule is effected. The walls of the tubule in this region are very permeable to all the solutes except urea and the nitrogenous excretory products, so that the filtrate has extracted from it most of the glucose, amino-acids and mineral salts, together with most of the water. In the case of sodium and water, it is as much as 85 per cent of the total quantity entering the capsule. The resorption process is described as "active" to distinguish it from a passive diffusion mechanism. The tubule walls are very turgid, implying an energy output. The filtrate in the proximal tubule is isotonic with the blood plasma.

The passage of the weakened filtrate continues through the loop of Henlé but there is no evidence as to the precise function of this region of the tubule. In the more proximal end of the distal tubule, further "active" resorption takes place and up to 15 per cent more resorption of the filtrate, except urea and similar compounds, can occur. Sometimes the resorption of substances can total over 99 per cent. In the more distal parts of this tubule, the character of the wall changes. It becomes more or less waterproof and consequently further resorption is prevented. The distal tubule, at a point beyond its convolutions, is in intimate contact with the afferent arteriole, and surrounding both the tubule and the arteriole is a thickened pad of cells, the *juxta-glomerular complex*. The cells are sensitive to the Na^+ concentration in the blood, and when this concentration falls below a certain critical level, the hormone *renin* is secreted by the cells. It acts on a plasma globulin *hypertensinogen* converting it into *hypertensin* (*see* p. 288). This now acts as a hormone, causing vasoconstriction and stimulating secretion of mineralocorticoids from the adrenal cortex. These cause a rise in the amount of Na^+ resorption from the distal tubules. It has been suggested that these cells of the juxta-glomerular complex, surrounding both the arteriole and tubule, are able in some way to detect the difference in concentration of Na^+ ions in the blood and in the urine at this point.

It must also be remembered that there is overall control of the resorption process by the hormone *vasopressin*, a neurosecretion from the hypothalamus (*see* p. 280). Apart from causing vasoconstriction, it plays an important part in controlling the resorption of water and Na ions from the uriniferous tubules.

In the region of the juxta-glomerular complex, the filtrate can be regarded as having the same composition as urine, while in the more proximal regions of the distal tubule its tonicity with the blood plasma is variable. The fluid in the more proximal regions may be isotonic while that in the waterproofed region may be hyper- or hypotonic.

Thus the renal tubules and the associated capillaries actively and selectively resorb substances from the crystalloidal filtrate produced by the Malpighian bodies. The unwanted excretory products are passed out of the body in the urine via collecting ducts, ureters, bladder and urethra, which are all more or less impermeable to water; the water and salt content of the blood are carefully regulated, so that excess of either is eliminated in the urine. The hormonal factors which control various aspects of the process are described on pp. 210 and 288. It *must be stressed that this osmoregulatory function of the kidney is of equal importance to the excretory function.*

Much work has been done on the physiology of frog kidney since the tubules are more easily dissected than those of the mammal. One or two differences in function are worth mentioning. Whereas mammalian urine is generally hypertonic to the blood plasma (i.e. contains less water in proportion to dissolved substances), amphibian urine is hypotonic, indicating a lower rate of water absorption. Some physiologists have pointed out the absence of the loop of Henlé in the frog, and have ascribed to it the role of extra water absorption in the mammal. It is true that under amphibian conditions, there is no advantage gained by conserving water to the extent desirable in a land mammal and thus there has not been evolved the greater complexity of the renal tubule.

Also, in the frog, ammonia is formed in the walls of the distal parts of the tubule and enters into the composition of the urine there. It is presumed to be formed from some precursor present in the cells.

The table below shows the substances found in 24 hrs. urine produced by a resting man on a standard diet, and some of the substances found in glomerular filtrate. These data cannot be compared quantitatively since they were derived from kidneys working under different conditions.

	24 hours urine	24 hours glomerular filtrate
Water . . .	1·5 dm³	180 dm³
Urea . . .	35·0 g	60 g
Uric acid . .	0·75 g	
Creatinine . .	1·0 g	
Hippuric acid . .	0·5 g	
Ammonia . .	0·75 g	
Glucose . . .	trace	200 g
Chlorine . . .	10·0 g	
Phosphates . .	2·5 g	
Sulphates . .	2·5 g	
Sodium . . .	6·0 g	600 g
Potassium . .	2·0 g	35 g
Calcium . . .	0·2 g	5 g
Magnesium . .	1·5 g	

Other Excretory Processes

So far no mention has been made of excretory processes other than by specialized organs. Other processes which may directly or indirectly assist in excretion do occur, and the following are worthy of mention.

It is probably the case that the cells of the skin of most animals serve some excretory function but nowhere is this so obvious as in the mammals. The sudoriferous glands secrete large quantities of water containing dissolved substances, chiefly urea and sodium chloride. Whilst the rate of perspiration is carefully regulated as part of the mechanism of heat conservation, sweating undoubtedly serves a useful excretory function in mammals.

Excretion through the gut lining or glands associated with the gut may also occur in many animals. This is the case in echinoderms, where there are no other special excretory organs. Uric acid crystals have been reported in the mid-gut of insects and crustaceans. We have noted that bile pigments in mammals are excretory products and it is known that insoluble salts which could clog the renal tubules are exclusively removed by the gut lining in mammals. Salivary glands are reported to secrete minute quantities of HCN in their juices.

Excretion by deposition in organs may occur, comparable with what happens in plants. The fat bodies of insects often show accumulation of uric acid, whilst the calcium carbonate deposits in animals such as the earthworm may also be excretory.

The gills of the crayfish are also known to be actively engaged in excretory activity and the gills of teleost fishes eliminate salts.

It is not wise to elaborate the possibilities further, since this can lead to misconceptions regarding the purpose and nature of many animal substances. For example, the formation of a perfectly good, protective cuticle may sometimes be described as a method of excretion.

EXCRETORY METHODS IN PLANTS

There are no specialized excretory organs or systems in plants. What excretion there may be, is carried out by one or more of three methods. Unwanted metabolites may be secreted into intra- or intercellular spaces. The aerating system of the plant makes possible the ready diffusion of respiratory carbon dioxide into the surrounding water or atmosphere. Parts such as leaves, seeds and fruits may be shed and carry with them accumulations of unnecessary substances. It is to be noted that such elimination is purely secondary to the true purpose of leaf-fall and seed and fruit dispersal.

CHAPTER 16

SENSITIVITY AND RESPONSE TO STIMULATION: BEHAVIOUR

SENSITIVITY, or irritability, may be defined as that capacity of an organism to perceive and respond to changes in external or internal conditions. The capacity resides in the protoplasm and is a manifestation of life, which in animals at least, is one of the more obvious to the casual observer. Any change in conditions which is pronounced enough to produce a change in the activities of a whole organism or any of its parts, can be termed a *stimulus*. The change in activity on the part of the organism, constitutes its *response*. Thus the *behaviour* of an organism, defined as the sum of all its activities, continuous and otherwise, can be affected profoundly by the responses made to stimulation. The responses constitute changes in behaviour.

Stimuli are many and varied but never provide any of the energy which may be used by an organism in making its responses. Changes in conditions which are known to evoke responses by living things may be summarized as below, but it must be remembered that they do not apply necessarily to both plants and animals.

Stimuli External to the Organism. Changes in water conditions; light; temperature; gravitational force; chemical conditions such as oxygen and carbon dioxide concentration, the presence or absence of food substances, irritants and numerous other chemicals; pH and mineral salt balance; physical conditions such as density of the medium, contact, pressure and sound changes, and electrical forces.

Internal Stimuli. Changes in metabolic conditions such as rate of metabolism, excess or lack of nutrients, production or accumulation of waste, excess or lack of oxygen; disease conditions; changes in secretory activity; turgor of parts; body equilibrium; sex urges; parental urges; memories; habit formation and many other unexplained inherent factors.

Thus the organism is at the centre of a large number of forces all tending to modify its behaviour. To maintain itself in adjustment with whatever environmental conditions obtain at any time, it must be able to detect or perceive the changes in these forces, and to make suitable co-ordinated responses if it is to be assured of survival. It must, therefore, possess parts fitted to perceive the changes, and a means of transmitting perception to the parts which can make the adjustment.

Thus we speak of *receptor parts* capable of receiving the stimulus, *transmission systems* capable of conducting the effect of the stimulus on the receptors to other parts, and *effector parts* capable of bringing about the necessary response.

In animals, the structures and mechanisms associated with perception of stimuli and transmission of the direct effects of stimulation are much more clearly defined than is the case in plants. Both the nervous and hormone systems are involved and often work in conjunction with one another. There are discrete structures forming the sensory and endrocrine systems and all are linked, through the central nervous system and body fluid circulation, with one another and with the effector parts. In Chap. 8 the co-ordinating function and modes of operation of these systems have been described.

By contrast, plants do not possess parts in any way comparable with the sensory and nervous systems in animals and although in a few cases, certain cells may be specialized as receptor parts, e.g. trigger mechanisms in some insectivorous plants, they are not linked by specialized tissue to the effector parts. Most of the reception of external stimulation by plants seems to reside in the protoplasm of young cells such as occur at root and shoot apices or other young growing parts. Transmission of the effect of the stimulus takes the form of a changed pattern of distribution of auxins to parts which are capable of making responses. Except in freely motile plants where, as in animals, the responses can be locomotive, they are most commonly observed as alterations in the growth and developmental pattern of the plant.

There is far more known about the perception of, and response to, changing external conditions than to changing internal conditions, chiefly because the former can be controlled much more readily for experimental purposes. We shall deal chiefly with the former in this chapter.

SENSITIVITY IN PLANTS

This is manifested by responses on the part of the plant which can be summarized as below—

1. Movements of entire free-swimming plants (ciliate or flagellate plants), of entire free-swimming parts (motile spores and gametes), and of cell protoplasmic constituents (protoplasmic streaming, chromosomes).
2. Movements of parts of fixed plants (roots, shoots, leaves, etc.).
3. Changes of developmental pattern.
4. Changes in metabolic activity.

Of these, some are responses to well-defined external changes and are said to be *paratonic*, and some to less clearly defined or often unknown internal changes and are said to be *autonomic*. These latter include such occurrences as protoplasmic streaming, the movement of chromosomes and other cell inclusions, some of the developmental changes, and many of the changes in metabolic activity. Our lack of knowledge concerning these, will not allow of much further profitable discussion but some will be mentioned briefly later. The changing developmental pattern in response to external conditions has already been mentioned in Chap. 13 and will be discussed further in Chap. 17. This section will be chiefly confined to the paratonic movement responses of plants.

Plant Movements in Response to External Stimuli—Paratonic Movements

These are broadly of two kinds, namely, those which are purely *mechanical responses* by plant parts such as the dehiscence of sporangia and fruits, and the scattering of reproductive parts, and the *movements induced by stimulation of irritable protoplasm*.

Mechanical Movements

Many of these have been described in Vol. I and need not be further treated here except for a brief summary. We may classify such movements as follows—

(*a*) *Hygroscopic Movements*. These are shrinkage movements of dead cell walls in response to drying-out by the atmosphere. Such examples as capsules of liverworts and mosses, sporangia of ferns, cones of gymnosperms, anthers and dry fruits of angiosperms have been described in Vol. I. The movement is generally one to open the spore or seed container. Associated with it, there may be similar movements of specialized dead cells such as the elaters of liverworts and peristome teeth of mosses, to ensure the most favourable dissemination of the capsule contents. The cohesive properties of water may be used in conjunction with the shrinkage properties of the dead cells as exemplified by the annulus of the fern and special cells of the anther wall.

(*b*) *Turgor Movements*. These are movements due to variation in turgor of living cells and are similarly often associated with spore and seed dispersal. Excellent instances are afforded by the squirting cucumber (*Ecballium*), and balsam (*Impatiens noli-me-tangere*).

Movements of Irritability

These may be summarized as *tactic*, *tropic* and *nastic*.

Tactic Movements. A tactic movement is one made by a whole organism or freely locomotive part of an organism, such as a gamete, in response to a stimulus; the direction of action or source of the stimulus has a direct bearing on the direction of the movement elicited. Such movements are of a locomotive nature and are effected by organelles such as cilia or flagella, by protoplasmic streaming or by extrusion of cell substances.

Taxes are generally described according to the nature of the stimulus which evokes the response, by the addition of a prefix to the word "taxis."

Phototaxis: response to variation in light intensity and direction.

Chemotaxis: response to variation in concentration of chemical substances.

Aerotaxis: response to variation in concentration of oxygen (special chemotaxis).

Rheotaxis: response to variation in direction of flow of liquids.

Osmotaxis: response to variation in osmotic conditions.

Phototaxis can best be studied by using cultures of free-swimming algae such as *Chlamydomonas*, *Volvox*, or the flagellate *Euglena*. A

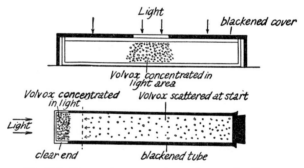

Fig. 16.1. Phototactic responses by motile green flagellates.

culture of such organisms on a window-sill, usually shows a gathering of the organisms on the illuminated side of the dish. If the light is very strong, the aggregation may be on the darker side of the dish, the organisms having moved away from the very high light intensity. Previous treatment of the organisms may affect what they will do under particular conditions. If the organisms have been cultured in a very high light intensity, they will search out a higher light intensity than corresponding organisms cultured in weak light. Thus there is no absolute intensity of illumination to which such organisms respond. The phototactic response may be induced by either a directed beam of light or by variation in light intensity. In some cases, the organisms will swim along a light beam either towards or away from a light source, or they may merely concentrate in a well-lit area (*see* Fig. 16.1).

Observations on different organisms have indicated various types of reaction. The purple sulphur bacterium, *Chromatium*, does not appear to be able to move in the direction from light to darkness. If a cell is swimming in a darkened area on a slide and enters a light patch, it shows no reaction at all but, on passing back into the darkened portion, it suddenly checks, reverses direction and swims back into the

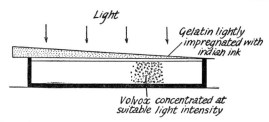

Fig. 16.2. *Volvox* plants congregating in a suitably lit position under a light screen.

light. It thus appears to be a negative reaction to darkness, or in other words a change from a high to a low light intensity which causes the response (*see* phobotaxis, p. 572).

Members of a *Volvox* culture, if subjected to the conditions shown in Fig. 16.2, will swim about until a suitable intensity is found and there

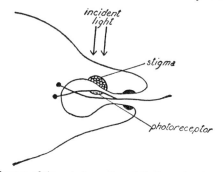

Fig. 16.3. Diagram of the anterior region of *Euglena*, showing the position in which the photoreceptor is completely shaded by the stigma.

they will remain without further movement. This seems to indicate that the organisms only make phototactic movements if removed from the most suitable light intensity.

A euglena will respond to a directional stimulus by swimming towards a light source. If the position of the source of light is changed, the euglena will change the direction of its movement. An individual swimming along a light beam proceeds in a straight line but with its anterior end rotating about its axis. If the specimen is then darkened, its immediate response is to stop swimming forward and to rotate its forward end in ever-increasing sweeps. If it succeeds in detecting a new light source, it promptly swims in this direction. In many cases, the light-sensitive part of the organism has been located as the red eye-spot. Phototactic responses are greatest to light of 400 nm to 500 nm wavelength.

Chemotactic response is probably shown by the great majority of free-swimming organisms, but not many have been studied in any detail. Most of the work has concerned the chemotactic movements of

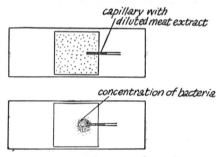

Fig. 16.4. Chemotactic response by motile bacteria.

antherozoids. It is not difficult to show experimentally that organisms will congregate in definite concentrations of particular chemical substances, having reached the desired concentration by swimming along the concentration gradient. Capillary tubes containing given concentrations of materials under test are slipped under cover slips covering the organisms (*see* Fig. 16.4). If the substance is acting as a stimulant, the organisms will quickly concentrate around the open end. Many motile putrefactive bacteria will concentrate in meat-extract juices. As a result of such tests on plant antherozoids, it seems that each is responding to a particular concentration of one of a variety of organic compounds. Some ferns, *Selaginella* and *Equisetum* species show positive reaction to malic acid, *Lycopodium* species to citric acid, mosses to sucrose and liverworts to proteins. It seems certain that these substances are secreted by the respective archegonia. Sometimes there are exceedingly fine powers of discrimination between chemically similar substances. For example, malic, maleic and fumaric acids are closely similar, yet antherozoids can discriminate without fail. In the case of malic acid, it is known that it is the malic ion and not the whole molecule which is the stimulant. The concentration of the substances acting as stimuli is often incredibly low. *Bacillus termo* will react to 0·001 per cent peptone, 0·0018 per cent K_3PO_4 and 0·1 per cent glucose.

Aerotaxis is largely confined to the aerobic bacteria, which will swim from a low oxygen concentration to a higher one.

Rheotaxis is characteristic of some naked plasmodia of the slime fungi (Myxomycophyta). If a plasmodium is placed at the foot of a strip of wet filter paper which passes upwards and dips into a dish of water, the organism will creep slowly up the paper against the current, which is siphoning downwards. On reaching the lip of the dish, it will continue over the edge downward against the upward current.

Osmotaxis depends on the tonicity of the solution with respect to the protoplasm of the organism. Very dilute solutions usually repel; strong solutions may either attract or repel, according to the organism. Most organisms appear to favour conditions of isotonicity.

Tropic Movements (Tropisms). A tropic movement is one made by a fixed part of a fixed plant, the direction of the movement being determined by the direction from which the stimulus originates. For a tropic response to occur, the stimulus must be acting from a particular direction, that is to say, the stimulus must be from one point only or be greater on one side of the stimulated part than any other. Tropic movements are the result of growth curvatures and are therefore only executed by such parts of the plant as are capable of making them. A growth curvature is due to an increase or decrease in the amount of growth made by the side of the organ nearest the stimulus with respect to the opposite side. We use terms to describe the position taken up by an organ with respect to the direction of the stimulus when a movement has been completed. A *positive* tropic curvature is said to occur when the organ grows more rapidly on the side furthest from the stimulus, so that the organ bends itself towards the stimulus. The opposite occurrence is referred to as a *negative* curvature. When an organ, as a result of either negative or positive curvature, sets itself in the line of action of the stimulus, it is said to be *orthotropic*. If the organ sets itself across or at an angle to the line of stimulus, it is said to be *plagiotropic*. A special case of this in which an organ such as a rhizome sets itself strictly at right angles to the vertical may be described as *diageotropic*. Tropisms are also described by the addition of a prefix according to the nature of the stimulus.

Geotropism: response to gravitational forces.
Phototropism: response to unidirectional light.
Chemotropism: response to a concentration of chemical substance.
Hydrotropism: response to a water source.
Haptotropism (Thigmotropism): response to contact.
Traumotropism: response to wounding.
Thermotropism: response to heat source.

Galvanotropism: response to electric current.
Rheotropism: response to water current.

Tropic responses are the result of the ability of the plant to perceive a stimulus and to respond. Since the regions of perception and response are almost invariably separated by a distance, there must be some form of conduction within the plant. A knowledge of the special characteristics of the plant hormones (auxins) has aided in elucidating the complex nature of some tropic movements. In fact, much of our knowledge of the auxins has come from experimental work on tropisms.

GEOTROPISM. Many plant parts show the ability to make geotropic responses. Below, the usual types of geotropic response made by various plant parts are summarized.

Plant part		Response
Higher plants	Primary roots	Positively orthogeotropic
	Secondary lateral roots	Plagiogeotropic but directed downwards
	Tertiary and lesser roots	No response or ageotropic
	Main stem	Negatively orthogeotropic
	Lateral branches	Plagiogeotropic but directed upwards
	Leaves	Plagiogeotropic but at right angles to the gravitational pull
Fungi	Stalks of fructifications	Negatively orthogeotropic
	Vegetative hyphae	No response or ageotropic
	Gills of some Basidiomycete fructifications	Positively orthogeotropic
	Liverwort sporogonia setae	Negatively orthogeotropic
	Moss stems	Negatively orthogeotropic
	Algal holdfasts	Sometimes positively orthogeotropic

These responses have obvious biological advantages in allowing the roots to tap the greatest possible volume of soil and the shoot system to obtain adequate light and air. In the fungi, the advantage is to aid in spore dispersal so that the maximum number of spores are given a free fall under gravity when set free of the fructification by some dispersal mechanism. There are, of course, exceptions to the general rules outlined above and organs may vary in their responses during their different developmental stages. For example, flower buds of the poppy are subtended on drooping stems which gradually become erect as flowering and fruiting follow. This is said to be due to changing geotropic response. Primary stems of several woodland plants such as moschatel, wood anemone, herb paris are at first negatively geotropic, then later positively geotropic, so that they re-enter the soil layer quickly and then become diageotropic, to grow horizontally under the soil surface as rhizomes. Injury or the removal of parts may change the type of response made by a lateral organ. For example, the removal of the apical sections of some stems results in the change from plagio- to negatively orthogeotropic responses in nearby lateral branches. Changes in external conditions such as exposure to near freezing tempera- tures brings about a change in response in some stems such as dead- nettle from negatively orthogeotropic to plagiotropic. Changes in light conditions may also cause responses to vary from those normally ex- hibited. Under varying conditions of exposure to particular light wavelengths and intensities, it has been shown that the geotropic response by *Avena* coleoptiles varies also. For instance, red light of 661 nm wavelength and 7×10^{-4} joule per square centimetre intensity promotes the response whilst blue light at 479 nm of about half that energy content suppresses it by altering the growth rates of the two sides of the coleoptile.

A tremendous amount of experimental work has been done in attemp- ting to explain geotropic responses. Much of the earlier work succeeded only in establishing a few of the facts. A brief résumé of some of the more important experiments is given here. Roots were first shown to be sensitive to gravitational forces by Knight in 1809. He did so by

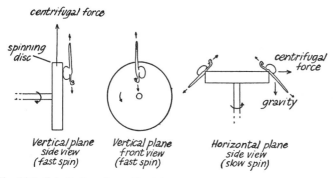

Fig. 16.5. Substitution of centrifugal force for force of gravity and its effect on growing bean radicles.

substituting for the force of gravity, a centrifugal force which he caused to be effective on seedlings by spinning them on rotating wheels. His results are depicted in Fig. 16.5.

By his experiments, Knight at least eliminated the suggestion that roots grew downwards under their own weight. Technique of the same kind can be used to study the effects of changes in direction and strengths of the forces applied.

The modern clinostat is an instrument on which similar results can be achieved by causing the root to rotate about its own axis in any selected plane, so that the force of gravity is never applied continuously to any side of the root, with the result that the root continues to elongate without bending (*see* Fig. 16.6). Sachs first used the instrument in

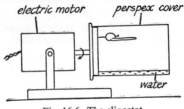

Fig. 16.6. The clinostat.

1879. By creating conditions of an ineffective all-directional gravity, the clinostat makes it possible to study the effects of exposure of an organ to a unilateral gravitational or centrifugal force over a given time.

It can be shown that decapitated roots cannot respond to gravitational stimuli, *see* Fig. 16.7, and when the apices of coleoptiles are removed,

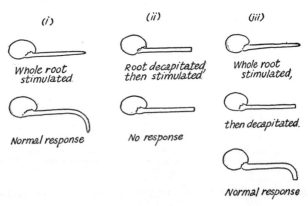

Fig. 16.7. The effect of decapitation on geotropic response given by bean radicles.

the lower regions lose the power to respond to gravity also. For som roots, e.g. maize, it has been claimed that when the root cap is dissecte away without damage to the apical meristem they will continue t elongate but will not curve downwards when placed horizontally. Th indicates that perception could be located in the root cap in such case

Czapek, using an ingenious device of glass covers for the root tips, was able to show that as long as the root tip was out of the vertical, response would occur no matter where the rest of the root was (*see* Fig. 16.8).

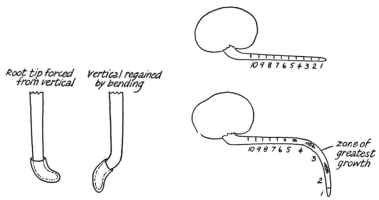

Root tip forced from vertical   Vertical regained by bending

zone of greatest growth

Fig. 16.8. Czapek's glass covers.

Fig. 16.9. Bean radicle marked to show zone of greatest response to gravity.

Sachs, in 1874, first showed that the region of response is always in the most actively growing zone of the root, a millimetre or two behind the tip. This was demonstrated by marking of equal spaces along the root from the tip backwards, allowing a geotropic curvature to occur, and studying the spacing of the marks in the region of the bend (*see* Fig. 16.9). The response can also be shown to be one of growth, since in conditions which will retard growth or even prevent it, such as lack of oxygen or too low a temperature, a root will not bend however much it is stimulated. Further, microscopic examination of the cells in the region of the bend, shows that those on the upper side elongate to a much greater extent than those on the lower side. The bending, therefore, is undoubtedly due to differential growth in size of the cells on the upper and lower sides. Generally a minimum stimulation is required to evoke a response in an organ. This can be expressed as the *presentation time* for a given force applied. Below this level, no response is elicited. When a stimulus above this "threshold" value has been received, there is a time lag before any perceptible curvature results. This period has been called the *reaction time* or *latent time*. The bending response is, therefore, the result of a series of events forming a reaction chain in which one event triggers the next.

It is only since Boysen-Jensen's discovery of auxins and F. W. Went's experiments on the effect of auxins in bringing about phototropic responses that any reasonable explanation of the curvature phenomenon has been possible. Went's work on auxins in oat coleoptiles is described more fully in the next sub-section, but from a knowledge of the effect of

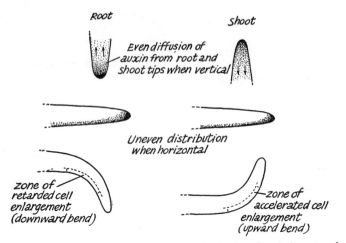

Fig. 16.10. Diagram to illustrate the uneven distribution of auxin in root and shoot tips when horizontal, and the growth curvatures resulting therefrom.

auxin on cell enlargement, which has already been explained, it will be understood that different concentrations of auxin on opposite sides of the root would have the effect of causing a bend in the root. In the case of oat coleoptiles, Went was able to show that where the auxin was in greatest concentration, there cell enlargement was greatest. But experiment has shown that when a root is stimulated by being placed horizontally, the greatest concentration of auxin occurs on the under side. If the auxin acted in a root in the same way as it does in a coleoptile, this should cause an upward bend. On the contrary, it causes a downward bend. Cholodny offered a suitable explanation based on the two following conditions: (1) that there is a greater accumulation of auxin on the underside of a root due to a lateral displacement whilst it is diffusing through the tissues from the tip backwards, and (2) that stems and roots are different in the manner in which they respond to given concentrations of auxin; in the shoot, high auxin concentration accelerates cell enlargement, in the root high auxin concentration retards cell enlargement (*see* Fig. 16.10). Dolk's work on oat coleoptiles largely confirms these suggestions. He has been able to prove that there is a redistribution of auxin in the tip of an oat coleoptile when the tip is placed in a horizontal position, and that the higher auxin concentration in the lower portion, whilst of a concentration sufficient to cause acceleration of cell enlargement in the coleoptile, would cause retardation of cell enlargement in a root. His methods followed those of Went and need not be described here. Whilst these suggestions may account for the bending of roots and shoots, there is so far no satisfactory explanation of the lateral displacement of the auxin when diffusing along a horizontal root or shoot.

Similarly, whilst differential auxin distribution may be cited as the cause of the reaction effect, it does not give any indication of the nature of the perception of the stimulation. Theories regarding the registration of perception have been quite numerous, ranging from strains set up in a displaced organ, movement through the cytoplasm of loose particles such as *statoliths* of starch or microsomal bodies, to a geo-electric effect allowing the polarized movement of auxin so that it always concentrates on the lower side of an organ. Only the starch statolith (amyloplasts) hypothesis seems to have any real basis for acceptance, for there is a strong correlation between the appearance of statoliths in the cells and the sites of sensitivity in organs. Sedimentation rates of starch grains or amyloplasts are commensurate with measured presentation time. But again the issue is clouded by the fact that when all the statolith starch in a root has been physiologically removed, it may still show geotropic curvature. It is also the case that cold-treated *Sphagnum sp.*, not showing any starch grains in its cells, will show geotropic responses. In the absence of much more concrete information it has to be recognized that there is as yet no fully acceptable explanation of all the facets of geotropic response. It should be noted that some of what has been observed under this phenomenon is not unlikely to be a combination of epi- or hyponastic curvatures, see p. 620, with geotropic responses.

PHOTOTROPISM. The types of phototropic response made by various plant parts are summarized below.

Plant part	Response
Higher plants { Roots	In general, no response or aphototropic
Stems	Positively orthophototropic
Leaves	Plagiophototropic, at right angles to the incident light
Fungal fructifications	Positively orthophototropic
Bryophyte and pteridophyte rhizoids	Negatively orthophototropic
Algal holdfasts	Negatively orthophototropic

These responses also have obvious biological advantages. The photosynthetic areas will be exposed to the maximum light. Leaf mosaics of many plants show that this is the case. In roots, the lack of response is of no significance, since they would be growing in darkness normally. Roots of such climbers as ivy are negatively orthophototropic and will therefore tend to grow inwards towards the supporting structure, with obvious advantage. When gravitational and light stimuli are antagonistic to one another, the light usually proves the stronger, although in many cases the final situation of leaves and stems may be due in part to both responses. In organs that show dorsiventrality, such as leaves, the final positioning of a blade may also be partly due to epinastic or hyponastic curvature, *see* p. 620. As with geotropism, some plants show interesting changes in response during their developmental stages. For example, the ivy-leaved toad-flax, *Cymbalaria muralis*, produces flowers on stalks which are at first positively phototropic. Immediately after fertilization, however, they become negatively phototropic and carry the ripening capsules back towards the darkness of the support on which it is growing, thus giving the seeds a chance to take root in a suitable position.

As with geotropic response in roots, phototropic responses in shoots have been the centre of a great deal of experimental work. There is no history as to who first demonstrated the fact that shoots respond to light stimuli; it seems to have been known from time immemorial. To show that it is the tip of the shoot which is sensitive to the stimulation, Darwin used grass seedlings and demonstrated, among other findings, that decapitated seedlings or those with their tips covered with black paper caps could not normally perceive the light stimulus and consequently did not bend appreciably towards a laterally placed source of light (*see* Fig. 16.11). In fact, when the top 5 mm of an oat

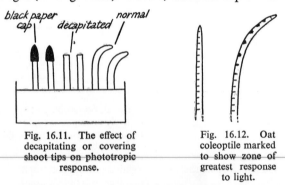

Fig. 16.11. The effect of decapitating or covering shoot tips on phototropic response.

Fig. 16.12. Oat coleoptile marked to show zone of greatest response to light.

coleoptile are covered and it is then stimulated from the side by light, it will show some curvature but very much less than if the tip were clear. The lower part of the coleoptile is thus sensitive to some degree and this indicates the presence of a photoreceptor substance. The best attempts to localize the most photosensitive regions of the oat coleoptile have shown it to be in a region about 0·1 mm behind the apex, although there is some sensitivity in the whole length to the base. This means that if there is a photoreceptor substance present in the coleoptile, then it is distributed along its whole length but not necessarily evenly. It is the case that the yellow pigments, either carotenoids or flavins (FMN), that might be regarded as the most likely photoreceptor substances, are to be found in higher concentration near the apex than elsewhere. There appears to be a real correlation between the distribution of plastids and light responsiveness in the coleoptiles. Plastids, pale yellow in etiolated material, greenish in coleoptiles that have been exposed to red light, are to be found in the cells immediately surrounding the vascular strands. The plastids are absent from the extreme apex, most concentrated just behind it, and fall off in numbers from here towards the base. The pigments appear to be carotenoids and protochlorophyll. Such findings point to the possibility that there is some photoreceptor mechanism at work through the agency of special pigments. The fact that shoots with apices covered or even decapitated shoots are capable of a little response is understandable in the light of such observations.

By marking the upper portion of the shoot with equally spaced marks, it was also possible to show that the region of response was in the growing region, a few millimetres behind the tip (*see* Fig. 16.12). More careful studies of internal growth changes verify this. The nature of the response has been shown conclusively to be one of a growth curvature similar to geotropic growth curvatures. The cells in the growing region of the stem or coleoptile farthest from the source of light enlarge more rapidly and to a greater extent than normal, whilst in those nearest the light source, the reverse is the case with a consequent bend towards the light. If conditions are not favourable for growth, there can be no response. There must be oxygen available and the temperature must be suitable.

There have been a number of explanations put forward to explain the curvature phenomenon, but only that of Went and Cholodny, concerning the role of auxins, need be mentioned. Went's work was done on the responses by oat coleoptiles, and he was first able to show that it is the tip of the coleoptile which is rich in auxin, and that preceeding farther and farther downwards, the auxin concentration grows less and less (*see* Fig. 16.13). He was able to do this by perfecting a method of extracting the auxin from portions of the coleoptile into agar blocks, by allowing the auxin to diffuse out of the coleoptile section into the agar. The agar blocks, thus charged with auxin, when placed on the cut stumps

of other coleoptiles, acted as normal tips and curvatures could be induced in the stumps as though they had never been decapitated.
Further, if these charged agar blocks were placed eccentrically on the

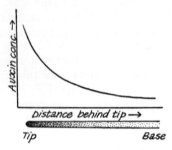

Fig. 16.13. Graph of auxin concentration against distance from coleoptile tip.

cut stump, curvature of the stump was induced in the absence of
stimulation (*see* Fig. 16.14). Went was able to show that the amount of
curvature induced by a charged agar block was directly proportional to
the number of coleoptile tips which had discharged their auxin into it.

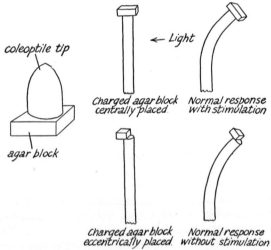

Fig. 16.14. Collection of auxin in an agar block and the effect of placing the
charged block on the stump of a coleoptile.

Therefore he was able to base his estimate of the quantity of auxin in
any agar block on the amount of curvature the eccentrically placed
block would induce in a given time. As a result of a great deal of such
quantitative work, a unit quantity of auxin was defined. This was

known as the *Avena-Einheit*. It was the quantity of auxin contained in a block of agar 2 mm × 2 mm × 0·5 mm, i.e. 2 mm³ which would produce a 10° curvature in 2 hours at a temperature of 22–23°C in 95 per cent humidity.

Having proved that the source of the auxin was in the tip of the coleoptile and that it diffused from the tip downwards, so that its

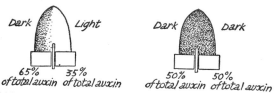

Fig. 16.15. The effect of unilateral light on the distribution of auxin in a coleoptile tip.

concentration decreased with distance from the tip, Went next tried to determine the effect of unilateral light on the distribution of auxin in the tips of the coleoptiles. Using tips which had been illuminated strongly from one side only, and control tips which had not been illuminated at all, he allowed the auxin in the illuminated side to diffuse out separately from that in the darkened side, by inserting a mica slip into the coleoptile to separate the two halves (*see* Fig. 16.15).

He compared the amounts of auxin in the two agar blocks under a tip previously illuminated from one side only, with the amounts in two agar blocks under a tip which had not previously been illuminated at all, by seeing how much curvature they would induce in a decapitated coleoptile. He proved indisputably that the effect of the unilateral light was to cause a displacement of auxin from the lighter to the darker side, as it diffused down the coleoptile. (*Note*: his actual results showed also that the unilateral illumination had destroyed some auxin since the total amount of auxin received from the unilaterally illuminated tips was only about 84 per cent of the total auxin produced in the darkened tips.) Went's explanation of the uneven distribution was that the effect of the unilateral light was to polarize the auxin laterally across the coleoptile. The alternative possibility that auxin is photolysed in any appreciable amount on the illuminated side has been shown not to be the case, but so far there is no accepted explanation of the asymmetrical distribution of auxin under the influence of unilateral illumination.

In view of the known effects of auxin on cell enlargement, it can now be readily appreciated that the excess auxin diffusing down the darkened side of a laterally illuminated coleoptile, will cause greater growth on that side and a consequent bending of the coleoptile towards the light (*see* Fig. 16.16). An acceptable explanation of phototropic response, at least in oat coleoptiles, embodying all the reasonably well authenticated and confirmed observations, would be as follows. A yellow pigment, probably a carotenoid located in plastids, acts as a photoreceptor substance and, upon the absorption of light, triggers off an unknown (as yet) sequence of events that affect the permeability properties of the cells containing the pigment. As a result, the transport of auxin in the apex to base direction is affected. The asymmetrical distribution of auxin brings about a curvature through its effect on cell enlargement, *see* p. 270.

OTHER TROPIC RESPONSES. None of the other tropic responses known to occur in plants appear to have been as deeply studied as geotropism and phototropism. Consequently, little more than the fact that they do occur, is known about them.

Some plant parts are undoubtedly influenced in their direction of growth by the presence of chemical substances. Such chemotropic

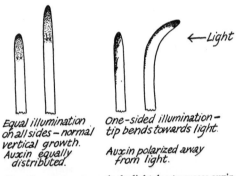

Equal illumination on all sides – normal vertical growth. Auxin equally distributed.

One-sided illumination – tip bends towards light. Auxin polarized away from light.

←Light

Fig. 16.16. Growth curvature towards the light due to excess auxin on darkened side.

responses may be positive or negative. Pollen grains allowed to germinate on gelatine show that the pollen tubes are positively chemotropic to portions of the appropriate stigmatic tissue buried in the gelatin. The pollen tubes of *Scilla* have been shown to react to the presence of sugars, particularly sucrose. The same tubes show negative responses to inorganic salts such as phosphates and ammonium compounds. Some roots have been shown to be negatively chemotropic to strong concentrations of substances in the soil. A whole root will turn aside from a block of copper placed in its path, whilst a decapitated root is unable to make such a response. In general, roots are said to respond positively to the presence of salts, particularly salts of calcium and negatively to alkaline or acid concentrations. Fungal hyphae exhibit positive reactions to concentrations of food materials. Phosphates, proteins, peptones, amino-acids and sugars often elicit positive responses from the hyphae of moulds. They show equally strong negative responses to concentrations of acids and alkalis.

Hydrotropism can be considered as a particular case of chemotropism. Some plant roots are able to respond to differences in water content of adjacent soil masses and show positive orthohydrotropism by growing towards the wetter region. This can conveniently be demonstrated by germinating peas on saturated porous porcelain blocks as shown in Fig. 16.17. Such experiments show that the response is a growth curvature and that the water stimulus is even greater than that of gravity. Perception is localized in the root tip, but the stimulus must be strong for it to occur at all. Some fungal hyphae appear to be negatively hydrotropic and will grow away from moist surfaces into the atmosphere. The exact causative factors in any case are unknown.

Aerotropism may likewise be regarded as a particular case of chemotropism in which the organ is responding to the presence of oxygen.

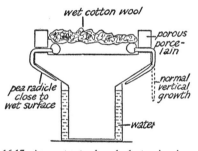

Fig. 16.17. Apparatus to show hydrotropism in roots.

Roots are often positively aerotropic, growing away from anaerobic conditions. Pollen tubes have been stated to be negatively aerotropic, but the evidence is not conclusive.

Haptotropism or thigmotropism is exhibited by many plant parts, but is particularly a characteristic of tendrils and other organs by which the plant secures support. The tendril of the marrow plant has been the subject of much investigation. The young tendril is coiled in a spiral with the lower side outwards. An increased growth rate on the upper side results in the straightening of this, until the tendril stands out stiffly from the main stem. Growth in length then occurs at the base, whilst the tip starts to execute sweeping circles in the vertical plane (see Fig. 16.18). If the tendril is not stimulated by contact, it coils in the opposite direction to that when miniature and soon withers. If, however, the tendril contacts a hard surface, and it seems that the morphologically lower surface is much more sensitive than the upper, it will attempt to twine itself around the support offered, and if this is of reasonable dimensions, it will twine through several coils until adequate support is achieved. The stimulus must be one of contact with a solid body. A jet of water, directed on the tendril, elicits no response. A glass rod, covered with gelatine, has no effect either. Thus it appears that the tendril cannot respond to a watery system. When the first reaction is completed, a secondary reaction occurs in the

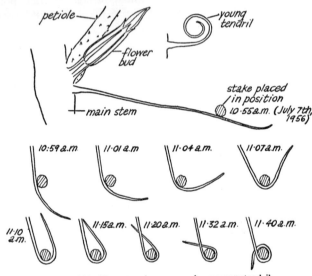

Fig. 16.18. Haptotropic response by marrow tendril.

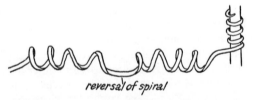

Fig. 16.19. Marrow tendril showing completed response.

basal part of the tendril. This coils into a corkscrew and draws the main stem nearer to the supporting structure (*see* Fig. 16.19). Finally the tendril undergoes great thickening in its lower regions. In order to cause the complete cycle of events, the stimulus must be sufficiently strong. With small doses of stimulation, i.e. stroking for short periods a bending of the tendril may commence, but straightening-out will occur later. For the secondary reaction to occur, continuous contact must have been obtained by the tip.

The bending is certainly a growth reaction and the cells on the outward side of a bend are elongating much more rapidly and to a greater extent than those on the inside. The nature of the contact stimulus is not understood, but since a liquid causes no reaction, it may be that unequal pressures on the cell surfaces by a solid object set up deformation of the protoplasm.

Haptotropism is not restricted to tendrils only. Some young growing shoots, particularly etiolated ones, may show similar responses. Neither is the response always one of bending. The Virginian creeper responds to contact at its tendril tips by the development of sucker-like discs.

Traumotropism has been demonstrated in several plant parts. Roots have been seen to execute curvatures as a result of damage on one side,

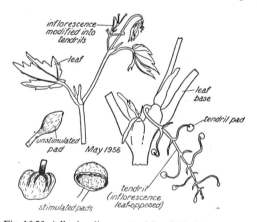

Fig. 16.20 Adhesive discs on tendrils of Virginian creeper.

the curvature always being away from the injured side. In some seedlings, the amputation of a cotyledon may cause a bending of the hypocotyl or epicotyl. Removal of part of a leaf lamina may induce curvature in the petiole. Complete removal of leaves may lead to curvatures in stems. Such phenomena have never been precisely explained.

Thermotropic responses have been recorded in several instances. A difference in temperature on opposite sides of an organ, may constitute a directional stimulus and the organ bends towards the warmer side. This phenomenon has been used to explain the movement of certain flowers such as anemone and tulip, in following the sun. Others have explained such phenomena as being particular cases of phototropism in which the organ is responding to the longer infra-red wavelengths of the spectrum.

The occurrence of galvanotropism has been described by some investigators who have claimed to show curvature by some roots in response to a current of electricity. It is not unlikely that such curvature is in response to the movement of mineral nutrients under the influence of the current and consequently a chemotropic response.

Rheotropism has also been observed as a response by young roots to a water current in which the root tips come to be directed against the current. It is said that a current of distilled water will not cause the reaction but that a solution of nutrient salts will, so that it may be in reality a chemotropic response.

Nastic Movements (Nasties). A nastic movement is one made by a part of a fixed plant in response to a non-directional or diffuse stimulus and such being the case, the direction of the movement is always related to the nature of the responding organ and not to the stimulus, cf. tropism. The movements are varied in nature and they may be the result of growth curvatures or of sudden changes in turgor pressure of some cells. The former are then called growth movements and the latter variation movements. Nasties may be described according to the stimulus evoking them.

Nyctinasty: response to changing day and night conditions. These are usually light intensity and temperature changes.

Photonasty: response to change in light intensity.

Thermonasty: response to change in temperature.

Haptonasty: response to contact.

Seismonasty: response to shock.

Hydronasty: response to humidity changes.

Chemonasty: response to presence of specific chemical substances.

The terms *epinasty* and *hyponasty* refer to the tendency for a dorsiventral structure such as a leaf or petal to curve away from (epinasty) or towards (hyponasty) the axis as a result of differential growth rates at the upper and lower faces. When the adaxial side grows faster than the abaxial, the curvature is said to be epinastic; when the reverse is the case, it is hyponastic. There is some evidence that these curvatures are connected with differential distribution of IAA in the organ concerned but how this is influenced is not known. Such nastic curvatures can be compounded with orthogeotropic responses to produce plagiogeotropic positioning. The oblique setting of a lateral branch may well be the resultant of an epinastic curvature and a negative geotropism. The opening and closing of some flowers due to alternate epinastic and hyponastic curvatures seems to be controlled by a rhythm from within, e.g. the night-flowering cactus, *Cereus grandiflorus*, and evening primrose, *Oenothera biennis*. For reference to endogenous rhythms, *see* Chap. 6.

NYCTINASTIC MOVEMENTS. These are related to the rhythmic changes in day and night conditions. The most noticeable are those performed by flowers, leaves or leaflets, and are sometimes referred to as "sleep-movements." It can be shown that the response is almost always due to changes in atmospheric temperature or light intensity or both. As night falls, both temperature and light intensity become less and the flowers or leaves make their characteristic movements. Nyctinastic movements are therefore thermonasties, photonasties or combinations of both.

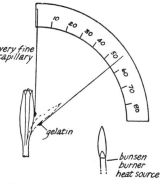

Fig. 16.21. Movement of crocus petals in response to changing temperature.

The opening and closing of crocus flowers affords a good example of such movements resulting from growth differences between the upper and lower surfaces of the petals. This can be demonstrated very clearly, using the apparatus shown in Fig. 16.21. As the temperature of the surrounding atmosphere or the light intensity increases, the upper surface at the base of the petal grows faster than the lower surface and the flower consequently opens. A reversal of the growth rates of upper and lower surfaces causes the closure as temperature or light intensity decrease. In the tulip, there is evidence that the cells at the inner and outer surfaces near the base of the petals have different optimum temperatures for growth, the outer cells growing faster than the inner at lower temperatures and vice versa at higher. There are numerous other examples of sleep movements in flowers and inflorescences, but the manner in which the stimulus is detected is not known.

Similar movements in leaves and leaflets are due to turgor pressure changes in the cells of special structures at the bases of the leaves or leaflets. These structures are known as *pulvini* and are thickened cortical regions made up of soft parenchyma with large intercellular spaces. When the cells of the pulvinus are all fully turgid, the whole structure serves to keep the main axis of the leaf or leaflet rigid and more or less erect. When all are flaccid, the leaf or leaflet tends to droop under its own weight. Curvature of the pulvinus can occur if there is differential turgor on its opposite sides. Sometimes changes in length of the surface of a pulvinus are increased by folding of the surface tissues. Under the stimulation of a lowering of temperature or light intensity, the turgid pulvinus becomes wholly or partly flaccid by losing water to the intercellular spaces and the leaf or leaflet executes a movement accordingly (*see* Fig. 16.22). Leaflets of many of the Leguminosae such as clover, acacia, mimosa and *Phaseolus* show such movements, as do the wood sorrel (*Oxalis acetosella*) and *Desmodium gyrans*, the telegraph plant. Nyctinastic movements presumably have some biological significance in that the closing of flowers by night would tend to protect delicate inner parts, whilst the folding of leaves or leaflets may prevent excess cooling or protect against excess water loss. Some flowers, such as the night-scented stock, open only at night. They are normally pollinated by nocturnal moths.

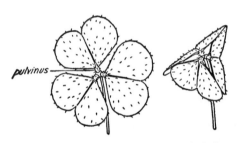

Fig. 16.22. "Sleep" movement of *Oxalis* leaf.

It is of interest to note that many of the nyctinastic movements are continued by the plant even when exposed to unchanging temperature and light intensity. There seems to be some inherent periodicity of the movements. There is no real understanding of the causes of the turgor changes that occur to bring about nyctinastic movements. Attempts to associate the movements with IAA and gibberellin distribution have not clearly established a connexion in most instances. Another suggestion that the movement may depend on the presence of phytochrome in the tissues requires fuller confirmation. Changes in permeability of the cell membranes appear to occur, leading to rapid in and out passage of electrolytes, but under what influence is not known.

HAPTONASTIC MOVEMENTS. Certain plant parts will execute nastic movements when stimulated by the right kind of contact. The most noticeable are those of floral parts operating in pollination mechanisms and the movements of leaves of insectivorous plants.

Typical examples of the former are the movements of stamen

filaments in response to contact stimuli. *Centaurea* filaments contract violently to pull the stamen tube quickly downwards, so exposing the stigma, and at the same time, the pollen previously shed inside the tube

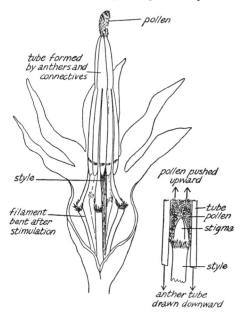

Fig. 16.23. Movement of *Centaurea* filaments to effect expulsion of pollen from anther tube.

(*see* Fig. 16.23). If the filaments of the barberry flower are touched, the stamens immediately spring inwards towards the middle from a previously spread-out position. This would tend to dust pollen on any insect causing the contact.

The most startling of the movements executed by insectivorous plants is that by the Venus fly-trap, *Dionaea* (*see* Fig. 16.24). When the trigger hairs are suitably stimulated by being touched, large "motor" cells along the midrib respond by losing water rapidly, and the trap is sprung, the leaf shutting with the spines interlocking.

SEISMONASTIC MOVEMENTS. These are movements in response to some shock, such as jarring or injury. The best known example is undoubtedly *Mimosa pudica*, the sensitive plant which not only executes ordinary sleep movements of its leaves by turgor changes in pulvini, but exhibits very rapid folding responses when properly stimulated. Petiole and leaf structure and a section through a pulvinus (tertiary) appear as in Fig. 16.25.

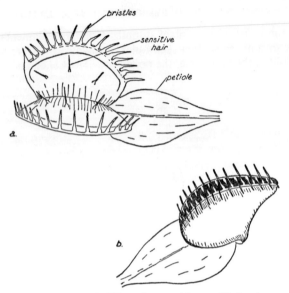

Fig. 16.24. Leaf of Venus fly-trap, (*a*) open, (*b*) closed.

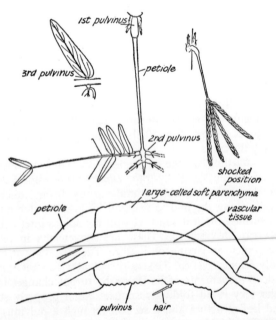

Fig. 16.25. Seismonastic movement and pulvini of *Mimosa pudica*.

The folding movements of the pinnules and the drooping movements of the leaflets and the whole leaf, are due to changes in turgor pressure on the upper and lower sides of a pulvinus. The lower side loses water quickly into the intercellular spaces and the whole structure supported on the pulvinus droops. The normal erect position is regained later, due to the gradual increase in turgor of the pulvinar cells. Stimulation repeated at too-frequent intervals, ceases to bring about a response. The stimulation may be conducted to all parts of the plant from a single source and the extent to which it is conducted, depends upon its intensity. The order in which the pulvini collapse, depends upon the location of the stimulus. When a terminal pinnule is injured, the response usually travels along the pinnule pulvini (1) to the leaflet pulvini (2) and thence down the petiole to the main pulvinus (3). From there, it may travel up and down the stem to the other leaves which respond in the order 3–2–1. If the stem is injured first, the sequence of response is 3–2–1 pulvini. The rate of the conduction of the stimulus is very great comparatively and can be as rapid as 1 to 2 m per minute. Much work has been done to try to establish the presence of a nervous system in *Mimosa pudica*, but the contrary is undoubtedly the case. However, the transmission of exitation is certainly accompanied by changes of electrical potential in the tissues through which it passes, somewhat comparable with nervous conduction in animals. If a water gap is interposed in the petiole, the whole leaf can still respond to injury, thus indicating that the conduction of the stimulus is due to the movement of a diffusible substance. A possible hormonal substance, not fully identified chemically, seems to be effective at concentrations as low as one part in five hundred million. It has been shown that this hormone, carried in the transpiration stream, can account for some of the features of seismonastic response in *Mimosa pudica*, but certainly not all of them, particularly speed of conduction of the stimulus in some cases. Such rapid conduction seems to be confined to living cells in the xylem although special tube cells associated with the phloem have been held responsible. The sensitive plant also responds to the change from day to night, by folding its leaves, but the mechanism of the nyctinastic movement seems different from that of the seismonastic.

HYDRONASTIC MOVEMENTS. Some flowers, for example the dandelion, are said to open and close with changes in humidity. Dandelion inflorescences will close in very moist conditions. Some nyctinastic movements have also been associated with changes in atmospheric humidity as night follows day.

CHEMONASTIC MOVEMENTS. These are executed in response to the presence of chemical substances. Among those most fully investigated are the movements of the tentacles of the sundews (*Drosera*). Charles Darwin investigated these particular movements in great detail. He found that the placing of a drop of liquid, containing a nitrogenous substance such as ammonium phosphate, urine, urea, egg albumin or meat, on the leaf, resulted in a curvature of all the tentacles towards the middle of the leaf. Solutions of sugary substances or starch or oily substances had no effect. It has been shown since that the movements are due to unequal growth on the inner and outer sides of the tentacles and there is need for only one tentacle to come into contact with the nitrogenous substance. The stimulus, when strong enough, is transmitted eventually to all tentacles through the leaf, possibly by some hormone, but the precise nature of this is not known. There is good reason to suppose that some of the movements executed by the tentacles of the sundew are haptonastic movements, since brief contact of a solid body with the tip of a tentacle may result in its curvature towards the middle of the leaf.

Plant Movements in Response to Internal Stimuli—Autonomic Movements

Both freely motile plants such as algae, or plant parts such as spores and gametes, and fixed parts of fixed plants such as stems, may execute autonomic movements.

The freely motile structures make locomotive movements more or less continuously, even when all external conditions are maintained constant and evenly applied, so that the movement cannot be associated with external stimuli. It is as though the organism possesses some inherent ability, controlled from within, to make its cilia or flagella execute their beating movements slowly or quickly or not at all. What the controlling factors are, or how they operate, are unknown. The same applies to streaming movements of protoplasm, considered by some to be a feature of most living plant cells. Under varying external conditions, the movements can be made to proceed more slowly or more quickly or be completely inhibited, but beyond a little knowledge of the effects of some external conditions, the phenomenon is not at all understood. The movements of chromosomes are equally difficult of explanation.

Of the autonomic movements of fixed plant parts, the *nutation* is the best known example. When a shoot elongates, its apex does not move in a straight line, but follows the path of a helix. It is said to *circumnutate* and the circumnutation can be demonstrated using the apparatus as shown in Fig. 16.26 with a convolvulus plant. The rotation of the tip is due to the fact that one part of the apical meristem is growing more rapidly than the rest, and this point of most rapid growth moves slowly around the apex so that there is always a slight curvature of the young stem successively in every plane. A similar circumnutation occurs in many young uncurved tendrils and is also to be observed in roots, flower stalks and the sporangiophores of some fungi. The nutating stems of many plants enable them to twine around suitable supports e.g. the bindweed, wistaria, etc. In these cases and in those of nutating tendrils, the phenomenon is very marked, and the young apex may sweep out a very wide spiral until a suitable support is found. It is interesting to note that in most species, the direction of the coiling is constant for the species.

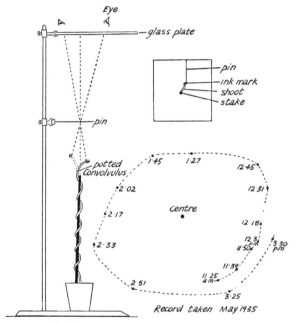

Fig. 16.26. To demonstrate circumnutation.

SENSITIVITY IN ANIMALS

As in plants, sensitivity in an animal becomes apparent when it makes responses to changing conditions. The responses may be summarized thus—

1. Movements of whole animals (locomotive responses).
2. Movements of parts.
3. Secretory activity.
4. Colour changes.
5. Changes in metabolism, growth and development.

The effector parts which bring about these responses are chiefly contractile tissues such as flagella and muscle and secretory tissues of the glands. Occasionally electric organs, light-producing organs and pigment cells are involved. The modes of action of these have been described elsewhere and need not be treated here, except in the case of pigment cells (chromatophores). In this section we shall deal almost exclusively with the nature and mode of functioning of the receptor parts of the animal.

The Special Senses of Animals

Most animals possess a wide variety of receptor cells, often localized in sense organs, by which they detect variations in conditions. Receptors which are known to be associated with specific stimuli may be summarized as chemoreceptors, photoreceptors, vibration receptors, pressure-change receptors, touch receptors, temperature-change receptors, equilibrium-change receptors and proprioceptors (stimulated by mechanical movements of the body).

The sensation of pain cannot be associated with special cells, nor is it localized in any part of the body. The five senses of higher animals, taste, smell, sight, sound and touch, embrace the activities of most of these types of receptors. If the receptor is located to receive stimulation from changes outside the body, it is called an *exteroceptor*. If the receptor is exposed to stimulation from within the body but outside the receptor itself, as for example in the gut, it is called an *enteroceptor*. Proprioceptors may be regarded as being in neither of these categories; being buried deeply in the muscle, tendon or joint tissues, they receive stimulation from changes in tissue tensions. Every receptor is linked by a series of nerve fibres with the central nervous system into which is passed an electrical impulse generated by the receptor or sensory cell as a result of the stimulation. From the central nervous system the impulse is relayed through other neurones to the effector parts and a suitable response to the stimulation may be induced.

Chemoreception

Animals are able to make use of their sensitivity to the chemical constituents of the environment in finding food, in placing themselves in the correct position for the reception of food, and in regulating their feeding habits by selecting the correct foods. The majority use the same chemosensitive mechanism for the avoidance of unfavourable chemical environments, and many use it as an important part of their reproductive behaviour, as in the attraction of male to female.

The sense organs concerned with chemoreception have not been fully identified in all groups of animals, but there seem to be three clearly distinguishable types at least. In man we may recognize the *olfactory* sense organs, the *gustatory* sense organs and the sense organs of a *general chemical activity*. In applying these terms to other organisms, there is likelihood of confusion, since it is difficult to distinguish between smell and taste organs in aquatic animals. In land animals, the smell organs are stimulated by air-borne substances, whilst the taste receptors are stimulated by substances in solution. In aquatic animals, both receptors are stimulated by substances in solution and there would appear to be no significant difference in their activities. An attempt to distinguish between the olfactory and gustatory senses on the grounds that the former acts as a receptor of distant stimuli and the latter as contact stimuli, also breaks down when it is found that the so-called taste buds on the surfaces of fish, such as the

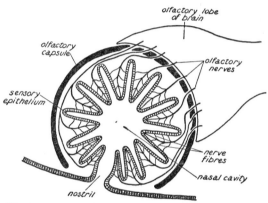

Fig. 16.27. V.S. through an olfactory organ of the dogfish.

catfish, are able to detect the presence of food at a distance from the body. Despite these difficulties, it is better to adhere to the terms in common usage and so avoid confusion.

There is no space to discuss chemoreception and consequent behaviour of the invertebrates in this section. References to chemoreption by these animals are made in the types described in Vol. I.

Chemoreception in the Vertebrates. There is a very high degree of structural and functional specialization of groups of cells to give the senses of smell and taste and the general chemical sense. In all vertebrates, it is probable that the two former are essential to survival, whilst the third may have survival value in that it may work in a way analogous to the pain sense, to serve as a warning under certain circumstances.

OLFACTORY SENSE. The cells associated with smell are located in special cavities of the anterior part of the head. In fishes these are the olfactory or nasal pits which in some cases have openings into the mouth. Jacobson's organ is also well developed in some fish and may be an accessory olfactory organ. In the amphibians the olfactory organs are similarly located and always open externally as well as internally into the mouth. In reptiles and birds the sense of smell is not well developed, the nostrils being narrow, encased in horn, and dry. Both depend to a very large extent on vision rather than on chemical sense. In mammals, the olfactory organs are located in the upper nasal cavities and cover large or small areas of the mucous membrane surface according to species. In man, the area is about $2 \cdot 5$ cm^2 per nostril. The sense of smell in mammals varies in acuity. Rodents have a poorly developed sense, whilst dogs exhibit exceptional powers, being able to detect minute traces of chemical substances. This is probably an adaptation to predatory activities.

Compared with other sense organs, the olfactory organ is simple in construction. Histologically, in mammals, the olfactory epithelium consists of neurones embedded in columnar epithelial cells, some of which contain a yellowish-brown pigment, giving the organ a yellow colour (see Fig. 16.28). The neurones are bi-polar with the dendrite

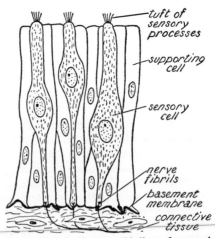

Fig. 16.28. V.S. olfactory epithelium of mammal.

forming the receptor proper and the cell body performing the function of relaying the impulse to the unmyelinated fibre which passes into the olfactory nerve. The olfactory epithelium also contains free nerve endings of the trigeminal nerve which function as part of the common chemical sense receptor. Stimulation probably occurs through the hair-like processes, and the sense cell, serving as receptor and conductor, transmits this to the brain. It is assumed that odorous substances act on the hair-like processes to initiate nerve impulses in the neurones, but the manner in which this is accomplished is unknown. It is not possible to correlate chemical structure of the substance with its odour, since isomers and stereoisomers often have quite different smells. One primary requisite of all odorous materials is that they must be volatile; they must also be readily adsorbed and they must be relatively fat-soluble. It may be that the ready adsorption of odorous substances, coupled with their high fat-solubility, facilitates concentration on, and rapid penetration into, the olfactory neurones.

It is generally recognized that the sense of smell is much more acute than the sense of taste. Man can detect the odour of methyl mercaptan at an approximate concentration of 9×10^{-13} mol dm^{-3}. He is even more sensitive to skatol. Not only is the olfactory sense in man the most acute of the chemical senses, but it gives rise to the greatest variety of sensations. It is reckoned that a man may discriminate up to four thousand odours, but all attempts to reduce man's olfactory sensations to a system of classification have met with failure. Subjectively, odours have been divided into nine classes such as ethereal, aromatic, nauseating, etc., but such a method is of little value in studying the physiology of the subject unless there can be shown to be correlation between the class of odour, the chemistry of the odorous substances and the existence of groups of neurones primarily concerned with each class of odour.

GUSTATORY SENSE. The special cells associated with taste are most usually located in and around the mouth where they play their chief role in feeding habits. In the fish's mouth they occur on the soft palate region and in a ring around the opening to the swim-bladder, where this occurs. In some fish, they are much more widely distributed than this and may occur on barbels, over the entire surface of the body, and on pectoral and dorsal fins. Some fish undoubtedly locate food by means of gustatory organs and if they lost their sense of taste, they would starve in the presence of food. The amphibia have taste buds in the mouth. The reptiles have little sense of taste in general and this is true of birds also. The tongue papillae of the latter, do contain taste-sensitive nerve-endings, but they are poorly developed. In mammals, the sense of taste is very well developed in most species, but there is little definite comparative information.

In adult man, the gustatory cells are restricted to the sides and top of the tongue and to the epiglottis; in children, they are also present in the cheeks. The tongue shows four types of papillae. *Filiform papillae* are conical projections covering the whole upper surface and tip and edges of the tongue; they are not concerned with taste, but in cats are highly developed for scraping and rasping purposes. *Fungiform papillae*, which resemble button mushrooms in shape (*see* Fig. 16.29), are not so numerous and are scattered over the front upper surface; they contain taste buds. *Circumvallate papillae* are seven to ten in number in man and are larger than the two mentioned previously. They lie towards the hinder upper surface. Each papilla is a flat-topped mound surrounded by a groove, both sides of which contain taste buds (*see* Fig. 16.29). *Foliate papillae* are vestigial, forming ridges at the edge of the back of the tongue.

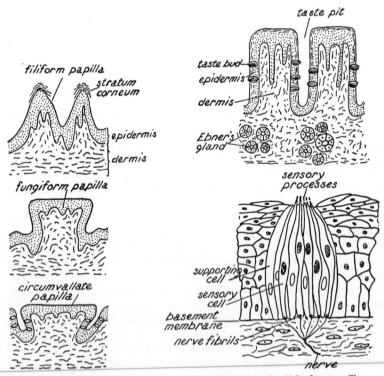

Fig. 16.29. Types of papillae from human tongue, V.S.

Fig. 16.30. Taste pits, V.S., from papilla foliata of rabbit, and taste bud, V.S.

Each taste bud consists of a group of a dozen or more elongated cells, some of which have minute hair-like processes, and the cells are embedded in the sides of the papillae, where they are slightly sunk below the surface, opening to the exterior by the gustatory pore (*see* Fig. 16.30). The hair-like processes of the sense cells project through the pore into the cavity between the papillae.

It is generally accepted that stimulation of the chemoreceptors which cause the sensation of taste in mammals, requires the contact of the stimulating substance with the sensory cell. Exactly how the nerve impulses are generated is not known, but it is assumed that penetration of the substance into the cell or adsorption on to the cell membrane is essential. The nature of chemical structure and stimulating effects have been closely studied, but as with the olfactory sense, there seems to be no correlation whatsoever. In some cases in man, there is indication that stimulation may be a simple matter of pH changes. This seems to be so for the inorganic acids in the pure state.

Physiologists agree that there are four fundamental tastes, namely, sweet, sour, salt and bitter, and that the taste buds associated with these, are more or less localized in certain regions of the tongue. The taste buds sensitive to sour tastes are on the back lateral surfaces, those for sweet and salt tastes near the tip, and those for bitter tastes towards the back. This distribution is irregular and there is much overlapping, but the wider separation of the receptor cells for taste, allows for greater experiment. Sour taste, which is associated with acid foods, is definitely a function of pH. The most typical of salt tastes is that of sodium chloride, but many other salts such as potassium chloride and ammonium chloride are also salty. Others are salty and bitter, e.g. potassium bromide, and others are chiefly bitter, e.g. caesium chloride. The sweet taste is given by a wide variety of chemical substances. Salts of beryllium and lead elicit the sweet flavour reaction and this is attributed to the metallic portion of the molecules. Organic substances such as sugars, dihydroxy- and polyhydroxy-alcohols, saccharin, amino-acids and some esters also do the same. Many substances may give a bitter-sweet effect. The bitter taste is elicited also by a wide variety of substances; some salts such as caesium chloride, the iodides, and calcium, ammonium and magnesium salts are bitter. The alkaloids are the most bitter of all compounds.

If there are only four tastes, then it should be possible to duplicate any complex taste by using a mixture in the right proportions of four suitable substances, for example, sodium chloride, sucrose, oxalic acid and quinine hydrochloride. It is possible to do this and some investigators have claimed to be able to duplicate any taste by mixtures of this sort.

When experiments on taste are being carried out, great care must be taken to control odours. It is easy to show that the olfactory sense interferes with the sense of taste. Many substances taste alike when the olfactory organs are not functioning.

There is no record of a person being completely deficient in any one of the four tastes, but complete loss of taste may follow damage to the nervous system. However, there is variability in ability to taste certain substances. For example, it is found that roughly two-thirds of a population can taste *p*-ethoxyphenylthiourea whilst the remainder cannot. This ability to taste the substance is an inheritable character.

Many land vertebrates possess Jacobson's organs. They consist of a pair of pockets ventral to the nasal cavities, opening into the mouth. These organs contain patches of sensory cells similar to those in the nasal cavities and they appear to have the function of smelling food in the mouth. In snakes, where they are best developed, the forked tongue enters their openings on the roof of the mouth. Thus substances collected on the tongue when protruded, are smelt in Jacobson's organs. They are absent from many mammals, including man.

GENERAL CHEMICAL SENSE. In mammals, all mucous surfaces exposed to external influences, such as those of the respiratory tract, mouth, pharynx, and anus, have developed a general chemical sense which is stimulated by irritation. The sense is active only in fairly high concentrations of the substances and gives rise to protective reflexes such as coughing and sneezing. Most terrestrial vertebrates have much

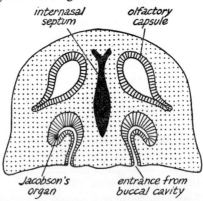

Fig. 16.31. Jacobson's organ of a lizard, T.S.

the same distribution of this sense as man, but in fishes and amphibians, sensitive areas are present all over the external surface. The stimuli are probably perceived by bare nerve-endings, which are derived from the facial nerve in the head, and from the spinal nerves elsewhere. It is to be noted that these nerve-endings are to be distinguished from touch-endings and pain-endings.

Photoreception

Of all the sense organs which influence the activity of animals in their external environments, the light sensitive organs or photoreceptors are the most widely represented. When present, they show a great diversity of structure, but not all animals which show light sensitivity possess the discrete structures by which it is manifested. There appear to be two forms of light sensitivity in animals. The first we may call a *diffuse sensitivity* in the sense that no discrete structure of the body seems specially adapted for this function. This is the case in some of the protozoa such as *Amoeba*, which respond to changes in intensity of illumination without possessing any light-sensitive structures at all. The echinoderms can detect changes in light intensity and it seems that pigment cells of the integument are responsible; no other structures are discernible. In the earthworm, we find light sensitivity all over the body, due to the occurrence of light-receptor cells in all segments, with the anterior end, particularly the prostomium, most plentifully supplied. The second form of light sensitivity may be termed *localized sensitivity*, for the reason that the body possesses discrete photoreceptor organs at fixed points only. Most animals belong to this category, and the structures they have evolved, display a very wide variety of composition and function. They range from parts which show intensity perception alone, to the light-sensitive organs more truly designated eyes, which are capable of intensity perception, pattern vision and form-perception, with sometimes colour vision as well. Intensity perception alone is found in the light-sensitive organelles of flagellates such as *Euglena*, and in the simple ocelli of coelenterates, flat-worms and some annelids. Some of the molluscs, such as the snail, are probably capable of pattern vision as well as intensity perception. Eyes seem to have several evolutionary origins. The complex eyes of the cephalopods resemble the eyes of the vertebrates in structural and functional complexity; in the arthropods, the compound eyes serve for pattern vision. These three types of eyes appear to have independent evolutionary origins. Reference to the structure and functioning of some invertebrate light-sensitive organs has been made in Vol. I. Here we shall confine our description to the vertebrate eye. Throughout the vertebrates, its structure is remarkably constant, and the human eye, which is very generalized, is here taken as a representative example.

Photoreception in the Vertebrates; the Eye

In its mode of functioning, the eye is similar in many respects to the camera. The cornea and lens form an inverted image on a photosensitive screen, the retina, fully in accordance with geometrical optics (*see* Fig. 16.33). The aperture of the lens can be varied by the iris and the eyelids serve to exclude light completely and to prevent damage to the corneal surface. The retina is backed by a layer of black pigment which reduces internal reflection and makes the image sharper. Part of the focusing mechanism, the lens, is controlled by the ciliary muscle,

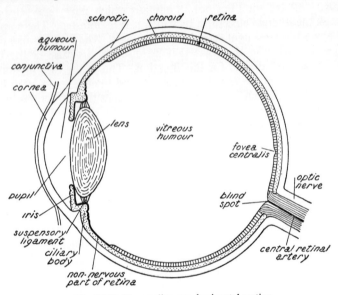

Fig. 16.32 Mammalian eye, horizontal section.

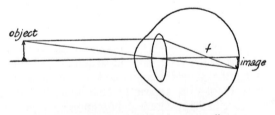

Fig. 16.33. Image formation by the mammalian eye.

so that objects at differing distances may be brought sharply into focus in their turn. Focusing of light on the retina is achieved partly by this mechanism and partly by the curvature of the cornea. The refractive index of the cornea is 1·376 and that of the lens is 1·42, whilst the refractive index of both humours is 1·33. It follows that because the greatest refractive index difference is between the air and the cornea this interface must be the most important in image formation. The lens acts as a fine adjustment, giving very delicate and accurate control

The range over which the lens can change the focus of the eye, is known as the *accommodation range*. This range varies with age and may be as much as 20 dioptres in young children down to an average value of about 4 dioptres at the age of about 40 years. (If the focal length, f, of a lens is expressed in metres, the quantity $(1/f)$ is called the *power* of the lens and the unit, that is, the power of a lens of which the focal length is one metre, is called the dioptre.) When the normal eye is relaxed, it is focused at infinity and accommodation is accomplished by the changing shape of the lens, when the ciliary muscles are contracted. The lens is an elastic structure and the decrease in focusing power with age is due to hardening, so that it becomes progressively less able to become rounder and thicker on contraction of the muscle. This causes the normal *presbyopic* condition of the eye with advancing age and is known as "old" sight. The common abnormalities of the eye which may be corrected with spectacles are nearsightedness or *myopia*, in which an elongated eyeball causes an image to be formed in front of the retina, farsightedness or *hypermetropia*, in which a shortened eyeball causes an image to be formed behind the retina (*see* Fig. 16.34), and *astigmatism*, due to the non-uniform curvature of the lens or cornea,

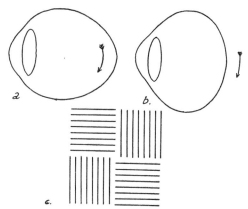

Fig. 16.34. Defects in vision. (*a*) Myopia, (*b*) Hypermetropia. (*c*) Test for astigmastism.

causing the occurrence of different focal points for beams of light in different planes. A symptom of this is the blurred appearance of vertical lines when horizontal lines at the same distance are quite clear, or vice versa.

The structure of the retina is shown, much simplified, in Fig. 16.35; it is known as an *inverted retina*. The light enters the retina through the layer of nerve fibres which forms its front and then successively passes through a layer of ganglionic neurone cell bodies, a layer of synapses, a layer of secondary neurone cell bodies, another synaptic layer and finally through the cell bodies of the rods and cones, which are the primary neurones or sensitive cells. It is the portion of the rod or cone corresponding to the dendrite of the primary neurone which contains the photosensitive substance. This is *rhodopsin* in the rods and *iodopsin* in the cones. Only the photosensitive substances absorb much light as it passes through the retina, and any light not absorbed here, is absorbed by the choroid coat.

There are complicated connexions between the neurones in the retina. The figure shows only the basic connexions. Some cones connect by way of the secondary and tertiary cells directly into the brain and thus have exclusive use of a single fibre. Some cones are connected to fibres which may receive impulses originating in other cones or in rods. Some cones which have individual fibres, may also be connected through secondary cells to nerve fibres which receive impulses from many rods and cones, and it is thought that these connexions are important in colour vision. All the nerve fibres which pass across the

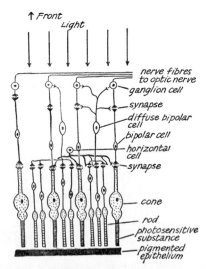

Fig. 16.35. Diagram representing structure of the retina. Much simplified with some types of retina cell omitted.

inner face of the retina enter the optic nerve, and at this region there are no rods or cones and this constitutes the "blind spot." Around the optical axis of the eye is the area called the fovea centralis, or "yellow spot," where there are abundant cones but few rods.

The two distinct forms of sensory cells, the rods and cones, function differently by being adapted to work most effectively over widely-differing light-intensity ranges. The rods, containing rhodopsin, are much more sensitive to light than the cones, which contain iodopsin. Rhodopsin is very rapidly bleached in high light intensities and so the rods can function only when the light intensity is low. It takes upwards of half an hour in the dark for the rods to become maximally sensitive. Iodopsin is not nearly so photosensitive and so the cones are able to function at much higher light intensities. The combinations of rods and cones enables the human eye to be sensitive over a very wide range of intensity.

It is known that illumination of the photosensitive cells causes an increase in negativity in the electric potential at the apices of these cells. This is called the retinal action potential. This action potential is conducted via the optic nerve to the visual centres of the brain, where it is translated into the sensation of seeing. Thus there are two physiological problems of interest, namely, the action in the photosensitive cells of the retina and the action in the central nervous system.

The first step in the process of vision is a photochemical one. The rhodopsin or visual purple of the rods is a conjugated protein possessing a carotinoid prosthetic group which seems to play a part in light sensitivity. Rhodopsin possesses a characteristic magenta colour. If extracted from a retina which has been in darkness for some time, it shows a very low yield of free carotinoid substance. If the extracted rhodopsin is now exposed to light, it bleaches rapidly, giving rise to orange and yellow products. Rhodopsin extracted from a retina after a short period of illumination, yields a quantity of a stable yellow carotinoid known as *retinene*. If extraction is made after a long period of illumination, there is a high yield of a carotinoid identified as vitamin A. Both retinene and vitamin A are of low molecular weight, indicating that they have been split off the conjugated protein.

If extracts of rhodopsin are subjected to light at very low temperature, an orange decomposition product is produced. When its temperature is raised, it changes colour rapidly to yellow. This *transient orange* is believed to be the first product of rhodopsin breakdown, in which the carotinoids are not actually split from the protein. At ordinary temperatures, transient orange quickly develops retinene. The formation of transient orange is believed to be the only step which requires the presence of light, all other transformations being carried on in the dark. The bleaching effect of the light on the rhodopsin leads to the development of action potentials in the rod fibres. These can only be short-lived in daylight, since all the rhodopsin will quickly become bleached and remain so, as long as the rods are brightly illuminated. However, in very low light intensities or in darkness, the vitamin A is reconverted to visual purple, and sensitivity of the rods is maintained at an equilibrium by the balanced operation of events represented in this equation.

$$\text{Visual purple (rhodopsin)} \underset{\text{darkness}}{\overset{\text{light}}{\rightleftharpoons}} \text{Bleached products}$$

A fuller sequence of events is given in the cycle below—

$$\text{Rhodopsin} \underset{\text{darkness}}{\overset{\text{light}}{\rightleftharpoons}} \text{Transient orange} \underset{\text{vitamin A}}{\rightleftharpoons} \text{Retinene}$$

The great preponderance of rods in the eyes of nocturnal animals can be understood if this is the true picture of the sequences in the rods. In diurnal animals, daylight vision must be associated with the cones only, since the rods will be inactivated for most of the time as a result of the continued bleaching of the visual purple. At levels of brightness corresponding to full moonlight, there is a transition effect when both rods and cones are functional. For light brighter than this, when only the cones are functional, the eye has the ability to discriminate different light wavelengths as colour sensations. In most vertebrates, vision is said to be *duplex*, being mediated by rods and cones according to light intensity.

The substance iodopsin, so named because of its violet colour, has been extracted from the cones of domestic fowls. It shows similar light absorption properties to rhodopsin, but the photochemical reactions are not known. It is assumed to be less sensitive to light than rhodopsin, hence its functional value in full daylight.

Although rhodopsin has been extracted from all vertebrate retinae tested, other light-sensitive substances have been found on several occasions. Porphyropsin has been found in the retinae of fresh-water fishes. It is similar to rhodopsin and is included in the collective name visual purple, but its breakdown products are slightly different, being known as retinene 2 and vitamin A_2.

In the case of the neurones forming the rods, they are of definite and constant construction, being composed of a cylindrical part or rod proper, a conducting fibre and a nucleus or rod body. It is in the rod proper that the photochemical events occur. It is divided into an outer segment and an inner segment; the former probably contains the photosensitive substance whilst the latter seems to be concerned with its reformation after bleaching. The outer segments, on examination with polarizing and electron microscopes, can be shown to be alternate layers of protein and lipoid substances.

The physiology of colour vision is to a large extent theoretical. It was suggested as long ago as 1807 by Young and later expanded by Helmholtz in 1852, that colour vision was fundamentally due to the ability of different receptors to respond to different light wavelengths of the spectrum. To the normal human eye, the spectrum appears as a series of colours varying from red through orange, yellow, green, blue, to violet. All these colours can be duplicated by mixtures of three lights, a red, a green and a blue light, provided each can be independently adjusted for intensity. Two of the three are not sufficient, but a fourth may make colour matching more perfect. A normal person is said to be *trichromatic*, since he can duplicate the spectrum with three colours.

The Young–Helmholtz theory, for vertebrate eyes, has, to some extent, been substantiated by Granit. He used a technique of placing micro-electrodes on single tertiary neurones of the retina before they enter the optic nerve. Upon stimulation of the sense cells, in the area connected to this fibre, with light of known wavelength and intensity, he was able to establish the sensitivity of individual visual units to various wavelengths. He found that all the visual units of the retina were not the same. Some gave normal rod and cone sensitivity curves, indicating that some tertiary fibres receive impulses from both rods and cones. Other tertiary fibres received impulses from cones only and these seemed to be of several types. Some were most sensitive to blue light, some to green, some to yellow, and some to red. Such a result can be explained if it is assumed that there are cones with narrow sensitivity bands in blue, green, yellow and red wavelengths and that these can be connected both singly and in combination to tertiary neurones.

Abnormal colour vision in human beings, usually known as colour-blindness, can be explained in terms of the above discoveries, assuming that some of the colour-vision units are absent or non-functional. Colour-blindness is known to exist in three forms. Some individuals require lights of three colours to match any given colour but have to use an intensity adjustment different from normal. They are said to have a "colour weakness" or a weak sensitivity to certain colours. Others can match the spectrum, as they see it, with only two colours. They are said to be dichromatic or colour-blind. They exist as several types. *Deuteranopes* can distinguish yellow from blue, but not green from red; they see greens and reds as shades of grey. *Protanopes* cannot differentiate red from green, and are much less sensitive to reds, which they see as dark grey or black. *Tritanopes* are rarer; they can distinguish between red and green, but not between yellow and blue. The third form is complete colour-blindness in which the person is

unable to detect any colour differences in the spectrum. The deu-
teranope and protanope conditions are known to be inherited as sex-
linked characters and are about twenty times as common in men as in
women. The phenomenon of nightblindness is a different condition.
It is due to the inability of the sufferer to manufacture visual purple.
Deficiency of vitamin A produces the condition.

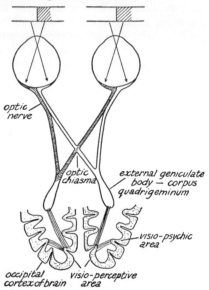

Fig. 16.36. Interconnexions of the eyes, the visio-perceptive and visio-psychic
areas.

THE CENTRAL NERVOUS SYSTEM VISUAL MECHANISM. Nerve fibres
from the retinae lead to the thalamus, and thence, after synapses,
impulses are conveyed along other fibres to the *visio-perceptive areas* at
the back of the occipital cortex. Each point on each retina is represented
in this region of the brain. However, there is no recognition or interpretation of visual impulses without reference to the *visio-psychic areas*,
just in front of the visio-perceptive areas (*see* Fig. 16.36). Here, it seems,
the memory of things seen previously is situated, and thus the nature of
an object is identified. This, however, is not the complete explanation.
Lateral connexions lead forward from the visio-psychic areas to several
regions of the temporal lobes; their function is not known, but there are
certain strange effects caused by injury to these forward areas. For
example, a person who could formerly read well may be able to read
individual words, but not sentences. There are many other examples of
such partial deficiencies in other senses as well as sight.

VERTEBRATE VISUAL ADAPTATIONS. It is not surprising that vertebrate eyes, whilst fundamentally similar in structure, show many
adaptations according to the conditions under which their possessors
live. These adaptations are of two main kinds. In the first place,
animal behaviour varies with respect to day and night. Some are
diurnal, chiefly active by day, and some are *nocturnal*, chiefly active by
night, whilst others may seldom rest or sleep at all. These last may be
described as *arhythmic* in behaviour.

The eyes of diurnal animals may be adapted in several different ways
to give a great degree of visual acuity. There is a marked increase in
the number of cones over rods and in some areas of the retina, there may
be few if any rods at all. Such areas are known as foveae, and in man
there is the fovea centralis. The general effect is to increase sensitivity
at high light intensities but the shape of a fovea may also serve to
spread the image over a larger number of cones, thereby again increasing acuity. A depressed fovea would have the effect of a local concave
lens, since the refractive index of the vitreous humour is less than that
of the retina (*see* Fig. 16.37). This is most noticeable in birds. A
larger image may also be achieved by enlargement of the eye as a whole,
thus increasing the lens-retina distance. The diurnal animals have a
greater lens-retina distance than nocturnal and arhythmic animals.
Another method of increasing acuity is to use colour filters which
absorb the blue wavelengths and thus decrease chromatic aberration
at the retina. Yellowish lenses as in some reptiles, some mammals and
in man, red or yellow droplets of oil in the cones as in some reptiles
and birds, a yellowish cornea as in some fishes,
a yellow fovea as in man, all serve this purpose.

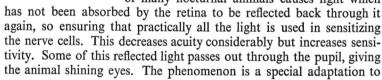

Fig. 16.37. Effect of
fovea in increasing
visual acuity.

In nocturnal animals, there is the general
tendency for rods to preponderate over cones,
thereby increasing sensitivity in a dim light. The
slit pupil of the nocturnal animals is an advantage over the round pupil in that it can close
more tightly in bright light and thus prevent
insensitization of the rods more efficiently. The
slit can also be more widely extended in dim
light, thus allowing more light to enter the eye.
The presence of a reflecting layer in the choroid
of many nocturnal animals causes light which
has not been absorbed by the retina to be reflected back through it
again, so ensuring that practically all the light is used in sensitizing
the nerve cells. This decreases acuity considerably but increases sensitivity. Some of this reflected light passes out through the pupil, giving
the animal shining eyes. The phenomenon is a special adaptation to

nocturnal habits and is common in vertebrates. It can be seen in the cat.

In the arhythmic animals, a good development of both rods and cones gives sensitivity over wide ranges of light intensity. There may also be the condition of the pigment cells migrating into the sense-cell layer of the retina, or the sense cells migrating into the pigment layer to guard against very high light intensities. This is found in some fishes. There may also be very great mobility of the pupil, as in higher vertebrates, whereby the eye aperture can be very quickly adjusted.

Other adaptations of the eye may be correlated with the fact that vertebrates inhabit two distinctly different media from the point of view of eye function. Aquatic and terrestrial animals show differences in their eye structure accordingly. Since the refractive index of the cornea is close to that of water, the image is formed very largely by the lens in aquatic eyes, and so the cornea may be almost any shape without affecting the functioning of the eye. This allows the cornea to fit into the general streamlining shape of the fish. In air, the cornea must be heavily protected against drying out. In man and many other animals, this is accomplished by the spreading of a fluid from lachrymal glands across the eyeball by blinking the lids. In birds and many mammals, there is the additional structure, the nictitating membrane, which serves a protective function. In birds, it is probably covering the cornea most of the time, to prevent rapid drying out by the bird's slipstream. Perhaps the most striking difference between air and water vision are the two fields of view. A fish in water will have a composite field of view due to the total internal reflection at the water–air interface. This is represented in Fig. 16.38.

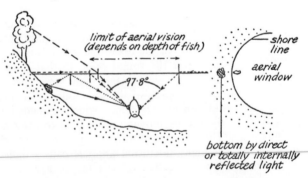

Fig. 16.38. Diagrams representing a fish's-eye view.

Animals which live in both media must be adapted to both. They must have a very great range of accommodation. The fish, *Anableps*, is curiously adapted for seeing both in air and in water. It has two pupils for each eye, and two retinae, one for use in air and the other in water. Fish living in very deep water, where light intensity is very low, often have large eyes with relatively larger pupils and lenses, a great concentration of rods, a tubular eye shape which gives concentrated central vision, and they illuminate objects by their own bioluminescence.

Any image formed by an eye should be capable of being translated by the brain into a number of qualities. For instance, relative sizes, shapes and positions of objects should be readily discernible; brightness, colour, and movement should be detected. There are many eye adaptations which are all aimed at improving this perceiving ability. *Binocular vision* is a great aid in giving a better sense of the distance and relative positions of objects. The single or monocular field of vision of the vertebrate eye is about 170°, lower in deep-water fishes and greater in some land animals like the cat. The visual angle of the human eye is 150°. The situation of the eyes in the head will fix the extent of binocular vision. A few animals, such as large whales, have no binocular vision at all (*see* Fig. 16.39 (*a*)). Both eyes can never be focused on the same object. Most fish and birds have some degree of binocular vision. In man, the visual field is almost entirely binocular as is the case in the carnivores, such as the cat and dog, where it is an advantage to see directly ahead (*see* Fig. 16.39 (*b*)). Animals whose

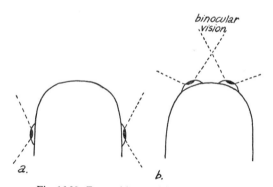

Fig. 16.39. Eye positions and binocular vision.

main concern is to escape enemies profit by having a much wider total field of view, hence binocular vision is sacrificed to a great extent. Hares and rabbits have two monocular fields of about 190° each, overlapping slightly to form binocular fields both anteriorly and posteriorly. Due to the independent muscular control of the eyes in the chameleon and the wide visual field in either eye, the animal is fully adapted for both conditions.

Movement of the eyes to maintain the visual field constant, and to ensure that during binocular vision the images are simultaneously focused on the same parts of the two retinae, is effected by the extrinsic eye muscles.

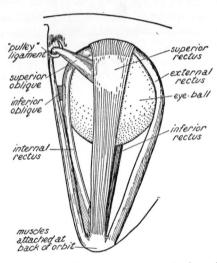

Fig. 16.40. Mammalian eye, extrinsic muscles from above.

The movement of these muscles may be voluntary or involuntary. The involuntary movements are due to automatic reflexes which maintain the constancy of the visual field during locomotion or movements of the head. When voluntary movements are made, the two eyes may move independently without co-ordination, or there may be varying degrees of co-ordination. In most reptiles and birds, the movement is independent. In some fishes and in the chameleon it can be either. In man, it is never independent, but is always co-ordinated in such a way as to ensure that a single object is always focused simultaneously in the same corresponding positions of the two retinae. Birds which feed in flight, such as swallows, are peculiarly adapted in that they have two foveae in each eye. One of these in each eye is used monocularly, whilst the other is used binocularly with the corresponding fovea in the other eye. Thus, such a bird can have in focus at the same time, three objects, two monocularly and one binocularly.

A sense of depth and distance may be obtained either monocularly or binocularly. Sensations of depth can be gained from perspective, parallax, size of image, and several other characteristics. It is not possible to say how any particular animal is able to use any of these characteristics in helping it to fix distances between objects.

Perception of movement is particularly complex. Normally a small moving object can be detected more easily than if the same object is stationary. Very rapid movement results in a blurred visual image; the successive images are partly superimposed. On the other hand, an object moving very slowly may not be seen at all. For most creatures, there is an intermediate speed at which the object is most clearly visible.

Colour vision in animals serves a purpose in that it increases the ability to distinguish more clearly between one object and another. From the behaviour of numerous animals in response to colour, it is fairly safe to assume that many possess it. Those possessing it with greatest certainty are the bony fishes, some reptiles (turtles and lizards), birds, and the primates among the mammals.

Sound Reception

Waves, consisting of the vibratory movement of particles of a medium, are always emitted from a vibratory source. For example, when a tuning-fork is struck, the backward and forward movements of the prongs set up alternate compressions and rarefactions in the

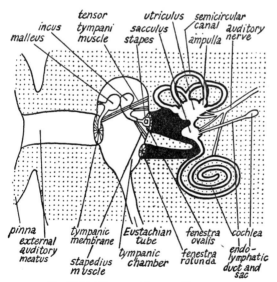

Fig. 16.41. Diagram of mammalian ear structure.

adjacent air. Because of the elasticity of the air, these alternate changes in pressure are propagated outwards in all directions in wave form. Such waves may be propagated through gaseous, liquid, or solid media. They constitute sound-waves when the frequency of vibration is within certain limits known as *audiofrequencies*. Some animals are capable of detecting such vibratory movements of a medium and have specialized receptor organs for the purpose. The study of sound perception or hearing, in animals, has been confined almost entirely to the vertebrates and to the insects among invertebrates. It must not be assumed, however, that no other groups of animals can perceive sound vibrations. Here, we shall confine our account to the vertebrate sound-detector, the ear, which has, in addition another function, that of perceiving changes in body equilibrium. It is to be noted that many animals have special sensory organs by which they can detect pressure changes of much lower frequency, and also mass movements of the medium. Such are the lateral line organs of fishes: the ear appears to be a highly specialized part of the lateral line system.

Sound Reception by the Mammalian Ear. The structure of the mammalian ear has been described in Vol. I, Chap. 29. Its mode of functioning is best described with reference to Fig. 16.42. There is a sequence which commences with the impinging of sound waves on the tympanic membrane and ends with stimulation of the sensory cells in the organs of Corti. In many mammals, but not in human beings, the pinna may be instrumental in concentrating the waves into the auditory meatus. The ear-drum, having no frequency of vibration of its own, vibrates in sympathy with the air. It is able to do this because the Eustachian tube is in free communication with the pharynx, and thus the air-pressures on both sides of the membrane are equalized. The vibrations are communicated to the auditory ossicles, and through them to the fenestra ovalis. This oval window thus vibrates in unison with the

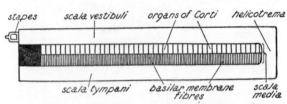

Fig. 16.42. Diagram of cochlea, unrolled.

tympanic membrane. The two small muscles attached to the ossicles, the tensor tympani and stapedius, may act reflexly to deaden the vibrations, or they may adjust the mechanism so that only certain vibrations pass satisfactorily.

From the oval window, the vibrations are transmitted to the perilymph of the vestibule. They are then passed up the fluid of the spiral scala vestibuli and into the scala tympani via the helicotrema. The vibrations pass down the spiral to the fenestra rotunda which bulges outwards synchronously with the bulging inwards of the fenestra ovalis.

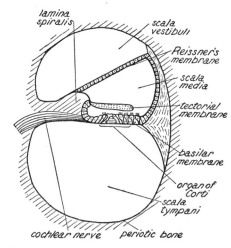

Fig. 16.43 Portion of cochlear canal, enlarged, V.S.

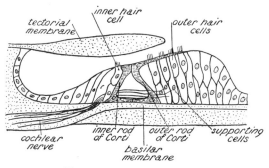

Fig. 16.44. Organ of Corti, V.S.

The cause of stimulation of the sensory cells of the organ of Corti is not clear. It could be that vibrations passing up the scala vestibuli are transmitted through Reissner's membrane. This would cause oscillations of the tectorial membrane which might disturb the hair-like processes of the sensory cells sufficiently to stimulate them. Alternatively, vibration of the basilar membrane would cause movement of the processes against the tectorial membrane. Experiment has shown that when the processes are moved against the tectorial membrane, the sensory cells become negative from their free ends to their bases. Presumably, this is the initiation of the action potential sent out along the fibre to the auditory nerve. Some physiologists however, consider that the effect of distortion of a sensory cell is to cause it to release some chemical substance which then causes the action potential to develop.

The process of hearing goes beyond the mere ability to detect sound waves. In human beings, at least, the sound waves are distinguished by their frequencies, by the relative intensities of the frequencies, and by the directions from which they are coming. Furthermore, the human ear is able to distinguish all these simultaneously. In some animals, for example the fishes, the process has advanced no further than sound detection. In the bat, there is an improvement on man, in that it can deal with a much wider range of frequencies.

Discrimination between frequencies by the mammalian ear is almost certainly due to the fact that a given frequency stimulates only a specific small part of the organ of Corti. The base of the cochlea is sensitive to oscillations of high frequency, whilst farther and farther towards the apex the cochlea becomes sensitive to lower and lower frequencies. In man, this frequency range lies between about 16 and 20 000 hertz (Hz) or cycles per second, but many mammals can hear sound of much higher frequency than this. Both the dog and the bat can detect vibrations which are supersonic to man.

The manner in which the ear perceives loudness or intensity of sound may be effected by either or both of two mechanisms. The single nerve fibres from the organ of Corti may carry an increasing number of impulses in a given time under the influence of louder and louder stimulation or more sensory cells may respond so that in either or both cases more impulses reach the brain. The human ear is responsive to a wide range of intensities. Intensity or loudness is commonly expressed in the unit *bel* (B), or a submultiple, *decibel* (dB). This unit is not recognised in the S.I. (Thunder overhead about 120 dB, pneumatic drill about 80 dB, whisper 15 dB.)

Fixing a source of sound is done by the human ear with great accuracy. This is due to the functioning of both ears at once, which causes two effects. In the first place, one ear may not hear the sound at the same intensity due to "shadowing" by the head and secondly one ear may hear the sound slightly out of step with the other. Thus there will be a phase difference in the sound waves reaching the two ears. Both these effects can aid in pinpointing a sound source.

The sensation of sound in the brain is interpreted in the tempora lobes of the cerebral cortex in the mammal. The pathway of nerve fibres from the organ of Corti to the brain is made up of the neurones which innervate the hair cells. These are bipolar and have their cell bodies in the ganglion of the modiolus. Their axons form the cochlear branch of the VIIIth cranial nerve which enters the brain at the dorso-lateral wall of the pons Varolii where it meets the medulla oblongata.

The structure of the mammalian ear is remarkably constant for all but the completely aquatic species. The cochlea is always highly developed and coiled, but not of the same size. The horse has two turns to the spiral, for example, whilst the guinea pig has four. There is also some variation in the range of frequencies which can be detected by different mammals. Many can detect frequencies well above those audible to human ears. The so-called "silent" whistle for dogs has a frequency of about 25 000 Hz. Bats are credited with being able to hear sounds with a frequency of 98 000 Hz or more vibrations per second. Furthermore, it has been conclusively proved that the bat is able to emit sounds of high frequency and to detect echoes from surrounding obstacles, so that it can avoid them even in total darkness. This device is strictly comparable with modern radar in which high frequency electro-magnetic waves are emitted and then received as reflections from obstacles in their path. Both range and direction of the obstacle can be found from the reflected wave and the bat seems quite able to use reflected sound waves for similar information.

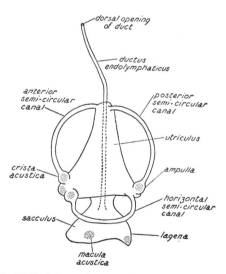

Fig. 16.45. Left membranous labyrinth of the dogfish.

Sound Perception in Other Vertebrates. It is well known that the hearing mechanism of lower vertebrates is far less efficient than that of the mammal. From the fishes to the amphibians, reptiles and birds, the auditory organ shows an increasing complexity of structure. All appear to possess a more or less similar organ of balance but the auditory apparatus is very poorly developed in the lower classes. Fish can almost certainly hear sounds propagated through the water, but may be unable to detect sounds made in air, since the air–water surface may prevent the sound entering the water. In some fish, hearing may be very acute and this is achieved by using the swim-bladder as a means of picking up vibrations in the water and relaying them to the auditory organ through a series of bones, the Weberian ossicles, which correspond functionally to the middle ear ossicles of higher vertebrates. In some fish, the lateral line organ is known to take part in reception of sounds of low frequency. Some amphibians show an improvement, in possessing a middle ear with a columella transmitting vibrations from an ear drum to the inner ear. Frogs are known to have a very acute sense of hearing, but may not often respond to sound. It has been demonstrated that during the early development of the tadpole, the round window of the inner ear is connected with the lung sac by means of a fibrous cord. This cord becomes cartilaginous and functions as a columella which transmits vibrations to the inner ear. The cord disappears at metamorphosis.

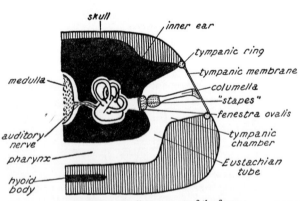

Fig. 16.46. Auditory organ of the frog.

The development of a structure which may be regarded as a primitive cochlea is a noticeable advance in reptile sound-organ structure. The lagena is elongated and attached to two sides of the surrounding bony cavity, thereby forming the characteristic three ducts of a cochlea.

all the reptiles except snakes, there is a middle ear and this is traversed by two bones, the outer of which touches the tympanic membrane. Snakes have no middle ear and the outer end of the columella is in contact with the quadrate bone of the skull. They are thus very sensitive to vibrations through the ground but insensitive to sound vibrations in air.

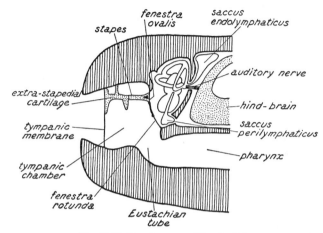

Fig. 16.47. Ear structure of lizard. T.S.

Birds possess ears very similar in structure to those of reptiles, but in general, their sound-detecting capacity and frequency discrimination are very highly developed, even comparable to those of mammals.

Detection of Low Frequency Vibrations and Movements of the Medium

Among the vertebrates, these senses are most strongly developed in fish and the larval stages of amphibians, where the surrounding medium is water. The terrestrial vertebrates, during their evolution, have lost also the apparatus developed by their aquatic ancestors for these purposes, namely the lateral line system.

The Lateral Line System of Fish. This takes the form of a network of canals just below the skin, opening to the exterior by pores. The two main branches of the system are the lateral lines one on each side of the body extending the whole length of the trunk to the tail region. At the anterior end, these branch into an anastomosing network over the head and snout (*see* Fig. 16.48). In a sectional view at right angles to the axis of the canal can be seen the organs of special sense which detect vibrations and water movements. These are the "neuromast" organs (*see* Fig. 16.49). Each consists of a group of sensory cells set on a cushion on the inner wall of the canal. Each sensory cell has a fine cytoplasmic tip projecting into the fluid-filled canal and from the base of each a nerve fibre passes inwards, where all eventually form the lateral-line branch of the vagus nerve. The sensory cells are stimulated by any vibratory or mass movement of the fluid which disturbs the cells. Fish can thus detect pressure waves and ripples originating from a distant source. Tests have shown that they are best able to perceive low-frequency vibrations. The cells are also stimulated by any movement of the fluid in the canal which may cause their distortion. Thus, fish are able to detect the movement of other objects in the surrounding water without seeing them, and the presence of still objects from which the vibrations set up by the moving fish may

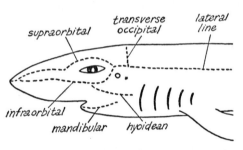

Fig. 16.48. Lateral view of the anterior region of dogfish to show lateral line system.

be reflected. This is of undoubted significance in keeping shoals of fish together and in aiding the fish to avoid one another and to avoid other obstacles. Further, movement of the fish's own body is sufficient to distort the neuromast cells and they may thus act as proprioceptors to assist in co-ordination of swimming movements. Amphibian possessor

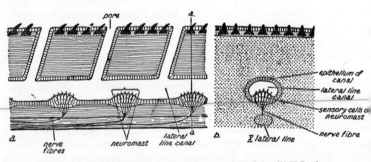

Fig. 16.49. (a) Longitudinal V.S. of lateral line of dogfish. (b) T.S. plane a—a in (a).

of lateral line systems are known to use them in a similar manner, being able to localize moving objects with great accuracy and to detect fixed objects by reflected vibrations.

Detection of Changes in Hydrostatic and Atmospheric Pressure

Such pressures are applied continuously and uniformly over all parts of the body and may change with extreme rapidity. Thus it is not surprising to find that the effects of change in pressure of the outside medium are to be noticed more quickly in the respiratory system than in a more specialized sensory system.

The Effect of Changing Hydrostatic Pressure on Fish. It is noticeable that fish tend to remain within a certain pressure range to which they are best adapted. Fish which possess swim-bladders apparently do this by varying the amount of gas in the swim-bladder. A sudden decrease in pressure (by as little as a small fraction of an atmosphere) is sufficient to cause an increase in the gas content of the bladder. In fish where there is no link between the swim-bladder and the exterior, this gas must be secreted by the blood and presumably this is the case also for fish in which the swim-bladder is connected to the oesophagus. Where the Weberian ossicles are present, they are known to transmit pressure changes in the swim-bladder to the inner ear which would thus become sensitive to changing hydrostatic pressure as well as to sound vibrations. It is known that changes in swim-bladder pressure in some fish causes a strong response of all fins and other reactions. Thus the bladder must be acting as a sensory mechanism which detects the changing hydrostatic pressure. Fishes, such as the elasmobranchs, which possess no swim-bladder, may be able to detect hydrostatic pressure changes through the ductus endolymphaticus which opens on the surface of the fish.

The Effect of Changing Atmospheric Pressure on Land Vertebrates. Within the limits of the normal vertical range which land animals usually inhabit, atmospheric pressure cannot change a great deal nor can it change abruptly. Thus there is no sensory system known which is responsive to the small slow changes which may naturally occur. Animals, subjected to violent pressure changes in the atmosphere, show effects on the respiratory system which could ultimately lead to death if such conditions were applied over long periods. Under natural conditions, animals would not tend to subject themselves to such conditions and the need for adjustment has never arisen. Only man has ever attempted to live outside the natural vertical limits to which he is adapted, and he has been able to devise artificial aids to assist him to accomplish his purpose.

The physiological effects of rapid ascent by man into the atmosphere are due to three causes. In the first place, lack of oxygen will lead to gradually increasing enervation and eventually unconsciousness. Man has learnt to combat this by carrying a supply of oxygen in cylinders. Secondly, the reduction of pressure with altitude will cause bleeding from capillaries because the body fluids are at a higher pressure than the surrounding atmosphere. In high-altitude aircraft this is overcome by the use of pressurized cabins. However, man is capable of some adjustment to both conditions by gradual ascent and periods of acclimatization at the successive higher levels. The third physiological effect is due to falling temperature; suitable clothing will to some extent compensate for this.

Rapid increase of pressure is only experienced naturally by descent in water. Here slow descent is essential to avoid the crushing effect due to higher external pressure. The higher pressures cause more gas to go into solution in the blood and therefore the ascent has to be slow to avoid liberation of gas bubbles which give the dreaded "bends" of divers.

The Detection of Mechanical Contact

The ability to detect and respond to the contact of the body with other objects is referred to as the tactile sense or sense of touch. All animals seem to be capable of detecting such stimuli and in some the sense is very highly developed. The organs concerned in animals above the protozoa and sponges are usually special receptors or sensory endorgans located at the surfaces of the body.

The Tactile Sense of Mammals. This is comparatively highly developed and the ability of a mammal to locate a contact stimulus is very exact. The organs concerned are the sensory nerve-endings in the skin and there are several varieties of these, including tactile corpuscles

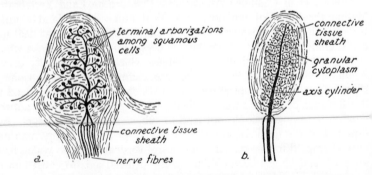

Fig. 16.50. (*a*) A tactile corpuscle in dermis of skin. (*b*) An end bulb from connective tissue.

end bulbs, Pacinian corpuscles and organs of Ruffini. It is known that between them they are responsible for the sensations of touch, heat, cold and pain but the particular function of each is not clearly understood.

The tactile sense has been studied most usually by the use of flexible

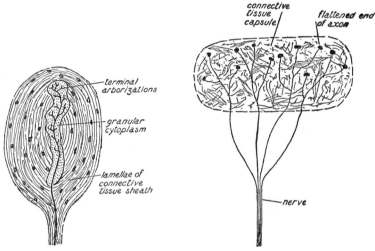

Fig. 16.51. A Pacinian corpuscle from mesentery of cat.

Fig. 16.52. Organ of Ruffini from dermis of skin.

fibres which can be applied to the skin with variable pressures in very localized areas. Two aspects of the phenomenon have been studied in detail in man, namely, the sensitivity of different parts of the body and the ability to distinguish between a single stimulation and two stimuli close together. For man, it has been shown that sensitivity varies greatly according to the locality on the body. The greatest sensitivity is associated with the presence of short hair. At the base of each hair is a nerve ending and this is stimulated very readily by the lightest touch of the hair. Where hair has not grown or where it has been shaved off, the sensitivity of the skin itself varies as follows; greatest sensitivity to lowest sensitivity in this order: tongue, forehead, nose, finger-tip, back of hand, abdomen, back of forearm, buttock. The ability to resolve two stimuli close together does not follow the same order. This ability is at its greatest in the finger-tips, then palm of the hand, back of the hand, upper arm, thigh and back, in that order. This is accounted for, when it is realized that the sensory unit seems to be one sensory neurone together with all the sensory endings which make up the fibre of that neurone. Thus one fibre may be serving many square millimetres of surface area of the skin and receiving impulses accordingly. The sensation finally obtained would be the same from anywhere inside that area.

Temperature Perception

For poikilothermous vertebrates, it has been shown that there are special temperature receptors in the skin. In fishes, it seems certain

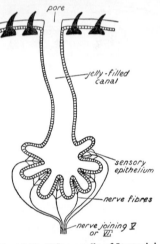

Fig. 16.53. V.S. ampulla of Lorenzini.

that the lateral-line organs and also the ampullae of Lorenzini are stimulated by temperature changes. The amphibia and reptiles possess receptors for both heat and cold in the skin. With all these poikilotherms, rise of temperature leads to a higher metabolic rate, and lowering of temperature to the reverse. With decreasing temperature, they may be gradually reduced to a state of increasing torpidity.

The birds and mammals are homoiothermous and further, they have in the brain a temperature-sensitive centre. This can adjust retention and production of heat and so act as a thermostat for the whole body. In addition, there are vasodilator and vasoconstrictor reflexes which regulate heat loss from the body surface. These reflexes are elicited in response to perception by the thermal receptors of the skin. The temperature-sensitive brain centre appears to be stimulated by changes in blood temperature, which is partly dependent on the thermal gradient in the skin.

In man, the warmth-receptors are the organs of Ruffini which lie deeper than the bulbs of Krause, which appear to be the cold-receptors. Both are located in the dermis.

Pain Perception

Pain, in human beings, results from excessive stimulation of nerve endings or even of nerve trunks. It is not certain whether any animals except man are sensitive to pain in the sense we understand it. Nevertheless, the majority react rapidly to violent stimuli. In man, pain acts as a protective mechanism, informing the individual of the location of the damage, and sometimes of its extent; it has probably had considerable survival value in man's evolution. The child pricked by a pin or cut with a knife will be more circumspect in his future use of those articles.

It is not known which particular types of receptors perceive pain, if indeed there are particular types. Several lines of evidence indicate that pain impulses are carried along different fibres from tactile or temperature impulses. In some diseases of the nervous system such as locomotor ataxia, pain may be perceived when mere touch is not. Some parts of the body such as the intestine, which cannot perceive touch, can register pain. Tactile impulses travel along relatively large fibres with high velocity; pain impulses travel along small fibres with low velocity. Stubbing his big toe will convince the reader of this fact; the pain is felt a fraction of a second later than the tactile sensation. In the teeth, there are large fibres stimulated by pressure and smaller ones by pain.

An interesting phenomenon is that of "referred" pain. Stimulation of a nerve trunk, causes the pain to be "referred" to the peripheral endings of that nerve. Thus a blow on the ulnar nerve at the elbow gives a sensation of tingling in the fingers. Pains which are really located in the intestines may be felt as if they arise from an area of skin.

Equilibrium Perception

Maintenance of equilibrium results from the power of co-ordination of the muscles. This is achieved in vertebrates as a result of sensations perceived by three types of sense organ; they are the eyes, the proprio-ceptors of muscles and joints, and the specialized organs of balance located in the ampullae, the utriculus and sacculus of the inner ear. The power of orientation with respect to gravity is very important in animals which have achieved symmetry, and even more so in those which have not only symmetry but longitudinal differentiation along an antero-posterior axis. The mode of locomotion, the method of feeding, the position of sense organs and protective devices, these and many other features, depend upon this power of orientation.

Some Protozoa such as *Paramecium* show well-defined gravity reactions, though the mechanism is not known. In many invertebrate groups, statocysts are the organs of perception concerned. They have been described in Vol. I for coelenterates (Chap. 10) and for the crayfish (Chap. 22). Insects rarely possess statocysts, but it is known that the halteres act as gyroscopic mechanisms and perceive any departure from normal body position. In all these cases, reflex adjust-ment of body position is made.

In vertebrates, especially mammals, the eyes play a part in main-tenance of equilibrium. The animal tends to orientate its body in relation to objects around it. A man with locomotor ataxia can walk only by purposeful concentration on placing his foot at a point determined by his eyes. If he closed his eyes, he would at once fall. The giddiness which some people experience on high cliffs or towers when looking down, is due to unfamiliar visual conditions; closing the eyes will end the sensation temporarily.

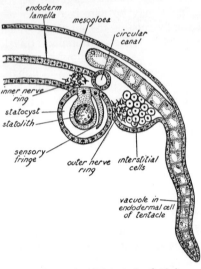

Fig. 16.54. V.S. through an ad-radial tentacle of *Obelia* medusa showing statocyst.

Streams of impulses from proprioceptors in muscles and joints are constantly flowing in to the central nervous system. It may be said that they are informing the control centres in the cerebellum of the degree of stretch of muscles, tendons and ligaments.

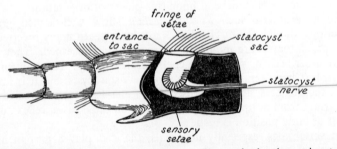

Fig. 16.55. Statocyst of *Astacus* in antennule. The proximal podomere is cut longitudinally.

There are, however, specialized organs in vertebrates for perceiving any departure from equilibrium. They are patches of sensory epithelia located in the ampullae of the semicircular canals, in the utriculus and sacculus. The description applies to the mammal, and the locations of

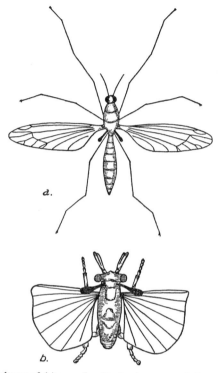

Fig. 16.56. Halteres of (*a*) cranefly, *Tipula oleraceae* and (*b*) a strepsipteran, *Stylops*.

these special sense organs in the membranous labyrinth are shown in the figure on p. 662. On the inner wall of each ampulla, projecting into the cavity, is a crest called the crista acustica; in the utriculus is a similar patch, the macula utriculi, and in the sacculus, the macula sacculi.

A single ampullary sense-organ consists of the crista, which is enveloped except at its base by a gelatinous cupule. The sensory cells of the crista bear long stiff hair-like processes projecting into the gelatinous mass (*see* Fig. 16.58). Movement of the head, for instance to the left, will cause greater pressure of the endolymph on the cupula and hence on the hair-like processes. The disturbance sets up an action-potential in the nerve-endings which lie in or between the sensory cells. The impulse is propagated along the VIIIth nerve into the cerebellum where the necessary reflex adjustment is made, and the head is moved into the correct position again.

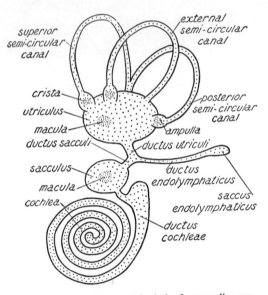

Fig. 16.57. Membranous labyrinth of mammalian ear.

These organs have been called "organs of the sense of space." It seems certain that they give us our conception of three-dimensional space, and in the past it has been customary to relate this with the positions of the semi-circular canals which are in three planes at right-angles to each other. Whether these three canals, in the three planes of space, give us our appreciation of space, is not known. What is certain is that the cristae play an important part in the sense of balance.

Proprioceptors. Some nerve cells situated in cranial and spinal ganglia have afferent fibres leading from many of the deep tissues of the body. In these tissues, the free end of the fibre takes the form of very fine fibrils which are in close contact with the cells of those tissues: a condition of free nerve-termination. These nerve endings in an internal epithelium are enteroceptive in function but those commonly found in muscle, tendons, connective tissue and joints are proprioceptive. Some are found even in bone marrow. Such nerve endings may sometime be directly related to specialized proprioceptor structures such as the muscle and tendon spindles. Muscle spindles occur in striated muscle, most often close to the insertion of tendon. Each has the shape of a long, thin spindle (fusiform) and is surrounded by a thick sheath of connective tissue, within which is a small bundle of muscle fibres.

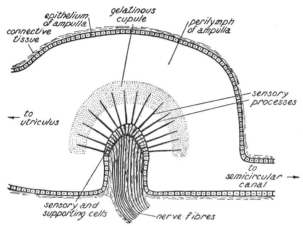

Fig. 16.58. V.S. single crista acustica.

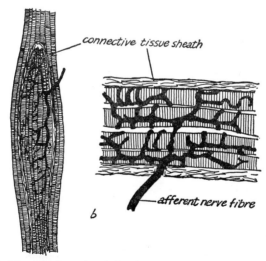

Fig. 16.59. Diagram of muscle spindle. (a) Whole spindle. (b) Part enlarged showing nerve ending.

These fibres, known as intrafusal muscle fibres, are generally shorter and narrower than the normal and the striations show only weakly. They enter the spindle from one end as prolongations of small fibres outside. Entering a spindle is one (or more) afferent nerve fibre which ends freely on the intrafusal fibres. The ending may form spirally coiled branches around the fibres or merely break into a system of diffuse branching fibrils (*see* Fig. 16.59). One end of the sheath is attached to the tendon, and the afferent nerve ending presumably registers changes in the contraction of the muscle. Entering the spindle, there are also efferent nerve fibres which end as motor end-plates on the muscle fibres.

Another kind of afferent nerve-ending can also be found in striated muscle. On the rounded ends of some of the fibres, where they are attached to tendon tissue, may be seen terminal arborizations (*see* Fig. 16.60). These also would appear to be proprioceptive.

The organs of Golgi, found in tendon, are of comparable function. These organs, sometimes known as tendon spindles, resemble in general form the muscle spindles, but, of course, they contain tendon fibres, not muscle. These structures are regarded as tension receptors.

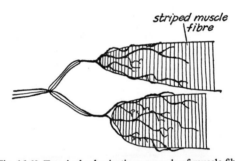

Fig. 16.60 Terminal arborizations on ends of muscle fibres.

The proprioceptors serve to give an animal some awareness of the state of its muscles, tendons and other connective tissues, i.e. whether under tension or otherwise. In so doing, they do not cause the registration of any conscious feeling in the animal, but undoubtedly serve in the co-ordination of movements made by the animal and in assisting it to maintain the correct tissue tensions consistent with correct equilibration. In this connexion, the animal is sometimes said to possess a *kinaesthetic sense*.

Colour Change Responses

The possession of special pigment cells or chromatophores by many animals was referred to in Chap. 6. Such cells are usually located in the skin, but may be found in deeper-lying tissues. They can be seen very easily in the stretched web of a frog's foot, for example. Within the chromatophore, pigment granules are developed and these may be of various colours. Melanin is a black pigment and is the commonest, but browns, reds, yellows and whites are comparatively frequent. Several pigments may be contained in one cell. The pigment cells influence the general coloration of the animal in one of two ways. A *physiological colour change* is effected when the cell either disperses the pigment over a wider area or concentrates it into a very small space. The wider the area covered by the pigment, the more will it tend to impart its own colour to the skin. When the cell contains pigment in a concentrated state, it is referred to as *punctate* and when expanded as *reticulate*; an intermediate condition is *stellate* (*see* Fig. 16.61).

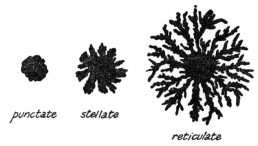

punctate   stellate

reticulate

Fig. 16.61. Punctate, stellate and reticulate chromatophores.

Chromatophores may also influence the skin colour by varying the amount of pigment which they contain and by being developed in greater or smaller numbers. A colour change produced by such means is known as a *morphological colour change*. Generally, both physiological and morphological changes occur in the same animal.

The most usual form of chromatophore is that occurring typically in crustaceans and vertebrates. It is made up of a single cell or small syncytium with a highly irregular outline as shown in Fig. 16.61. By streaming into or out of the branched cell processes, the pigment is distributed or concentrated. The pigment cells of cephalopods, e.g. octopus and squid, are distinctly different. Each is composed of a central cell containing the pigment substance enclosed by a very elastic membrane. Radiating from this, in a plane parallel to the skin surface, are a number of uninucleate smooth muscle fibres. By contracting, these fibres stretch the central cell out to a greater size, so increasing the area covered by pigment, and a darker colour is produced (*see* Fig. 16.62). When the fibres relax, the colour density is reduced as the pigment cell regains its smaller dimensions. The contractions and relaxations of the muscle fibres seem to be under nervous control and the changes in colour are generally much more rapid in the cephalopods than in any other animals.

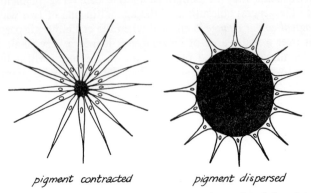

pigment contracted   pigment dispersed

Fig. 16.62.   Contracted and dispersed chromatophore of cephalopod (*after Bozler*).

A number of factors of the environment are known to influence colour changes in animals. Temperature, touch, humidity and psychical stimuli are all known to produce responses in certain cases, but light seems to have the greatest influence of all and affects more types than any other stimulus. The light, in most cases, seems to be effective through the eyes, involving therefore the peripheral and central nervous systems. The efferent pathway from central nervous system to chromatophore effector may be directly nervous, or less directly hormonal, or in some cases both. Colour changes which are known to be controlled through the eyes are known as secondary responses to differentiate them from primary responses made under the influence of light received by any means other than the eyes. In most adult animals, the secondary colour changes dominate any others. In the shrimp, *Crangon*, it has been clearly shown that unless the eyes are present and functional, no colour changes will occur under any stimulus, the chromatophores remaining fully expanded. In this case, it has also been shown that the mechanism controlling the chromatophore changes is hormonal. The eye stalks are the sites of production of a hormone which when relayed to the chromatophores via the blood stream, causes them to concentrate the pigment. A hormone produced in the rostral region of the animal will cause the reversal of this process. In the general case, it is not the total amount of light falling on the eye which is the factor concerned in evoking the response, but rather the ratio between the amount of light falling on the eye from above and the amount of light entering the eye by reflection from the animal's background. On equally illuminated dark and light backgrounds, the animal tends to become darker or lighter accordingly, so that the response is conducive to a good degree of protective coloration with respect to the background. Animals which normally live closely applied to the bottom of the sea, or the soil, or are more or less permanently attached to other structures, tend to show the response to a well-marked extent. But often the response is not only one of becoming lighter or darker in terms of white and black. Many animals have the capacity to match coloured backgrounds. The prawn, *Hippolyte*, will match reds and yellows, and the prawn, *Palaemonetes*, is said to be able to match black, white, red, yellow, blue or green backgrounds. Such adaptations are due to differential movements of the various pigments of the chromatophores, the colour in question being dispersed while the others remain concentrated. Perhaps one of the most striking of such colour-matching adaptations is to be seen in the flat fishes, which not only develop the right colours, but in the right places to match the chequered pattern of the background on which they rest, so making themselves almost invisible. The colour-adaptive capacities exhibited by a wide variety of animals indicate that the eyes, in so far as they are concerned with chromatophore activity, are not only able to detect how much light comes from above or below, but also are able to distinguish between the wavelengths of the light which may be reflected from the background.

Of the other factors known to affect pigment disposition in animals, temperature has been mentioned in Chap. 6, where it was suggested that pigment dispersal may serve a thermoregulation function altering the heat-absorbing and reflecting properties of the skin. This is not a universal condition among animals, however, and only in some reptiles does there appear to be any true correlation between colour and temperature. In the range 35–45°C, the tendency is towards pigment concentration, thus lightening the animal, whilst in the range 1–5°C, darkening occurs. Among the invertebrates, the position is not at all uniform. For example, in the crustaceans, both high and low temperatures may cause darkening or they may both cause lightening. There is great variation among species as to which effect is exhibited.

Touch stimulation is not effective in producing colour changes generally, but it has been recorded that stroking of suctorial discs of the appendages of some cephalopods, will cause the chromatophores to disperse their pigment. Atmospheric humidity changes likewise only produce colour change responses in isolated cases. The stick insect, *Dixippus*, is an example. The brown variety is normally dark by night and light by day and this has been proved to be due to changes in humidity and not to variation of light intensity. If the posterior part of a pale insect is enclosed in a damp chamber whilst its head protrudes into dry air, the whole body darkens from the head backwards. When returned to dry air, the whole animal blanches in an hour or two. In this case, the response has been proved to be due to detection of the stimulus by nerve endings in the skin, followed by neurosecretion of a hormone from the brain into a gland (*see* Vol. I, Chap. 23). This then secretes a hormone into the blood. On conduction of the hormone to the chromatophores, they disperse their pigment and the animal darkens.

When excited by the presence of enemies or possible mates, some animals make very distinct colour change displays. Squids, cuttlefish and some lizards are particularly noteworthy examples. When preparing to mate or fight, there may be colour changes which sweep over the body from end to end in rapid succession giving the appearance of a wave-like series of changes. Such colour alterations may be rightly ascribed to stimulation of the senses merely by the presence of the mate or enemy, and they reflect the state of excitement which is produced in the animal.

It was earlier stated that the co-ordinating link between the stimulus receptors and the chromatophore effectors could be nervous, hormonal or a combination of both. The mechanism varies in the different animal groups. Some annelids, particularly leeches, show a condition where the chromatophore responses are wholly under nervous control. Those sea urchins which make colour changes could be within the same category, although there is evidence that light-intensity changes may directly affect the chromatophores without any part of the nervous system being involved. The cephalopods have been mentioned previously as possessing a type of chromatophore peculiar to themselves. The chromatophore muscle fibres receive nerve endings and it is through these that they are stimulated to contract or relax, thus dispersing or contracting the pigment. Among the insects, colour changes are nearly all morphological. One of the few exceptions, *Dixippus*, has been mentioned. The destruction or increased production of pigment by which morphological colour changes are characterized, is influenced by illumination, background colour and sometimes humidity, but response to changes is a slow process and the mechanism involved is little understood. Among crustaceans, the co-ordinating mechanism is undoubtedly a combination of the nervous system and hormones. The eyes seem to be responsible in most cases for initiating nervous impulses under the influence of changing light intensity and background colour conditions. These impulses, on being relayed to the brain, stimulate neurosecretion via the *sinus glands* in the eye-stalks, and the *post-commissural organs* just behind the brain, of two hormones called *chromatophorotropins* (*see* Vol. I, Chap. 22). These have antagonistic effects, one causing pigment in the chromatophores to expand, and the other causing it to contract, thus having lightening and darkening effects respectively. These hormones are discharged into the haemocoel by way of the sinus glands and the post-commissural organs and circulated in the blood. This type of chromatophore has been shown to occur in prawns and other large decapods. However, in some Crustacea, which appear to lack both the sinus glands and the post-commissural organs, there is probably direct neurosecretion from the brain and possibly also from some of the ganglia, the chromatophorotropins having the same effects. Control of the chromatophore responses, in all the vertebrates which possess them, seems to be only hormonal in many cases, but direct innervation of the pigment cells is the case in some fishes and lizards. In such cases, the pigment movements are certainly under the influence of substances released at the nerve endings in contact with the chromatophores. Some fish species, e.g. *Fundulus*, show an apparent double innervation of the pigment cells and it may be the case that one nerve causes pigment dispersion whilst the other inhibits it. In some other bony fish and in all the cartilaginous species investigated, the control is at least partly hormonal and the pars intermedia is the site of hormone production. In the dogfish, *Mustelus canis*, blanching is caused by stimulation of the nerves and darkening is caused by a hormone. In *Scyliorhinus*, the chromatophores have no innervation and their regulation is entirely hormonal. In the amphibians, it is certain that it is the pars intermedia of the pituitary gland which is the source of the substances which cause pigment dispersion and contraction. There is no innervation of the chromatophores. The concentration of the hormones in the blood depends almost entirely upon stimulation of the eyes by changing light intensity and background conditions, since blinded animals make practically no responses. There appear to be three hormones concerned with colour changes in frogs. One substance from the pineal, *melatonin*, causes lightening of the skin. Two others, termed α- and *β-melanocyte-stimulating hormones* (MSH), produced in the pars intermedia, respectively cause lightening and darkening. Purely hormonal responses are slower than those caused by nervous stimulation, and animals possessing this kind of co-ordinating mechanism, lack the ability to reproduce contrasting colour patterns to match a mottled background, except in so far as the chromatophores on different areas of the body may be different colours. Frogs cannot reproduce a dark and light chessboard pattern, as can some flatfish. Among the reptiles, only lizard forms seem able to make secondary colour-change responses and these may be controlled in several ways according to species. In the chameleon, the pigment cells are directly innervated by the sympathetic system and the suggested scheme of control is as follows. The chromatophores in darkness are kept in a state of contraction by autonomic impulses through the fibres. An afferent impulse caused by light falling on the skin, or by one from the retina when the animal is on a dark background, causes an inhibition of this autonomic impulse, so that the pigment is dispersed. Another afferent impulse from the retina, when the animal is on a light background, cancels any effect of direct light falling on the skin. This scheme accounts for the usual colour changes of the animals which are very pale in total darkness and dark when blinded. But in the light, they darken when on a dark background and blanch when on a light background. Studies on the horned toad, *Phrynosoma* (a lizard), have shown that the control is partly nervous, partly hormonal. Lightening in colour can be caused either through the nervous system or by adrenalin in the blood, whilst a second hormone from the pituitary gland can cause pigment dispersion. In the iguana, *Anolis*, the nervous system, other than that of the eyes, plays no part in colour changes. Once the stimulus has been received through the eyes, control is entirely hormonal and the site of hormone production is in the pituitary gland.

In mammals, the pars intermedia produces α- and β-MSH, which affect the formation of melanin granules in chromatophores, and may also affect their distribution. Light stimuli perceived by the eyes are mediated by both the pineal and the hypothalamus (*see* p. 280), the former secreting melatonin, and the latter producing a releasing factor inducing the production of MSH. It should be noted that changes of colour produced by blood supply to the skin do not affect the chromatophores. Sun-tan is due to production of greater quantities of melanin, after stimulus by MSH.

ANIMAL BEHAVIOUR

In the broadest sense, all the activities, continuous and otherwise, of an organism, contribute to its behaviour. Any animal is carrying out some activities continuously and therefore it exhibits some pattern of behaviour as long as the external and internal conditions are favourable to life. For example, the cells are always respiring; there is movement of some kind proceeding all the time; the cells continually rid themselves of waste products, and so on. Even when an animal is to all appearances completely at rest, the fact that it is alive at all, is the manifestation of behaviour of its parts. But superimposed on this background of continuous activity, there are actions which are specific and clear-cut and can be related to definite changes in conditions both internal and external. An animal makes a response to a stimulus, and in so doing, exhibits a particular pattern of activity which can be directly associated with the stimulation. This is the more restricted meaning of behaviour and is the one that is usually employed. By restricting the meaning, the physiologist draws a line between the phenomena of metabolism, growth, development, regulation and co-ordination, and the phenomena of response.

There are many complications which arise in the study of behaviour; some lie in methods of approach and others in the interpretation of observations. In the first place, it is not easy to be completely objective, that is, to view the organism in a completely detached way. It is too easy to attempt to construe a particular activity on the part of one animal in terms of similar activity in others, particularly man. It is very difficult to avoid this tendency toward *anthropomorphism*. Even after strictly objective observation and recording of the patterns of an animal's behaviour, it is very difficult to explain these *ethograms* without being subjective, without crediting the animal with conscious ness and thought.

Then there is often great difficulty in working out the details of the complicated stimulus-response processes which underlie behaviour. Knowledge of the structure and mechanism of the parts concerned, is well advanced in some cases, but the understanding of the intricate activity of a complicated nervous system is so far not within full grasp. There is the inevitable clash between vitalistic and mechanistic inter-pretations. What cannot be explained in terms of physics and chemistry is referred to a life-force, as yet not clearly defined.

To the beginner, the subject cannot be easy, since it can be approached properly only when there is a deep knowledge and understanding of all the other aspects of physiology, coupled with some appreciation of psychology and sociology. We can do little more to help the beginner than to set out the generally recognized content of the subject, indicating and exemplifying the types of response which may be encountered.

There is no doubt that an animal's behaviour, like its anatomy and physiology, is a product of evolution. It is rarely possible to describe how it has evolved, but it certainly has inherited components. Super-imposed on these are the "learned" components acquired as a result of experience, and often it is almost impossible to separate the two. Several categories of inherited behaviour are described below.

In recent years, the main tendency in *ethology*, the study of behaviour, has been the insistence on a much more rigorous approach to the subject,

especially bearing in mind a main axiom of science, that any phenomenon which can be observed can be explained. We might not yet be in
possession of enough facts to explain a particular example of behaviour,
and then we must have recourse to the formulation of hypotheses and
the testing of them, one by one. However, there is always value in
accurate recording of objectively studied behaviour, as long as we do
not attempt to give attractive and apparently reasonable explanations
which are not based on a full knowledge of the facts.

There are several questions which are pertinent in any attempt to
explain a particular example of behaviour. Firstly, we have to inquire
how it is brought about, what structures and mechanisms are concerned
and what are the relationships between them. Secondly, it is necessary
for us to know the function of the behaviour, of what use it is in the life
and survival of the animal. If we can answer these satisfactorily, we
then need to know how and when the ability to behave in a particular
manner developed in the creature's life and, what is much more
difficult, how it evolved in the species. Finally, since much of behaviour
is species-specific, we should try to investigate how much of the behaviour is innate and how much acquired by experience. It can be
stated categorically that on these criteria, no single, full explanation of
any animal's behaviour, or even of one aspect of it, has been completed.
The results of all the patient and careful work of ethologists are necessarily fragmentary.

Some of the difficulties may be exemplified by trying to solve an
apparently simple problem. For what reason *do wood-lice congregate
in groups in damp and dark crevices*? By maintaining constancy of all the
environmental factors we can think of, we can show by laboratory
experiments that they prefer certain types of food, that they are
negatively phototactic, that they like dorsal and lateral as well as
vertical contact with solid objects (*thigmotaxis*) and that they prefer
humid to dry air. In addition, it can be shown that they secrete a
pheromone which attracts them to each other. If we provide the wood-
lice with what we consider to be optimal conditions, some of them,
possibly all, will occasionally leave this environment and move for a
time into drier air, or brighter light, or higher temperature, etc. Perhaps
they have a homeostatic "drive" to lose water, or to find different food;
we do not know the answer. The problem would be less difficult if all
the wood-lice always stayed in such an environment; unfortunately,
they do not. Our experiments have ignored variation in their physiology, and also the influence of internal factors on their behaviour.
Nevertheless, every piece of experimental work and every observation,
helps in the gradual building-up of the full explanation.

Not very long ago, it was the usual practice to classify forms of behaviour into three groups. In the first group all the so-called simple reflexes were placed; in the second group, sequences consisting of innate stages of behaviour and called instincts; and in the third group, learned behaviour, due entirely to the individual's experience. The term "instinct," as a name for behaviour patterns, has now been abandoned, mainly because it has been used to describe a variety of phenomena, and also because much of the behaviour formerly considered "instinctive" has been shown to have both innate and learned components.

We can do little more to help the beginner than to set out the generally recognized content of the subject, indicating and exemplifying the types of behaviour which may be encountered.

Orientation in the Living Space

Animals tend to place themselves in advantageous positions in the regions which they inhabit. To accomplish this, they make movements; some of these movements are not directed with regard to the source of a stimulus, and some are directed. Those movements in which the *rate*, but not the direction, varies with an environmental stimulus, are called *kineses*. The locomotive response bears no relation to the direction of the stimulus. *Orthokinesis* occurs when the speed of movement is under the influence of some diffuse external factor such as intensity of light, temperature or degree of humidity. For example, the wireworm larva of *Agriotes* is very sluggish in saturated air, but becomes more active as the air begins to dry. In the soil, it responds to the moisture content and eventually, without any directed response, comes to moist conditions and, because of the considerably lower rate of movement, tends to remain there. Wood-lice (*Porcellia sp.*) are similarly more active in moist air; their rate of movement can be plotted and it is found that they move faster in dry air than in moist air, where they tend to remain because of the lessened rate of movement. Animals which show orthokinetic responses, tend to aggregate in optimal conditions, because they move more slowly in these conditions and thus tend to remain there. Laboratory experiments on orthokinesis can be carried out in *choice chambers* of various patterns (*see* Fig. 16.63). The apparatus can be adapted for use with a variety of stimuli, and for aquatic as well as terrestrial animals.

The aggregation of animals in optimal conditions may also come about by changes in the rate of random turning which an animal makes as it moves. This is termed *klinokinesis*. Planarians in a half-shaded choice chamber will eventually cluster in the darker part. If conditions are favourable, the animals change direction infrequently; if they encounter less favourable conditions, the rate of turning increases. If a sheet of squared paper is placed beneath the glass floor of the choice chamber, the direction changes can be plotted against time on a similar sheet of paper.

A more efficient method of arriving at a favourable area is to move directly towards the source of a stimulus or directly away from it, or at a fixed angle to the direction of the source. Such directed movements are called *taxes*. The movement may be towards or away from the source of stimulation, i.e. positive or negative, and there are a number of possible stimuli, such as chemicals, water, sound, touch, light, water-currents, gravity and heat. The responses are described by adding a suitable prefix to the word "taxis"; for example, chemotaxis, hydro-taxis, audiotaxis, thigmotaxis, phototaxis, rheotaxis, geotaxis and thermotaxis. An organism exhibits *klinotaxis* if it attains and maintains orientation by continuous oscillation of the whole or part of the body (e.g. the antennae or the head) about the line of stimulus. For example,

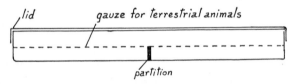

Fig. 16.63. Choice chamber in section; this can be circular or rectangular.

the larvae of the common house-fly move away from a single source of light, swinging the head from side to side as they move. The light-sensitive organ is a cluster of photoreceptor cells above and behind the mouth; this response is *negative photoklinotaxis*. The opposite condition, positive photoklinotaxis, can be observed in a euglena, which maintains a position in which its anterior-posterior axis is along a beam of light; any deviation from this will cause a shock reaction which automatically swings the axis back into the path of the beam. Similar klinotactic response seems to be the general case for organisms in which there is a single stimulus-intensity receptor.

Organisms which show *tropotaxis* move directly towards, or away from, the source of a stimulus. They have paired receptors and main-tain orientation, without deviation, by comparing the intensity of stimulation on both sides of the body. If confronted with two sources of a stimulus such as light, such an organism will orientate itself on a line between the two sources so that the left and right receptors are exposed to equal intensity. When an animal is under the influence of two different stimuli, to each of which it responds tropotactically, then the two stimuli are treated as being of the same kind, and the animal will move along a resultant between the lines of action of the two stimuli. For example, the slug *Agriolimax*, shows negative phototaxis and negative geotaxis. When placed on a slope and illuminated from one side, it will move upwards at an angle away from the light. In an animal which shows tropotaxis, if one of a pair of receptors is put out of action, e.g. the left eye covered or painted over, the animal will circle, in this case to the right.

Telotactic behaviour is exhibited by animals which, when exposed to two sources of stimulation, orientate themselves directly with relation to one of them, not intermediate between the two. The response occurs chiefly with light stimuli, and an animal may switch orientation unpredictably from one source to the other, thus exhibiting irregular zig-zag movements. Many insects (e.g. the honey bee) and crabs (e.g. *Eupagurus*) exhibit telotaxis. A honey bee will walk a beam of light towards its source even if one eye is painted over. If an animal which shows positive phototelotaxis is confronted with two lights of equal intensity it will move directly towards one or other of the lights; this obviously involves selection. On the other hand, an animal which shows klinotaxis or tropotaxis will move along a line equidistant from both sources of the stimulus.

An animal is said to exhibit *phobotaxis* if it responds to an unfavourable stimulus by making an avoiding movement. For example, if a paramecium encounters an obstacle, strong light, high or low temperature or irritating chemicals, it will respond by reversing the beat of its cilia, slowing down the rotation of its body about its longitudinal axis, and pivoting its anterior end towards the aboral surface through an arc centred on its posterior end. The net result is that it moves away from the source of the irritating stimulus for a short distance and then proceeds forward in a different direction from the previous forward movement. This may occur repeatedly until eventually a direction is found in which the stimulus is no longer effective. Because of these phobotactic reactions, many Protozoa and motile algae tend to remain in the most favourable conditions available to them. Some tactic and kinetic responses have elements which can be ascribed to phobotaxis or shock reaction.

There are a number of biologically interesting examples of orientation behaviour where an animal's movement is directed at an angle with the direction of the stimulus, and not along its line of action. Insects such as bees and ants, returning to their nests, can orientate themselves by moving along a line at a fixed angle to the sun's rays. A homing young ant, if covered with a box for several hours so that the angle of incidence of the sun's rays has changed, will continue to move along its original path and will miss the nest. It will usually find it after searching the area. A mature ant allows for the sun's apparent movement and proceeds in the same compass direction after liberation. This shows learning from past experience. This *light-compass* orientation is not a fixed feature common to a whole species but must be learnt by each individual and for each journey; furthermore, return orientation is obviously the opposite of that on the outward journey. Lindauer brought some Indian bees across the equator to Munich; after a short period, they were able to navigate by the sun, then south of them instead of north. If hive bees are not exposed to the sun in their early (imago) life, they cannot develop light-compass navigation; this underlines an essential part of much learned behaviour in animals, that of a sensitive period in early life. Bees use sun navigation even when the sky is completely clouded; this is achieved by their perception of the plane of polarization of light from the sky.

Some species of bees navigate by means of scent trails; they have been observed daubing body secretions on objects along their flight paths. Most, if not all, species of bees, ants and social wasps can steer by landmarks (*pharotaxis*); this, of course, involves learning.

The great majority of animals orientate themselves at right-angles to the incident light from the sky. This *dorsal-light reaction* is important with regard to the position of the body in space. Some few animals which swim upside-down, e.g. the brine shrimp *Artemia*, show *ventral-light reaction*. A fish with its ears destroyed will still swim with its dorsal surface facing the light; under experimental conditions, lateral light evokes swimming on the side, but still with the dorsal surface facing the light. Though postural orientation in vertebrates is mainly determined by impulses from sense organs in the inner ear, nevertheless, this may be reinforced by a dorsal-light response.

Migration

This may be considered as orientation on a grand scale. In this connexion, migration is defined as the movement of whole populations from one region to another and their return to the original habitat. In many species of birds and butterflies there is a *circannial* rhythm (about a year between the beginnings of successive migrations). Such movements range from relatively short journeys such as those of many mountain species, e.g. chamois, *Rupicapra rupicapra*, which move down the mountain slopes as winter approaches and back up in the spring, to the spectacular migration of the Arctic terns. These birds breed in the Arctic in summer and during September and October they fly 17 500 km to the Antarctic pack ice where they spend the southern summer (northern winter) and then fly back to the Arctic during March and April. Thus they have four months northern summer, four months southern summer and two periods of two months each for the southerly and northerly journeys. The greater part of their stupendous flight lies over the sea.

A great many species of birds migrate shorter distances. A very interesting example is that of the European white stork, *Ciconia ciconia*. These birds breed in central and southern Europe in summer and spend our northern winter in equatorial and southern Africa. It is a strange fact that those which nest roughly west of a line of longitude through Denmark, fly first south-west, cross the straits of Gibraltar, and then proceed overland to west central Africa, while those which nest east of this line fly south-east, over the Bosphorus, and winter in the east central and South Africa. On their migration routes, none of the storks is ever out of sight of land. In an experiment, eggs of "eastern" storks were hatched in West Germany and the birds were not allowed to migrate until all the "western" storks had gone. When freed, they flew south-east, over the Bosphorus to East Africa. This is obviously innate behaviour of a fixed species-characteristic type. We cannot yet explain it, but we can say that the animals had no possible opportunity of learning the direction and route.

Many other examples of animal migration pose what are as yet unanswerable questions. Salmon hatched in streamlets in Western Europe spend a long period (1 year in England; up to 7 in northern Scandinavia) growing and maturing in fresh water and making their way down to the sea. At the smolt stage they arrive in the open sea and take 1 to 6 years reaching sexual maturity. Tagging experiments prove that they return to mate and spawn in the actual streamlet where they were hatched.

Eels, born in the depths of the Sargasso sea, swim as leptocephalus larvae across the Atlantic Ocean to the western coasts of Europe, taking 3 years. At the elver stage, they swim up rivers to their higher reaches, often journeying overland to ponds and lakes. In their fresh-water habitats, they feed, grow and mature, the males taking 4 to 8 years and the females 7 to 12 years. Once mature, the eels journey down to the sea and back to the Sargasso where they breed and never return.

Green turtles (*Chelonia mydas*), which live in shallow waters off the eastern Brazilian coast, travel 2500 km across the open Atlantic to the isolated Ascension Island where they mate and lay their eggs. The young turtles are hatched at night and they make straight for the sea, the directional stimulus being the quality of light above the water. Those that survive the numerous hazards eventually arrive at the Brazilian feeding grounds where they grow and mature.

The study of the migration behaviour of even a single species is a formidable task. In a great many cases there are accurate and objective accounts of the observed behaviour, but we are very far from explaining it. In the case of many bird migrations, there is undoubtedly sun-compass navigation which is often coupled with pharotaxis, but cases such as that of the white storks cannot be thus explained. We can often see some advantage in a particular type of migration, e.g. swallows cannot find their food in northern climates in winter; salmon eggs and young will not develop in salt water; the purer water of the open ocean is a better medium for the growth of young turtles than the variable coastal waters. We are still left, in all these cases, with the necessity of explaining the behaviour in terms of the animal's nervous and endocrine systems.

Exploration

All the movements so far described have, in general, been responses to external stimuli, coupled in some cases with adaptability attributed to previous experience. It is a matter of common observation that animals make searching movements in their living space; these movements are in many cases necessary to supply bodily needs such as food, water or a suitable temperature. Such movements have a homeostatic function, and nowadays they are generally termed "*appetitive behaviour*". These ranging movements are usually unpredictable, especially on first occupation of a living space. Often, however, an animal will develop a preference for certain feeding and drinking sites and will move directly to them; it is thus that appetitive searching has led to learning.

Apart from appetitive behaviour, many animals range round their environment for no reason which we can discern. They seem sometimes to be activated by what in human terms we would call curiosity. Careful observation has shown that bees in a hive spend a great deal of time in apparently aimless movement about the hive; this occurs especially in the early imago period. If laboratory rats are placed in a new enclosure, they will, even if hungry, neglect their food in favour of exploration of their new quarters. The same sort of behaviour can be observed in many animals in most of the big groups. It seems that there is a "drive" to explore, to experience a variety of kinds of stimulation, especially in the young. Many experiments have shown that ability to solve "problems" in later life is to some extent dependent upon the diversity of stimulation present in early life. Observation of mammals in their native environments shows clearly this tendency of the young to explore. Experiments with animals deprived of such variety in sensory experience when young show that in later life they are less capable of solving problems than are those with a multifarious experience of sensory stimuli.

Animal Societies

Few animals live alone all their lives without some contact with their own species, if only for the purpose of sexual reproduction. Some, however, form well-established aggregations for part, or all, of their lives. Thus collective nouns have come into common use to signify such groupings, e.g. herd, shoal, school, pack, swarm, etc. The functional value of such groupings may be mutual protection, or defence of a feeding area, which would tend to restrict numbers to such as the habitat can support. There are possibilities of division of labour and more successful rearing of young.

Some of these social groupings have fixed and rigid caste systems; the best examples are found in the ants, bees and termites. The members of the group, in these cases, are kept together by the passing of food from one to another (*trophallaxis*) and by pheromones which give a common odour to the colony. Thus, a strange bee attempting to enter a hive is stopped, thoroughly investigated by guard bees and repulsed. The tasks of providing food, building and repairing the cells, tending the queen and the larvae, etc., are carried out by the workers. The queen alone lays eggs and one of the drones will mate with her. The bees communicate by odour, by touch and by species-specific movements known as "bee-dances." The dominance of the queen is maintained by the secretion of certain pheromones (*see* Chap. 8); the workers lick these secretions when they tend the queen, and the chemicals are quickly distributed round the whole colony. Indeed, if the queen is screened off by a meshwork, the organization will proceed quite smoothly as long as the workers can touch the queen.

In these insect caste systems, there are "reproductives" and "non-reproductives"; the distinction, in the case of the females, depends on the type of food supplied to the larvae. In bees, any female larva can be raised to the queen status by being fed with "royal jelly" (*see* Vol. I, Chap. 23).

The termites have kings, queens, workers and soldiers of both sexes. There are also secondary female reproductives with rudimentary wings; if the queen dies or becomes infertile, one or more of these secondaries will develop fertility after special feeding. The workers forage for food, build and repair the mounds, tend the larvae and feed the soldiers, which are unable to feed themselves. The latter act as guards against their arch-enemies, the ants. Both termites and ants, in some species, cultivate fungi.

The general pattern of organization in the social insects is species-specific and almost entirely rigid. Yet, in all cases, there is some ability to learn; all mature workers exhibit pharotaxis.

In contrast to the social instincts, there are large-scale but transitory groupings formed in the same species at certain particular times in the life. Examples are seen in the aggregations of frogs in ponds in early spring, the swarming of cephalochordates and certain marine poly-chaetes in the surface waters, in each case for breeding purposes. Swarms of locusts are of a different nature; they occur spasmodically and years may pass without any swarming. All locusts begin life as solitary grass-hoppers and may continue thus for the whole life-span. If, however, environmental conditions concentrate egg-laying in a few favourable sites, crowding occurs, and the young grass-hoppers begin to march, devouring all green plant material in their path; as they mature, they form the dreaded flying swarms. Gradually, predators and unfavourable weather reduce the numbers drastically and the survivors revert to the solitary life.

Somewhat more permanent aggregations are found in packs of wolves and wild dogs, herds of cattle, deer and sheep, troops of monkeys and flocks of some birds. In all of these communities, there is a type of hierarchy, which reaches its height in the "peck order" system char-acteristic of domestic fowls. In this case, each member of the group has a position or rank which it maintains by pecking any birds below it and refraining from pecking any above it in rank. The number one in the order has first choice of food and can peck all the others. Weaken-ing of any individual fowl, from any cause, will force its descent in the peck order.

The phenomena of dominance and submission occur commonly in avian and mammalian groups. In a pack of wild dogs, each member adopts a characteristic aggressive or submissive posture depending on which other member it is faced with. In some cases, e.g. sea-lions and stags, dominance is only achieved after a series of contests. They are usually trials of strength, the antagonists rarely inflicting serious injury on one another; death in such hierarchy contests is very rare. In the monkeys and apes, dominance is achieved less rarely by fighting, but by menacing postures and ferocious sounds. Each mammalian social species has evolved its own series of signals, which evoke counter signals from other members of the group and from other groups. Again, it is worthy of note that though this type of species-specific behaviour is partly innate, it is not completely so. Young animal which are deprived of normal group contacts in early life are late unable to take part in the life of the group.

Another line of evidence which points to the same conclusion is th careful examination of bird song. Many people can identify the char acteristic song of a few birds such as a sparrow, a blackbird or a thrusl and it is a prevailing impression that the song of every blackbird identical with that of every other. Careful analysis, especially in wor on chaffinches, has shown that there are what one might call "dialects" in the song. Chaffinches in adjacent valleys have slightly different song patterns. If young chaffinches are deprived of hearing the typical adult song, then in later life they will never develop it completely and perfectly. They may sing the same length of song, but the full melodies and correct sequences will never be achieved. In all the types of social behaviour which have been thoroughly investigated, there are evident innate and learned components.

Territorial Behaviour

In this connexion, a territory is defined as an area which is defended. The practice is common among many groups of animals; in some, a territory is established and defended only at breeding times; in others, the same territory is occupied throughout life. *Nereis pelagica* inhabits a small rock crevice which it defends against intrusion by others of its species. Fiddler crabs (*Uca* species) live in burrows in sand on the coasts of Panama and Brazil. At low tide, the crabs emerge and immediately commence signalling, waving the brightly coloured, enlarged right cheliped. Each male defends his territory, which is a small area, by vigorous contest with any intruding male. Sticklebacks normally live in shoals, but in spring each male becomes solitary and builds a nest in the sand beneath water-weeds. This is defended vigorously against other male sticklebacks; presumably attack is provoked by the red belly of another male, since it has been shown that models with ventral red colouration will incite attack. On the other hand, females with uniform silver are not attacked but courted. After egg-laying, the male guards the nest and maintains a current of water over the eggs by fanning with his fins. After hatching, the male continues to defend the young for some time, any tendency to eat them being suppressed.

Herring gulls establish small territorial nesting areas on rocky cliffs or islands, defending them with loud-squawking, wing-flapping, threatening postures and even direct attack. Each pair can identify their own site among a thousand others; this must entail learning and memory of small landmarks.

A male robin will reconnoitre a territory and gradually establish it as his own, by singing first from the ground, then from higher points, and finally from the highest points in the area. He will defend this territory throughout mating, nesting and brooding the young, by exhibiting his red breast in aggressive posture. If these fail, there will be actual combat with any intruding robin. Other species are, however, tolerated within the area, even if they nest there.

Among the mammals, there are numerous examples of territorial establishment and defence; in fact, it is almost the rule and not the exception. A troop of the ring-tailed lemurs of Madagascar will establish a territory among the trees by marking branches with exudations from the scent glands on the shoulders. A herd of hippopotami have a singular type of organization in their area, which is usually a mud- or sand-bank fringed with vegetation. The herd consists of 20 to 100 animals; the central area is tenanted by the females and young, while each male has a guard station on the periphery. Furthermore, each male marks his own feeding path with faeces and urine. At breeding time, there is violent fighting among the males and, though serious wounds are inflicted, they are rarely fatal.

The musk deer of Siberia, *Moschus moschiferus*, like most ungulates, are strongly territorial. Each herd occupies a territory which is about 3 square kilometres in area. There is usually a rocky outcrop somewhere near the centre of the territory: this acts as a look-out point and a central easily defended position. The male deer mark the peripheral regions with musk from glands under the tail, the marks being made on low branches and dead wood. The males defend their territory with vigorous combat, using their four-inch tusks to inflict severe wounds on intruders; each area is occupied permanently by the same small herd.

The red howler monkeys of the Brazilian forests average about 18 in a troop. The males are dominant and have a definite hierarchy with a "peck order". The females have a separate but similar order. The territory varies somewhat in shape and size according to the available food supply. The monkeys advertize their presence, and warn of intruders with their penetrating howls. Observers state that they begin their day very early in the morning with the most frightful cacophony. Within the territory, the members of the group occupy a fixed number of favourite sleeping trees. Close observation of several adjacent groups suggests that there may occasionally be some regrouping among the troops.

Most cases of territorial defence involve a system of signalling and of recognition of the signals of other individuals. The signals may be based on colour, posture, odour or sound, or any combination of these. All the signals, including the pheromones, are species-specific. Within each species there is variation between groups in small details of posture and sound, as much as one might expect in behaviour variation, just as it exists in anatomical and physiological variation. It is also significant that in laboratory experimental work, deprivation of contact with the species shows that the full territorial behaviour is never afterwards attained. The value to the species of territorial establishment and defence may be twofold. First, it may set a natural limit to the density of population in a given area and thus there will be less demand on the available food supply. Secondly, there is likely to be less disease and "emotional" disturbance than there would be in an overcrowded area. Under completely natural conditions, there is an optimal population for all the species in a given area; if this varies widely, it is soon adjusted by regrouping or by natural checks on population increase.

Homes of Animals

In this connexion, the word "home" is used to denote a more or less permanent centre, to which an animal returns after excursions into the environment, or in which it may stay and feed by extending movable structures out into the environment, or by creating inward currents of water from which it extracts its food. The home may provide one or more of these advantageous factors; shelter from severe environmental fluctuations, protection against predators, a relatively safe place in which to rear offspring. There are many common and well-known examples; the tunnels and chambers of earthworms, the burrows of rabbits, moles and voles, the nests of birds, the form of the hare, etc.

Some of the structures elaborated by animals are extremely complex; for example, the nests of bees and wasps, the pillars and mounds of termites, the webs of spiders and the nests of many birds. Others may be very simple; such are the scrape on the ground used as a "nest" by seagulls, a rock crevice used by a fox, the flattened grass of the hare's form and the depression in a rock to which a limpet always returns after feeding. The limpet enlarges the depression by erosion with the edges of its shell so that it always fits exactly. Many animals will use a burrow vacated by some other type of animal, others have highly characteristic burrows.

Different species of spiders have webs of different structure, each characteristic of its group. But there are obvious variations even among a single species, e.g. the webs of the common house spider, *Tegenaria domestica*. There is first a search for a suitable site, and then the main scaffolding of the web may be roughly triangular or rectangular according to the supports available. The webs of young spiders are much more rudimentary than those of adults; there is improvement. The common European swallow, *Hirundo rustica*, builds a nest of mud strengthened with pieces of dry grass. The mud has to be of a suitable consistency and the swallows will search widely to find it; then the structure is fashioned pellet by pellet. There is no standard place in which to build; the nest may be on a rafter, under an angle of the eaves, or even fastened to the vertical surface of a brick wall beneath a projecting roof. The pairs return to the same nesting places year after year.

Some interesting examples of homes are those of the hermit crabs. In *Pagurus bernhardus*, the young crabs occupy winkle shells, then graduate to top shells and finally to whelk shells. *Pagurus prideauxi*, which is commensal with the sea anemone, *Adamsia palliata*, starts by inhabiting a small shell and remains there for life, the overgrowth of the base of the anemone always providing sufficient shelter. In this case, it is interesting to note that *A. palliata* is never found anywhere else, but always in association with *P. prideauxi*. Many tropical hermit crabs use any available cover; they have been found in hollow bamboo canes, in broken glass jars and in a variety of molluscan shells. *Pagurus typica* uses no shell but relies solely on the cover provided by *Anemonia mammillifera*.

Caddis fly larvae may be divided into two ecological groups, those which live free and are at least partly carnivorous, and those which build portable cases or tubes and are herbivorous. In the latter group, the larvae of various species build a variety of tubes, all of which show careful and delicate workmanship. The tube is always open at the head end of the larva and the posterior end of the tube is sealed with a silken mesh; hooked appendages enable the animal to cling to the tube, from which it does not emerge until metamorphosis. Members of the genus *Phrygania* cut pieces of leaves and stick them together with silk. The most familiar tubes are those of the genus *Limnophilus*; they are made of tiny stones, pieces of plants or broken snail shells. The limnophilids will also use any material supplied to them experimentally, e.g. glass beads and small pieces of substances such as nylon. *Stenophylax sp.* use sand grains to make a straight tube, while *Heliopsyche sp.* use the same material to make helical tubes like snail shells. Cases made of heavy materials, such as small stones or sand, have their weight reduced as the larva grows by the trapping of a bubble of air. In all the tubicolous species, the tube is extended at the head end as the larva grows.

These examples, together with a great many others, pose difficult problems for the student of behaviour. While some aspects of home-making are innate, others are not. In all cases there must necessarily be a search for a suitable site or for suitable materials; this involves inspection, rejection and selection. In the limnophilid caddis flies, there appear to be only two criteria involved in selection of materials, firmness and size of particle; in the shell-inhabiting hermit crabs the criteria are size, shape and weight. The limpets select only rocks with smooth surfaces for they cannot adhere satisfactorily to rough surfaces. Formerly, all examples of home-building would have been ascribed to instinct, but it is certain that some aspects of the behaviour are not innate.

Courtship and Mating

Throughout the great animal groups, from some annelids to mammals there are numerous examples of ritualized courtship behaviour, some simple and some complex. These rituals depend on the giving and receiving of signals by the two partners concerned. The stimuli range through colour, pattern, touch, posture, specialized movements, odours and even in the emission of light, as in the fireflies, family Lampyridae, in which the flashing signals enable male and female to locate each other. A few examples of more elaborate rituals are described briefly here.

The fighting fish (*Betta splendens*) of Thailand are very pugnacious towards each other, especially the males; patches of scales and pieces of fin are bitten off, and sometimes one of a pair of antagonists dies after a contest. In the breeding season, the males become heightened in colour, the general background being a brilliant metallic blue, with flowing red fins. A male swims around a female with his fins spread and performs a series of dancing and embracing movements. Before this performance the male has already built a nest consisting of bubbles of air, each enclosed in sticky mucus; he makes these, one by one, in his buccal cavity. The male eventually turns the female on her side and wraps himself round her; in this position, the female lays 3 to 7 eggs and the male fertilizes them. He then stations himself beneath her and catches each egg in turn in his mouth, coats it with mucus and sticks it on the underside of the bubble raft. After all the eggs are thus treated, the male loops himself again around the female and the same procedure ensues. This continues until several hundred eggs are fastened beneath the bubble raft. Then he drives the female away and guards the eggs until the young hatch (24 to 30 hours later), and then leaves them to fend for themselves. Sometimes a female will prove unreceptive and will drive the male away; sometimes she will choose one of a pair of antagonistic suitors. It is of interest that some of the behaviour of the male may be stimulated by using a model of the female.

Fruit flies of the genus *Drosophila* will be familiar to most students in connexion with their genetical studies (*see* Chap. 19); less familiar, perhaps, is the important work which has been carried out on their behaviour. Among the various species, a female can recognize a male of her own species, by sight, hearing or smell, or by a combination of all three. A courting male will circle around a female, vibrate one or both wings, lick the female and finally mate. If the female is unreceptive, or the male is of a different species, the female will buzz violently and kick him away. If she is receptive (third to tenth day of imago life), she will open her genital plates and allow mating. Most remarkable is the fact that a female is only "tuned in" to frequency of buzzing of her own species. This buzzing is so faint that it can be recorded only by placing the flies on the actual diaphragm of a microphone in a sound-proofed box.

In these two examples, as in all cases of ritual courtship, the behaviour of the participants is not only complex, but species-specific. The whole sequence depends on successive signals, each of which elicits a particular response in the other partner. It is evident that much of the behaviour is innate, but there are always doubts as to how much of the behaviour depends on normal contact of an animal with its own species. Studies on rhesus monkeys have shown that if a male is isolated from birth, it is inept and sometimes impotent in sexual matters, even though it is sexually mature at the time of observation.

Some other Patterns of Behaviour

All animals with a sexual process make some provision for their offspring, even if only to the extent of providing the necessary genetical equipment, protoplasmic organelles and a sufficient store of food material in the egg. Many oviparous animals deposit their eggs in favourable positions for the protection and development of the offspring; a number of such examples will be found in Vol. I. The highest degree of parental care is found in the birds and mammals. The latter not only feed their young for long periods after birth, but also actively instruct them in various activities. Varying patterns of this parental care can easily be observed in domesticated animals.

In all the Eutherian mammals (see Vol. I, Chap. 4), the birth process takes place in essentially the same manner, with some variation in specific details. It has, therefore, been customary to state that the whole sequence is instinctive. This statement is, however, open to doubt. A female rat, isolated from birth and constrained by a type of collar which prohibited self-licking and grooming, proves to be quite inept at cleaning the young after birth and, in some cases, quite incapable of doing this. The experience of self-grooming and of licking vaginal and uterine secretions may thus, in some mammals, be an essential part of youthful experience and necessary to produce a good mother. There are a number of reports of the effects of the mere presence of a strange male on pregnant rats. Up to a certain stage of pregnancy, the female is quite likely to miscarry if confronted by a strange male rat.

One must be careful in interpreting laboratory experiments in which animals live in completely artificial conditions and, indeed in many cases, they have been bred under these conditions for many generations. Nevertheless, observation based on such conditions are valid as long as we are careful not to extend the interpretation of the results to the whole field of behaviour. The great need in modern ethology is the study of behaviour in natural environments.

Rhythms

It is probable that the great majority of animals exhibit rhythmic behaviour. A full account of this phenomenon, together with that of "biological clocks," is given in Chap. 6. There it is pointed out that periodicity in certain functions appears to be built-in and yet, in many cases, the natural rhythm can be altered under experimental conditions.

Drive

An animal in a resting state, apparently uninfluenced by external stimulation, will start moving, find food, start eating and finally stop eating. The same is true of drinking and of salt-licking. In the latter case, an animal may have to travel a considerable distance. The value of such behaviour is concerned with aspects of homeostasis, but we cannot state exactly what causes an animal to start eating and then to stop eating. It is no explanation to say that the animal is hungry and after a certain time it is not hungry; these are terms which describe the observed behaviour. It is certain that one of the factors governing feeding is the time that has elapsed since the last meal. Experiments based on locating fine electrodes in certain regions of the hypothalamus show that stimulation of one area will lead to non-stop eating; in another area, stimulation will lead to non-eating.

A house-fly will move around, apparently aimlessly, on an undirected course. Ultimately, this random movement may bring it into a position where sense organs on the antennae are stimulated by the odour of food. The animal will then move with directed flight towards the source of the stimulus. When it alights on or near the food, signals from taste organs on the antennae stimulate the animal to extend its proboscis. Other receptors on the end of the proboscis are next stimulated by contact with the food, saliva is secreted and the animal begins feeding by suction. It has been discovered that when the material in the insect's crop reaches a certain level, minute enteroceptors initiate impulses which travel in fibres of a prominent dorsal fore-gut nerve to the brain, and the animal stops feeding. If this nerve is cut, the fly will over-eat. This is a simple negative feed-back control; the survival of the huge numbers of flies is a measure of its efficiency. By analogy with similar work on mammals, it is presumed that enteroceptors are stimulated by the state of dilation of the crop in its anterior region. Impulses from these enteroceptors affect certain regions of the brain and two processes operate to stop feeding. Firstly, there is cessation of a neurosecretory hormone which activated salivation and, secondly, redirected impulses through motor nerves activate the complex muscles concerned with cessation of feeding and the proboscis is withdrawn. This account is obviously simplified; the contraction of one set of muscles involves the relaxation of others, also the receptors on the proboscis tip, etc., must either be ignored or temporarily inactivated. The example illustrates the difficulties of ethological study; a discovery such as that concerning the fore-gut nerve merely leads to further problems.

Work with laboratory rats shows that signals from the stomach wall influence the onset of feeding, but a rat will both start and stop feeding if all the nerves to and from the stomach are cut and will even do the same if the stomach is completely removed. There is some evidence that signals from the pharynx and buccal cavity play some part in both the commencing and the cessation of feeding. It is also a common observation that animals, for no detectable reason, will "go off their food." In man, this condition is attributed to emotional disturbance and it is called *anorexia nervosa*. Often, an animal, though "hungry" in the sense of the time that has elapsed since the previous meal, will refrain from eating in a strange environment or even if the food is placed in a strange type of container. Warehouse rats will ignore poisoned bait placed along their runs for several days; when the novelty wears off, they will consume it. It is evident that other parts of the brain, apart from the hypothalamus, are involved in the process of eating.

In the past, the term "drive" has had various meanings. Apart from its association with the hypothalamus in aspects such as hunger, thirst and temperature regulation, etc., it has been used to denote sexual phenomena, care of the young, territorial behaviour and a host of other behavioural phenomena. Nowadays, it is usually limited to aspects of homeostasis, and ethologists attempt to explain such behaviour in terms of the control systems of the body, nervous and endocrine.

Learning

It is difficult to give a comprehensive definition of learning; as in all attempts to explain aspects of behaviour, we are dealing with processes which we do not completely understand. Possibly an animal may be said to have learnt when it exhibits adaptive change in its behaviour apparently due to its past experience. Thus defined, learning will include at least some parts of all aspects of behaviour except reflex actions.

If a group of snails is placed on a board and allowed to feed or range around freely, a sudden bang on the board will cause the snails to withdraw into their shells immediately. Continuation of the stimulus at frequent intervals will result in the snails apparently ignoring the noise and vibration, at least as shown by the fact that they will not withdraw into their shells. Perhaps we can say that the snails have learnt that the noise and vibration are irrelevant to their well-being. It is to be noticed that this explanation is in human terms; it is what we would possibly do under similar circumstances. We do not know what processes in a snail's nervous system have led to the change in behaviour. After a fairly long period of twelve hours or so, during which no further noise stimuli are applied, the snails have "forgotten" and they go through the whole process of "conditioning" again; we say that they do not remember very well. This ability to ignore persistent disturbing stimuli is very widespread. People and animals can live in the shadow of a gasworks and eventually they do not appear to notice the smell of gas; cattle feeding in meadows close to a railway or an airfield learn to ignore the sight and sound of trains or aircraft.

Many experiments on "conditioning" have been based on the classical experiments of Pavlov, in which dogs learnt to salivate at the stimulus of a sound or a light formerly associated with the presence of food. Fish in an aquarium can readily be conditioned to rise for food at a certain corner of the tank, by tapping on the glass, or ringing a bell, or shining a light. Such conditioned reflexes make up a large part of an animal's total behaviour. A great deal of work with laboratory animals is based on endless varieties of conditioning experiments. For example, rats, cats, pigeons and monkeys can easily learn which of two choices gains a reward. The choices may be concerned with levers, maze paths, patterns or colours, but the same type of behaviour is being investigated. Other "choice" experiments differentiate between punishment and no punishment, e.g. electric shock.

Many animals can learn by "trial and error" how to escape from confinement; an ape can learn to open a door which has a number of varied devices securing it. The modern emphasis on such experiments is not so much to test whether an animal can learn choices, but whether its training in puzzle-solving will enable it to cope with new problems more easily than untrained animals. All the work in this connexion has led to the formulation of the conclusion that the greatest possible variety of sensory and motor experience is of great value to any animal, and that it has had some importance in natural selection. In general, it has also been shown that there are critical periods in the life of animals, especially in the early life, when variety of experience is of the greatest value.

Memory

Learning, as defined and exemplified above, involves memory, which is here defined as some trace or mechanism remaining in the nervous system as a result of each experience. We do not know the material basis of memory, or even if there is such a thing, and we do not know where it is located. There is a great deal of experimental evidence which shows that memory becomes more fixed or fortified by the frequency of an experience. Indeed, much of our human education is based on this precept. Conclusions based on repetition experiments have led to the formulation of the *theory of facilitation*. The theory states that the more frequently a nervous circuit or series of circuits is used, the easier does impulse transmission become on stimulation of the particular circuit or circuits involved. For reflex actions, facilitation is said to be innate. From our knowledge of nerve-impulse transmission and of the passage of impulses across synapses, there does not seem to be any material basis for the theory of facilitation, though it has been claimed that there is some change in a synapse whenever learning occurs.

The student will probably be familiar with experiments on planarians which purport to show that memory can be transferred from trained to untrained individuals. Planarians were trained to negotiate a simple Y maze in which one direction led to a food reward. Eventually the animals learned to make the correct turn. Two series of experiments were then performed on the trained worms. In one, worms were cut up and the pieces allowed to regenerate; the new flatworms were tested and it was claimed that they "remembered" the training. In the second series of experiments, portions of trained worms were fed to untrained worms, and again it was claimed that they "remembered." Furthermore, the experimenters stated that there was justification for believing that the "memory" was stored in RNA molecules and that the transfer of these from worm to worm conveyed the "memory." This is a very attractive suggestion and gives rise to a fascinating vista of possibilities. Unfortunately, the experiments have not been successfully repeated, and their validity and interpretation are both open to grave doubt.

Modern Study of Behaviour

One of the main trends in modern ethological study is to avoid the use of vague terms which themselves bear overtones of explanation. Such are instinct, thought, consciousness, insight and intelligence. There is increasing emphasis on careful and accurate recording of observation and on rigorous analysis of behaviour in terms of the nervous and endocrine systems and of changes which take place in them. There is less tendency to classify behaviour but rather to think of graded evolutionary series, in which advances in anatomical complexity are paralleled by complexity of behaviour.

At the lowest levels are the Protozoa, the Porifera and young metazoan stages. Some groups of Protozoa possess conductive threads, neuronemes, which affect locomotor organelles such as cilia and flagella, and there are many examples of sensory organelles, e.g. eye-spots and sensory cilia. In some ciliates there are even "control" centres such as the motorium of Paramecium. In some work on protozoan behaviour, there is evidence of "ignoring" persistent stimuli, for example an amoeba's reaction to a periodic pencil of light (*see* Vol. I). In general, however, protozoans show mainly reflex behaviour, in the true sense of the word "reflex"; the behaviour reflects the stimulation from the environment.

In the Porifera, there is no sign of a nervous system, yet the animals respond to stimuli. Unfavourable stimuli lead to contraction and often closure of the pores. In our present state of knowledge of these lowly animals we think of their behaviour as being unorganized and unco-ordinated; this is merely an expression of our inability to explain the behaviour.

In young metazoans, and especially in embryos, there is obviously co-ordination or the animal would not develop into its specific form. We know that there are hormonal controls given the names of organizers and evocaters and we can observe and verify their effects, but we do not know how they invoke such phenomena as cell movements or the localized development of particular tissues and organs. We have to be satisfied, at present, with descriptions of the observed effects. Coelenterates and some of the lowly metazoan groups exhibit largely reflex behaviour, and yet, even at this level, there are aspects which involve learning.

With the increasing size and complexity of the nervous and endocrine systems, behaviour becomes more adaptable; there is more power of learning and of combining sensory stimulation with the results of previous experience. In the "highest" animals, the brain is more self-exciting and can be more free of domination by the senses; the brain has complex autonomous activity, with the capacity to store the results of past experience and the ability to organize these results. The highest abilities in behavioural contexts are measured by the storage capacity of the brain, the number and variety of connexions between its parts, and the rapidity with which stimuli from the environment are combined with organization of the requisite portions of stored information to deal with any situation. If "intelligence" is to mean anything in an ethologist's vocabulary, it is a position on a graded scale of increasing complexity.

Structural and Physiological Basis of Behaviour

In most cases, the behaviour patterns of animals have never been correlated in detail with the structure and physiology of the animal concerned, but there are some broad generalizations which may be made. For example, in the majority, the structure and function of the parts concerned with reception of the stimulus and of those making the response are well known. But there are gaps even in this knowledge, particularly with regard to the protozoans, many of which do not seem to possess parts specially associated with perception of stimulation, or specialized effectors. An amoeba appears to possess a "diffuse irritability," in that any part of it may detect a change in conditions and the whole animal may respond without any apparent lines of communication between one portion of the protoplasm and another. On the other hand, the structure and mode of functioning of receptor organs such as the eyes of vertebrates, and the mechanisms by which muscles and glands make their responses are well known.

There is no doubt that increasing complexity of behaviour can be correlated with the complexity of the nervous and endocrine systems of an animal. It has been shown, in developing embryos, that behaviour patterns develop all of one piece, as it were, and only when the embryo has reached a stage of development where the nervous connexions are detailed enough to permit them, or where the endocrine organs are sufficiently formed to be functional. The unco-ordinated responses of simple animals such as sponges, where there are not even protoplasmic bridges between the cells, much less a nervous system, are due to independent responses on the part of individual cells which then collectively bring about the animal's behavioural display. In the simpler metazoans such as the coelenterates, which have no central nervous system, there is much independent behaviour of the parts similar to that of sponges. But in the case of the coelenterates, there is a nerve net which can convey impulses about the body from specialized receptor cells in the surface layers to effectors such as the muscle tails. Weak stimulation at one point produces only local response, stronger stimulation may lead to more widespread activity of the parts. A small piece of food on one tentacle of a hydra, causes it to bend towards the mouth, a larger piece may elicit the response from all the tentacles. A light touch on one tentacle may cause it to contract independently of the other tentacles. A vigorous prod anywhere, usually results in the contraction of the whole animal. But nowhere does there seem to be a central control of responses; the nerve net serves only to conduct impulses and nothing more. Of course, some activities of a hydra appear to originate from within. The animal is capable of quite complicated locomotive movements and as far as is known, these are initiated internally, each act becoming a stimulus for the performance of the next, a kind of chain behaviour in which, for example, the attachment of the mouth end of the animal automatically leads to release of the substratum by the basal disc and this to flexure of the body and so on.

No radially symmetrical animals have developed any part of the nervous system as a centre of control except perhaps some echinoderms which possess an oral nerve ring, but in all the bilaterally symmetrical animals there is some form of central nervous system with a "brain." In the most lowly, e.g. the planarians, the "brain" is chiefly concerned with receiving impulses from the head region, but since many planarian activities involve the whole of the body, some co-ordinating function of the central nervous system must be envisaged, making it a much more efficient system than a nerve-net. Further, the planarian, *Leptoplana,* is capable of learned behaviour, such as not moving when exposed to light, a response which it normally makes, but nothing seems to be known about the structural basis of this, beyond the fact that it is possible only when the nervous system is properly developed.

In annelids, the "brain" is better developed, and there is a higher degree of control over the rest of the body. The brain can initiate behaviour and control muscle tone, types of control which are much more highly developed by higher animals. Among the invertebrates, the arthropods, particularly insects, show the most complicated of centrally controlled behaviour, coupled with a much more elaborate central nervous system.

Some description of the vertebrate central nervous system and its correlation with behaviour has been given in Vol. I, Chaps. 26, 27, 28, 29. Reference to the types there described, will show that there is a graded series from fish to man, wherein the greater complexity of the brain and its correlation and co-ordination centres is mirrored in the behaviour of the animal.

CHAPTER 17

REPRODUCTION AND SEX

USED biologically, the term reproduction means the formation of new, separately existing individuals of a species from members already in existence. This capacity to reproduce is universal and the significance of the characteristic is obvious, for in its absence, a species must sooner or later die out. Moreover, in the eternal competition for space which all living things experience, any advantage gained by more efficient reproduction and dispersal of offspring leads to greater success of a species, where success is measured in terms of numbers and their distribution. Perpetuation and increase in numbers of a species depend on successful reproductive processes. It is little wonder therefore, that there have been evolved reproductive mechanisms of the utmost complexity to suit particular conditions and that all living things, at some period in their lives, direct all their energies into reproductive activity, sometimes even at the expense of their own continued existence.

In an analysis of the reproductive processes adopted throughout the plant and animal kingdoms we can distinguish clearly between two methods. These are called the *sexual* and the *asexual* methods of reproduction. A tremendous amount of study has been made of all forms of both kinds and a terminology has grown up which can lead to bewilderment, unless all the terms are clearly defined and their customary usages understood. The following summary may assist the student in the avoidance of misunderstanding even though the sight of so many terms in so short a compass may be frightening.

Sexual reproduction involves the formation of two types of specialized cells, called *gametes*, neither of which singly, in the ordinary run of events, is capable of giving rise to a new organism. The two types of gametes are most often complementary to one another, and when two such gametes of the same species meet and fuse together, a unified *zygote* is formed. This is a unicellular structure capable of developing into a new individual when given suitable conditions. When a species produces gametes identical in structure, it is said to be *isogamous*. Such is the case in some algae, fungi and protozoa. A *heterogamous* (*anisogamous*) condition exists when a species produces two distinctly different types of gametes, which are designated *male* and *female*. This is the case for most plants and animals. The differing gametes may be

694

developed on the same individual, in which case it is *hermaphrodite* or *bisexual*, or on separate individuals of the same species, in which case it is *unisexual*. Most plants conform to the former condition and most animals, the latter. The terms *homothallic* and *heterothallic* are also used to describe algae and fungi, the former indicating production of both types of gamete on the same thallus, and the latter, segregation of gamete-producing organs to separate male and female thalli. Sometimes these thalli and sex cells are indistinguishable and the species is said to exist as *strains*, e.g. the + and − strains of *Mucor* (*see* Vol. I, Chap. 33).

The term hermaphrodite is also used to describe single flowers of seed-bearing plants, when pollen and ovules occur together. *Monoecious* is used to describe the condition in which separate male and female flowers are borne on the same plant, and *dioecious* describes the plant in which male and female reproductive organs are developed on separate plants. The term *heterogamy* is also used by zoologists to describe the condition in an animal when there is an alternation of two forms of reproduction in successive generations, as for example, in aphids, where normal gamete fusion and parthenogenesis (*see* p. 696) regularly follow one another.

Where a species is represented by distinct male and female individuals, there is nearly always divergence of morphological, physiological and, in higher animals, psychological characters between them. The phenomenon is known as *sexual dimorphism*. It is clear that nearly all individuals inherit their potentiality for developing as one or other of the sexes, in a Mendelian fashion. Sex determination is dealt with more fully in Chap. 19, and will be mentioned again later in this.

Slightly differing forms of sexual reproduction may be distinguished by the adoption of descriptive names. *Amphimixis* describes the method of sexual reproduction in which two sex cells not closely related to one another participate in the zygote formation. This is the condition obtaining in most animals as well as most plants. When the formation of the gametes is restricted to specialized parts of the parent body, reproduction is said to be *merogamous*, whilst it is *hologamous* when the gametes are formed from two whole mature individuals. The latter is the case in some protozoa, e.g. *Polytoma* and *Copromonas* and in some fungi, e.g. *Polyphagus*.

Automixis is said to occur when there is self-fertilization involving closely related sex cells or nuclei; this includes *parthenogamy* which describes the process of fusion between two cells of the female sex organs, and *autogamy*, when there is fusion between two nuclei within a single cell of the female sex organs. Such mechanisms occur in some fungi where sexuality is very reduced, e.g. *Psalliota*. In the protozoa a comparable condition exists in which a fusion may occur in one organism between two nuclei derived from the same parent nucleus, e.g. *Paramecium*.

Pseudomixis is used to describe the case in some fungi when a fusion occurs between two vegetative cells, the resultant then acting as a zygote. Some yeasts reproduce in this way.

Apomixis describes the case when a gamete can develop into the next generation without the fusion with its counterpart. Sometimes a male gamete may merely enter an ovum without a subsequent nuclear fusion. The stimulation of entry may be sufficient to cause the ovum to develop. This is known to be the case in some nematode worms and a few higher plants. The process is described as *pseudogamy*. In other instances, an ovum may develop without even the penetration of a male gamete. This is known as *parthenogenesis* and occurs in both animals and plants, sometimes with great regularity and as part of the normal life cycle. Such is the case in rotifers, some species of insects, e.g. aphids, some wasps and sawflies, and the cladoceran, *Daphnia*. In the gall wasps and aphids, there is regular alternation between normal sexual processes and parthenogenesis (*see* heterogamy above). In bees, the queen controls parthenogenesis in her eggs by allowing sperm to come into contact with some, but not with others. *Endomixis* is a special form of parthenogenesis occurring in the protozoan *Paramecium*, in which the micronucleus of a single individual may, without nuclear fusion, give rise to micronuclei and meganuclei of fresh individuals. This is distinct from autogamy mentioned above. In some plants, parthenogenesis occurs regularly, as for example in species of Compositae, the dandelion among them. The special case of parthenogenesis in larval forms of animals is known as *paedogenesis*. The name should not be confused with *paedomorphosis* which is the prolongation of an early stage of development of a part of the body into sexually mature life, e.g. the continued development of juvenile down feathers on some birds. Parthenogenesis may be described as haploid or diploid depending on whether the ovum is in the haploid or diploid condition when it starts its development into an embryo. Ova are normally haploid, but diploid ova are produced in some cases when the meiosis which should precede their formation is omitted from the life cycle. More information on naturally occurring parthenogenesis is given in Vol. I where specific instances are described. The term apomixis sometimes includes the condition in which cells of the nucellus of an angiosperm ovule penetrate the embryo sac and there develop as a normal embryo. These cells are, of course, diploid like all others of the sporophyte. Such offspring are diploid and perfectly normal, but neither kind of gamete has entered into their formation. Strictly, this process is asexual.

The well-defined process by which gametes, under normal circumstances, blend to form the zygote, is known as *syngamy*. *Karyogamy* describes the nuclear fusion, and *plasmogamy*, the merging of the cytoplasm. *Plastogamy* describes the process which occurs in slime fungi and in *Amoeba diploidea*, in which two protoplasmic masses may fuse without fusion of the nuclei. *Fertilization* is used synonymously with syngamy but is sometimes used to mean the phenomenon by which one gamete initiates the further development of another by fusing with it. Fertilization may have still another shade of meaning when it describes the active process of fusing, whereas syngamy always embraces the whole fusion.

When it is the case that two organisms become linked or coupled together so that their gametes may intermingle, they are said to *copulate* or take part in *coition*. The term *conjugation* is sometimes used synonymously with both syngamy and copulation, but it is more correct to use it specifically for the coming together of two whole morphologically identical cells, which, acting as gametes, fuse to form a zygote. Such is the case in members of the Conjugales among the algae, e.g. *Spirogyra*. The term may also include the process seen in some ciliates, e.g. *Paramecium*, in which the fusion is partial and only nuclei are exchanged between conjugants.

When an animal becomes sexually mature at an early stage of development, that is to say, the development of its reproductive organs is accelerated in relation to the rest of its body, it is said to exhibit *neoteny*. The axolotl, already mentioned in Chap. 13, is a typical example. The term neoteny has sometimes been used synonymously with both paedogenesis and paedomorphosis, defined earlier, but should be avoided when either of these is meant.

Asexual implies the negative of sexual, i.e. non-sexual, and can be used to describe any reproductive processes not involving gametic cells. Botanists generally distinguish between cases involving production of specialized but non-gametic cells such as spores, and a variety of other cases in which unspecialized body parts such as roots, stems and leaves may become reproductive structures. The former is more generally called *asexual reproduction*, and the latter, *vegetative reproduction* or *propagation*. Zoologists generally make no attempt to differentiate between asexual and vegetative reproduction. Asexual reproduction can be described under various names such as *fission, sporulation, fragmentation, budding*, etc. The meanings of these terms will be made more clear in the later paragraphs devoted to this mode of reproduction.

SEXUAL REPRODUCTION

There are comparatively few organisms which do not employ this method of reproduction. The essential feature of the process, the formation of a zygote, involves a set sequence of events which may be summarized as—

1. The attainment of sexual maturity by the organism.
2. A process of gametogenesis in which the gametes are made ready for their function.
3. The liberation of at least one type of gamete (sometimes both) and the coming together of a complementary pair.
4. The fertilization process in which the zygote is formed by the syngamy of the two protoplasts.

Sexual Reproduction in Animals

There are innumerable variations in the ways in which sexual reproduction may be brought to completion. Many detailed accounts of the reproductive processes in animal species are given in Vol. I; most emphasis here will be placed on the sequence of events and the influences believed to control them as they occur in higher animals.

Attainment of Sexual Maturity

Sexual maturity is said to have been reached by an animal when its gamete-producing organs, the gonads, and its accessory sex organs, the genitalia, are fully functional. Except in rare instances, they are the last organs of the body to mature.

In vertebrates, gonads develop from cells of the mesoderm which form paired genital ridges along the inner border of the dorsal body wall. In all the females and in most males, they remain in this position throughout life, connected to the exterior by ducts. In mammalian males, in most cases, the testes descend into special outgrowths of the body cavity, the scrotal sac (or sacs) which is situated between the hind-limbs. Accompanying the development of the gonads is the growth of the characteristic male and female genital organs. As the gonads reach functional maturity, the animal is said to have reached puberty. At the same time, in most cases it develops secondary sexual characteristics peculiar to its sex. Very marked sexual dimorphism may become apparent.

Development of the gonads and the stages in the fulfilment of their function are controlled ultimately by hormones. It must, however, again be emphasized that over-riding control of the endocrine system lies in the hypothalamus. There, a variety of stimuli concerning internal state, and environmental conditions such as day length and temperature are modulated. If the state of the body and external conditions are suitable, the hypothalamus secretes releaser factors which are carried in the portal system (*see* p. 280) to the pars anterior. These factors stimulate production of gonadotropins which control the development and maturation of the gonads and at sexual maturity induce the secretion of gonadial hormones. Lack of gonadotropins in young animals will cause arrested gonadial development; lack at maturity will cause atrophy of the gonads and the cessation of gamete production. The same gonadotrophins are responsible for the correct functioning of the gonads, and ovum and sperm production cease in their absence.

Regulation of the development of male and female secondary sexual characteristics and accessory sexual organs, and control of the reproductive impulse and behaviour during reproductive activities, are exercised by the gonadial hormones which are formed in the mature gonads and other places. These hormones include the androgens and oestrogens. The androgens stimulate the development and activity of the male accessory reproductive organs and are also partly responsible for the differences in form of the male body from the female body in such features as the structure of the sound-producing parts and the distribution of hair. In lower vertebrates, they control such things as formation of the breeding pad in male frogs, the cock's comb-form, the male newt's crest and certain colour differences between the sexes in many animals. The oestrogens stimulate growth and function of the female accessory reproductive parts and control the development of the other typically female characteristics.

A vertebrate animal, having reached sexual maturity, can continue to reproduce over much of the rest of its life span in most cases, but senility is usually accompanied by loss of reproductive powers, particularly in mammals. The balance of hormones gradually changes, so that in the female, the shedding of eggs ceases, and in the male, sperm are not produced. In the case of the male, the change is spread over a much longer time. Women may suffer considerable mental and emotional disturbance during this period, which is known as the menopause, and complete stability is not achieved until a new balance of hormones is established.

Gametogenesis

Gametes are produced, sometimes in gigantic numbers, only in the specialized gonads. The ova are products of the ovary and the spermatozoa of the testis. Both types of gametes are produced by divisions of cells of a germinal epithelium of the appropriate gonad. Meiotic divisions always occur during their formation so that a gamete contains

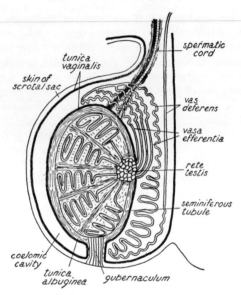

Fig. 17.1 Median L.S. of rabbit testis in scrotal sac.

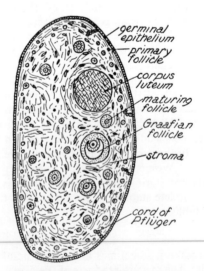

Fig. 17.2 Median L.S. of ovary of rabbit.

only the haploid number of chromosomes. In general, an ovum is a comparatively large cell (exceptionally big in reptiles and birds), non-motile, and containing food storage materials. By contrast, the spermatozoon is very tiny, motile by means of a flagellum, and it lacks any obvious storage materials. In the formation of spermatozoa (spermatogenesis) and in the formation of ova (oogenesis), a comparable

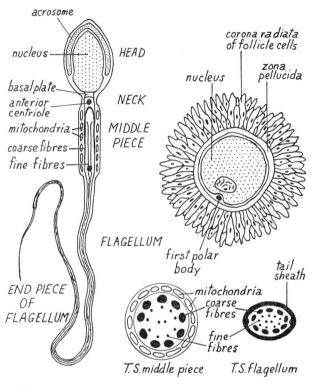

Fig. 17.3. Diagrams representing mammalian sperm and ovum.

sequence of events occurs. First, any cell of the germinal epithelium may undergo repeated mitotic divisions, i.e. a phase of multiplication, then these new cells increase in size, i.e. a phase of growth, and finally by a maturation process they become functional gametes.

Spermatogenesis. The sequence of events is diagrammatically illustrated in Fig. 17.4. The parent cell in the germinal epithelium, known as the *primordial germ cell*, undergoes repeated mitotic divisions. Since this cell contains the diploid number of chromosomes, all its offspring, known as *spermatogonia*, will be diploid likewise. Each spermatogonium comes to lie in a more superficial region of the epithelium, that is, towards the inside of the seminiferous tubule, and, under the nourishing influence of the cells of Sertoli, enlarges. In this condition, it is known as a *primary spermatocyte* and is ready for its maturation phase. It is during this further development that the first and second meiotic divisions occur. As a result of the first division,

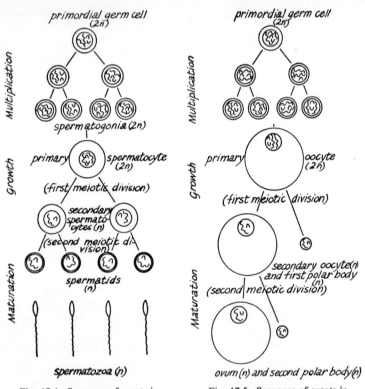

Fig. 17.4.　Sequence of events in spermatogenesis.

Fig. 17.5.　Sequence of events in oogenesis.

each primary spermatocyte produces two haploid *secondary spermatocytes*, and as a result of a second division by each of these, four haploid *spermatids* in all are formed from each primary spermatocyte. Each spermatid now changes into a mature *spermatozoon*, complete with head, middle piece and flagellum, and lies in the fluid secreted inside the seminiferous tubule, awaiting transfer to the ova of the female. It is important to note that each primary spermatocyte gives rise to four functional gametes and that each of these is a haploid cell.

Oogenesis. Each ovum originates from a primordial germ cell of the germinal epithelium at the surface of the ovary and the sequence of events is comparable to that described for spermatogenesis. A collection of cells derived by mitosis from a *primordial germ cell* passes inwards into the stroma of the ovary, there to develop further. Each of the group is a potential *oogonium* but only one is singled out for further change into an ovum. The remainder form a layer around this cell to constitute a *primary follicle*. Within its follicular layer, which serves a nutritive function, the oogonium enlarges considerably to become the *primary oocyte*. When the oocyte is fully enlarged, the follicle is known as a *Graafian follicle*. The primary oocyte now undergoes its maturation divisions, the place and time of occurrence varying with species. In some cases, the ripe follicle may move to the ovary boundary, rupture, and extrude the primary oocyte before any divisions occur; in others, the first maturation division may have occurred before the oocyte is extruded. In any event, the primary oocyte undergoes a first meiotic division to produce two haploid structures. One of these is comparable in size with the parent cell and is known as the *secondary oocyte*, whilst the other is very small and known as the *first polar body*. The secondary oocyte, to complete its maturation, now undergoes the final division, but this may not occur until a spermatozoon has penetrated its membrane. As a result of the division, the secondary oocyte gives rise to the mature *ovum*, a large cell, and the *second polar body*, once more of insignificant size. In the case here described, the result has been the production of three cells, one ovum and two polar bodies. Sometimes the first polar body may also undergo a division to produce two very tiny cells, in which case the usual number of four cells resulting from a meiotic division is achieved. The polar bodies are not known to function in any way during subsequent development of the egg, but can be seen adhering to the mature ovum. They represent unwanted nuclear material without cytoplasm.

Liberation of the Gametes

Before syngamy can occur, the male and female gametes must come into contact with one another. In some animals this seems to be a matter of chance, in others, it is controlled with great certainty. Where both spermatozoa and ova are released into water, motility of at least one gamete is essential. A few aquatic animals appear to have little direct control over the meeting of their gametes, releasing them indiscriminately into the water as they continue other activities, e.g. *Hydra*. But in most animals, particularly land animals, there are many mechanisms set in motion which ensure with great certainty the meeting of gametes. Such assurance comes from the fact that most animals show a definite mating process during which gametes are released synchronously under the influence of distinct stimuli. The mating process may vary between mere deposition of eggs and spermatozoa in proximity to one another in the water, as in many fishes and amphibians, to the actual placing of the sperm by the male inside the female body during copulation. Fully ripened sperm are then ejaculated only as a response to stimulation, to meet fully receptive ova. In reptiles and birds, this is followed by the laying of fertilized, protected eggs which develop outside the parent. This is the *oviparous* condition. The reproductive process has reached culmination on land in the mammal, where male and female copulate, fertilization being effected inside the female genital ducts, followed by implantation and development of the embryo, nourished through the mother's blood-stream up to the time of birth. This is the *viviparous* condition. The *ovoviviparous* state occurs in some snails, insects, snakes, fish, etc., when fertilized ova, protected by egg-membranes, develop within the mother until hatching takes place.

In most cases, the mating activities of animals are initiated by perception of environmental conditions (*see* p. 708). In other cases, for example, the mouse, rat and man, there is a rhythm which is quite independent of seasonal fluctuations (*see* p. 557). The whole sequence of events is initiated internally, in both the above cases, by hypothalamic secretions. In many species of birds, increasing day length causes enlargement of the gonads and stimulates the pre-mating behaviour. This effect of the photoperiod has been discussed on p. 205. The extraordinary courtship behaviour in some animals provides a series of signals which are evidently recognizable and meaningful to members of the same species. The whole sequence stimulates both partners through their senses; their bodies respond internally through the medium of hormones; these external and internal sequences build up towards a state of readiness, so that there is exact synchronization of action at coitus. There is undoubtedly a large inherited component in these courtship behaviour patterns since they are species-specific.

Fertilization and the Zygote

Apart from the placing of sperm near eggs during mating, the final contact between gametes is effected by the sperm, since only they are motile. Within the testes, sperm are generally mature but do not commence to swim until they are under the stimulus of the medium in which fertilization occurs. This may vary between sea-water as in the case of echinoderms and the secretion of the prostate gland in a mammal. It appears that the medium contains specific activators which cause the sperm to commence its locomotion. The hormone thyroxin has been shown to be effective in some cases. There is certainly a connexion between the thyroid gland and sexual activity in the mammal. The effect of the stimulant could be to raise the metabolic rate of the sperm, since, in general, sperm are unable to absorb food material from the surroundings. Thus each sperm can live only for as long as its food reserves hold out. Inactivity until just before fertilization is therefore necessary, otherwise most sperm would die before contacting an egg.

Temperature also affects the rate of activity of sperm, and at higher temperatures they are more active, but die more quickly. They can be preserved for long periods in the frozen state (artificial insemination). During movement, oxygen consumption depends on the rate of activity and it has been shown that a given quantity of sperm, whatever the life span, will consume the same quantity of oxygen. This also points to the fact that each sperm has a limited quantity of stored food of which it can make use. The life span for most sperm seems to be about 24 hours after leaving the male, but there are a few cases in which the female may store sperm for long periods, e.g. the earthworm,

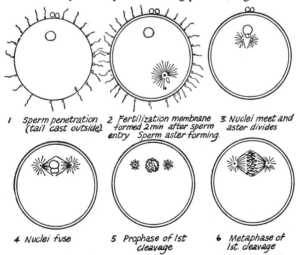

1 Sperm penetration (tail cast outside).   2 Fertilization membrane formed 2 min after sperm entry Sperm aster forming.   3. Nuclei meet and aster divides

4 Nuclei fuse   5 Prophase of 1st cleavage   6 Metaphase of 1st. cleavage

Fig. 17.6. Diagrams representing fertilization in sea urchin.

honey bee, cockroach. Animal sperm are not known to execute clear-cut chemotactic movements as is the case for many plants, but there must be some similar phenomenon in those numerous cases in which they swim along female ducts always in the right direction to meet the egg. When eggs are contacted, many sperm may make efforts to enter the cytoplasm by burrowing into protective jelly, but only on rare occasions does more than one sperm succeed. This seems to be due to the immediate reaction by the egg when one sperm has penetrated it.

For long it has been considered that some form of fertilization membrane is produced by the egg that prevents further penetrations, but recent work indicates that the process controlling sperm penetration is much more complex and subtle than just the formation of a physical barrier. It is, to some extent at least, associated with the fertilization reaction itself. When sea urchin sperm are placed in sea water that has previously contained sea urchin eggs, it is seen that the sperm quickly clump together or agglutinate and lose motility, something that they do not do when in pure sea water. From this it is concluded that the eggs release into the water an agglutinating substance. This is known as *fertilizin* and has been identified as the material forming the gelatinous covering of the eggs which goes into solution as the eggs stand in the sea water. On the sperm surface is another substance, *antifertilizin*, and it is the combination of this with the egg fertilizin that causes the sperm to agglutinate. In conditions where the sperm are allowed to contact eggs, it is this combining reaction that encourages the sperm to become attached to the eggs. In natural conditions, enormous numbers of sperm attempt to make contact with each egg. In sea water surrounding an egg there will be some dissolved fertilizin and thus most of the sperm will quickly be inactivated in absorbing this excess fertilizin and hence will never actually reach the egg surface. However, owing to the numbers, at least one sperm with its antifertilizin still uncombined is likely to reach the egg. Since the fertilizin–antifertilizin system is very highly chemically specific to each kind of egg and sperm it is clear that no sperm other than that of the same species as the egg is likely to make union with it. The system has been likened to the antigen–antibody reaction seen in body fluids (*see* p. 883).

The precise mechanism of sperm entry to the egg has been studied with the electron microscope for the marine worm, *Hydroides*. The sperm is seen to respond to the presence of the egg by forming a process at the tip of the acrosome and it is this that attaches it to the egg's surface. When the sperm has made proper contact by its acrosomal process, enzymes, released from the acrosome, dissolve the vitelline membrane. When this is achieved, the ovum is stimulated to further development; there is a rapid rise in its metabolism; a whitish colour spreads through the ovum from the point of attachment, and the larger cortical granules disintegrate. The acrosome wall breaks down further and the ovum plasma membrane forms a cone that projects slightly into the aperture in the vitelline membrane; the two plasma membranes now fuse together. In mammals, it is known with certainty that pores develop in the acrosome surface and the enzymes are released through these. It is the action of the enzymes which dissolves a pathway for the sperm through the corona radiata, zona pellucida and vitelline membrane.

In mammals, the ovum reacts rapidly to contact by a sperm. Changes in the zona pellucida and on the surface of the vitelline membrane tend to prevent penetration by more than one sperm. These changes are not always fully effective and several sperm may penetrate; this

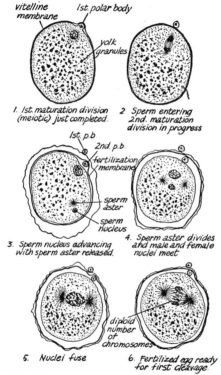

Fig. 17.7. Diagrams representing fertilization in *Amphioxus*.

polyspermy causes early death of the embryo. A structural change, which may be related to sperm repulsion, is the extrusion of well-marked cortical granules from the periphery of the ovum cytoplasm into the perivitelline space. These cortical granules possibly induce changes in the zona pellucida, which repel further sperm.

Once a sperm has penetrated, it moves across the egg cytoplasm towards the nucleus, which in many cases will be undergoing its second maturation division as a response to the stimulus of penetration.

Precise details of sperm entry, and subsequent direction of movement within the egg, vary with species, but in all cases, the entry is followed by the blending of the two sets of haploid chromosomes, one possessed by the sperm and the other by the egg. The diploid nucleus of the zygote is usually reconstituted on a spindle formed under control of the sperm centrosome. On this spindle, the zygotic nucleus immediately undergoes its first mitotic division preliminary to the first cleavage. Figs. 17.6, 17.7 represent the sequence of events which occurs during fertilization of eggs in the sea urchin, *Arbacia*, and in *Amphioxus*.

The zygote is unique among cells; it should not be described as a "generalized animal cell." Its potentialities are enormous and from it are derived all the cells of a new organism. As a result of their differing modes of differentiation, these cells build up the wide variety of tissues and organs which make the adult complete. The sequence of events which leads to the formation of the next generation from a zygote, comes under the heading of embryology, to which much space has been given in Vol. I.

Reproductive Rhythms

It was stated earlier that in some cases the coming together of the gametes seems to be almost accidental, while in others, it is very carefully controlled. In either case, there is advantage to be gained if the liberation of sperm and eggs from the gonads is exactly synchronized. In many cases, this is reasonably assured by the occurrence of *reproductive rhythms* which may affect both sexes. In species where eggs and sperm are liberated into water, the parents are influenced in the timing of the liberation of their gametes by the time of the year. The reproductive capacity in each sex reaches its climax at a particular season, and the gametes are freely mixed in the water at the same time. In animals where sexual intercourse between the parents is necessary, there are also clear-cut reproductive rhythms which may also be seasonal. Thus we speak of the nesting period for birds and the rutting season of many mammals. The *menstrual cycles* of women and the periods of heat of bitches are also outward signs of a definite rhythmic cycle of events. Such rhythmic cycles are most clearly marked in mammalian females. To follow this *oestrous cycle*, as it is called, knowledge of the mammalian ovary structure and function must be recalled. The product of the ovary is the Graafian follicle containing the ovum bathed in its liquor folliculi. *Ovulation* is said to occur when the follicle wall ruptures at the surface of the ovary to release the ovum into the body cavity. The number of follicles rupturing together in each ovary is variable; in the rabbit about six from each ovary at the same time, in the human being, one only, alternately from each ovary.

After the ovum is extruded, the remains of the follicle pass back into the ovary and change form by cell enlargement to become the corpus luteum. The total time taken for the development and degeneration of a follicle governs the length of the oestrous cycle. In some mammals, this period is one year, and they are said to be *monoestrous*, e.g. the fox. Most are *polyoestrous*, although this does not necessarily mean that they mate more than once a year. The breeding season for most mammals is spring in the northern hemisphere and autumn in the southern, i.e. the period when day-length is increasing. Such animals, if moved far across the equator, will change their mating time from one season to the other to suit the new conditions.

In each oestrous cycle there are four main phases. First is a latent period, called *dioestrus* or *anoestrus* during which there is no visible sexual activity. This may be quite prolonged in animals with a long cycle. The second phase, known as *pro-oestrus*, is the phase during which the follicle develops in the ovary. *Oestrus*, the "period of heat," is the third phase and this is often accompanied by changes in the external and internal genital parts. Only at this time is the female willing to receive males. It is at this time that ovulation occurs in most mammals. The rabbit, cat and ferret are exceptions, since they ovulate only after copulation. Finally, there is the period of *post-oestrus*, which is a time merging into *pregnancy* if copulation is successful, or *pseudopregnancy* where it is not. If the union has been successful, the egg or eggs will have been fertilized at the upper end of the Fallopian tube, cleavage will have commenced and the young embryo (or embryos) will have been implanted in the uterus wall. The mother is then said to be pregnant. Changes which have commenced in the endometrium of the uterus soon after ovulation, become very marked at this stage, and a placenta is formed where tissues of the embryo and mother come into close contact. As the embryo develops further, the uterus increases in size to keep pace. Mammary glands develop in the later stages of pregnancy and the female eventually adopts certain behaviour patterns according to species. She may make a nest, or a burrow, or merely retire into seclusion to await the birth of her offspring. If fertilization does not occur, then pseudopregnancy results. The early changes seen in pregnancy occur, but are much less pronounced. The unfertilized egg passes out of the body and is lost for ever.

The events of such a cycle are known to be initiated and controlled by hormones. The development of a follicle to the Graafian state does not determine the other changes which occur. Oestrus occurs when all follicles are destroyed, but does not occur if both ovaries are wholly removed. Thus an oestrus-causing substance must be produced in the ovary stroma. This hormone, an oestrogen, can be extracted from female urine in greatest quantity at or about the time of ovulation and during pregnancy. Such a substance, if injected into a mouse from which the ovaries have been removed, will induce oestrus within 48 hours. Several substances with similar properties have been isolated from males as well as females; they are oestrone, oestriol, oestradiol, equilin and equilenin.

The change from oestrus to pregnancy is clearly marked by the development of the corpus luteum and it is from this that another hormone, progesterone (progestin), emanates. The hormone stimulates the uterus to prepare for embryo implantation, causes the changes in the mammary glands and reproductive organs, and also inhibits further ovulation. Oestrus can only be repeated when the effects of this hormone have ceased. In most mammals, corpora lutea remain active throughout pregnancy, but in some, such as the horse, rat, cat and woman, they may quickly disappear, their function being assumed by the placenta.

Whilst the effects of oestrogens and progestins may be the immediate cause of oestrous and pregnancy changes, they cannot account for the origination of the oestrous cycle in young females nor for its rhythmical reappearance. Other hormones are known to be responsible and these have been traced to the anterior lobe of the pituitary gland. Extracts of pituitary gland cause ovulation, which is followed by the formation of corpora lutea (luteinization). The hormones concerned in these two processes are sometimes known as FSH (follicle-stimulating hormone) and LH (luteinizing hormone) respectively. It is known that in some mammals, possibly all, prolactin is necessary in addition to LH for the maintainance of the corpora lutea and for causing them to secrete. Prolactin is therefore known as luteotropic hormone LTH.

The whole series of events making up the breeding cycle in females under hormone control can now be summarized with reference to Fig. 17.8.

1. A combination of internal and external factors, modulated in the hypothalamus, stimulates secretion of releasing factors. These induce production of the gonadotropins FSH and LH from the pars anterior.

2. FSH stimulates development of the Graafian follicles and induces production of oestrogens in the ovary. LH activates the ovulation process, and the feed-back of oestrogens to the hypothalamus cuts off the supply of FSH, so that no further follicles become mature and there is no further ovulation.

3. LH stimulates luteinization of the empty follicle, whilst the LTH component causes the formation of progesterone by the corpus luteum.

4. Progesterone from the corpus luteum (or later the placenta)

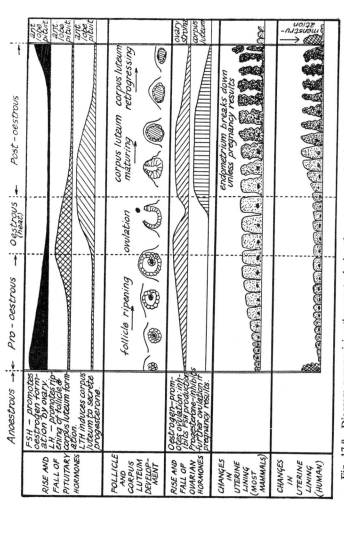

Fig. 17.8. Diagram summarizing the events during the oestrous cycle of a polyoestrous mammal.

initiates the pregnancy changes and also inhibits further ovulation, so that in the case of a successful mating, another oestrous cycle cannot commence until the effects of progesterone cease.

The menstrual cycle in woman does not at first appear to fit into this cycle of events, but clearly can be made to do so. In the menstrual cycle, there is first a period when the follicle is developing up to the time of ovulation. This corresponds to the periods of pro-oestrus and oestrus, when the oestrogen hormone is dominant. During this time the uterus lining is thin and unchanged. After ovulation, the corpus luteum is formed, progesterone dominates the cycle and the uterus wall thickens and becomes highly vascular. This is the pre-menstrual period and corresponds to post-oestrus. If no fertilization is effected, no placenta is formed, the corpus luteum degenerates and progesterone is not produced for long. This results in endometrium breakdown which causes the menstrual discharge of blood and cells. This period and the short post-menstrual period which precedes the next development of a follicle corresponds to anoestrus.

In man, FSH, or its corresponding male hormone ICSH (*see* p. 287), controls the maturation of sperm. The hormone is continuously produced and thus the production of sperm proceeds uninterruptedly in adult life. LH controls the secretion of androgens from the interstitial cells of the testes. In some male mammals there is rhythmic sexual behaviour, in others there is not. A cycle of events comparable with oestrus is not necessary for the success of the reproductive processes. As long as the female will only accept males when she has just ovulated or is about to do so, and the male can at that moment supply fertile sperm, fertilization is reasonably well assured.

Care of the Young

Development of the mammary glands is promoted by oestrogens which induce nipple and duct growth. Later, oestrogens in conjunction with progesterone induce vigorous growth of the lobule-alveolar system; the latter is further stimulated by prolactin, which finally brings the glands into a secretory state. The actual liberation of the milk is promoted by oxytocin within ten to fifteen seconds of the stimulus of suckling.

Many animals, especially mammals and birds, show varying degrees of care of the young. Such aspects as defence and feeding (apart from mammals) are undoubtedly instinctive and are brought into action by stimulation of the parents' sense organs. Mammals, birds, some insects and a few examples from other groups, have evolved very complex instincts related to care of the young.

Sexual Reproduction in Plants

The elucidation of problems associated with sexual reproduction in plants has lagged behind the understanding of this process in animals. Most of the work that has been done, applies to the flowering plant.

Attainment of Sexual Maturity or Onset of Flowering

Some of the facts concerning the onset of flowering have been stated in the account of the developmental cycle in Chap. 13 and it has been pointed out that both internal and external conditions may have some influence. No angiosperm seems to be capable of flowering until a certain amount of vegetative growth has been made. This indicates that some internal factor is at least partly controlling the process. Also, both temperature and light conditions, sometimes interacting, are known to affect the time of flowering in some plants; the phenomena of vernalization and photoperiodism have been well-established. They can be thought of as the mechanisms, more often acting separately but sometimes in combination, through which the developmental processes leading to flower development are triggered to operate. The one is a response to alternating periods of light and darkness and the relative duration of these and the other is a flowering requirement supplied by a chilling treatment. Some plants need a period at low temperature before they will flower in response to the appropriate day-length conditions, this being a means of ensuring that flowering in the winter is a very unlikely event. This is made so because it is during the winter that they receive the first of the two requirements, without which they might respond to the day-length conditions of autumn and be flowering during the least favourable season.

Although photoperiodic responses are generally thought of only as flowering responses others are known, and a given timing of alternating periods of light and darkness may act as the stimulus to such events as the formation of tubers, the development of branches and breaking of dormancy. It is in fact the case that even in some mammals, birds and insects, photoperiodism is the mechanism underlying the control of reproductive behaviour and development according to the season of the year, *see* p. 205. In no case, plant or animal, is it ever a question of light being required to energize the processes involved so that the whole phenomenon is clearly one of an initial activation only. Much work has been carried out to elucidate the real nature of this triggering process in relation to flowering and plants are now commonly described as being in one of three categories according to their responses to different photoperiodic treatments. Some plants are quite unaffected by variations in the ratio of natural day-length to night-length. They are said to be *day-length indifferent* and common examples include the tomato and pea. Some are said to make a short-day response or are *short-day plants*. This is somewhat misleading since it may be taken to mean that these plants will flower only after being subjected to alternating short periods of light and long periods of darkness. What it really describes is the requirement for the plant to receive *less than* a certain time period of illumination per 24 hour day. That is, for each plant in this category there is a critical light period per day or *critical day-length* (cdl) that must not be exceeded if the flowering response is to occur. For some the period may be as short as 8 hours, for others it may be as long as 15 hours. By the converse, plants described as making a long-day response or as *long-day plants*, require a period of illumination *longer than* their critical day length. To emphasize the distinction between the so-called short-day and long-day plants, the cocklebur, *Xanthium sp.*, will flower only when subjected to light periods of shorter than $15\frac{1}{2}$ hours, its cdl, but the henbane, *Hyoscyamus sp.*, will flower only when it has experienced light periods of longer than 11 hours, its cdl. The former is categorized as a short-day plant whilst the latter is referred to as long-day. To complicate matters still further, there are some plants known to require short-day and long-day conditions in the proper sequence and duration before they will flower.

Within each of these main categories there are now recognized still further differences between individuals. Some are clearly *obligate* short or long-day requirers, meaning that no other conditions will induce flowering. Obligate short-day species include soy bean, *Glycine max*, cocklebur, *Xanthium pennsylvanicum*, and red goose-foot, *Chenopodium rubrum*, whilst henbane, *Hyoscyamus niger*, ribwort plantain, *Plantago lanceolata*, bearded darnel grass, *Lolium temulentum*, and barley, *Hordeum vulgare*, are all obligate long-day forms. Other plants may be said to be *conditional* short-day or conditional long-day plants because if they are subjected to certain other conditions then their photoperiodic requirements need not be fulfilled. For example the otherwise short-day plants, morning glory, *Ipomoea purpurea*, and a variety of tobacco, *Nicotiana tabacum*, do not need less than their cdl treatment if subjected to chilling. The duckweed, *Lemna perpusilla*, another normally short-day plant, likewise becomes day-length in different if grown in an abnormally high level of copper ions, it i reported. Some plants categorized as short-day are not really so at all because they will flower eventually even when permanently given greater than their cdl treatment. They do flower much more rapidly however, when given less than their cdl conditions and have been referred to as *quantitative* short-day plants. Conditional and quantitative kinds of long-day plants are also known, for example spinach, *Spinaci oleracea,* will flower most rapidly if given long-day treatment, but this can be substituted by a chilling treatment. But even if kept at shorter than its cdl without chilling, it will eventually flower. An example of a plant needing first less than a cdl followed by greater than a cdl treatment is the wheat, *Triticum aestivum,* whilst a species of *Bryophyllum* shows the reverse requirement.

Irrespective of the precise requirements of individuals there seem to be three stages in the sequence of events that make up the photoperiodic control of flowering. There are the *perception* of the photoperiodic conditions leading then to a state of full *inducement to flowering* after which no further stimulation is required, culminating in the *initiation and subsequent development of floral structures* at the appropriate apices, provided all nutritional and other external physical requirements are met.

It has been shown many times that photoperiodic conditions are perceived by the leaves of a plant most commonly. For example, if an obligate short-day plant, grown in conditions of light periods longer than its cdl, has all its leaves removed prior to return to short-day conditions, no flower initiation will follow and all its apices remain vegetative. But if one leaf only is given short-day conditions while all the others receive long-day conditions, normal flowering frequently occurs. This might indicate that if the appropriate day-length conditions are experienced by leaves, inducement to flowering is accomplished by the formation in the leaf tissues of a substance in the nature of a hormone, which when translocated to the stem apices stimulates them to form flowers. Such a substance, called *florigen,* has been postulated but never isolated chemically and despite a great deal of investigation into the effects of the three major photoperiodic factors, namely light, darkness and the times of application of these, there is no positive evidence that one separately or more of them in combination is instrumental in bringing such a substance into existence.

One question that still has not been wholly resolved is the relative significance of the light and dark periods—which is really of greater importance, the day length (light period) or the night length (dark period)? It is known that if the dark periods of a photoperiodic treatment are interrupted by light periods, i.e. a light break is applied, then this may interfere with the effects that the treatment would otherwise have. For example, typical short-day plants illuminated at below their cdl will not flower if the dark periods are broken by further illumination. On the other hand, long-day plants given light for less than their cdl, will flower if their dark periods are interrupted. Corresponding interruptions in the light period by darkness has no equivalent effects, thus such events were initially interpreted as indicating that it is the dark period that counts and that the short-day plants are really long-night plants with their inducement to flowering controlled by a requirement for an uninterrupted period of darkness. But unfortunately the situation does not seem to be as simple as that. It is the case that the timing of the dark period interruption by light, the light break, can have some significance. Whereas it might be thought that a light period interposed at the midpoint of the dark period must always be most effective in altering the outcome of a photoperiodic treatment, this is not so and an early interruption may have a much more drastic effect than a later one.

Another area of study related to the interruption of the dark period by a light break has concerned the photoreceptor substance, *phytochrome, see* p. 542. So far it has been indicated that the concentration or level of P_{730} during the dark period may be of significance in photoperiodic responses and that it might be the length of the dark period that influences this level. There certainly seem grounds for supposing that phytochrome may be involved because light breaks are greatly more effective if made with red light of the wavelength known to convert P_{660} to P_{730} (P_r to P_{fr}) and their effects can be more or less nullified if a further break at far-red wavelength, at which P_{730} becomes P_{660}, is given immediately after. From this it might be inferred that the presence of P_{730} in the leaves of short-day plants, if it is produced during the dark period, results in the failure of such plants to flower but that the reverse holds good for long-day plants. The general conclusion then might be made that the dark period is really the effective one in controlling the level of P_{730} in the receptive organs, this having different flower-inducing properties in short-day and long-day plants. But, once more, this would be to over-simplify the situation because there is strong evidence that a far-red exposure given to a long-day plant just before it enters its critical dark period will promote flowering rather than subdue it, a negation of the hypothesis that long-day plants require a high level of P_{730} during the dark period in order to induce flowering.

Investigations into the effects of the light period, other than through photosynthesis, have taken into acount both intensity and wavelength as well as the duration of the exposure to light. Some unexplained situations have been disclosed. For example, in addition to the known effects of light breaks with red and far-red wavelengths, some plants are known to have a blue light requirement. The henbane is reputed to be photoperiodically sensitive to blue-violet or far-red wavelengths only and is insensitive to the green, yellow, orange and red parts of the spectrum. Thus there has been postulated another photoreceptor pigment, different from phytochrome, one that mediates the photoperiodic responses through high energy wavelengths only. It has never been demonstrated.

It seems that events neither of the dark period nor of the light period, separate from one another, can be held responsible for the photoperiodically regulated flowering responses but that the two must interact in some way and any acceptable explanation must be in these terms. One such is the suggestion that the duration of the dark period has its effect through the conversion of P_{730}, synthesized during the light period (when under natural conditions all wavelengths of the sun's radiation would be acting), to P_{660} and that in the continued presence of this, the so far unidentified flowering hormone, florigen, or at least some flower-inducing stimulus, is produced eventually to reach a level at which it can promote the initiation of flowers at an otherwise vegetative apex.

For this to be true some such substance as florigen must exist and there are some conditions of photoperiodic responses that cannot easily be explained by any other means. For example, it is possible to graft parts of one photoperiodic response type, say a short-day type such as cocklebur, to another response type such as the long-day flea-bane, *Erigeron sp.*, or the long-day henbane to a short-day variety of tobacco. When this is done, long-day scions (parts grafted on to stocks) will flower under short-day conditions, provided their stocks are short-day plants and are kept under short-day conditions. The same scions under short-day conditions will not flower if the short-day stocks are kept under long-day conditions. Similarly, a short-day plant can be made to flower on long-day treatment when grafted to a long-day stock receiving long-day treatment and the short-day soy bean will flower when grafted to a day-length indifferent stock. Although such experimental findings point strongly to the existence of a flower-promoting hormone, there is still some doubt. The fact that no substance with its properties has ever been isolated or synthesized makes it difficult to accept as a reality, however conveniently it may fit the requirements of a hypothesis. An alternative explanation to flower-promoting effects, through the mediation of a hormone, is that these effects are really due to the removal of some influence that inhibits flower production. The possibility that a gibberellic acid could be the florigen concerned has been mentioned on p. 274.

Another aspect of photoperiodic response is that concerning the extent to which endogenous circadian rhythms, *see* p. 226, may be involved. This is a field that is complicated in the extreme. A brief reference has been made to "biological clocks" in Chap. 6 and will not be pursued further here.

Vernalization, or chilling treatment, is also known to affect developmental processes other than flower initiation. Among these are dormancy of seeds and buds but it is common to refer to the phenomenon in terms of its effect on flowering only, since by far the most study has been made in this field.

There have been many investigations into the effects of temperature treatment in inducing flowering since 1918, when Gassner of Germany showed that winter cereals planted in spring would fruit in the same year if subjected to periods of low temperature. Gregory, Purvio and others, using the variety of rye known as Petkus, elucidated the nature of the vernalization response in such plants. They found that onset of flowering at the normal period is due to the effect of temperature on the subsequent development of parts initiated at the apex in what may be called a "labile" condition. The first seven initials at the apex invariably develop as leaves under all the conditions. Following these are about eighteen "labile" initials which are double structures; one part is bract-like and the other, in its axil, is a flower primordium. Which of these two structures develops, depends upon external conditions. In the one case, at higher temperatures all the time, the bracts develop into leaves but the flower primordia do not mature; in the other case, where the plant is subjected to low-temperature treatment, the bracts do not develop but the flower primordia do. After the twenty-fifth initial, all give rise to flowers in any conditions and the vernalization requirement is said to be quantitative as distinct from absolute essential or qualitative requirement. This is shown by some plants like *Hyoscyamus niger* that will not flower at all unless chilled for a sufficiently long period. The low-temperature treatment thus results in the reduction of the number of leaves, after the seventh, produced by the labile initials, and a corresponding increase in the number of flowers. It is the case that the longer the chilling period, up to a limit, the fewer initials remain vegetative, although even short chilling periods of three or four days can reduce the vegetative development. Spring rye will produce flowers after the seventh leaf without the low-temperature treatment.

It is also known, however, that these phenomena occur only under long-day conditions. In permanent short-day conditions neither form of rye will produce flowers until the twenty-fifth leaf is produced, even if the low-temperature treatment is applied to the winter variety. There is evidently some interaction of temperature and light conditions, and this applies to other plants such as the beet. Whilst there may not be any direct connexion between the vernalization requirement and a particular form of photoperiodic response, it happens that many long-day plants need to have their vernalization requirement satisfied before they are able to respond to the long-day treatment. There seems not to be the same necessity in short-day plants. There are many other puzzling features associated with the whole process and so far it has not been possible to find a convincing comprehensive explanation. For example, in most cases where vernalization is a requirement, there is a complex relationship between the temperature at which the chilling occurs and the length of time for which it is applied. It is also the case that in some plants of the biennial kind they must reach a certain size before vernalization is effective in inducing them to flower. This may be bound up with the fact that vernalization seems to be an energy-consuming process in that a respiratory substrate in sufficient quantity must be present in the tissues and oxygen must be available.

Another peculiar feature is the possible reversal of vernalization or devernalization. This can be achieved in rye by drying previously chilled seeds and keeping them at room temperature for about two months. The effect of this on the subsequent plants is to increase the number of branches they produce from the base (tillers), that is, to increase vegetative activity instead of reproductive. High temperatures applied over a period are known to have a devernalizing effect on a number of plants and various conditions of illumination are effective in others.

Attempts to localize the site of the vernalization process in the tissues points to the meristematic apex and once this has been subjected to chilling all the tissues developed from it require no further treatment. Work with *Lunaria biennis*, honesty, has shown that plants developed from leaves taken from a previously vernalized plant will flower without a further chilling but those developed from leaves taken from an un-vernalized plant will not do so until given low temperature treatment. From these cases it is clear that a vernalized vegetative apex need not be present to ensure flowering in its later unvernalized derivatives but that the products of an unvernalized apex will themselves need chilling individually.

There are a number of confusing circumstances to be found when photoperiodic response requirements are examined in relation to vernalization. For example, if spinach seeds are vernalized, their products will flower without the application of the long-day treatment that they normally require. In perennial rye grass, a chilling require-ment coupled with long-day treatment, its usual requirements, can be replaced by a short-long-day treatment. In other words, rye grass is a plant normally requiring chilling prior to long-day illumination, but it can be brought to flowering if given short days in place of low temperature before the long-day treatment. It can be converted from vernalizable long-day to short-long-day. In *Chrysanthemum sp*, vernalization changes the photoperiodic response requirement from long-day to short-day.

Still further complications arise when it is recognized that some plants such as tomato and chrysanthemum appear to make what can be described as *thermoperiodic responses*, that is, they flower more quickly and abundantly than normal if subjected to alternating periods of low and high temperature. In both cases low night and higher day temperature is effective. It is not known whether these are just variations of the normal vernalization requirement or whether they indicate the necessity for a special kind of low temperature requirement to be met as in the stock, *Matthiola incana*. It is reported for this plant that it cannot be made to initiate flowers at all unless it is kept at a temperature below 19°C for at least 21 days and even then it must be kept in the chilled condition for the primordia to complete their differentiation. A break in the chilling sequence, at any time before this full differentiation is accomplished, inhibits the flower forming process. But once some full flower formation has occurred at an apex, removal from cold to a warm temperature does not halt the process further. This would not appear to be a usual vernalization phenomenon although presumably related in some way. There are many others in the same undefined category, lacking clarification, so that for the moment there is no real understanding of the nature of the chilling requirement or how it acts to promote flower formation.

One possibility, not so far mentioned here, is that the correct photoperiodic and vernalization conditions are not in reality positively promoting the formation of flowers in an otherwise inactive system but that instead they are reversing a real inhibitory effect brought into being by the opposite non-flowering conditions. In the case of photoperiodism there is some evidence that this could be so because inhibition of flowering in the strawberry has been shown to be linked most strongly with the movement of an assumed translocatable substance in the tissues. In this case, it has been well established that the plant has a less than critical day-length requirement or is a short-day plant with respect to flowering and when subjected to long days, flowering does not occur but the formation of runners is very strongly promoted. The indication that a flower-inhibiting substance is involved came from work carried out on plants developed in succession along the same runner, clearly genetically identical and in other ways equivalent. It was found that if one of the plants was given long-day treatment whilst the other was given short days then neither flowered and the one given short days showed enhanced vegetative development. This was even more pronounced if external and other conditions were arranged so that conduction between the plants was more likely to be from the long-day to the short-day plant. This result clearly points to the existence of an active flower-inhibiting substance that promotes vegetative growth whilst doing nothing to support the hypothesis that flowering is due to the presence of a flower-promoting substance such as florigen.

Studies of the effects on flowering of chemicals, both naturally occurring and otherwise, have been made but have yielded little of an explanatory nature. The gibberellins seem to have the greatest effect as flower-inducing substances, as illustrated by their action on rosette plants under otherwise non-flowering conditions such as no chilling for those requiring vernalization or permanent short days for those needing long-day treatment. It does not seem likely, however, that a gibberellin is the hypothetical florigen or a component of an equivalent chemical system. At least, one has never been identified acting in this way despite extensive efforts to do so. Other plant hormones, auxin and the cytokinins, seem very indirectly connected with flowering in response to photoperiodic or vernalization treatments. Ethylene chlorohydrin has a very pronounced effect on cocklebur, bringing it to flowering quite rapidly under its normally non-flowering conditions. Others, including acetylene, naphthalene acetic acid and 2:4 D, have been shown to have similar effects on various plants but studies of them have yielded no significant information so far.

There seems at present no real grasp of the nature of the physiological processes that lead to flowering at previously vegetative apices and likewise very little seems to be known of the effects of day-length and temperature on the development of sex organs in lower plants.

It should be noted that the distribution of plants over the earth's surface must be greatly influenced by the ranges in light and dark periods and in temperature, which occur at different latitudes.

Gametogenesis

There is no need here to dilate on the details of gamete production in plants. Space is devoted to this in Vol. I under the descriptions of the different plant types. One important difference from the case in animals needs stressing. As the reader should be aware, in the majority of plants, the gametes are produced only by the gametophyte generation. This applies equally well to the seed-bearing plants, where the gametophyte is extremely reduced and retained on the sporophyte, as to the more lowly forms, where the two generations lead independent existences. The gametophyte generation is haploid, and is so because of a reduction division which occurs immediately prior to the production of the spore from which it develops. This meiosis is comparable in all respects with that occurring in animals, and normally four haploid cells result from the division of the parent cell. In plants in general, therefore, the division immediately prior to gamete formation is mitotic, but it must never be overlooked that because the gametophyte is already haploid, the gametes also will still be haploid.

The physiology of gamete production by plants has not been the subject of much study and there is little information regarding the controlling factors.

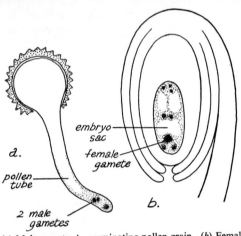

Fig. 17.9. (*a*) Male gametes in germinating pollen grain. (*b*) Female gamete in embryo sac.

Liberation of the Gametes

Except in the higher gymnosperms and all the angiosperms, there is always one motile gamete. In some of the lower algae, both gametes are freely motile. Thus there is, in many cases, the necessity for liberation of the male gametes. The antherozoid, or spermatozoid, as it may equally well be called, can find its partner only in the presence of water through which it can swim. Because of the sedentary nature of plants, no mating processes can occur and the possibility of gamete-pairing can only be assured if the motile male gamete can reasonably easily find its non-motile female counterpart. Some assurance of this meeting of the gametes is afforded by the fact that plant male gametes make chemotactic responses to specific substances secreted by the female gametes or the organs which contain them. For example, fern antherozoids respond positively to malic acid; other examples are quoted in Vol. I.

In the angiosperms, where neither gamete is freely motile, there is not a comparable liberation of the gametes. The male, inside the pollen grain, is first transferred to the vicinity of the female either by air currents or by insects and then conveyed by an outgrowth of the pollen grain (siphonogamously) to the female within the ovule. Pollen tubes are known to exhibit chemotropic responses to substances secreted by stigmas and ovules, and this can be demonstrated by germinating them in a suitable sucrose solution which contains small pieces

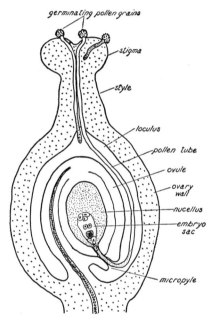

Fig. 17.10. Diagram of the fertilization process in an angiosperm.

of the corresponding stigma or ovules. What guides the tip of the pollen tube so precisely to the egg apparatus is not known with certainty, but it may be some effect of the two synergidae which flank the oosphere.

By such responses on the part of the male structure and by such transference methods in flowering plants, the chances of gametes coming into contact are much enhanced, but as in animals, whereas the female gametes are usually relatively few in number per organism, the male produces prodigious numbers, reducing the odds against failure.

It is noteworthy that in the evolution of terrestrial plants, there has been a gradual trend towards increased chances of gametic union by the development of the siphonogamous habit, which dispenses with the necessity for water films.

Fertilization and the Zygote

As with animals, the fusion of the material of the male and female gametes constitutes the act of fertilization. Despite the fact that the possession of a wall is almost universal among plant cells, the gametes are almost invariably naked. This allows the fusion to occur with the greatest efficiency. In some cases where both gametes are shed into water, as in *Fucus*, there is penetration of the oosphere membrane by the antherozoid comparable with the case described for animals. The oosphere becomes impervious to others and the two nuclei fuse. Where only the male gamete is shed, the female gamete must be exposed to the exterior by an opening in the sex organ containing it. Such is the case in the oogonia of *Oedogonium* and the archegonia of liverworts, mosses and ferns. In a few cases, a definite conjugation tube is formed which unites the male and female organs so that the passage of gametes is readily effected. In flowering plants, the female gamete is exposed to the male only when the pollen tube effects entry into the embryo sac and there disrupts to discharge its contents.

THE ADVANTAGES AND SIGNIFICANCE OF SYNGAMY

A zygote is formed when the nucleus and cytoplasm of one gamete are fused with those of another. By this fusion, the diploid condition of the nucleus is reconstituted, and to some extent, there is a blending of cytoplasm from both parents. The nuclear substance undoubtedly carries factors or genes which control the pace and pattern of the development of the new organism resulting from the fusion cell. Cytoplasmic constituents probably represent more of these controlling influences as plasmagenes. The new generation thus inherits its characters from both parents, a condition which ensures, except in the case where parental constitutions are genetically identical, that it will not be like either parent in precise detail. Variation is introduced, and here lies an advantage of great significance when sexual reproduction is compared with asexual processes. If succeeding generations are all variants, then natural selection can act to weed out the least fitted for survival under the particular environmental conditions. The position of the species in the battle for existence, which must occur in over-crowded conditions, is considerably strengthened by such means, since only the strongest survive to hand on their special characters to new generations. An asexual process cannot introduce such possibilities of variation since blending of material from two organisms does not take place. Such offspring cannot inherit fresh potentialities; they are, genetically, exact replicas of the single parent which gave rise to them. The same is of course true in the case of sexually produced offspring where they result from self-fertilization. Both animals and plants have evolved along lines which tend to exclude self-fertilization from the sexual process.

Some biologists have marshalled evidence from the protozoan animals to indicate that the blending of protoplasm during sexual processes not only results in variation among the progeny but also induces a rejuvenescence in the zygote. It has been shown convincingly in some species, that by asexual reproductive methods alone, the race can continue only for a limited time. In the absence of conjugation, the fission rate slows down and the individuals pass slowly into a decline ending in death. Such an end is prevented only by conjugation between individuals, the resulting offspring then being able to enter a new phase of full activity and reproduction by asexual means once more. That such rejuvenation is in reality a necessity for the continued vigour of the protoplasm in all living things seems very doubtful, since many higher plants in cultivation can be propagated only by vegetative means and have successfully been so propagated for very long periods.

There is, of course, one obvious disadvantage in the sexual process. It is attended by the risk of not being achieved at all, since two structures which might never meet are involved. Many and varied mechanisms have been evolved by plants and animals to reduce this risk and to avoid wastage of material and energy. The fact that the sexual process is almost universal among living organisms is a measure of its success.

ASEXUAL REPRODUCTION

Asexual methods of reproduction cannot bestow biological advantage upon a species in terms of rejuvenation and possibility of evolutionary advance, but they possess another advantage in that any such process can be accomplished without such grave risk of failure. Consequently there is considerably less wastage of material and energy in the asexual process as compared with the sexual.

Asexual Reproduction in Animals

Whilst all methods of reproduction which are not sexual are called asexual by the zoologist, it is usual to differentiate them into types.

Except in special instances they all correspond more closely to what botanists call vegetative propagation, than to the distinctive asexual spore-producing process exhibited by many plants by which these link the two alternating generations at one point in the life cycle. Only in some sporozoan protozoans among animals is a haploid body produced by spore production. This is comparable with the plant form of life cycle. This special case is pointed out in the summary of asexual processes which follows.

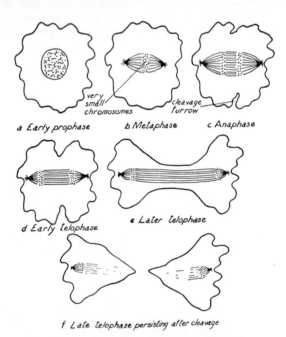

Fig. 17.11. Stages in binary fission of *Amoeba*.

The process known as *fission* (splitting) is strictly only applicable to a method of reproduction encountered in protozoan animals. The parent body divides up into more or less equal parts and thus ends its own existence as a single individual. When the fission is into two daughter organisms only, the process is *simple* or *binary fission*, e.g. *Amoeba*, *Trypanosoma*, *Euglena*. In *Vorticella*, such binary fission may be repeated rapidly by successive generations of cells before the indivi duals are free to lead separate lives. When, within the parent, there is repeated division of the nucleus before the cytoplasm divides to form separate new animals, the process is called *multiple fission*. This occurs in such forms as *Nosema apis*, parasitic in the bee, and *Sarcocystis lindemanni*, parasitic in man. The process may also be termed *sporulation*. In the cases of some sporozoa, e.g. *Plasmodium*, the spores are direct products of a zygote by meiotic division and the process corresponds to the condition in plants mentioned above. The special case of plasmotomy occurring in some protozoa, e.g. *Opalina*, involves the division of the cytoplasm of a multinucleate organism independently of the nuclei, so that several multinucleate pieces are formed.

Fragmentation describes the breaking up of a metazoan animal body into two or more parts when there are no special cells associated

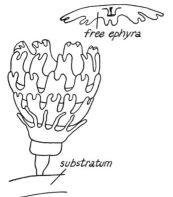

Fig. 17.12. Scyphistoma of jelly-fish.

with the process. It is characteristically the method adopted by some aquatic annelids, e.g. *Lumbriculus*. Other processes which may rightfully be called fragmentation, include the strobiloid development of medusae by successive transverse division of the scyphistoma of jelly-fish. The longitudinal break into two parts of the sea-anemones and the coral polyps is a comparable process.

Budding or *gemmation* is characteristic of other coelenterates. It may be distinguished from fragmentation by reason of the fact that the bud or gemma is derived from a small group of initial cells laterally placed on the parent. It is more of an offshoot than a fragment of the parent. The parent retains its identity, whereas in fission and in fragmentation it does not. Such lateral buds occur on a hydra, initiated in the interstitial cells of the body wall. When it is fully formed as a perfect but immature hydra, the bud enforces a break away from the parent. Under conditions of starvation the buds are not produced, and if they have already commenced development, they may even be absorbed into the parent body. The production of new hydroids, which do not separate but remain to form a colony, is also a budding process, e.g. *Obelia*.

It is characteristic of the higher animals that they exhibit no forms of asexual reproduction whatsoever.

The factors affecting asexual reproduction in animals, other than those associated with growth conditions such as temperature and nutritional influences, have not been extensively studied. The part played by hormones, if any part is played at all, is completely unknown.

24

There may conceivably be some connexion between the regeneration processes and those of asexual reproduction in animals. Both are largely confined to the lower groups.

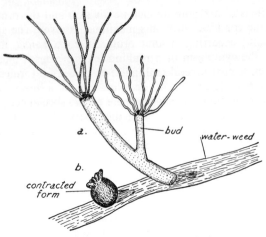

Fig. 17.13. *Hydra fusca* with a bud.

Asexual Reproduction in Plants

We have already indicated that asexual processes in plants may conveniently be described in two groups. There are the numerous instances of spore-production in which specialized cells are formed by special structures, and there are the instances in which a parent plant may give rise to new individuals by some budding process, the offspring eventually becoming separated in a more or less fully formed state.

The term spore is derived from the Greek "speiro," to sow, and thus need not be associated with asexual processes only, since it can be used to describe any single- or several-celled reproductive body, however formed, which becomes detached from the parent to develop into a new individual. Generally a prefix to the word "spore" indicates something about its formation and possibly its behaviour. For example, an oospore is the resting cell derived from a zygote and is obviously not asexually produced. In the fungi, the spores formed in basidia and asci are known as basidiospores and ascospores respectively, indications that they arise in special organs. In both cases, the organs are formed only as a result of some sexual process, even though this may be very reduced. Zoospores are those liberated in a motile condition into water; conidiospores are fungal spores formed exogenously on a conidium and liberated into the atmosphere in a non-motile state;

aplanospore can describe any non-motile spore produced within a sporangium, i.e. endogenously. The reader will undoubtedly have encountered a number of these spore forms during his systematic study of plants.

Spores are produced asexually by a very wide variety of plants, and in those exhibiting alternation of generations, spores form the resting and dissemination stages between the sporophyte and the gametophyte. In this case, the spores result from meiotic division of cells on the sporophyte and are therefore haploid. Typical examples are the spores of mosses, liverworts and ferns, and these are truly asexually produced, for no fusion of protoplasm has immediately preceded their formation. In some species of algae and fungi, where there may be no indication of an alternating life cycle, spores of various kinds may be produced mitotically by asexual means to propagate the species as exact replicas of the parents. We see examples in *Chlamydomonas*, *Ulothrix*, *Vaucheria* and *Oedogonium* among the algae and in *Mucor*, *Pythium*, *Eurotium* and *Erysiphe* among the fungi. In the colonial genera of the green algae, e.g. *Pandorina*, *Eudorina* and *Volvox*, the spores unite in colonies before liberation from the parent. Spore-production by the seed-bearing plants is by far the least obvious of all cases, but nevertheless, it occurs immediately before the gametophyte generations, however reduced these may be. The pollen grain is the microspore and the megaspore develops within the ovule.

Vegetative propagation by plants involves the separation of comparatively large multicellular masses which have been produced by a budding process. The separated part is usually a specialized portion of the vegetative body of the parent and may be a stem tuber (potato), root tuber (dahlia), corm (gladiolus), bulb (tulip), bud or bulbil (lily). Comparable structures are the gemmae of liverworts and moss gametophytes. Horticulturists and agriculturists make great use of this method of propagation and have extended the process beyond its natural occurrence by such methods as layering, taking of cuttings, budding, dividing and grafting.

Conditions affecting spore-production by plants have not been studied extensively. There is no general information which relates sporing to external conditions or hormones over a range of plant types. Isolated facts have emerged, as for example, that temperature of the water determines the time of spore-production in the fungus *Monoblepharis*, that light conditions do not appear to affect spore-production in the fungi, and that members of the liverwort genus *Marchantia* spore effectively only under long-day conditions followed by cold temperatures. On the other hand, considerably more is known about the factors affecting the vegetative processes. This is because they are of great economic importance. The formation of some tubers and bulbs is related to day-length. In general, a shortening of the light period in proportion to the dark, tends to favour tuber formation. With some varieties of potato and artichoke, no tubers will form during the long-day period of summer, but can be induced to form if the light periods are artificially shortened. The runner bean can be induced to form large tuberous roots under short-day conditions. Bulb-formation in the onion is known to be favoured by a long-day treatment coupled with high temperature. Short days and low temperatures together, almost completely suppress it.

The commercial use of plant hormones, natural and synthetic, is becoming more widespread and there are many preparations on sale which help to ensure the successful development of cuttings, grafts, layers, etc.

Artificial Parthenogenesis

Parthenogenesis has previously been defined and is known to occur naturally in animals and plants, but the greatest physiological study has been made on its occurrence when produced by artificial means. Most work has been done on animal eggs, particularly those which are shed into water in the unfertilized state. At some time or another, eggs from all animal groups have been under survey, and in very few cases only has it been found impossible to bring about parthenogenesis. One interesting feature is the very wide variety of methods of treatment to which eggs will respond. Often within a single species, several treatments may be effective. In the case of the sea-urchin, *Arbacia*, eggs have been activated by exposure to both hypotonic and hypertonic salt solutions, increases and decreases in pH of the water, the presence of fat solvents and some alkaloids, radiation with infra-red, ultra-violet and radium emanations, and even puncture with a fine needle. Generally one or other of these treatments will cause any egg to respond, but so far there seems to be no single treatment which affects all eggs. Several theories have been advanced to explain cell stimulation by natural as well as artificial means, but none is satisfactory. Some attribute a special rôle to calcium ions, in the absence of which no agent seems to operate successfully. Special interest has been shown in the formation of the spindle which precedes the first cleavage after the egg is activated. In natural conditions, it is the centrosome in the middle piece of the sperm which controls this spindle formation. In artificially-produced parthenogenesis, it has been found that the centrosome of the egg takes over the function in some cases, whilst in others, new centrosomes appear as a result of the treatment.

INTERSEXES AND SEX REVERSAL IN ANIMALS

Normally, animals of different sexes in the same species are clearly distinguishable in one or more of several ways. There are differences in form and structure of gonads; the accessory sexual apparatus, such as ducts and glands, and the external genitalia are distinctive. There are often variations in other structures such as skeleton and epidermis and there are usually physiological differences. This distinct unisexuality is known to be determined at the time of fertilization, when the chromosome complement of the zygote is made from the two sets of chromosomes carried by sperm and ovum. If the zygote contains a complete set of paired homologous chromosomes, then the embryo develops as one sex; if one of the chromosomes is not exactly paired, or as in some instances has no counterpart at all, then the embryo develops as the other sex (see Chap. 19). The chromosomes which control sex determination are known as the sex chromosomes and are of two kinds, labelled for convenience X and Y. In most cases, if XX chromosomes are brought to the zygote, then this leads to femaleness, but if XY chromosomes (or X alone), are brought, then maleness will follow. In butterflies and moths (lepidopterans) and birds, the reverse is the case.

During the development of the zygote, sex-determining factors on the chromosomes cause the sexual characters of the adult. Sometimes however, this translation fails to be complete in varying degrees and animals which are termed *intersexes* are produced. The classic investigation into intersexuality was carried out by Goldschmidt, who by crossing Japanese and European varieties of the gypsy moth (*Lymantria*) was able to produce every grade of intersexed moth, from full male to full female, at will, and was able to predict with accuracy the result of any cross he chose to make. From his experimental results, he was able to draw certain conclusions with regard to the occurrence of intersexuality in this insect. The sexual differentiation which the zygote will pursue is fixed at fertilization, but each sex possesses the potentialities of the other, since either kind of zygote can become intersexual. Normal development of one sex or the other, not mixed, is bound up with the nature of the sex chromosomes, that is, whether XX or XY, but even when there is nothing abnormal about these chromosomes, intersexuality is not prevented. Therefore the mere presence of sex chromosomes is not sufficient to ensure normality; it is their effect on development which is important. In an intersex, the observable features are those of a mixture of sexual parts in which some are male and some female. Those which are male are linked together by time of development, as though they had come under some influence all at the same time. Those which are female are likewise linked by their time of development but at a different time from the male parts. The mixture can be very variable, as though the relinquishing of control by one influence and the handing over to another can occur at any time. Thus we may see cases of intersexuality where only the sex parts which are the earliest to be differentiated are of one sex, whilst all the later differentiated parts are of the other sex, and from this by grades, through every possible mixture of sexual parts. It is as though the gypsy moth intersex is an individual which has developed as male (or female) up to a certain point in its life history and then has continued to develop as a female (or male). There are never cases of moths developing as male and female at the same time. The degree of intersexuality is determined by the time of the switch-over of the one controlling influence to the other. The controlling influence has been explained in terms of sex-determining substances produced under genic control. In a full male, it is supposed that during development, a male-determining substance is always produced in greater quantity than female-determining substance, and thus there is never any sign of femaleness. In a full female, the reverse is the case, but if it should occur that the sex-determining substances are produced at different rates, and that some genes have the power of causing more production of substance in a given time than others, then this can lead to a switch from male-substance preponderance to female-substance preponderance or vice versa, at some stage in development, and thus mixed sex in the same insect is possible. The final result, i.e. the degree of intersexuality, depends on the actual amount of each sex-determining substance produced at any moment, since this bears on the time when the switch is made from preponderance of one to the preponderance of the other.

In vertebrates the final sex-determining substances or hormones are postulated as being products of the gonads. The kind of gonad developed is the direct expression of sex-determining factors carried on the sex chromosomes, and the substances which they produce, act during the embryonic stages to control proper development of the accessory structures in keeping with the gonads. In other words, the chromosome make-up governs the development of gonad differentiation, and then hormones produced by the gonads govern the development of the other sexual parts one way or the other. Evidence for this comes from the occurrence of free-martins in cattle. A free-martin is the female of a pair of twin calves of opposite sexes. In this calf, the ovaries develop poorly or are even testis-like and the female ducts are degenerate, with male ducts sometimes showing partial development. The explanation offered is that during the embryonic period, when the blood streams of the twin calves are linked in the embryonic membranes, substance secreted by the male gonads, which develop at a very early stage, permeates the developing female tissues. This male-producing substance inhibits development of the female gonads and ducts, and encourages male structures. Not all conditions of intersexuality in vertebrates can be explained on these lines, and there is evidence from work on other mammals that the explanation of the free-martin is open to doubt. In the opossum, for instance, removal of the gonads from embryos of both sexes has no effect on the later development of the corresponding male or female ducts. It should be remembered that in many vertebrates, both male and female ducts are initially formed.

Intersexuality in vertebrates can be as widely variable as in the invertebrates. In mammals, including man, abnormalities of all kinds may be found. The gonads may consist of one testis and one ovary, with complete or partial development of the appropriate ducts on opposite sides; the gonads may be complete and correct but ducts of the opposite sex accompany them; male-looking animals may occur, lacking testes and having female tracts, and likewise the reverse of this. This parallels the conditions found in the lower animals and it is quite possible that in all animals, the genetic composition, i.e. the quantitative balance of male- and female-determining factors on the chromosomes, is the primary basic condition responsible for sex differentiation in the first instance.

Sex reversal, or change from fully functional male to fully functional female or vice versa, is quite normal in some animals. The oyster regularly changes its sex. It starts life as a male, then, when about two years old, it becomes a female and produces ova. Within a month or so of shedding the ova, i.e. becoming "white sick," and while still carrying embryos, it can revert to the male condition once more and continue to alternate from male to female at regular intervals. Similar occurrences are seen in the slug, *Limax maximus*, and the starfish, *Asterina gibbosa*. There are many well-authenticated accounts of other naturally occurring complete sex reversals in higher animals, including teleost fish, amphibians and birds, long after they have reached maturity. In the mammal, however, such complete sex reversal cannot occur in adults because of the differences in mode of development of the internal and external sex organs.

There are several possible causes of sex reversal. External agencies, which influence the development of a zygote, or even an unfertilized egg, may override the effects of the chromosome constitution. In frogs, if fertilization is delayed after the eggs are ripe, or if the eggs are exposed to a high temperature (27°C) before fertilization, all the embryos develop into males. This means that the 50 per cent of the total offspring which should have been females, develop as males, despite their chromosome make-up. Sex reversal can also be due to a disturbance of the general physiology of the animal during the embryonic period and in some cases during the post-embryonic period. If the ovary of a salamander (*Amblystoma*) embryo is replaced by a grafted embryonic testis of the same age, the remaining intact ovary gradually becomes modified into a testis and eventually produces fertile sperm. Castrated young male toads, on a diet rich in fat and lecithin, have become fully functional females. Hens which have produced normal eggs for some period have been recorded as becoming fully functional males.

It should be remmebered that in all these cases where the process of sex differentiation appear reversible, the chromosome constitution of the changed animal is not affected, and thus the form and function of a gamete are not affected by the chromosomes it carries but by the gonad which produces it. This means, for example, that a male toad made to function as a female, produces ova with the XY chromosomes just as the sperm would have possessed had the toad developed normally.

The *gynander*, or *gynandromorph*, is an animal which shows male characteristics in some parts of the body and female characteristics in others. The distinction between the gynander and the intersex lies in the fact that in the former the different parts are fundamentally dissimilar in that the cells contain unlike sex-chromosome complements, i.e. some XX, some XO, whereas in the intersex all cells contain the same sex-chromosomes, i.e. all XX or XY (XO).

Gynandromorphism is common among insects, and in the fruit fly, *Drosophila*, the genetic conditions which underlie the development of gynanders have been carefully analysed. In this fly, the gynanders generally begin development as females with two X chromosomes. During development, one of these X chromosomes is lost by a cell so that all its descendants will have the XO composition which leads to maleness. The result is a patch of tissue with male characteristics, amongst female tissues. Male regions of gynanders may be quite distinct from the female regions. The right half may be male and the left female; the anterior may be female and the posterior male; there may merely be islands of male tissue scattered among the female tissue. It depends upon the place of the accident and the stage of development of the embryo. Occasional cases of gynanders among vertebrates have been reported, as in some species of birds, but whether these are in the same category as those occurring in insects is doubtful.

CHAPTER 18

VARIATION

DEVIATION from a standard, or variation, is an outstanding feature of living things. Even the most untrained eye can discriminate between the majority of plant and animal classes. No one is likely to confuse an elephant with a grasshopper, a kangaroo with a scorpion or an elm tree with an aspidistra. Such distinction between main groups of organisms is obvious to all. Less obvious at first sight, but nevertheless equally present, is the distinction between individuals of the same group. No two elephants, if examined in detail in terms of the numerous qualities that each possesses, are found to be identical. The chances of their weight, shape, volume, colour or linear dimensions alone being all alike is so remote as to be virtually impossible. Similarly the chances of two oak trees having precise similarity in every respect is beyond computation. To go further, there exists the same degree of improbability of similarity between say, the trunks of two elephants or even between two leaves from the same oak tree. In fact no detailed comparison between two organisms, corresponding parts of two organisms or even corresponding parts of the same organism has ever shown complete similarity. "As like as two peas in a pod" will not bear literal translation.

Such variation has long been of interest to biologists, who have studied it and made use of it for many centuries. Their studies have been along three main lines. First, investigations into the properties in which organisms differ have been made from earliest times. Details of structure and function of plants and animals have been recorded and are still being investigated. Secondly, experiment and observation of natural variation have enabled the systematists to compile a logical classification of all known living things, in accordance with the observed differences. The third matter of interest to the biologist is one of more profound depth, namely, a study of the causes of variation. Whilst establishing what sort of variation may exist, and making use of such knowledge in classification may be matters of some difficulty and by no means satisfactorily complete, to state the causes of variation in precise terms is to reach the ultimate limit of attainment in biological knowledge. The fact that variations may be produced artificially at will in the laboratory has taken us nearer to the achievement of this aim but the story is by no means complete.

735

The painstaking systematists have produced from their variation studies an acceptable classification into which any known or newly discovered organism can be fitted. The whole range of living things has been grouped into kingdoms, phyla, classes, orders, families, genera and species such that in ascending order from species to phylum the broader phylogenetic relationships are clear, but never can the groups be linked by intermediates in a completely smooth or continuous way. There are no clear transitions between mammals and insects or between ferns and the algae and such groups can only be linked together in a discontinuous way. They do not merge gradually one into the other. Such variation is known as *discontinuous variation* and such differences exist between all groups of organisms right down to species. Each species is represented by a number of individuals all of sufficient likeness to warrant identifying them as the same kind of organism. If, however, a large number of individuals of the same species are examined in detail and their qualities such as shape, colour, weight and linear dimensions are compared, as has been said, never within the group will be found two individuals of exactly identical characteristics in every respect. It will be found often though, that for each such quality, the variations can be arranged so that they exhibit a continuous series from the one extreme to the other with every possible gradation between. Such variation is known as *continuous variation*. It must not be thought that discontinuous variation cannot be found within a species. Species may be further subdivided into varieties. For example, the tall and dwarf varieties of garden peas and the coloured varieties of domestic rabbits are well defined by their discontinuity. Furthermore, it was early recognized that by selective breeding such discontinuities could be more heavily accentuated, as may be witnessed by the existence of the many varieties of domestic plants and animals not occurring in nature. It has also been found that whilst some can be, many of the continuous variations within a species cannot be selectively bred and thus must be considered to be due to an influence outside the parentage of the individual. Living things do not exist, cut off from their surroundings, in conditions of complete constancy, so that the fluctuating conditions under which they live can be expected to exert an influence upon their development. Thus variation, whether continuous or discontinuous, has long been recognized as the result of one of two influences, or the interaction of both. These are environmental influences and the influences of parentage. How much of each, and how much of their interaction, is responsible for any particular variation is often a matter difficult to elucidate. There are those who would say that it is impossible to distinguish between them.

MEASUREMENT OF VARIATION

A detailed analysis of the variability of a group of organisms for a particular character may be made by the application of *biometrics*. The biometrician applies mathematical methods to the study of small but measurable differences between individuals. He is often able to deduce valuable information concerning the variability of a given population of organisms. It is possible to deduce for instance whether or not a population is homogeneous, that is to say, whether all the members are related by the same parentage and by the same environment. It is also possible to state the degree of variation for a particular character in a population in mathematical terms and, as it were, measure it. Further, from a given set of variation observations, the biometrician may be able to demonstrate that at least some part is due to inheritance and some to environment.

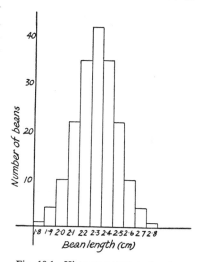

Fig. 18.1. Histogram representing the frequency distribution of the length of beans.

Where variation in a character is continuous, what is known as *frequency distribution* for that character can be found by recording the numbers of individuals which fall within given ranges of measurement. For instance, the frequency distribution of length in beans may be obtained by selecting at random a large number of beans and measuring the length of each to the nearest millimetre. Then, by placing into classes all those whose length is the same to the nearest millimetre, and plotting as a histogram the numbers in each class, i.e. the frequency of individuals in a class, as ordinates, against the length in millimetres of the classes as abscissae, a frequency distribution of the length of the beans may be obtained. The histogram in Fig. 18.1 represents such a result. The general outline is one indicating the gradual falling away in numbers of beans with lengths successively smaller or greater than the commonest length. If such were not the case, suspicion might be justified that the material under consideration is not homogeneous. If, for instance, the histogram had two clearly marked maxima as in Fig. 18.2, we might deduce the fact that the measured individuals consisted partly of a small variety of beans and partly of a large variety, or if care had not been taken in selecting the material, we may have chosen two distinct age groups of beans. Returning to the histogram in Fig. 18.1, quite obviously the outline shows a discontinuity since the heights were measured to the nearest millimetre. If we were to repeat the whole experiment but measure the bean lengths to the nearest one-tenth of a millimetre, we should find fewer beans in each class, but the outline of the histogram would be a much closer approximation to

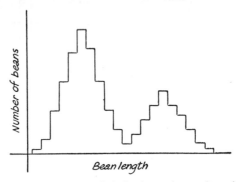

Fig. 18.2. Histogram with two peaks indicating a mixture of two clearly-defined varieties of bean.

a continuous line, provided sufficient beans were included in the experiment. If an infinite number of beans from a homogeneous population were measured to within infinitesimally small gradations in length, the curve would be perfectly smooth. From the characteristics of such a smooth curve, it is possible to deduce a method of extracting from any set of observations, information which will lead to a measurement of the variation of a character within a population. Such a curve is superimposed on the histogram in Fig. 18.3, and is known as a *normal distribution curve* or Gaussian curve (after Gauss, who applied it originally to the arrangement of errors occurring in physical experiments). In such a distribution, the commonest measurement, i.e. that which is applicable to most individuals is called the *mode*, and exactly coincides with the average measurement for all the individuals which is known as the *mean*. To obtain such a perfect curve in practice is highly unlikely, but a study of variations in living things shows that they conform, at least, to a good approximation to this normal distribution which has a definite mathematical form, and thus a statistical treatment may be applied to any such set of variation data. Two constants peculiar to the normal distribution curve are of value in variation studies. For instance, we may calculate the *mean value* for a character and also mathematically express the *amount of variation* of that character within the population.

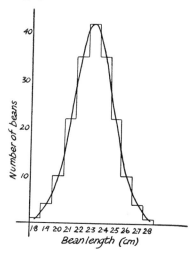

Fig. 18.3. A normal distribution curve.

The general formula for a normal distribution curve is given by—

$$df = \frac{1}{\sigma\sqrt{(2\pi)}} \cdot \exp\left(\frac{-(x-\mu)^2}{2\sigma^2}\right) \cdot dx$$

where,

df = the number of individuals within the infinitesimally small range of the measurement taken, dx.

π = the ratio of the circumference to the diameter of a circle.

exp = the natural log base.

μ = the mean value of all the measurements.

σ = the standard deviation displayed by the population.

The variable part of this function, specifying the normal distribution curve, is $-(x-\mu)^2/2\sigma^2$. σ^2 is the *variance* or "spread" of the distribution and provides a measure of the variation present in the population. If a number of distribution curves plotted on the same scale are examined, it can be seen that some tend to be wide and squat, indicating a wide range of variation, whilst others tend to be narrow and sharply peaked, indicating a narrow range of variation. The square root of the variance, σ, is known as the *standard deviation* of the distribution.

From any sample of observations, in order to describe the population in terms of its variation in the character chosen, it is necessary to obtain estimates of the two constants of the curve, μ and σ. Any such estimate is termed a *statistic*. It can be demonstrated that the best obtainable estimate of the quantity μ is found by calculating the arithmetical mean of the measurements or observations made, i.e.

$$\mu = \frac{\Sigma(x)}{n}$$

where n is the number of measurements and $\Sigma(x)$ denotes the total of all the measurements. The best estimate of the variance, σ^2, is then given by,

$$\sigma^2 = \frac{\Sigma(x - \mu)^2}{(n - 1)}$$

The standard deviation, σ, is therefore the square root of this value and is given by,

$$\sigma = \sqrt{\left(\frac{\Sigma(x - \mu)^2}{n - 1}\right)}$$

Below is set out in tabular form the manner in which estimates for the constants μ and σ can be calculated from the data from which the histogram in Fig. 18.1 was constructed. It is stressed again that the data are purely imaginary and have been chosen for simplicity.

STANDARD DEVIATION IN LENGTH IN A SAMPLE OF RUNNER BEANS

Bean length (cm)	Number of beans with this length, i.e. frequency	Frequency × length	Deviation from mean length	Deviation ^{2}	Frequency × deviation ^{2}
x	f	$\times x$	$(x - \mu)$	$(x - \mu)^2$	$f.(x - \mu)^2$
1·8	1	1·8	− 0·5	0·25	0·25
1·9	4	7·6	− 0·4	0·16	0·64
2·0	10	20·0	− 0·3	0·09	0·90
2·1	22	46·2	− 0·2	0·04	0·88
2·2	35	77·0	− 0·1	0·01	0·35
2·3	42	96·6	0·0	0·0	0·0
2·4	35	84·0	+ 0·1	0·01	0·35
2·5	22	55·0	+ 0·2	0·04	0·88
2·6	10	26·0	+ 0·3	0·09	0·90
2·7	4	10·8	+ 0·4	0·16	0·64
2·8	1	2·8	+ 0·5	0·25	0·25
	$n = 186$	$\Sigma fx = 427\cdot8$			$\Sigma f(x - \mu)^2 = 6\cdot04$

Mean bean length, $\mu = \dfrac{427\cdot8}{186} = 2\cdot3$ cm.

Standard deviation, $\sigma = \sqrt{\left(\dfrac{\Sigma f(x - \mu)^2}{n - 1}\right)} = \sqrt{\left(\dfrac{6\cdot04}{185}\right)} = 0\cdot18$

By filling in a table of this kind from data obtained in an actual experiment, the student will realize that the procedure is simple, if somewhat tedious, arithmetic. Lengths or widths of seeds, fruits and leaves may be measured quickly and easily and afford material for such exercises. As a long-term exercise the same data can be collected from succeeding generations of a species in a natural population, and the change in variability of a particular character examined. Biometricians use this method to establish whether or not a character is being acted upon by natural selection, and if so, in what direction the selection is acting.

As has been previously stated, it is found in nature that a great many characters of living things show variation from the normal pattern such as might be due to pure chance, and in general, where such normality is shown, we are tempted to conclude that such variation is being influenced by the chances of environment only. Note that some characters attributable to pure chance may produce a variation which is not normal. The curve is said to be "skew," meaning that it is not symmetrical about the modal ordinate. The modal value differs more or less widely from the mean of the measurements. But normal variation does not necessarily mean that environment only is influencing the character, and a further use of biometrics, linking the variation in a character between parent and offspring, may show that inheritance is playing a part. If it can be shown that deviations from the mean in a parent's characters are preponderantly accompanied by deviations from the mean in the offspring's characters *in the same direction*, then it may be concluded that the parent is influencing the character possessed by the offspring. To do this, what is known as the *covariance* between the characters in parent and offspring must be calculated. To quote an example, suppose it is required to study the effect of inheritance on height in human beings, say, between father and daughter. We should first measure the heights of numerous fathers and for each one, the height of his daughter. For each group, i.e. fathers and daughters, the variance (σ^2) would be given by—

$$\text{Variance for fathers} = \sigma_f^2 = \frac{\Sigma(x_f - \mu_f)^2}{n-1}$$

$$\text{Variance for daughters} = \sigma_d^2 = \frac{\Sigma(x_d - \mu_d)^2}{n-1}$$

The covariance of fathers' and daughters' heights is given by—

$$\frac{\Sigma[(x_f - \mu_f)(x_d - \mu_d)]}{n-1}$$

It should be noted that whereas variances must always be positive because they are derived from the sums of squares, covariances may be either positive or negative because they are derived from the sums of cross-products of deviations. The magnitude of the value for the covariance between fathers' and daughters' heights indicates the degree to which one is following the other. The sign of the covariance shows in which direction one is influencing the other. A positive covariance indicates that deviations from the mean in fathers' heights are being followed by deviations from the mean in daughters' heights in the same direction, i.e. tall fathers, tall daughters. When the sign of the covariance is negative, it indicates that deviations in the two distributions are preponderantly in opposite directions, i.e. tall fathers, short daughters.

The value of this calculation is now plain. If the variation is to some extent under parental influence, we should expect tall fathers generally to have tall daughters and short fathers generally to have short daughters. A positive covariance between height of fathers and daughters does in fact exist, and it was first demonstrated by Galton who therefore showed that stature in man is in part inherited. If inheritance plays no part, the covariance would have been zero.

The size of the covariance when compared with some standard, gives a measure of the degree to which father influences daughter. The standard taken is that given by the variances of the two separate distributions. We may either compare the covariance with each variance separately and so establish the *regression coefficient*, or compare the covariance with both variances together and establish the *correlation coefficient*.

Regression coefficient of daughters on fathers =

$$\frac{[1/(n-1)] \cdot \Sigma[(x_f - \mu_f)(x_d - \mu_d)]}{\sigma_f^2}$$

Regression coefficient of fathers on daughters =

$$\frac{[1/(n-1)] \cdot \Sigma[(x_f - \mu_f)(x_d - \mu_d)]}{\sigma_d^2}$$

Correlation coefficient,

$$r = \frac{[1/(n-1)] \, \Sigma[(x_f - \mu_f)(x_d - \mu_d)]}{\sqrt{(\sigma_f^2 \cdot \sigma_d^2)}}$$

or,

$$r = \frac{\Sigma[(x_f - \mu_f)(x_d - \mu_d)]}{\sqrt{[\Sigma(x_f - \mu_f)^2 \cdot \Sigma(x_d - \mu_d)^2]}}$$

The correlation coefficient will always have a value between -1 and $+1$. If the value is $+1$, it would signify a complete association between father and daughter in the matter of height, meaning that the height of the father fully determined the height of the daughter. If the value is zero, then it indicates that there is no relationship whatsoever between fathers' and daughters' heights. Values between 0 and +1 and 0 and −1 show partial determination in the positive and negative directions respectively. It should be noted that where the correlation coefficient is small, the relationship between the characteristics is very slight and in view of the extent to which, in actual observations, the measurements may be effected by disturbances which have no relation to the matter under examination, e.g. errors on the part of the measurer, no great significance can be attached to the occurrence of correlation coefficients of small magnitude. The correlation coefficient found by Galton between heights of parent and offspring in human beings was about 0·5, indicating quite definitely that our height is to some extent dependent on the height of our parents. By means of similar correlation studies, it has been shown that a number of characters showing continuous variation in many plants, animals and man are in part inherited.

It is a fact that many biological characters, both morphological and physiological, are correlated in some degree, depending on the closeness of their association in the organism. The study of correlation often gives valuable information about the make-up or workings of an organism which cannot be obtained by simple observation or experiment. Its use has sometimes proved to be of great practical importance as for example in forecasting a future event. Date of ripening of crops is often strongly correlated with the earliest date of opening of the flowers, and if during the year the flower-opening data is noted, the date at which the crop may be harvested may be predicted with some accuracy from previous experience.

The manner in which a correlation coefficient can be calculated is set out in tabular form on p. 744. Once the data in the first two columns have been obtained, the rest is a matter of simple arithmetic once more.

Correlation coefficient

$$r = \frac{\Sigma(x_w - \mu_w)(x_l - \mu_l)}{\sqrt{\Sigma[(x_w - \mu_w)^2 \cdot \Sigma(x_l - \mu_l)^2]}}$$

$$= \frac{0·45}{\sqrt{(0·56 \times 0·60)}}$$

$$= 0·776$$

This is a high positive correlation.

Correlations can also be represented graphically. This can be done by plotting the two measured quantities against each other on suitable

CORRELATION BETWEEN WEIGHT AND LENGTH IN A SAMPLE OF RUNNER BEANS
(Weighings taken to the nearest decigram and lengths to the nearest millimetre)

Bean weight (g) x_w	Bean length (cm) x_l	Deviation from mean weight $(x_w - \mu_w)$	Deviation from mean length $(x_l - \mu_l)$	$(x_w - \mu_w)(x_l - \mu_l)$	$(x_w - \mu_w)^2$	$(x_l - \mu_l)^2$
1·7	2·7	0·3	0·5	0·15	0·09	0·25
1·5	2·6	0·1	0·4	0·04	0·01	0·16
1·6	2·2	0·2	0·0	0·0	0·04	0·0
1·4	2·1	0·0	− 0·1	0·0	0·0	0·01
1·6	2·4	0·2	0·2	0·04	0·04	0·04
1·4	2·3	0·0	0·1	0·0	0·0	0·01
1·0	2·0	− 0·4	− 0·2	0·08	0·16	0·04
1·1	2·0	− 0·3	− 0·2	0·06	0·09	0·04
1·2	2·1	− 0·2	− 0·1	0·02	0·04	0·01
1·1	2·0	− 0·3	− 0·2	0·06	0·09	0·04
$\Sigma x_w = 13{\cdot}6$	$\Sigma x_l = 22{\cdot}4$			$\Sigma(x_w - \mu_w)(x_l - \mu_l) = 0{\cdot}45$	$\Sigma(x_w - \mu_w)^2 = 0{\cdot}56$	$\Sigma(x_l - \mu_l)^2 = 0{\cdot}60$

Results—

Mean weight, $\mu_w = 1{\cdot}4$ g (to nearest decigram)

Mean length, $\mu_l = 2{\cdot}2$ cm (to nearest millimetre)

$n = 10$

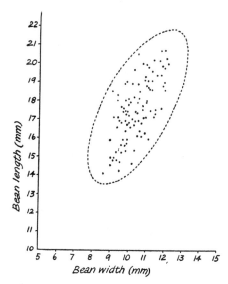

Fig. 18.4. Dot diagram of bean length against width showing a high positive correlation.

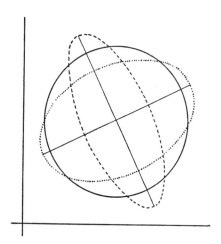

Fig. 18.5. Forms of dot diagram.

Full line (circle)—no correlation. Dotted line (ellipse with major axis sloping upwards —slight positive correlation. Dash line (ellipse with major axis sloping downwards)— high negative correlation.

squared paper. As an example such a dot diagram is shown in Figs. 18.4 compiled from measurements of bean lengths and breadths. For every measured length of a bean, a dot is placed against the corresponding breadth of the bean, breadths being ordinates and lengths abscissae. The shape enclosed by a line surrounding all the points and the direction of its slope, if it has one, indicates the degree and kind of correlation between the two characters. If the two quantities have a complete direct correlation then all the dots will fall on a straight line sloping upwards. A similar line sloping downwards indicates complete negative correlation. A circle indicates no correlation at all, whilst intermediate ellipses indicate greater or lesser degrees of correlation according to the ratios of their major and minor axes. The shapes in Fig. 18.5 show the possibilities.

INHERITED AND NON-INHERITED VARIATION

It was stated earlier in this chapter that variation between individuals could be due to one of two kinds of influence. The subjects of comparison may each possess a different inward constitution inherited from their parents or alternatively, they may have been subjected to different external conditions under which to develop. Thus observed differences may be due to "nature," that is, an inherited potentiality for developing in a particular way, or due to "nurture," the set of environmental conditions under which the organism has been allowed to develop. For example, as we shall see in the next chapter, pure-breeding tall peas always tend to produce tall offspring when growth conditions are good, but the progeny could not be expected to be of the greatest stature if they were starved or deliberately stunted by some other means. It is not always an easy matter to distinguish between the variations which result from these two causes, and to do so, some kind of breeding test must be devised in which all the subjects of the experiment are grown under controlled conditions of both "nature" and "nurture." The results of such an experiment can be fully interpreted only when subjected to biometric analysis.

The first experiment of this kind to be recorded was performed in 1900 and later years by Johannsen of Denmark, who set out to determine whether the average weight of seed produced by a plant at maturity was in any way influenced by the weight of the seed from which that plant developed. He started with nineteen seeds, one from each of nineteen different dwarf bean plants, *Phaseolus vulgaris*. This plant reproduces unfailingly each season as a result of self-fertilization so that all the descendants of a particular plant will have identical inherited potentialities. He called such a succession of descendants from a common starting point a *pure line* (*see* p. 760). He started, then, with nineteen seeds each representing a different pure line. These he planted and cultivated in equivalent environmental conditions and harvested seed from each parent separately. For each line the seeds were weighed and the weights recorded. The following season only the lightest and heaviest seeds of each line were planted, and the same procedure of harvesting and weighing repeated. Again, only the lightest and heaviest seeds were selected by which to propagate, and so on for several more generations. Johannsen was practising selection for seed weight within each line. When sufficient generations had been harvested, he examined his results to see if he had been able to separate new varieties according to seed weight within any of the pure lines. The results achieved for Line XIX are shown in the following table. All other lines showed corresponding results.

Line XIX	Average weight of selected parent seeds		Average weight of daughter seeds	
Year	Lighter seeds	Heavier seeds	Lighter seeds	Heavier seeds
1902	300 mg	400 mg	360 mg	350 mg
1903	250	420	400	410
1904	310	430	310	330
1905	270	390	380	390
1906	300	460	380	400
1907	240	470	370	370

An examination shows that there was a remarkable retention of weight averages within each line, generation after generation, no matter whether the line was reproduced by the heaviest or the lightest seed. In fact, in the sixth generation in Line XIX, the smallest (240 mg) and the largest (470 mg) both produced daughter seeds averaging 370 mg. Johannsen had demonstrated clearly that within a pure line of beans, selection for weight has no effect. The variations in the size of the beans were not due to some internally controlled potentiality, but to slight differences in the conditions to which they were subjected during their development. Such factors as position in the pod or position of the pod on the stem, differences in warmth, light and moisture, are in control.

But this is not the whole of the story. When Johannsen's results are examined from the point of view of comparison of the average weight of beans produced by the different pure lines, then it can be seen that quite a different state of affairs is presented. From among the pure lines with which he started, Johannsen was able to distinguish between some that tended on the average to produce small beans, and some that tended to

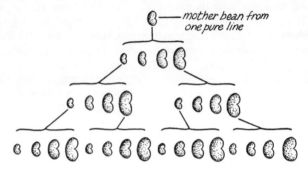

a. In many succeeding generations, lightest & heaviest beans always produced the same average weight of daughter beans

mother beans from different pure lines

b. Different lines showed different weight averages

Fig. 18.6. Diagrams representing Johannsen's work on pure lines.
(a) The effect of selection for weight within a line.
(b) The effect of selection for weight within several lines.

produce heavy beans, outside of the influence of external conditions. Some of his results are shown in the table on p. 749. They indicate clearly that his original nineteen pure lines did in fact differ from one another in their ability to produce light or heavy seeds, even though selection for seed weight within the lines was ineffective. The differences between the lines are obviously internal, that is, due to parentage and not to environment.

Weight of mother bean to nearest 10 mg	I	II	V	VII	Line XII	XIII	XV	XVIII	XIX
200 mg	—	—	—	460	500	—	470	410	—
300	—	—	—	—	—	480	—	410	360
400	—	570	530	500	—	450	—	410	350
500	—	550	490	—	450	450	450	—	—
600	630	570	—	490	440	460	450	—	—
700	650	560	500	—	—	—	—	—	—
Average of line to nearest 10 mg	640	560	510	490	460	460	450	410	360

Whilst the table above gives a reasonably clear indication that a pure line of heavy beans tends to produce, on average, larger beans than are produced by a pure line of less heavy beans, thus pointing to inherited variation in the bean weights, a distinction between this and any environmental influence on bean weight variation is not always so readily disclosed. In many instances it would be necessary to subject the data to some statistical analysis from which it could be assessed with what degree of certainty it might be assumed that any variation shown was really inherited, that is, under control of some internal hereditary factor or alternatively due to chance, meaning the random vagaries of growth conditions. Such biometrical work would necessitate the evaluation of the statistics, means, variances (standard deviations) and covariances as previously described. It is beyond the scope of this work, however, to develop this further but it should be borne in mind that it becomes an essential tool of the plant breeder who is dealing with continuous variation in quantitative characters.

Johannsen's experiment made several things clear to those of his day who were interested in the causes and inheritance of variation. The subject had been of great importance since the first theories of evolution were put forward, and was further brought into prominence by the rediscovery in 1900 of the work on inheritance carried out by Mendel some forty years previously. The work on pure lines showed that both heredity and environment can influence the same character at the same time, but that it is the effect of the former only which is transmitted to succeeding generations. Further, a character which shows continuous variation can be under the influence of internal as well as external factors, a condition previously denied by many biologists.

In the next chapter we shall give an account of the work of Gregor Mendel, who by the application of biometrics, made the first major contribution to the solution of one of the most fundamental of all biological problems, namely, the way in which characters are inherited.

CHAPTER 19

GENETICS

REFERENCE was made in the previous chapter to the conformity to a pattern within a species despite individual variation; lettuces reproduce lettuces, and the offspring of cats are cats. The general form of an individual always agrees so closely with that of its parents that we suppose it is in fact inherited from them. Genetics, a term coined by Bateson, is the biological science which seeks to account for the resemblances and differences which are exhibited among organisms directly related by descent.

That like, in general, begets like, has been realized all through the ages. The irregularity of the process has also been apparent ever since attention was paid to the breeding of domesticated plants and animals. The ancient use of the terms "sport" or "lusus naturae" to indicate a new variation in a group of progeny, illustrates this. But to such early breeders the whole processes of variation and inheritance were completely obscure. This was due to the fact that nobody logically worked out a series of carefully controlled experiments to investigate them. Such work is vitally necessary to the understanding of these phenomena. Little insight into the nature of inheritance had been gained by plant breeders up to the first half of the nineteenth century, and it was not until 1865 that the foundations of modern knowledge were laid by Johann Gregor Mendel. He planned and executed successfully a series of experiments from which he could make sound postulations as to the manner of inheritance of characters between parent and offspring. His brilliance lay in his method of approach to the problem, not in any special technique. It is by his name that we perpetuate the scientific theory which explains the fundamentals of organic inheritance. *Mendelism* is the theory relating to the manner in which characteristics of parents are distributed throughout succeeding offspring; it attempts to interpret what Mendel was able to reveal by his own plant breeding experiments. Born of a peasant family in 1822, he became a monk and eventually an abbot in the Augustinian monastery at Brunn in Moravia. His plant breeding was carried on in the garden of the monastery, where he died in 1884.

THE WORK OF GREGOR MENDEL

Mendel selected for investigation the problem of the manner in which true-breeding variations within a species are related to one another.

He had noticed that in a number of garden plants it was possible to find within a species, here defined as an interbreeding group, sharply circumscribed variations. For example, the garden pea *Pisum sativum* exhibited both tall (2–2·5 m) and dwarf (0·2–0·4 m) varieties; their seeds could be smooth and round, or wrinkled in appearance; their unripe pods could be green or yellow, and so on. He set out to find how these variable characteristics were distributed numerically throughout successive generations and to explain the mechanism whereby each variety inherited its special characteristics. His approach was fundamentally biometric in that his results could only be interpreted by mathematical analysis. His knowledge of the life histories of the material used in his experiments was complete since the story of sexual reproduction in plants and animals had long been worked out. He knew that the garden pea was normally self-fertilized and that he could therefore use for his original parents, plants which he knew to be in succession from lines which had bred true, i.e. showed no gross variation from the parent pattern, for many generations. He proceeded as follows. He selected parents for seven pairs of contrasting characters as set down in the Table on p. 753. He then effected cross-pollination between parents exhibiting the contrasting characters by first removing the stamens from the flowers (emasculating) on one plant before they were mature, so ensuring that self-fertilization could not occur. Then, when the ovaries on the emasculated flowers were fully formed, he transferred pollen from the other parent of the pair to the stigmas of these ovaries. Between the emasculation process and the artificial pollination, he carefully protected the young stigmas so that they could not be contaminated by foreign pollen. He thus had male and female parent peas of his own choosing, and as a check on his work he made a cross-pollination in the other direction. For example, in one cross-pollination he used a tall father and a dwarf mother, and as the check he used a tall mother and a dwarf father. We say that he made *reciprocal crosses*. In due course, the pollination resulted in fertilization and the female parents set seeds. These he carefully harvested, counted and stored separately until the following season, when he planted them and watched their development. He had produced this generation from his first cross. While they were maturing, and at maturity, he examined each plant carefully and noted its characteristics with respect to the pairs of contrasting characters which he had selected for observation. All these plants were allowed to become self-fertilized and in due course each set seeds which were again harvested, counted, and stored as before. In the next season, each was planted and allowed to mature. This was now the second generation from his original parents, and once more he carefully examined each plant for the selected

TABLE OF CONTRASTING CHARACTERS AND FREQUENCIES OF
OCCURRENCE IN MENDEL'S EXPERIMENTS

Contrasting characters of the parents	Frequency of each in the first generation	Frequency of each in the second generation	Ratio of one to the other in the second generation
1. *Seed form:* round *v.* wrinkled	All round	5474 round 1850 wrinkled	2·96 round 1 wrinkled
2. *Colour of reserve food in the cotyledons:* yellow *v.* green	All yellow	6022 yellow 2001 green	3·01 yellow 1 green
3. *Ripe pod form:* inflated, smooth *v.* constricted, wrinkled	All inflated, smooth	882 inflated 299 constricted	2·95 inflated 1 constricted
4. *Seed coat and flower colour:* whitish seeds, white flowers *v.* greyish seeds, purple flowers	All greyish seeds with purple flowers	705 grey, purple 224 white, white	3·15 grey, purple 1 white, white
5. *Colour of unripe pods:* green *v.* yellow	All green	428 green 152 yellow	2·82 green 1 yellow
6. *Flower position:* axial *v.* terminal	All axial	651 axial 207 terminal	3·14 axial 1 terminal
7. *Stem length:* tall *v.* dwarf	All tall	787 tall 277 dwarf	2·84 tall 1 dwarf
		Totals: 14949:5,010 (74·9%) (25·1%)	Mean ratio: 2·98:1

characters, noting down the frequency with which each character appeared. Once more he allowed each plant to become self-fertilized and harvested the seeds from each. The following season he examined the third generation, counting the frequency with which his contrasting characters appeared.

The selected contrasting characters and the frequency with which they occurred in the first two successive generations are given on p. 753.

These results can best be interpreted if we set out the facts of the experimental procedure and the frequencies of one of the pairs of contrasting characters thus:

Tall female parent artificially pollinated from dwarf male parent produced—

All tall plants in the first filial generation, or F_1, which when self-fertilized produced—

787 tall plants and 277 dwarf plants in the second filial generation, or F_2.

Ratio	2·84 tall plants : 1 dwarf plant
Approximate ratio	3 : 1

From an examination of all the frequencies and their ratios, Mendel was able to draw very definite conclusions. First, in all the descendants (F_1) of the original cross, one of a pair of contrasting characters under consideration always failed to appear, as if it were subjugated by the other. To the overriding character, Mendel applied the term *dominant*, and to the subjugated character, the term *recessive*. In the second place, both characters appeared in the second generation, and in every case, very closely approximating to a 3 dominant to 1 recessive ratio. It did not matter in which direction the first artificial cross had been made; both sets of results gave this ratio. In the third generation, the frequencies indicated that all plants showing the recessive character in the second generation bred true for this character, and that of those showing the dominant character in the second generation, one-third bred true and two-thirds again segregated into two groups, three times as many showing the dominant character as those showing the recessive character.

From these numerical results Mendel formulated a law of inheritance concerning the gametes which he knew had to fuse to form the zygote from which a new plant could develop. He knew nothing of their cytology but he suggested that they carried something, which when able to exert its influence, was responsible for the appearance of a particular character, and for each of a pair of contrasting characters he supposed a *germinal unit* or *factor* in the gamete. He maintained that these germinal units remained unaltered in the gametes through successive generations, even though the characters they represented did not always appear. They would not be able to exert an influence, for example, if they were paired at fertilization with other germinal units which could nullify their presence. In the first generation of his crosses, one of Mendel's chosen pairs of contrasting characters, such as dwarfness or wrinkledness, always disappeared, hence his terms dominant for the tall and round characters, and recessive for the dwarf and wrinkled characters. They would be able to exert their influence in later generations if paired with germinal units of their own recessive nature. In the second generation, some did so, and the dwarfness and wrinkledness, etc., reappeared.

Mendel's first law of inheritance stated that *of a pair of contrasted characters only one can be represented in a single gamete.* This is sometimes known as the law of the purity of the gametes.

We can more easily interpret Mendel's results and understand what the law means if we use symbols to represent the germinal units. Let T stand for a germinal unit for tallness, and t for a germinal unit for dwarfness. Now, if Mendel's original parents were pure-breeding tall and dwarf varieties, then they must be represented as TT in the one case and tt in the other, since each resulted from the fusion of two gametes each carrying a single germinal unit. That is to say, the zygote from which the tall parent was formed had brought to it T from one gamete and T from the other, whilst the zygote from which the dwarf parent was formed likewise arose from two gametes each carrying a t. When these parents produced their own gametes, according to Mendel's law, those gametes could carry only one germinal unit each. Thus from the tall parent, male and female gametes, each carrying a single T, were formed, and from the dwarf parent, male and female gametes, each carrying a single t, were formed. When Mendel effected his artificial cross, he deliberately eliminated some of those gametes and arranged it so that all new zygotes formed were produced by a male gamete carrying T and a female carrying t or vice versa, so that his first filial generation plants can be represented by Tt. When these reached maturity, they were all tall plants because the germinal unit T dominated the unit t. When this generation produced gametes, each plant could give rise to male and female gametes each of two kinds, namely males carrying T or t and females carrying T or t, and if self-fertilization were permitted, then by the laws of chance there should be equal probability of four kinds of zygotes being formed. A male gamete carrying T could combine with a female gamete carrying either T or t and a male gamete carrying t could combine with a female gamete carrying T or t. Thus the zygotes could be of the four kinds TT, Tt, tT and tt. The resulting plants carrying TT were like the original tall parent and were tall, those carrying Tt or tT were like those of the first filial generation, and because T dominates t, were also tall, whilst those carrying tt were like the original dwarf parent. Thus three-quarters of the plants in the second filial generation were tall and the remainder dwarf. Mendel's figures showed a very close approximation to this ratio. In the third filial generation, the plants constituted TT and tt would again breed true for tallness and dwarfness respectively, whilst those constituted Tt or tT would produce offspring again in the ratio 3 tall to 1 dwarf. The scheme below summarizes this symbolic representation of Mendel's results and his first law of inheritance.

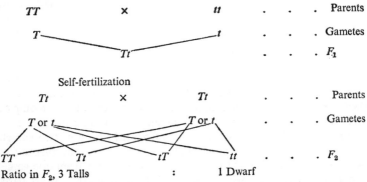

It is important for the student to note that the proportions of peas showing tallness and dwarfness in Mendel's experiments are characteristic of the assortment or segregation of two characters when influenced only by the laws of probability and chance. For instance, if a bag contains equal numbers of red and black marbles identical in every other respect, and pairs of marbles are withdrawn and the frequency of the colour combinations noted, after a large number of pairs have been withdrawn, the frequencies will closely approximate to the ratio 1 (red and red):2 (red and black):1 (black and black). The greater the number of pairs withdrawn, the nearer to these exact proportions will the frequencies become. Similarly, if two pennies are tossed simultaneously, the frequencies of two heads, one head one tail and two tails, show the same ratios. The parallel between the assortment of colour in marbles, the heads and tails with pennies, and Mendel's germinal unit assortment, should be plain.

Mendel's experiments were not ended by the investigation of the behaviour of one pair of contrasting characters, for he went further and included experiments to test the assortment of two pairs of such characters. For instance, in one case, he chose seed form and seed colour as his two subjects of study, using a pure-breeding plant with round and yellow seeds as one parent, and a pure-breeding plant with wrinkled and green seeds as the other. His mode of operation was the same as that previously described, but he recorded the frequencies with which each character appeared with the other, instead of their separate frequencies. His results showed that in the first filial generation all the plants had round and yellow seeds, indicating the dominance of round over wrinkled, and yellow over green characters respectively. When these offspring were selfed, of a total of 556 seeds, the second filial generation produced the following frequencies: 315 round and yellow-seeded plants, 101 round and green-seeded plants, 108 wrinkled and yellow-seeded plants and 32 wrinkled and green-seeded plants. The calculated proportions are $9.9:3.1:3.4:1$ or approximately $9:3:3:1$.

From these results, Mendel deduced a second law of inheritance concerning the free or independent segregation of the germinal units. He stated that *each of a pair of contrasted characters may be combined with either of another pair.* This may best be understood by the use of symbols for germinal units once more. His original parents may be denoted by $RRYY$ and $rryy$ respectively, where R = round, Y = yellow, r = wrinkled, and y = green. The gametes from these parents can carry only one germinal unit representing each character and may be denoted as RY in the case of one parent and ry in the other. This was stated in Mendel's first law. When Mendel deliberately brought these gametes together at cross-fertilization, the first filial generation were made to carry the germinal units $RrYy$. The plants all possessed round and yellow seeds because those characters are dominant. They were the parents of the second filial generation when allowed to self. From each there are four possible kinds of male and female gametes, namely, RY, Ry, rY and ry. The laws of chance again dictate that any one of the male gametes may unite with any one of the four possible female gametes, making altogether sixteen possible combinations. Thus, using symbols, we have—

RR YY	×	rr yy	.	.	.	Parents

RY⟍　　　　　　⟋ry　　. . . Gametes

　　　　⟍$Rr Yy$⟋　　　. . . F_1

Self-fertilization

Rr Yy	×	Rr Yy	.	.	.	Parents

$RY\,Ry\,rY\,ry$ ·　　　　　　$RY\,Ry\,rY\,ry$　. . . Gametes

The possible combinations are shown best by the following method—

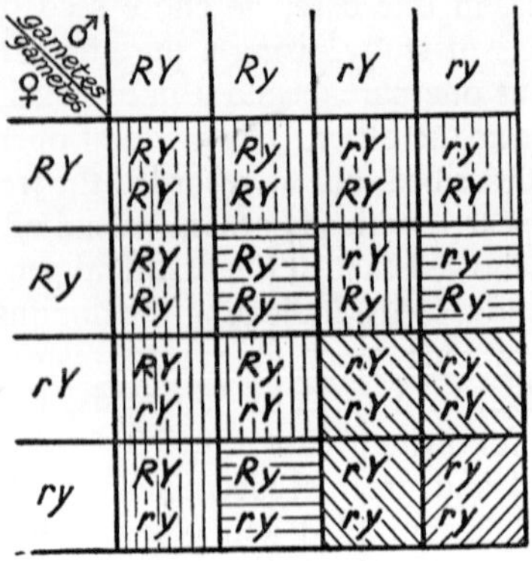

Allowing for the dominance of roundness and yellowness, if each of the sixteen possible combinations is studied, the population of the second filial generation will appear in the proportions of nine plants with round and yellow seeds, three with round and green seeds, three with wrinkled and yellow seeds and one with wrinkled and green seeds. The shaded squares indicate these proportions. Whilst only four different-looking types appear in the second filial generation, the germinal units which they carry are much more varied. Only four of them will breed true on selfing, namely, *RRYY*, *RRyy*, *rrYY* and *rryy*. The remainder, if allowed to self, would produce once more a mixed population following the laws formed by Mendel.

Mendel's two laws can now be combined and stated in another way. We may say that *each inheritable character is represented in a gamete in some form; for each pair of contrasting characters, only one can be represented in the gamete, and that for two or more pairs of such characters each representative in the gamete acts independently of the others.*

This was the extent of Mendel's work published in 1866 and to him must go great credit for discovering the two major requirements for success in any experiment on inheritance, namely, that crosses must be made between sharply differing parents with respect to the character under study, and that these parents must come from lines of which the ancestry is completely known.

Copies of his papers were passed to many learned societies including some in Great Britain, but for some unknown reason, no one attached any importance to Mendel's work and he died an obscure monk with little recognition of his brilliance. Then, in 1900, his papers were unearthed almost simultaneously by the biologists de Vries, Correns and Tschermak and quickly recognized at their true worth, since each was able to substantiate Mendel's findings from experiments with other plants. Within the next ten years, Mendel's discovery that the factors determining hereditary characters segregate into the gametes according to definite and quantitative laws, became the core of a biological science which has expanded probably more rapidly than any other. Men like Bateson, Morgan, Punnett, Saunders, and their pupils, who extended Mendelism to many plants, animals, and man, placed on Mendel's foundations an edifice of remarkable structure and complexity. The science is known as genetics and it embraces all other biological sciences in its scope.

TERMS USED IN MODERN GENETICS

With the growth of the subject, a vocabulary of specialized terms has arisen. Some of these and their definitions are given below—

Gene: a unit of inherited material, first called "gen" by Johannsen in 1909. It corresponds to Mendel's germinal unit or factor which he supposed was the cause present in the gamete for the appearance of a character in the adult.

Allelomorphs (Alleles): may be applied to the characters or genes which follow the Mendelian laws. Allelomorphic characters are those like the tallness and dwarfness of stem for instance, or the greenness and yellowness of seeds, which can be paired up from the point of view of inheritance. Allelomorphic genes are the representatives in the gametes of these characters, and they produce their different effects on the same developmental process.

Homozygous, heterozygous: terms applied to the genetical constitution of an individual with respect to the possession of a particular pair of allelomorphic genes. Mendel's original parent plants are said to be homozygous, those of the F_2 generation which possessed two factors for tallness or two factors for dwarfness are also homozygous, whilst those of the F_1 generation possessing one factor for tallness and one for dwarfness are said to be heterozygous.

The homozygous individual receives like genes for the same character from both its parents, the heterozygous individual receives unlike genes for the same character from its two parents.

Hybrid: the result of a cross between two parents showing unlike characters. Mendel's F_1 generations were hybrids. They did not breed true.

Monohybrid ratio: the ratio between the numbers of individuals possessing different genetical constitutions in the F_2 as a result of crossing a single pair of contrasting characters. Mendel's F_2 proportions, 1 homozygous tall:2 heterozygous tall:1 homozygous dwarf, represent the monohybrid ratio in which there are always four

25

possibilities. Figures concerning an experiment which show this 3:1 ratio indicate that only one pair of genes is involved.

Dihybrid ratio: the same applied to the crossing of two pairs of contrasting characters. Mendel's F_2 proportions, 9:3:3:1 represent the dihybrid ratio in which there are always sixteen possibilities. Figures concerning an experiment which show these proportions indicate that two pairs of genes are involved.

Polyhybrid ratio: the same applied to the crossing of more than two pairs of contrasting characters. The number of possible combinations depends on the number of characters involved, but may be derived by substitution in the formula $(2^n)^2$, where $n =$ the number of characters. For example, in the trihybrid ratio, three pairs of characters are involved, and the number of combinations is given by $(2^3)^2 = 64$. The proportions of types resulting in the F_2 would then be 27:9:9:9:3:3:3:1.

Genotypic: a group of organisms all having the *same genetical constitution.* Mendel's F_2 plants were of three genotypic groups, homozygous tall, heterozygous tall and homozygous dwarf.

Phenotypic: a group of organisms all having the *same observable characters* irrespective of genetical constitution. Mendel's F_1 plants and all the talls of his F_2 plants were phenotypic. They all *looked* alike, but some were homozygous and some heterozygous for tallness.

Clone: descendants produced vegetatively or parthenogenetically from a single original parent, and therefore having identical genetical constitutions.

Pure line: a succession of generations of organisms homozygous for all genes. Continued self-fertilization will rapidly lead to a homozygous condition in a species, since all factors already homozygous cannot change, whilst of the heterozygous allelomorphs, half become homozygous at each generation so that the number is steadily reduced.

Reciprocal cross: crosses between parents in both directions. The whole of Mendel's results were based on such crosses.

Back cross: the crossing of a heterozygous individual of the F_1 back to either of the homozygous parents. A back cross to the homozygous recessive parent is known as a *test cross* and can be used to test for dominance of one character over another, or to test for the independent segregation of two characters (*see* pp. 762 and 772).

Polyploid: the condition of chromosome sets increased above the diploid ($2n$) state, e.g. triploid ($3n$), tetraploid ($4n$), etc.

Gene pool: the total of all the allelomorphs for a character at all the loci on the chromosomes of a whole population of organisms, taking into account their frequencies. To describe a gene pool accurately it is necessary to specify the gene frequencies as well as which different allelomorphs are present (*see* p. 807).

Supergene: a gene combination within which crossing-over and recombinations are rare due to the extremely close linkage on the chromosome of the smaller component genes.

Genome: the entire genotype; total genetic information; complete set of genetic material for an individual.

APPARENT DEVIATIONS FROM MENDELIAN INHERITANCE

It was perhaps fortunate for Mendel that he chose the garden pea as his object of study and that his chosen characters behaved as they did. Otherwise he might have had considerably greater difficulty in interpreting his results. The experimenters of the early 1900's found many examples of complications which did not appear to agree with the simple Mendelian hypothesis. Some of these are outlined below and serve to illustrate the fact that genes are frequently more complex in their behaviour and influences than the simple Mendelian laws might lead us to believe.

Incomplete Dominance or Blending

This is the apparent failure of one allelomorphic character to dominate the other in the F_1 generation. It is found commonly in both plants and animals. One good example is that seen in a variety of domestic fowl. It is possible to obtain a pure strain of black Andalusian fowls in which the black colour is due to the development of melanin, and likewise a pure strain of white Andalusian fowls in which the absence of melanin leads to lack of pigmentation of the plumage. If a cross is effected between these strains, all the offspring of the first generation are neither black nor white, but an intermediate colour called "blue" or "splashed white." This is due to partial development of melanin, giving the less pronounced dark colouring. If these F_1 fowls are allowed to breed among themselves, the F_2 generation shows a segregation into blacks, blues and whites in the proportions $1:2:1$ as follows—

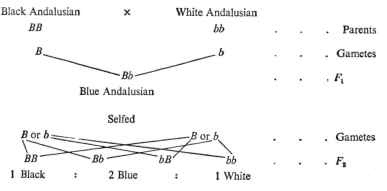

Black Andalusian   ×   White Andalusian

BB bb .   .   .   Parents

B b .   .   .   Gametes

Bb .   .   . F_1

Blue Andalusian

Selfed

B or b B or b .   .   .   Gametes

BB Bb bB bb .   .   . F_2

1 Black   :   2 Blue   :   1 White

From these results, we may conclude that the gene for melanin production in the Andalusian fowl is only partially effective when present as a single gene, or on the other hand, that there has been a blending of a gene for blackness with a gene for whiteness. This can be put to the test by making a "back cross," i.e. crossing the F_1 heterozygous "blends" with the homozygous parent whites. If a true blending had occurred in the F_1 we should expect offspring which again showed an intermediate colouring between blue and white. In fact, the results of the back cross show a segregation of 1 blue : 1 white. Such a result can only prove that in fact there has been no real blending between genes but that for full development of colour, a fowl must possess a gene for melanin production from both parents. The absence of one of these is sufficient to curtail melanin production severely, and the absence of both of them will prevent it altogether.

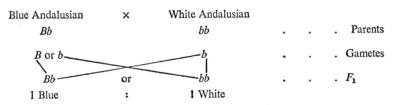

Many other examples of incomplete dominance occur. Flower colour in plants such as the pink varieties of *Mirabilis jalapa* and *Antirrhinum*, is due to hybridization between red and white strains. Roan shorthorn cattle are heterozygous for allelomorphic genes in which the corresponding homozygotes are red and white. A rare allele in man producing a very severe anaemia (thalassemia major) in the homozygous condition, causes only mild anaemia (thalassemia minor) when heterozygous.

Multiple Factors, Polygenes and Multiple Allelomorphs (Poly-alleles)

Mendel was fortunate in selecting material in which single genes determined the difference between his chosen characters. It is well known now that in many instances there may be two, three or more genes all affecting the same character. We know, for instance, that at least ten genes affect coat colour in mice and that thirteen genes affect pollen tube growth in the cherry. When such a series of genes occur, they are referred to as *multiple factors*. When they have separate, small, cumulative effects and together control continuous variation (*see* p. 736) without dominance over one another they are more usually called *polygenes*. Characters influenced by such genes can be described in a quantitative way only and will therefore be concerned with some measurable feature such as length, weight, volume, that is, they are *metrical characters*. Such were those described by Johannsen in his work on pure lines of beans, (*see* p. 746). Another example is to be seen in bantam size inheritance. The difference in size between the Hamburgh and Sebright bantam fowls appears to be due to four genes of which the Hamburgh possesses three, *A*, *B* and *C* and the bantam one, *D*, all in homozygous form. They are all incompletely dominant. When crossed, the F_1 offspring are intermediate between the parents and uniform in size. The F_2 is very variable and includes genotypes *AA*, *BB*, *CC*, *DD* and *aa*, *bb*, *cc*, *dd* which are larger and smaller respectively than either parent. Polygenic characters are inherited in the same manner as others controlled by single genes but it requires a complicated biometrical treatment to analyse their effects so as to distinguish them from the effects of chance environmental fluctuations.

In *Primula sinensis*, a gene A^1 suppresses the yellow eye of the flowers, while the gene *A* merely restricts it. Both these genes are dominant to a third, *a*, which allows its full development. This is a case of multiple allelomorphs or poly-alleles and differs from the previous case. Poly-alleles are variants of the same gene, situated at the same locus on homologous chromosomes, whereas multiple genes or polygenes occur at different loci and are in no way related except in their effects.

Lethal Characters

The development of characters which lead to the early death of individuals in the progeny of a cross may also interfere with the Mendelian monohybrid ratio. This may be exemplified by crosses with yellow mice, where it has been shown that the homozygous dominant for yellowness is never born. Any yellow mouse must therefore be a heterozygote for yellowness, and in a cross between such mice, the ratio of yellows to normals can be worked out as below. The homozygous

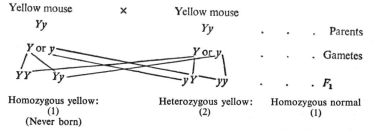

dominant for yellowness never appears, thus the ratio of yellows to normals is 2:1 and not 3:1. It is often the case that there is no sharp distinction between lethal and other genes, since many recessives when homozygous have a harmful effect. The individuals are less viable, particularly when conditions are not very favourable for their survival. Lethal genes are common and may affect the gamete as well as the zygote.

Complementary Genes and Other Gene Interactions

Some genes appear to operate in a manner complementary to one another, i.e. one must be present for the other to have an operative effect. There is a factor C, which appears to control the production of a basic substance in flower pigment in wild peas. Another factor R controls the conversion of this basic substance into a purple pigment. If either of these factors is absent, no colour can be developed. This condition was first shown in a breeding experiment carried out by Bateson who crossed two true-breeding white strains of pea and produced coloured F_1 offspring, the F_2 from which contained coloured and white flowered peas in the ratio 9:7. When further tests were made, three different forms of the white flowered peas were found in the proportions 3:3:1. Thus the 9:7 was really a dihybrid 9:3:3:1 ratio, the modification being caused by the fact that the colour is produced only when the two different dominant genes, C and R, are present together in the same pea. One of the parent white strains lacked colour because it lacked the gene C, that is, had the constitution $ccRR$. The other parent was $CCrr$ and also lacked colour. In the F_1, the offspring were $CcRr$ and thus were coloured. In the F_2, those containing C and R were also coloured but any $CCrr$, $Ccrr$, $ccRR$, $ccRr$ and $ccrr$ were all white because they were without one or both of the *complementary genes*.

There are other forms of this kind of interaction between two genes. There is the case in which a character shows only when both recessive genes are present. That is, only genotypes $aabb$ can show the character, none containing one or both of the dominants A or B can do so. Hence only one in sixteen can be of this kind, the other fifteen must be without the character. The F_2 will therefore show the ratio 15:1 or really (9 + 3 + 3):1, the 15 being made up of either $AABB$, $AaBB$, $aaBB$, $aaBb$, $AaBb$, $Aabb$, $AABb$ or $AAbb$ but all being alike phenotypically in not showing the character in question. Such interacting genes are said to be *duplicate*. Another less common case occurs when the character is developed due to one dominant and one recessive gene operating together, that is, in the genotypes $AAbb$ or $Aabb$. In such case the character appears in the F_2 three times in sixteen, with a ratio 13:3. One of the genes is said to be a *recessive suppressor* or inhibitor of the other.

A still different form of gene interaction is exhibited by the inheritance of coat colour in mice. The colour of wild mice is called agouti and such coloured mice differ from black, non-agouti mice due to a single gene A, the agouti condition being dominant. But the presence or absence of pigment (albinism) is also governed by a single gene C with the pigmented state dominant. Thus colourless mice can arise in two ways: the ones that would be agouti if they had the gene C for any colour at all, that is they are $AAcc$ or $Aacc$, and those that would be black if they were able to possess pigment, that is they are double recessives, $aacc$. Now if a black $aaCC$ is bred with a pure albino of the type $AAcc$, this gives an F_1 of $AaCc$ and these are pigmented agouti because they possess C and A genes. In the F_2 there will be 9 agoutis, that is either $AACC$, $AaCC$, $AACc$ or $AaCc$, 3 blacks either $aaCC$ or $aaCc$, 3 albinos carrying agouti but unable to show it, being $AAcc$ or $Aacc$ and 1 albino, double recessive $aacc$ that would have been black if it could develop pigment. Thus there appears a $9:3:(3+1)$ or $9:3:4$ ratio of agouti:black:albino mice. This kind of interaction between genes is referred to as *epistasis* in which the gene for colour at all is the primary gene, epistatic over the kind of colour that is said to be the hypostatic or secondary gene. There is a variation of this in which instead of albinism being recessive to coloured, it is dominant. This is so for plumage colour in White Leghorn fowls. Representing dominant white by W, colour by w, blackness by B and brownness by b, then any containing both a W and a B or one W and bb would be white, those with ww and BB or Bb would be black and those with the double recessives, $wwbb$, would be brown. Hence in a cross such as above the heterozygous F_1, $WwBb$, would all be white whilst the F_2 would consist of the proportions 12 white to 3 black and 1 brown, that is a $12:3:1$ ratio with the 12 really being $(9+3)$.

The Inheritance of Comb Form in Domestic Fowls

Among the comb forms of domestic fowls, two are easily recognized as the "pea" and "rose" combs (*see* Fig. 19.1). Both breed true, but when crossed, the resultant is neither "pea" nor "rose" but one quite different and described as "walnut." If fowls with "walnut" combs are allowed to in-breed, then in the next generation, four comb forms appear. These are "pea," "rose," "walnut" and the fourth called "single," different again from any of the others. The proportions in which they appear are 9 "walnut":3 "pea":3 "rose":1 "single," indicating a case of dihybrid inheritance. The four comb forms can be explained as follows. There are two factors operating which affect comb form. We can call them P for "pea" and R for "rose." When a fowl possesses one or two factors for pea comb but none for rose comb, it produces a "pea" comb. When a fowl possesses one or two factors for rose comb but none for pea comb, it produces a "rose" comb. When it possesses one or two factors for both pea comb and rose comb, it produces a "walnut" comb. When it has no factors at all for either pea comb or rose comb, it produces a "single" comb. Letting the symbols P stand for the presence of the factor for pea comb and p for its absence, and letting R and r represent the same respective conditions for presence or absence of the factor for rose comb, we can illustrate the inheritance of comb form as on p. 767.

Fig. 19.1. Comb forms in domestic poultry: (a) walnut, (b) single, (c) pea, (d) rose.

Such a mode of inheritance illustrates again the fact that more than one gene may affect a single character. The genes for pea comb and rose comb, when separate, act in one way, when together, they act in another. It further leads us to consider also the nature of the dominance of one gene over another. From the simple Mendelian inheritance in peas, the inference is that the dominant and recessive characters of an allelomorphic pair are each represented by a gene. This comb-form mode of inheritance indicates that dominance may be due to the presence of a gene and recessiveness merely to its absence, that is, only one factor is involved in the inheritance of one or other of a pair of allelomorphic characters, not two. The single comb represents the one double recessive in a dihybrid ratio and arises only in the absence of pea and rose comb genes, thus indicating that the pea

Pea comb	×	Rose comb			
PP rr		*pp RR*	.	.	. Parents
Pr		*pR*	.	.	. Gametes

	Pp Rr		.	.	. F_1
	Walnut comb				

Inbreed

PR Pr pR pr   *PR Pr pR pr*   .   .   . Gametes

♂ gametes / ♀ gametes	*PR*	*Pr*	*pR*	*pr*
PR	*W*	*W*	*W*	*W*
Pr	*W*	*P*	*W*	*P*
pR	*W*	*W*	*R*	*R*
pr	*W*	*P*	*R*	*S*

.   .   . F_2

9 Walnut : 3 Pea : 3 Rose : 1 single

and rose conditions, both dominant over single, are due to the super-imposition of something on a condition which was already in existence.

Whilst such peculiar modes of inheritance appear complicated and not comparable with the results obtained by Mendel in his pea-breeding experiments, not one of them shows a real departure from the Mendelian segregation laws. Their peculiarities can be explained in terms of the way in which the genes act after segregation, not during it.

THE MATERIAL BASIS FOR MENDELIAN INHERITANCE: THE CHROMOSOME HYPOTHESIS

During the time that Mendel's work lay unrecognized, attention was being centred on the precise behaviour of the cell nucleus during division. The phenomena of mitosis and meiosis were clearly described in their general essentials in the period 1876–1900 by Strasburger, Fol, Flemming and others. A connexion between the laws of Mendelian inheritance and the behaviour of the nucleus in cell division was not long in being established. Details of nuclear divisions are given in Vol. I, Chap. 2, but a brief summary of the events may serve a useful purpose here.

In mitosis, a number of clearly recognizable double-structured chromosomes become apparent. Each is composed of two chromatids which separate from one another and migrate to opposite poles of the spindle. Each new nucleus consists therefore of the same number of chromatids as the parent nucleus had chromosomes. During the interphase, each chromatid is duplicated to form another double-structured chromosome which clearly reappears at the next division.

In meiosis, a number of chromosomes become apparent but they have no double structure. If examined closely, they will be found to

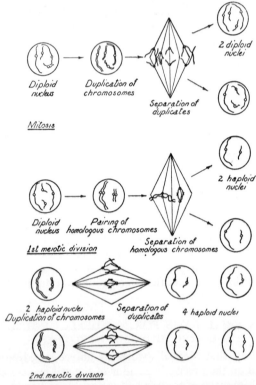

Fig. 19.2. Diagrams comparing mitosis with meiosis.

form a number of pairs of similar chromosomes (homologous pairs). These homologous pairs lie together and each member of each pair develops a double structure. Then they separate, one member of each pair proceeding as a double structure to one pole, and the other member to the other pole. Each new nucleus possesses only one half the total number of chromosomes of the parent nucleus, one of each of the homologous pairs. Each nucleus then proceeds to divide again as in mitosis, so that eventually four new nuclei are formed, each with half the original chromosome number.

When an organism is developed as a result of sexual fusion, the nucleus of the male gamete always fuses its substance with that of the female gamete. Each gamete contributes half the chromosome material of its parent to form the full chromosome complement of the zygote, which must therefore be made up of half maternal and half paternal chromosomes. All body cells of the new individual are derived from this zygote, which is specific in that it will always give rise to the same species as that from which the gametes were derived. It follows therefore that the factors controlling the future development of the zygote must have been brought to it by the gametes and, since in most cases, the only common feature of the gametes is that they contain a nucleus, it can be fairly concluded that the location of these hereditary factors or genes lies within the nucleus. The chromosomes appear to be the most likely structures for bearing the genes, since the separation of maternal and paternal chromosomes into different gametes during meiosis provides the means of segregation of the genes as postulated by Mendel in his first law of the purity of the gametes, i.e. they could carry only one germinal unit for a particular character. The diagram opposite links the behaviour of the chromosomes carrying the factors for tallness and dwarfness in Mendel's parent pea plants with the results he obtained in the F_1 and F_2 generations.

Working with the common fruit fly, *Drosophila melanogaster*, Morgan and his pupils did much work in the early 1900's towards establishing proof of this chromosome hypothesis. The fruit fly is a convenient subject with which to work, since it reproduces rapidly and has only eight chromosomes in the body cells and four in the gametes, i.e. $n = 4$. Their breeding and cytological studies showed clearly the mechanism of Mendelian inheritance; the chromosomes are the carriers of Mendel's germinal units, or in the modern term, the genes. The hypothesis has been generally accepted and it forms the basis on which many otherwise inexplicable inheritance phenomena can be explained. One of these is the linkage of inherited characters. In addition, it makes clear one significant feature of sexual reproduction. The blending of two sets of nuclear material from different sources makes it possible for new genetic constitutions of cells to arise. When an organism reproduces by vegetative means, the cell divisions leading to the formation of the offspring are all mitotic and consequently all new cells are identical in genetic constitution with those of the parent. The offspring cannot be expected to vary from its parent in any major way, as it might if its genetic constitution were different. No major variations can be introduced into a line of vegetatively produced organisms as could happen when they are sexually produced. It should be remembered also that one of the occurrences during meiosis is that there is a reshuffling of the maternal and paternal chromosomes of the parent nucleus when the members of a homologous pair separate. The movement of the maternal and paternal partners follows no fixed

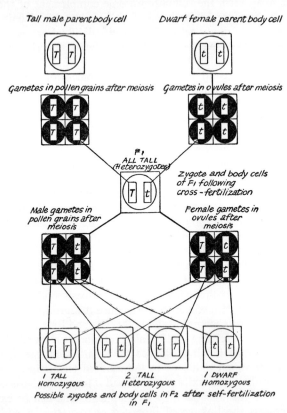

Fig. 19.3. Diagram linking behaviour of the chromosomes with simple Mendelian inheritance.

direction along the spindle. Each might go either way, but always one to each pole, so that according to the total number of chromosomes present, there can be many different mixtures of the maternal and paternal chromosomes in the new nuclei. A gamete thus formed will not necessarily have precisely the same nuclear material as either of the gametes from which its parent cells were formed. This also leads to a continuous blending of different nuclear material as generation succeeds generation. Fresh mixtures of genes could lead to variation, and as will be seen in a later chapter, this may form the basis of organic evolution when natural selection operates on the variants produced.

LINKAGE AND CROSSING-OVER

Morgan and his co-workers left little doubt that the behaviour of the chromosomes at cell division provides the material basis for the numerical proportions of offspring in breeding experiments; that the genes are carried by the chromosomes. Sutton and Bateson pointed out something which must follow from this. They reasoned that if the chromosomes carry the genes, and there are only a comparatively small fixed number of chromosomes but a large indefinite number of inheritable characters, then each chromosome must carry a number of genes. Further, if the chromosomes behave as interpreted from observation, then there will be whole aggregates of genes which can never segregate from one another except by accident. That is, they are *linked*, or fall into *linkage groups*. If this is the case, the genes of any organism should be arranged in a number of linked groups equal to the haploid number of chromosomes. The theory has been put to the test many times and has been found to be borne out for all subjects so tested. In *Drosophila*, which has been investigated most thoroughly of all, the number of linkage groups is four, corresponding to the four pairs of homologous chromosomes. The term *linkage* is used to describe the two kinds of behaviour of non-allelomorphic characters with respect to one another. In the one case such characters remain together always, and in the other case they are permanently separated. The former case is called *coupling* and the latter *repulsion*. Coupling will occur between two non-allelomorphic characters when they are represented by genes carried on the same chromosome of an homologous pair, so that in theory they cannot be separated. Repulsion will occur between two non-allelomorphic characters when each is represented on one of an opposite pair of homologous chromosomes, i.e. one on the paternal and the other on the maternal chromosome of the pair. Theoretically, such characters should never appear together, since during meiosis the homologous chromosomes separate into different gametes.

When genes are linked, the linkage reduces the number of recombinations of genes that should be obtained in a particular cross if they were obeying Mendel's second law of independent assortment. Linkage of genes therefore makes Mendel's second law invalid in such cases. It must be remembered though, that he had no knowledge of cytology and was fortunate in using characters which showed no linkage. A simple example will illustrate the segregation which may be expected to occur between coupled genes. Consider the behaviour of a pair of genes A and B carried on one of a homologous pair of chromosomes and their alleles a and b carried on the other of the pair. The heterozygote for A and B can be symbolized AB/ab. This is the usual way of writing the genetic constitution of an individual when dealing with linked genes. It indicates that the genes AB were received from one parent coupled on one chromosome and the genes ab were received from the other parent coupled on the other chromosome of the homologous pair. The homozygote for AB would be written AB/AB, and one of its gametes $AB/\ldots$ Suppose that this heterozygote, AB/ab, is crossed with the homozygote recessive for both genes, ab/ab, and that the genes are linked as we have said. The cross can be represented thus—

$\dfrac{AB}{ab}$	$\times$	$\dfrac{ab}{ab}$	.	.	.	Parents
$\dfrac{AB}{..}$ or $\dfrac{..}{ab}$		$\dfrac{ab}{..}$ or $\dfrac{..}{ab}$	.	.	.	Gametes
$1\dfrac{AB}{ab}$	:	$1\dfrac{ab}{ab}$	.	.	.	F_1

This means that only two kinds of F_1 offspring should appear and be in equal numbers. This happens because at gamete formation in the parents, whole chromosomes segregate and not the genes. If the genes did segregate quite independently of one another irrespective of the chromosomes on which they are borne, then we should expect four kinds of offspring in the F_1 as below—

$Aa\ Bb$	$\times$	$aa\ bb$	.	.	.	Parents
$AB\ Ab\ aB\ ab$		ab	.	.	.	Gametes
$1\ Aa\ Bb:1\ Aa\ bb:1\ aa\ Bb:1\ aa\ bb$			.	.	.	F_1

In actual breeding experiments, the results of back crosses show that linkage occurs, but that the case of total linkage, such as has been described, is rare. In fact, the offspring of such a cross are more likely to appear as four types, but the four types are not in equal proportions. The linkage between genes is nearly always broken to some extent. The reason for breaks in the linkage must be sought in the precise behaviour of the chromosomes during gamete formation, i.e. during meiosis. It will be recalled that during a meiotic division, there is frequently an interchange of material between bivalents. We say that *crossing-over* occurs and the phenomenon of the formation of chiasmata accompanies the occurrence. We may illustrate crossing-over in

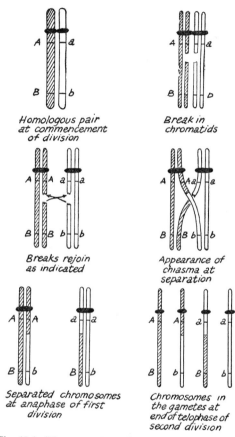

Fig. 19.4. Diagrams representing crossing-over during meiosis.

terms of genes by the diagrams above. Had there been no crossing-over the gametes resulting from a meiosis would have appeared as in Fig. 19.5.

These simple diagrams illustrate the single cross-over case between one pair of chromatids. During meiosis a bivalent may show more than one chiasma and all four chromatids can be involved. Crossing-over, in breaking linkage groups, has the effect of increasing the number of different genetic constitutions which may appear in the gametes, and so cancels to some extent the lessening in gamete variability which must arise because of linkage and the consequent non-independent segregation.

As a result of much work with *Drosophila*, Morgan came to realize that the extent to which two genes crossed over could be used in working out the relative positions of genes on chromosomes. He visualized

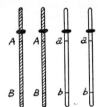

a linear arrangement of the genes at fixed distances apart along the chromosomes, and made the guess that crossing-over would occur more frequently between genes which were far apart than between genes close together. He postulated that the relative frequency of crossing-over would be a measure of the distance apart of genes. The frequency by which two genes cross over is known as the *cross-over value* or *C.O.V.* He crossed red-eyed, yellow-bodied flies with white-eyed, grey-bodied flies. Red eyes and grey bodies are the normal wild-type conditions and are the dominant

Fig. 19.5. Result of division when no crossing-over has occurred.

characters. The sign + represents any such normal wild type character in *Drosophila*. His cross can be represented—

$$\frac{+y}{+y} \qquad \times \qquad \frac{w+}{w+} \qquad \qquad . \quad . \quad . \quad \text{Parents}$$

$$\frac{+y}{..} \text{ or } \frac{..}{+y} \qquad \qquad \frac{w+}{..} \text{ or } \frac{..}{w+} \qquad \qquad . \quad . \quad . \quad \text{Gametes}$$

$$\frac{+y}{w+} \qquad \qquad \qquad . \quad . \quad . \quad F_1$$

The F_1 progeny were heterozygotes for both characters but looked normal. When these were backcrossed to the double recessive wy/wy, he found four types of flies in the offspring, namely $+y/wy$, $w+/wy$, $++/wy$ and wy/wy. The last two of these could only have occurred if there had been crossing-over between the genes w and y. Their proportions amounted to 1·5 per cent of the total and he said that this could be interpreted as indicating that the loci of w and y were 1·5 units apart

$$\frac{+b}{+b} \qquad \times \qquad \frac{w+}{w+} \qquad \qquad . \quad . \quad . \quad \text{Parents}$$

$$\frac{+b}{..} \text{ or } \frac{..}{+b} \qquad \qquad \frac{w+}{..} \text{ or } \frac{..}{w+} \qquad \qquad . \quad . \quad . \quad \text{Gametes}$$

$$\frac{+b}{w+} \qquad \qquad \qquad . \quad . \quad . \quad F_1$$

on the chromosome. He then crossed flies with red eyes and bifid wings with white-eyed, normal winged flies. The F_1 progeny were heterozygotes for both characters but looked normal. When these were back-crossed to the double recessive wb/wb he found that 7 per cent of the offspring had white eyes and bifid wings, indicating that crossing-over had occurred between w and b. On the same reasoning, the loci of w and b were judged to be 7 units apart.

If this reasoning was correct, Morgan knew that by simple arithmetic, the loci of y and b should be 5·5 units apart. He tested this by crossing yellow-bodied flies with normal wings with grey-bodied flies with bifid wings. His results showed 5·49 per cent crossing-over, so that the experimental data fitted his supposition perfectly.

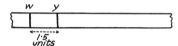

Fig. 19.6. Loci of w and y, 1·5 units apart on *Drosophila* chromosome.

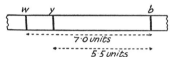

Fig. 19.7. Loci of w and b 7·0 units apart on *Drosophila* chromosome. Loci of y and b 5·5 units apart.

Since Morgan's original work much has been done to map the chromosomes of *Drosophila*, and this has shown that the problem is not as simple as the early work suggested. First, the occurrence of double cross-overs must be allowed for. If for example, cross-overs had occurred between w and y and between y and b simultaneously, then the effect would be to cancel the crossing of w and b but to separate y from both of them thus—

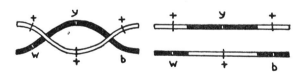

Likewise, owing to the mechanics of crossing-over, two chiasmata cannot arise infinitely close to one another. The prevention of the formation of a chiasma at a particular point on the chromosome is known as *interference*. Interference can be expressed as a constant by dividing the actual double cross-over value obtained in an experiment by the expected double cross-over value. The expected double cross-over value between these points depends on the product of the distances between the first and second, and second and third points. This constant representing interference is known as the *coincidence*, and varies considerably from point to point on a chromosome.

In addition to mapping chromosomes by the calculation of cross-over values, much work has been done more directly by examination of the effects of structural changes in the chromosomes which occur by accident, or such as can be induced by short-wave irradiation of the chromosomes. The changes can be of several types, such as complete loss of a fragment of chromosome, or the attachment of one bit of a broken chromosome to a non-homologous chromosome, or the inversion through 180° of a middle portion of a chromosome which has been broken in two places. By analysing the genetics of flies in which these accidents happen, the precise piece of chromosome affected and the genes which lie on it can be determined. Dobzhansky performed much detailed work of this kind and from direct cytological evidence was able to construct more accurate maps of the chromosomes of *Drosophila*. They bore out Morgan's original supposition, that is, that the genes have a linear arrangement on the chromosomes, and the two sets of maps, although not agreeing as to distances apart of the genes, at least show them in the same order. Results of this kind, produced experimentally, leave little doubt that the chromosome hypothesis is correct.

SEX INHERITANCE AND SEX LINKAGE

There are few species of organisms which do not reproduce by the sexual method. Most animals are unisexual and each must mate with another of different sex. Most plants are hermaphrodite and can produce gametes of both kinds, but most of them have evolved the means whereby the union of gametes from the same individual is very unlikely to happen, even if not entirely prevented. In either case, animal or plant, the process at its most advanced level prevents inbreeding and consequently tends towards the production of greater variation in a species.

Where the sexes are separate, the cause of the separation in a particular case can be one of two. Environmental conditions are known to be directly responsible for the sexual development of a few animals. For example, the larvae of an annelid worm, *Bonellia*, which develop in close contact with the female parent, become males, whilst those which are more widely dispersed, become females. The second cause of unisexuality is genetic. It was soon realized by geneticists that sex inheritance could be interpreted as Mendelian, since the mating of male with female always results in male and female offspring in approximately equal numbers, a result which would be achieved by crossing a homozygote for one character with its heterozygote. An inspection of the chromosome complement of males and females of the same species gave the solution. They are different. The exact replica of one of the homologous pairs in a female cannot be found in a male or vice versa. The homologous pair in question exists as an identical pair in one sex but not in the other. Of the whole chromosome complement these are known as the *sex chromosomes*, the remainder forming the *autosomes*. In most animals and dioecious higher plants, the male sex chromosomes are not alike in size or shape, whilst in the female, the sex chromosomes are identical and correspond to the more normal chromosome of the male pair. Note that there are exceptions to this in some animals, e.g. birds and some insects, the female possessing the unlike sex chromosomes. The sex possessing the like pair is called the *homogametic sex*, whilst the other is called the *heterogametic sex*. The homogametic sex is said to carry two X-chromosomes, whilst the heterogametic sex carries one X-chromosome and one Y-chromosome.

If we represent these chromosomes by the symbols X and Y we can show their segregation in the normal Mendelian manner.

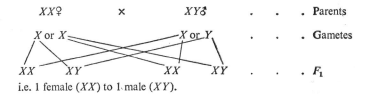

i.e. 1 female (XX) to 1 male (XY).

The homologous segments of the X- and Y-chromosomes that can pair and exchange genetic information by crossing over are called *pairing segments.* Other parts that are non-homologous and where the sex-determining factors are located are called *differential segments.* When these latter parts of the X- and Y-chromosomes are different in form and size the sex chromosomes can readily be identified from all the autosomes in cytological preparations.

The X-chromosomes do carry genes that are not sex-determining but clearly, since genes occur in linkage groups, these must be coupled to the sex-determining genes. This coupling of certain inheritable characters to the sex of the offspring is known as *sex-linkage.* To consider its effect in inheritance it is useful to illustrate with actual examples.

In the case of *Drosophila*, sex appears to be dependent upon the more common case of XX giving female and XY giving male, although by accidents at fertilization some flies may occasionally appear as XO males and XXY females, this clearly indicating that the Y chromosome is playing no part in sex determination. Femaleness is really being determined by the presence of three pairs of autosomes plus two X-chromosomes and maleness by three pairs of autosomes plus one X-chromosome.

Now, a male fly can transmit its X-chromosome only to a daughter, since the other gamete forming the daughter zygote can bring only one X-chromosome from the female parent, and this paternal X-chromosome can reach another male only through a daughter. A male fly can inherit an X-chromosome only from its female parent. Any other characters that accompany the sex-differentiation are the ones said to be *sex-linked*, as stated above, and genes for such characters are carried on the X-chromosome.

Consider the reciprocal crosses between flies and the progeny in the F_1 and F_2 generations shown in Fig. 19.8, where w represents an eye-colour variant recessive known as "white-eye" lacking pigment in the eye cells. Wild-type or normal flies have red eyes, denoted by $+$.

In cross 1: of the F_1, all are normal in both sexes (white-eye is recessive)

in the F_2, of all flies $\frac{3}{4}$ are normal and $\frac{1}{4}$ white-eyed, but when eye colour and sex are correlated it is seen that all females are normal whilst half the males are normal and half are white-eyed.

In cross 2: in the F_1, of all flies $\frac{1}{2}$ are normal and $\frac{1}{2}$ white-eyed, but when sexed it is seen that all the normal flies are female and all the white-eyed are male

in the F_2, of all flies $\frac{1}{2}$ are normal and $\frac{1}{2}$ white-eyed, but when eye colour and sex are correlated it is seen that half the females are normal and half white-eyed, and in the males the same equal proportions of normal to white-eyed occur.

There is obviously a connexion between the appearance of the white-eyed condition and the sex of the fly. Note that corresponding reciprocal crosses not involving sex-linked characters give the same results at both the F_1 and F_2 no matter which way round the cross is made. In this case of inheritance of white-eye, if the female parent is white-eyed all its male offspring are white-eyed. If the female parent is normal for eye colour, all her offspring are normal and white-eye is seen again only when the offspring are inbred and then only in some males.

If it is postulated that the gene w and its wild-type allele is carried on the X-chromosome this would fit perfectly. The males at F_1 in both crosses are receiving their X-chromosomes from their mothers and the

Cross 1: Homozygous normal female x white-eyed male

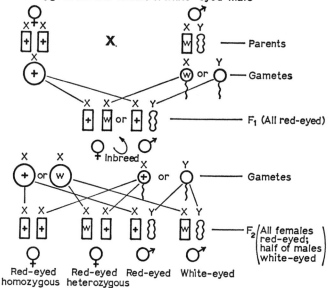

Cross 2: Homozygous white-eyed female x normal male

Fig. 19.8. Sex-linkage in *Drosophila*.

male parents are transmitting their X-chromosomes to their daughters. In cross 1 above, the F_1 males are red-eyed since their mothers produced gametes with X-chromosomes carrying the normal allele of w. In cross 2 above, the F_1 males are white-eyed because their mothers produced gametes all with X-chromosomes carrying the w mutant gene. Since the wild-type red-eyed gene is dominant, the females in both crosses at F_1 are normal for eye colour. They received one X-chromosome from a normal parent and one from a mutant parent. It follows also that because the males in the F_1 of cross 2 are white-eyed, the Y-chromosome does not appear to carry an allele of w. In fact this chromosome in *Drosophila* is genetically inactive with respect to most of the gene loci in the X-chromosome and it is usual to refer to male *Drosophila* as *hemizygous* for sex-linked genes, since there are no alleles in the Y-chromosome.

At this point it is worthwhile noting that with the inclusion of the reference to sex linkage, three items of evidence in support of the chromosome hypothesis have now been presented. One, the events at meiosis are strongly correlated with the segregation of a pair of allelomorphic characters. Two, the random movement to the poles of the spindle of the partners in the bivalents during meiosis metaphase of the first division clearly reflects the phenomenon of independent assortment. Lastly, sex linkage clearly points to the fact that some genetic material is transmitted from generation to generation according to the mode of transmission of a particular chromosome, thus establishing a direct connexion between the inheritance of a specific character with the possession of a specific chromosome. This last is the only really direct evidence of the three kinds so far put forward.

There are several conditions in man which are known to be inherited as sex-linked characters. Colour-blindness and haemophilia are among them. Note that a sex-linked character carried on an X-chromosome cannot be transmitted directly from father to son. This is the case in the two conditions mentioned above.

With regard to the inheritance of sex, it must be realized that the qualities of maleness and femaleness exist in a wide range of degrees. The possession by zygotes of XX or XY chromosomes does not necessarily mean that all individuals will exhibit the same degree of either maleness or femaleness. In human beings and in many other animals, intersexes are well known (*see* Chap. 17). They combine, partly or wholly, the structure and functions of both sexes. Indeed, it has been stated that at least 1 per cent of human beings are intersexes. Their parents were normal but they are usually sterile. Castration has long been recognized as affecting the secondary sexual characteristics; if carried out at an early age it will prevent their appearance completely.

It is not only the gamete-producing power that is lost, but also the tissue largely responsible for production of the male hormones. Removal of the ovaries in female birds and grafting testes in their place, causes the hens to assume gradually the appearance and behaviour of cockerels.

In the final analysis, the ultimate appearance and potentialities of any individual are due to the interplay of three factors; the genes, the hormones and the environment. Of these, the genes play the prime, the initial, part. This applies to sex as well as to other characteristics of the individual.

GENETICAL EXPLANATION OF THE ORIGIN OF VARIATION WITHIN A SPECIES: MUTATIONS

Before Darwin, it was believed, even by biologists, that all the various kinds of living things had been created as they were and had undergone no change. Darwin postulated evolutionary change and ascribed it to the effect of natural selection on small variations within a species. He could only assume that all variations were due to an accumulation of small changes in individuals, called forth by differences in environment, and that the ones most fitted to the environment survived to hand on these beneficial characters to their offspring. In the light of modern cytology and genetics, we can account for the origin of variation, both continuous and discontinuous, in several ways.

There can be the sudden appearance of a recessive character after many generations. Such variation is not real, however, since it has merely been hidden and could have arisen at any time if the already existing genes chanced to come together. On the other hand, there do arise quite new forms which can be distinguished from the above by genetical tests. Such new forms are the result of some sudden change in the genetical constitution of the individual as distinct from a rare form of recombination of already existing genes. Such a change in an inheritable character is known as a *mutation*, a term applied by de Vries, who encountered several such abrupt departures from normal form in the evening primrose, *Oenothera* (*see* p. 893). We now know that mutations can arise in one of two main ways. First, the change may be due to alterations from normal in the number of chromosomes, or the genes on one or more chromosomes may become rearranged with respect to one another due to some accident. Both these departures are known as *chromosome mutations*. Secondly, a new character may arise as a result of some change in a single gene. Such a mutation is known as a *gene mutation*, and it should be noted that where the term mutation is used in an unqualified way, it refers to this kind of mutation.

Chromosome Mutations

The term *polyploidy* is used to describe the condition in which whole sets of chromosomes are present in more than the diploid number. The prefixes tri-, tetra-, penta-, etc., indicate the actual number of sets, and may be written $3n$, $4n$, $5n$, etc. The term *polysomy* indicates that individual chromosomes may be present in excess of an otherwise diploid set. The condition is written $2n + 1$, $2n + 2$, etc. Polyploids can arise in one of two ways. The chromosome number of an otherwise normal organism may suddenly become doubled, or alternatively, a hybrid organism may appear in which the two basic sets of chromosomes have been multiplied up. The former is a case of *autopolyploidy* and the latter *allopolyploidy*. Autopolyploids may arise owing to some accident in a cell division. At gamete formation, all the chromosomes may go to one pole of the spindle to give a diploid gamete. If this fuses with a normal haploid gamete and the offspring does not die, then a triploid organism is produced. In some plants, for example, the tomato and *Datura*, a cell in a bud apex may have a similar accident. It becomes tetraploid as are all its descendants. If these eventually give rise to gametes, they will be diploid. Fusion of such diploid gametes gives rise to tetraploid offspring. Autopolyploids in general are usually enlarged versions of the diploid varieties and possess greater vigour. A typical allopolyploid is *Primula kewensis* which arose from a cross between *P. floribunda* and *P. verticillata*, both of which have a haploid number of 9. The cross results in a hybrid containing 9 chromosome pairs, but these are not homologous in origin. At pollen grain or ovum production, these 18 chromosomes will not pair as in normal meiosis. Instead, one set of 9 splits into 18 which separate to opposite poles of the spindle and the other set of 9 does likewise, so that the gametes each contain 18 chromosomes. Fertilization results in an allotetraploid variety of *Primula*. Allopolyploids generally possess different characters from either of the parents. There are a few authenticated cases of polyploidy in animals but they seem always to be autopolyploids. In plants, the condition is much more common, and in fact, most of our cultivated plants can be placed in a series according to chromosome number, which indicates the way in which they may have evolved. Einkorn wheat, a wild variety now cultivated only by a few primitive peoples, has 7 chromosomes as the haploid number. The Emmer wheats with $n = 14$ seem to have been the varieties most widely cultivated by neolithic man, and evidence from the pyramids indicates that the ancient Egyptians used them exclusively. Durum or macaroni wheat, also $n = 14$ was used by the Romans and is still widely used because of the special properties of the grain. The most commonly cultivated wheat today is the bread wheat, *Triticum aestivum*, which possesses 21 chromosomes in the haploid condition. The Elizabethans knew only the wild type strawberry, $n = 7$. In 1712, a new variety was planted in Kew and gave us the modern strawberry varieties. It was hexaploid. The species *Fragaria chiloensis* is octoploid, i.e. $n = 56$.

Polyploids of these kinds have arisen with the interspecific hybridization that results from the crossing of two species. It is almost always so that when two species give rise to a third by breeding together then the new species is an allopolyploid.

Changes in the structure of a chromosome which alter the arrangement of genes or even the number of genes on it, do occur naturally, but are more readily produced by artificial means. X-ray irradiation, the exposure to sudden large temperature changes, and some chemicals, may all cause fragmentation of chromosomes. The parts may later

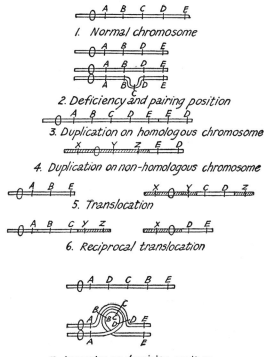

1. Normal chromosome

2. Deficiency and pairing position

3. Duplication on homologous chromosome

4. Duplication on non-homologous chromosome

5. Translocation

6. Reciprocal translocation

7. Inversion and pairing position

Fig. 19.9. Summary of chromosome structural changes which alter arrangement of the genes.

join up in a variety of ways or may be lost completely. Such chromo-somal changes are described in terms of the nature of the change. A *deficiency* is said to occur when a part of a chromosome carrying one or more genes breaks away and is lost during a nuclear division. Most frequently, the effect is lethal. When a part of one chromosome is duplicated and attached to another chromosome, the condition is said to be one of *duplication*. A *translocation* is said to occur if a part of a chromosome is broken away from its normal position and attached either to another chromosome or to a different region of the same chromosome. *Inversion* occurs if a segment of a chromosome becomes detached and then joined up once more, but the opposite way round. These occurrences are summarized in Fig. 19.9.

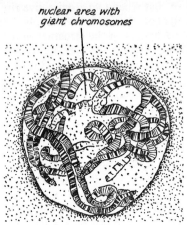

Fig. 19.10. Giant chromosomes of *Simulium*.

Most of the cytological investi-gations into the structural changes in chromosomes have been made on the giant chromosomes of the salivary glands of *Drosophila* and other flies. Fig. 19.10 indicates the general appearance of these. They are extremely large chromosomes called *polytenes*, formed when successive replicate strands of chromosome substance fail to separate from one another, and even minute changes in their structure can be detected.

Gene Mutations

If cultures of normal wild type *Drosophila* are kept going indefinitely, sooner or later there will appear among them, suddenly and spon-taneously, new forms. Close investigation of the chromosome comple-ment of the new forms will show no readily apparent departure from normal. The cause cannot be explained in terms of chromosome mutation. The change in genetic constitution of the mutant forms can be explained only in terms of the genes on the chromosomes. A gene representing a particular character has individually undergone some change from normal, which is reflected in the bearer of the abnormal gene as a change in that particular character. For example, among the commonest of such changes in *Drosophila* is the sudden appearance of individuals with white eyes. The red eye colour of the normal flies is absent. The white-eyed flies are mutants and continue to produce white-eyed offspring when inbred. A careful examination of the giant salivary gland chromosomes shows that in one of them, one of the dark bands is missing. This indicates a link between the absence of eye-colour and the absence of the band. The inference is that the removal of the band destroys the gene which controls pigment production in the eye.

There is evidence that genic mutations occur regularly in all classes of organisms but that any individual gene mutates only rarely. The frequency of mutation seems to vary with the gene and may be as high as once in 50 000 cells or as low as once in 1 000 000 cells or more. But considering all the genes in all the chromosomes of an individual organism as complicated as man, where there may be as many as 10 000, then if every gene can mutate at, say, an average rate of one in 100 000, it means that in every ten gametes, one carries a freshly mutated gene. This may seem a very high estimate but is not improbable. Even this rate of mutation can be speeded up considerably. Among the agencies which have been proved to cause gene mutations, i.e. are mutagenic, in organisms including *Drosophila*, wasps, mice, fungi and some flowering plants, are X-ray, α-ray, β-ray, γ-ray, neutron and ultra-violet light irradiation, high temperature treatment and the presence of some chemicals. The pioneer of research into the cause and nature of gene mutations was A. J. Muller, who discovered that if he subjected *Drosophila* individuals to doses of X-rays, then he could induce mutations at a rate up to 150 times as fast as their spontaneous occurrence. Among the chemicals known as mutagenic are mustard gas, mustard oil, caffeine and formaldehyde; for others *see* p. 799.

As a result of the work of Muller and other workers certain general observations on the phenomenon can be made. There is no difference between the types of gene mutations which occur as a result of artificial treatment and those occurring spontaneously. All ionizing radiations are mutagenic, as are certain chemicals. It is therefore safe to assume that spontaneous mutations are due to the same causes as the artificial ones, since all organisms are subject to a certain amount of ionizing radiation from cosmic rays and radioactive material, and could come into contact with the mutagenic chemicals which occur in nature. When a mutation is produced, it is much more likely to be harmful to the bearer than otherwise. This can be explained by saying that existing organisms have for long been subject to the same mutant changes and thus all the "good" ones will be already "selected in" and the "bad" ones "selected out." Thus the possibility of an advantageous mutation arising is less than that for a disadvantageous mutation. Although mutations can occur in any cell, they do not occur at the same rate in all genes. In general, the artificially produced mutations occur in about the same proportions as the spontaneous. A mutation in one direction may be followed by a later mutation in the reverse direction, and thus a new mutant gene may revert to the normal type. This indicates that the mutation has arisen through some alteration to the existing gene and not through its extinction, since the change can occur in both directions. There is direct proportionality between the quantity of ionizing radiation used and the number of mutations it can bring into being. This indicates that the mutation is the result of an ionization occurring in the chromosome carrying the gene, or in its near neighbourhood. A weak dose of radiation for a long period has the same chance of producing such ionizations as a strong dose for a short time. No dose of radiation is so low that it cannot cause mutations sooner or later. This is of significance in modern times when radioactive fall-out, however small, produced by atomic and hydrogen-bomb explosions, can only increase the number of possible mutations.

The nature of a gene mutation, and the exact mechanism which causes it, are to be sought in the chemical constitution of the genetic code carried by the DNA molecules in the chromosomes. A mutation may be defined as an inherited change due to an alteration in the nucleotide sequence in such a molecule (*see* pp. 789–798).

PLASMAGENES

So far in this chapter, we have regarded the chromosomes with their genes as having the sole prerogative of control over hereditary potential, considering the cytoplasm as merely the agent of the nucleus. It is, however, becoming increasingly evident that there are variations in species which do not depend upon chromosome differences; sometimes these differences can be attributed to definite discrete particles in the cytoplasm, and sometimes they appear to be due to the whole chemical composition of the cytoplasm.

There are a number of cases of self-perpetuating bodies in the cytoplasm. Examples are the plastids, mitochondria, basal granules, parabasal bodies and centrosomes. These have precise arrangement and all arise by division of previously existing structures. Some contain RNA but it is not known if this is coded. Some cases are known of chloroplasts and parabasal bodies which contain replicating DNA. Altogether, over one hundred cases of cytoplasmic inheritance are now recognized.

In a variant of *Primula sinensis*, if a yellow-leaved female parent is crossed with a green-leaved male parent, the F1 offspring all show yellow leaves. But if a green-leaved female parent is crossed with a yellow-leaved male parent, then the F1 offspring all show green leaves.

It seems obvious that the leaf colour is determined by the plastids, which are carried into the zygote by the maternal cytoplasm. The

1. Yellow ♀	×	green ♂	.	.	.	P
	all yellow		.	.	.	F_1
2. Green ♀	×	yellow ♂	.	.	.	P
	all green		.	.	.	F_1

green-and-white striping of leaves has a similar hereditary basis. A number of such cases of plastid-borne inheritance are known, and hence the term *plastogenes* has arisen.

Several examples of cytoplasmic inheritance via the mitochondria have been discovered. The "petite" strains of yeast lack the enzyme cytochrome oxidase which is known to occur in the mitochondria. In the mould, *Neurospora*, there is a "poky" or shrunken form. There are two types of gametes, one with substantial cytoplasm and one with very little. The "poky" character is passed on to the offspring only in conjunction with the larger gamete. Here again, the mutant form is due to an abnormality of respiratory enzymes in the mitochondria.

Some types of trypanosomes show remarkable resistance to drugs. This faculty is inherited and has been shown to be associated with changes in the structure of the parabasal body.

The existence of other determinants, not yet identified with discrete particles, is inferred from various crossing experiments. In the willow-herbs, if *Epilobium hirsutum* as the female parent, is crossed with *E. luteum* as the male parent, the F1 offspring are all dwarfed plants with reduced flowers. But if the cross is made with *E. luteum* as the female parent and *E. hirsutum* as the male, the F1 offspring are all large and bushy, with well-developed flowers. In *Funaria* spp., the form of the capsule is determined by the cytoplasm of the oosphere.

It seems that we must recognize the existence of cytoplasmic determinants and it is possibly in these that we may also find the explanation of differentiation in development. Unequal distribution at cytoplasmic cleavage would then lead to different potentialities in the daughter cells. Equal distribution would lead to similar potentialities and hence a homogeneous tissue. Some connexion between plasmagenes and virus particles has been suggested, the basis of the suggestion being that viruses represent plasmagenes which have dissociated themselves from cytoplasmic control and can exist separately but can only multiply in a suitable cytoplasmic medium. It is certain that both plasmagenes and viruses are subject to mutation.

Our conception of heredity must be widened. The nuclear genes have the major measure of control, expressing themselves through the cytoplasm; plasmagenes occur linked with nuclear genes, and finally there are plasmagenes which seem to function independently of nuclear control. The cell, and hence the organism, is the expression of all these factors in conjunction with the environment.

HYBRID VIGOUR—HETEROSIS

A variation of a quantitative nature sometimes arises in the hybrid offspring of crosses made between two normally inbred varieties of a species. The hybrids of the F_1 generation often show a considerable increase in capacity to grow. They are bigger and more vigorous and frequently produce a higher yield of vegetative material and reproductive parts. The phenomenon is known as *hybrid vigour*. There are two possible theories to explain it. The first assumes that the heterozygous nature of the hybrids as compared with the homozygous condition of the inbred parents is responsible. That is, if two alleles A and a each have an effect on growth then the combination Aa produces greater vigour than either of the homozygotes AA or aa. Hence the name *heterosis*, shortened from heterozygosis, is used as an alternative to hybrid vigour. The second theory assumes that each parent (homozygous if inbred for many generations) will possess some of, but not necessarily all, the possible genes dominant for vigorous growth, and that when two such parents each possessing different genes of this sort are crossed, then the offspring will inevitably carry a greater number of such genes than either of the parents. For example, in a cross between parents genetically constituted $VVwwxxYYzz$ and $vvWWXXyyZZ$, where capitals represent dominants, the progeny must be $VvWwXxYyZz$ where all five dominant genes for greater vigour appear as opposed to two in one parent and three in the other. Under favourable conditions therefore, they will develop to greater proportions than either of the parents.

As an increase in bulk and vigour (although theoretically it can be a decrease or negative heterosis), heterosis is generally easily recognizable since its greatest effects are to be seen in the F_1 generation only, with a gradual decline in succeeding generations when these are inbred. This falling-away can be explained by the first theory in terms of the gradual increase in proportion of the homozygous condition among the succeeding offspring, i.e. the heterozygotes will segregate as homozygotes and gradually become fewer in proportion throughout the total number of individuals. This reduction in the extent to which hybrid vigour persists into succeeding generations, particularly in self-fertilized plants, is therefore automatic. For example, suppose an individual shows hybrid vigour due to the heterozygous pairing of genes Aa. When self-fertilized the offspring will appear 1 AA:2 Aa:1 aa, i.e. only half show the heterozygous condition necessary for hybrid vigour. On continued breeding more homozygotes will be produced from Aa types and so the proportions of those showing Aa will gradually be reduced.

On the second explanation of heterosis it should be possible, in some cases at least, to perpetuate or fix the condition in succeeding generations. This would involve the production of individuals which are homozygous dominants for all genes favouring vigour. From the example given above, if the vigorous $VvWwXxYyZz$ F_1 progeny were selfed they should produce some offspring which are constituted $VVWWXXYYZZ$, and these should breed true. However, in view of the fact that there are undoubtedly many more than five genes involved (some estimate over two hundred) and that they do not necessarily freely segregate because of linkage, it is highly improbable that such a breeding experiment could be successful until countless millions of such offspring were bred.

A knowledge of hybrid vigour can have practical value. For example, crosses can be made between maize plants which result in offspring showing from 100 to 200 per cent increase in yield of grain weight over either parent, but, if this maximum vigour is to be maintained, then it is necessary to repeat the original parental cross for each new crop. When a plant can be vegetatively reproduced, the hybrid vigour can be retained indefinitely by cuttings, grafts, bulbs, corms, tubers, etc., and many of the high-yielding plants used in horticulture have been produced in this way. Pears, apples, strawberries and raspberries are among them.

Heterosis may also occur when crosses which result in polyploidy are made between different species. These species hybrids, as distinct from the variety-hybrids described above, may breed true, since the separate diploid sets of chromosomes from the two parents may act separately during meiosis, and so any extra vigour in the hybrid is passed on to future generations.

MODERN CONCEPTS OF THE GENE AND GENE ACTION

To understand the more recent concepts of the gene and how genes may act to produce their effects in organisms, some knowledge of the genetic material, that is, of the chemical structure of chromosomes is necessary, for it seems abundantly clear that they are the material basis of inheritance. In Chapters 2 and 4 (pp. 36 and 111), this has been briefly treated in references to the nucleic acids, desoxyribonucleic acid (DNA) and ribonucleic acid (RNA), but further consideration of the nature of these macromolecules will be necessary before the significance of the content of this section can properly be appreciated.

Although the nucleic acids were identified as long ago as 1897 their tremendous genetic significance was not demonstrated until about 1944 when it was realized that the substance responsible for "transforming" bacteria of one genotype into another was DNA (non-virulent, non-capsulated R-type *Diplococcus pneumoniae* into virulent, capsulated S-type—*see* Vol. I, p. 1240). It was here clearly demonstrated that hereditary changes, that is, alterations in the genotype, can be produced in bacteria when DNA molecules from cells of one genotype are taken into cells of another. Similar evidence came from work with bacteriophages, in which it was shown that it was the DNA portion of the bacteriophage only which was necessary to the formation of more bacteriophage within the bacterial host cell. In the case of the tobacco mosaic virus it has also been shown that it is the RNA portion of the virus that is essential to the reproduction of more virus bodies, the protein covering playing no part in the virus activity within the tobacco cells.

There has been brought forward no direct evidence parallel to this to support the idea that the nucleic acids are genetic materials of higher organisms, but there are some very strong pointers to this being the case. In the first place, of all the cell constituents, DNA is the one substance that appears to remain constant in quantity from cell to cell and from generation to generation, whereas all other chemical cell constituents are constantly varying. This matches the idea that if DNA is the genetic material it should be transmitted from cell to cell and from parent to offspring in unchanging quantity and state, appearing always in the same amount per cell of a given organism from generation to generation. Actual measurements of DNA in the cells of a particular plant or animal and its offspring show this to be so, except during the period of nuclear division when the DNA, located in the chromosomes, is doubled in quantity. From this kind of evidence and other indications, which alone may not "prove" anything but which collectively fit together to make a sensible hypothesis, there seems little doubt that DNA is the material substance of the genes in higher organisms.

The structure and composition of DNA have been described in earlier chapters, according to the Watson-Crick model, and need not be described here except in outline. These workers proposed that the DNA molecule is composed of two helically coiled polynucleotide chains bound together in a specific way, namely by hydrogen bonding between a purine base (adenine or guanine) on one chain and a pyrimidine base (cytosine or thymine) on the other. Since hydrogen bonding between adenine and cytosine and between guanine and thymine is very difficult to attain and hence highly improbable in such a uniform structure as indicated by X-ray crystallography, in their model the two polynucleotide chains are bound by hydrogen bonds between the pairs of bases so—adenine (purine) and thymine (pyrimidine), two bonds, and guanine (purine) and cytosine (pyrimidine), three bonds. The model proposed allows perfectly for the exact replication of the structure and this is believed to occur by the separation of the two chains by breaks at the hydrogen bonds, each chain then acting as a template for the construction of its partner chain from the necessary units of sugar, phosphate and purine and pyrimidine bases previously manufactured in the cell. At the end of a nuclear division, each original DNA molecule (chromosome) will have become two (chromatids).

This Watson-Crick proposal has been substantiated in a number of experimental ways, working with DNA from a wide variety of sources. One of the most convincing was the work of Meselson and Stahl using

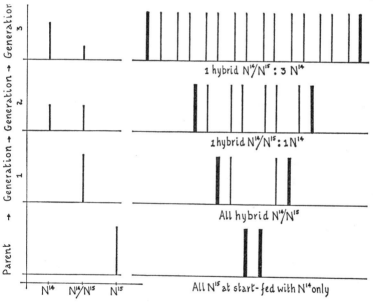

Fig. 19.11. Distribution of ^{15}N in successive generations of *E. coli*.

the heavy isotope of nitrogen, ^{15}N, to trace the distribution of the nitrogenous bases from generation to generation in the bacterium *Escherischia coli*. They showed that the DNA in which all the nitrogen of the bases was initially ^{15}N, changed its density from all "heavy" to

26

intermediate or "hybrid" density to increasing proportions of "light" DNA exactly as would be expected in successive duplications of the DNA on the Watson-Crick hypothesis, when supplied with ^{14}N only during the duplication periods. Their results are represented graphically in Fig. 19.11.

There is no doubt that the mass of evidence is in favour of the correctness of the Watson-Crick model of the DNA molecule and that this substance is indeed the material basis of inheritance, that is, represents the genotype. The question still remains—what properties does it possess that are recognizable as the means whereby it is responsible for the development of a given phenotype? The more acceptable answer to this appears to be that the DNA molecule represents only "information" required for the building of a phenotype structure, that is, it is a molecule carrying a message, that when passed in some way to a "manufacturing" site and translated, controls the formation of the individual structural units necessary to the complete, finished article. It does not itself play any part in the manufacturing process, merely giving instructions as to what shall be processed and how this must be done.

The suggestion is that the DNA molecule carries its messages in some form of *code*, this being represented by the sequence of the bases in the polynucleotide chain. At first this idea may be difficult to accept since there are only four bases and obviously many messages to convey. But if the four initial letters of the bases, A T C G, are considered as letters of a four-letter alphabet from which words can be built, it is not difficult to see that many messages can be spelt out according to the sequences of the letters, even from such a restricted alphabet. Each message, or sequence of messages, might then be thought of as being equivalent to the gene. But before the probable nature and meaning of such a code can be discussed further it will be necessary to examine the possible way in which the inheritance messages are translated into building processes. Evidence has accumulated to suggest that there may be some relationship between genes and enzyme production by cells and in some cases, even to relationships between particular genes and particular enzymes.

Biochemical mutants are the chief sources of this evidence, such a mutant being an organism differing from the normal in that it is unable to carry out some particular stage in a metabolic process or carries it out in a different way. When such a mutation is not lethal, as it most probably would be in many cases, the mutant variety of the organisms is recognizable from normal forms because the intermediate and/or end products of its metabolism are different and this can be detected by chemical means.

One well-known case occurs in humans as the inherited disability known as phenylketonuria. In addition to other symptoms of the condition, sufferers excrete phenyl-pyruvic acid in the urine, whereas normal persons do not. The reason why the urine contains this substance is because the affected cases are unable to convert the amino-acid phenylalanine into tyrosine. It is known that this chemical change is dependent on the presence of a single enzyme, an amino-acid oxidase. Thus it is reasoned that the condition of excreting the abnormal substance phenyl-pyruvic acid in the urine is due to the inability of the subject to form the oxidase enzyme at all or the ability to form it only in a changed state. When the genetics of this situation are studied by careful analysis of family pedigrees, it is found that the sufferers are homozygote recessives for one gene. Thus a link between a gene and its expression in a biochemical reaction via an enzyme seems probable.

The most detailed knowledge of the possible relationships between genes and enzymes has been produced during studies of microbes, in particular, the fungus *Neurospora*. The geneticists, Beadle and Tatum, have been largely responsible for the work done with this material. To understand the significance of one of their experimental findings it is necessary to be aware of the events that occur during the biosynthesis of a complex substance from simpler materials. In nearly all such cases there is no direct construction of the larger molecule. Instead, the synthesis proceeds by a series of stages, each step in the process being catalyzed by the appropriate enzyme. For example, it is known that during the synthesis of the amino-acid arginine, the last steps are from—*ornithine precursor* to *ornithine* to *citrulline* to *arginine*, each step being under the influence of a different enzyme.

Now, in wild-type *Neurospora*, the synthesis of arginine proceeds along this pathway, provided the fungus is supplied with a nutrient containing all the basic raw materials necessary for it to do so. However, there are several mutants of this wild type that cannot manufacture arginine under these raw material nutrient conditions. It appears as though the mutation is being expressed as a block in the synthesizing system or, in other words, the inability to produce the appropriate enzyme for a particular step in the synthetic pathway. In fact there are at least three such mutants each apparently blocked at a different step from the others and when crossings are carried out back to the wild type it is found that there are three non-allelic gene loci concerned, each segregating in a one-to-one manner with the normal locus. It is possible to discover at which point the path is blocked for each of the mutants by feeding the fungus with a nutrient containing one of the intermediate substances known to occur along the pathway, e.g. ornithine precursor, ornithine or citrulline, and then finding out whether or not the mutant is capable of producing arginine. When this is done the following is disclosed: one mutant will grow only when fed with arginine, implying a block between citrulline and arginine, since none of the earlier intermediates nor the raw materials will support growth. Another mutant will grow when fed with arginine or citrulline, but not with the others, implying a block at the ornithine-to-citrulline stage. A third will grow if supplied with arginine, citrulline or ornithine, indicating a block at the ornithine precursor-to-ornithine stage.

From this it seems reasonable to suppose that the mutants are varieties of the fungus in which a gene change has occurred, this being expressed as an inability to produce a particular enzyme, or at least as an ability to produce it only in a non-functional form. The relationship between the genetic message and enzyme production in cells seems reasonably well established in this case, for it seems a valid inference that gene mutation has brought about a change in the genetic control of a biosynthetic step and this is expressed in the organism as an inability to perform one operation in a metabolic activity.

As a logical extension of the inference to be drawn from the above discoveries it may now be further suggested that since enzymes are proteins, the expression of genes is initially through the formation of these substances in the cells and that such substances form the intermediate stages between the genetic code and its ultimate expression as the phenotypic characters. Much will need to be done before this can be accepted as fully representing the case, but already there is mounting a good deal of information that it may be so. An acceptable description of what may be happening in cells, leading to the formation of proteins, is given below.

Protein molecules are extremely complex, being made up of chains of large numbers of amino-acids (of which there are twenty different kinds) arranged in specific sequences and united by peptide linkages (see p. 95). These chains of amino-acids are crosslinked and folded to form the most complicated of all organic molecules. It is obvious that the synthesis of such intricate systems of particles is not accomplished in any haphazard fashion. The deciphering of the "genetic code" has brought a much better understanding of the way in which the "messages" within the code are translated into cell activity, that is, how the cell "factory" converts the messages into finished products such as proteins.

The cell carries an extremely detailed "blue-print" for each kind of protein required, the "blue-print" being in the DNA molecule. This is described on pp. 112–113. This DNA is separately stored within the cell as chromosome material but is capable of passing copies of itself, as a form of RNA, to other cell regions. Such "messenger RNA" molecules carry information from the genetic code to the sites where protein is to be manufactured. There is a difference between the nucleotide bases of the DNA and those of the messenger RNA, uracil being found in the latter in place of thymine. The sites or "work benches" are the ribosomes (see p. 29). The materials to be worked with are, of course, amino-acids, and these are conveyed to the ribosomes attached to "carriers," each carrier being capable of picking up one only of the twenty kinds of amino-acids. The carrier substance is one of another series of RNA forms known as "transfer RNA". Each transfer RNA molecule carries a triplet of bases or *anticodon* that recognizes its counterpart triplet sequence of bases or *codon*, representing a specific amino-acid on the messenger RNA by the base-pairing rule cytosine and guanine only and adenine and uracil only. The carrier molecules are thought to enter the "factory", each with its amino-acid, and become attached to the messenger RNA which is already there, the point on the messenger RNA at which a carrier can become attached being governed by the necessity for its anticodon to match a messenger RNA codon. Thus, although the messenger RNA will have many points of attachment for carrier molecules, a particular type of carrier, with its particular amino-acid, can be attached at its specific site only and no other.

This sequence of carrier acceptance points is determined by the messenger RNA which in its turn represents the message dictated by the DNA of the chromosomes. Thus when all the points of attachment on the messenger RNA are filled with the appropriate carrier RNA molecules, each with its amino-acid, there becomes lined up a sequence of these which can be joined by peptide bonds to form a chain. As a chain is completed and a protein molecule portion is formed, this is released, as are the carrier RNA particles which can then pick up their appropriate amino-acid loads once more. At the same time, the messenger RNA is freed to become the template for another similar sequence of amino-acids. Clearly, the finished product at the ribosome is fashioned according to the genetic message embodied in the chromosome DNA and the genes are apparently finding their expression as specific sequences of amino-acids.

The process outlined above is represented diagrammatically in Fig. 19.12.

Note that the "work benches", the ribosomes, are each considered to be made up of two joined particles, together measuring about 20 nm across. In the test-tube they can be separated by removing magnesium ions from the solution. Each particle consists, in about equal proportions, of RNA and protein. The RNA in ribosomes does not reflect the DNA coding and it has been shown that ribosomal RNA from many different species is very similar whereas the DNA from the same range of species is very different. The role of ribosomal RNA is not fully clear but it is thought to present a surface to which various parts of the protein-making machinery can form a temporary attachment. Obviously those points of attachment must follow some precise pattern because it is essential that different parts of the machinery are positioned in the right sequence and correctly aligned both in space and time if the finished product is not to be malformed. There is evidence that the ribosomes act together in groups (threes, fours, fives or more), which accounts for their appearance as ribosomal clusters or *poly-ribosomes* in the cell. When clustered they are held together by an RNA strand many times longer than a ribosome and this could be the messenger RNA mentioned above. If this is so then the following could be a possible interpretation of what may be happening in the cell.

When a ribosome begins to synthesize an amino-acid chain, it is at first attached to the "leading" end of the messenger RNA and proceeds to move along this, compounding the sequence of amino-acids as indicated by the messenger RNA. As one ribosome moves away from the "head" of the messenger RNA molecule, another ribosome becomes

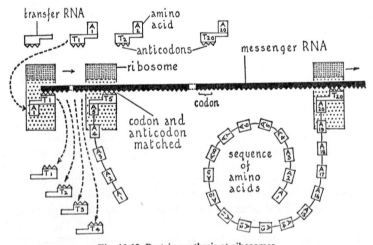

Fig. 19.12 Protein synthesis at ribosomes.

attached and commences its journey in the same direction. As preceding ribosomes move along, more become attached, one behind the other. Each is building the same protein component and when each reaches the "tail" of the messenger RNA, it is released. It drops its load and is ready for service once more.

To return now to the "coding" of messages in the light of what has been said, it seems certain that this is the means of determining the relative positions of the different amino-acids required to build a particular protein. Several forms of code have been suggested to account for the manner in which four bases can be sequenced to represent twenty amino-acids.

The most acceptable is that resulting from the work of Leder, Nirenberg and others which indicates clearly that the code by which proteins are synthesized is represented in the DNA molecule by triplet sequences of the bases adenine (A), thymine (T), cytosine (C) and guanine (G). Each such triplet or codon represents one amino-acid and the code is "non-overlapping". This latter concept has arisen from work with viruses in which mutants are unable to produce a protein covering layer identical with that of the normal virus. The mutants vary from the normal generally by a change in one only of the constituent amino-acids of the protein, rarely in two, and in this latter event the amino-acids concerned are never adjacent in the protein molecule. This indicates that the code is not over-lapping, because if this were so, a

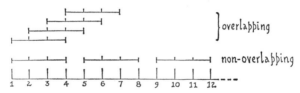

Fig. 19.13. Overlapping and non-overlapping codes.

change in one amino-acid due to a change at one point of the code would have to be followed by changes in others close to it in the protein molecule. To compare the effects of changes in overlapping and non-overlapping codes see Fig. 19.13.

The code is also clearly "degenerate", that is, several different codons or triplet sequences can represent the same amino-acid. This follows from the fact that there are sixty-four possible codons to be made up from the four DNA bases but only twenty amino-acids for these to represent. It has now been worked out which codons represent which amino-acids in terms of the sequence of bases on the messenger RNA where, it should be remembered, that uracil (U) appears in place of thymine of the DNA bases. The following are examples—

Glycine —GGU or GGC or GGA or GGG
Lysine —AAA or AAG
iso-Leucine—AUU or AUC or AUA

It seems clearly the case that most of the amino-acids are sufficiently well coded by the first two bases of a triplet sequence and if these appear in the correct order it does not matter what the third may be. This seems certain for all but three of the amino-acids, but in the cases of arginine, leucine and serine, in which the coding has been worked out as follows—

> Arginine—AGA or AGG or CGU or CGC or CGA or CGG
> Leucine— UUA or UUG or CUU or CUC or CUA or CUG
> Serine— AGU or AGC or UCU or UCC or UCA or UCG

they are represented by either of two sets of codons differing from one another in the first two bases of a triplet.

Two triplets, UAA and UAG, do not appear to be employed for the coding of specific amino-acids at all. It is possible that their appearance in a long sequence of codons represents the starting or termination points of a polypeptide chain. This would mean that a particular sequence of triplets starts from a fixed point in the whole DNA structure and thus if somewhere along the ensuing sequence one base is missing, duplicated or substituted, the code then becomes meaningless for the remainder of the sequence. Alternatively, base changes in the code could lead to gene mutations.

What is a Gene?

Apart from a vague reference to the gene as the "unit of inherited material" on p. 759, there has been no further attempt to define it in more precise terms. If the reader has fully comprehended the significance of the sections dealing with chromosome behaviour, including linkage and crossing over, gene mutation and what has been written so far in this section concerning possible gene action, he should be able to understand that there are, therefore, three possible bases on which the gene may be more clearly defined.

If the chromosome is visualized as a linear series of inheritance units, like beads on a string, each such unit or gene can be considered as the smallest quantity of genetic material, occupying part of a chromosome, that is capable of being separated from the others in its linkage group by cross-over processes and hence capable of being recombined with any equivalent unit of genetic material at any other point in the series. This follows from the recognition of the existence of linkage groups among the genes, a knowledge of the fact that cross-overs between the parts of chromosomes do occur and that genetic material at the same locus or point in the linear series on homologous chromosomes is allelic and not capable of recombination by cross-overs, whereas two units of genetic material at different loci can be so recombined. In this sense the gene can be defined, therefore, as the *smallest unit of genetic material capable of being separated from the rest and of being recombined with it*. The exact limits of each such piece of genetic substance along the linear series cannot always be determined precisely in practice, but the many chromosome maps that have been made are clear demonstrations that this conception of the gene has at least some tangible foundation even if this cannot be brought sharply to view.

Another possible way of defining the gene comes from a knowledge of gene mutations. These are changes at molecular level in the structure of a chromosome leading to a transmissible alteration in the genotype. A good example is white-eye of fruit flies which is the direct result of a fly's inability to synthesize the normal pigments. It would appear to be the case that the mutant white-eyed fly lacks, within its DNA structure, the information necessary for the production of the colouring substances and the visible effect in the phenotype represents a change, in this case a loss, of a genetic function. The gene can therefore be defined as the *smallest amount of genetic substance capable of undergoing a change or mutation resulting in an alteration in an inheritable character*. Such a mutation could be the result of changes in the sequence of the purine-pyrimidine base pairs in the DNA molecule due to errors in pairing during replication or to the substitution of one or more of the bases by some other suitable molecular structure. In practice it is possible to replace some of the DNA bases by analogous substances and also, by the use of chemicals, to bring about changes in the structure of the bases normally occurring. For example, 5-bromouracil is a *biological analogue* of thymine (i.e. a substance whose molecular structure is very similar to a natural compound but which acts antagonistically to it). Bromouracil is known to cause mutations in bacteria and bacteriophages. Similarly, an analogue of adenine, 2-aminopurine, is said to be mutagenic in algae. A substance that is known to act mutagenically by chemically altering the bases that normally occur in DNA is nitrous acid. This is known to have mutagenic activity in fungi, bacteria and bacteriophage. The mutagenic effect in all these cases is believed to be due to an ultimate alteration of the base sequence in DNA. Clearly, the smallest unit capable of mutating must lie in the DNA structure somewhere and could be as small as a single pair of nucleotides.

The third conception of the gene can be derived from an understanding of the probable way in which genes act, that is, the gene may be considered as the unit of function. The proposed role of the genetic material in protein synthesis incorporates the idea that sequences of base pairs in the DNA represent the order in which amino-acids are brought together to form protein molecules. If this is truly the situation, and there is much evidence to substantiate it, a gene can be considered as the *smallest unit of base sequence representing a specific biosynthetic activity in the cell.*

The question that now must clearly arise is: do these three definitions or concepts of the gene all describe the same thing?

From the work on the arginine-requiring mutants of *Neurospora* it may appear, at least superficially, that they do. In that particular case, the three mutants all arose independently of one another and in crosses with the wild type each mutant gives a one-to-one segregation. An analysis of cross-overs, that is, of recombination between any one of the mutant genes and the genes in the same linkage group shows that each mutant has a specific locus in a particular chromosome, as does its wild-type allele. Thus a genetic test for recombination has demonstrated the separation of pairs of allelomorphic genes from each and all others. Further, each mutant can be shown to be functionally different from the others in that it is blocked at a different point in a biosynthetic pathway. Hence we seem to have, embodied by the same entity, all three properties by which we might attempt to designate the gene, namely, recombination, mutation and function.

However, not all cases are as clear-cut as this appears to be, and there are many other situations which call for at least a modification of this view in order to explain them satisfactorily. For example, breeding experiments involving multiple allelomorphs clearly indicate the existence of a wild-type gene with a number of mutant alleles all at the same locus and the mutants behave with one another when crossed exactly as any pair of alleles do. Functionally, each mutant shows a difference from the others in being responsible for a different effect on the same character in the phenotype and it can be concluded that each allele has been produced by a mutation that affects the overall function of the wild-type allele in a different way. That is, there are a number of places in the same gene locus at which a mutation can occur, i.e. there may be more than one unit of mutation within a gene locus, as defined in terms of recombination, and mutations at different points within this single locus lead to differences in the expression of the gene functionally.

A further case that must also be considered complicates the issue still further. There is evidence that functionally identical units of genetic material can be shown to be separable by tests for recombination, i.e. that two mutants, arising independently of one another and clearly spatially separated from one another along a linear series are functionally alike. They are known as *pseudo-alleles* and are best described as units of genetic material, apparently concerned with the same function, always so very closely linked that cross-overs between them occur at very rare intervals, but nevertheless do occur. The genetic constitution of heterozygotes carrying these alleles can be represented as

$$\frac{\text{pseudo-allele 1} \qquad + \qquad}{+ \qquad \text{pseudo-allele 2}} \text{ or } \frac{\text{pseudo-allele 1} \qquad \text{pseudo-allele 2}}{+ \qquad +}$$

In the first case the heterozygote is said to be in the *trans*-form and in the second in the *cis*-form. The former show the mutant effects in the phenotype whilst the latter show wild-type characters. This *cis/trans* effect is shown only between gene loci (defined on the recombination basis) that are very closely linked and of which the functions are similar. It obviously has an important significance in the question of how the gene can be defined as a functional unit, and this definition can then be only in the following terms. Any two pieces of genetic substance that, in a diploid, show the *cis/trans* effect must be considered to be the same gene functionally even though they may be separable on a recombination basis.

To summarize briefly, there are three possible ways in which we can conceive the gene: as the smallest piece of inheritance material capable of segregating from and recombining with other portions of the same chromosome; as the smallest portion of the DNA molecular structure capable of undergoing chemical change or substitution to transmissible changes in the genotype; as the smallest portion of inheritance material that can be associated with one specific function of the cell. The evidence from investigations into the nature of the gene do not appear to indicate that these are the same thing in all cases.

However, the work of Benzer, in which mutant strains of a bacteriophage were used, has assisted greatly in building a more comprehensible presentation of gene nature. His view is one of the existence of basic regions of genetic substance along the chromosome, separable as units of function. He has called them *cistrons*. Each cistron may be a comparatively large unit since each may contain a number of points or *mutons* at which a mutation can take place as well as sections, called *recons*, between which recombination can occur.

GENES AND DEVELOPMENT

The processes by the completion of which the development of an organism is accomplished must be gene controlled since the organism is the product of its own genetic potential operating in an external environment. But there must also be brought into play some programme or sequence of gene action as part of this control mechanism, for not all the genes can be active at the same time or equally operative in different parts of the same developing plant or animal. The totipotency (*see* p. 539) shown by cells, that is, the ability of each to produce offspring that can develop into all the others under tissue culture or regeneration conditions, cannot be allowed full freedom or there would be no differentiation of parts. In other words, since the nuclei of different cell forms do not show any inheritable distinctions, there must be some very precise mechanism for calling into play some genes and suppressing the activities of others at carefully timed periods during the development of an organism so that the whole series of events proceeds smoothly.

More recently geneticists have turned their attention to elucidating how genes may be "switched on and off," as it were, and the nature of the control over the switching sequence. Most information concerning genetic control of cell activity has come from studies of bacteria and thus it should not be assumed that this necessarily has any relevance to the condition in higher organisms. What has been suggested as an explanation in the case of the microbes follows from the relationship known to exist between genes and enzymes with the genes determining the sequence in which amino-acids can be put together as proteins (*see* pp. 794–798). Now, in bacteria there appear to be three enzyme forms from the point of view of their appearance and permanence in the cell. One kind, called *constitutive enzymes*, are present all the time whilst another kind, called *inducible enzymes*, are produced only when the substrates with which they are associated are present. An example of an inducible enzyme is β-galactosidase, produced in any quantity by *Escherischia coli* only when a galactoside substance (galactose sugar + non-sugar group) is present in the culture medium; removal of galactoside from the medium results in the cessation of formation of the enzyme. The third form of enzyme, by an appearance criterion, contrasts with this and whilst the induction process operates for some there is a reverse effect known as *repression* for others, that is, the enzymes may be referred to as *repressible enzymes*. For example, when *E. coli* is grown without histidine in the medium, the enzymes necessary for histidine synthesis are quickly formed but as soon as histidine is added to the medium, production of the enzymes stops. This can only mean that something, in this instance presumably the metabolic product histidine, acts as a *repressor* of the gene(s) concerned with histidine formation. This can be referred to as "end-product repression" but similar enzyme repression can be effected by substances other than an end product.

The phenomena of enzyme induction and repression have been used by Jacob and Monod as the basis of their hypothetical explanation of control of gene activity in bacteria. This in general terms is as follows.

It is recognized that a specific enzyme structure (protein) will be decided by the genetic code in a section of DNA in a chromosome.

Such a sequence of DNA material that determines the structure of an enzyme protein can be thought of as a *structural gene*. To convey the message carried in the DNA code to the cytoplasm, messenger RNA is required and this is translated into an amino-acid sequence at the ribosomes. It is theorized that the synthesis of messenger RNA occurs only under the control of given regions of DNA strands, referred to as *operators*. An operator may control several structural genes in the copying or transcription of DNA to the messenger RNA form and any section of DNA under such control of a single operator is referred to as an *operon* (*see* Fig. 19.14). In conjunction with each operon is a

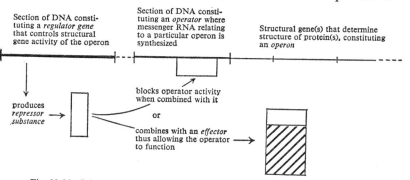

Section of DNA constituting a *regulator gene* that controls structural gene activity of the operon

Section of DNA constituting an *operator* where messenger RNA relating to a particular operon is synthesized

Structural gene(s) that determine structure of protein(s), constituting an *operon*

blocks operator activity when combined with it

produces *repressor substance* →

or

combines with an *effector* thus allowing the operator → to function

Fig. 19.14. Scheme representing the Jacob and Monod hypothesis of control of gene activity.

section of DNA, known as a *regulator gene*, functioning to regulate the transcription activity of the structural genes by causing to be produced in the cytoplasm, substances that can repress the structural gene's ability to produce messenger RNA. A *repressor* formed under the influence of a given regulator is presumed to have an affinity for a specific operator and when bound to it chemically prevents the production of messenger RNA by the whole operon associated with that particular operator. This is effective in preventing particular proteins coded in that operon from being synthesized.

A repressor is also presumed to have the property of being able to combine with and dissociate from another molecule, referred to as an *effector*, so that in a given system there may be the reversible situation so:

free repressor + free effector ⇌ repressor–effector combination.

The difference between inducible and repressible systems can now be explained in these terms. In the former, the free repressor molecule alone is active in blocking the action of the operon by combining with the operator. When an effector (or inducer) is present this renders the repressor inactive by combining with it and the required messenger

RNA for enzyme synthesis can be produced at the structural part of the operon. In the latter, the repressible system, only the combined repressor–effector substance is active (the effector possibly being the end product of metabolism, e.g. histidine) in blocking the operator and synthesis of messenger RNA by the operon, which would otherwise proceed in the absence of the effector (or repressing metabolite), is prevented.

Some ideas have been put forward with regard to the nature of the product of the regulator gene. It has been conceived possibly as an allosteric protein, that is one with two reactive sites, one able to react with the operator and to block it and the other able to react with another compound that alters its structure in such a way as to prevent it from reacting with the operator. The operator is then in effect unrepressed.

Whilst there is some evidence that such a system of gene control may be operative in bacteria, there is no indication that the same applies to other higher organisms. In fact, a rather different mechanism has been suggested in which the histone protein component of the chromatin material has the chief gene-regulating function. Chromatin, the whole chromosome substance, is known to be composed of DNA, histone proteins, small quantities of other proteins and probably an RNA-polymerase enzyme, responsible for linking the nucleotides when messenger RNA is formed. Work carried out on the possible role of histones as repressors in higher plants has given some indication that they could be involved but it seems not likely that they constitute the whole control mechanism because while there are very many genes to be repressed there are comparatively few different histone structures. It would seem more likely that each gene would require the activity of a specific repressor substance and there are not enough histones to fill this requirement. Also against the hypothesis is the fact that bacterial chromatin contains no histone but shows very pronounced gene repression and activation conditions.

Other features of development are almost certainly related to activity of genes. The fact that development follows always an orderly sequence of stages with smooth transition from one to the next implies accurately timed, successive periods of repression and activation of particular genes. The phenomenon of "puffing", seen in the giant chromosomes of insects, gives some indication that the chromosomes are active in different sections at different times and hence this "puffing pattern" may be the chromosomal demonstration of the developmental sequence. Each distinguishable section on a giant chromosome could conceivably coincide with an operon and "puffs" when it swells with RNA transcribed from DNA, meaning that the puffing represents gene activity at specific loci. It has been observed that different tissues show different puffing patterns and one interesting observation concerns the effect of treating a larva with ecdysone, the moult hormone. When the hormone is administered, the puffing pattern quickly changes as moulting commences. It seems possible therefore that the successive phases of development are the outward manifestations of some kind of internal "programme selection" mechanism, the manner of operation of which is little understood.

In the same sort of way, as individual cells (particularly of plants) are produced and reach maturity there must be some control of the sequence of internal metabolic activities that occur to bring about the observable changes. In different tissues with different functions there must be varying sequences of activity at different genetic sites to account for the range in mature cell form and function. When mature, the various kinds of cells with their diverse metabolic operations must be subject to dissimilar gene activation and repression patterns. What this biochemical differentiation between cells may be controlled by, in genetic terms, is not yet elucidated.

It has been considered that hormones may in some way be concerned in the control of gene action. In plants, there are examples of hormones promoting activities that might be thought of as resulting from the switching on of a new gene complex. For example, auxin stimulates the production of adventitious roots by stems and cytokinin promotes bud formation in tobacco pith, in both cases from cells not previously active in these ways. In animals, hormones appear to have an effect on rate of enzyme synthesis to increase it. The hormone effects are visualized as being less likely due to direct switching on or off of genes but more in the nature of indirect control through some gene repression process. The interaction of hormones and genes is undoubtedly exceedingly complex, no less so because the same hormone seems to have the property of stimulating different tissues to react in different ways.

What has been said so far concerns the control of gene action in terms of genes associated with DNA in the nucleus. But, in some plants at least, cytoplasmic inheritance is undoubtedly involved and furthermore, some cell constituents such as plastids and mitochondria appear to have their own DNA coding so that they may operate independently of the nucleus. What factors are concerned in the regulation of the activity of such inheritance material is comparatively undisclosed, but occasionally some insight into the nature of the condition is gained from isolated investigations. One such is that involving the alga *Acetabularia spp.*, of the coenocytic Siphonalean form. A plant is composed of a basal rhizoidal part, a middle stalk portion and a distal, segmented cap. The whole protoplast is enclosed by a single continuous cell wall and there is a single nucleus located most commonly in the rhizoidal portion. It will regenerate a whole new plant very easily from any part containing the nucleus, much less readily from the remainder. For instance, when the stalk is severed above the rhizoid, the latter gives rise to a whole new plant that survives normally, whilst the stalk portion, lacking the nucleus, if it has not already developed one, will produce a normal cap but such a plant has no rhizoidal portion, neither does it survive for as long as the other. The grafting of parts shows more interesting conditions. When the apical portion of one young, uncapped stalk is grafted to another rhizoidal portion with its nucleus, a normal plant results. When the young enucleate part of one species, such as *A. mediterranea*, is grafted to the nucleate base of another, *A. crenulata* (or vice versa), there develops a new plant mainly but not entirely with the characters of the species supplying the nucleus. This last result would seem to indicate that whilst control of gene action, as manifested by developmental or regeneration processes, is mostly attributable to nuclear DNA, it may not be so entirely. The regeneration of a cap from an enucleate stalk seems to indicate that there may be long-lasting effects of the nucleus on the cytoplasm that can operate to maintain to completion a developmental process even when the nucleus is removed. It could be that the cytoplasm already contains sufficient messenger RNA produced by activated operons to control the process through to its conclusion.

So far, control of gene action, whilst a subject engaging the attention of many geneticists, is still very largely an unexplained phenomenon.

GENETICS OF POPULATIONS

In the preceding sections consideration has been given only to the way in which inheritable characters are transmitted through succeeding generations, commencing with a single pair of parents of known genetic constitution. The laws of Mendelian inheritance indicate what proportions of genotypes and phenotypes can be expected to appear in each successive generation. Under natural conditions, however, the control of mating between individuals which is exercised during experiment, i.e. when a single mating pair are isolated, does not exist. Pairings may be made between any two of a large number of potential parents so that the distribution of inheritable characters throughout the succeeding generations in a population is a much more complex state of affairs. It is with the distribution of inheritable characters through a freely breeding population of organisms that the study of *population genetics* deals.

To create the kind of situation that has to be dealt with and considering only the simplest degree of complexity initially, suppose in a given area there is not one pair of insects capable of breeding but many pairs and that phenotypically they vary, that is, the insects vary, say, in body colour, some being red-bodied and some being yellow-bodied, with red colour dominant to yellow. Let males and females be equal in number in both cases. The issue of what body colours will appear in successive generations, now becomes complicated because there are three possible matings that can occur: red-bodied can mate with each other, yellow-bodied can mate with each other and red-bodied can mate with yellow bodied. The problem to be considered is: how can the body colour of the offspring of such a population through successive generations be predicted?

To be successful in arriving at the correct conclusion it is important at the outset to understand that *it is genes, not whole organisms, that have to be studied when making such predictions.* The basic concept of the genetics of populations is that of *gene frequency.* This should be evident from the understanding that when the act of fertilization occurs it is the genes of each of the parents that are brought together and it is the genes that determine the characteristics of the new generation.

Returning to the situation cited above, with this in mind, when gametes are produced in an insect each contains one of a pair of alleles for a given character (body colour in this case) and can thus be divided into two kinds on this basis. The genotype of the insects can be represented by a pair of genes whilst a gamete can be represented by one of the pair. Now, if in the population there are, say, 100 pure-breeding red-bodied insects, this can be represented as 200 genes responsible for the red-bodied character. Similarly, with the yellow-bodied insects, 200 genes can be said to be responsible for yellow-bodiedness. The total number of genes contributing to the production of either red or yellow body colours in the insect population can be represented as 200 red plus 200 yellow. This constitutes the *gene pool* for body colour in the population. It is this pool of genes that must be the subject of study if the body colours in a future population are to be predicted correctly.

To illustrate what can happen by chance in the way of possible pairings between these genes in the pool, the analogy of pairings between different-coloured balls drawn from containers can be used. Suppose red and yellow balls (representing the genes) are placed in equal numbers in each of two containers (representing the sexes) and the balls are drawn at random in pairs, one from each container at a time. This will represent how the genes in the pool may be paired by chance. If a large number of pairs are drawn (if this is tried in practice it is advisable to use not less than 400 balls of each colour) and the proportions of pairings are counted, it will be found that they occur in the ratio 1 red/red: 2 red/yellow (i.e. 1 red/yellow and 1 yellow/red): 1 yellow/yellow. Carrying the analogy back to the insects where the genes are representing body colours and remembering that red is dominant, the following body colours in the offspring can be expected from the possible three kinds of matings—

3 red-bodied: 1 yellow-bodied, composed genotypically of 1 homozygote red (red/red): 2 heterozygote red (red/yellow): 1 homozygote yellow (yellow/yellow).

Assuming that nothing happens to disturb this condition, i.e. that all the offspring grow up to mate with one another and that the sexes

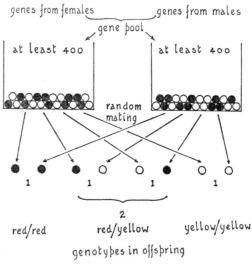

Fig. 19.15. Chance pairings of coloured balls.

are still produced in equal numbers, what will happen in the next and following generations? To find a solution, extend the analogy of the red and yellow balls. Suppose, therefore, half the number of balls representing each genotype placed into two separate containers representing the sexes and pairs are drawn, one from each container at a time, on a random basis. If this procedure is repeated many times, each repetition simulating what happens at the formation of succeeding generations in the insects, it will be found that the proportions of the genotypes within the whole population, i.e. their frequencies, will not vary at each successive mating. This is illustrated in Fig. 19.15. The result is fully in accordance with the following reasoning: in each generation, the gametes will contain the allelomorphic genes in the proportions in which they are found in the genotypes, thus at each generation the gene pool will contain the same proportions of alleles. Since the gametes fertilize at random, the same combinations are possible in the formation of each generation, and it follows that the proportions of the genotypes in each successive generation must remain constant.

It is possible to express the situation in mathematical terms by representing the frequencies of genes in a pool by algebraic symbols.

Let the frequency of the gene for redness (R) in the pool be a and that for yellowness (Y) be b. Then of the male gametes produced by all the male insects in the population, a will contain R and b will contain Y and of the female gametes produced by all the female insects in the population, a will contain R and b will contain Y. Insects of the genotypes RR and YY can be produced only by the union of male and female gametes, each containing an R in the first case and a Y in the second. Insects of the genotype RY can be produced either by the union of male R and female Y or vice versa, i.e. in two ways. Thus the relative frequencies of genotype in a generation can be represented as

frequency of genotype $RR = a \times a = a^2$
frequency of genotype $YY = b \times b = b^2$
frequency of genotype $RY = 2 \times a \times b = 2ab$.

If it is now considered that the total of all frequencies making up the whole gene pool can be represented by unity, it follows that

$$a^2 + 2ab + b^2 = 1.$$

From this equation it is possible by a substitution process to predict (i) frequencies for each genotype in a population, even though they may not be observable through differences in the phenotype, as when homozygotes are indistinguishable from heterozygotes due to dominance of one gene over another, and (ii) the constitution of future generations.

The expression is sometimes known as the *Hardy–Weinburg equation* after the men who first formulated it. Most clearly it refutes the fallacious argument that dominance of one gene over another will lead to an increase in the numbers of organisms exhibiting the dominant character at the expense of those carrying the recessive allele.

Now, it must be noted that what has been said so far applies only to those populations where no factors are operating to disturb the theoretical values of gene frequencies in the pool. If something is operating to change from equality the numbers of mating types, then the frequencies of genotypes in successive generations will not remain unchanged.

In natural surroundings there are four possible things that can happen to disturb this equality. In the first place, *selection* can occur. It may be that the possession of a particular character gives a survival advantage (or disadvantage) over its non-possession. For example, red-bodied insects may be more difficult to find against the background than yellow-bodied insects, hence predators would be more likely to eat more of the latter than the former before they were mature enough to breed. Such an occurrence would seriously disturb the equality of the mating types. Secondly, there may be *immigration* or *emigration* of one only of the mating types from or to the area. This also would affect the relative numbers of possible pairings. Thirdly, there could be a *mutation* in an inheritable character that could bring about a similar imbalance. Finally, if the population is small, the matings which occur will tend not to approximate to the equal numbers of pairings of all kinds that are assumed to occur by chance in a large population. This means that there can be differences between what is predictable on chance alone and what happens in an actual case. These "random sampling errors" are reflected in fluctuations and irregularities from the theoretical in the numbers of phenotypes actually observed. This condition is sometimes referred to as *drift*.

The effects of the first three of these possible disturbances on the proportions of genotypes in successive generations are all somewhat similar in that they will tend to create a depletion (or excess) of one of the mating types with respect to the other, with the consequent effect on the gene pool. As an illustration of what this effect is likely to be, consider the following. Suppose a population commencing with two strains of a species, represented genotypically by AA, BB, with B dominant to A, in equal numbers and the sexes equal in both strains. Because of some special feature in the phenotype of AA suppose that suddenly 20 per cent of this strain is killed off at each generation before mating has occurred. How does this affect the genotypes in later generations?

The proportion of the organisms being selected out will decrease markedly in successive generations whilst the proportions of the homozygote dominant allele will increase. It follows also that the proportion of heterozygotes will also tend to decrease. After several generations AA may become "extinct" but the gene A will still be represented in the AB genotypes so that it is likely that the homozygote recessive will occasionally reappear. Thus, although it is possible that selection operating against a homozygote can result in the eventual extinction of a gene, the process is not a rapid one. The task of eradicating deleterious genes in a population by selection is by no means simple and it is necessary for the selecting influence to act for a long time.

The procedure outlined below illustrates the case just described in terms of the random selection of coloured balls when twenty per cent of those representing one genotype are removed between successive chance pairing operations. It should be noted that the number of balls in the containers are always kept high, but in the appropriate proportions of one colour to the other to simulate the effect of the selection, thus allowing for the fact that it is gametes carrying genes that are being represented, not the numbers of individual organisms.

Procedure to investigate the effect of selection using coloured balls to represent the gene pool

Assume the genotype *AA* (represented by red/red balls) to be selected out at the rate of 20 per cent at each generation and genotypes *AB* and *BB* (represented by red/yellow and yellow/yellow balls respectively) to be unaffected.

For the initial mating, represent the genes in the pool by equal numbers, keeping the numbers high (not less than 400 total) to represent the abundance of genes (gametes).

In containers representing the sexes of the parents place the balls in equal numbers as follows—female, 100 red + 100 yellow; male, 100 red + 100 yellow.

Draw the balls in pairs, one from each container, at random and record the pairings. These represent the genotypes in the first generation and should be roughly in the proportions—1 *AA* (red/red): 2 *AB* (red/yellow): 1 *BB* (yellow/yellow).

Simulate selection by removing as nearly as possible 20 per cent of the red balls in the pile representing the genotype *AA*, since this is the genotype being selected out.

Bring the total number of balls back to over 400 by adding balls in the proportions, red to yellow, that now exist between these colours.

In the containers representing the sexes, place equal numbers of red and yellow balls from the total balls now in use.

Draw in pairs as before and record the pairings. These represent the genotypes in the second generation (or the first after selection has been allowed to act). The proportions of genotypes will be different from those above.

Again allow selection to act by removing as nearly as possible 20 per cent of the red balls in the pile representing *AA* and bring the total number back to over 400 as before in terms of proportions, red to yellow.

Make another draw and record the genotypes.

Repeat for about ten generations.

Plot the proportions of all genotypes, *AA*, *AB*, *BB*, against generation and look for rates of change.

Selection clearly has its influence by changing the proportions of the genes in the gene pool. Consequently, the proportion of genotypes and hence phenotypes is changed at each generation; they do not remain constant as they do when no selection is operating.

A study of differences from the expected due to "accidents of sampling" in a small population, coupled with the above opposing systematic variations due to selection, etc., will involve the reader in statistical processes well beyond the scope of this book.

All the above relates to autosomal genes, not sex-linked, that is. In the special case of sex-linked genes, one sex (usually the male) has only one chromosome transmitted to half the gametes it produces while the other sex has the normal pair of chromosomes, equally distributed at gamete formation. As males inherit sex-linked genes only from their mothers while females inherit equally from both parents, the condition may be represented thus—

$$q_m = q'_f \quad \text{and} \quad q_f = \tfrac{1}{2}(q'_m + q'_f),$$

where q_m and q_f represent frequencies of a sex-linked gene in gametes of offspring males and females respectively and q'_m and q'_f represent the frequencies of the gene in the parent gametes. The effective gene frequency of the population is given by

$$q = \tfrac{1}{3}(q_m + 2q_f) = \tfrac{1}{3}(q'_m + 2q'_f) = q'$$

Although, therefore, the frequencies of sex-linked genes tend to remain constant from generation to generation in the population as a whole, the frequencies in males and females separately do not come to an equilibrium at once. There are fluctuations about the mean which tend to die out over a period. The same holds for the frequencies of the various genotypes in polyploid organisms.

From all the foregoing it should be clear that the characteristics which may be expected in a future generation are not determined merely by the features observable in the parents. They are determined by the nature of the complete gene pool of the population.

A SUMMARY OF THE MAIN PRINCIPLES OF GENETICS

The main principles of genetics appear to be universally applicable to the whole range of living things and the basic premises, or propositions from which inferences can be drawn, have been demonstrated many times in one of two ways—by breeding experiments and by observations on chromosomes during cell division, this being the event common to all growth, developmental and reproductive processes. The behaviour of chromosomes during cell divisions and the transmission of inheritable characters from parent to offspring are closely correlated and it is really the case that the chromosomes are not obeying the laws of inheritance, they are dictating them, with, of course, the exception of those cases known to be related to the distribution of cytoplasmic factors. Thus the two methods of study are complementary to one another and when the basic premises of genetics are tested in both ways, the results should always be seen to match. Evidence from breeding should be corroborated by cytological evidence.

One genetical premise is that the genotype will remain constant from generation to generation as long as sexual reproduction is not involved. This is demonstrated during the formation of a *clone*, a collection of individuals of a species produced vegetatively or in some other asexual manner, during which meiosis does not occur. Provided all the individuals of a clone are developed in uniform environmental conditions, the members cannot be phenotypically distinguished from one another or from their parents and their breeding behaviour is the same in all cases. The cytological evidence supports this in that it can be observed during mitosis that daughter nuclei are formed with identical chromosomes, also identical with those of the parent cell.

Another of the fundamentals is embodied in Mendel's first law of segregation, namely that a heterozygote produced by the union of two gametes differing in one respect only, will itself produce gametes equivalent to those of which it was formed and in the same one-to-one ratio. Mendel drew his conclusion indirectly by inferring from the relative frequency of the two types of offspring the natures and numbers of the gametes which produced them. Cytological evidence derived from a study of meiosis which precedes gamete formation matches the breeding results.

Mendel's second law of independent assortment of factors (determinants or genes) is now known to be true only for those carried on different chromosomes. This is a third basic premise. Observations on the behaviour of bivalents at the first meiotic division and the separate chromosomes at the second meiotic division show that they separate in random fashion to opposite poles of the spindle, thus if genes are on different chromosomes they will also recombine at random. If genes are on the same chromosome they should be linked in heredity, meaning that they can recombine but not at random. The fact that the number of linkage groups revealed in breeding experiments is equalled by the number of bivalents at the first meiotic division (or the haploid number of chromosomes) verifies this conclusion. Theoretical considerations concerning the recombination of genes on the same chromosome indicate that this can be at random only when an exchange of chromosome material occurs between two chromosomes and the probability of this happening is dependent on the distance between the loci of the genes in question. Breeding experiments rarely show complete coupling of all genes in a linkage group and the recombination frequencies between genes in the same linkage group have been used to map the order of genes on particular chromosomes. The genetical cross-overs demonstrated in breeding experiments have their cytological equivalents in the chiasmata visible during meiosis.

The more recent studies concerning the chemical basis of inheritance have produced a great deal of evidence to indicate that DNA is the genetic material or at least is very closely associated with it. Watson and Crick's molecular model, based on information gained in a number of ways (chemical analysis, electron micrographs, X-ray diffraction studies and physicochemical dimensions) describes the chemical structure of this substance. It is considered that it is the sequence of the two purines, adenine and guanine, and the two pyrimidines, thymine and cytosine, which embodies or "codes" the genetic message. It is considered to be the case that the two intertwined helices of sugar-phosphate are held together by hydrogen bonds between the purine and pyrimidine pairs at corresponding levels. This means that the double helix can be regular in its form only if a purine on one polynucleotide chain pairs with a pyrimidine on its partner. Two purines would be too large to bridge the fixed distance between the sugar-phosphate chains and two pyrimidines would be too small. All this is borne out in fact by chemical evidence which shows that in a DNA molecule the amount of adenine is balanced by the thymine and the guanine by the cytosine.

An essential property of any genetic material is that it must be capable of replication. This can be seen to have happened during cell divisions, and in terms of the DNA structure the replication process is considered to take place by the separation of the two polynucleotide chains due to breaks at the hydrogen bonds, each then serving as a template for the construction of its complementary chain. The new double chains are identical with one another and with the original double helix. This corresponds with what is seen to happen during mitosis and at the first prophase of meiosis.

Finally, there has for long been the concept that the inheritance material exists in units that have been called genes, each gene being considered responsible for the appearance of a particular inheritance character in the phenotype. Attempts to state precisely what such units are composed of are not straightforward undertakings since there are several criteria by which such units might be delimited from one another. When the different criteria are applied, it is seen that the units so delimited do not always come to the same thing. The smallest piece of genetic material that can be recombined with any other piece rarely, if ever, coincides with the smallest piece capable of mutation and there are often a number of separable and recombinable parts within the unit of material which is known to be in control of a particular biochemical function and, within each of these, more points at which the genetic substance can change its molecular structure. Until more is known about the coding of genetic messages in a much greater number of different organisms and how the code is translated and converted into cell activity in these, the concept of the gene as some kind of unit cannot be further clarified.

CHAPTER 20

LIVING THINGS THROUGH THE AGES

FROM our common experience, we all know that organisms live and die, each generation in turn giving rise to the next. For how long this has been happening on earth has been a matter of speculation since man first became interested in his own ancestry. Originally, those who established beliefs conceived a single creation of all living things within a time scale which they could comprehend, a few thousand years at the most. A more real conception of the time during which organisms have lived and died on this planet has only been formulated within the last hundred years or so. Reasonably accurate measurements of the age of the earth, the ages of the different rocks which make up its crust, and the dating of the remains of past floras and faunas which occur in them, have only been possible by application of more recent scientific discoveries and techniques.

Palaeontology, comprising knowledge of the ancient origins of living things, involves the study of their remains, called *fossils*, which are found on or in the earth's crust; *palaeobotany* and *palaeozoology* are subdivisions of it. For its fullest study, a knowledge of biology must be coupled with geology. First, some comprehension of the morphology and anatomy of modern living things is required, so that the fossils can be described and linked with modern representatives of the same or allied groups. Secondly, an understanding of the processes of fossilization is necessary for the proper comprehension of the nature and structure of a fossil, so that its story can be most fully read. Thirdly, there must be some appreciation of the geological time-scale, so that fossils may be dated. Lastly, it is necessary to understand the nature of the rocks in which fossils are found, so that there may be some conception of the condition of the earth in a particular period.

A comprehensive knowledge of palaeontology is unnecessary at this level, but some introduction to this fascinating subject is essential for the proper understanding of evolution.

THE AGE OF THE EARTH AND THE GREAT GEOLOGICAL TIME-SPANS

The earth has a crust of rocks, many thousands of feet thick in parts enclosing a molten centre. Most of the crust is hidden by water, which has filled the depressions in its corrugated surface. The crust was formed by the cooling of molten surface materials, and the oceans and seas by precipitation of condensed water vapour formed in the atmosphere as the intense heat diminished.

There was no real appreciation of the vast extent of geological time until the eighteenth century when James Hutton (1726–97), a Scottish geologist, recorded observations which led to a modern acceptable evaluation of the facts. In his *Theory of the Earth*, he advanced the view that much of the earth's outer crust had been formed by disintegration of the original rocks by erosion, followed by deposition of small particles under the sea and then consolidation into new or secondary rocks by the enormous pressures prevalent on the sea-bed. Convulsions beneath the crust caused elevation of some parts and depression of others, and those parts which were thrust above the surface were again subjected to erosion, followed by deposition and consolidation into new rocks.

Hutton postulated that by repetition of these processes, the crust of the earth had had a changing history far older in time than any previous conception. His followers based their time estimates on the incredible slowness of the eroding, sedimenting and consolidating processes. At the present time, geologists estimate that there have been at least ten cycles of major importance involving the same sequence of events. Each such sequence has taken many millions of years to complete.

There have been numerous attempts to measure the age of the earth. One of the first to be based on scientific principles, was due to Joly, who in 1898, reviving an older suggestion, tried to fix the age of the oceans by measuring the sodium content of sea-water, assuming that it had increased quite regularly over the period since the oceans first came into being. He found that the total amount of sodium in the sea could not have been built up in a period of less than 80–90 million years. It is now understood that his major premise, i.e. the regular rate of sodium deposition, was incorrect and his time estimate far too small. With the discovery of radioactivity (*see* p. 944), a much less irregular chronometer became available. The valid use of such a time-measuring device depends on the regularity of the rate of decay of radioactive elements into others; the rate, for instance, at which uranium and thorium are converted by their own radioactivity into helium and lead. From a knowledge of the rates at which such substances decay, and by measuring the amount of lead or helium produced in rocks by such radioactive decay of the parental elements, it has been possible to calculate the age of any rock in which a radioactive element occurs. Use of such a method of measurement is justifiable since it fulfils the necessary requirements of accuracy. It is based upon a natural process which must have gone on throughout the whole history of the earth; the present rate of the process can be accurately measured and it can be shown that there has been no variation in the past.

By the use of methods based on the radioactive properties of uranium and thorium, modern estimates of the age of the earth have been given as between three and five thousand million years. To a man, whose life-span is reckoned in tens of years, the idea is somewhat bewildering. The record of life on earth does not extend throughout the whole of this time, for the earliest clearly recognizable fossils can be dated between 500 and 1000 million years ago. The greatest wealth of fossil specimens occurs in relatively recent rock formations.

As geologists found more and more fossils, they became aware that the rocky strata which bore them were distinctly characterized by types of fossils. Other strata in the neighbourhood, above or below that under investigation, showed each its own forms of ancient plant and animal life. Gradually, it became recognized that the fossils were so characteristic of the various rocks that they could be used to identify the strata, wherever they might occur; that rocks bearing identical fossil forms must have been laid down at the same time and in continuity with one another. The classification of rocks having the same origin, and the charting of the changes in the outline of land masses with the ages in terms of the fossils, became a possibility. This is a branch of geology known as *stratigraphy*. When a certain group of fossils in one part of the earth's surface can be found in exact replica in other parts, it is safe to say that the rocks which bear them were once all part of the same sea-bed or land mass on which the organisms lived. For example, the agreement between the fossils found in the Kent coalfields and those in coal seams of the continent, is so great that it is fair to assume that England was once part of the larger European land mass. Similarly, the occurrence of remains of marine organisms in rocks far inland indicates that whole areas which are now land were once sea-covered.

In addition to using the fossils as evidence of relationships between rocks, geologists, assuming evolution as a fact, worked on the principle that the deeper below the surface a fossil-bearing rock is found, the earlier in the earth's history was it formed and thus the more ancient are its fossils. In an analysis of the results of such stratigraphical studies, geologists found it convenient to divide the whole series of rocky strata deposited in the past according to the forms of life which existed when they were formed. The larger geological time-spans, the *eras*, have been named on the basis of the history of life in the past.

Five eras have been delimited, each of the three most recent being further divided into *periods* according to the rock formations which originated during their time. The earliest era is sometimes known as the *Azoic* or *Archaeozoic* era, meaning the era of life's beginning. It includes all the time between the beginning of rock formation and about 1100 million years ago. No fossils have ever been found in the rocks dating from those times. Following this is the *Proterozoic era* or the era of early life. Very few fossils are found dating back to this time, which ended somewhere in the region of 500 million years ago. Then comes the *Palaeozoic era* or era of ancient life, in which fossils are much more plentiful. The era is divided into six main periods named according to the nature of the fossil-bearing rocks which were formed within its time span. The earliest period, lasting about 100 million years, is the *Cambrian*, so-named because rocks typical of the age are found in Wales. It ended approximately 400 million years ago. The *Ordovician* period, named from a Celtic tribe (Ordovices) who once lived where some of these rocks are found, lasted about 70 million years. Then is recognized as the *Silurian* period, also named from a Celtic tribe, of about 30 million years duration. The next period of the era, named *Devonian* because of the occurrence of rocks of the time in Devon, extended over 45 million years and passed into the *Carboniferous* or coal-producing period which lasted about 60 million years. The *Permian* period, named from the province of Perm in the Ural mountains, extended over a period of about 25 million years to close the era somewhere near 200 million years ago. The era lasted for more than 300 million years and its rocks contain many fossils of the early forms of life both in the sea and on land. The first three eras so far named are sometimes collectively referred to as the *Primary era*, and any rock formations occurring in times earlier than the Palaeozoic era are called *pre-Cambrian*.

The next more recent large time-span is the *Mesozoic era*, or the era of middle life-forms. It lasted about 120 million years and is subdivided into three periods. The oldest, the *Triassic*, named from a rock system known as Trias because of its three-fold division, lasted about 25 million years. The middle period, the *Jurassic*, named from the Jura mountains, lasted about 30 million years, and the era ended with the *Cretaceous*, or chalky period, of about 70 million years' duration. The era is sometimes called the *Secondary era*.

The next more recent era is the *Cenozoic* or time of recent life. It is often referred to as the *Tertiary era*. According to the relative numbers of fossils representing the more recent life forms which are found in its rocks, it is divided into four periods. The oldest is the *Eocene*, or dawn of recent life, lasting about 25 million years. Next, lasting about 10 million years, comes the *Oligocene* or period of few recent life forms, followed by the *Miocene* of about 20 million years and the *Pliocene* of about 15 million years, each named in terms of

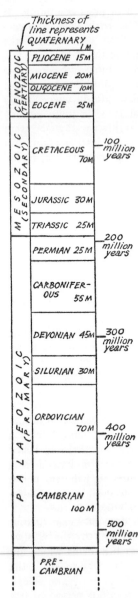

Thickness of
line represents
QUATERNARY 1M

CENOZOIC (TERTIARY)	PLIOCENE 15M	
	MIOCENE 20M	
	OLIGOCENE 10M	
	EOCENE 25M	
MESOZOIC (SECONDARY)	CRETACEOUS 70M	100 million years
	JURASSIC 30M	
	TRIASSIC 25M	200 million years
PALAEOZOIC (PRIMARY)	PERMIAN 25M	
	CARBONIFEROUS 55M	
	DEVONIAN 45M	300 million years
	SILURIAN 30M	
	ORDOVICIAN 70M	400 million years
	CAMBRIAN 100M	500 million years
	PRE-CAMBRIAN	

Fig. 20.1. Geological time scale.

their increasing yield of more recent life forms. The Cenozoic era ended about one million years ago and gave place to the fifth main time-span known as the *Quaternary*. This is divided into two periods, the *Pleistocene* and the *Holocene*. The former means literally, most recent, and it included the Great Ice age; the latter means wholly recent, and it includes the post-glacial period up to modern times.

It is interesting to note that man's presence upon the earth, as far as his fossil remains tell us, can be reckoned in a fraction of a million years.

A geological time scale indicating the sequence and duration of the great eras and periods is shown in Fig. 20.1.

FOSSILS

The term "fossil" originally included any unusual object unearthed by digging, but nowadays it is applied only to the preserved remains of living things or to objects which can be used as evidence of their former existence. Fossils may range in form from whole, perfectly preserved bodies with tissues intact, to mere imprints of plant and animal parts in hardened rock.

The process of fossilization can be any process which preserves in some form the original structure or evidence of its passing. Generally it is one which has prevented or delayed decomposition of the original organic matter or its secretions for a period long enough to allow its place to be taken by some hard and durable material. We may summarize and exemplify the forms which fossils take and the processes involved as follows.

Preservation of undecayed organisms in a more or less intact state sometimes occurs. It can be achieved only under conditions which exclude bacterial and fungal action completely, and under conditions where the organic matter cannot be dissolved away. Imprisonment in ice gives the most complete preserved specimens, of which the woolly rhinoceroses and mammoths dug out of the Siberian ice are the best examples. Such specimens in which even the softer tissues are not decomposed are very rare and do not date back much beyond 25 000 years. Nearly perfect preservation, at least of the hard parts of whole organisms, is achieved by their imprisonment in globules of durable substance such as amber. The amber found in the Oligocene rocks of the Baltic coast is the hardened resin of coniferous trees. Insects which were trapped in the resin during its formation, are preserved in a state in which the minutest details of the exoskeleton can be made out. Small parts of plants, such as flowers and leaves, have retained their external form equally well over millions of years. Other natural preservatives are oily and tarry substances. The "asphalt lakes" of California frequently yield teeth and bones in an almost unchanged condition. Softer parts of bodies have not survived so well. Wolves, sabre-toothed tigers and their unfortunate prey are amongst the better known fossils from that source. Very acid bog water may prevent decay, at least of harder parts, and in such conditions, woody parts of plants and skeletons of animals a few thousand years old can be found in the peaty areas of Northern Europe. Perfectly preserved antlers of the comparatively recent Irish Elk illustrate this, whilst the peat itself may be considered as fossilized plant remains some thousands of years old. Mummified remains of organisms of comparatively recent times can sometimes be found in places of extreme dryness.

Impressions or prints of plant and animal remains, both hard and soft, are frequently found on rock surfaces. The organic material has completely disappeared, leaving only an outline of its shape. The best impressions are made in rocks in which the grain of the sediment, from which they were formed, is of the finest. The outlines of leaves and the intricate details of the venation are found in a wide variety of rocks dating back many millions of years. Perhaps the most outstanding of impressions or traces of soft organisms are those of jellyfish of the Cambrian period found in rocks of British Columbia. Sometimes an impression may be coated with a thin black carbonaceous film which represents all that is left of the organic matter after the more volatile substances have disappeared. In the Carboniferous rocks, fossils of this kind are in abundance, and indeed, coal itself is made up of the carbonized remains of countless millions of plants compressed into seams by colossal pressure from above.

Among the more frequent of fossils are those produced as moulds or casts of the original structures. Again, fineness of detail depends upon the grain of the rocky substance of which it is formed. If an organism became buried in clay, sand or ooze, which became compacted before appreciable decay of the body set in, and then slowly the organic matter became dissolved out of the compacted mass, it would leave behind a space or mould exactly in the form of the original. If later, the hollow mould became filled with mineral particles which became hardened under pressure, a perfect cast would be made inside the natural mould. Many such moulds and casts exist in a wide range of rocks. Fossil plants of numerous kinds exist as casts, particularly from the Carboniferous period. The casts of giant horsetails (calamites) are usually those of natural cavities formed within the trunk. When the trunk fell, if not already hollow, its pith rapidly decomposed leaving a space, whilst the outer rind was more resistant. Inside this pith cavity, sediment collected and gradually hardened, thus carrying on its outer surface an exact impression in reverse, of the inside of the rind. Flattening of the trunk by pressure during the formation of the cast could cause much distortion. Many of the fossil shells of invertebrate animals are similarly castings of moulds formed by the empty shells. An internal cast of a shell is produced by the filling up of the cavity formed when the soft body of the animal was decomposed, the space acting as a mould. Casts of molluscs often show detail of the muscle attachments which controlled the opening and closing of the shell when the animal was alive. Casts of the shells themselves are produced when the material of the shell is later replaced by other sediment settling in the space once occupied by the shell material. The outside surface of such a cast shows precise detail of the external form of the animal. On rare occasions, when sediment has filled the alimentary canal of an animal, an excellent mould of this structure is produced.

Petrified fossils remain where organisms have literally been turned into stone. Petrifaction can occur where there is abundance of mineral-containing water, and the process is one of deposition of mineral matter such as silica, pyrites or calcium carbonate to replace the organic molecules washed out by the water. By such means an exact stony replica of the original structure may be formed, complete with detail of tissue construction. The "coal balls" found in English coal fields contain specimens preserved in this way and from them microscope preparations can be made. In these, the anatomical details of 250-million-year-old plants can be studied. The petrified forests to be found in various parts of the world provide examples of the same phenomenon. Among the more famous plant remains of the petrified kind are those of very primitive land plants of the Devonian period. These were found in the chert beds near Rhynie in Scotland. They represent practically the whole of our knowledge of early plant life on land, but in some cases the petrifaction is so perfect that we can see the detail of their internal anatomy. Petrifaction of animal remains is not so common but there are some notable examples. Sponges and corals in silicified form show exact detail of external structure. The skeletal parts of the extinct graptolites have been replaced by iron sulphide, pyrites, so that they can be found in almost perfect form.

As has been pointed out, a fossil need not represent any part of an organism; it can be anything which substantiates its existence. Among such fossils are included impressions in soft mud of foot-prints and trails left by limbless creeping animals. Footprints of the reptiles of the Triassic period are not uncommon. They yield a good deal of information about the size of the animal concerned and the nature of its gait. Many worms and molluscs made burrows in the soft mud or sand, and casts of these were made when they filled with sediment which eventually hardened. Coprolites are fossilized faecal pellets. They can indicate sometimes the feeding habits of the animals concerned, particularly when they are found to contain undigested matter such as scales and teeth.

It is quite obvious that of all the millions of organisms which have existed, only an infinitesimally small number have been preserved in any way at all, and of these only a small proportion have ever been unearthed. In general, it is only the more robust of living things which have persisted long enough after death to become fossilized. Very tiny and soft plants and animals could not possibly have made good material for any of the fossilization processes to work upon. Despite all this, palaeontologists have managed to piece together much of the jig-saw pattern of life through the ages. Their success is due partly to their own skill in finding, preparing and reading the fossil story, and partly to the fact that enormous quantities of material are not always essential. One well-preserved specimen will tell the story of all the others of its kind. Nevertheless, there are still gaps to be filled and the search for new material goes on incessantly. Most of the fossils which have been found can be classified within the major classes or orders of existing plants and animals. Their resemblance is close enough to warrant this and at the same time is direct evidence of the descent of modern forms from ancient ancestors (see Chap. 21). On the other hand, some fossils have no modern counterparts, thus indicating the extinction of a line, due possibly to the advent of new environmental conditions to which the representatives of that line were not adapted. There is no doubt that the geography of the earth and its climatic conditions have undergone periodic variations sufficiently drastic to cause such extinction. Detailed studies of the fossil floras and faunas tell us much about the climatic conditions under which they existed and aid in the composition of the story of the earth's history. In the next section, a brief account of the succession of life forms through this long history is given.

Fossils are named in Latin and there is sometimes an apparent confusion of names because parts of the same organism are given different titles. This is not quite as stupid as it appears to be at first sight, since it must be remembered that in most cases whole organisms are rarely dug up, but mostly isolated parts. When these parts are first found, if there is no evidence for connecting them together, then they must be described and named individually. Thus it has come about, for example, that parts of a Carboniferous lycopod now understood to belong to the same species, have received three different names, *Lepidodendron* for trunks and leaves, *Stigmaria* for roots and *Lepidostrobus* for reproductive cones.

THE SUCCESSION OF LIFE THROUGH THE AGES

There is no indication of the origin of life in the fossil record. The first living things could not have had forms comparable with the majority of those existing today since the environment could not have supported them. There is evidence that in the beginning the earth's atmosphere was almost lacking in oxygen, so that the earliest recognizable living things could only have been comparable to the few micro-organisms of the present day which can manage to exist in its absence. The origin and nature of the earliest forms of life are still mysteries and they may always remain so. We have found no fossils older than about 500 million years and thus there is a gap of at least 2500 million years in the history of the earth. About this vast span of time we have practically no knowledge relating to living organisms.

The fossil record commences in the Proterozoic era, that is, in pre-Cambrian times (before 500 million years ago), but fossils are rare and point only to the existence of soft-bodied organisms which could have lived nowhere other than in watery surroundings. The rocks of the succeeding eras yield fossils in ever-increasing abundance, telling in outline the story of the gradual advance in complexity of the aquatic forms and the slow but sure conquest of the land by both plants and animals.

Pre-Cambrian Eras (Azoic and Proterozoic)

In the rocks of the early part of this time there are no fossils at all, and even towards its close, the evidence for life is still very scanty. There are good reasons for believing that any living things which may have existed at the beginning were only very minute structures, or if macroscopic, they were of an extremely delicate nature. Such forms are not readily fossilized. However, the presence of the crystalline form of carbon, graphite, dating back to about 1000 million years ago, could indicate the presence of early forms of aquatic life.

Direct evidence of distinct plants and animals is found only in the later rocks of the Proterozoic era. These were formed at the bottom of seas, and in them, occasional relics of plant members, possibly allied to the present algal forms, have been unearthed. They take the form of calcareous nodules such as might have been formed by algae living in colonial association. Similar structures are reputed to be of bacterial origin. No indication of plant life on land during this era has been found.

Evidence of animal life is no less sparse. The more recent pre-Cambrian rocks do contain occasional imprints of coelenterates, brachiopods (lamp-shells) and worm-like creatures, and these, coupled with more abundant fossils of similar kind in the Cambrian period, indicate an animal population of the seas of varied nature and some degree of organization. The fossil, *Xenusion*, of the late pre-Cambrian rocks of Sweden, has been interpreted as an intermediate between a worm and an arthropod.

From the fact that the seas supported a reasonable variety of invertebrate animals, it follows that there must have been an equivalent variety of sea plants to support them.

The Succession of Post-Proterozoic Floras

Palaeozoic Era

During this great division of time, there is an abundance of evidence of an enormous increase in complexity of all forms of life and in the range of conditions which could support their existence. Plants became well established on land and most of the major groups of present forms came into being.

Simple plants are clearly traceable from the Cambrian period. The earliest are of doubtful form but have been allied to the blue-green algae. Later forms, of the Ordovician, Carboniferous and Permian periods, may be early ancestors of the Chlorophyceae. One, *Dimorphosiphon*, very closely resembles a member of the modern Siphonales, *Codium*. Bituminous shales indicate the presence of oil-storing algae in the Carboniferous period, and at the same time, members of the red algae are known to have lived. The fossil, *Nematophyton*, of the Silurian and Devonian rocks, is very similar to the stipe of a seaweed. It measures over half a metre in diameter, indicating considerable size of the whole plant. The fossil, *Algites*, found in a Devonian swamp, is considered to be a close relative of the modern stone-wort, *Chara*. In the later rocks of the era, there is evidence that fungi allied to the modern Phycomycetes were living either as parasites or saprophytes in the tissues of other plants. The record of bryophytes is very sparse. Their delicate structure is not easily preserved. However, liverwort and moss-like forms have been unearthed in the Upper Carboniferous rocks. Their form is very comparable to that of modern representatives and seems to have changed little in 200 million years.

Fossils of vascular plants inhabiting the land are found in ever increasing numbers from the middle of the Silurian period. Among the earliest and most famous of them all are the specimens found in the Rhynie chert beds of Scotland. They have no modern counterparts and are placed in the fossil order Psilophytales of the Pteridophyta.

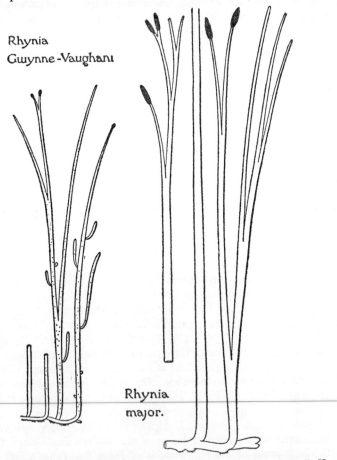

Rhynia Gwynne-Vaughani

Rhynia major.

Fig. 20.2. Restorations of Rhynie plants; about half natural size. (From D. H. Scott's *Extinct Plants and Problems of Evolution* (Macmillan & Co. Ltd.).)

They became extinct towards the end of the Devonian period. Three genera from Scotland have been named, *Rhynia*, *Asteroxylon* and *Hornea*. They were very simple plants, showing little sign of organization into root, stem and leaf, but each possessed a vascular cylinder made up of clearly defined xylem and phloem elements. Reproductive structures, which may be called sporangia, occurred at the ends of some of the erect branches (*see* Fig. 20.2).

Other pteridophytes, some of tree-form, were abundant in the later

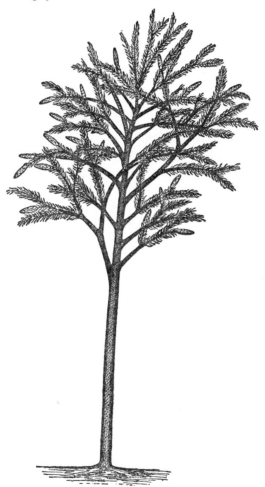

Fig. 20.3. Restoration of *Lepidodendron elegans*, a large tree bearing cones. (From D. H. Scott's *Studies in Fossil Botany* (A. & C. Black, Ltd.).)

Palaeozoic. We must imagine enormous forests of them growing in the warm, wet atmosphere of the Devonian and Carboniferous periods. Some had characters in common with the smaller lycopods of present times. One of the commonest of these was *Lepidodendron* (*see* Fig. 20.3). This was a tall tree of some 30 m, surmounted by a tuft of branches bearing scale-like leaves which, when they fell, left characteristic scars indicating the presence of a ligule and parichnos strands (*see* Fig. 20.4). The trunk was anchored to the soil by a dichotomously

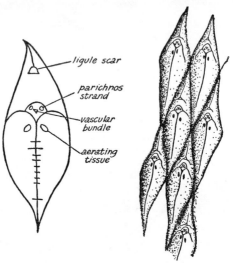

Fig. 20.4. Leaf bases of *Lepidodendron*

branched system of root-like structures (fossil name—*Stigmaria*). The aerial branches bore cones (called *Lepidostrobus*) which were similar to those of a modern *Selaginella* in that they were heterosporous. Many other similar forms of lycopods are known to have existed and they ranged over very large areas of the existing land. Certain of them bore reproductive structures approaching seed-form. One such fossil, *Lepidocarpon*, is found as isolated cones and sporophylls which have attached megasporangia, each containing only a single megaspore. This was enclosed, when mature, by outgrowths of the sporophyll, except for a narrow opening along the apex which served as a micropyle (*see* Fig. 20.5). It cannot be considered as a true seed, however, since it did not develop prior to shedding from the parent plant. These ancient, large lycopods all started to die out in Permian times and were extinct early in the Mesozoic era. The genera *Lycopodium* and *Selaginella* are small extant representatives of the group. Similar small fossilized forms can be traced back to the times when their giant relatives flourished.

Other pteridophytes which flourished alongside the large lycopods were the giant horsetails. Fossils of the vegetative parts of these plants are placed in the genus *Calamites* (*see* Fig. 20.6). During the Carboniferous period, these plants reached heights of over 30 m and

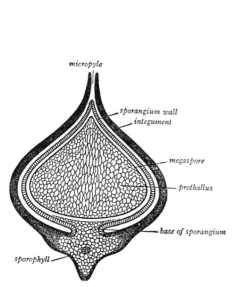

Fig. 20.5. Diagrammatic section of seed of *Lepidocarpon lomaxi*. (From D. H. Scott's *Studies in Fossil Botany* (A. & C. Black, Ltd.).)

Fig. 20.6. Carboniferous horsetail; a small *Calamites* abundant in the coal-measures of Europe and North America.

(By permission of the British Museum, Natural History.)

diameters up to 1 m. They developed from rhizomes of similar dimensions, and the stems had a jointed appearance with prominent ribs alternating in position from joint to joint. An exactly similar condition is found in present horsetails of the genus *Equisetum*. The hollow trunks of these ancient plants produced many excellent fossils in the form of pith casts (*see* Fig. 20.7). Fossils of the reproductive cones show that most of the representatives of this class of plant were homosporous. Other fossil genera of similar nature to the calamites are quite common in rocks ranging from Devonian to Permian times. They include *Sphenophyllum*, *Calamophyton* and *Hyenia*.

Among these larger forms, many other pteridophytes flourished, the ferns in particular. The oldest known fern fossil seems to be that of *Iridopteris* from the Devonian rocks, but in those times ferns were not plentiful. They became among the most abundant of spore-bearing plants in the Lower Carboniferous period and their descendants have continued to exist up to the present day. Most of the fossils are impressions of stems and leaves, which show that in many forms, branching of the frond was in several planes, giving the plant a bushy appearance. Most modern ferns show fronds branched in one plane only, with

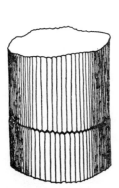

Fig. 20.7. Piece of *Calamites* pith cast.

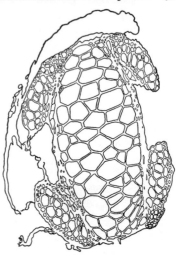

Fig. 20.8. T.S. xylem of *Etapteris* showing anatomical detail.

a characteristic flatness. Some ferns of the Upper Carboniferous and Permian beds were tree-ferns. They occur as impressions or casts of large stems named *Caulopteris* and *Megaphyton*. A few comparable forms exist today, notably *Dicksonia*. Petrified remains of ferns indicate much of their internal anatomical structure right down to detail of tissues. Among the better known fossil genera of this kind are *Ankyropteris*, *Botryopteris* and *Etapteris* (*see* Fig. 20.8). Fronds bearing sori of sporangia are not uncommon and indicate that the manner of spore production was similar to that of modern ferns. Among the best fertile structures preserved are the fronds of the fern *Asterotheca*.

The true seed-bearing plants, the spermatophytes, also appeared in the Palaeozoic era. None of them were flowering plants (angiosperms) but primitive gymnosperms were quite common. Side by side with the pteridophytes lived the pteridosperms or seed-ferns. They are classified in the gymnosperm order Cycadofilicales [Pteridospermae]. All are now extinct. The pteridosperms and ferns were much alike vegetatively and unless the fronds bear either sori of sporangia or true seeds, it is very difficult to distinguish between them. Impressions of such plants as *Pecopteris* were for a long time considered to be ferns and it was not until seeds were found in association with the fronds that their true nature was understood. There are many such impressions

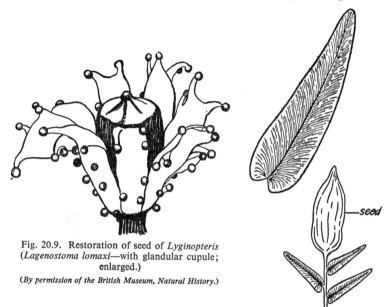

Fig. 20.9. Restoration of seed of *Lyginopteris* (*Lagenostoma lomaxi*—with glandular cupule; enlarged.)

(*By permission of the British Museum, Natural History.*)

Fig. 20.10. *Neuropteris* pinna and seed.

of fernlike plants which have never been satisfactorily assigned to their proper groups. Among the most completely known pteridosperms is *Lyginopteris oldhamia* of the Upper Carboniferous period. It is common in the coal-bearing strata around Oldham in Lancashire. Its seed was enclosed in a lobed husk bearing easily identifiable glandular hairs which occur also on the leaves (*see* Fig. 20.9). Other moderately well-known pteridosperms of similar times include *Neuropteris* (*see* Fig. 20.10) and *Alethopteris*, which are vegetative structures, whilst *Potoniea* and *Trigonocarpus* are their respective reproductive parts. The widespread appearance of the seed-ferns in Palaeozoic times indicates clearly that the seed habit was a comparatively early evolutionary tendency in the history of plants.

The fossil order Cordaitales is also made up of primitive gymnosperm plants, none of which survive today. The best known is the form *Cordaites* from the Upper Carboniferous rocks (*see* Fig. 20.11). It was a lofty tree, branched only near the apex, and its stem was not unlike the modern monkey-puzzle, *Araucaria*, in structure. The leaves were large, linear and parallel-veined. The seeds were developed in cones, but not on cone-scales as in the modern conifers. Pollen grains

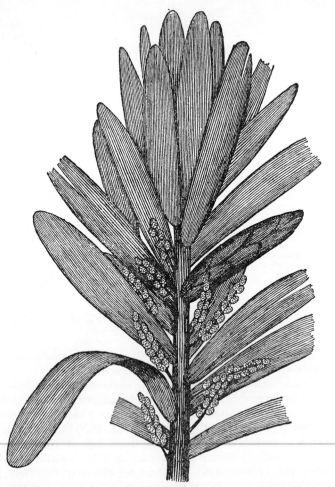

Fig. 20.11. Restored branch of *Cordaites laevis* with female inflorescences and a large bud; reduced. (From D. H. Scott's *Studies in Fossil Botany* (A. & C. Black, Ltd.).)

were produced in microsporangia attached to long, delicate stems arising from between small scale leaves.

Three other gymnosperm fossil forms are found in Palaeozoic rocks. They are the fossils referred to the orders Bennettitales, Ginkgoales and Coniferales. All are only scantily represented. The first assumed greater importance in the next era before finally dying out. The other two also became more abundant and have not yet disappeared from among the earth's flora. The maiden-hair tree, *Ginkgo biloba*, is the only remaining species of the order to which it lends its name, whilst our modern conifers, of greater variety and abundance, still hold a significant place in the earth's vegetation. Representatives of these three orders will be described later under the periods in which they assumed greater abundance.

Of great importance to the ordinary man are the plants which flourished in the Carboniferous period. Coal is a comparatively modern discovery, but it was from the plants growing in the forest swamps upwards of 200 million years ago that it was formed. As the generations of plants shed their parts and finally died and fell into the water, they built up great thicknesses of peat. These were compressed and changed into bituminous coal as subsidence of the ground caused deep flooding and sedimentation above them. The seams of coal at different levels represent repeated alternating peat-forming and coal-forming changes in the same area. A seam of coal in one locality can be identified with another, many miles away, solely by comparison of the fossil remains which it yields. Not least among these are the spores and pollen grains of the pteridophytes and early seed-bearing plants. No happier hunting ground for fossils is to be found than the waste tips in the neighbourhood of coal mines. Coal formation is probably still going on in some parts of the earth. In the Ganges delta, there are being built up beds of peat formed of tropical vegetation alternating with sediments as the delta is periodically flooded. In the distant future, the same vegetation may be unearthed as coal. It is interesting to note that all the energy released when coal is burnt originated as light energy from the sun about 250 million years ago. It was absorbed by green plants and used in the construction of their body components which, in the fossilized state, can be made to produce the power to keep man's machinery working or alternatively yield useful substances ranging from dyes to aspirin.

Mesozoic Era

Towards the end of the Palaeozoic and continued into the early Mesozoic era, there seems to have been a time when there were great changes in geographical and climatic conditions. Gradually, the environment became unfavourable to the continued expansion of many of the plant forms making up the luxurious vegetation of the Carboniferous and Permian periods. Indeed many of them ceased to exist early in the Mesozoic era; the Pteridospermae and Cordaitales were among them. But from the Palaeozoic forms, new types arose and many of them are represented in one form or another at the present time. The vegetation began to look more like that of today by contrast with the ancient types which constituted the earlier land floras.

Algal forms continued to flourish in the seas, of course, and rocks of the Triassic period have yielded abundant evidence of calcareous seaweeds of a type now common in tropical waters. The fossil *Diplospora* has a stalk several centimetres long, bearing clusters of small branches. A group of the red algae, the Corallinaceae, is represented in marine deposits of the Cretaceous period. One genus, *Lithothamnion*, still has living representatives, indicating a continuous history of over 70 million years. Fossils of the stone-wort, *Chara*, are known from the same period. Fossil diatoms begin to appear in the Jurassic period.

As is the case in all the rocks so far investigated, fossil bryophytes are very poorly represented, but it has been pointed out that this does not necessarily mean that they were rare. One Jurassic fossil impression, *Marchantites*, shows a flattened, forking thallus, possessing a distinct midrib, not unlike that of the modern genus *Marchantia*, from which it was named.

During the Mesozoic era, considerable changes in the pteridophyte flora occurred. The large lepidodendroids began to lose their dominance and most became extinct. One or two forms survived into the Triassic period. Among them are *Lycostrobus*, represented by cones and spores similar to those of *Lepidodendron*, and *Pleuromeia*, an unbranched shrubby plant with needle-like leaves, and rootlets growing from a base similar in character to the much smaller modern quill-wort, *Isoetes*. Fossils of small lycopods, named *Lycopodites* and *Selaginellites*, have been found in Mesozoic rocks and are considered by some to be the ancestors of the modern genera *Lycopodium* and *Selaginella*.

Although the form of the horsetails changed somewhat with time, they did not recede as completely as did the lepidodendrons. The giant calamites are not much in evidence, but their place was taken by similar smaller plants. The fossil-form *Schizoneura*, which first appears in Carboniferous–Permian time, seems to have been widespread in the Triassic period. It and its close relative *Neocalamites* seem to have arisen from the Palaeozoic calamites. The nearest approach to the present genus *Equisetum* was the fossil form *Equisetites* which seems to be different only by being rather larger.

The ferns continued to be among the dominant forms of the flora well into Jurassic and Cretaceous times, but unlike their earlier counterparts, they had more of the characteristics of the modern fern (*see* Fig. 20.12). Indeed, many of them are still represented in the earth's vegetation. The fossil genus *Marattiopsis* closely resembles the modern *Marattia*. Other past and present forms which can be similarly closely coupled are *Laccopteris* and *Matonia*, *Dictyophyllum* and *Cheiropleura*, *Osmundites* and *Osmunda* (the royal fern), *Todites* and *Todea* and

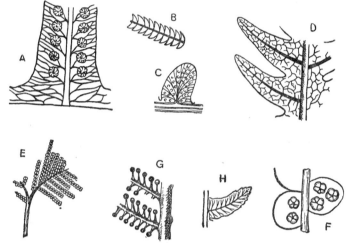

Fig. 20.12. Mesozoic ferns.

A. Laccopteris polypodioides; B. Todites williamsoni; C. Todites williamsoni, fertile, frond; *D. Dictyophyllum nilssoni; E. Gleichenites delicatula; F. Gleichenites gracilis*, fertile; *G. Coniopteris arguta*, fertile; *H. Cladophlebis denticulata*. *A.C.F.* enlarged; *B.D.E.* reduced; *G.H.* natural size. (After Seward and Nathorst.) (*By permission of the British Museum, Natural History.*)

Gleichenites and *Gleichenia*. Tree-ferns also continued to exist and one form found in the Cretaceous rocks, *Protopteris*, is very similar to the modern tree-fern, *Dicksonia*.

Whilst many of these primitive pteridophytes were dying out, the dominant position in the land flora was being taken by seed-bearing plants, some of which were already in existence during the Palaeozoic era. Of these, the pteridosperms and cordaitalean forms reached their zenith in early Mesozoic time and passed into extinction. Their place was taken by the expansion of some of the old gymnosperm seed-bearers and by some entirely new forms. Among the gymnosperms represented earlier is the fossil order Bennettitales, allied to the modern Cycadales. Representatives have been found in all the Mesozoic strata. *Williamsonia*, of the Jurassic rocks of Yorkshire, is one of the most completely preserved of the earlier forms. It was a stumpy plant bearing long narrow leaves in a tuft at the crown. The reproductive parts may be termed flowers. They were formed at the ends of shoots covered with overlapping scales, and consisted of an axis with whorls of organs bearing ovules or pollen sacs enclosed by bracts, which may have been protective only or, alternatively, coloured and attractive. They

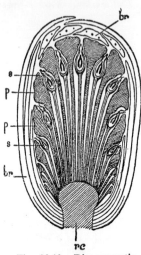

Fig. 20.13. Diagrammatic longitudinal section of cone of *Cycadeoidea* [*Bennettites*] *gibsoniana*; *br.* bracts; *s.* seeds; *p.* interseminal scales; *rc.* receptacle. (From D. H. Scott's *Studies in Fossil Botany* (A. & C. Black, Ltd.).)

Fig. 20.14. Sectional restoration of *Cycadeoidea* flower showing hairy bracts, compound stamens, and central seed-bearing cone. (After Wieland.)

(*Redrawn, by permission of the British Museum, Natural History.*)

were not comparable with angiosperm flowers, however, since the ovules were not enclosed in a carpel. Such flowers were unisexual, but a comparable form, *Williamsoniella*, possessed hermaphrodite structures. This particular flower is probably the nearest approach to the form of an angiosperm flower yet discovered. Also occurring in Jurassic rocks are the remains of the bennettitalean form *Cycadeoidea* [*Bennettites*]. This was another cycadean plant bearing flower-like reproductive parts. A specimen named *Cycadeoidea gibsoniana*, shows the characteristic cycad vegetative form, bearing flowers of very striking pattern. The axis terminated in seeds borne on long stalks between which were sterile scales whose ends were enlarged over the seeds to leave only a small opening above each one (*see* Fig. 20.13). The whole structure was enclosed by sterile protective bracts. There were no male structures in this flower but similar species found in America show the presence of much-branched stamens inserted below the receptacle. Such a reproductive structure, known as the Bennettitalean or Cycadeoidean flower (*see* Fig. 20.14), has for long been of great interest to botanists who see in it the early beginnings of the angiosperm flower form.

Bennettitalean plants came to a sudden end at the close of the Mesozoic era and there are no modern representatives, but alongside them, arising early in the era, are many remains of plants which agree closely in structure with the modern order Cycadales. The vegetative form of these orders was very closely similar, so much so that the two are often referred to collectively as cycadophytes. The Mesozoic cycads are represented among the fossils chiefly as vegetative parts. There are no sure indications that these bore reproductive cones similar in nature to the modern cycads, but some detached sporophylls bearing megasporangia make it seem probable that they did. The fact that cycads very similar to the modern forms were plentiful in early Cenozoic times, points to the probability that they were already abundant in the later periods of the Mesozoic era.

The other gymnosperm orders which had beginnings in the Palaeozoic era, namely, the Ginkgoales and the Coniferales, must be mentioned again. The former order is most abundantly represented in the Upper Triassic floras. The genera *Baiera* and *Ginkgoites* were among the commonest. The living genus *Ginkgo* first appeared in Jurassic rocks, and one of its species, *G. biloba*, has survived to the present day. Members of the Coniferales were also abundant and widely distributed in Mesozoic periods. One genus, *Voltzia*, already present in the Palaeozoic, is found in late Triassic rocks. It may have been the ancestral stock of the fossil genus *Araucarites*, which itself apparently gave rise to the modern genus *Araucaria* (monkey-puzzle). During the Cretaceous period, there arose many forms closely similar to modern gymnosperms; they include the giant red-woods, pines, firs and yews, so that these can be said to date from that time.

Whilst much of the Palaeozoic flora was dying out and the typical Mesozoic flora was expanding, there occurred an event which has had the greatest influence of all on the form of modern vegetation. Early in the Mesozoic era, the angiosperms arose from ancestors which are entirely unknown. There is little evidence of their existence prior to the Cretaceous period, but their abundance in the older Cretaceous floras and the rapidity with which they spread during the whole of that period to all parts of the earth, suggests an earlier origin. By the end of the Mesozoic, they dominated the vegetation of the earth. The lower Cretaceous rocks of western Greenland yield some of the earliest angiosperm fossils as leaf impressions and these include magnolia, persimmon, fig, poplar, plane and other broadleaved trees. By the latter part of the Cretaceous period, angiosperms were sufficiently widely distributed and of so many forms as to give a decidedly modern aspect to the vegetation. Representatives of the modern families Magnoliaceae, Caprifoliaceae, Juglandaceae, Betulaceae, Fagaceae, Ulmaceae, Alismaceae, Gramineae and Liliaceae, together with many others, are to be found in the later Cretaceous rocks. The phenomenal rise of the angiosperms has never been fully explained, but there is little doubt that the mutual adaptation of pollinating insects and honey-bearing flowers was a factor which greatly contributed to their success.

One other group of plants of the Mesozoic era should be mentioned; they make up the fossil order Caytoniales. Although possibly arising at about the same time as the angiosperms, they were never destined for the same success. A few species only became widespread and all became extinct in early Cretaceous times. Fossils of the foliage are called *Sagenopteris* and consist of about four small, net-veined leaflets. The reproductive parts are mid-way between those of pteridosperms and those of angiosperms. The fossil genera *Caytonia* and *Gristhorpia* are small berry-like fruits. These appear to have been formed from closed ovaries but not of a form corresponding to those of modern angiosperms.

Cenozoic Era

The earth's vegetation continued to possess its modern characteristics. No new forms of major significance arose to displace the angiosperms, whilst the ranks of some of the older forms became thinned out still further, leaving only scattered representatives in many cases.

Primitive plants are still only occasionally found in the fossil record, but there is evidence that blue-green algae played a major role in the formation of oil shales during this era. Lime-secreting forms of green and red algae, similar to present types, built up marine limestone deposits in many parts of the world. *Chara*, continuing down to the present time as one of the most persistent of all plants, was instrumental in developing freshwater limestone beds. Diatoms, particularly abundant during the Miocene period, formed thick strata of diatomaceous earth from their siliceous shells. Various types of fungi have been found, including *Fomes*, one of the more resistant bracket fungi.

The bryophytes are once more but rarely encountered as fossils. There are occasional liverwort forms of modern appearance and various types of fossilized mosses in deposits of the Pleistocene period.

Of the pteridophytes, the once abundant lycopods and horsetails show up only occasionally in their modern form. The ferns were very much subordinate to the seed-bearing plants. Occasional tree-ferns of the Cyatheaceae are found in Cenozoic floras but the majority of ferns were smaller and more fragile. The most widespread modern fern family, Polypodiaceae, was represented by the genera *Adiantum*, *Dryopteris*, *Asplenium*, *Pteris* and *Polypodium*, among others. *Osmunda* occurred in somewhat similar proportions as it does today. A few of the tropical ferns now in existence such as *Lygodium*, *Marattia*, *Gleichenia* and *Matonia*, are clearly traceable from early Cenozoic times.

The only abundant gymnosperms were members of the Coniferales. The pteridosperms, Cordaitales, Bennettitales and Caytoniales were by now extinct and the cycads and ginkgos very reduced in numbers. Several modern cycads, such as *Cycas*, *Dioon* and *Zamia*, were represented but showed only very limited distribution, whilst the solitary genus *Ginkgo*, managed to maintain a precarious position. Most of the modern families and genera of the Coniferales were established early

Fig. 20.15. Twig with cone of *Taxodites europaeus* (natural size). (From A. C. Seward's *Fossil Plants* (Cambridge University Press).)

in the era; they include *Araucaria*, *Sequoia*, *Taxodium*, *Taxus*, *Abies*, *Picea*, *Pinus*, *Cupressus*, *Juniperus* and *Thuja*.

During the Cenozoic era, another gymnosperm order came into existence. This was the order Gnetales, represented by the genera *Ephedra*, *Gnetum* and *Welwitschia* at the present time. The fossil history of the order is very meagrely recorded, but twigs and pollen grains of *Ephedra* are found as fossils in several Cenozoic strata. There is little evidence from the rocks to substantiate the theory that the Gnetales may be the link between gymnosperms and angiosperms.

By the close of the era, the latter were firmly established in their present dominant position and fossils representing about 260 families of the class, as it stands today, have been found. The most primitive are considered to be the Magnoliaceae and the Ranunculaceae, and it is from such primitive forms that we attempt to derive the more advanced floral structure of the others. Monocotyledons are never so abundantly represented among the fossils as the dicotyledons. There is no real evidence showing when they became a distinctive group, but members of the grass family were becoming more numerous and widespread in the Miocene period, foreshadowing their eventual rise to their present eminence.

Quaternary Era

Little need be said of the succession of plants during this time since our modern vegetation had already been well established at its commencement. The distribution of fossils formed during the era does,

Fig. 20.16. Twig of the dwarf birch (*Betula nana*), about natural size.
(*By permission of the British Museum, Natural History.*)

however, indicate changes in climatic conditions. A great glacial period obviously exterminated many forms or forced them into warmer areas. This accounts for the few present relics of some of the more ancient Cenozoic plants such as the tree-ferns, cycads and *Ginkgo*, which can live only in tropical conditions, and for the occurrence of present arctic plants such as *Betula nana* (the arctic birch) as fossils in England (*see* Fig. 20.16).

The chart in Fig. 20.17 indicates the probable times of the rise and fall of the major plant groups mentioned, and the area enclosed with each group name gives some idea of the relative proportion of the total vegetation of the time which that group represented, as far as can be determined from the incidence of their fossil remains.

The Succession of Post-Proterozoic Faunas

Palaeozoic Era

Although the record of plants, upon which animals must have depended, is meagre in early Palaeozoic times, the latter are much more numerously represented as fossils. This is largely because many

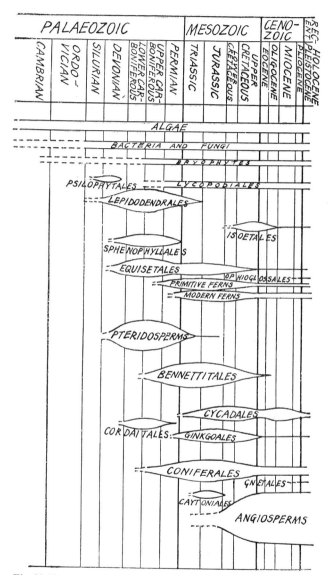

Fig. 20.17. Chart showing the rise and fall of the major plant groups.

of them possessed hardened parts much more likely to be preserved. By the end of the Cambrian period, all the main groups of invertebrates were in existence; by the end of the Permian period, all the major

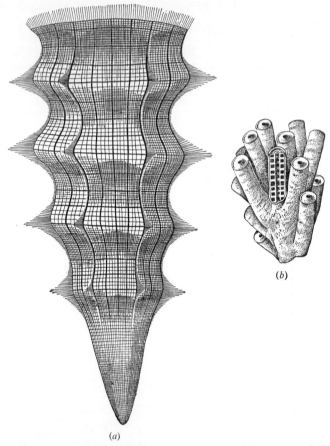

(a)

(b)

Fig. 20.18. (a) Devonian glass-sponge, *Hydnoceras*. (b) Cretaceous calcareous sponge, *Barroisia*.
(*By permission of the British Museum, Natural History.*)

forms except birds and mammals had appeared, and some of the earlier types had become extinct.

Fossils of shell-making protozoa, i.e. radiolarians and foraminiferans, are to be found in the Cambrian rocks and evidence of their existence extends all through the Palaeozoic strata. The largest of these, the fusulinids, as big as wheat grains, occurred in the Carboniferous period. The sponges, represented sparsely early in the era as spicules of siliceous type, gradually assumed local abundance up to the Devonian period and beyond. *Hydnoceras*, a glass-sponge, like the Venus' flower basket of today, is first represented in Devonian rock (*see* Fig. 20.18). Coelenterates, although rarely preserved from Cambrian times, were undoubtedly present, as is witnessed by the natural casts of jelly-fish interiors found in British Columbia. In the Ordovician period, the lime-secreting corals are well repre-

sented. Some indicated an individual existence, whilst others grew in colonies. Similar coral-fossils are found in many rocks of the whole era. From Silurian times to the end of the era, the corals have left their traces in the form of chalky limestones which must have been marine coral reefs of the time. Throughout most of the Palaeozoic era, the dominant corals belonged to the group known as the Rugosa. They differ somewhat from typical corals, particularly in bilateral symmetry, and their affinities are some-what obscure. They died out in the Permian period.

Fig. 20.19. A carboni-ferous compound rugose coral, *Lonsdaleia*.
(*By permission of the British Museum, Natural History.*)

Several groups of invertebrates, which are either no longer in exist-ence or only very poorly represented at the present time, flourished in the seas. Among those now extinct were the graptolites, which are sometimes considered to be relatives of the primitive chordates (*see* Vol. I, Chap. 24). They were colonies of minute animals enclosed in a

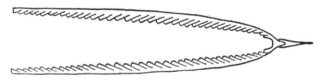

Fig. 20.20. *Didymograptus protobifidus*, an Ordovician graptolite.
(*Redrawn by permission of the British Museum, Natural History.*)

common sheath of horny substance and they floated in the surface waters of the sea. Net-like or dendroid graptolites appeared first, followed by branched forms such as *Didymograptus* (*see* Fig. 20.20), in the Ordovician period. These ancient animals died out early in the Devonian period. The brachiopods, or lamp-shells, rarely represented in modern faunas (e.g. *Lingula*), gradually rose to dominate the shallower waters of the seas during the Silurian and later. They con-sisted of a short fleshy stalk surmounted by a body enclosed in two bilaterally symmetrical horny shell valves of unequal size (*see* Fig. 20.21). During Devonian times, enormous numbers of them existed, and from one form, *Conchidium*, rock-like deposits were built up. They continued in large numbers right into the Mesozoic era. Polyzoans, still represented today by a few forms, were very plentiful and of varied form during the Ordovician and Silurian times.

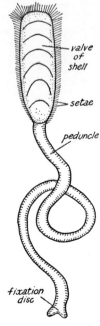

Fig. 20.21. A modern brachiopod, *Lingula*.

Of worms there is little fossil evidence, but molluscs, echinoderms and arthropods have left many traces. In Cambrian times, there existed a few sea-snails and bivalves, whilst the Ordovician rocks contain fossils of a primitive octopus-cuttlefish form, i.e. cephalopods. These continued to exist into the Silurian period, by the end of which, bivalves very similar to freshwater mussels had appeared. In Devonian rocks is the first hint of the ammonoid molluscs which become so common in later strata. From the Carboniferous times to the end of the era, bivalves and gastropods very similar to those of the present day, appeared in increasing numbers. A large swimming snail, *Bellerophon*, came into existence, but later, it died out before the era ended. Cephalopods of the nautilus type arose, and the slug-like chitons, which appeared in the Ordovician, have similar existing forms today. Through the Carboniferous and Permian times, land snails appeared, but undoubtedly the predominant molluscan fauna was marine and fresh-water.

The echinoderms are also represented among the fossils of Cambrian rocks, and remains of starfishes, sea-cucumbers and the more common crinoids have been identified. These last are the so-called sea-lilies. They were fixed, stalked creatures not like the present free-swimming crinoid, *Antedon*. Fossils of sea-urchins are found in the Ordovician rocks and all of these echinoderm forms continued to be well represented right through the whole era.

Of the arthropods, there seem to have been none very similar to modern types, but about half of all the fossils found in Cambrian rocks are those of a now extinct arthropod group, the trilobites (*see* Fig. 20.22). They were the dominant animals of the seas and were mostly small, averaging about 25 mm in length, but a few larger specimens up to half a metre have been found. Some of them resembled the modern woodlouse in external appearance. A few were swimmers but most spent their lives crawling on the sea floor. These ancient animals maintained a dominant position among the inverte-brates for many millions of years, but during the Carboniferous period, they began to wane in numbers and few fossils of them are found later than the Permian. Other arthropods are represented in the early rocks; crustaceans of shrimp-like form in the Cambrian, bivalve crustaceans or ostracods, and eurypterids or sea-scorpions, in the Silurian. In rocks of Devonian times are to be found some land-living arthropods. These were similar in form to millipedes, spiders and woodlice. Also some wingless insects are in evidence. In the Carboniferous rocks are the first signs of winged insects, including very primitive cockroach and dragonfly forms (*see* Fig. 20.23). More air-breathing arthropods appeared, including land scorpions, millipedes of several kinds, centipedes and spiders. Among the arachnids, a water-living form of king-crab appeared (*see* Fig. 20.23 (*b*)). An

Fig. 20.22. A Carboniferous trilobite, *Callavia*.

(*By permission of the British Museum, Natural History.*)

animal of similar form, *Limulus*, seems to be the only survivor today. By the end of the Permian period, most of the modern forms of insects had begun to appear and some good fossils of bugs and beetles are in existence.

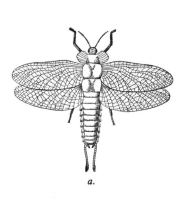

a.

Fig. 20.23. (*a*) *Homoioptera*, a primitive insect from the Carboniferous period. (*b*) *Belinurus*, a Carboniferous king-crab.

(*By permission of the British Museum, Natural History. (b) redrawn.*)

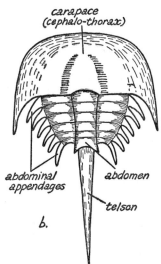

carapace
(cephalo-thorax)

abdominal appendages

abdomen

telson

b.

The earliest signs of vertebrate animals are the occasional remains of the armoured scales of the fish-like ostracoderms found in Ordovician rock (*see* Fig. 20.25). Perhaps the most primitive of fossil vertebrates ever to be found is the specimen named *Jamoytius kerwoodi* unearthed from the Silurian rocks of Lanarkshire (*see* Fig. 20.24). It measured

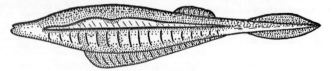

Fig. 20.24. *Jamoytius*, the most primitive known fish-like vertebrate.
(*Redrawn by permission of the British Museum, Natural History.*)

about 25 cm in length, had a finned, streamlined body not entirely unlike *Amphioxus* and probably possessed a cartilaginous endoskeleton. In Devonian times, the fish-like vertebrate form became very common and the period is sometimes known as the Age of Fishes. There must have been rapid evolution and spread of vertebrates, for in these times, ostracoderms (*see* Fig. 20.25) were abundant, and in

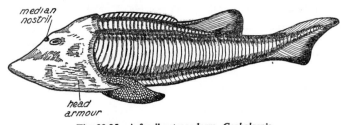

median
nostril

head
armour

Fig. 20.25. A fossil ostracoderm, *Cephalaspis*.

Fig. 20.26. A lungfish, *Ceratodus*.

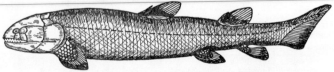

Fig. 20.27. A typical crossopterygian, the fossil *Osteolepis*.

addition, dipnoan lungfishes with paired fins such as *Ceratodus* (*see* Fig. 20.26), crossopterygians like *Osteolepis* (*see* Fig. 20.27), and sharks (*see* Fig. 20.28), have left their traces. It is from the crossopterygians that the land vertebrates probably evolved. Early in the Carboniferous period, the ostracoderms died out, but fish became more and more plentiful, and in the Permian fresh-waters, primitive bony fish appeared. Amphibian fossils are found as far back as the Devonian rocks, and during the Carboniferous period, salamander-like forms such as *Eryops*,

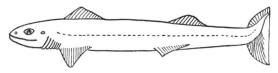

Fig. 20.28. *Cladoselache*, a Devonian shark.

(*see* Fig. 20.29) up to 3 m long, were in existence. In rocks of the same period are found a few small skeletons of primitive reptiles. From remains in the Permian rocks, it is clear that during this time a great variety of reptiles emerged. Many were ponderous and slow creatures, but some showed the more agile lizard form. The more advanced types already showed some similarities to mammals. Nearly all of them were carnivorous. The reptiles represent the highest form of animal life evolved during the Palaeozoic era. During the following hundred

Fig. 20.29. *Eryops*, a Permian amphibian about 3 m long.
(*By permission of the British Museum, Natural History.*)

million years, they were to reach their zenith and then dwindle in the face of changing conditions and competition from new forms.

Mesozoic Era

This era saw the dawn of many new animal forms both in the seas and on the land, but not all pre-existing representatives of the main groups died out. The most lowly of invertebrates, the protozoans, poriferans and coelenterates, continued much as before; calcareous sponges and corals were so abundant in many places as to form reefs in the waters. Some of the corals were similar to those which exist today, for in the Cretaceous rocks are found some of the alcyonarian forms which are represented by the genus *Heliospora*. The brachiopods still survived, but by the Jurassic period, only two types were common, the rhynchonellids and the terebratulids. Bivalve molluscs became increasingly common during the Triassic period and the ammonites are among the commonest of all fossils to be found in the Jurassic rocks. They are seen as large snail-like animals with coiled shells (*see* Fig. 20.30). They had obviously declined to extinction by the end of

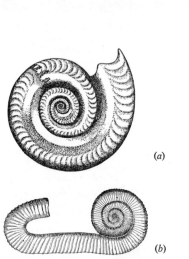

(a)

(b)

Fig. 20.30. (a) *Hildoceras*, a common Jurassic ammonite. (b) *Macroscaphites*, a Cretaceous ammonoid.

(*By permission of the British Museum, Natural History.*)

Fig. 20.31. Restoration of a Jurassic belemnite

(*By permission of the British Museum, Natural History.*)

the Cretaceous period and in some forms the shell was no longer coiled (*see* Fig. 20.30 (*b*)). As the brachiopods declined, the molluscs became the dominant shell-forming animals and some modern types emerged. Oyster-like forms were plentiful in the Jurassic period and *Trigonia*, a small bivalve still extant in Australasia, appeared. Some of the cephalopod fossils are very distinctive of the Jurassic period. These are the belemnites, a type of cuttlefish which abounded in the seas of that and the following period (*see* Fig. 20.31). These animals had an internal shell shaped like a bullet, and casts of these are of great frequency. They do not appear in rocks later than those of the Cretaceous period.

Echinoderms, particularly crinoids, also flourished, but forms of modern kind began to appear; sea-urchins, starfishes and ophiuroid brittle-stars were among them. In the Cretaceous rocks, sea-urchin fossils are very common, and *Micraster*, the burrowing heart-urchin, has a continuous fossil record extending over five million years in the chalk deposits of the period (*see* Fig. 20.32).

Early in the era, land arthropods were on the increase; they included scorpions and insects of some wide variety. By the Jurassic period, these latter included grass-hopper, dragonfly, beetle, termite and dipteran forms. In the water, crab- and lobster-like crustaceans were

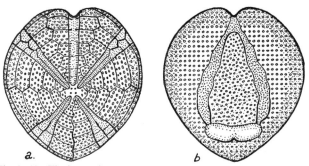

Fig. 20.32. The burrowing heart-urchin, *Micraster cor-anguinum*, from the Cretaceous rocks, (*a*) dorsal surface, (*b*) ventral surface.

appearing in significant numbers, but the greatest expansion took place among the vertebrates, particularly the fish and reptiles. In the sea, primitive bony fish abounded, and cartilaginous sharks and rays, not unlike those of modern times, were no rarity in Triassic times. Lung-fish appeared in fresh-water. They were similar to the modern *Ceratodus*. By the Cretaceous period, the fish all had a much more modern aspect and types related to herrings and other bony fishes were evolving. Coelacanths, represented at the present day by *Latimeria*, etc., are known to have existed from this time (*see* Fig. 20.33). Meantime, preying on the fish, the fish-like ichthyosaurian reptiles roamed the seas and continued to do so as the dominant marine animals until well into Jurassic times. The very large amphibians began to dwindle, but smaller frog and toad forms arose.

Among the land vertebrates, the reptiles became the dominant fauna, their remains appearing in ever increasing numbers from Triassic rocks upwards. Among the earliest were the tuataran forms of today and some small mammal-like reptiles, the earliest of the dinosaurs. In Jurassic rocks may be found a wide range of now extinct reptilian forms, the large dinosaurs and the flying pterodactyls together with lizard-like crocodilians and turtles. The large dinosaurs must have been among the most spectacular of all animals which have ever existed. The giant *Diplodocus* was in the region of 30 m long from snout to tail and stood about 6 m at the shoulder. *Brontosaurus* (*see* Fig. 20.34), was of similar dimensions and its weight has been estimated at over thirty tonnes. *Stegosaurus* was smaller, about ten tonnes. Its brain weighed about 85 g of this total. These were all vegetable-

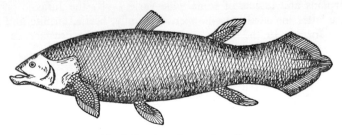

Fig. 20.33. A modern coelacanth.

(a)

(b)

Fig. 20.34. (a) *Brontosaurus*, a Jurassic dinosaur about 24 m long. (b) *Triceratops*, a Cretaceous dinosaur about 6 m long.

(*By permission of the British Museum, Natural History.*)

eaters, but in Cretaceous rocks a greater proportion of remains of carnivorous dinosaurs are to be found. These include *Antrodemus*, about 9 m long, and *Tyrannosaurus*, 12 m long and 6 m high with 15 cm fangs and vicious claws. Also in Cretaceous times lived *Triceratops* (*see* Fig. 20.34), a plant eater, and *Iguanodon*, which had an upright stance and bird-like, three-toed, hind-feet. It is very probable that all these land reptiles laid eggs. Clutches of dinosaur eggs have been found as casts in the rocks of Mongolia. The icthyosaurs of the seas, however, are known to have been viviparous.

Fig. 20.35. *Rhamphorhyncus,* a small Jurassic pterodactyl with a wing-span of 75 cm.

(By permission of the British Museum, Natural History.)

The pterodactyls were extraordinary creatures (*see* Fig. 20.35). Whilst in all main respects they were reptiles, they could fly. Their remains, found in Cretaceous rocks, show that some attained a wing-span of six metres or more. The wings were not feathered but were membranes extending as out-growths of the body from fore- to hind-limbs, somewhat as in modern bats. They had toothed jaws and seemed to have spent most of their time preying on other animals in the shallower waters of the coasts.

Another flying animal, which can be considered the first bird, also evolved from reptilian ancestors during the Mesozoic era. This was *Archaeopteryx* (*see* Fig. 20.36). It had feathers instead of scales but retained the reptilian characters of toothed jaws and jointed tail. There is no evidence of its existence after the Jurassic period, but rocks of the Cretaceous period have yielded fossils of a large flightless bird form, *Hesperornis* (*see* Fig. 20.37). From its structure it seemed to have lived on the coastal waters, swimming and diving for prey which it caught in its spiny toothed jaws. There is no doubt that the birds evolved from reptiles during the Mesozoic era but did not achieve the heights to which they later rose.

It is likely that the mammals had emerged before the end of the Triassic period, but their remains are not conspicuous; only a few fossil teeth have yet been found. In the Jurassic rocks of Somerset, a fossil reptile, with some features akin to those of mammals, has been unearthed. This fossil skeleton has been named *Oligokyphus*; it was an animal about the size of a rabbit, possessing hair and living a burrowing existence with others of its kind (*see* Fig. 20.38 (*a*)). Most of the larger mammal-like reptiles, such as *Kannemeyeria* (*see* Fig. 20.38 (*b*)), seem to have disappeared by this time. During Cretaceous times, the mammals remained inconspicuous, but fossil evidence shows that ancestors of the marsupials had spread over a wide area including

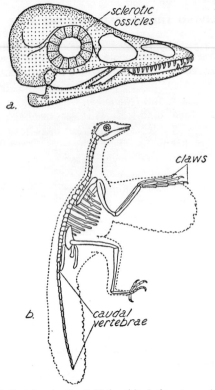

Fig. 20.36. Earliest known fossil birds. (a) *Archaeopteryx* (skull, side view showing teeth). (b) *Archaeornis* (skeleton, side view, showing long tail).

Fig. 20.37. *Hesperornis*, a Cretaceous toothed sea-bird.
(*By permission of the British Museum, Natural History.*)

South America, and they had penetrated into Australia via south-east Asia. By the end of the Mesozoic era, there are signs that the placental mammals had emerged, but in terms of size and numbers, the hundred

Fig. 20.38. (a) *Oligokyphus*, a Jurassic mammal-like reptile about 60 cm long. (b) *Kannemeyeria*, a Triassic mammal-like reptile, about 2 m long.
(*By permission of the British Museum, Natural History.*)

million or more years through which it lasted is rightfully known as the Age of Reptiles.

Cenozoic Era

As with the flora, so the fauna began to be transformed into something approaching the modern condition. The end of the giant reptiles and the advent of new forms of placental mammals were the chief changes. The invertebrate fauna underwent no great alteration in general pattern, but the insects rose to new heights in variety of form and numbers.

The time from the Eocene to the end of the Miocene saw the rise and fall of a new form of foraminiferan protozoans. These were the nummulites, which possessed flat, round, chalky shells which in some species were up to 70 mm in diameter. With these were other foraminiferans and radiolarians which all seemed to flourish in the warmer seas. Sponges of modern form have been preserved in the Oligocene rocks, together with reef-forming corals. The fossils of calcareous tube-like structures from Eocene times indicate some forms of tube-living worms. The largest of all the brachiopods, *Terebratula*, ceased to exist after the Pliocene period. Among the molluscs, cockles, mussels, oysters and wood-boring forms like *Teredo* the ship-worm, marine gastropods like cowries, and cephalopods of the nautilus and cuttle-fish forms, have left their traces from the same period. Land and fresh-water snails date from the Oligocene time. Among the echinoderms, heart-urchins were abundant. Aquatic arthropods included crabs, lobsters and acorn barnacles of the present *Balanus* form. On the land, the insects were rapidly evolving parallel with the rise of the flowering plants. Among the most famous of all fossils are those of insects, trapped in amber, which come from Oligocene rocks of the Baltic regions. Perfectly preserved in the hardened gum, are insects of all kinds, comparable with those of today. They include butterflies, bees, ants and sawflies. Later, in the Miocene period, there is evidence of the existence of beetles, bugs, crane-flies, tsetse flies, dragon-flies and ants.

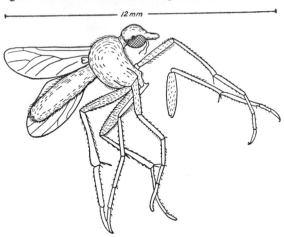

Fig. 20.39. Insect in amber: Diptera: Platyura.
(Drawn from a specimen in the British Museum, Natural History, by permission.)

The Cenozoic era saw no tremendous alteration in the aquatic vertebrates; sharks and bony fishes abounded in the seas. It was on the land that the most significant changes started to occur during the Eocene period. The dinosaurs and other large land reptiles, which were so plentiful in Cretaceous times, gradually receded. Only the smaller reptiles such as crocodilians and turtle forms remained. All the toothed birds became extinct and their places were taken by a few flightless kinds. More and more birds appeared during the succeeding periods and there are fossils of wading ducks, pelicans and ibises in the Miocene rocks. Gradually, the mammals assumed the ascendancy and from small, five-toed animals of primitive mammalian form, early in the Eocene period, there slowly arose the wide variety of present types. By the end of the Eocene period, all the main groups of mammals were in existence. They included pouched opossums, rodents, insectivores, primates, ungulates, cetaceans (whales) and sirenians (sea-cows). The remains of the ancestors of pigs, horses, cattle, rhinoceroses, tapirs and elephants are clearly distinguishable. Modern horses had their origin in the fossil *Eohippus* or *Hyracotherium* of Eocene times (*see* Fig. 20.41). This was a small animal, standing no more than 45 cm and possessing four-toed feet. Its teeth were low-crowned and adapted for browsing on shrubs, etc. The evolution of the horse is described in Chap. 21. The herbivorous and insectivorous mammals were first on the scene, the creodonts or carnivorous forms not becoming plentiful till the

Fig. 20.40. *Smilodon,* a Pleistocene sabre-toothed "tiger," a metre high at the shoulder.

(By permission of the British Museum, Natural History.

Oligocene, in which rocks the ancestors of cats, dogs and bears have left their traces. Flying mammals of bat-like form were not uncommon. There were gibbon-like primates related to the higher apes and man.

In the Miocene and Pliocene periods, the mammals assumed an even more modern appearance and included antlered deer and camel-like types. The sabre-toothed tiger (*see* Fig. 20.40) and many other carnivorous mammals became frequent. Elephants became more widespread than at any other time; the now extinct mastodons were alongside them. By the end of the Pliocene period, herds of horses grazed the plains. *Hipparion* was the commonest, and by this time, the horse had evolved as a much larger animal with only a single toe on each foot touching the ground. Long and short-necked giraffes have been preserved, some very similar to the existing okapi of Africa. Man-like apes began to appear and a primitive monkey-like ape, *Proconsul,* dates from the Miocene. The fossil ape-man of South Africa, *Australopithecus,* found in Pliocene rocks, shows that it was of upright stance, hinting at the future of the primate animals.

The Cenozoic era is most often known as the Age of Mammals.

Fig. 20.41. (a) *Eohippus*, (b) *Mesohippus*, (c) *Merychippus*, (d) *Equus*.

The Quaternary Era

During this fourth era of life, glacial and warmer periods alternated. Each successive change in conditions had a considerable effect on the distribution of living creatures. For example, during cold periods, arctic animals such as the reindeer and arctic fox lived in southern England. In the interglacial periods, the lion hunted in Yorkshire and the hippopotamus inhabited the river Thames. In Pleistocene times, the woolly rhinoceros and the mammoth, both adapted to cold conditions, roamed about England. But the majority of animal forms were similar to those of the Cenozoic era, and from them arose the present genera and species, many of which are represented as fossils in Pleistocene rocks. A few new forms appeared and some of them had a relatively transient period of existence. One such was the giant sloth, *Megatherium*, which was a huge South American animal, over 6 m high (*see* Fig. 20.42).

Undoubtedly the greatest development of the Pleistocene period was the emergence of Man, the animal destined to rise to the greatest heights of all. The fossil record of the Hominidae is incomplete, but there is evidence that ape-like forms, which appeared at the end of the Cenozoic era, gave rise to more intelligent creatures. The evolution of the Hominidae, culminating in modern man, *Homo sapiens*, is bound up with enlargement of the neopallium, perfection of the erect posture, reduction of the face and jaws, and loss of the power to oppose the big toe to the remaining four.

There has been considerable difficulty in devising a reasonably accurate time-scale to cover the Pleistocene period. Though there is still controversy, it is generally considered that there were four major ice-ages, named after localities in Switzerland; Gunz, Mindel, Riss and Wurm. These were separated by warmer interglacial ages and

Fig. 20.42. *Megatherium*, a Pleistocene giant ground-sloth, about 6 m high.
(*By permission of the British Museum, Natural History.*)

succeeded by a postglacial age which is still with us. In tropical and sub-tropical parts of the world, where the actual polar ice did not penetrate, there were pluvial or wet periods, interspersed with inter-pluvial or dry periods. The pluvial periods coincided with the ice-ages and the interpluvial with the interglacial periods. Hominid fossils and relics are dated mainly with reference to the glacial time-table even though that is not fully established. Where they exist, associated animal and plant remains give further clues, and in some cases, it is possible to make more accurate estimates by finding the amount of radiocarbon present in particular specimens. Finally, in the progress of the Hominidae, we find increasing evidence of culture, manifested by the use of tools, burial of the dead, the use of fire, the domestication of animals, crop cultivation and the establishment of settled communities. By the degree of cultural development, we are able to estimate the relative age of the remains.

The first hint of the primate stock (*Proconsul*) appeared in Miocene times and probably gave rise to two divergent lines, the Pongidae and the Hominidae. To the Pongidae belong the anthropoid apes, represented now by the gorilla, chimpanzee, orang-outan and gibbon. *Homo sapiens* is the only extant species of the Hominidae and its origin is still not apparent. A goodly collection of more or less fragmentary remains has been amassed, and among these, the more important are assigned to the following species: *Australopithecus africanus*, *Pithecanthropus*

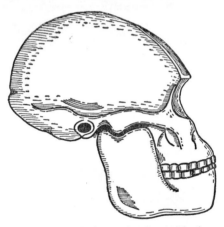

Fig. 20.43. Skull of *Australopithecus*, side view.

erectus, *Sinanthropus pekinensis*, *Homo heidelburgensis*, *Homo rhodesiensis* and *Homo neanderthalensis*. None of these lines appears to have given rise to the present human stock, but *Homo sapiens* or his precursors probably existed contemporaneously with them.

In 1924, at Taungs in Bechuanaland, South Africa, the fossilized skull of an ape-man was unearthed. It shows a number of features intermediate between those of apes and men. The brain capacity is estimated at 900 cm^3 whereas the largest modern ape has about 650 cm^3 and the average for modern man (white races) is 1500 cm^3. The brow ridges are less projecting than those of modern apes, and the teeth agree closely with those of modern man (*see* Fig. 20.43). Placed in the genus *Australopithecus* (southern ape), the fossil is dated at about 1 000 000 years ago, either at the end of the Pliocene period, or early in the Pleistocene. While the creature existed too late to be an actual ancestor of man, it certainly lies near the beginning of the hominid line. By 1947, remains of half-a-dozen other specimens, including the skulls of three, had been discovered, also in South Africa. One is stated to be another species of *Australopithecus* and two others have been named *Plesianthropus* and *Paranthropus*. All these are closely akin and they exhibit a line of evolution between the apes and man. There is no definite evidence of culture and none are placed in the genus *Homo*.

Probably contemporaneous with *Australopithecus* were Java man, *Pithecanthropus erectus*, and Pekin man, *Sinanthropus pekinensis*. Remains of the former were first discovered in Java by the Dutchman, Du Bois, in 1892. They show that the animal had a definite hominid, erect posture and a brain capacity somewhat larger than that of

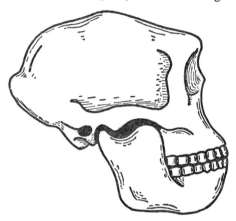

Fig. 20.44. Skull of *Pithecanthropus*, side view.

Australopithecus, but there were still many ape-like features (*see* Fig. 20.44). There are indications that tool-making had begun and thus, in archaeological terminology, *Pithecanthropus* is placed in the lower Palaeolithic culture period. These remains of Java man are more than 500 000 years old and the rock formations in which they occur, are either middle or lower Pleistocene. In 1937, there were several additional discoveries of *Pithecanthropus* in Java.

About three dozen fossils, discovered in China between 1925 and 1936, have been assigned to the species *Sinanthropus pekinensis*. They have been dated to the middle of the Pleistocene period and are thus about 500 000 years old. Pekin man had the human posture and gait, and there were less ape-like features than in Java man (*see* Fig. 20.45). He had developed a more advanced culture; crude stone tools are found and fire was used. The brain capacity ranged from 950 to 1200 cm³. It is considered that Pekin man was closely related to Java man.

The best known of sub-human species is Neanderthal man, so-called from the skull discovered in the Neander valley of western Germany in

1856. Since then, numerous and more complete remains have been found in many localities in Europe, in the middle East, in north Africa, in Asiatic Russia and in south-east Asia. From these relics, we learn that *Homo neanderthalensis* had a brain capacity ranging from 1300 to 1600 cm^3, but was specialized in cranial structure, in dentition and in

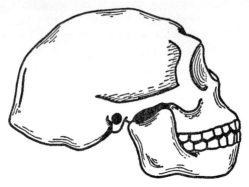

Fig. 20.45. Skull of *Sinanthropus*, side view.

the limbs, and is quite sharply differentiated from *Homo sapiens*. The skull bones were thick; there were well-developed brow ridges and the face was chin-less (*see* Fig. 20.46). There is clear evidence of workmanship in stone, of fire-making and of burial rites. Neanderthal man was probably related quite closely to *Homo rhodesiensis*, a skull of which was discovered during mining operations at Broken Hill in Rhodesia. This Rhodesian man is considered by some experts to be related to the

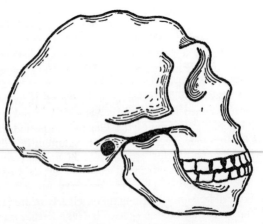

Fig. 20.46. Skull of Neanderthal man, side view.

ancestors of the present bushmen. *Homo neanderthalensis* had disappeared by the end of the Pleistocene period.

The Piltdown skull, unearthed in Sussex in 1911, provided something of a puzzle. Its brain capacity of 1350 cm³ is well within the limits for modern man. A jaw discovered near it, has been shown to be that of an ape. Modern tests have shown that the parts were "planted" by a hoaxer. Skull fragments found at Swanscombe in Kent are associated with the lower Palaeolithic hand-axe culture.

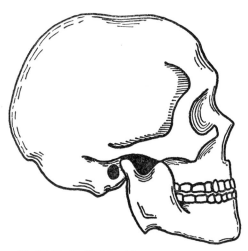

Fig. 20.47. Skull of Cro-Magnon man, side view.

Towards the end of the last glacial age (Wurm), men of our own species entered Europe. Their place of origin is somewhat doubtful but is considered by many experts to be in the region of the Iranian plateau. They are dated between 40 000 and 13 000 B.C. and they reached a much higher state of cultural development than did Neanderthal man. The majority are assigned to the Cro-Magnon group, though there were probably related stocks ranging over a wide area, from China to France. They belong to the Upper Palaeolithic culture and are characterized by their parallel-sided stone blades, their use of fire, their burial rites and the piles of bones containing remains of the mammoth, the woolly rhinoceros, the reindeer, wolf and fox. A number of different cultural groups developed in different localities, one of the most famous cultures being the Aurignacian, which left us the cave art of southern France and northern Spain. Cro-Magnon men showed several racial peculiarities. They were tall and had large, massively-built heads. There were no brow ridges on the face, which was essentially of modern proportions, with the nose prominent and narrow (*see* Fig. 20.47).

Modern man is undoubtedly descended from Cro-Magnon man and kindred groups, all of which may be classified as *Homo sapiens*. There is as much difference between some extant races as there is between Cro-Magnon man and modern man. The later stages in human

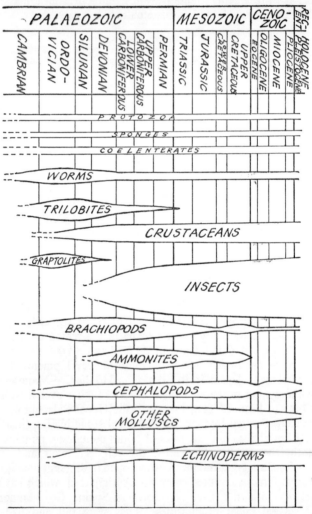

Fig. 20.48. Chart showing the rise and fall of the main invertebrate animal groups.

evolution come within the scope of history and are classified in terms of culture as Neolithic, Bronze age and Iron age. It may be considered that we are still living in the Iron age. Intermixture and migration have led to hopeless confusion of races. Indeed, it is no easy matter to define the term "race" as applied to human beings. With regard to the definition of "race," a committee of experts, organized by the United Nations, has submitted a report which has now been adopted. The

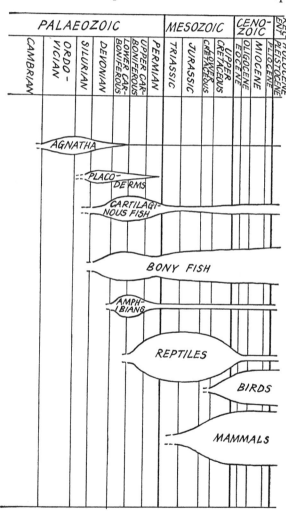

Fig. 20.49. Chart showing the rise and fall of the vertebrate animal groups.

report states that there are three ethnic groups, the Mongoloid, Negroid, and Caucasoid, and that the main differences between these groups are due to the history of their cultural experience. All have equal educability and there is no justification for prohibition of intermarriage. This seems to dispose of the idea of racial superiority which, throughout history, has been claimed by the dominant power of the moment.

The early ancestry of *Homo sapiens* is still not clear. All the other types mentioned are probably parallel species and it seems that we must look for our early precursors as far back as the Lower Pleistocene or even earlier.

More recent discoveries, in the period 1959–1965, by Dr. and Mrs. L. B. Leakey, tend to confirm this view. After many years of digging in the Olduvai Gorge, Tanzania, E. Africa, they have discovered at least two other types of ancient man. The first, represented by several teeth and fragments of a skull with powerful jaws, they named *Zinjanthropus* or Nutcracker Man, who was evidently a vegetarian. Later finds revealed numerous skull fragments of a more slightly built hominid who made simple tools and used fire. These fragments were assigned by Leakey to a species *Homo habilis* or Handy Man. Both these types have been dated about 1 000 000 years ago, and Leakey suggests that they are closer to the direct line of modern man's ancestry than are any of the later Australopithecines. However, it is only fair to state that certain other anthropologists do not agree with Leakey. They consider that both *Zinjanthropus* and *H. habilis* are other types of Australopithecines. There is a great deal of digging now taking place in this East African area and further finds may resolve the problem more clearly.

The charts in Figs. 20.48, 20.49, shows the times of origin, expansion and decline of the main animal groups. The writers realize that textbook description and illustration can do little to convey to the reader the immensity of time which has elapsed since the Earth began, or to conjure up satisfactory pictures of the variety of life it has supported. It is urged that the student should make use of the galleries of museums, in which the fossils themselves represent the sentences, paragraphs and chapters of the story in a much more realistic way.

CHAPTER 21

EVOLUTION

STUDY of the selected examples of free-living plants and animals in the sequence given in Vol. I, that is, in ascending order of complexity, coupled with the general information concerning the history of modern forms to be gained from the previous chapter, should have brought the reader a long way towards the realization that the examples are not wholly unrelated individuals, but that each forms part of a complete system of living things which has slowly been developed by gradual and persisting changes over many millions of years. Such a conception is embraced by the theory of Organic Evolution, which postulates that all present representatives of the plant and animal kingdoms have themselves arisen from pre-existing forms by a gradual process of change.

Whilst there are few biologists who would attempt to refute this theory, it is only within the last hundred years that it has gained general acceptance. Prior to that time, the conception was vague and had never been seriously considered although mentioned as a possibility by several writers.

In the development of a sound concept as to the origins of the numerous forms of life, the first step most certainly had to be the establishment of a firm understanding of, and belief in, the reproduction and permanence of species. Records dating back to Aristotle (384–322 B.C.) and Theophrastus (b. 372 B.C.), show that the earliest naturalists attempted to accumulate information about plants and animals concerning their specific characteristics, but the writings indicate that none showed an understanding of what a species is or how it originated. Most believed that many organisms could be generated spontaneously from inorganic material. Such a belief continued to hold the attention of many eminent writers of the middle ages, and there are in existence recipes for generating forms such as flies, bees and even mice from dust. Such beliefs die slowly, and it was not until the middle of the seventeenth century that Redi was able to prove to the satisfaction of all that flies come from eggs and maggots. Finally Louis Pasteur (1822–95) was able to put an end to ideas of spontaneous generation by his experiments showing that even micro-organisms had a living ancestry. He thus established an irrefutable notion that each and every living thing could arise from pre-existing organisms of the same kind only.

A modern conception of the species began with John Ray (1627–1705)

and reached preliminary heights of importance in the system of classification of Linnaeus (1707–78), who catalogued a wide range of organisms
into genera and species. Whilst he could not conceive that all the
material he handled might have had some common starting point, he
did make a suggestion that cut across the currently accepted principle
that all forms of life had been specially created and had remained
unchanged from the day of their special creation. He suggested that
the genera which he outlined were the separately created forms and the
species were variants of them. To this extent he admitted some change.
He was supported by some later eighteenth and early nineteenth century
biologists, who pointed to the implication of a continuity of descent in
the natural system of classification of living things, which was gradually
being built up. Some even postulated methods by which the transformations could have come about. Buffon (1707–88) suggested that the
external environment might have a moulding effect. Erasmus Darwin
(1731–1802) and J. B. Lamarck (1744–1829) both advocated evolution
through the inheritance of characters acquired by adaptation to particular environments. Others such as George Cuvier (1769–1832) were
bitterly opposed to such theories and tried to explain the accumulating
evidence for evolution by a series of catastrophes which wiped out all
existing forms and which were followed by new creations. Geoffroy
Saint-Hilaire (1772–1844) tried to combine the two ideas of discontinuity of form and continuity of descent by the hypothesis that new
species of a greater complexity did arise from the occasional appearance
of "monstrosities" which found an environment in which they could
flourish at the expense of their ancestors who thus died out.

So the discussions and sometimes bitter arguments continued
throughout the early half of the nineteenth century, with no one really
convinced one way or the other until Charles Darwin (1809–1882,
grandson of Erasmus Darwin), forestalled in fact by A. R. Wallace
(1823–1913), put forward the theory of natural selection as a simple
logical explanation, with wider scope and greater credibility than any
previous, to account for the mounting evidence that evolution was a
fact. His *Origin of species by means of natural selection, or the preservation of favoured races in the struggle for life*, published in 1859, sets
out in detail his long and arduous work, and his ideas on the manner in
which evolution could have been brought about. The publication
immediately created a controversy and there waxed a bitter struggle
between the pro- and anti-evolutionists. The public quarrel between
T. H. Huxley and the Bishop Wilberforce of Oxford at the British
Association meeting of 1860 was one of many clashes.

Gradually, however, the fact of evolution became widely accepted,
and at the present time, few would dispute it. The precise manner in

which it has been brought about is still elusive, and even today there are differing opinions on this score.

In the following sections we shall outline the evidence on which we may base our theory that evolution has occurred, and then give a brief account of the possible mechanisms by which it may have been achieved.

EVIDENCE THAT EVOLUTION HAS OCCURRED

This evidence may be grouped conveniently under four categories.

1. Direct evidence from the existence of fossils and the sequence in which they occur (geology and palaeontology).

2. Deductions from the recognition of homologous structures and the consequent morphological and anatomical relationships presumed to exist—

    (i) between present groups of plants and animals;
    (ii) between present groups and fossil forms;
    (iii) between the early developmental stages of present types;

(comparative morphology, anatomy and embryology).

3. Deductions from similarities between the chemistry and methods of functioning of present types (comparative physiology).

4. Deductions from the geographical distribution of plants and animals both present and fossil (plant and animal geography).

Evidence from Palaeontology

It has already been pointed out in Chap. 20 that the fossilized remains of plants and animals have been discovered in large numbers dating back to times estimated at many millions of years ago. At the time that Cuvier, who founded the study of vertebrate palaeontology, proposed his theory of catastrophes to account for the distribution of extinct forms as fossils in the rocks, creation was considered to have occurred only a few thousand years previously. From the records left by man over this period, it was obvious that he had not changed a great deal since his apparent time of creation and it was difficult to see how he or any other form of life could have evolved from a common starting point in so short a time. It needed a very drastic change in the appreciation of the earth's history from the point of view of time, to make any hypothesis of evolution compatible with the currently accepted facts.

More penetrating studies into rock formation brought about this change of view. Careful examination of thousands of feet of sedimentary rocks, showed characteristics in them which could only be accounted for if they had been deposited extremely slowly, and estimates of their age, based on probable rates of deposition, were in millions of years instead of hundreds. With the help of the physicists, geologists have since been able to make generally acceptable estimates of the age of many deposits of rock-containing fossils, and thus to date the appearance of practically all known forms of life.

If any particular deposit is examined in great detail from the point of view of relative situation of fossils, it is found that wherever sufficient material is available, any changes in structure observed in a particular fossil form always follow a time sequence. There is a gradual changing of character from the oldest to the youngest, which coincides exactly with their relative positions in the rock. There is never any indiscriminate mixing of the varying forms. To draw a parallel, there are no Bren guns to be found in caches of mediaeval fire-arms.

An excellent example of such complete harmony between age and change in form is afforded by the fossil remains of the sea-urchin, *Micraster*, which occur with great frequency in the Cretaceous chalk deposit of southern England. This chalk, in places, is up to 460 m deep and took upwards of ten million years to be deposited. The conditions prevailing during its deposition appear from its construction not to have changed to any great extent, and through most of this time, conditions on the bottom of the sea continued to be so similar that large numbers of the same kinds of organisms are represented as fossils almost throughout its whole depth. *Micraster* can be found persisting through about 160 m of this chalk, equivalent to four or five million years. Hundreds of thousands of fossils can be collected and a gradual change can be traced from the lower strata upwards, that is from the oldest deposits to the youngest. The changes, largely in shape, are nowhere abrupt, merging gradually into one another. There is a slow alteration in shape from flattened to arched and from elongated to almost equal length and breadth. The mouth position moves gradually forward from a point about one-third of the body length from the front of the lower surface, to a position about one-sixth of the body length from the front, a distance of about 1 cm. Gradually appearing is a low ridge along the under part of the upper surface. This ridge, from total absence, first appears and then increases in height. The grooves from which the tube feet emerge grow longer, and from a smooth surface they develop a sculptured one. The mouth becomes increasingly overhung by a protruding lip of the hard skeleton. In some regions, the sculpturing on the skeleton gradually becomes more prominent.

All these changes are extremely minute and each appears to have no great significance, but nevertheless, they are there, and furthermore are absolutely continuous, the sea-urchins at any one level merging into those above and below. There is no gainsaying the fact that these animals had their existence and reproduced in continuous succession so that their remains illustrate a line of descent extending over at least four million years.

Another oft-quoted example of a continuous series of fossil forms showing a full pedigree, as it were, of a present genus, is the series of fossils of the horse. These extend over more than forty million years.

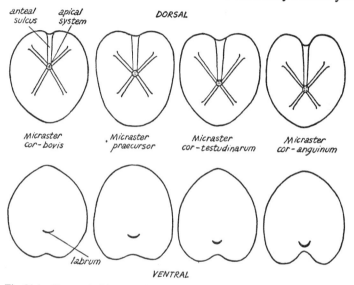

Fig. 21.1. Changes in *Micraster*, in successive zones of the Cretaceous chalk, the earliest on the left.

Present horses are very distinctly characterized by their possession of only one toe to each foot and teeth distinctive in three ways, namely, in being all alike instead of being differentiated into molars and premolars, in having a complicated surface pattern caused by the irregular ridges of enamel, dentine and cement laid down in their development, and in exhibiting continuous growth (hypsodont) over the first six to eight years of the horse's life. The peculiar structure of the leg and foot is in keeping with fleet-footedness, a necessity when the animal has to exist in regions of flat plains with little cover. The structure of the teeth shows admirable adaptation for chewing tough leaves and stems of grasses.

Fig. 21.2 shows the structure of the modern horse's limb. The visible "leg" of the horse corresponds to the middle digit, the wrist or ankle, and the fore-arm or shank bones of an ordinary five-toed animal. Only the last joint of the digit touches the ground. The hoof is equivalent to the finger- or toe-nail. Above this digit is the region corresponding to the palm or sole, represented by the single cannon-bone on which are two small splint bones representing the second and fourth digits. The horse's "knee" or "hock" joints mark the position of the wrist or

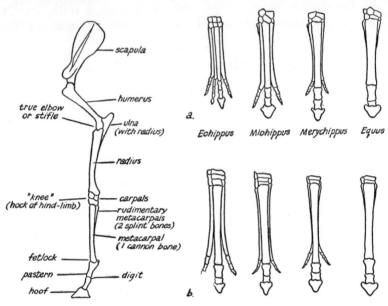

Fig. 21.2. Skeleton of fore-leg of modern horse.

Fig. 21.3. Lower parts of (*a*) fore- and (*b*) hind-limbs of ancient and modern horses to show disappearance of all but one digit. (Not drawn to scale.)

ankle. The true elbow or knee is well above this and is known as the stifle.

If we examine the fossils of mammals as far back as the earliest period of the Cenozoic era, there are none showing such distinctive peculiarities of either limbs or teeth. They all appear to be small five-toed animals with short-crowned teeth lacking grinding ridges. There was no evidence of obvious carnivores or herbivores and clear distinctions between these are not noticeable until the later stages of the Eocene.

The oldest known link with the modern horse was described by Owen in 1856 under the name of *Hyracotherium*, now known as *Eohippus* or the dawn horse. This was a fossil of the Lower Eocene and may be described as having been about the size of a fox, with four complete toes on the fore-foot, three on the hind-foot and possessing short-crowned, lightly ridged or bunodont molars, and distinct, smaller, simple premolars. Between this and the modern genus, *Equus*, eight other genera totalling close on three hundred species have been found, and this series bridges completely the gap between the earliest and most modern horse in four main stages.

In the Middle and Upper Eocene periods, fossils of the genera *Orohippus* and *Epihippus* are found. The lateral digits of the feet are somewhat reduced although the outer fifth digit of the fore-feet is functional. The premolars, except the first, have become very much like molars in form and the molar crowns are higher with distinct crests.

In the Oligocene, the horse is represented by the fossils *Mesohippus* and *Miohippus*. There are three toes on fore- and hind-feet with the lateral digits reduced, whilst the middle digit is enlarged and the fifth digit of the fore-foot is reduced to a short splint. The teeth, except the first premolar, are all of molar form.

Parahippus of the Lower Miocene and *Merychippus* of the Middle and Upper Miocene show fully unguligrade characteristics, i.e. running on one toe. The side toes are very much reduced and do not touch the ground; the central digit is still more pronounced. The teeth are much longer than any previous form and show development of cement on the crown, giving the first indications of the tooth pattern found in modern horses.

In the Lower Pliocene, *Pliohippus*, and in the Upper Pliocene, *Plesippus*, both bridge the last gap in possessing teeth which are all long-crowned and molar-like, and in having feet with the lateral digits reduced to splints, whilst the middle digits are very pronounced. *Equus*, the modern horse, appears early in the Pleistocene rocks.

Accompanying these genera are numerous others not on the direct line of descent. They may be related to other modern animals such as the rhinoceroses and the tapirs. The two examples quoted, the sea-urchin and the horse, serve to show evolutionary trends within two widely separated genera. There are also, however, numerous examples of fossils which undoubtedly link classes together and give some indication as to the possible changes by which one gave rise to the other. They are "missing links," as it were, in the story of descent.

The primitive bird fossil, *Archaeopteryx*, found in the Bavarian rocks of the Jurassic period, occurs when reptiles were undoubtedly the most successful of animals. Examination of the almost perfectly preserved remains shows that this early bird form was in reality a reptile specialized for aerial life, thus confirming the long held suspicion that birds are descended from reptiles.

The fossilized remains of plants may yield similar information

There are no living representatives of plants which bridge the gap between the seedless, spore-bearing pteridophytes and the seed-bearing spermatophytes, but flourishing alongside the spore-bearing forms of the later half of the Palaeozoic era, are remains of seed-bearing plants, the pteridosperms or "seed-ferns" as they are called, which combine the vegetative characteristics of the ferns with the seed habit of the flowering plants. There is no evidence here that flowering plants descended directly from ferns through the Pteridospermae, but at least there is the indication that the pteridosperms and flowering plants have some common ancestral form, possibly some fern-like plant.

Many other examples could be cited from known fossil material, and no doubt in the future, new fossil discoveries will serve to fill more completely the jig-saw pattern which at the moment has many gaps. The fossil evidence is direct in that the fossils are real and can be said to depict truly the extinct forms of life in the sequence in which they occurred. When the sequence is complete for any particular form, it tells an undeniable story.

Deductions from Comparative Morphology, Anatomy and Embryology

The building of a single natural system of classification has been based almost entirely on the comparison of structures. Similarities have been used to indicate relationship between one organism and another. There are however two types of similarity. Animal or plant parts may be similar by *analogy*, that is to say, they serve only the same purpose. The wings, legs and eyes of the fly are analogous to the same parts of the bird in that the two may fly, walk and see respectively with them. But there the similarity ceases for they have no common structural feature. Similarity by *homology* on the other hand, implies structural and developmental likenesses which go far beyond the coincidence that the parts concerned may perform the same function. In fact many animal and plant parts which are undoubtedly homologous structures have long since ceased to perform the same functions for the organisms concerned. That the wing of a bird and the fore-limb of a mammal are based on a common pentadactyl limb pattern would be difficult to deny, but they no longer serve the same purpose. Homology implies a common origin and it is on this type of similarity only that our classification has been based. This supposition, that only homologies between structures can show relationship, is amply justified when we examine and compare enough organisms. We are forced to the conclusion that throughout both plant and animal kingdoms, the structural plans common to groups within them can provide us with the only logical means of cataloguing the almost infinite variety of living things. Each

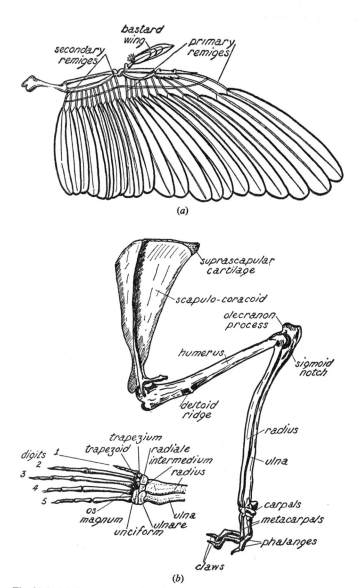

Fig. 21.4. (a) Arrangement of remiges on wing of pigeon. (b) Fore-limb of rabbit, side view, natural position.

large or small group is characterized by a common form that underlies the structure of its members. That is undeniable, but why should it be so unless all members of a group had their origin in a common ancestor?

There are countless examples of conformity to a pattern within the existing animal and plant groups. The flowers of the angiosperms all exhibit a fundamentally similar pattern, even though within the group may be seen unlimited diversity of detail. How can we account for this unless we recognize that all angiosperms are descended from a plant form which possessed the basic characters from which all the variations could be derived?

The jaws and mouth parts of the insects, each designed for a particular mode of feeding, provide similar variations on a common theme. The sucking tube of the mosquito, adapted for piercing hardened surfaces, the coiled proboscis of the butterfly, designed for extracting nectar from long corolla tubes, the chewing parts of the ant, all may be reduced to a simple plan consisting of an upper lip, a pair of strong mandibles and two pairs of maxillae, the posterior pair of which are fused to form a single lower lip. Such a fundamental plan is displayed by the cockroach.

The basic pattern of the brain throughout the vertebrates, the structure and mode of formation of xylem vessels within the angiosperms, the characteristic structure and life history of the mosses and liverworts, the conformity to a common plan in the appendages of the crustaceans, and countless other examples, serve only to subscribe to the idea that all members within each of these groups are in some way related one to the other.

The homologues of a single basic type, showing differences through adaptation to different environmental conditions and modes of life, are said to show *divergent evolution* or *adaptive radiation*. By contrast, the tendency of basically different types in similar environments to show analogous similarities is known as *convergent evolution*.

Not only may we deduce relationship between members of a group of present organisms, but in some cases, we may even postulate relationship between two groups from a study of the anatomy of an existing organism not wholly in either. The worms and the arthropods may be linked by the caterpillar-like *Peripatus*. In general appearance, this creature even looks half-way between a grub and a worm. It has a worm-like body bearing appendages, but these are flexible without being jointed. Internally, it shows arthropod features, e.g. its body-cavity contains blood, but it also resembles the worms in having a series of nephridia, one pair per pair of legs, similar to those of worms. Its central nervous system is formed of two cords one each side of the body and interconnected by cross-strands, very reminiscent of some worms. In development, it shows in its embryonic stages, an ordinary blood system with veins and arteries and an ordinary body cavity with no blood in it, once more characteristic of the worms rather than the arthropods.

Balanoglossus similarly links chordates and sea-urchins. Although little like vertebrates in gross structure, it possesses one or two indisputable chordate characters. For example, it has gill slits which are respiratory in function, and the front part of the body is somewhat stiffened by the presence of a structure comparable with a notochord, lying, as in chordates, ventral to a central nervous system on the dorsal side of the body. Despite this structure of the adult body, in the embryo, *Balanoglossus* has a structure very much like the larvae of sea-urchins.

In the study of comparative anatomy, on numerous occasions, there have been discovered structures which are inexplicable except on an assumption that evolution has taken place. These are broadly described

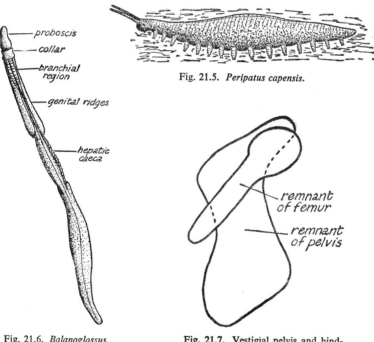

proboscis
collar
branchial region
genital ridges
hepatic caeca

Fig. 21.5. *Peripatus capensis.*

remnant of femur
remnant of pelvis

Fig. 21.6. *Balanoglossus.*

Fig. 21.7. Vestigial pelvis and hind-limb of *Balaenoptera musculus*, the blue whale; the right side, seen from beneath.

as *vestigial structures*. They often seem to be rudimentary or undeveloped and of no value to their possessor, but yet are clearly homologous with structures of great use in other creatures. Perhaps the
most striking of all such vestigial structures are the hind-limbs of
whales. Whales have well-developed fore-limbs in the form of flippers
which show a structural plan comparable with the fore-limbs of land
mammals, but they have no external signs of hind-limbs at all. However, dissection discloses the fact that the hind-limbs are represented
by a few bones embedded in the flesh, and these may sometimes take
the form of a vestigial pelvic girdle with thigh bones attached. The
presence of such vestiges in the whale can only have significance if we
believe that the whale is descended from land mammals, which returned
to the sea and evolved the fish-like form, but retained all the essential
mammalian characteristics.

Similar limb vestiges may be found in snakes. In the pythons, for
instance, the remnants of hip-girdles and hind-limbs are clearly formed,
showing resemblance to the four-legged reptiles. Such resemblance is
hard to explain as pure coincidence.

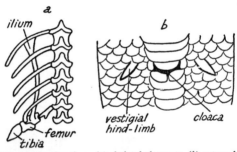

Fig. 21.8.   Hind-limb of python, (*a*) skeletal elements, (*b*) external appearance.

Vestigial structures are to be found in plants as well as animals.
The prothallial cell of a *Pinus* pollen grain appears to represent the link
between this gymnosperm and an ancestor which produced a more
definite male gametophyte. In flowers of families usually displaying
symmetry of parts in fives, sometimes there will be found a bilateral
arrangement of the stamens in two pairs with the odd one reduced to a
vestige never forming pollen. Such is the case in the figworts, *Scrophularia*.

Not only do homologies exist between the parts of different present
types, but the fossil record shows that numerous extinct organisms
possessed parts which were undoubtedly built on the same plan as
many existing today. There is no doubt that the vestigial splint bones

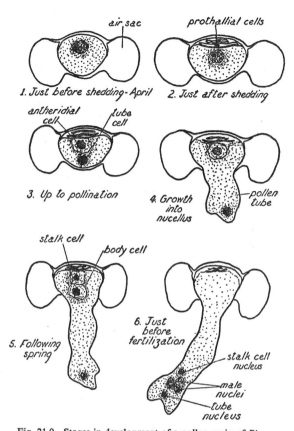

Fig. 21.9. Stages in development of a pollen grain of *Pinus*.

Fig. 21.10. Corolla of *Scrophularia nodosa* opened out to
show four stamens and staminode.

of the modern horse are homologous with the digits of an Eocene fore-runner. The ear ossicles, malleus and incus, of the modern mammal, can be traced to their origin in fossils of the theromorph reptiles. Here they were at first part of the structure of the jaw, but

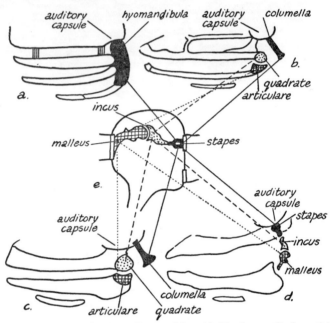

Fig. 21.11. Evolution of the auditory ossicles; (a) side view, cartilaginous fish; (b) side view, amphibian; (c) side view, theriodont reptile; (d) side view, developing mammal; (e) adult mammal, the ossicles in position in the tympanic chamber.

gradually became incorporated into the mechanism of the middle ear. They are undoubtedly homologous with the quadrate and articulare bones associated with jaw suspension in some fossil reptiles. Thus a link is established between the reptiles and the mammals in still another respect. The chain is considerably lengthened when we trace the jaws of reptiles, amphibians, bony fishes and cartilaginous fishes from the gill-supporting structures of jawless primitive chordates. The ligule of *Selaginella* is almost assuredly homologous with the ligule of the fossil lepidodendrons of the Carboniferous period. The archegonia of the *Pinus* ovule are homologous with the archegonia found in fossil seeds.

The study of comparative embryology provides numerous examples of structures common to whole groups of plants and animals in their developmental stages. Logical explanation of such common structures cannot be given except on the basis that evolution has occurred. Von Baer was the first to record observations on the similarities of the embryos of reptiles, birds and mammals. Later, Haeckel suggested that every individual, during its embryonic period, goes through the stages by which it evolved from a primitive form. This is tantamount to saying that during its development, every organism "climbs its own genealogical tree." Haeckel's view was somewhat exaggerated, but nevertheless it is considerably substantiated. It is true to say that even experts in animal embryology could not distinguish between very young mammalian embryos. All are very unlike the adults but are so similar to one another that identification becomes well-nigh impossible. Further, these young mammalian embryos are very similar to those of birds and reptiles, and yet neither class of embryo has much resemblance to the adult form. In fact, at comparable stages, all these embryos have common fish-like characteristics. Each develops visceral clefts in the pharynx, and a vascular system with a single circulation. In each, the heart is not divided into right and left halves, and the visceral clefts, though never functional as gill clefts, are served by arteries undoubtedly homologous with those of an adult fish. As each embryo completes its development, changes characteristic of its group take place in these basic structures, and, at the end of the developmental period, each emerges as an unmistakeable reptile, bird or mammal. Study of the embryology of any terrestrial vertebrate will illustrate

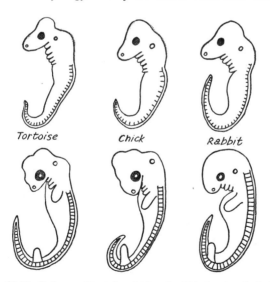

Fig. 21.12. Embryos of tortoise, chick and rabbit showing similarity.

many other similarities, and we cannot but infer that these land animals had their origin in aquatic ancestors.

To carry the argument to its logical conclusion, the zygote which begins a new organism's existence would compare with the unicellular stage of its phylogeny, and the gastrula with the diploblastic stage. Unfortunately, even if a gastrula is two-layered, these are germ layers, and include in the case of higher animals, the third layer blended with the second. Besides, gastrulation is carried out in several different ways in different groups. Whereas in coelenterates it occurs by delamination, in *Amphioxus* it occurs by invagination. The existence of an embryonic two-layered condition does not necessarily imply that its possessors had coelenterate ancestors. Indeed there is some evidence that the coelenterates themselves arose by reduction from the triploblastic Platyhelminthes.

The modern view is that the theory of recapitulation cannot be endorsed in its entirety. It is probable that no living organism during development passes through the complete series of its evolutionary ancestors. It would be more correct to say that in general, at various times in its development, it *resembles the embryos of its ancestors*, and that these recapitulatory phases are very much shortened and often modified in adaptation to particular environments. An embryo reptile, for instance, does not resemble an adult fish, but it is very much like an embryo fish.

Nevertheless, when we find within a large group such as a phylum, that the same structure arises in a similar way, we are entitled to infer that this is strong evidence for a common ancestry for all members of the phylum. In the Chordata, all animals from *Amphioxus* to man, develop the notochord, the visceral clefts and the nerve cord, in the same manner, and this indicates strong relationship between them and wide separation from non-chordates. The lancelet retains these structures in a condition which is partly primitive and partly specialized and therefore we say that the vertebrate line did not evolve from *Amphioxus* but from an ancestry which possessed these structures in the primitive unspecialized condition. We have looked for and possibly found such an ancestor in *Jamoytius*. We can trace the line, somewhat tenuously, through the crossopterygian fishes, the stegocephalian amphibians and the theromorph reptiles, to the mammals. In the individual vertebrate's development, we see a brief retracing of the major evolutionary steps, blurred considerably by the specializations characteristic of its particular type.

Animals which have undergone extensive modification in evolution, and especially some parasites, have young stages which are without counterpart in other animals of the same phylum. Thus the larval stages of a liver-fluke bear no resemblance to any stages of the free-living planarians. A developing earthworm never has the remotest resemblance to the typical annelid trochophore larva.

Among the plants, there is comparable evidence of relationship between groups but none so spectacular as that which can be found in animals. The developing embryos of plants cannot be said to trace the evolutionary path along which the adults have come, as clearly as do those of many animals, but a similar kind of link with the past may be illustrated by the processes which lead up to the formation of an embryo. Fertilization in the lower plants, Bryophyta and Pteridophyta is nearly always effected by a swimming male gamete. In most representatives of the Spermatophyta, the male gamete is transferred to the female organ in a pollen grain and delivered to the female oosphere by a tube which penetrates the outer protective tissues. The gamete has lost its free-swimming characteristic. In *Cycas*, we see the combination of the primitive and the more advanced. The male gametes are transferred in wind-borne pollen grains. These develop pollen tubes which penetrate the tissues surrounding the oosphere, but as the gametes mature, they show the link with the primitive in that they develop cilia and may swim along the pollen tube during the fertilization process.

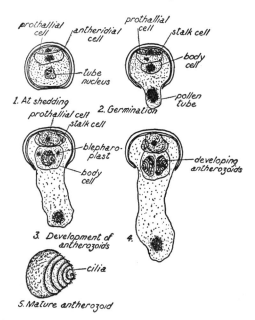

Fig. 21.13. Stages in development of pollen grains of *Cycas*.

We can only account for the vegetative ascendancy of the diploid sporophyte over the haploid gametophyte in the higher terrestrial plants, by assuming that it was more adaptable to a drier environment.

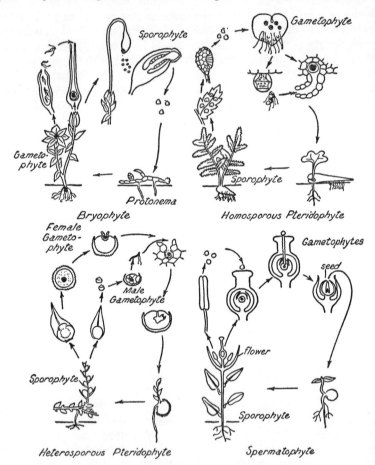

Fig. 21.14. Diagrams of the life cycles of a bryophyte, a homosporous pterido-phyte, a heterosporous pteridophyte and an angiosperm.

We can trace the same type of life cycle, that of alternation of generations, from the most primitive of algal forms through the very simple terrestrial bryophytes to the more thorough-going land plants, the pteridophytes and spermatophytes. The basic pattern of events in the life cycles of all is the same, pointing to a common origin, but the gradual switch of vegetative activity from the gametophyte to the sporophyte is undoubtedly bound up with the adoption of a land habit. We can only believe that land plants commonly evolved from aquatic ancestors. Why else should they be linked together by a similar life cycle which varies only in the extent to which it is adapted to the new conditions?

Innumerable examples from the field of comparative structure can be used to illustrate relationship between existing forms on the basis that they have a common ancestry. The evidence is indirect, but it cannot be explained by any other logical argument.

Deductions from Comparative Physiology

If we are able to deduce relationship between individuals by reason of the fact that they display a conformity to pattern in structure, the same should equally apply if they conform to pattern in function and chemical structure. Needless to say, physiological comparisons must be restricted to those of existing types. It is not possible to support deductions made from living animals and plants with those from fossils, since it is impossible to reconstruct the fossil as a living creature. Nevertheless, logical arguments in favour of evolution can be put forward purely on the evidence derived from the chemistry of living organisms.

Perhaps the most conclusive of these is that based on comparative serology which compares the nature of the proteins contained in the body fluids of organisms. It is a fact that the proteins of species invariably show specificity and yet it may be shown by serological methods that the proteins of the body fluids of species closely related from a structural standpoint, have much more in common than have those of species only distantly related. The method of testing similarity is based upon the effect of the proteins from the species concerned upon the body fluids of another species. If a small quantity of the protein from one of two species under comparison is injected into a third, usually a rabbit, the blood of the rabbit reacts to its presence by developing the power to precipitate it. The rabbit's blood is able to manufacture an antibody, a precipitin, which will precipitate the foreign protein, the antigen, and so render it innocuous. This precipitin, once manufactured in the rabbit's blood, may remain there for some time and retain its property of destroying any of the antigen which caused its first appearance. The important fact in comparative serology is that the precipitin is fully effective only on proteins identical in structure with the original antigen. Its effect on distantly related proteins is negligible, and on totally unrelated proteins, there is no effect whatever. If therefore, the second protein under test is not precipitated by the blood sensitized by the first, then it is considered that there is no relationship between the two.

If a rabbit is made to develop antibodies against human blood, the effectiveness of the antibodies can be demonstrated by withdrawing from the rabbit some blood serum and mixing it in a tube with some human blood. The human blood proteins are immediately precipitated as a visible greyish cloud. With the blood of say a horse, cow or pig, no such effect would be observed. The anti-human rabbit blood is apparently fully effective against human blood only, but if blood from an anthropoid ape is used instead of that from a horse, cow or pig, a precipitate is formed although less dense than with that of human blood. A precipitate is also formed in the anti-human serum with blood of the old-world monkeys and with blood of the South American monkeys but in each case the effect is progressively less striking. No precipitate at all is formed with the blood of lemurs.

By this method of testing, the human species seems to be most closely related to the anthropoid apes and less so to the monkeys. This is precisely the same relationship as can be deduced from structural comparisons. Attempts have been made on a large scale in both plants and animals to make classifications based on these methods. Wherever this has been completed, as for example in the monocotyledonous angiosperms, there has always been remarkably close agreement with the classification based on morphological comparisons.

Comparison of the physiology of the major groups of plants and animals undoubtedly shows distinctions from one another but there are always similarities between members of the same group which may add weight to the pro-evolution argument. The almost universal occurrence of cellulose in plants and its corresponding absence in animals, sharply divides the two but welds the separate groups the more closely. The same reasoning applies to the presence of starch in plants and of glycogen in animals. The epidermal cells of vertebrates characteristically develop keratin, whilst the corresponding substance in the arthropods is chitin. Arginine phosphate plays a part in crustacean respiratory processes analogous to that played by creatine phosphate in vertebrates. The iron-containing haemoglobin is invariably the oxygen transporter in vertebrates, whilst the copper-containing haemocyanin is much more common amongst invertebrates. Haemoglobin does however occur in some molluscs, annelids and arthropods, but this apparently erratic appearance can be explained when it is understood that its prosthetic group, based on porphyrin, is found in oxidizing enzymes of almost universal occurrence (cytochrome, cytochrome oxidase, catalase) and also in the chlorophyll in green plants.

Glucose is found universally in plants and animals. Even so, it is remarkable that the glucose is always the dextro-rotatory form and never its stereoisomer.

The analysis of the body fluids of both aquatic and terrestrial animals shows that they are fundamentally similar. This may not have much significance until it is realized that both are strikingly similar in ionic composition to sea-water. Such facts require explanation. We can only interpret them logically if we assume the ancestry of the land animals to be common with that of the marine aquatic forms. The sea, in composition and stability, appears to be well constituted as a physiologically balanced solution to which cells became accustomed. It is reasonable to suppose that the early invaders of the land could survive only if their body fluids were likewise maintained at the same composition as the fluids to which their cells were already adjusted. The higher animals have continued to make use of the conditions which were found favourable by their early ancestors.

But the land habit has posed problems in the maintenance of constancy of the body fluids; they have been solved by the evolution of new structures with new functions. The ridding of the body of unwanted substances likely to interfere with the local physiological balance is simple for organisms whose bodies are readily permeable to the surrounding fluid environment. The problem is much more difficult for the animals whose bodies are cut off from such watery media. The evolution of such organs as nephridia, Malpighian tubules and kidneys, has been the solution.

Deductions from Geographical Distribution of Plants and Animals

A study of the distribution of plants and animals over the earth's surface presents some peculiarly striking facts. Different regions differ, often very markedly, in respect of whole groups of plants and animals. Often, these marked contrasts are not explicable in terms of climate. In the southern hemisphere, we may take for comparison the enormous land masses of South America, Africa and Australia. All have comparable climates ranging from tropical to temperate, and yet each is most clearly characterized by its inhabitants.

South America is the natural home of the edentates such as the sloths, ant-eaters and armadillos, the primitive monkeys such as the spider monkey with a prehensile tail, and rodents such as the guinea-pig. None of these occur naturally in Africa or Australia. Africa is characterized by the wide variety of antelopes, the zebras, the giraffes, the lions, the baboons, and by the rarer gorilla, okapi and aard-vark. Of these, the giraffe and the aard-vark families are to be found nowhere else in the world. Australia is even more unlike any other land mass in possessing no natural placental mammals of any kind except the bat, some mice and the dingo. The presence of the first is explained by its possession of wings, and the other two seem to have accompanied the earliest human colonizers of the continent. The characteristic mammals are the marsupials such as the kangaroo, and the peculiar egg-laying mammals or monotremes, the duck-billed platypus and spiny ant-eater.

The occurrence of the different types of animals in the three different areas might be explained by saying that each is supporting the type of

Fig. 21.15. Probable distribution of land during Cretaceous period superimposed over map of modern world.

animal which is best able to survive under those conditions. But this is patently not the case, as is borne out by the overwhelming success for example, of such animals as the rabbit, after introduction into Australia, and the red deer, introduced into New Zealand by white men.

The explanation must be sought elsewhere and the only logical interpretation is in terms of the origin of the different types of animals and their past migration and distribution over the earth's surface when the land masses were not as they are at the present time. Geology provides us with the means of plotting the land masses as they have varied through the millions of years. Fig. 21.15 shows the probable distribution of land and sea about half-way through the Cretaceous period, and from it we may deduce the reason why Australia possesses no natural mammals other than the bat. The true placental mammals did not appear before the late Cretaceous period, but previous to this, the monotremes and marsupials were in evidence all over North America and Europe. Some time during the Cretaceous period, they spread over the land bridges via Asia into Australia. This fact is established from the fossil records of America, Europe and Australia. There they developed from somewhat small, primitive forms into the present types, but they were never followed by the placental mammals (with the exception of the bat) because the land bridge between Australia and Asia was broken before the latter could spread from their point of origin which was almost certainly North America. The marsupials and monotremes have flourished in Australia because they had no competition from the much more vigorous placental mammals. Upon their advent, the latter have, in large measure, been able to oust the indigenous mammals. Geographical barriers, not climate, prevented the spread of placental mammals into Australia.

South America and Africa were connected to North America during the periods which followed the evolution of placental mammals, and these animals spread into both areas. In the case of South America, the early connexion lasted for a relatively short time, and for many millions of years only the more primitive of placental mammals, none of which were carnivorous, lived there and were sheltered from the competition of the more vigorous, advanced mammals. Later, however, South America once again became joined to North America, as it is today, to let in carnivorous mammals. They have gradually reduced the types of primitive placental mammals to very few indeed. The fossil record shows that whilst they were protected by the sea barrier, the primitive mammals flourished abundantly and developed into forms seen nowhere else on earth.

Africa has not the same characteristic of being populated by primitive types of mammals as have South America and Australia. It would appear that the land links between Europe and Asia and Africa were barriers as effective as the sea. The dry northern deserts would prevent movement of animals from the north into the south. Africa was not populated by land mammals until after the marsupials and the very early placental mammals had long since disappeared from Europe and Asia. Thus Africa's first mammals did not include marsupials and monotremes but were among the more advanced. These have remained there still, cut off from the north, and have evolved more or less along their own paths.

Thus, in the case of these three land masses, we can account for their individual and characteristic mammalian fauna only by supposing that each is supporting descendants of extinct primitive forms which radiated from their points of origin and became cut off. The marsupials and monotremes were cut off in Australia, never to be linked again with the point of origin of the placentals. The primitive placentals were cut off in South America and were only superseded by the more advanced in relatively recent times. Only the more advanced mammals ever got

29

into Africa. In each case though, the earliest colonizers gave rise, through an evolutionary process, to types adapted to the prevailing conditions. Hence, in every case, the fossil history shows the rise and fall of numerous types derived from the early ancestors.

Another striking fact which becomes clear from a study of animal and plant distribution is the incidence of obviously closely related types at widely different points of the globe. For example, tapirs occur in South America and Malaya only; lung-fish are found only in localized regions of tropical Australia, tropical South America and tropical Africa; camels and llamas, which are close relatives, are found only in North Africa and Asia and South America respectively. Once again, fossil studies show that such discontinuous distribution was not always the case. Each was once very widespread but has become exterminated in all save the places mentioned. We can only suppose that later forms eliminated them from most of the wide areas which they once inhabited.

An examination of more local floras and faunas may also provide facts which cannot be explained logically save on the basis that evolution has occurred. This is particularly so when oceanic islands are the objects of investigation. Darwin, who spent several years in studying the flora and fauna of many different parts of the world (*Voyage of the Beagle*) was extremely influenced in his thoughts on evolution when he surveyed the Galapagos Islands. These lie in the Pacific Ocean about 1000 km from the nearest land which is the west coast of South America. The inhabitants correspond to those found on the mainland as far as general types are concerned, but by far the greater majority are, as species, quite distinct from the mainland plants and animals. Darwin could only suppose that the islands, being volcanic in origin, had been populated by organisms from the nearest land but that these earliest inhabitants had evolved along lines different from those along which the common ancestors of the mainland had evolved. There has arisen therefore some divergence in the species of mainland and island. A similar state of affairs exists in the Cape Verde islands off the African coast. These islands are very similar in character to the Galapagos islands from the point of view of climate and the two might therefore be expected to support an identical population. This is far from the case however. The inhabitants of the Cape Verde islands are in general type similar to those of the African mainland, but once more, as species, they are endemic to the islands.

From these lines of evidence we must draw our conclusions. It is noteworthy that nowhere are there any conflicting or confusing facts which may cloud the issue. The present living creatures have undoubtedly descended from the living creatures of the past. Slowly, over millions of years, changes have come about; these changes have persisted and they have accumulated. Organic evolution may be defined as a process of cumulative change, and all the evidence clearly indicates that it has occurred. That it is still taking place, there can be no doubt when the influence of man on domestic plants and animals is taken into account. That it will continue to occur is also certain. How it has happened is still a riddle, to which as yet there is no single clear-cut answer.

HOW EVOLUTION MAY HAVE OCCURRED

Accepting evolution as a fact on the foregoing evidence, for want of some other logical explanation, we must now pass to the more difficult task of appreciating the way in which it may have been brought about.

Lamarckism

The need for such an explanation had become apparent with the realization that the Linnaean natural system of classification implied a continuity of descent among living things. Buffon's suggestion that the environment of an organism might play a part in directly moulding it into a new form, which could persist and thrive in those conditions, could hold good only if directly impressed characteristics were handed on from parent to offspring. Erasmus Darwin and Lamarck both advocated this idea that changing form was brought about by the handing-on of characters, acquired by one generation, to the next. An acquired character may be defined as one that has developed in the course of the life of an individual in the somatic or body cells, usually through some external change in the environment or through excess use or complete disuse of a part of the body. The inheritance of such characters means the reappearance of them in some of the individuals in the following generations.

Whilst Lamarck was not the first to expound such a theory, he was the first to enlarge upon it and quote specific examples in support of it. Thus the conception of the inheritance of acquired characters to account for evolution is often referred to as Lamarckism.

Lamarckism may be summarized as follows. Changes in conditions create new needs. The new needs lead to new methods of behaviour involving fresh uses (or disuses) of existing organs of the body. These fresh uses (or disuses) lead to changes in their structure or method of functioning. The resulting bodily changes are inherited. Lamarck gave rather crude examples to illustrate his points. He suggested, for instance, that if seeds from a marsh plant fell on high ground and germinated there, they might become adapted to the drier soil, and change by the acquisition of new characters more fitted to their fresh surroundings. This new type, by perpetuating itself, would establish a distinct species. In animals, a new environment, calling forth new needs, causes the animal to seek to satisfy them by making some kind of effort. Thus new needs lead to new habits and hence to modification of old structures. The giraffe, inferred Lamarck, endeavouring to feed on foliage growing higher and higher above ground, stretches its neck and legs. As a result of this stretching through many generations, the giraffe has developed into an animal with long front legs and neck. Similarly, birds which have to come to rest on water to find food, have developed webbed feet through the constant spreading of the toes and stretching of the skin. Flat-fishes have arisen by comparable means from fish with the habit of turning on one side in shallow water.

Such was the reasoning of Lamarck, but in the light of modern knowledge it cannot be said to fit the facts. It must be remembered that Lamarck was not in possession of any of the facts of inheritance as we understand them today, neither had there been built up a mass of evidence all tending to indicate that acquired characters such as he visualized are not in fact inherited. It would be fairer to Lamarck to remember him as a man of many positive achievements in zoology rather than as a protagonist of a theory of evolution which was to be discredited. There have been supporters who have adhered to Lamarckism in principle, to account for certain otherwise unexplained phenomena. A school of geneticists under Lysenko, following the teaching of Michurin in Russia, have sought to point out that, under certain circumstances, new characters may be impressed on an organism by its physiological environment and that these are to some extent inherited. Since, however, the experimental evidence seems overwhelmingly against the possibility of the inheritance of characters acquired in the way Lamarck envisaged, few modern biologists would subscribe to his views.

Darwinism

On his travels with the *Beagle*, Charles Darwin had collected a mass of evidence indicating that evolution had occurred. In his *Origin of Species*, he sought to give an explanation of the process. The theory which he put forward is known as Darwinism and may be summarized as follows.

The potential offspring from an individual reproductive process are almost always relatively enormous in numbers and all organisms tend to increase their numbers in a geometrical ratio. For example, a female cod may carry as many as five million eggs in one season; a giant puffball may produce seven million million spores; a single poppy capsule several thousand seeds. If each potential new cod, puff-ball or poppy were successful and reproduced itself at the same rate as its parents, in a few generations there would be countless numbers of cod, puff-balls and poppies.

Despite this enormous potentiality, in natural conditions, the numbers of a species tend to remain constant. The seas are not solid with cod, nor the land piled high with puff-balls and poppies. By far the majority of potential offspring die before they can reproduce.

The death of many such individuals can be attributed to their loss of a battle for survival between themselves and surrounding organisms. There is continuous competition between individuals for space, food and all the other requirements of living things. Therefore, Darwin's first major conclusion was that there is a *struggle for existence*. If a large number of individuals of a species are examined, it is abundantly apparent that they exhibit *variation*. No two of them ever look or behave exactly alike. This was Darwin's second point.

Since there is a struggle for existence, and since all the combatants are different from one another, some are sure to possess characters which are more advantageous in the struggle, and others will possess characters which are less advantageous in this respect. It is reasonable to suppose that on the average the best equipped, that is, those showing the more favourable characters, will survive the struggle and reproduce at the expense of those showing less favourable variations. Darwin's next point therefore was that only *the fittest survive*.

He then postulated that a great deal of, if not all, variation is transmitted from parent to offspring, so that new generations will increasingly tend to be composed of individuals most fitted to survive. The other variants would have been selected out, by losing the battle for existence. By this means, from any starting point, therefore, over many generations, it is conceivable that a species may change by becoming increasingly adapted to its surroundings and ultimately become a new species. *Evolution may thus be brought about by the agency of natural selection on a wide range of inheritable variations.*

Darwin made no attempt to distinguish between continuous and discontinuous variation and assumed that any or all variation could be transmitted from generation to generation. His inheritable variations also embraced acquired characters and to that extent he subscribed to Lamarckian principles.

Darwin's work did two things. It finally forced the issue with regard to the acceptance of evolution as a fact and it also focused attention on to its weakest point. Some of the more penetrating and critical minds, whilst impressed with the mass of detail so carefully prepared by Darwin, and with his conception of natural selection, pointed out that, although he may have accounted for the *survival of the fittest*, he had done no more than Lamarck *to account for their arrival*. The manner of inheritance of variation continued to be a point of much speculation, but for want of more positive information in this branch of biological knowledge, Darwinism became widely accepted as a theory of evolution. Darwin did attempt to vindicate his views on the inheritance of variation in his theory of "Pangenesis," which postulated that particles were carried by the body fluids from adult organs to the reproductive cells, where they had their effects and thus influenced the next generation.

Weismannism

The first major criticism of a theory of evolution involving the inheritance of all somatic or body variation as envisaged by Lamarck and Darwin, came from the German biologist, Weismann, who between 1868 and 1876 published a series of papers on the inheritance of variation. He contended that acquired characters could not be inherited. Although he had carried out a series of crude experiments such as the removal of mouse tails for many generations, and had shown that every generation always produced mice with tails whether the parents had tails or not, he did not make much use of this as evidence. But more from hypothetical reasoning than from direct evidence, he stated that there was a distinction between the body of an organism, the *soma*, and the cells concerned only with reproduction, the *germ plasm*, and so formulated his *Theory of the Continuity of the Germ Plasm*. He postulated that only the protoplasm of the germ-cells could affect inheritance, and that the soma could play no part. In each generation, special reproductive protoplasm was set aside at an early stage of development. This germ plasm could not in any way be influenced by what happened to the rest of the body. Weismann even went so far as to develop a theory of inheritance which was based on chromosome behaviour, the details of which had already been described for mitosis. Weismann's theory was a serious blow to Lamarckism which was thereby largely discredited. Since that time, the Lamarckian views have had few whole-hearted supporters.

Neo-Darwinism

A few years later there was a revolt against orthodox Darwinism. In 1894, William Bateson published his *Materials for the Study of Variation*. In it, he deliberately drew attention to variations between related species which were of the discontinuous kind, in contradistinction to the continuous variations which had been so carefully studied by Darwin. It was suggested that evolutionary advances could have

been made as a result of the appearance of these discontinuous variations rather than as a result of the variations of the continuous kind.
This departure from Darwinism received very considerable support a
few years later from the personal observations of Hugo de Vries on the
plant *Oenothera*, which had been introduced into Holland from
America. In his cultured plants, de Vries discovered a few completely
new forms which appeared among the numerous ordinary forms, when
all had been growing under the same conditions. The variants were
quite distinctive. There was no question of their being included within
a range of continuous variation for the character in question.

De Vries gave the name "mutation" to this phenomenon of the
production of new forms by sudden change. It was quite sharply
defined as an abrupt change and could not be attributed to a gradual
process due to accumulation of continuous variations under the
influence of selection. It was a totally new concept of the origin of
variation. It could account for the appearance of discontinuous
variation.

De Vries' account of his observations was published in the period
1900–1903 as *The Mutation Theory*, and in it he suggested that evolution
found its raw material in sudden, spontaneous changes, rather than in
the continuous variations under the influence of natural selection. This
publication almost coincided with the rediscovery of Mendel's work on
inheritance, which had remained unappreciated since 1866. The rapid
development of genetics, especially by Bateson and Morgan, made it
possible to replace many of the speculations concerning inheritance by
deductions from soundly based experimental evidence. The conception
of the gene as the vehicle of inheritance, and the early supposition that
continuous variation is not inheritable, led to the formation of a new
concept of evolution. It may be called *the mutation theory*, since its
more extreme exponents completely abandoned all Darwinian ideas
and assumed that all new forms that had ever existed arose only as a
result of some abrupt change in chromosome or gene composition. So
thoroughly did the idea of mutation colour the evolution picture that
even the fundamental principle of Darwinism, natural selection, failed
to find any place in the current theory of the time.

It is a fact that the mutation theory has the merit of being the only
one borne out by laboratory work. Early in the work on genetics, it
was discovered that mutations are of fairly frequent occurrence. They
can be produced artificially in the genes of *Drosophila* under the
influence of X-rays. To some extent they may be predicted. The rates
at which they occur can be measured. Changes in genic composition do
give rise to inheritable discontinuous variation. So also do changes in
chromosome composition and arrangement. A new species of plant,

Raphanobrassica, was produced by Karpechenko from an artificial cross between *Raphanus sativus*, the radish, and *Brassica oleracea*, the cabbage. In nature, a similar phenomenon has been known to occur. *Spartina townsendi* was first found in 1870 on the mudflats in southern England. By 1900 it had multiplied enormously. Its chromosome count of 126 indicated that it was a hybrid between the European *S. stricta* with 56 chromosomes, and the American *S. alternifolia* with 70 chromosomes. In a habitat which had previously been sparsely occupied, its adaptation was astonishing and its success phenomenal.

There is little doubt that numerous species of flowering plants have arisen by some comparable means. De Vries' mutations in *Oenothera* have since been shown to be rearrangements of chromosomes. There are probably similar examples in animals, but the almost universal occurrence of biparental reproduction in these would tend to prevent polyploidy from occurring very frequently.

Despite the apparent strength of the mutationists' case, not all attention was diverted from study of continuous variation. The school of biometrics which had been founded by Galton, following the publication of the *Origin of Species* by his cousin, had for long been engaged in such studies and continued to do so under Pearson. They did not forsake the Darwinian principle that continuous variation may be inherited, and by mathematical methods applied to variation within large populations, they were able to show that in some cases it is. Galton demonstrated conclusively by the use of statistics that height in human beings is such a case, and his school of workers developed a mathematical theory of selection of such inheritable continuous variation. Johannsen's work on pure lines showed experimentally that such variation can, in fact, be inherited.

Gradually, over the last thirty years, as genetics has expanded, as all the many facets of biological study have yielded more and more information concerning variation and inheritance, and as mathematical methods have been adapted to test evolutionary postulates, there has been a merging of ideas into a wider and more readily acceptable hypothesis. Despite the apparent eclipse of Darwinism by the early mutationists, this "modern synthesis" has been unable to exclude Darwin's original conception of natural selection. A modern acceptable theory of evolution has been presented by Gaylord-Simpson and may be summarized as follows.

Variation is the fundamental raw material necessary for evolution to proceed. The source of this variation must be in the genotype since purely phenotypic variation alone is not inherited and therefore cannot effect changes in succeeding generations. Lamarckism is not an acceptable hypothesis and the Russian Michurinism is not sufficiently convincing. Changes in the genotype must be accompanied by changes in the phenotype, since it is upon the latter only that the process of evolution can work by selection.

There are two sources of variation known to geneticists. Mutations of genes and chromosomes are known to give rise to new inheritable characters. The recombination of genes at crossing-over during meiosis, and the recombination of whole chromosome sets at the conjugation of gametes, may also produce variation, but in such cases there is really no new genetical material. This can only be provided by mutations. Recombination can only bring variation out of storage, as it were, by providing new sets of conditions in which old genetical constitutions may act. It cannot of itself produce new genic material but is a necessary adjunct to the evolutionary process in that it brings about the spread of a new inheritable variation through an interbreeding population. The clone, a population regularly reproducing by asexual means alone, may be regarded as an evolutionary dead end.

It was once thought that only mutations of a considerable size, which produced large discontinuous variations, were really inheritable and thus able to influence evolution, but it is now known that some continuous variation follows Mendelian laws of inheritance. There is a distinction now drawn between the so-called "major genes" and the "polygenes." The former are single-acting entities and have their effect on the phenotype as large discontinuous variations such as might be exemplified by the tallness and dwarfness of peas. The latter are collections of genes acting together and affecting variation in a continuous but nevertheless inheritable way. Inheritance of stature in human beings may be under the influence of such a collection of polygenes. It seems that genic construction must be regarded as being very variable over a graded range of structure from something large and complex to something small and relatively simple (*see* p. 798). There is to be found a great store of variation among polygenes, particularly when numerous polygenic alleles are heterozygous.

Mutations vary considerably in size, from the large, easily recognizable mutations of genetical experiments down to small modifications. The former could be responsible for the creation of completely new patterns, whilst the latter might be regarded as effecting a *microevolution*, as it were. But both, and all grades between, are equally involved in evolution. Any size of mutation may act independently to bring into existence a new form at any evolutionary level from phylum to subspecies. It is exceptional for a single large mutation alone to initiate any considerable evolutionary change, since it rarely persists, and therefore does not come to characterize whole populations. But when accompanied by the more numerous smaller mutations, its effects may be modified and integrated into variations which persist and lead to fresh forms. It might be said that from the point of view of evolution, leaving out occasional cases which do not fit the rule, the smaller mutations seem to be of greater importance.

Mutation rate is obviously of some importance to rate of the evolutionary progress, but this aspect is difficult to study, and there is no conclusive work to indicate just how mutation rate affects the problem.

The effects of mutations can only become operative in populations which are interbreeding, and the new study of population genetics indicates that the result of genetic changes is influenced by several factors. The size and structure of the population, the type of reproductive process, and whether the interbreeding groups are localized or not, all modify the effect of the initial genetic change.

None of these ideas can be said to have been taken into account by Darwin because he had no knowledge of the facts of inheritance, but nowadays no one would suggest that his conception of natural selection can be discarded as easily as his theory of the inheritance of variation. To this extent, Darwinism is reborn as Neo-Darwinism and evolution thus regarded as the effect of natural selection on the continuous appearance of mutant forms.

That selection must be effective in some cases is obvious. Any organism must be able to reproduce in its available environment in order to persist, and the survival of the fittest rule must inevitably eliminate at least those forms which are grossly unadapted. But the role of selection in cases where there is less distinction in form, is debatable, and points of view vary between considering selection as operative only on the obvious cases just quoted with no effect elsewhere, and considering selection as the only really essential factor in evolution.

Those who argue from the former viewpoint, stress the facts that many real evolutionary advances have been non-adaptive and are therefore inexplicable if selection is an important factor, and that many evolutionary phenomena are so minutely adaptive that selection could never have been efficient enough to produce them. The antlers of the Irish elk are in the former category and the suture patterns on ammonites in the latter. Mimicry in animals is of no use until very well developed, and in many cases it has been developed beyond the point where any greater degree of protection is afforded by a further approach to perfection.

The general concensus of opinion in this argument can best be summed up by Gaylord-Simpson's own words.

Whatever the proportion of adaptive or non-adaptive characters in any given case, it is certain that a high degree of adaptation is universal in nature. This relationship between organisms and their environment cannot now be considered as due primarily to the environment (Lamarck-ism), nor to the organisms (mutationist) nor yet to some supernal plan, as the finalists say. It must be that adaptation results from material interaction of organism and environment. The mechanism of the inter-action is selection, and the role of selection in evolution is the production of adaptation. This does not mean that selection is all-powerful, so that all evolutionary change is adaptive. Change, unguided by selection, can also occur; selection may even produce non-adaptive changes as a secon-dary effect, although its primary effect is necessarily adaptive in some way. It is also probable that some adaptive change may arise without benefit of selection, although it is improbable that this can persist or lead to significant evolutionary movement without involving selection.

In conclusion, it might be said that the foregoing summary of modern thought on evolution must not necessarily be treated as the last word on this subject. The disclosure of new facts from new methods of study and the re-interpretation of old-established facts in the light of the new knowledge, will undoubtedly recolour or even recompose the picture from time to time. This summary is therefore intended to do no more than give the student the basic composition of the evolutionary picture as it appears to be at the present time.

CHAPTER 22

ECOLOGY

ECOLOGY means literally "the study of home conditions." The term was coined by Haeckel in 1869 from the Greek "oikos," meaning home or dwelling. He spelt it "oecology," and since then, with its modern spelling, it has become used to describe the study of living things as they have their existence in their own natural surroundings, or as it were, in their own homes. Such study is one of accurate and precise natural history, and it may be more briefly defined as the interrelationships between living things and their environments. It is less a branch of biology than the embodiment of all things biological. In the ultimate analysis of the relationships between an organism and its surroundings, knowledge of every other facet of biological study must be utilized. Field studies must be properly linked with laboratory work involving investigations into every aspect of morphology, anatomy, physiology and genetics. Further, the ecologist needs to have at his command the necessary knowledge of physics, chemistry and other sciences, to enable him to analyse the inorganic as well as the living part of the environment. He will need to employ statistical procedures. To make his study complete he must be the complete scientist.

Plant ecology and animal ecology are most often treated as separate studies. In reality no such distinction should exist, since all living things, including man, are so intimately in association with one another that it is impossible for a biologist to pay undivided attention to one or other of them and at the same time read into his observations their fullest significance. All the branches of ecology are parts of a general study which embodies the conception that all living things and their environments are inseparable, and that they form collectively an indivisible whole.

Such a system in which organisms live as communities related in an inseparable way to their environments is known as an *ecosystem*. It may be a pond or a forest, a fresh-water lake or a salt marsh, but in every case the nature of the environment and the changes in it have an influence on every aspect of the lives of the organisms that inhabit it and they in their turn exercise some control over the nature of the conditions in which they live. Each ecosystem must be treated in its entirety of all living and inanimate material together with the energy flow and transformations taking place within it. All the ecosystems ultimately merge

898

and combine to form the *biosphere* or *ecosphere*. Any such system with an energy flow through it will reach a condition of stability by the operation of self-regulating mechanisms, i.e. it has the nature of a cybernetic system (*see* p. 146). In order to see how this stability is achieved or how it may be affected, all parts of the system must be investigated.

The correct approach to the study of any ecosystem must therefore take into account every one of its four main kinds of components. These are, the *non-living* or *abiotic substances*, for example, water and all that may be dissolved in it; the *producer component* of the living organisms found in the system, namely the photoautotrophes, chiefly green plants; the *consumer element*, mostly animals; the *decomposers*, the fungi and bacteria. Such study can be achieved properly and with meaning for the student if he has been trained as a biologist rather than as either a botanist or a zoologist. The modern trend in teaching is towards the integration that will make ecological study a more fruitful part of the biologist's work.

There are two fairly well defined approaches to ecological study; one deals with the individual, the other with the community. If we are concerned with a single plant or animal species throughout its life history in relation to its living conditions, we are making an *autecological study*. On the other hand, if we concern ourselves with the plants and animals which occur together in common surroundings, we are making a *synecological* approach or an approach through sociology.

Autecology, the detailed study of the individual, its complete life story including form and physiology, its methods and rate of spread, its adjustments and behaviour under different habitat conditions, is a long and difficult study. It may be the ideal approach to the wider subject of sociology, but can hardly be achieved successfully by the student at this level. It is more convenient for him to embark on synecology, the ultimate end of which in any case will lead to autecology, since he will always have to consider the individual in making a complete study of a community.

The first step in any synecological study, whether it be of a large or a small area, is the making of a survey involving two clear-cut practical projects. First, a list of all the plants and animals which may be found in the area must be compiled, so that the community can be described in terms of its constituents. Then a detailed analysis of the habitat factors must be made in order to reach an understanding of the conditions in which the inhabitants associate together. In connexion with the first, notes on the habits of the animals listed are very necessary, so that a picture may be formed of the behaviour of each animal in the area, whether feeding, breeding or merely passing through or over it.

Great skill and cunning are usually needed to observe the habits of the larger animals, but patience is often rewarded with information vital to the ultimate analysis of all the findings. Although the listing of species may call for a good deal of painstaking effort, it is, in general, more straightforward than is the task of analysing the habitat factors. This calls for a knowledge of geological formations, soil conditions, climate and meteorological data, and the precise effect of one organism on another, among other things. Resort to good topographical and geological maps may be necessary, a local meteorological station may have to be consulted or a station may have to be set up in the area. A good deal of work on soil analysis must be carried out in the laboratory.

In the following sections, we shall attempt to outline suitable general procedures which may be found useful in making an ecological survey of a given area. It will be found that it is most advantageous for the beginner to concentrate first on a study of the vegetation of the selected area, for several reasons. Plants associate in static and well-defined units or communities which in themselves form a background against which other work may be undertaken. For instance, animal communities may be studied in relation to the vegetational units, i.e. the fauna of woodland, heath, moor, pond or lake. Further, the plants of an area are much more easily found, counted and identified, so that much more information can be compiled in a given time. The feeling of achievement goes a long way towards encouraging the beginner to maintain an interest in his work.

It is the case that most of the general survey work described below tends to be qualitative or descriptive rather than quantitative. To be otherwise, that is to express the nature of a community in measured terms and to say in other measured terms how that nature may vary in accordance with measured differences in habitat factors, calls for the application of methods that may be beyond what can be expected of students at this level. It is not so much the collection of numerical data that creates the difficulties, but its proper interpretation. The skill and patience required in the field needs to be matched by an equally expert application of statistical analysis to the data there collected if it is to have any real meaning. It is considered that to involve the reader here in such matters is well beyond the scope of this work but we advise reference to an appropriate text dealing with statistical processes applicable to the analysis of ecological data. There are many situations where detailed intensive study is required and the fully quantitative approach is essential. In this introduction to a field of biology of ever-widening importance we can do no more than outline its main features and suggest some very general practical approaches to its study.

SURVEY PROCEDURES

We shall assume here that a biology class is proposing to undertake the study of the living population of a selected area within reasonable range of the school. This last condition is requisite because such work can only achieve any measure of completion if repeated visits are made to the same area. For the country or seaside school, the choice of a suitable area usually presents no great difficulty. The same cannot be said of the city school and it is an obstacle which with the best will cannot always be overcome. However, very careful application of the first survey procedure, even in a city, sometimes leads to unexpected results which can be followed up with some success. Open spaces, however small, generally support some kind of life, and it is on these that attention should be focused.

The survey can be divided broadly into two parts. What may be described as extensive but cursory study of the locality must first be made, followed later by the second part, intensive study of smaller parts as they present themselves within the main area. The extensive study is really a preliminary reconnaissance aimed at obtaining a general idea of the variation in the vegetation which the area has to offer. At its completion, no more than this need have been achieved. To carry it out, the area must be explored systematically but not in detail. The party must traverse the area in several directions, noting the salient features of the plant life and suitable routes by which smaller localities can most easily be reached. Each of the well-marked natural or semi-natural vegetation units need only be briefly examined but each should be noted as prospective future material and its position roughly indicated on a map. Reference to a one inch to the mile ordnance survey map will help in the construction of a sketch map showing the rough outline of the major units encountered. A successful reconnaissance is obviously dependent to some extent on a knowledge of common species, but assistance is usually obtainable where difficulties of this kind arise, and practice at plant recognition soon makes for proficiency. The map of the area, when completed, should show the relative positions of any woods, heaths, moors, bogs, lakes, streams, sand-dunes, rocky shores, cliffs, etc., in the neighbourhood. Less natural vegetational units such as arable fields, meadows and plantations can be included, since they may afford opportunity for work of a more specialized kind, if that is considered desirable.

Once this overall picture of the vegetation within the area has been sketched, intensive study of selected communities may now be made. Such study involves four major processes to make it complete. The first is a careful examination of the floristic composition of the chosen community. Next comes a full investigation of the habitat factors.

These are followed at intervals by similar work to discover what changes, if any, are occurring in the vegetation and the environment, and lastly, the findings concerning one community must be correlated with the findings about others so treated. As each community is dealt with, large scale maps, which more nearly suit the needs of the work, may have to be constructed. It is advisable to make such maps in any case, since their making leads to a much more accurate appreciation of the extent and character of a community or part of it, than would otherwise be obtained. Further, reference to a position on a map facilitates any discussion concerning that position.

Mapping the Area

Maps need not be elaborate and can be constructed for reasonably sized areas by the baseline and offset method, the principles of which will be readily understood by reference to Fig. 22.1 (*a*). The method is

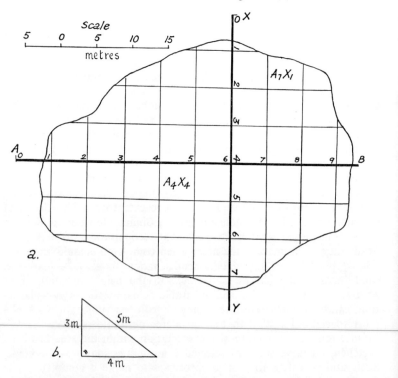

Fig. 22.1. (*a*) Baseline and offset method for mapping an area. (*b*) Cord right-angled triangle.

not very practicable for large, irregular areas in which more complicated survey apparatus would be necessary. Since, however, a school party would be ill-advised to undertake detailed work involving a large area, it is the only method which will be described here.

A perfectly straight line is run out across the area, preferably along high ground, where all points on it can be seen from everywhere within the area. The baseline AB on the diagram represents this line. It can be made straight by sighting along three vertical poles spaced one at each end and one in the middle. A second line XY is laid across the area to intersect the first at right angles near its mid-point. It is necessary to ensure that the intersection is exactly at right angles and this can be achieved well enough by use of a cord triangle measuring 3 m × 4 m × 5 m as in Fig. 22.1 (b). From the point of intersection of the base lines, pegs are driven in along each in both directions at equally spaced intervals until the edge of the area is reached (1, 2, 3, 4, etc., on Fig. 22.1). Intervals should be chosen to suit the size of the area. Five metre intervals are useful for small areas of about fifty metres across. It is from these pegs on the baselines that the offsets to the edges of the area are measured, care being taken to see that each offset is perpendicular to the baseline. If all the accumulated data are plotted to scale on squared paper, a serviceable map of the area is quickly made. It is a great advantage during later work to pinpoint exact spots within the area, and this can be done very easily, if, during the offset measurement, pegs are driven into the ground along each offset as it is measured, at the same intervals as those used along the baselines. The effect is to divide the area up into squares each outlined by pegs. Reference to the baseline pegs will quickly give the relative position of any square so marked out. The squares outlined in the diagram would be known as A_7X_1 and A_4X_4 respectively. The exact position of any point within a square is simply a matter of measurement from two sides of the square. If it is intended to visit the area at different times, it will be found an advantage to leave these pegs in position. If this is the case, they need to be durable, easily visible and well driven so that they are not lost or dislodged.

Floristics

The area must be examined systematically and the plants encountered must be listed. Here arises one of the greatest difficulties, that of identification, since little progress can be made until plants are readily recognized. Use of a flora is usually essential since accuracy is of paramount importance. Not infrequently, advice of an expert must be sought. Access to a good herbarium is useful, whilst the construction of a herbarium, in association with the work being undertaken, will prove an invaluable guide to others who may follow. In such a herbarium, the species should be displayed to show the general form of the plant and any special identification features. If several communities are tackled, the herbarium plants should be arranged according to the communities in which they are found.

The floristic composition can be used to describe the area under study in terms of the dominant or codominant species. This can be done only if the relative frequencies of the species are known. A counting system may have to be used in many communities before the limits of a particular plant association can be fixed. Two methods of counting are usually adopted, the *quadrat* and *transect* methods. The quadrat method involves the construction on the ground of a metre square marked out by cord. Detailed study of plants within the square is then made. The list quadrat is merely a list of the plants in the square with the numbers of each. The more instructive chart quadrat is made by plotting the relative positions of all the plants in the square (*see* Fig. 22.2). The position of every quadrat made should be fixed on

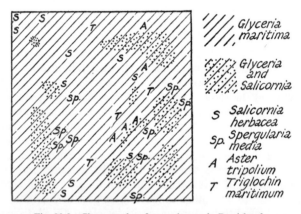

Fig. 22.2. Chart quadrat from salt marsh, Portishead.

the large scale map and if possible, permanently marked on the ground with pegs, so that it can be listed or charted again in the future. The transect method involves the use of a tape or string line marked at convenient short intervals. This is stretched across the area and the plants encountered at the interval marks are recorded. This is a random sampling of the flora and counts made give a fair estimate of the relative frequencies of each species over a comparatively wide area. When all the plants have been listed, and numbers estimated, each species can be given a symbol as below, to indicate its frequency of appearance—

d = dominant	f = frequent
$v.a.$ = very abundant	$l.f.$ = locally frequent
a = abundant	o = occasional
$l.a.$ = locally abundant	r = rare

The Habitat Factors and their Measurement

The habitat factors include—

1. Factors peculiar to the precise location of the area with respect to the rest of the district. They may be called *physiographic factors* and embrace the general topography of the area, its altitude, slope of land, drainage conditions, mobility of the soil due to tidal action, erosion, silting, etc., and tidal frequency and depth where applicable.

2. *Factors due to climate.* They include general climatic conditions such as temperature, rainfall, prevailing wind, light intensity variations and evaporating power of the air.

3. Factors due to the physical and chemical properties of the soil. These are called *edaphic factors.* Points to be considered include soil texture, water content and availability, organic content and pH.

4. Factors due to the influence of other living things or *biotic factors.* These are difficult to define, for if the community is looked upon as being controlled by factors, it becomes difficult to decide which organisms constitute the community and which are operating as controlling factors. Nevertheless, certain broad effects may be assessed, not least, those of man on natural communities.

Physiographic Factors

The general topography and altitude of an area can be studied from an ordnance map. Direction and degree of slope within a small area, are usually only found by measurement on the site. The simple water level is convenient to use in conjunction with surveyor's poles or other suitably marked rods (*see* Fig. 22.3). Contour maps of the area can be constructed quickly from readings taken. Of the remaining physiographic factors mentioned, some can be assessed by eye only. Drainage conditions become reasonably apparent after the ground has been walked over a few times, whilst the movement of soil, if occurring in appreciable amounts, is equally apparent. The extent of silting of a particular point can be roughly estimated against marks on a rod driven sufficiently deeply. Information concerning tides is always readily available and the high and low tide marks are easily found.

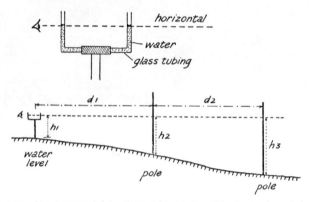

Fig. 22.3. Simple water level. From a knowledge of h_1, h_2, h_3, d_1, and d_2, the slope of the ground can be calculated.

Climatic Factors

Records of the prevailing weather conditions should be made during the period of the work undertaken. They should be continuous to be of any real value, since readings taken at odd intervals may lead to false conclusions about the weather in general. Continuous records can always be obtained at the local meteorological station, but it is more instructive if the readings are taken by the class engaged on the work. Daily maximum and minimum temperatures can be plotted in the form of a graph over as long a period as required. Rainfall by weekly amounts can likewise be recorded. Information regarding conditions of light intensity is essential when comparisons are to be made between small localized areas, as for example, the inside and outside of a wood or the north and south sides of an east-west ridge. These can best be made with a light intensity meter graduated in foot candles. Alternatively a photographic exposure meter, suitably calibrated, can be used. Failing the availability of either of these, a simple bee-meter can be employed. The light intensity is expressed as the time taken for a piece of light sensitive paper to acquire a standard colour. Seasonal variations in light intensity and duration may be recorded in terms of a "light sum." Intensity readings are taken at half-hourly intervals throughout the day and plotted on a graph against the time of day. An estimate of the total light for the day may be arrived at by measuring the area between the outline of the curve and the base line (*see* Fig. 22.4). Such graphs compiled during the winter, spring, summer and autumn periods, are very instructive when compared with one another.

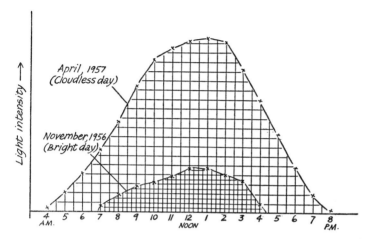

Fig. 22.4. Light sum diagrams.

The evaporating power of the air depends upon many conditions such as temperature, relative humidity and air currents, but these need not be measured separately since the atmometer will give directly all the information that is required for comparative purposes. The atmometer is represented in Fig. 7.5 and its construction explains its mode of working and its usage. It is exactly similar to the potometer in principle and it is very instructive if the two instruments are used simultaneously in the field and their readings compared. Relative conditions of exposure to a drying atmosphere in open and sheltered positions and at different levels in a stratified vegetation, can be studied by means of the atmometer. If it is found necessary to measure relative humidity at specified points, a hygrometer is required. Simple patterns combined with thermometers are obtainable and are quick and easy to use.

Edaphic Factors

Measurement and analysis of the soil factors are of great importance, because if properly investigated, they will show that the distribution of plants within an area is greatly influenced by them. Most of the soil analysis can be achieved only in the laboratory, so that it is necessary to collect soil samples. These should be taken at representative points in the area, due importance being given to the depth at which each is taken. This depth should not be arbitrary but should be selected with respect to the active parts of the root systems of the plants constituting the vegetation at that point. It is instructive to sketch the side of the excavation at which each soil sample is taken, showing the positions of roots of named plants in the soil and indicating on the drawing the exact position from which the sample was taken. Each soil sample should be placed in an air-tight container and suitably labelled as to place and date of taking. Soil may be taken from any depth from surface to about three feet by use of a soil auger. If the auger contents which are removed at each boring are arranged in succession, a good indication of the soil profile may be obtained. Alternatively, a narrow straight-sided trench can be dug to any depth which is practicable.

Analysis of a soil sample in the laboratory should consist of first an estimation of its texture. This can be achieved fairly well by sifting oven-dried soil through a series of suitably graded meshes, so that the proportions of the differently sized particles can be ascertained. Alternatively, Hardy's Texture Index can be calculated from "sticky point" measurements (see p. 171). More simply still, the differently sized particles can be separated out if a suspension is allowed to settle in a tall glass jar.

Water content may be measured by subtracting the weight of an oven-dry (105°C) sample from its fresh weight. If the sample, weighed fresh, is first air-dried to a constant weight and then oven-dried to a constant weight, the total water content may be expressed as capillary water, that which can be lost by evaporation to the air, and hygroscopic water, that which is retained by the much greater hygroscopic forces of colloids. Note that air-dry and oven-dry weights have no validity unless they have remained constant for several weighings.

The organic content of a soil containing little or no free carbonates can be approximately determined by simple ignition of weighed samples. The soil reaction can be tested in the field to within 0.25 pH, by use of a soil indicator and colour chart. The British Drug Houses, Ltd., make a transportable outfit of a suitable kind; it gives sufficient accuracy for the purpose.

Other soil factors should also be measured. Lime content may be estimated roughly from the amount of fizzing produced when hydrochloric acid is added to a sample. If no such fizzing occurs, the lime deficiency may be estimated in the laboratory using Comber's thiocyanate test. A saturated solution of alcoholic ammonium thiocyanate is added to a small quantity of dry soil in a test-tube. If there is a deficiency of lime, a red colour is produced, and the deeper the colour the greater the deficiency. The test is really one for ferric ions and the inference is that only where there is a deficiency of bases will there be any excess of ferric ions able to react with the thiocyanate to give the red colour.

A careful study of the three main edaphic factors, namely, water content, organic content and pH, and comparisons of these between different places will often explain the distribution of vegetation. Water content is very important and may be the limiting factor in some cases. Organic content needs careful consideration also, because it can exist in the form of humus, which makes the soil richer in available nutrients, or in the form of undecayed raw peat of no particular value in that state.

Biotic Factors

As has been stated, these are the most difficult of all to assess. The effect of plant on plant, e.g. the shade produced by some over others, the competition for space between species and any special modifications associated with this, possessed by some plants and not by others, the interrelations of plants and animals with particular reference to pollination and seed dispersal mechanisms in plants, the depredations of feeding animals, the incidence of parasitism and symbiosis are but a few which should be taken into account. The activities of fungi and bacteria in the soil are of considerable importance. Little appreciation of the biotic factors can be gained until a reasonably good survey of the animal population of the community has been made. The microscopic animals of the soil or water should not be overlooked in the hunt for larger species. But even when a complete knowledge of all the inhabitants of an area has been gained, it is still not easy to estimate the full effect of any one of them upon the others in many cases. Some observations can be made more easily than others. For instance, it is possible to exclude the grazing or herbivorous animals from a patch of land by fencing, and to compare the effect of this protection with other land not so protected. Most of the work concerning biotic factors calls for a great deal of time spent in close and accurate observation. Although a class may not have all the time necessary to achieve the fullest results, they will have lost something by completely disregarding the biotic factors of a habitat.

Plant Succession

Every plant community has developed in a once uninhabited place. Starting from the colonization of bare rock or soil surface, freshly-formed pool or ditch, or a newly silted river bank, so the inhabitants to be found there at any time thereafter have followed in succession on one another. The culmination of this succession is the vegetation of a stable community consisting of life forms of the highest kind which the habitat is capable of supporting. During the succession, changes have occurred in the habitat factors, especially edaphic and biotic, with each successive vegetation form to appear. These changes allow of further changes in the floristic composition of the area, and eventually the point is reached when the area supports what is called vegetation of the *climax* type. Sometimes, one or more factors are of sufficiently strong influence to arrest the succession before the real climax is attained. A community of plants developed so far and then ceasing to develop further, is said to be of the *sub-climax* type. The stage of development may be far from or near to the climax according to the conditions prevailing. A completed succession right through to the climax is called a *sere* and the various stages in its completion are called *seral communities* as opposed to *climax communities*. A sere which originates in an area never before inhabited within recent geological times is a *primary sere*. Such would be the colonization and subsequent development of vegetation on a rocky mountain face. A sere which originates in an area which has previously borne vegetation, but from which this has been removed by some agency, is called a *secondary sere*.

With this simple conception of plant succession in mind, it becomes apparent that no ecological survey can be complete unless some knowledge of succession within the area is gained. In closed communities, those in which all the available space is taken up by plants, succession may be a very slow process and no appreciable amount may be seen over long periods. In open communities, those in which there is still space for new arrivals, succession is usually much more rapid and can become the subject of a short-term study. Check on succession in an apparently stable community can be kept only by returning to quadrats and transects made in previous years and noting what changes, if any, have occurred. The need for dating all observations is obvious when such work is contemplated. In some areas, much better opportunity arises for watching succession. Colonization of a naked surface may be studied in sand dune, salt marsh or scorched heath as it occurs naturally. Alternatively, areas may be deliberately denuded of vegetation for the sole purpose of watching colonization. When a wood is coppiced or felled, work of great interest may be undertaken, and it is very instructive to follow the series of events which occur on soil after a bonfire.

The Animal Population

It would be pointless to pretend that an ecological survey of the fauna of a given area can be achieved by a school class on as broad a basis as that outlined for the flora. Nor is there ever likely to be quite the same feeling of completion when the time available is exhausted. One of the secrets of achieving anything at all in animal ecology lies in not attempting too much or tackling any project of so nebulous a character that it seems to have no limits. Small, but digestible portions give the greatest satisfaction. If a survey of the local vegetation has already been made, this will save a great deal of time and wasted effort, since in general, it is wisest to select for study the animal population of a particular plant community or part of it. If such a survey has not been carried out, then some comparable work must be undertaken, so that likely areas can be selected.

There are certain difficulties encountered in animal ecological studies which are either not present at all in plant studies, or not present to the same degree. Animals do not usually make their presence conspicuous, and any disturbance due to the observer may cause many to conceal themselves. Their hiding-places and even their presence may not be detectable without undue destruction of the natural conditions, and this will render later observations inadmissible for future surveys of the same area. Thus, in some types of survey, especially of larger animals, some form of hide may have to be constructed.

The great majority of animals move about from place to place. These movements may vary from small and random searches for food to mass migrations over hundreds of miles. Some animals have a fixed home for shorter or longer periods, whereas others lead a nomadic existence for the greater part of their lives. Thus, an animal may be present in an area only because it is passing through the area. Some animals rest in one type of habitat and feed in another. Also, there are those which are active in daylight and others which emerge at night, while yet others are most active at dawn and dusk. Daylight surveys may reveal no trace of nocturnal prowlers, unless the observer is skilled enough to recognize the significance of owl pellets, scratch-marks on bark, foot-tracks, etc.

For animals in certain groups, identification of species may be difficult, and this is especially so with larval forms and adults which show pronounced sexual dimorphism. Although excellent keys to most groups are obtainable, it may often be necessary to obtain the verdict of an expert. Hence, it is very desirable that specimens should be collected and housed in suitable containers so that they may be despatched to an expert for identification. Full details of the exact locality should be included with the specimen.

Possible Projects

It is important that the project selected for study is within the capabilities of the class, and that every student or group shall have a clearly defined task which will form an integral part of the total survey. The following types of project are suggested—

1. A survey of the animals of one group in one particular locality, e.g. the birds in a wood, the insects on a particular plant species, the protozoa in a ditch, the crustaceans or molluscs in a pond.

2. Animals in a particular locality at different times of the year, e.g. summer and winter inhabitants.

3. Animal succession in a new area, e.g. a freshly-dug ditch or a newly-made pond.

4. Animals of marginal habitats, e.g. the sea-shore, the edge of a wood, a hedgerow, the bank of a river or pond.

5. Animal parasites on and in one particular species, e.g. a tame rabbit, rat or mouse.

Survey Procedure

The following data, where applicable, should be recorded in a field notebook and where necessary, specimens should be collected. (*Caution* —species which are few and likely to be rare should not be removed but drawn and described *in situ*.)

1. The exact locality should be defined and its extent stated.

2. The date and time of the observations.

3. Details of climatic factors at the time of the survey, e.g. temperature, humidity, wind, insolation, etc.

4. The species of animals found, with the numbers of each species, the sex and the exact situation within the survey area.

5. The total number of species.

6. The behaviour of the animals, whether resting, feeding, mating, constructing nests, etc.

7. The food materials of the animals included in the survey, including details of predators, etc.

The actual survey should be followed by correlation and recording of all the results, together with habitat diagrams and other types of suitable illustration, e.g. chart quadrats, transects.

Interpretation of the Results

Any ecological survey, however limited in scope, if carefully carried out will yield results of value, and in conjunction with other surveys may well give a fairly complete picture of the life of the area. So that the results may not be mere collections of numbers and names, some suggestions for their interpretation are given here.

With regard to the number of species, many careful surveys have shown that it seems almost always to lie within 60 and 140. Although a survey of the same type of area in another geographical location will give approximately the same number of species, they will not necessarily be the same species. If further surveys yield figures of the same order, then it may be possible to draw certain conclusions. In the first place, a locality does not necessarily contain all the species which are adapted to live there. Secondly, the number of species which inhabit a new area will quickly reach saturation point. Though the types may change until the climax is reached, they will then present a fairly steady number of species.

Some caution must be exercised in drawing conclusions about the number of animals of a particular species. There is no easy or standard method of counting the numbers of animals, since first they have to be found and it is rarely possible to ensure that all within the survey area are found. Hence, for many species, the method of sampling is used. It is obviously impossible to count the total number of individuals of a single protozoan species in a ditch, but if small sample areas are taken in different localities in the ditch, then the average may be used for estimating the total number. From this final figure or from the average, the number per square metre may be estimated. Similar difficulties arise with counts of larger animals. Results of many surveys show that a species rarely reaches the optimum density of population for a particular locality. Further, the numbers show great fluctuation. Not only is there seasonal fluctuation, there are effects due to the vagaries of climate and to parasites. But even in ideal and constant conditions, there are population fluctuations which are due to no obvious cause. For some species, it has been shown that there is some periodicity in these fluctuations. If any data concerning number changes have been obtained, then some attempt should be made to account for the changes.

Counts of the sex numbers for one species will often show that there are far more males in the population than the number necessary to impregnate the females. In most groups, approximately equal numbers of male and female offspring are usual because of the method of sex determination, though there are notable exceptions, e.g. social bees. Hence, there is often competition among the males leading to some form of sex display in mating seasons.

Behaviour studies not only reveal many interesting facts about the particular species, but also link it with other species and thus give considerable help in an attempt to understand the interrelationships of the members of a community. In this connexion, a knowledge of the food materials of a species will be of great importance. All the creatures in a particular community are linked together in a number of food chains. Some examples of such chains are—

1. Green plant → aphis → lady-bird → insectivorous bird → hawk.
2. Marine alga → copepod → small fish → large fish → man.
3. Green plant → rabbit → fox.

It will be seen that the food chains above begin with the solid basis of the green plant and end in a large animal which has no predators relying on it for food. The lower members of every such predator chain have to support the higher members, and thus they have to reproduce at a rate greater than that necessary to maintain their species numbers because they must provide a margin to feed those above them in the chain. Thus we can construct a pyramid of numbers; those at the base must be plentiful, while those at the top are few (*see* Fig. 22.5). Any serious

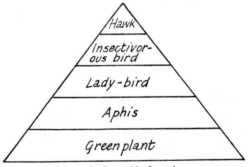

Fig. 22.5. Pyramid of numbers.

fluctuation in numbers at any point in the chain will have repercussions affecting all. Man is slowly learning that unconsidered interference with natural populations might have serious results. Other kinds of food chains are those associated with parasites and consist of smaller animals preying on larger ones. In saprophytic chains dead organic matter is converted into the substance of micro-organisms. In any ecosystem and in the ecosphere as a whole, the vast number of individual food chains are never isolated sequences but are always interconnected to form a complex *food web*.

Some attempt should be made to link the food chains with one another. The higher members of a chain do not usually confine themselves to one particular food species. The topmost carnivore or scavenger will usually take a variety of species for food.

Correlation of Ecological Studies

This phase of the work is essential if the previously described practical work is to have any real meaning. The information collected about several communities, the more the better, must be brought together and the communities considered in relation to one another. The aim in view is to explain as far as possible, the nature of the differences between communities as they are found to exist, in terms of the controlling habitat factors. Completely satisfactory conclusions cannot often be drawn from work carried out at school level, but this is not a reason for failing to undertake ecological work. The important lesson to be learnt is that it is only when trustworthy data have been collected about each community that the way is opened to comparisons between them. The practical work outlined indicates, at least for plants, accurate and methodical ways by which reliable information can be accumulated, and is therefore in itself an essential part of a scientist's training, irrespective of its ultimate use. In any event, attempted analyses of sets of data will always lead to the formation of possible explanations for similarities and differences, and the crystallization of problems requiring solutions, either of which can lead to further original research.

PLANT COMMUNITIES

By the application of field studies, it has become quite clear that plant communities can be regarded as groups of plants living together under the same habitat conditions and affecting one another in ways comparable with human or other animal societies. They may be mutually interdependent or in eternal conflict, but in any case, they always exert considerable influence on one another. A particular community will exist only when a certain set of controlling factors is in operation. If the same set of factors operates in different places, then we can expect to see similar types of vegetation in those places.

The term community does not take size into account. A community may be large or small and a small one may be within a larger one. Ecological study must always commence with small communities, but it is as well for the young ecologist to know from the beginning how his small patch of vegetation fits into the larger scheme of things. The largest of all plant communities are the gigantic divisions of the earth's vegetation. These may be classified as follows—

1. Tropical forest.
2. Tropical grassland.
3. Hot desert.
4. Temperate forest.

   (*a*) Warm temperate.
   (*b*) Cool temperate.

   (i) Deciduous summer forest.
   (ii) Coniferous forest.

    5. Temperate grassland.
    6. Tundra and ice desert.
The locations of these are largely determined by climatic conditions
and thus the vegetation of the earth shows a distinct zonation according
to the factors which control the climate (*see* Fig. 22.6). These include

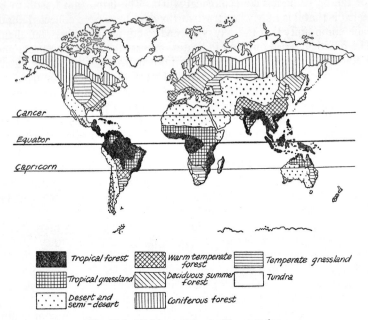

Fig. 22.6. Divisions of the earth's vegetation.

latitude and altitude, size and relative positions of land and water
masses, ocean currents and winds. Within the main divisions of
vegetation are smaller ones, varying from one another chiefly because
of soil differences. The vegetation of the British Isles comes into the
deciduous summer forest category and the climax of the greater
part is the oak forest, except in parts of northern Scotland where this
gives way to pines. The most important smaller communities within
the vegetation of Great Britain may be summarized—
    1. Woodland communities: oak, birch, ash, beech and pine woods.
In very localized areas, yew and alder woods may be encountered.
Woods are not, however, simple communities and should be considered
as consisting of several habitats. The members of the tree canopy, the

undergrowth and the ground flora have distinctly different sets of conditions in which to live.

2. Grassland communities: grass heath and chalk down.

3. Peat communities: heath, moor, bog and fen.

4. Aquatic communities: marsh, swamp and the true aquatic vegetation of lakes and rivers.

5. Maritime communities: salt marsh, sand dune, rocky shore and shingle beach.

These groups are sometimes termed *natural communities*, the inference being that they have not been subject to interference by man or his domestic animals. In a country such as Great Britain, where the greater part of the land has been cultivated for a very long time, there is little of the natural vegetation left, and even where this occurs, it has been determined by man and is, at best, semi-natural. Such are the woods which are regularly felled or coppiced, the heaths and moors which are periodically burned and then pastured, marshes which are drained, and grasslands which, although not sown, are cut or grazed. The only truly natural communities are those of the coasts and the remote mountainous parts. At the other extreme are the communities completely under human influence. These include sown pasture land, land under cultivation, hedges, plantations, canals, reservoirs and so on. But, bearing in mind the fact that human interference can be looked upon as another ecological factor, these last mentioned types of vegetation may be given equal status with the others, and from the point of view of field work can prove to be equally as instructive.

We have so far used the term community to describe any collection of plants found growing together, however small or extensive an area they may cover, however relatively important or unimportant they may be. We have used it in the same sense as we use it in human affairs to describe the units of population forming cities, towns, villages or single families. To simplify and classify plant communities, a number of terms are employed with special meanings. Apart from the large divisions determined by climate, the largest community now generally recognized is the *association*. We may define it by saying that it is of uniform nature as far as the life form of the commonest species is concerned, and the habitat factors which govern its existence are also generally constant over the area. Such associations are the deciduous and coniferous forests, the heaths and the maritime communities. They have reached an equilibrium with the habitat and represent the climax state. Strictly speaking, the term association should not be applied to communities which represent a stage in successional development of an area, but it often is. Most associations have more than one dominant plant. They are co-dominant. This is because the larger association may be made up of several smaller communities, each with its own dominant. The deciduous forests of Great Britain are really composed of smaller woods of oak, beech, ash and so on. Such smaller communities, each with its own dominant, form *consociations*. Within these, there will be numerous minor communities governed by the many small localized differences in habitat factors. Each of these will have its own particular dominant species and is known as a *society*.

ANIMAL COMMUNITIES

It is possible to divide animals into ecological groups inhabiting the major plant divisions of the world. Thus there are animals characteristic of tropical forest, of tropical grassland, of hot desert, of temperate forest, of temperate grassland and of tundra. In general, the denizens of these regions show interesting adaptations, as for example the various arboreal animals of the tropical forests which range from primates to tree-frogs among vertebrates alone. However, such a system of major ecological groupings is not favoured by zoologists, for several reasons. In the first place, the law of over-production in animals leads to active struggle for existence rather than the passive competition found in plants. Thus while plants are dispersed more or less capriciously, animals tend to migrate from their birthplace and their range is extended eventually to the limits of suitable territory. In the past,

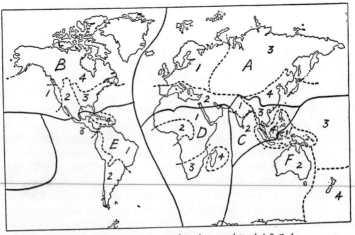

Note: each region A, B, C, D, E, F has subregions numbered 1, 2, 3, 4

Fig. 22.7. Zoogeographical areas of the earth.

there have been certain major highways of migration which do not now exist. Such, for example, were the northern passage across the Behring Straits, and the Asia-Australia passage via a land link now broken up into the East Indies. Secondly, natural barriers have arisen, either by great earth movements or by climatic changes, so that certain animal stocks have been isolated and the entry of further animals prevented. Examples of such barriers are ranges of mountains, e.g. the Himalayas, large bodies of fresh or salt water, swift deep channels, and deserts. One particular isolated stock has given rise to many specialized groups adapted to the various niches in the environment. An outstanding example is seen in the evolution of the mammals of Australia, which has many forms comparable with the mammals of Asia and Africa, but whereas the former are all marsupials, the latter are not. Both groups evolved from the same stock, but since isolation, the end-result is quite different.

Alfred Russell Wallace first suggested that natural barriers have divided the earth into major zoogeographical areas. His original scheme has been modified in certain details but the main outlines are unchanged. These regions are indicated in the table below and illustrated in Fig. 22.7. Each of the six regions has been divided into four sub-regions which are indicated on the map by the broken lines.

ARCTOGAEA	Palaearctic	A
	Nearctic	B
	Oriental	C
	Ethiopian	D
NOTOGAEA	Neotropical	E
	Australian	F

It is obvious that there will be some overlapping on the fringes of these regions, and that for flying animals or small wind- and water-borne creatures, the natural barriers offer little obstacle. Within each sub-region there is a wide variety of habitats. For example, sub-region 2 of the Ethiopian region is mainly tropical forest, but within it, there are rivers, lakes, ponds, swamps, clearings and uplands. Again, within each of these, there are progressively smaller niches until we arrive at the parasites on a single species or the protozoa in a handful of soil.

Each vegetation association has a typical animal community and during development to its climax the nature of the community changes. In the British Isles, man's interference with natural animal communities has been even more castastrophic than with plants. Very few areas

30

remain where truly natural communities may be found. We may, however, for ecological purposes, recognize the following larger groupings—

1. Woodland communities.
2. Moorland communities.
3. Bog and fen communities.
4. Various aquatic (fresh-water) communities, e.g. pond, lake, river.
5. Various maritime communities, e.g. intertidal sandy, muddy or rocky coast, estuarine waters.

With animals, there is rarely any question of a dominant species. In each community, there are a number of pyramids; the topmost species may be a large carnivore, a large scavenger or a large herbivore. In the British Isles, the great majority of the larger native wild animals have been destroyed. The wolf and the boar exist only in fable; the wild cat, buzzard and golden eagle are great rarities.

THE VALUE OF ECOLOGICAL STUDY

If a student is fortunate enough to be able to take part in the elementary survey procedures involving a number of areas, certain facts about the distribution of plants and animals and the kinds of communities they form, must begin to emerge. He should see that it is logical to argue that the natural community expresses perfectly the physical environment; that each different set of environmental conditions, however great or small the differences, possesses its own characteristic community of plants and animals, and therefore that by these communities sets of conditions can be classified. It follows that if we know our communities well enough, we can use our knowledge of them to solve many problems which cannot be approached by direct measurement. It has been truly said that in problems of use of land and in its management when under use, ecology occupies a position comparable with the use of physical and chemical sciences in the solution of industrial problems. This brings us to the major purpose of field studies. With the intelligent application of knowledge gained from ecological study, man can go a long way towards establishing a more healthy and long-lasting balance between both wild and domestic plants and animals and their environments. He can know before he starts what disasters he may cause through uncontrolled interference with natural populations and he can safeguard against such events. Ecological knowledge could have been used to prevent many of the difficult situations which have arisen in our own times, and must be used on an increasing scale as the only method of finding satisfactory solutions to many of the problems which now exist. Man must first learn above all to make the best use of the great variety of organisms with which he has been provided, and also of the fertile soils, forests and grasslands which their activities have produced. He must not destroy them without thought to the future since they are no less useful to him than his own cultivated fields. They yield many of his raw materials. Through ecology, man can learn to extract his requirements from natural communities whilst still allowing them to maintain the healthy balance of growth and decay on which they have normally existed for so long. Too often has he stripped them of all value and finally abandoned them. Where this has occurred, he can learn to restore them to a state properly suited to the conditions of the locality. Another intelligent application of ecological knowledge lies in making the most of the rich harvests of the seas. Marine ecology is no less important than land studies. Finally, the spread of diseases and pests among man, domestic animals and plants, and stored agricultural products, can be better controlled if applied ecology is used in finding methods.

When the student is engaged in learning the rudiments of field methods during practical exercises on small local areas, he must not lose sight of the great significance of his work. He is merely learning how to use the tools of a trade. He can be compensated for what may appear to be a certain pointlessness if he carries in his mind the thought that economic discoveries of the greatest importance have been made by the use of similar tools, and that he may one day be in a position to assist in such great things when he himself has mastered the general principles.

CHAPTER 23

MAN: HIS SUCCESS AND HIS PROBLEMS

No study of biology would be complete without some discussion of the success of man, his effects on other biological groups, and the problems he has created. The family Hominidae comprises the genera *Australopithecus*, *Pithecanthropus*, *Sinanthropus* and *Homo*; some anthropologists would include others. Here it is proposed to discuss *Homo sapiens*, modern man, a species somewhat difficult to define, since most of his characteristics apply in some degree to other Anthropoidea. Man is a bipedal primate with very characteristic dentition; he has learnt to use tools which he has fashioned for particular purposes, and he has evolved the power of speech.

MAN AS A SUCCESSFUL ANIMAL

Evolutionary success may be judged by certain criteria, though admittedly these criteria have been established by man himself. He can only judge by his own standards, since it is impossible for him to be completely objective. A primary requisite for success is the ability of a species to maintain its numbers and if possible to increase them. Secondly, a successful species must occupy a dominant position with regard to other living organisms, and thirdly and perhaps most important of all, continued success necessitates some degree of adaptability, so that changing conditions or different environments do not entail extinction. On all these counts, man may be judged the most successful of living organisms; indeed, there is quite a body of expert opinion which maintains that man is the only species of animal capable of progressive evolution. Some of the reasons for his outstanding success are pointed out here.

Man is a very generalized animal, conspicuous for his lack of specialized features; he is not well adapted for any particular environmental niche. Whereas other animals have evolved a host of delicate adaptations which enable them to fit perfectly into a particular habitat, man has become generalized, fitting no environment perfectly but able to adapt himself to a great variety of surroundings with success. He is able to live in all climates from extreme cold to extreme heat. He is not well adapted to running, jumping, swimming, or climbing, but can achieve a fair degree of proficiency in all of them.

The perfection of the erect position of the body has left the hands free from participation in locomotion and therefore able to perform other activities. Retention of the opposable thumb, enabling grasping, has been an indispensable adjunct in the use of tools, and the ability to adopt the pronate or supinate positions has meant that the hands can be used both for collecting food and for conveying it to the mouth.

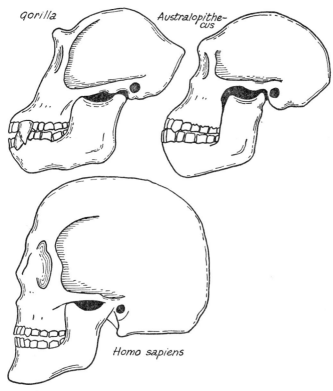

Fig. 23.1. Side view Gorilla, *Australopithecus* and modern man.

Elaborate muscle equipment of the fingers has made delicate and skilled movements possible. It is to be noted that the evolution of the upright stance has been coupled with loss of opposability of the big toe, a faculty which must have been of great use to man's arboreal ancestors and which is still possessed by other primates.

Correlated with the use of the hands as food-collecting organs, there has been reduction of the front of the face and particularly of the massive jaws associated with other Anthropoidea (*see* Fig. 23.1). The use of the hands for food capture led to the use of weapons, and later, implements were used for division of the food and conveyance to the mouth; the latter trend has culminated in the use of knife, fork and spoon, or chopsticks!

Perhaps the most important reason for man's success has been the continuation of enlargement of the cerebral hemispheres and particularly the perfection of the neopallium. This most important part of the brain has been folded into numerous sulci (*see* Fig. 8.13 (*b*)) and thus very large surface area has been achieved. On this large area, it

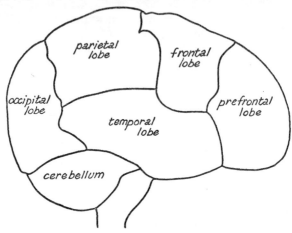

Fig. 23.2. Lobes of the human brain.

has been possible to develop an extensive mosaic pattern of correlation centres. Three particular regions of the neopallium show greater enlargement than that found in other Anthropoidea. They are the prefrontal, parietal and temporal areas (*see* Fig. 23.2) and it is noteworthy that they are the last parts of the brain to form in the embryo, evidence that they are recent evolutionary acquisitions. Specifically associated with the enlargement of these areas are greater proficiency in the tactile sense and co-ordination of the eyes (prefrontal), sight (parietal) and hearing (temporal).

Greater proficiency of sight, hearing, and tactile sense were undoubtedly necessary prerequisites for the development of the higher mental powers. Sight has been of particular importance; binocular vision, found in all the anthropoids, gives estimation of distance, comparative size and the relationship of objects to one another. Thus perception of irregularity of posture has been largely taken over from the semicircular canals. The adaptations in the eyes and optic nerves which have made these improvements possible are two. Both eyes focus the same image simultaneously on their respective maculae, and fibres from both optic nerves pass via the optic thalamus to both prefrontal and parietal lobes. Acute sense of hearing appears to have been a necessary adjunct to the development of speech. The delicate and complicated musculature of the fingers, together with increased development of the tactile co-ordination centres in the prefrontal lobe, have resulted in the uncanny accuracy of such craftsmen as the wood-carver, the watchmaker and the miniature portrait painter.

The change of diet from that of his vegetarian forbears to that of a semi-carnivore was undoubtedly a great step forward in man's evolution. The richer diet meant that less time had to be spent on eating and this left more time for other activities. Hunting for meat was probably an important factor in developing the use of weapons and that, of itself, encouraged search for better materials, and thus man progressed from sticks and stones to shaped flints, to bronze and finally to iron.

Life in social groups eventually led from nomadic to settled existence, though some tribes have retained the wandering way of life to the present day. Settled communities were hardly possible without agriculture; the growing of crops and domestication of animals, both as food and as beasts of burden, may be regarded as important steps towards civilization, which may be defined as life in cities. Producers of food supplied the excess to the city dweller, who compensated the farmers and hunters with tools, weapons and clothing. Man's cultural achievements are largely due to the settled life, and they range from cave-art and crude pottery to the multifarious products of the present time.

With tools, weapons and clothing, man has been able to extend his environment far beyond the range possible for any other animal species. Gradually he has evolved the peculiar facility of taking his environment with him. Provided with clothing, weapons and tools, he has achieved considerable independence of his surroundings, and thus is more adaptable though less adapted than other animals.

Man's greatest achievements are the result of development of the higher mental powers. He has sought to come to terms with his environment, to understand it, and thus constantly to improve the conditions of his life. A huge body of scientific knowledge has been accumulated, and at present, this mass of data is already beyond the comprehension of any one man. He has tried to understand his own nature and to find a purpose for living. From propitiating the forces of nature, he has formulated moral laws and promulgated religions, which all have a basic common core. He has sought to project his personality into the future in the shape of work for posterity; this thought for the future betterment and happiness of mankind is perhaps the greatest characteristic of the finest of men.

THE EFFECTS OF MAN ON HIS ENVIRONMENT

Two main trends have been of paramount importance in elevating man to his present dominant position. They are the vast increase in his numbers and the rapid development of means of transport. The population of the world three hundred years ago was less than 500 million; now it is 3500 million and still increasing rapidly. The development of transport makes a fascinating story marked by several important stages, of which the invention of the wheel, the sail, the steam engine, the internal combustion engine, the jet and atomic engines, were outstanding landmarks. Nuclear power promises new possibilities. Gradually, distance has come to be less and less of an obstacle and no place on earth is more than a few hours journey from any other. Oceans, mountains and deserts, have become less important as barriers, and apart from the abyssal depths of the sea, no place is inaccessible. The whole world is now man's geographical range and the conquest of outer space may not be entirely fantasy. He is able to exploit the natural resources of all regions and often to mould large areas to his own purposes. This wide-ranging and rapidly increasing population has had important and not always desirable effects on the soil, on plants and on animals.

Reference to Chap. 6 will recall to the reader the fact that soil is a complex mixture of constituents, all of which must be present in certain proportions if it is to support a healthy growth of plants upon which animals, including man, rely. Correct treatment of the soil demands adherence to the *law of return*. Whatever is taken from the soil must be replaced.

On the whole, man's treatment of the soil has been a sad story. Unless definite steps are taken in the matter of conservation, soil loses both its nutrients and its structure and eventually becomes a loose mass of particles which can be washed or blown away. Loss of top soil by both methods has occurred on a large scale, giving rise to dust-bowls and man-made deserts. Deforestation, especially of sloping ground, accelerates water erosion, and undoubtedly the removal of the tree cover has had an appreciable effect on rainfall. Lessening of the humus content is a prime cause of erosion; in many so-called virgin lands, man has grown quick crops for a few years and having exhausted the soil, has moved on to other virgin areas. Thus the dust bowl of North America has been created; much of what is now the Sahara desert was the granary of imperial Rome; the plains of the Euphrates and Tigris were once fertile land.

The loss of the humus sponge resulting from the despoiling of the mountain forests is a direct cause of the floods which now afflict lower river valleys, often with disastrous consequences. With accelerated drainage and run-off, trace elements in the soil disappear rapidly into the seas and thus deficiency diseases appear in plants and animals.

The total land surface of the world is about 145×10^6 km²; of this, only 66×10^6 km² are suitable for agriculture; 16×10^6 km² are at present under cultivation and it is estimated that about 8×10^6 km² have been destroyed beyond restoration in reasonable time. In man's treatment of the soil, either ignorance or the profit motive have predominated, and over the years a spate of books on the subject has appeared. Their very titles, e.g. *The Rape of the Earth*, *Our Plundered Planet*, are indicative of the gravity of the situation.

The picture is not however wholly bad. In some countries, careful husbandry has conserved the soil for centuries. Outstanding examples are found in certain parts of India and in China, where manuring and irrigation are devotedly practised and the soil is indeed "the good earth." There are indications that the problem of conservation is now being tackled scientifically in certain limited areas. It is heartening to see the reafforestation of uplands, new irrigation schemes, use of town sewage, invention of soil conditioners which will hold the moisture, the planting of tree shelter belts to lessen wind erosion, the use of marram grasses for binding sand, and the reclamation of land from the sea and from swamp.

The enormous increase in the population of human beings has necessitated clearance of large areas of natural vegetation so that crops can be grown in sufficient quantity to feed man and his animals. Increasing use of timber has led to wholesale destruction of forests; this has brought the soil conservation problems mentioned previously, besides the shortage of timber and especially of hardwoods. In addition, the tremendous amount of water transpired by forest trees is an important factor affecting rainfall. The amount of transpired water from one acre of beech trees amounts to about 1500 tonnes per season, but the bare slopes which ultimately result from deforestation and erosion will merely allow most of the water to run off, and thus there will eventually be less rainfall in such a region.

The extensive use of monocropping, by which is meant the growing of one single type of plant over a large area, tends to exhaust the soil of particular nutrients and to encourage plant disease. When vast acreages are under single crops, ideal conditions are provided for the spread of fungus diseases and insect pests. Examples which spring to mind are wheat rust, potato blight, turnip flea-beetle and bean aphis. Thus there has grown up an enormous industry for the manufacture of insecticides and fungicides. A well-known apple-grower has stated that to protect his trees adequately, they need annually ten different sprays as well as grease-bands.

Interference with plant associations has had severe effects on both flora and fauna with the result that some species have disappeared from the area affected. Unconsidered meddling with natural plant associations has often had serious repercussions. Examples of such are the introduction of cactus into Australia, of blackberry into New Zealand and of *Elodea* into Great Britain. The removal of suitable vegetation cover has driven away many small birds with consequent increase of insect pests.

On the credit side we have the enlightened use of modern crop rotations and the breeding of higher-yielding and more disease-resistant strains.

Apart from indirect effects of great magnitude due to clearing of natural vegetation, man has had drastic effects on animal populations. There has been wanton and thoughtless destruction of animals, often to the point of extinction. The Dodo of Mauritius, the Moa of New Zealand and the Solitaire of Rodriguez were all flightless birds which were easy prey for mariners. Steller's sea-cow (*Rhytina*), of the North Pacific, became extinct in like manner in the 18th century. The sea-otter (*Latax lutris*), of the North Pacific, was remorselessly hunted for its valuable pelt and is now either extinct or nearly so. In North America, the vast herds of bison were estimated to number over 50 million, but now relatively small numbers are preserved in national parks and zoos. The larger whales would undoubtedly have succumbed but for timely international agreement limiting capture.

Domestication of animals has resulted in selective breeding for particular results. One has only to consider the varieties of dog from toy terrier to great dane to realize the possibilities in a single species. Whether bred for hunting, drawing sledges, marshalling sheep or merely for companionship, they are bred for specific qualities. Similarly, cattle have been bred for beef (Herefords), for milk (Jersey) or for dual purpose (Friesian). With a host of other animals, from elephants and camels to bees and silkworm moths, there has been this selective breeding to obtain a desired result, and with our knowledge of genetics, there are great possibilities for future work of this kind. As with concentrations of crop plants, so with animals in domesticity; they provide ideal breeding-ground for disease. For example, foot-and-mouth disease, mastitis and Johne's disease are almost inevitably associated with herds of cattle, while hard pad and distemper are almost universal in domestic dogs.

Interference with natural animal populations has often caused serious trouble. The decimation of rabbits by myxomatosis resulted directly in attacks by foxes on domestic poultry. Now rabbits resistant to the disease seem to be on the increase. A classic example is shown in the introduction of rabbits into Australia. Their immediate success soon led to a situation which became almost uncontrollable until the advent of myxomatosis. It has however been shown most convincingly that study and experiment may make it possible to employ safe methods of biological control. Outstanding examples of this are seen in the destruction of Australian cactus by the larvae of the moth *Cactoblastis cactorum*, and the eradication of the citrus scale insect of California by ladybirds. In both cases, the biological controlling species was introduced after careful consideration of all the factors involved (*see* Vol. I, Chap. 23).

SOME PROBLEMS AFFECTING MAN'S FUTURE SUCCESS

The human race has advanced in knowledge, no less than in numbers, at an enormous pace. Progress cannot be stopped or even slowed down, and if man is to live a happy and useful life, he has to face many problems which demand solution. Though most of these problems are largely of his own creation, there seems little doubt that the accumulated knowledge and wisdom of human society can solve them satisfactorily.

Man is aware of his own evolution and knows that his greatest possibilities still lie in selection, though one can hardly apply the term "natural selection" here. His greatest achievement has been evolution of the mind and improvement lies in two directions. Since intelligence, as much as any other characteristic is the product of genetic inheritance and environmental influence, then it is possible at least to improve the nurture moiety. There is, in some responsible quarters, concern about alleged decline of intelligence. A post-war Royal Commission on population commented thus—

> We are not in a position to evaluate the expert evidence submitted to us to the effect that there is inherent in the differential birth-rate a tendency towards lowering the average level of intelligence of the nation, but there is here an issue of the first importance which needs to be thoroughly studied. We therefore urge that the Government should arrange for a thorough investigation of the problem.

The same commission has reported that the replacement rate per married couple is, for lower income families 2·36, while for higher income families it is 1·68. The birth-rate for all classes has been declining steadily (Great Britain) over the past sixty years, though there have been upward surges during the two world wars. In the past, the lower income groups constituted a vast reservoir of potential, but the trend towards equality of opportunity has gradually moved the more intelligent members upward. It seems almost a corollary that the higher level of social life thus achieved, results in fewer children. As the demand for highly trained scientists and technologists grows, is there not a possibility that there will be insufficient intelligent beings to satisfy it? A number of surveys carried out in England, Scotland and America show that a clear majority of the more intelligent children come from parents of above average intelligence. Some figures for a post-war Scottish survey are given below—

Possible	Children of one-child marriages averaged	42
score	Children of four-child marriages averaged	35·3
76 points	Children of eight-child marriages averaged	28·8

The average educational standard of the parents agreed closely with these results for the children. Is there more deliberate family planning in the higher income groups, due to economic or social factors? Do the demands of social life often make it undesirable for mothers to have children?

It must be pointed out that there is no final agreement about what we are measuring by intelligence testing and that no system yet devised is infallible as a method of predicting future performance. Yet, in view of widespread concern over this matter the question remains. Can it be shown that there is a decline in intelligence?

The world population is now about 3500 millions, and each year brings an increase of approximately 30 millions. Reduced to smaller proportions, these figures mean that every day, 82 000 new babies enter on the adventure of life, or approximately one per second. Over the last decade there has been a 12 per cent increase in population. The figures become very significant when it is realized that three-fifths live at starvation level or below. These starving hordes, mainly in the more backward countries, cannot be fed at the present level of world food production and distribution. In this enlightened age, there should be no question of burning surplus wheat, of turning sugar into alcohol, or of using oranges as fertilizer. Methods of distribution can be enormously improved; present economic conditions alone prevent it. No alleviation can be applied quickly or on a large enough scale; it must be tackled on an international scale, and here the work of the United Nations Food & Health Organization is making good progress. Is it possible to maintain the increasing world population by diligent application of scientific methods of food production and by sensible distribution?

There is widespread ignorance of biological phenomena, often even among the world's leaders. It is this ignorance which has allowed such catastrophes as land erosion, deforestation, well-meant but disastrous interference with natural populations, depredation of the fisheries, and costly but ill-advised large-scale cultivation schemes.

Even in the more advanced countries there is often little provision for human social biological education and thus many adolescents face life with little knowledge of the grave problems which beset the human race. Should we not endeavour to obtain more widespread knowledge of these problems? Here we may quote the geneticist, C. D. Darlington, who stated in his book *The Facts of Life*—

> Man is an animal. He is just as much a product of genes and environment as are fungi, insects, etc. The main task of the next century is a biological one. Man must begin to know where he is going and what kind of life he is likely to have thrust upon him. The guidance of our destinies is at present entrusted to supposedly educated men who are by no means ashamed of professing an ignorance of the discoveries which should determine their conduct. They are content to use the ideas of life, heredity and society which have done service since Old Testament times.

One of the topics which finds a prominent place in conversation today is the question of disease. It seems that this has been so for a long time, for one reads in novels of the eighteenth century of the "megrims," and the "vapours." The old herbals listed panaceas for all evils; some of them are now well-established drugs which have found a place in the British Pharmacopeia. The usual greeting between friends and acquaintances has become a polite inquiry into health. In 1851, the average expectation of life for male babies was forty years. One hundred years later, in 1951, it had increased to sixty-six years. Similar figures apply to all the more civilized nations, though in India for example, the male expectation of life in 1930 was twenty-five years, and today the position is little better.

Throughout human evolution, the three great causes of death have been famine, disease and war. The loss of life from the latter cause is almost negligible compared with the other two. In the past, famine has caused devastating loss of life, and even today there are large areas of the world where it is still a notable cause of premature death. Modern scientific agriculture, better distribution and storage of surpluses could eradicate famine for all time.

Disease is somewhat difficult to define. If we mean any departure from the normal condition of the body, then we have to define this normal condition. There are few people who have abounding and persistent health throughout their whole lives. It is however customary to define disease as some malevolent change in the tissues of the body, though this definition excludes certain mental disorders where no tissue change is demonstrable. Diseases may be divided into nine main groups.

1. Inherited abnormalities, e.g. haemophilia, one form of jaundice.
2. Dietary deficiency, e.g. scurvy, rickets, one form of anaemia.
3. Infection by pathogenic organisms; this is a vast group of diseases

ranging from virus to helminth-infection; it includes all the epidemic and pandemic diseases, e.g. poliomyelitis, influenza, tuberculosis, amoebic dysentery, malaria, sleeping sickness, hookworm disease, schistosomiasis, etc.

4. Physical injury, due to accident, from simple sprains to permanent physical distortion.

5. Poisons, e.g. carbon monoxide, chlorine, alkaloids.

6. New growth, ranging from benign to malignant carcinoma.

7. Degeneration with age, e.g. arteriosclerosis, heart failure of several types, cerebral haemorrhage.

8. Anxiety states, e.g. schizophrenia, depression, mania.

9. Hormone deficiency or excess, e.g. cretinism, diabetes mellitus.

In each group, there is a host of diseases and as medical knowledge advances, the number grows. It seems fairly certain that a great many diseases have increased in incidence because of three main causes. In the first place, the concentration of large populations in cities, makes for rapid spread of infection. Then ease of transport has spread many localized diseases all over the world. Examples of this are to be found in plague and tuberculosis. Thirdly, owing to the vast amount of medical attention now dispensed, it is often possible to cure diseases which would at one time have inevitably proved fatal. Thus there is, in most countries, a gradually ageing population, less resistant than the young, and this has stimulated the study of gerontology, the science which seeks to find the causes of ageing.

The accent in medicine is mainly on cure though there are signs that in epidemic disease, at least, there is considerable success in prevention by means of vaccines. A great deal of time and money is spent in research in chemotherapy, radiotherapy and dietetics. Should there not be some inquiry into the factors which constitute health? There is no doubt that in western civilization, at least, the percentage of unfit people is increasing. The cost of the medical service, with drugs and accessories, is already a serious item in national budgets, and as knowledge improves, it will inevitably rise. The whole question is largely bound up with increase in expectation of life. Is there not some likelihood that there may eventually be more pensioners than workers supporting them? Must we not re-orientate our ideas and organize health services so that all people are instructed in the major principles of health, and must we not ensure thorough medical inspection at regular intervals so that incipient disease may be detected where possible? At present, the cost is prohibitive, but there is hope that some of the vast sums spent in research on devastating weapons may be diverted to more profitable ends.

With the advent of atomic bombs and the realization that they may cause genetical damage, if not death, there has been world-wide concern over test nuclear explosions. It is certain that all types of ionizing radiation can cause chromosome and gene mutation as well as immediate damage such as radiation sickness. Apart from man-made sources of radiation such as X-rays and radioactive isotopes, we know that the human body is constantly exposed to radiation from cosmic rays and from natural radioactive substances. It has been shown in experiments on small animals, particularly on *Drosophila* and on mice, that inherited mutations can be caused by both natural and artificial types of radiation. It is also known that the mutation rate is directly proportional to the total irradiation, whether received in one large dose or in many small doses. Thus it seems that any increase in dosage to which human beings are subjected will increase the mutation rate. Nevertheless, it is reassuring to read in *Hazards to Man of Nuclear and Allied Radiation* (H.M. Stationery Office), that if tests of nuclear weapons continue at the present rate, the increase in radiation will not be more than 1 per cent of that which we receive already. Of more importance is the effect of direct radiation of the gonads for diagnostic purposes, and the production of radioactive strontium by hydrogen bomb tests. Strontium can replace calcium in bones and, if radioactive, it can have serious effects on erythro- and leuco-poiesis. With the increasing use of nuclear power stations, the problem of safe disposal of radioactive wastes might become a serious question.

Owing to the political feeling engendered in many countries by such terms as racial barriers and racial intermarriage, it is perhaps unwise to attempt a definition of race as applied to human beings. The report of a United Nations Committee of experts found that there are three racial groups of men, the Caucasoid, the Mongoloid and the Negroid. This report further states that there is no genetical difference of intelligence between these three groups and neither are there any genetical grounds for discountenancing intermarriage. Although it seems undoubted that natural selection has operated for example in the determination of black skin, yet we know that this particular characteristic, as in many other cases, is due to polygenic inheritance of a full number of dominants. Thus it is possible to obtain a large number of skin colours from black to white. Racial animosity has caused great trouble in the past and there are unfortunate outbreaks even today. Differences in outlook and in customs are due to the cultural history of a race and unless we wish to standardize culture of all kinds, we must expect men to differ in their philosophy, beliefs and practices. Is it not important that everyone should be educated to realize that biologically, man is a single species and that given suitable environmental conditions, white, yellow or black have the same potential?

We have attempted to outline a few of the major problems which demand solution. There are many minor problems; in fact, each individual has personal problems which he must learn to consider ecologically, in relation to his family, his social group, to the rest of mankind and to all living organisms. There has been and always will be survival of the fittest; man must strive to make himself physically, spiritually and mentally more fitted for survival.

APPENDIX

MATTER AND ENERGY IN RELATION TO LIVING SYSTEMS

THIS appendix has been compiled to assist those students whose previous experience in the physical sciences may be proving inadequate to meet their present needs. The many complex physical and chemical conditions and events that occur, both internal and external to living systems, can be understood only in the light of a reasonable knowledge of the nature and behaviour of matter and energy. What follows forms an elementary introduction to these concepts and is by no means to be regarded as a comprehensive treatment of all the physics and chemistry that a biologist should require to know. If it encourages the student at this stage to think along the lines dictated by modern trends in the subject it will serve at least one useful purpose. Where it is considered that a reader might benefit from reference to it, this is indicated in the main text.

It is emphasized that in its finest structure and in its energy relationships, the living system is not fundamentally different from a non-living system. Any apparent differences are due to the special arrangements of the components (atoms and molecules) of a living system which give it particular kinds of material exchange properties and energy relationships with its non-living surroundings that equate to being "alive".

Some Properties and Interactions of Matter

Everything that possesses *mass* and takes up space is composed of *matter*. The mass of a quantity is a measure of the amount of matter that composes it, but size is not itself an adequate measure of mass since a small portion of one substance, such as lead, may have a greater mass than a larger portion of another, such as aluminium. Whilst mass can be described as the quantity of matter contained in a body, it is not easy to formulate exactly what is meant by this. Because there is more concern with the effects of mass rather than with an exact definition of it, this is not of great importance here.

There are two distinctly different concepts of mass in terms of its effects, and a measure of the mass of a body may be obtained from either of them. One derives from the fact that an effort is required to change the velocity or rate of motion of a body in a given direction, i.e. give it an *acceleration*. Thus matter has a property known as its *inertia* by which it offers resistance to a change in its state of rest, or in its speed and/or direction of motion if it is moving. It has been observed that a force applied to a body produces an acceleration proportional to the force, and the constant of this proportionality represents the mass of the body. If two bodies are in direct impact and their resulting velocity changes are measured, it is found that the ratio between the changes is independent of the original velocities and is always constant for the given bodies. This ratio is known as the ratio of their *inertial masses*.

The other concept of mass stems from the fact that every portion of matter in the vicinity of the earth experiences a force of attraction towards the earth's centre. This force is called the *weight* of the body and can be measured directly in terms of the extension it imparts to a spring as in a spring balance. If the weights of two bodies are compared, as on a beam balance, it is found that the ratio of one to the other is always constant and this ratio is called the ratio of their *gravitational masses*.

As far as has been practically demonstrated the gravitational force between a quantity of matter or mass and the earth is proportional to the inertia of the mass, but it must be clearly understood that this does not mean that mass and weight are the same thing. The mass of a quantity of matter is constant, except at speeds approaching the speed of light (*see* p. 863), whereas its weight, being the force with which the earth attracts it, varies according to its location. A body's weight is the force that is required to support it stationary against the force of gravity. At a sufficient distance from the earth's centre and from all other bodies capable of exerting a gravitational force, a body may be regarded as weightless, but its inertial mass remains the same at all points in space provided all relativistic effects are ignored (*see* p. 994). The mass of a body is now properly regarded as a relative concept depending on the inertial system to which it is referred. If a body is moving with a speed v with respect to the reference system, its mass, relative to that system, is given by

$$m = m_0 \Big/ \sqrt{\left(1 - \frac{v^2}{c^2}\right)},$$

where m_0 is the mass of the body in its rest frame, known as its *rest mass*, and c is the velocity of light. It is really this rest mass that can be regarded as a measure of the amount of matter in that body. But weight and mass are obviously closely connected when a body is near the earth's surface, for the force of the earth's gravitational pull on a body, upon which weight depends, is proportional to the mass of the body, i.e. the greater the mass, the greater is the force by which the earth pulls it and hence the greater is its weight. Two gravitational masses can therefore be compared by comparing their weights at the same place on the earth's surface.

The SI unit of mass (*see* p. 5) is the *kilogram* (kg). The unit of weight or force is the *newton* (N). This is defined as the force required to give a mass of one kilogram an acceleration of one metre per second per second (m s^{-2}).

It should be mentioned here that the SI units metre, kilogram and second are sufficient for all physical quantities in mechanics one of which is force, already defined above. This definition stems from a form of Newton's Second Law of Motion, force = mass × acceleration, which assumes that the mass of a body remains constant with time.

The force exerted by the gravitational field of the earth on a body very near to its surface is an important physical quantity, not least to all living systems whose size and structure must of necessity be related to it. The measured value of this force acting upon a unit mass varies from place to place on the earth's surface due to the following factors: the earth's shape is not exactly spherical; the earth is rotating; there are local variations in the density of the earth. The force is about 9·78 N kg^{-1} at the equator and 9·83 N kg^{-1} at the poles. The definition of force shows that the acceleration, g, of a body falling freely is numerically equal to the value of the gravitational force at a particular point. Unless extreme accuracy is required, the value of g which it is necessary to use in calculations involving gravitational forces, is taken to be 9·81 m s^{-2}. It is necessary to include g in such cases since force has been defined in terms of unit acceleration.

For the purposes of the reader, who should be able to distinguish properly between mass and weight, all this can be summarized by saying that although the mass m of a body remains constant unless something drastic is done to it, its weight mg varies from place to place on the earth's surface because of the variation in g. This variation in weight can be detected with a sensitive spring balance but not by a chemical balance because the force with which the earth attracts the standard masses (usually incorrectly called "weights") alters in exactly the same way as the force with which the earth attracts the body that is being "weighed". If m_1 and m_2 are the masses in the scale pans on either side of a chemical balance, then at counterpoise, $m_1g = m_2g$ or $m_1 = m_2$. Thus weighing with a chemical balance really determines the mass and not the weight of a body.

As was stated earlier, the mass of an object is not necessarily proportional to the space or *volume* it occupies. Thus another factor, this volume, needs to be taken into account if there is to be a comparison of the masses of two separate bodies independently of their sizes. This can be done by measuring the *density* or mass per unit volume of each of the bodies. The unit of density is the *kilogram per cubic metre* kg m^{-3}. The density of pure water at 4°C is 10^3 kg m^{-3}.

Atoms

When matter is subjected to ordinary disintegrating forces there comes a time when it appears that the parts into which it has become separated are incapable of further division without their properties being changed. These apparently indivisible particles of matter are called *atoms* and it is recognized that the atoms of all the different elemental substances are different from one another. The evidence for the existence of atoms first came chiefly from work done on the nature of chemical reactions a long time ago, but much that has been discovered more recently substantiates this concept of the particulate nature of matter.

An early concept of atomic structure, primarily due to Bohr, was that of a small, dense central portion or *nucleus*, positively electrically charged, surrounded by *electrons* of very small mass and negatively electrically charged. The electrons were visualized as circling the nucleus in set orbits, rather like the solar system of sun and planets. These orbits are such that the angular momentum of the electron (product of moment of inertia, I, i.e. mass, m, times square of the distance, r, of the mass from the centre of rotation, and angular velocity or angle turned through in unit time, ω) about the nucleus is an integral multiple of $h/2\pi$, where h = Planck's constant (*see* p. 989) or $mr^2\omega = nh/2\pi$ where n is an integer. This was postulated by Bohr because, according to classical theory, an electron rotating in this way should radiate energy in the form of electromagnetic waves (*see* p. 1006). This could only be at the expense of the energy of the rotating electron which would eventually spiral into the nucleus. Bohr supposed that electrons can travel in certain permissible orbits as described above without radiating energy.

Some concept of the size and openness of atomic structure can be gained from a study of the approximate dimensions of its parts. The effective radius of an atom is about 10^{-8} cm and the effective radius of the nucleus about 10^{-12} cm or about one ten-thousandth of this. The effective radius of an electron is about 10^{-13} cm. Note that the use of the term radius is not intended to convey the idea that atoms or their parts are spherical lumps, like billiard balls.

It is now known that the nucleus is composed of more than one kind of particle, the first discovered being the *protons* of comparatively great mass and positively charged, and the *neutrons*, with slightly greater mass but possessing no electrical charge, all held together by *binding energy*. The proton is equally but oppositely charged as compared with an electron but its mass is about 1836 times as great. The mass of a neutron is about 1839 times that of an electron. More recently, other particulate structures have been shown to leave the nucleus such as *mesons* and *positrons*, but these need not concern us here. Because of the many similarities between the neutron and the proton, a common name, *nucleon*, is sometimes used to denote either of these nuclear constituents.

The more modern concept of the atom is not much like Bohr's original theoretical picture, particularly with regard to the disposition of the electrons around the nucleus. Whereas Bohr conceived the electrons as rotating in orbits around the nucleus as though they were concentrated, discrete particles, the present-day concept is more of the electrons occurring as standing electromagnetic wave patterns in what are known as *electron orbitals*. The position of an electron in an orbital can be given only in terms of probabilities. The electron charge and mass must be regarded as being "smeared out" over the orbital in proportion to the intensity of the electromagnetic wave at each point in the orbital. It should be noted that, strictly, the term orbital is used only in the wave-mechanical treatment of atomic structure and should not be used synonymously with orbit, in the Bohr sense. The reader at this stage may find it easier to visualize the Bohr atomic model.

All the different physical and chemical properties of different kinds of matter are expressions of differences in atomic structure. Each element has a different atomic structure from all others both in its nuclear components and in the numbers and distribution of electrons. The simplest atomic structure known to exist is that of the element hydrogen, consisting of a nucleus made up of a single proton orbited by a single electron. The next in order of complexity is the helium atom with two protons and two neutrons in the nucleus orbited by two electrons; the next is lithium with three protons and three neutrons in the nucleus and three electrons; and so on (*see* Table p. 940).

Electrons may travel distantly from the nucleus but if only those regions where they are most commonly moving are considered, a well-defined pattern of zones or layers of them appears. These are known as the *principal shells* and each shell can be known by its designated letter, K, L, M, N, O, P, Q, there being seven principal shells, or by its *principal quantum number*, 1, 2, 3, 4, 5, 6, 7. In each shell the electrons travel around the nucleus in complex, three-dimensional patterns (orbitals). The laws of quantum mechanics permit only certain types of these patterns. An electron following one of these paths possesses an amount of energy, its *energy level*, characteristic of that path. The fact that some shells may contain a number of different types of orbital need not concern us here. The shells can be regarded as more or less analogous to the planetary orbits postulated by Bohr, if a simple view of the structure is taken.

DATA RELATING TO ATOMS OF SOME ELEMENTS

Element and symbol		Nucleus Pro-tons	Neu-trons	Atomic number	Atomic mass of commonest isotope*	K	L	M	N	O	P	Q	Total electrons
Hydrogen	H	1	0	1	1	1							1
Helium	He	2	2	2	4	2							2
Lithium	Li	3	4	3	7	2	1						3
Beryllium	Be	4	5	4	9	2	2						4
Boron	B	5	6	5	11	2	3						5
Carbon	C	6	6	6	12	2	4						6
Nitrogen	N	7	7	7	14	2	5						7
Oxygen	O	8	8	8	16	2	6						8
Fluorine	F	9	10	9	19	2	7						9
Neon	Ne	10	10	10	20	2	8						10
Sodium	Na	11	12	11	23	2	8	1					11
Magnesium	Mg	12	12	12	24	2	8	2					12
Aluminium	Al	13	14	13	27	2	8	3					13
Silicon	Si	14	14	14	28	2	8	4					14
Phosphorus	P	15	16	15	31	2	8	5					15
Sulphur	S	16	16	16	32	2	8	6					16
Chlorine	Cl	17	18	17	35	2	8	7					17
Argon	Ar	18	22	18	40	2	8	8					18
Potassium	K	19	20	19	39	2	8	8	1				19
Calcium	Ca	20	20	20	40	2	8	8	2				20
Manganese	Mn	25	30	25	55	2	8	13	2				25
Iron	Fe	26	30	26	56	2	8	14	2				26
Cobalt	Co	27	32	27	59	2	8	15	2				27
Nickel	Ni	28	30	28	58	2	8	16	2				28
Copper	Cu	29	34	29	63	2	8	18	1				29
Zinc	Zn	30	34	30	64	2	8	18	2				30
Krypton	Kr	36	48	36	84	2	8	18	8				36
Silver	Ag	47	60	47	107	2	8	18	18	1			47
Iodine	I	53	74	53	127	2	8	18	18	7			53
Barium	Ba	56	82	56	138	2	8	18	18	8	2		56
Gold	Au	79	118	79	197	2	8	18	32	18	1		79
Uranium	U	92	146	92	238	2	8	18	32	21	9	2	92

* To the nearest whole number

Each shell has a certain maximum number of electrons that it can hold. The theoretical maximum possible number of electrons in the various shells has been given as K = 2, L = 8, M = 18, N = 32, O = 50, P = 72, Q = 98, but actually no elements are found having this theoretical maximum for the last three shells. It is found by examination of the elements that exist that the maximum numbers of electrons occurring in the shells are K = 2, L = 8, M = 18, N = 32, O = 32, P = 18, Q = 2. The table above shows the occupancy of these shells by electrons for some of the elements.

As stated earlier, the total electron component of the atom is arranged in a number of different energy levels and we can think of each electron as possessing a certain amount of potential energy (*see* p. 991), according to its distance from the nucleus. The energy possessed by an electron depends upon the energy level it occupies and each of these can be likened to successively higher rungs in a ladder. If the electrons are considered analogous to equal masses resting on the successive rungs from the ground upwards, from the knowledge that equal masses can be lifted to the higher rungs only by the expenditure of greater amounts of energy when compared with the lower, it can be understood that the greater the distance of the energy level (rung) from the nucleus (ground), the more potential energy an electron at that level possesses.

There is a force of attraction between the negatively charged electrons and the positive charges on the protons in the nucleus. This electro-static force is greatest at the shell nearest to the nucleus and decreases with distance from it. Hence electrons can more easily be lost from the more remote shells than from those closer in.

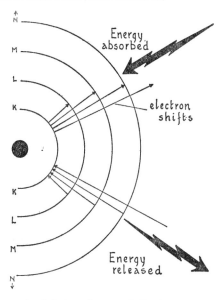

Fig. A.1. Movements of electrons between shells when energy is gained or lost by an atom.

Electrons can move from one energy level to another but, to effect such movement, the atom must absorb or lose energy according to which way the movement takes place. Because the attractive force between the nucleus and electrons is greatest at the closer shells, more energy is required and must be supplied to the atom to move electrons from the inner to the outer shells. Movement of electrons in the opposite direction is accompanied by the release of energy from the atom in the form of *electromagnetic radiation* (*see* Fig. A.1). Thus the shell closest to the nucleus is associated with the lowest energy state of the atom as a whole.

The electron distribution in the various shells for each element is known as its *electron configuration* and it is this that effectively governs the chemical properties of each particular element. This is so because, as will be seen, chemical reactions are primarily the results of interactions between electrons in the outer shells of atoms. Therefore the numbers and locations of the electrons in the outer shells are of great significance in determining what will happen when one kind of atom is brought into close proximity with another.

An atom in its *unexcited* or *ground state*, i.e. its most stable or lowest energy state, has its electrons placed in the available shells of lowest energy, the innermost first, working outwards. If energy is now supplied to such an atom, some of it can cause the movement of electrons from the inner to the outer shells. When this happens the atom is said to be *excited* and possesses energy above its ground energy state. If the electrons are to regain their original positions, they must now move inwards and in doing so will give up the same amount of energy as was supplied to the atom to excite it. Since the energies of the electrons in their different paths differ by discrete amounts, to move an electron from one to another will involve an exact amount of energy, either lost or gained by the atom. According to the *quantum theory* (*see* p. 989), energy exists in discrete units, only whole numbers of which can exist. Each such unit is called a *quantum* and the quantum of electromagnetic energy with which we are concerned here is called the *photon*. Exciting an atom therefore involves supplying it with a certain amount of energy for each outward jump of an electron. In regaining the unexcited state the atom releases the same amount of energy. The terms *quantum shift* or *electron transition* are sometimes used to describe the movement of electrons from one shell to another. It should be noted that an electron may jump from the N to the K shell in one step or it may move N–L–K or N–M–K or N–M–L–K or any of these in the opposite direction. The appropriate amount of energy must, of course, be supplied or abstracted for each jump.

If the energy supplied to an atom is great enough, it can result in the jump of one or more electrons from the inner shells right through the outer so that the electron escapes completely from the influence of the nucleus to which it originally belonged. This can be achieved by bombarding neutral atoms with fast-moving particles such as a stream of electrons. The bombardment is usually carried out with the element in the vapour state. The negatively charged electron that is removed from an atom in this way is a *free electron*, whilst the remaining portion of the atom will therefore carry one negative charge less, or looking at it the other way, one excess positive charge since it will now have insufficient electrons to balance the protons in its nucleus. An electrically charged atom of this kind is called a *positive ion* or *cation* because it would tend to move to the negatively charged cathode in an electric circuit. The least amount of energy required to free an electron from a particular atom is called its *ionization energy* or *ionization potential*, defined as the energy in electron-volts required to remove the most loosely bound electron from the normal atom. Note that an atom can also gain an excess of electrons in its shells and hence carry an excess negative charge, when it is then a *negative ion* or *anion* (*see* p. 963).

Ordinarily, when an atom is excited by having one or more of its electrons moved outwards into the higher energy levels it will regain its normal structure by the movement of the electrons back into their usual shells. As they do this they must release the energy that they gained. The release of this energy becomes apparent as the emission of radiation with particular wavelength such as X-rays or visible light according to the kind of atom and the energy levels involved. Atoms can be identified by the kinds of *emission spectra* that they produce when in the excited state.

If atoms are united to form molecules (*see* below), the energy levels of the individual atoms form molecular energy levels by their interaction with one another, and electron transitions or quantum shifts are possible within these just as they are within the energy levels of individual atoms.

Clearly, if each element is composed of atoms of different construction they can be distinguished from one another in terms of these atomic differences and there are three ways of identifying the elements by their atoms. The most straightforward way is to describe an atom of an element in terms of the number of protons in its nucleus. This is called the *atomic number* of the element, denoted by Z, and since the protons in a neutral (uncharged) atom are balanced by the electrons, the atomic number can give some indication of the chemical properties of that element. The atomic numbers of some elements are given in the table on p. 940.

Another way of describing the atom is to assign to it a *mass number*, denoted by A. This is arrived at by summing the number of protons and neutrons, i.e. nucleons, it possesses, each being given the mass unity, the mass of its electrons being neglected. Atoms of carbon which each possess 6 protons and 6 neutrons will then have atomic number 6 and mass number 12. The element carbon can then be symbolized $^{12}_{6}C$. The third way of describing atoms is in terms of *atomic mass*. This is arrived at by comparing masses of atoms relative to some standard atom. At present the system is to use carbon, with mass number 12, as the standard. On this basis the atomic mass of hydrogen is lowest at 1·00797 (note that no units are necessary because the value is a ratio between two quantities) and uranium is highest at 283·03, when we include the naturally occurring elements only. Note that this convenient method overcomes the problem of having to handle the cumbersome numbers that truly represent the absolute masses of individual atoms. For example, the mass of an oxygen atom is 0·twenty two noughts 266 g or $2·66 \times 10^{-23}$ g.

Mass numbers and atomic masses are closely similar numerically but whereas the former are whole numbers, because they refer to the number of nucleons in the nucleus of a particular atom, the latter are given to several places of decimals for accuracy. The mass number for a given carbon atom may be 12 but the atomic mass of the element carbon is actually 12·01115. This is because, in nature, nearly all elements exist in slightly differing forms or *isotopes* (called so because they occupy the same place in the Periodic Table (*see* p. 950)), and atomic mass is really the average mass number value for all the naturally occurring isotopes of an element. The isotopes are atoms of the same element that have the same atomic number but different mass numbers, i.e. have the same number of protons but different numbers of neutrons; the arrangements of their electrons are the same. For example, calcium exists in at least eight isotopic forms with mass numbers of 40 (20 protons + 20 neutrons), 42 (20 + 22), 43 (20 + 23), 44 (20 + 24) 45 (20 + 25), 46 (20 + 26), 48 (20 + 28) and 49 (20 + 29). In calculating the average of all these values the relative abundance of the isotopes, in nature is allowed for and the atomic mass of calcium is then given as 40·08. It should perhaps be mentioned here that the atomic mass of a particular atom is in reality always less than the sum of the masses of the electrons, protons and neutrons that compose it. This is because there is binding energy involved that shows up as a *mass defect*.

When atoms are combined as molecules, the *molecular masses* are given by summing the masses of the individual atoms in the molecule and it is usual then to use the atomic masses rounded off to the nearest whole number. The molecular mass of glucose ($C_6H_{12}O_6$) would then be given by $(6 \times 12) + (12 \times 1) + (6 \times 16) = 180$.

By establishing the nature of the electron, J. J. Thomson in 1897 showed that the atom is not an indivisible structure. Previously Becquerel had found that uranium salts were capable of emitting high-energy radiations that could penetrate paper and affect photographic plates. The name *radioactivity* was given to this emission of radiation and a number of heavy elements were found to possess the radioactive property. When the Curies isolated radium and other substances it became possible to examine the type of radiation emitted by them. The radiations were found to fall into three main types known as α-*radiation*, β-*radiation* and γ-*radiation*. The α-rays are known now to be positively charged helium atoms or really helium nuclei. The β-rays are fast-moving electrons with the properties of free electrons. The γ-rays are equivalent to very-high-energy X-rays. They are less particulate in their behaviour than α- or β-rays but they do show quantum behaviour ($E = h\nu$). They can be regarded as very high-frequency (short-wave-length) electromagnetic waves (*see* p. 1006), down to less than 10^{-10} cm. They are extremely penetrating and hence dangerous to living systems. Thus some atoms are capable of disintegrating, releasing smaller particles and large quantities of energy.

When the atom of a radioactive element undergoes such disintegration its structure cannot any longer be what it was previously; it must take on the form of some other element. Thus it was found that *transmutation* always accompanies the emission of α- and β-particles. Uranium, for example, when it emits an α-particle becomes thorium. This is called α-decay. Now, the place of helium in the Periodic Table is 2, that of uranium is 92 and that of thorium two places below it, and this fits the finding that an α-particle is a helium nucleus. In fact, whenever a substance undergoes an α-decay it becomes transmuted into the substance two places below it in the Periodic Table, e.g.

$$^{238}_{92}\text{U} \rightarrow {}^{234}_{90}\text{Th} + {}^{4}_{2}\text{He} \text{ or } {}^{238}_{92}\text{U} \,(\alpha,) \,{}^{234}_{90}\text{Th}$$

When a substance undergoes β-decay, that is, emits an electron it is transmuted one place up the Periodic table. For example, one form of bismuth upon emitting a β-particle moves up from place 83 to place 84 and becomes polonium, e.g.

$$^{214}_{83}\text{Bi} \rightarrow {}^{214}_{84}\text{Po} + \beta^- \text{ or } {}^{214}_{83}\text{Bi} \,(\beta^-,) \,{}^{214}_{84}\text{Po}$$

The emission of γ-rays does not involve a transmutation. It is a way in which an atomic nucleus can expend excess energy.

When the naturally occurring radioactive substances were subjected to tests to discover the rate at which they decayed it was found that the rate of decay was proportional to the amount of radioactive material present, i.e. the number of radioactive atoms present and to nothing else, not even temperature, pressure or any other physical or chemical treatments. What this really means is that the radioactivity falls away with regularly decreasing intensity in the following way (*see* Fig. A.2). Suppose the radioactivity of a piece of radium is given as 100 per cent at time zero and the time taken to reach 50 per cent of this activity is taken, then in a time interval following, equal to that just measured, the activity will fall to 25 per cent of the original value and in a further similar time interval, to $12\frac{1}{2}$ per cent of the original value, and so on.

The time taken for any piece of radioactive material to lose half its activity is known as its *half-life*. This is a constant for the substance and can vary from millions of years to a fraction of a second. Some half-life periods are given in the following Table.

THE HALF-LIFE PERIODS OF SOME RADIOACTIVE SUBSTANCES

Isotope	Half-life	Isotope	Half-life	Isotope	Half-life
^{14}C	$5 \cdot 6 \times 10^3$ years	^{35}S	$87 \cdot 1$ days	^{131}I	$8 \cdot 1$ days
^{24}Na	$14 \cdot 9$ hours	^{42}K	$12 \cdot 4$ hours	^{226}Ra	1,620 years
^{32}P	$14 \cdot 3$ days	^{45}Ca	164 days	^{238}U	$4 \cdot 4 \times 10^9$ years

Note that ^{15}N and ^{18}O are stable isotopes.

The intensity of the activity of a sample of radioactive material is expressed in terms of the number of disintegrations of atoms per unit time, chosen because this will be proportional to the number of radioactive atoms present. For example, one gram of radium exhibits an activity $3 \cdot 7 \times 10^{10}$ disintegrations per second ($3 \cdot 7 \times 10^{10}$ s^{-1}). It was this intensity of activity that was once the unit of radioactivity known as the *curie* (Ci). This unit is not included in the SI but its use is permitted, at least for the time being. For biological purposes quantities of material showing only about one thousandth to one millionth of the activity of a gram of radium are used, that is, $3 \cdot 7 \times 10^7$ to $3 \cdot 7 \times 10^4$ disintegrations per second. Note that for work in schools there are regulations concerning the use of radioactive substances. Permission must be obtained from the Department of Education and Science who will state the maximum quantities of radioactive materials that can be used in any one biological exercise or stored on the school premises. Relevant information concerning the granting of permission and the quantities and kinds of materials that can be used will be found in *Administrative Memorandum* 1/65 (8th Jan., 1965) obtainable from the Department of Education and Science.

The intensity of radioactivity of any given sample of radioactive substance can be measured with a Geiger counter. In biological work of an elementary kind it is often only necessary to detect the presence of a *tracer substance* and this can be done with photographic plates or papers.

So far, reference has been made only to those radioactive substances that occur naturally. Others can be obtained by artificial transmutation processes. Radioactive forms of carbon (^{14}C), phosphorus (^{32}P), sulphur (^{35}S), sodium (^{24}Na), potassium (^{42}K), calcium (^{45}Ca), iron (^{59}Fe) and iodine (^{131}I) are examples. Their half-lives are shown in Table 2. Radioactive phosphorus is obtained by neutron bombardment of sulphur and the form of carbon mentioned above is derived from nitrogen.

These radioactive isotopes must not be confused with *stable isotopes* that are also used as tracers in biological work, such as oxygen-18 and

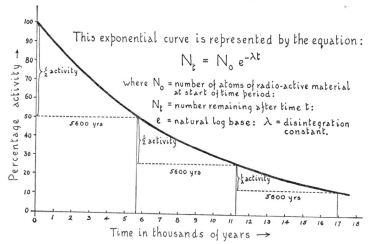

Fig. A.2. Curve representing the radioactive decay of carbon-14 with a half-life of 5,600 years.

nitrogen-15. These are prepared by fractionation of the naturally occurring mixtures of isotopes and are said to be stable because no change with time can be detected in their atomic structure, as opposed to unstable isotopes that show radioactive changes. Heavy isotopes such as those mentioned above, when used as biological tracers must be estimated by physical methods such as measurement of density, changes in refractive index or mass spectograph analysis.

Of all the elements in the Periodic Table there are about 340 isotopes occurring naturally and of these about 270 are stable. All the different isotopes showing differences in the composition of their nuclei really represent separate nuclear species. They are usually called *nuclides*. Nuclides having the same mass number (A) but with different numbers of neutrons (N) and protons (Z) are called *isobars*. Those having the same N but possessing different Z and A numbers are called *isotones*.

The use of subscript numbers for the Z value and superscript numbers for the A value with the element symbol, denotes which isotope is being referred to but there seems to be no universally obeyed convention. The use of the Z number is often omitted and the same isotope can be found symbolized as $^{12}_{6}C$, $_{6}C^{12}$, $^{12}C_{6}$, ^{12}C, C^{12}, C-12 or 12C. The first of these conforms to SI requirements.

Clearly, the use of both stable and unstable isotopes in biological work has been of the utmost value, making it possible to trace metabolic

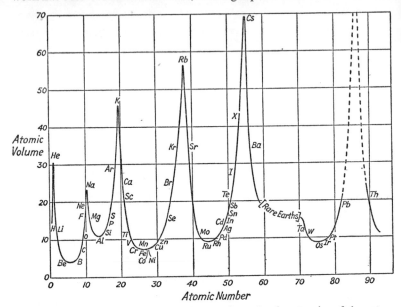

Fig. A.3. Graph of atomic volume against atomic number for a number of elements, to show periodicity.

pathways without interference with normal working of the living system. The occurrence of ^{14}C naturally in living things has enabled biologists to estimate the age of once-living organisms with great accuracy. ^{14}C is produced continuously in the atmosphere by the action of neutrons, released by cosmic radiation, on nitrogen. Living plants will absorb some of this along with the normal ^{12}C. Therefore all green plants will be slightly radioactive when they die. The half-life of ^{14}C is 5600 years and one gram of carbon from a living plant has an activity of 16 disintegrations per minute. Since, when the plant dies no more ^{14}C is absorbed and the quantity left in any dead material gradually decreases with the characteristic half-life of carbon, the activity of any dead specimen, when the weight of carbon in it is known, can be used to determine the age of the specimen. For example, a fossil containing 1 g of carbon showing 8 disintegrations per minute is 5600 years old; a similar specimen showing 4 disintegrations per minute is 11 200 years old.

When the variation in the properties of the elements is studied closely it can be seen that they vary in a regular or periodic way, involving regular repetition of certain characteristics. For example, if a graph is

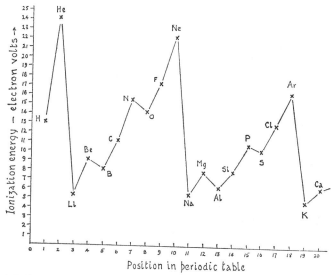

Fig. A.4. Graph of ionization energy (ionization potential) against position in the periodic table (atomic number) for some elements.

constructed by plotting atomic number (protons in the nucleus) against atomic volume in the solid state (atomic mass/density), the shape of the line joining the values for all the elements shows an unmistakeable regular periodicity of peaks and troughs (*see* Fig. A.3). Similarly, if a graph of atomic number against ionization energy or ionization potential (the energy required to remove an electron from the atom) is plotted, it is clear that there is not a smooth variation between the elements (*see* Fig. A.4). If every electron came out of the same orbit, merely making a small change in the energy of the orbit as it moved, we could expect a smooth variation in the graph. But instead we see very sharp breaks between pairs of elements such as helium and lithium, neon and sodium, argon and potassium and so on, with much less irregularity between lithium and neon, sodium and argon, etc. The

The Periodic Table of Elements

IA	IIA	IIIA	IVA	VA	VIA	VIIA	VIII			IB	IIB	IIIB	IVB	VB	VIB	VIIB	0
1 H 1·00797																	2 He 4·0026
3 Li 6·939	4 Be 9·012											5 B 10·811	6 C 12·011	7 N 14·007	8 O 15·9994	9 F 18·998	10 Ne 20·183
11 Na 22·990	12 Mg 24·312											13 Al 26·98	14 Si 28·086	15 P 30·97	16 S 32·064	17 Cl 35·453	18 Ar 39·95
19 K 39·102	20 Ca 40·08	21 Sc 44·96	22 Ti 47·90	23 V 50·94	24 Cr 52·00	25 Mn 54·94	26 Fe 55·85	27 Co 58·93	28 Ni 58·71	29 Cu 63·54	30 Zn 65·37	31 Ga 69·72	32 Ge 72·59	33 As 74·92	34 Se 78·96	35 Br 79·91	36 Kr 83·80
37 Rb 85·47	38 Sr 87·62	39 Y 88·91	40 Zr 91·22	41 Nb 92·91	42 Mo 95·94	43 Tc 99·00	44 Ru 101·07	45 Rh 102·91	46 Pd 106·4	47 Ag 107·87	48 Cd 112·40	49 In 114·82	50 Sn 118·69	51 Sb 121·75	52 Te 127·60	53 I 126·90	54 Xe 131·30
55 Cs 132·90	56 Ba 137·34	57–71 La* series	72 Hf 178·49	73 Ta 180·95	74 W 183·85	75 Re 186·2	76 Os 190·2	77 Ir 192·2	78 Pt 195·1	79 Au 196·97	80 Hg 200·59	81 Tl 204·37	82 Pb 207·19	83 Bi 208·98	84 Po 210·0	85 At 210·0	86 Rn 222·0
87 Fr 223	88 Ra 226	89– Ac** series															

	IIIA	IVA	VA	VIA	VIIA	VIII			IB	IIB	IIIB	IVB	VB	VIB	VIIB	IIIA
Lanthanide* series	57 La 138·91	58 Ce 140·12	59 Pr 140·91	60 Nd 144·24	61 Pm 147·0	62 Sm 150·35	63 Eu 151·96	64 Gd 157·25	65 Tb 158·92	66 Dy 162·50	67 Ho 164·93	68 Er 167·26	69 Tm 168·93	70 Yb 173·04	71 Lu 174·97
Actinide** series	89 Ac 227·0	90 Th 232·04	91 Pa 231·0	92 U 238·03	93 Np 237·0	94 Pu 239·0	95 Am 241·0	96 Cm 242·0	97 Bk 249·0	98 Cf 252·0	99 Es 254·0	100 Fm 253·0	101 Md	102 No	103 Lw

The periodic table above shows all the known elements classified into groups, O, IA, IB, IIA, IIB, etc., in vertical columns, according to their chemical characteristics. For each element, the symbols and names of which are shown separately on p. 951, the atomic number is given above and the atomic weight below. This latter is calculated as the average mass number value for all the naturally occurring isotopes of that element, the relative abundance of each natural isotope being taken into account. This explains the reversal of the order of increasing atomic weights for the pairs of elements, argon–potassium, cobalt–nickel, tellurium–iodine. The members of each group have chemical properties in common. In group IA, for example, the elements all have one valency electron (in the outermost shell); in group IIA, all have two valency electrons. The two series, lanthanide and actinide, have atoms with special electron distributions that give them common characteristics and are often treated as separate groups, although they are sometimes placed in group IIIA. For a fuller treatment of the chemical characteristics of the elements in each group, the reader is referred to an appropriate textbook of Chemistry.

Atomic number	Symbol		Atomic number	Symbol	
1	H	Hydrogen	53	I	Iodine
2	He	Helium	54	Xe	Xenon
3	Li	Lithium	55	Cs	Caesium
4	Be	Beryllium	56	Ba	Barium
5	B	Boron	57	La	Lanthanum
6	C	Carbon	58	Ce	Cerium
7	N	Nitrogen	59	Pr	Praseodymium
8	O	Oxygen	60	Nd	Neodymium
9	F	Fluorine	61	Pm	Promethium
10	Ne	Neon	62	Sm	Samarium
11	Na	Sodium	63	Eu	Europium
12	Mg	Magnesium	64	Gd	Gadolinium
13	Al	Aluminium	65	Tb	Terbium
14	Si	Silicon	66	Dy	Dysprosium
15	P	Phosphorus	67	Ho	Holmium
16	S	Sulphur	68	Er	Erbium
17	Cl	Chlorine	69	Tm	Thulium
18	Ar	Argon	70	Yb	Ytterbium
19	K	Potassium	71	Lu	Lutetium
20	Ca	Calcium	72	Hf	Hafnium
21	Sc	Scandium	73	Ta	Tantalum
22	Ti	Titanium	74	W	Tungsten
23	V	Vanadium	75	Re	Rhenium
24	Cr	Chromium	76	Os	Osmium
25	Mn	Manganese	77	Ir	Iridium
26	Fe	Iron	78	Pt	Platinum
27	Co	Cobalt	79	Au	Gold
28	Ni	Nickel	80	Hg	Mercury
29	Cu	Copper	81	Tl	Thallium
30	Zn	Zinc	82	Pb	Lead
31	Ga	Gallium	83	Bi	Bismuth
32	Ge	Germanium	84	Po	Polonium
33	As	Arsenic	85	At	Astatine
34	Se	Selenium	86	Rn	Radon
35	Br	Bromine	87	Fr	Francium
36	Kr	Krypton	88	Ra	Radium
37	Rb	Rubidium	89	Ac	Actinium
38	Sr	Strontium	90	Th	Thorium
39	Y	Yttrium	91	Pa	Protactinium
40	Zr	Zirconium	92	U	Uranium
41	Nb	Niobium	93	Np	Neptunium
42	Mo	Molybdenum	94	Pu	Plutonium
43	Tc	Technetium	95	Am	Americium
44	Ru	Ruthenium	96	Cm	Curium
45	Rh	Rhodium	97	Bk	Berkelium
46	Pd	Palladium	98	Cf	Californium
47	Ag	Silver	99	Es	Einsteinium
48	Cd	Cadmium	100	Fm	Fermium
49	In	Indium	101	Md	Mendelevium
50	Sn	Tin	102	No	Nobelium
51	Sb	Antimony	103	Lw	Lawrentium
52	Te	Tellurium			

very highly active substances, such as fluorine and chlorine, that combine so readily with hydrogen to form acids, occupy places on the graph just before the very inert gases, neon and argon. It can be taken for granted that such irregular variation in the properties of the elements are no accidental occurrences. There must be something in the structures of the various elements that governs their properties and there must be regular repetition of particular characteristics of atomic structure to account for the same regular repetition of the physical and chemical properties of the elements over the range of their atomic numbers. An understanding of atomic structure helps us to understand this periodicity. A modern simplified version of the Periodic Table of Elements is shown in the table on p. 950. It helps to substantiate the validity of the Periodic Law which states that "the properties of the elements are in periodic dependence of their atomic weights" first put forward by Mendeleev in 1869. A list of the known elements, in order of their atomic numbers, with their symbols, is given on p. 951.

Molecules and Chemical Bonds

Particles composed of more than one atom are called *molecules*. The numbers and kinds of atoms in a particular molecule can vary from the lowest limit of two of the same kind as in the combination of two hydrogen atoms to form the hydrogen molecule to very large numbers of atoms of different kinds to form the exceedingly complex molecules of proteins and other organic substances. The atoms are held together by energy relationships between them and we say that the atoms constituting a molecule are united by *chemical bonds*, where each bond can be conceived as representing a definite amount of *potential energy* (*see* p. 991). When such a bond is broken this potential energy is converted into *kinetic energy*.

A *chemical reaction* is said to occur when atoms come together to form molecules, i.e. bonds are built, or vice versa when molecules break down into constituent parts, i.e. bonds are broken. Thus all chemical reactions involve energy exchanges (*see* p. 999). The formation of chemical bonds is due to the interactions of electrons in the atoms concerned and in order for this to occur the atoms must be close enough together for their electrons to mingle. Ordinarily, there is a natural repulsion between the negative charges possessed by the electrons of two adjacent atoms and between their positively charged nuclei, thus for one atom to come close enough it must be moving with sufficient energy to overcome this repulsion. One way of assisting this is to raise the temperature of the mixture of atoms and in this way increase their energy of movement amongst one another (*see* p. 982). When two sets of electrons are brought close enough together rearrangements of the electrons in the outermost energy levels of each atom can occur. There are two possible rearrangements; one of the atoms may give up one or more of its electrons to the other atom or each of the atoms can share one or more of the electrons between them. Whatever happens, the result is that the total electric charge of one of the atoms will change with respect to the other and it is this electrostatic attraction between the atoms that constitutes the chemical bond. The readiness with which one atom will interact with another to form a molecule depends upon the stability or otherwise of its outer electron shell.

Now, with the exception of the K shell which is filled with two electrons, no matter which energy level is outermost in an atom it so happens that this level never contains more than 8 electrons. For example, in the M shell the maximum number of electrons possible is 18 but in the series of elements between atomic numbers 11 and 18, the M shell never contains more than 8 electrons. If another electron is added to this system it takes up position in the next outer energy level, the N shell. From this it must not be thought that the N shell will fill up successively as electrons are added. Argon has 8 electrons in the M shell. The next element, potassium, adds one electron in the N shell. The next element, calcium, adds yet another in the N shell to make 2. But the next element, scandium does not add a third electron in the N shell but a ninth electron in the M shell, while titanium has 10 electrons in the M shell and 2 in the N shell. This possession of 8 electrons in its outer orbit represents a *stable electron configuration* for an atom. Such stability occurs in the atoms of the well-known *inert* gases such as helium, neon, argon, krypton, xenon and radon and they were called inert because it was thought that they could never undergo chemical reactions with other atoms. It is true that these gases will form compounds with fluorine, e.g. xenon tetrafluoride, and this shows that the assumption concerning stability with 8 electrons in the outer shell is not wholly true but it certainly holds for most atoms. No atoms are known to have more than 8 electrons in the outermost occupied shell and all those with less than 8 will react with other atoms by losing, gaining or sharing electrons in an attempt to make up an outer shell containing 8 electrons.

The two major types of chemical bonds found in molecules are distinguished by the way in which stability is attained in the outer energy levels of the combining atoms. The one most commonly found is that in which one atom gives up its outermost electrons to one or more other atoms and is known as the *electrovalent* or *ionic bond*. By the transfer, the outermost energy levels of all the atoms concerned are made more stable. A good example of ionic bonding occurs between hydrogen and chlorine. A chlorine atom shows 2, 8 and 7 electrons in its K, L and M shells respectively. The hydrogen atom has one electron only in the K level. If it loses this to the chlorine it becomes stable but possesses a net positive charge. If the chlorine atom gains an electron it too becomes stable but now possesses a net negative charge. The attraction between these opposite electric charges will hold the two atoms together and a molecule of hydrogen chloride is formed.

When such ionic bonds are formed, the general rule is that those atoms with less than four electrons in their outer energy levels will tend to lose them and those with more than four will tend to gain electrons.

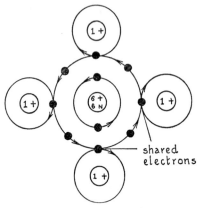

Fig. A.5. The covalent bonding of carbon and hydrogen atoms to form methane.

The former are called *electron donors* and include such atoms as boron, aluminium, sodium, potassium, hydrogen, calcium and iron, all these possessing three or fewer electrons in their outer shells. The latter are called *electron acceptors* and include atoms of chlorine, oxygen and sulphur each needing one or two electrons to make up the outer number to 8.

The other main type of chemical bond is known as the *covalent bond* and is exhibited by such atoms as carbon which has four electrons in its outer shell. Here the binding is achieved by the sharing of one or more pairs of electrons between atoms. The electrons in the outer shells of the atoms concerned undergo rearrangement so that neither actually loses or gains electrons but the electrons in the outer shells are shared between the nuclei of all the atoms in the combination. A good example of covalent bonding is shown by methane, CH_4. Here the molecule consists of a central carbon atom with four electrons of its own in the outer shell and four symmetrically placed hydrogen atoms, the electrons of which can enter the outer energy level of the carbon atom (*see* Fig. A.5). Such a molecule can be represented as

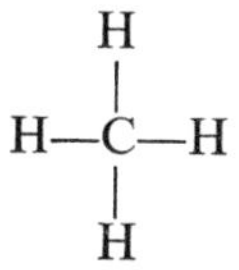

$$
\begin{array}{c}
H \\
| \\
H-C-H \\
| \\
H
\end{array}
$$

where each link between the carbon atom and the hydrogen atoms represents a pair of shared electrons. Another way of representing the molecule is

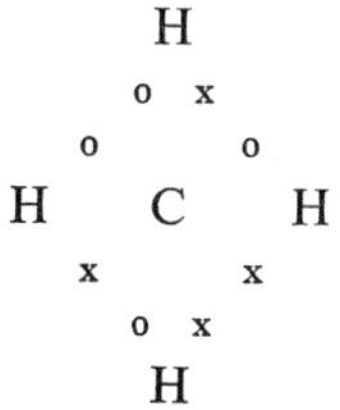

$$
\begin{array}{ccc}
 & H & \\
 & o\ \ x & \\
o & & o \\
H & C & H \\
x & & x \\
 & o\ \ x & \\
 & H &
\end{array}
$$

where o = electrons in the L shell of the carbon atom and x = the electrons in the K shells of the hydrogen atoms.

It is possible in the union between atoms of certain kinds to find an apparent unequal sharing of electrons between the atoms, that is, electrons spend more time with one atom than with the other. This is known as *partial ionic bonding* and is, as it were, a combination of ionic and covalent bonding.

The three kinds of chemical bonds so far mentioned, ionic, covalent and the intermediate condition, whilst accounting for a very large number of atomic unions to form molecules cannot account for all cases. Another type of bond, more difficult to visualize, but which occurs in some of the molecules of living substance is the *resonance bond*. This is the result of alternation between different possible electron configurations within the molecule. It can most easily be understood with reference to the hydrogen molecule and the covalent bond between its two hydrogen atoms. During the sharing of electrons two possibilities occur: each of the electrons can be near its own nucleus or alternatively, each can be nearest to the nucleus of the other atom. This should not appear to matter in the overall situation but it is the case that an exchange of electrons of this nature does not affect the binding energy between the atoms and an energy exchange is created between them. This is sometimes described as *resonance* and is exhibited by the well-known benzene molecule, C_6H_6. The structure of this molecule can be represented simply as

Each carbon atom has four electrons capable of making four covalent bonds. Of the total of 24 bonds, 6 are used in binding the 6 hydrogen atoms, leaving 18 for bonding between the carbons, i.e. 3 per atom. There are obviously two ways in which these bonds may be distributed with double bonding between different pairs of adjacent carbon atoms so

In fact, the benzene molecule is never wholly of one of these forms, it actually alternates or resonates between the two conditions and this results in an exchange of resonance energy which holds the carbon atoms together. The double bond can be thought of in a very simple way as the case where the electrons forming it spend half the time on one side of each carbon atom and half the time on the other side. By changing their positions between these two configurations the electrons pull the carbon atoms closer together than if single bonds only existed.

Another possible form of energy relationship between atoms is known as the *hydrogen bond*. This is produced by the electrostatic attraction between a positively charged hydrogen atom or ion, i.e. one lacking its electron, on one part of a molecule and a negatively charged atom of oxygen or nitrogen on the same or another molecule. It is the case that oxygen collects two electrons and becomes negatively charged to reach the nearest stable atomic configuration and hydrogen gives up one electron and becomes positively charged to reach the same state. Nitrogen also, in collecting three electrons to become stable, becomes negatively charged. In certain situations, as in some molecules where negative oxygen atoms are brought close to other negative oxygen or nitrogen atoms, it is possible that they could be bound by a positively charged hydrogen ion oscillating between them. This often occurs between $C=O$ groups on one part of a molecule and NH_3 groups on another part of the molecule or on a different molecule in which the binding hydrogen is one of the three normally associated with the nitrogen. The bond is a weak one and is so regarded for the reason that the hydrogen nucleus is many times more massive than an electron and its oscillations between the two bound units are much slower. However, some biological molecular structures are very large and there can be many places on the molecules where such bonding can occur. Thus large amounts of binding energy can be amassed between such molecules or between parts of the same molecule. It is believed that the two strands of the nucleic acid molecule (DNA) are held together entirely by such bonds between the purine and pyrimidine bases (see p. 111). There are other possible occurrences involving binding energy between atoms but these will not be touched upon here.

Of the two main kinds, the ionic and covalent bonds, the former is sometimes described as *unsaturated* for the reason that other particles can enter the binding relationship, which is due to electrical interaction, between the particles already engaged, whereas the latter is described as *saturated* because, in the sharing of electrons between atoms, there is no further possible involvement with other atoms unless the bond is first broken. This difference in the nature of these bonds is important when the likely effects of bringing other substances into contact with already chemically combined atoms have to be considered. The relative strengths of the chemical bonds mentioned above, allowing for variations in actual values, are given by

ionic and covalent $100 \times 4 \cdot 2$ kJ mol^{-1}
partial ionic 50,,
resonance 40,,
hydrogen 2–10 ,,

Attention is called here to the so-called "energy-rich" or "high-energy" phosphate bonds in adenosine triphosphate (ATP). In living systems the chemical bond energy of ATP (see p. 111) between two of the phosphate radicals and the rest of the molecule acts as the energy "currency" of the cell. Here energy is stored in small "packages" and can be turned to useful work by the severance of these "energy-rich" bonds. Measured in terms of heat energy the breaking by hydrolysis of each of the two "energy-rich" bonds to liberate the two phosphate radicals releases 8–$12 \times 4 \cdot 2$ kJ mol^{-1} whereas the breaking of the third phosphate bond releases only about 3–$4 \times 4 \cdot 2$ kJ mol^{-1}, according to various estimates. The term "energy-rich" or "high-energy" is therefore used purely relatively and is comparing one phosphate bond with another only. It is misleading in the sense that it tends to convey the idea that such phosphate bonds are very special in possessing more potential energy than any other kind. This is certainly not so and in living things there are many chemical bonds possessing more energy than the "energy-rich" phosphate bonds of ATP.

Valency

The combining power of one atom with another is referred to as its *valency* or *valence*. This is expressed as a number and has been defined as the number of hydrogen atoms that an atom will combine with or replace in a chemical reaction. For example, in water, oxygen combines with two hydrogen atoms and the valency of the oxygen is said to be 2. The valencies of different atoms vary from 1 to 8 (zero for the inert gases), but in some the valency depends upon the nature of the other atom in the combination. For example, hydrogen and chlorine normally have valency 1 (monovalent), oxygen has valency 2 (divalent), phosphorus and boron have valency 3 (trivalent), carbon and silicon have valency 4 (tetravalent) but some elements such as nitrogen and sulphur are polyvalent meaning that their valencies can vary. In nitrous oxide (N_2O) nitrogen is monovalent, in nitric oxide (NO) it is divalent, in ammonia (NH_3) it is trivalent, in nitrogen peroxide (NO_2) it is tetravalent and in anhydrous nitric acid (N_2O_5) it is pentavalent. In sulphuretted hydrogen (H_2S) sulphur is divalent, in sulphur dioxide (SO_2) it is tetravalent and in sulphur trioxide (SO_3) it is hexavalent. Manganese is normally divalent but in the oxide Mn_2O_7 it is

VALENCY VALUES FOR SOME COMMON ELEMENTS

1	2	3	4	5	6
Sodium	Calcium	Aluminium	Tin*	Phos- phorus*	Sulphur*
Potassium	Strontium	Iron*	Lead*	Nitrogen*	
Silver	Barium	Chromium	Carbon*		
Mercury*	Zinc	Nitrogen*	Silicon		
Fluorine	Magnesium	Phos- phorus*	Sulphur		
Chlorine	Manganese*		Nitrogen		
Bromine	Copper				
Iodine	Iron*				
Nitrogen*	Nickel				
	Tin*				
	Lead*				
	Mercury*				
	Oxygen				
	Sulphur*				
	Carbon*				
	Nitrogen*				

(Note that elements marked * have more than one valency)

heptavalent. Osmium in the oxide OsO_4 is octavalent. The valencies of some of the commoner elements are shown in the table on p. 958.

If what has been written earlier concerning atomic structure, its stability and the formation of molecules by the donation or sharing of electrons has been understood, it should not be difficult to grasp that the valency of an atom can be explained in terms of the number of electrons it loses, gains or shares during combination with another. It is perhaps more suitable to write the valency value as positive or negative when considering it from this point of view. A positive valency indicates that the atom donates electrons and a negative valency indicates the opposite. For example, a potassium atom, when it donates an electron to some other atom becomes a potassium ion with a valency of $+1$. By the same notation, since chlorine accepts an electron in combining with another atom its valency is -1.

If we know the usual valencies of atoms (or molecules) we can predict the proportions in which they will combine in chemical reactions. For example, the valency of potassium is $+1$ and that of chlorine is -1. To form an electrically neutral molecule from these two elements, one of each is required and we can write down the combination as KCl. In this case, the one-to-one match is perfect since chlorine needs one electron to fill its outermost M shell to become stable whilst potassium possesses a single electron in its outer N shell that, when donated, renders this atom stable. But the number of electrons that one atom can donate or may be needed by one atom to give it a stable outer shell is not always met by interaction with another single atom and the cases of magnesium chloride ($MgCl_2$) or aluminium oxide (Al_2O_3) show that several atoms of one element may be needed to satisfy the electron requirements of another. In the former, magnesium possesses two electrons in its outer M shell, chlorine possesses seven. To reach stability the magnesium therefore requires two atoms of chlorine, each capable of accepting one electron to reach stability, to accept its electrons, to form the molecule of magnesium chloride. In the reaction between aluminium and oxygen, each atom of aluminium has three outer electrons in its M shell to donate, whilst oxygen has six in its outer L shell and needs to accept two to become stable. The only satisfactory arrangement between these atoms is therefore two aluminium to three oxygen, the three latter being satisfied with two each of the six electrons available from the two aluminium atoms. A simple rule to remember and one that fits most cases, in constructing molecular formulae for the combination of two elements, is to write the symbols for the components and then to supply the subscript number for each element by quoting the valency number of the other, remembering, of course, that some elements may have more than one valency number.

The condition of polyvalency in some elements can be traced to the fact that in these the number of electrons in the outer shell can vary because one or more electrons can be moved between the outermost shell and the one beneath it. The iron atom, for example, can have the structure K2, L8, M14, N2 or K2, L8, M13, N3, because an electron can be moved between an outer shell and that next below it. This means that the number of electrons in the outer shell varies and hence there can be variation in the number of electrons that can be accepted or donated during a reaction. In the two cases for iron, one atom can donate two electrons, i.e. has valency 2, and become the ferrous ion (Fe^{++}) and the other can denote three electrons, i.e. has valency 3, to become the ferric ion (Fe^{+++}). Iron and some other substances such as cobalt, nickel and copper can thus have variable valency depending on the characteristic of being able to move electrons from one shell to another. Such substances owe their efficiency in electron transfer reactions in living things to this property.

It should now be clear that an important concept concerning chemical reactions is that the atoms and molecules that are involved combine or dissociate from one another always in definite and predictable proportions according to their particular structures and that the properties of the elements in their abilities to combine with one another depend essentially upon the configuration of the electrons in the atom, particularly those in the outermost shells.

Shapes of Molecules and Polarity

When two atoms combine they can always be visualized as lying together with their nuclei in one plane, but when more than two atoms combine their relationships with one another in space can be very complicated. In a multi-atomed molecule there is always a definite, three-dimensional, geometrical pattern of arrangement of the atoms with respect to one another and, according to the numbers and kinds of atoms concerned, this can be predicted.

When a water molecule is produced from one atom of oxygen and two of hydrogen, the atoms lie

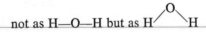

not as H—O—H but as H〈O〉H

where the angle between the lines joining the nuclei of the two hydrogens to the oxygen is 104°31′. This is called the *bond angle* and there are methods by which such angles between various combinations of atoms can be measured, so enabling the relative positions of atoms in a molecule to be established. The "flat" structural formulae for compounds often given in texts can be very misleading. For example, the methane molecule, usually shown as

$$H$$
$$|$$
$$H—C—H$$
$$|$$
$$H$$

is really a tetrahedron or four-faced figure contained by four triangles (pyramid with a triangular base) and is better represented as shown in Fig. A.6. The geometrical shape of the extremely complex DNA

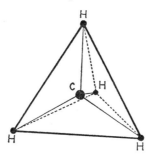

Fig. A.6. Representations of the methane molecule to show its three-dimensional form.

molecule, produced by the positioning of thousands of atoms is shown in Fig. 4.3.

Shapes of molecules are often important in determining the ways in which they will react with one another. This is particularly true of large molecules such as proteins, whose chemical properties can be drastically changed by the rearrangement of just a few atoms in the molecule.

One consequence of the particular shape of a molecule is the uneven distribution of electrical charges over its surface, that is, one part of a molecule can be positive or negative relatively to another part. When this is the case the molecule is said to exhibit *polarity*, i.e. acts as an *electric dipole* and the effect is to produce an arrangement that might be likened to a bar magnet (with a N pole at one end and a S pole at the other) except that the polarity is due to electrical and not magnetic conditions. The difference between the "poles" or charges in this case is due to an accumulation of electrons at one place and an equal deficit of electrons at another. The water molecule with the shape described above is a simple example of a polar molecule. The positioning of the hydrogen atoms with respect to the oxygen atom creates a molecule that is positively charged on the hydrogen side and negatively charged on the oxygen side so

$$+H \diagup \overset{\overset{=}{O}}{} \diagdown H+$$

As a whole it is electrically neutral but its opposite sides can act independently and effect electrical interactions. Thus water molecules can become interposed between the atoms of other molecules that are bonded by electrical forces. For example, sodium chloride is separated into sodium and chloride ions when placed in water (*see* p. 965). The water molecule is said to be *dipolar* but many larger molecules show *multipolar* properties. Glyceride fats are non-polar.

One important property of polar molecules is the fact that they tend to become oriented, that is, definitely positioned in space, with respect to other molecules. A good example of such orientation is afforded by the fatty acid molecules. Each is composed of a non-polar chain of

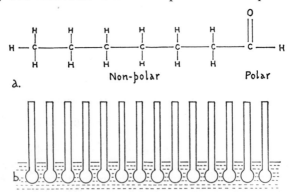

Fig. A.7. Polarity exhibited by the fatty acid molecule. (*a*) form of the molecule; (*b*) orientation of fatty acid molecules at a water surface.

carbon atoms with a polar carboxyl group (COOH) at one end (*see* Fig. A.7*a*). When these molecules are placed in contact with water, the non-polar ends are repelled by the water whilst the charged ends of the acid molecules are attracted to the water dipoles. The result is that the fatty acid molecules are all oriented with the carboxyl groups held in the water and the carbon chains extending outwards into the air (*see* Fig. A.7*b*). The distinctive structure of cell membranes is thought to be due, at least partially, to a similar action between phospho-lipides (combinations of fats and phosphate groups) and water. They form a boundary zone at the surface of the aqueous inner cytoplasm in much the same way as fatty acids do. In the sodium cholate bile secretions exist examples of polar molecules that play an important role in the digestion and absorption of fat. Sodium cholate forms a colloidal solution with water, containing aggregates of molecules in which their polar parts are all oriented outwards from the non-polar central portions. A comparable condition arises when the sodium cholate in aqueous solution comes into contact with a non-polar lipid. When this occurs the sodium cholate molecules in an aggregate separate and become oriented on the fat surface with their non-polar parts inwardly directed to the fat droplet in the same way as a detergent substance tends to stabilize the fatty material thus to facilitate the formation of a colloidal solution.

Within very large molecular structures, such as proteins, polarity is important in determining the geometry of the molecule and through this many of its chemical properties. Large protein molecules may possess many polar groups, i.e. points at which there is a build-up of positive or negative charges, and there may be attractions between these oppositely charged regions within the same molecule, resulting in very specific twisting and folding of the molecule as a whole.

Molecules in Solution: Ions and Radicals

It was stated previously that an ion is an atom or group of atoms that bears a net electric charge, that is, carries excess electrons (negative) or is deficient in them and hence has an excess of protons (positive). Many compounds are said to ionize when placed in water or other suitable ionizing agent. A simple example is that of hydrogen chloride, a compound that at room temperature is in the gaseous state. When molecules of this compound are placed with water they are said to go into *solution*, a state in which the *solute*, hydrogen chloride, particles are homogeneously distributed with those of the *solvent*, water, throughout the mixture. Under such conditions the hydrogen becomes separated from the chlorine in the hydrogen chloride molecule and the two are said to be *dissociated* but the dissociation occurs in such a way that the hydrogen atom does not take back the electron that it donated to the chlorine atom in creating the ionic bond between them. Thus the hydrogen atom is really a proton carrying a charge of $+1$ and the chlorine atom possesses an extra electron equivalent to a charge of -1. The hydrogen particle is now a *hydrogen ion* and the chlorine a *chloride ion,* written as H⁺ and Cl⁻ respectively. The ionization process may be thought of as being brought about by the tendency of the water molecules to interpose themselves between the parts of the hydrogen chloride molecule, and the chemical bonds holding the hydrogen to the chlorine are not actually broken (*see* Fig. A.8). When the water is removed from the system the oppositely charged hydrogen and chlorine atoms recombine to form the electrically neutral hydrogen chloride molecules once more. Many compounds are acted upon in the same way by water which is thus said to be a good *ionizing agent.* Any substance that dissociates into ions and so renders the solution a conductor of electricity is called an *electrolyte.*

Not all compounds are ionized by the separation of individual atoms. In some, one of the dissociation products is a combination of two or more atoms that carry an overall positive or negative charge. Such charged particles are known as *radicals* and are really collections of two or more atoms acting as single ions. For example, nitric acid (HNO₃) when placed with water, ionizes yielding hydrogen ions, H⁺, and

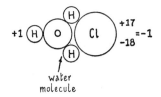

Fig. A.8. Diagram representing the dissociation of hydrogen and chlorine in a water solution.

nitrate radicals, NO_3^-, these being composed of one atom of nitrogen and three of oxygen.

This is a very simple way of expressing what happens in such a case. It is now considered that the hydrogen ion, H⁺, is such a small cation by comparison with others that it would have a very strong tendency to form a covalent bond with some other molecule or group. Thus a hydrogen ion in aqueous solution immediately enters into combination with a water molecule to form the *hydroxonium* (*hydronium*) ion, H_3O^+. On the postulate that it is the hydroxonium ion that confers acidity on a solution, the conversion of a pure acid, such as nitric acid, into aqueous solution is not an electrolytic dissociation, but a chemical reaction with water, so

$$HNO_3 + H_2O \rightleftharpoons H_3O^+ + NO_3^-$$

Nevertheless, the hydrogen ion is still commonly represented as though it actually exists in the uncombined state.

The dissociation of molecules into ions can be represented by ionization equations. For example, $HCl \rightarrow H^+ + Cl^-$ shows that, upon dissociation, one molecule of hydrogen chloride yields one positively charged hydrogen ion and one negatively charged chloride ion. The ionization of calcium chloride would be written as $CaCl_2 \rightarrow Ca^{++} + 2Cl^-$ and that of sulphuric acid as $H_2SO_4 \rightarrow 2H^+ + SO_4^{--}$.

Not all compounds when dissolved in water dissociate into ions equally readily. Molecules such as those of hydrogen chloride, sodium chloride and sulphuric acid are examples of almost complete ionization, that is, nearly all the molecules become dissociated. In the case of carbonic acid, H_2CO_3, only of the order of one in a hundred molecules are dissociated into hydrogen ions and the bicarbonate radical, HCO_3^-. Water is itself an example of a substance that ionizes only to a very slight degree.

In this case, the products of dissociation can be considered as hydrogen ions and hydroxyl radicals, OH^-, but as pointed out above the ionization of water is probably better represented as

$$H_2O + HOH \rightleftharpoons H_3O^+ + OH^-$$

It is calculated that at 25°C the concentrations of H_3O^+ or H^+ and OH^- ions are the same and equal to 10^{-7} mol dm^{-3} (*see* p. 64, concerning pH values).

The factor that controls the degree of ionization of a molecule in solution is the nature of the chemical bond holding its components together. Substances with ionic bonds tend to dissociate much more readily than those with covalent bonds. Whereas sodium chloride with its ionic bonding shows almost 100 per cent dissociation, methane with its covalent bonding shows virtually no ionization in water. The reason for the difference is to be found in the fact that in ionic bonding where one atom gives up one or more electrons to another to create stability, in each of the participants the outer electron shell of the atoms is satisfied and at ionization no further exchange of electrons is necessary, the ionization being accomplished by the overcoming of the force of attraction between the oppositely charged particles. This is not so in covalent bonding where the outer shells of the atoms are satisfied only by shared electrons. In this case the atoms are very closely set together and to force them apart, so that the outer electron shells become unstable, is very difficult.

Large molecules as well as small ones can undergo dissociation and some reference to this in connexion with amino-acids and proteins is made in Chap. 4.

One effect of dissociation of molecules into ions is to increase the number of particles per unit volume in the solution and this has a great influence on some of the properties of the solution (*see* p. 972). In order to calculate the number of particles per unit volume in a given strength solution of a dissociable compound, a factor, known as the *dissociation factor*, needs to be known. Dissociation factors for substances in given solvents can be found in appropriate tables.

States of Matter and Changes of State

We recognize matter existing in one or other of three interchangeable *states*, namely, *solid*, *liquid* or *gas*. (But some physicists suggest that a highly ionized gas is sufficiently different in its properties to constitute a fourth state of matter, namely, a *plasma*.) At the normally occurring temperatures (*see* p. 184) of the earth's surface most materials remain in the same state. For example, iron is a solid, mercury is a liquid and oxygen is a gas. When sufficient heat (*see* p. 1002), is added to or taken from a substance, however, changes in its state occur. Iron can become liquid when heated enough, mercury becomes solid and oxygen liquefies when cooled sufficiently. Quite ordinary heat exchanges between liquid water and the surroundings are sufficient to turn it into solid ice and vice versa.

The simplest descriptions of materials in the three states that we can apply are that (1) a solid is a piece of matter that maintains its own shape and volume independently of any attempt to enclose it, that is, it retains its shape and resists changes in volume; (2) a liquid will take up the shape of a container and fill it to an extent dependent on the quantity of liquid, that is, it alters shape to fit the container but will resist changes in volume; (3) a gas will spread itself evenly through the volume of a container no matter what the size of the container or the quantity of gas, that is, it resists neither changes in shape nor in volume. Generally speaking, solids and liquids with a definite volume are about one-thousand times denser than gases. Alternatively, it can be said that matter is solid or fluid according to whether or not resistance is offered to a shearing type of force, solids being in the former category; fluids are classified as liquid or gas depending on whether a free surface is displayed.

To find a simple but rational explanation of the natures of matter in the three different states with reference to the particles (atoms or molecules) of which it is composed we must apply the *kinetic-molecular theory*. We can visualize the solid as being composed of particles held closely together by forces of attraction strong enough to resist separate individual movements but kept apart by forces of repulsion strong enough to prevent the particles from collapsing together as a single mass. The result is a more or less rigid system of particles, each helping to hold all the others in place with little freedom to move in space, vibrations (or movements of the particles relative to one another) being present but limited to small oscillations about their permanent locations. In the liquid state the particles can be visualized as having greater freedom to move over and around one another through the volume of the liquid, but still held closely enough together so that the repulsion forces between them resist compression. In a gas the particles are quite separate and independent, distantly placed, with complete freedom to move relatively to one another through space making contact only when their chance movements bring them into collision, unless the temperature is greatly lowered or the pressure greatly raised.

This is borne out by what happens when energy in the form of heat is added to any system of particles. In the case of the solid, the energy added appears as an increase in the rate and amplitude of vibration of the particles about their more or less fixed points relatively to one another. There will be a tendency for the particles to vibrate more rapidly and to move further apart. The temperature, really the expression of the rate of vibration of the particles, rises and the volume increases, that is, the solid gets hotter and expands slightly. In the liquid, where the particles already have, in addition to their vibratory movement, freedom to move through space at a certain speed, i.e. translatory movement, the added energy increases this speed of movement and tends to separate the particles further from one another. Again, the liquid will get hotter, i.e. its temperature will rise, and expand, and the expansion will be to a much greater extent, in proportion to its volume, than in the case of the solid, because the liquid particles are initially more widely apart and have greater freedom of movement. In the gas, where the principal movement of the particles is at random through all the space occupied by the gas, the effect of adding heat energy to the system is to increase the speed of this random movement of individual particles. This will show as an increase in hotness (temperature) and if the gas is so contained that it cannot increase in volume it will mean that the number of particles per unit time colliding with the walls of the container will increase as will the speed at which any particle makes its collision. The overall effect is seen as an increase in the *pressure* of the gas within the container. If the container walls allow of being stretched, e.g. an air-filled balloon, this allows also of an expansion in the volume of the gas.

By adding or subtracting sufficient quantities of heat energy from a system of particles it is possible to cause the system to change from one state to another. This is purely a physical change and must not be confused with a chemical change (*see* p. 978). As a solid receives heat energy, the vibrations of its particles increase in speed and amplitude until they begin to approach the condition of separation and movement characteristic of a liquid. Just as they reach this point some energy is required to overcome the forces binding the particles into their fixed locations, characterizing the solid, and once this has been supplied the original solid takes on the form of a liquid by what is called *melting* or *fusing*. If a record is kept of the rise in temperature of the solid as the heat is supplied, it will be seen that at the point at which fusion occurs, although heat is added continuously, the temperature remains constant. The heat supplied during this period is not used to increase the speed of movement of the particles and so to raise the temperature of the particle system, but to overcome the binding forces between them. This heat energy is called the *latent* (*hidden*) *heat of fusion* because it is not manifested by a change in temperature. Each substance has its own particular *melting point*, at a given pressure, expressed as a temperature and its own particular specific latent heat of fusion expressed in joules per kilogram. Note that latent heat has the unit joule. If expressed per unit mass the word specific must be applied and the unit is then $J kg^{-1}$.

When all the solid is turned to liquid, if heat is still added, the temperature of the system will once more begin to rise as the particles go on increasing their speed of movement, that is, as their kinetic energy increases (*see* p. 991). They will not all necessarily possess exactly the same speed of movement at any one temperature, some possessing more kinetic energy and moving more quickly than others, within a range. If a fast-moving particle is near the surface of the liquid its movement can cause it to lose contact with all the other liquid particles by breaking through the surface and escaping into the surrounding space. In this space the escaped particle will move about freely as do those in the gaseous state. We say that the liquid is *evaporating*.

If the space above the liquid is confined, the escaping particles will gradually tend to fill it and some near the liquid surface may re-enter the liquid. We say they *condense*. Thus there will be a two-way exchange of particles between the liquid and the space. At first there will be more particles escaping (evaporating) than re-entering (condensing) but slowly the numbers will equalize and an equilibrium state will be reached at which, apparently, no more particles are escaping from the liquid. This is described as a *dynamic equilibrium* because there is still movement. We can measure the excess quantity of particles that have escaped by registering the pressure they exert on the walls of the container. For any given condition of temperature of the liquid this *vapour pressure* reaches a maximum, known as the *saturated vapour pressure*, but the maximum gets greater as the temperature rises.

If a liquid is heated with its surface open to the atmosphere there will come a time when the saturated vapour pressure of the liquid is equal to the atmospheric pressure. At this point, groups of particles moving at high speed in the liquid are forced apart and form bubbles of gas within the liquid. This is because the pressure of the atmosphere is not great enough to prevent this happening. The bubbling condition is known as *boiling* and the temperature of the liquid at which its saturated vapour pressure is equal to the pressure of the surrounding atmosphere is called its *boiling point*. Clearly, if atmospheric pressure changes, so the boiling point of a liquid changes. If the pressure is increased so the boiling point will be at a higher temperature and vice versa.

When a liquid particle receives enough heat energy to raise its speed of movement sufficiently for it to escape completely from its neighbours, some of the energy will be used to overcome the forces binding the particles together. Thus during evaporation and boiling some of the heat energy supplied will not be translated through the speed of motion of the particles, i.e. kinetic energy, as a rise in temperature but will be hidden or latent. This heat is referred to as the *latent heat of vaporization*. For all substances, the specific latent heat of vaporization is greater than the specific latent heat of fusion per kilogram of substance. This is because the change, liquid to gas, requires more energy than the change from solid to liquid.

Dispersion Systems: Solutions

Because living matter is composed entirely of some kinds of particles physically spread or *dispersed* through others to form *dispersion systems*, it is important that the biologist should understand at least the fundamental nature and properties of the different kinds of dispersion systems that are possible.

Dispersion systems can be formed between different kinds of matter in any of its three states. Air is a gaseous dispersion of nitrogen, oxygen, carbon dioxide and the inert gases. Soda water is a dispersion of a gas (CO_2) in liquid water. Brass is a dispersion of solid copper in solid zinc. Salt water is a dispersion of solid sodium chloride in liquid water. All can be referred to as *solutions* but the term is frequently applied only to those systems that are liquid. The real nature and properties of any whole dispersion system depend primarily on the relative sizes of the particles within the system. In any such system or solution there will tend to be one substance in excess, known as the *solvent*, and the other or others dissolved in it, the *solute(s)*.

Graham, in 1861, first distinguished what appeared to be two types of water dispersion systems, namely *crystalloidal* and *colloidal*, according to whether the solutes were capable of passing through parchment membranes from one zone of the solvent particles to another. We know

now that there is no such sharp dividing line and that what he regarded as a difference between types of solutions is due really to the sizes of the solute particles relative to the size of the solvent particles and has nothing to do with their chemical properties. If the solute molecules are small, comparable in size with the solvent particles, the solution is more or less homogeneous with no separation of the different particles under the influence of gravity. This dispersion system can be referred to as a *true solution*. If the solute particles are large by comparison with those of the solvent, the solution is a somewhat heterogeneous system, but so long as the particles do not settle out under the force of gravity when the system is not agitated, it is still a solution and is called a *colloidal solution*. If the particles are so large that they cannot remain dispersed through the solvent against the force of gravity unless the mixture is constantly stirred there is no permanent dispersion at all, but a *suspension* or *emulsion* is said to be formed, depending on whether the suspended particles are solid or liquid.

It must be noted that not all kinds of substances are normally capable of being dispersed through all other kinds. Each when in the liquid state is said to possess a degree of *miscibility* with another. For example, water and oil are immiscible, water and ethyl alcohol are completely miscible and will mix homogeneously in all concentrations, water and ether are partially miscible. The greater the tendency for one substance, the solute, to become homogeneously dispersed through a solvent the greater is its *solubility* in that solvent said to be. For example, oil is insoluble in water but is highly soluble in ether and acetone. The solubility of one substance in another at a given temperature and pressure is measured by the maximum amount of the substance that can be homogeneously dispersed through it.

Soluble substances are still sometimes classified as *crystalloids* and *colloids*, the former being those substances which form true solutions, molecular or ionic, the latter forming colloidal solutions. The distinction is not always valid, since some substances behave as crystalloids in one solvent and as colloids in another. Sodium chloride forms a true solution in water but a colloidal solution in benzene. Some substances, generally regarded as insoluble, can be obtained in colloidal solution by suitable treatment. For example, colloidal solutions of gold and other metals in water may be obtained by electric arc treatment. It is therefore advisable to refer to the *colloidal state* of a substance rather than to a substance as being a colloid.

It has been found that a convenient classification can be made with regard to the sizes of very small particles. They can be divided into three groups. *Amicrons* are below 1 nm in diameter; they form true solutions, molecular or ionic. Such, for example, are sodium chloride, urea and cane sugar. *Submicrons* have diameters ranging between 1 nm and 100 nm; they form colloidal solutions. Examples are clay, animal charcoal, albumin and gelatin. *Microns* have diameters over 100 nm and do not form solutions but suspensions or emulsions which gradually separate. For an appreciation of these sizes by comparison with macroscopic particles refer to p. 21. There is no absolutely abrupt line of division between these kinds of particles, but the submicrons merge into amicrons on the one side and microns on the other.

In the submicron range lie all the substances in the colloidal state, whether they are large molecules, e.g. starch and protein, or aggregates of molecules, e.g. NaCl in benzene. Such aggregates of molecules are termed *micelles*. The distributed substance in a colloidal solution is called the *disperse phase*, and the substance in which it is distributed is the *continuous phase*. When the solvent is the continuous phase, we have a *colloidal sol*, e.g. a hot solution of gelatin. When the solute forms the continuous phase, we have a *colloidal gel*, e.g. the same solution of gelatin, when cooled, sets into a jelly, with the water distributed through it. Colloidal sols are of two kinds, *suspensoids* and *emulsoids*. Suspensoid particles carry relatively strong electrical charges and thus tend to repel one another, but when small amounts of electrolytes are added, the charges are neutralized, the particles aggregate and are precipitated. Such precipitates cannot easily be brought into solution again, and for this reason, suspensoids are termed *lyophobic* (solvent-hating). Some examples are animal charcoal, clay and silicic acid. Emulsoid particles carry relatively small electrical charges. They also have great affinity for water and are hydrated in solution. For precipitation, high concentrations of electrolytes are required. Not only must the charges be neutralized, but dehydration must occur. Such precipitates are usually redissolved easily and hence emulsoids are termed *lyophilic* (solvent-loving). Some examples are starch, gelatin, agar-agar and proteins. In living cells, almost all the colloidal material exists as emulsoids.

The concentration or strength of a solution is given by stating the relative proportions of solute and solvent. There are various ways of doing this, each useful for particular purposes—

(i) as mass of solute per unit mass of solvent (kg per kg);

(ii) as mass of solute per unit volume of solvent (or solution) (kg per dm^3 or kg per litre);

(iii) as moles of solute per unit volume of solution (molarity) (mol per dm^3 or mol per litre);

(iv) as moles of solute per unit mass of solvent (molality) (mol per kg);

(v) the concentrations of the components of a solution can be described in terms of their *mole fractions*, the mole fraction of any component being the ratio of moles of that component to the total number of moles present; the sum of the mole fractions of all the component substances is equal to unity.

A *saturated solution* at a given temperature is that solution which is in equilibrium with undissolved solute. A *supersaturated solution* is one in which a metastable state is achieved by altering the solubility of a solute so that more of it enters into dispersion than otherwise normally would. The solute in excess can be deposited usually by stirring, shaking or by introducing a crystal of the solute. The supersaturated state can be reached by the evaporation or cooling of a solution.

The terms *soluble, insoluble* and *sparingly soluble* have no exact meanings. In general a substance is said to be soluble in another if a concentration of over 1 g per 100 cm³ of solvent can be attained. It is said to be insoluble if its solubility is less than 0·1 g per 100 cm³ of solvent and sparingly soluble if the concentration achieved is between these limits.

The solubility of a substance at constant pressure may vary with the temperature. The solubility of most substances increases with rise in temperature but some show a decrease. These conditions are associated with the fact that the process of one substance dissolving in another is accompanied by either the absorption or evolution of heat (*see* p. 997). Those solutes that absorb heat, i.e. have a negative heat of solution, on dissolving exhibit an increasing solubility with rise in temperature whilst the reverse holds for those that evolve heat, i.e. have a positive heat of solution, on dissolving. Most soluble substances belong to the former group with reference to water and are more soluble in hot solvent. Sulphuric acid is a good example of a substance with a positive heat of solution. Some substances such as sodium sulphate appear to exhibit a sudden change of solubility with variation in temperature. In this case, $Na_2SO_4.10H_2O$ shows rapid increase in solubility up to 32·383°C but above this temperature the solubility decreases. This is really because we are dealing with two different conditions of the substance. The temperature quoted represents a transition point at which $Na_2SO_4.10H_2O$ changes to Na_2SO_4. Below the transition temperature the solution, on cooling, would deposit $Na_2SO_4,10H_2O$, above it, it would deposit Na_2SO_4.

Many properties of true solutions are dependent upon the number and not the nature of the particles of solute present per unit volume of the solution. Such properties are commonly referred to as *colligative* or *osmotic properties*. They include the lowering of vapour pressure, the elevation of the boiling point, the depression of the freezing point and the osmotic pressure of solutions. For more detailed study of these phenomena the reader is referred to an appropriate textbook of chemistry and to Chap. 3 of this volume for an elementary treatment of osmotic pressure and diffusion.

Properties of Substances in the Colloidal State

To understand how the colloidal nature of protoplasm has a bearing on its structure and activities it is necessary to know something of the general properties of substances in the colloidal state.

Dialysis

The inability of colloids to pass through certain membranes has important consequences for living organisms. Thus it is that the colloidal protoplasm is retained within the plasma membrane, and thus colloidal food storage substances, such as starch and glycogen, cannot pass out of a cell as such, but must be reduced to smaller molecules.

The Tyndall Effect

If a beam of light, in an otherwise darkened room, is passed through a true solution, the solution appears to be perfectly clear and transparent when viewed at right angles to the path of the beam. But if a similar beam is passed through a colloidal solution, a cloudy pencil of light is seen where the beam passes. This opalescent effect is due to the scattering of the light by the tiny particles in the colloidal solution.

Brownian Movement

Very small particles in certain liquids, when viewed under a microscope with high magnification, appear to quiver without much change of position. The quivering is due to the bombardment of the particles by the moving molecules of the liquid. It is called Brownian movement. It is obvious that the particles must be small enough to be visibly

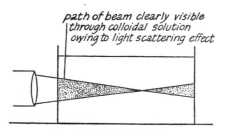

Fig. A.9. The Tyndall effect.

affected by this molecular bombardment. The effect can be demonstrated with a dilute suspension of Indian ink in water. All colloidal solutions exhibit this phenomenon and minute particles in cell sap often can be seen executing the movement.

Filtration

When a solution is filtered through a membrane, the particles which will pass, depend on the size of the pores in the membrane. When collodion membranes are used, the diameter of the pores can be controlled by washing the membranes in various strengths of ethyl alcohol. By this technique, the actual sizes of colloid particles can be measured.

Osmotic Properties

True solutions are characterized by high osmotic pressures (*see* Chap. 3). Lyophobic colloids have negligible osmotic pressures, and thus have extremely slight effects in elevation of the boiling-point and depression of the freezing-point. Lyophilic colloids have small but measurable osmotic pressures, and consequently they cause slight elevation of the boiling-point and depression of the freezing-point as compared with those values for the pure solvent.

Electrophoresis

Colloid particles are usually charged positively or negatively. Thus they respond to a type of electrolysis whereby charged submicrons move to the cathode or anode. Negatively-charged colloids are the more common. The technique has been used in the separation of the proteins contained in the muscle of various fishes, in an attempt to find chemical correlation with existing classification.

Precipitation

Precipitation of a colloid is sometimes termed coagulation; it applies more particularly to lyophobic colloids. The micelles come together and form larger masses which then settle under the influence of gravity. A positively-charged colloid will cause coagulation of a negatively-charged one, but precipitation will be complete only if the charges are exactly neutralized. Addition of an electrolyte will have the same effect, because the ions of opposite charge to that on the particles will be attracted to the colloid. Dilute acids and rennet will coagulate the casein of milk; on this process depends the making of cheese. Jam is made by the precipitation of pectin in the combined presence of acid, sugar and a fairly high temperature. Heat will irreversibly coagulate the albumin of eggs and the proteins of protoplasm.

Surface Properties

Owing to the small size of the submicrons and their huge numbers, they will present an enormous surface area to the surrounding medium.

The interface between the two phases is a type of membrane, and this ability of colloids to form membranes in contact with watery solutions is extremely important in protoplasm.

The surface energy at these interfaces is considerable; it is due to surface tension, for the molecules on both the micellar particles and the surrounding medium have free energy on the surface, whereas molecules deeper in both media are acted on by forces from other molecules surrounding them, and hence are in a more stable state. Many physical and chemical actions can take place at these free-energy surfaces.

A substance is said to *adsorb* when it collects upon its free surface molecules of another substance from the surrounding medium. Note that *absorption* involves the whole body of the absorber not only its surface.

Adsorption takes place readily on submicron surfaces. Ions carrying a charge opposite to that of the micelle, and certain dyes, are most readily adsorbed. Adsorption can be demonstrated by placing finely-powdered charcoal in a solution of methylene blue. The dye is adsorbed on the charcoal and the solution loses its colour. Another adsorption effect is the definite and orderly arrangement on submicron surfaces of large molecules possessing electrically-charged or polar groups. When one of the phases is a watery solution, there will be formed a regular monomolecular film on the submicron, with the water-soluble fraction in the water and the insoluble fraction in the submicron. This adsorption phenomenon can confer stability on certain emulsions which would otherwise separate slowly. The addition of egg stabilizes mayonnaise; soap stabilizes oil-based insecticides and tar-based winter washes.

Gel and Sol State

In the sol state, a colloidal solution is fluid; in the gel state it is solid. Starch dissolved in hot water forms a colloidal sol; when cooled, it forms a colloidal gel. Conversion from sol to gel is best regarded as a partial precipitation, not sufficient to give a separate solid phase. It can be effected in various colloids by different methods. Change of temperature or acidity, agitation, pressure increase and the presence of metallic ions are all effective in various cases. In most biological transformations it is probable that the effective agents are enzymes; the clotting of blood by the gelation of the protein fibrinogen is a familiar example. In the gel state, there is a structural framework holding together large quantities of water; in the sol state there is complete breakdown of this structural scaffolding.

Imbibition

Imbibition is the term used to describe the absorption of fluid by colloids. Gelatine will readily take in considerable quantities of water. It will also lose the water by slow drying, but its ability to take in water again is quite unaffected. The testa of a dry seed absorbs water by imbibition, and the cellulose walls of plant cells take up much water in the same way. The liberation of reproductive structures from their containers, in many cases, depends on the absorption of water by mucilaginous materials and their consequent swelling.

Hydrotropic Substances

Some substances have the very interesting property of making very sparingly soluble substances dissolve fairly readily in water without apparently altering the chemical nature of the dissolved substance. They are known as *hydrotropic substances* and good examples are the bile salts, taurocholates and glycocholates. The higher fatty acids are soluble in water to a degree of less than 0·1 per cent but this is increased to 0·5 per cent if bile salts are added to the mixture and still further increased to 4·0 per cent by the addition of lecithin. Other examples are benzoic acid, hippuric acid and soaps of the higher fatty acids. Among the substances rendered more soluble in water are fats, phospholipides and uric acid, the increase in solubility varying from ten to one hundred times normal.

This phenomenon has considerable biological importance when it is realized that substances so dissolved by the action of hydrotropic substances can diffuse through cell membranes whereas otherwise they could not, and thus these substances must play an important part in determining the readiness with which certain compounds can be absorbed by cells.

The precise nature of the hydrotropic action does not seem to have been fully explained but all hydrotropic substances appear to have one common property, that of decreasing surface tension, and it may be that their effect on solubility is through this characteristic.

Separation of the Components of a Dispersion System

There are various physical ways of separating two or more substances that form a solution or suspension. Evaporation of the solvent will leave the solute behind in those cases where the solvent is volatile and the solute non-volatile. Suspensions or emulsions of large particles will settle out under gravitational force. What is called "centrifugal force", applied through the agency of a centrifuge, can be used in effect to increase the pull of gravity on the particles and, in general, they can be made to settle out in order of their masses by varying the speed of revolution of the centrifuge. Thus it is possible to use the instrument to separate individual types of particles from a mixture in the same dispersion system. This applies particularly to the separation from one another of large molecules that occur in suspensions of biological origin.

Maintenance of Non-equilibrium between Cells and Non-living Surroundings

In non-living systems of particles in solution, it is the case that by the normal laws of diffusion there would be reached a state of equilibrium in which all the solute particles were evenly distributed throughout the solvent unless their movement was in some way restricted. That is, there would not exist concentration gradients between different parts of the system so far as each kind of particle was concerned. A living system, however, clearly shows permanent differences in concentrations of the same kind of particle on opposite sides of the membrane separating the protoplasm from the external solution and furthermore is capable of transferring particles through the membrane against such a concentration gradient. To maintain this composition, distinct from that of the surroundings, the living system must actively be performing work to prevent the equilibrium state from being reached. To move particles in the "up-hill" direction more work must be done. It is no doubt the case that when cells have lost the property of maintaining this state of non-equilibrium they are no longer alive.

It is possible with non-living systems to create conditions in which equilibrium is reached with different concentrations of particles of the same kind on opposite sides of a membrane. This may be done by separating different kinds of ions by a membrane that is not equally permeable to all of them. When an equilibrium is finally reached, known as the *Donnan equilibrium*, it is the case that there can be very steep concentration gradients across the membrane of ions of the same kind. For more detailed treatment of the Donnan equilibrium phenomenon, the reader is advised to refer to a suitable textbook of physical chemistry.

Whilst such a situation may be cited as an explanation of some of the ionic concentration phenomena associated with living cells, it by no means explains them all. There is no doubt that the conditions known to occur in living systems can only be the result of their active metabolism, some of the energy so released being used to do the work of maintaining the condition of non-equilibrium and to transfer particles against a concentration gradient. This kind of work is being performed at all times by active cells.

Chemical Reactions

There are two main aspects concerning the nature of the interactions between chemically reactive substances that should be taken into account. One has reference to the overall changes in the reactants and the products of these, including energy changes. The other concerns the mechanisms by which the reactions occur. In the first, concern is with what atoms and molecules enter and leave a reaction and whether energy is released or consumed. In the second, the movement or mobility of atoms and molecules, their geometric structures and the effects of such influences as temperature and the presence of catalysts must be taken into account.

There are four common patterns of chemical reactions or reaction types usually referred to, but not all chemical reactions can be classified within them. These are *combination reactions* or *additions*, *decomposition reactions*, *displacement reactions* or *substitutions* and *double displacement reactions*.

A combination reaction involves the chemical union of two substances to produce a third and can be represented as

$$A + B \rightleftharpoons AB$$

where A and B represent the two reactants that combine to form a product AB. For example, hydrogen and oxygen will combine to form water. During such a reaction there may be a release of energy or energy may be absorbed as heat. The formation of water by the union of its two component elements releases much heat energy, i.e. is *exothermic* and can be symbolized:

$$2H_2 + O_2 \rightleftharpoons 2H_2O + Energy$$

When sulphur and iron atoms unite to form iron sulphide energy must be fed into the system, i.e. the reaction is *endothermic* and can be symbolized:

$$Fe + S + Energy \rightleftharpoons FeS$$

A decomposition reaction occurs when there is a breaking of existing chemical bonds and this can be represented as the reverse of the combination reaction so: $AB \rightleftharpoons A + B$. An example would be the decomposition of potassium chlorate into potassium chloride and oxygen:

$$2 KClO_3 \rightleftharpoons 2 KCl + 3O_2$$

More than two kinds of atom may be involved and the products may be of more than two kinds. Energy as heat may be liberated or absorbed.

For example, the breakdown of water into hydrogen and oxygen demands that heat energy be supplied and is thus endothermic. The decomposition of carbonic acid into carbon dioxide and water is exothermic, releasing heat energy.

A displacement reaction occurs when one atom or group of atoms is replaced by another atom or group of atoms in a molecule and can be represented as

$$A + BC \rightleftharpoons AC + B$$

An example is the reaction that takes place when zinc is placed in contact with sulphuric acid:

$$Zn + H_2SO_4 \rightleftharpoons ZnSO_4 + H_2$$

Again, the heat energy exchange may be in one direction or the other according to the reactants. This kind of reaction is one of the commoner ones in living material in which the molecules are large and change chemically by the displacement of one atom or group of atoms by another.

The double displacement is said to occur between compounds that exchange parts, represented as

$$AB + CD \rightleftharpoons AD + BC$$

An example is the reaction between potassium chloride and silver nitrate:

$$KCl + AgNO_3 \rightleftharpoons AgCl + KNO_3$$

Once more, heat energy may be released or absorbed during the reaction.

The special cases of oxidation and reduction reactions are dealt with later in this section and more detailed considerations of the energy exchanges that occur during chemical reactions are treated in a separate section commencing on p. 995.

All chemical reactions are theoretically *reversible*, meaning that they can take place in both directions and the form of an equation indicating this is $A + B \rightleftharpoons AB$. Although in principle this is so, under normal conditions many reactions are less easily reversed than others. One way of indicating that a reaction occurs more readily in one direction than the other is to write it with unequal arrows so:

$$A + B \rightleftharpoons AB \qquad \text{or} \qquad A + B \rightleftharpoons AB,$$

according to which direction is most favoured. We say that the reaction is *shifted to the right or left* with reference to the favoured direction. In some cases the rate of reaction is so slow in one of the directions that for all practical purposes it is irreversible and the equation is written so: $A + B \rightarrow AB$. Such reactions tend to occur when, in the forward direction, a great deal of energy is released. To reverse the reaction the equivalent amount of energy would have to be supplied. The loss of one of the products of a reaction by either its escape as a gas or its being thrown out of solution as a precipitate may also cause irreversibility of a reaction. Some chemical reactions in living things appear to be irreversible for one or other of these reasons although nearly all biochemical reactions are really reversible.

The *rate of a reaction* is given by the amount of reaction, measured in terms of the change in concentration of reactants or products, in a given period of time. It can be expressed as

$$\text{Rate of reaction} = \frac{\text{Change in concentration}}{\text{Change in time}}$$

Under given conditions of temperature, etc., the rate of a reaction can be predicted for specific chemical systems, and such reactions can be described by their reaction rates as well as by their energy exchanges.

In a reversible reaction, there will be reached a point at which the rate of reaction in one direction is equal to the rate of reaction in the other; a state of *chemical equilibrium* is said to have been attained. From the definition of rate of reaction it can be seen that at equilibrium the concentration of the reactants does not have to be equal the concentration of the products, only the rates at which reactants are being converted into products and products into reactants. There is in fact usually a much higher concentration of products at an equilibrium than of reactants. Only the completely reversible reaction will show equal concentrations of reactants and products at the equilibrium point.

At any chemical equilibrium there will be changes occurring in both directions and so the state is said to be one of *dynamic equilibrium* (as in the case of evaporating and condensing particles, *see* p. 968). Since practically all biochemical changes are reversible to a greater or lesser degree, living systems must exist in this state. But should all metabolic changes in an organism reach their points of equilibrium the whole system would reach a standstill ending in death, since for "life" to continue it is necessary that individual syntheses and breakdown reactions should be going in one direction or the other, not more or less equally in both directions at once. In living things, the situation is made still more complicated by reason of the fact that metabolic changes occur in a series of steps and in each case, from reactants to products, there are a number of intermediate reactions and products. The problem of keeping a large series of reversible reactions going in one direction is solved in the living system by the use of the products of one reaction as the reactants for the next reaction and the removal from the site of reaction of the final product, thereby maintaining a continual shift in the required direction. The overall process of aerobic respiration in which glucose is broken down to carbon dioxide and water involves over thirty chemical reactions, nearly all reversible. Since one of the final products, carbon dioxide, is continuously removed from the site of its formation by diffusion, it cannot enter into the reverse series of reactions and the whole process continues indefinitely as if it were irreversible. This would not be so at the compensation point for green plants (see p. 359) but at the low light intensities at which this occurs no green plant could survive indefinitely.

There is no fully acceptable explanation of the causes of the interactions of atoms and molecules during chemical reactions but one theory, the *collision theory*, helps to clarify the nature of the phenomenon. This is based on the idea that all particles in a system are in constant movement and can come into contact with one another. The contact may be such that electron exchanges or rearrangements are made possible. Clearly, the likelihood of two negatively or two positively charged particles making such interacting contact is remote because of their mutually repellent charges but positively and negatively charged particles are much more likely to do so. The chances of successful collisions between such particles is enhanced if their velocities are high and each has more than its minimum kinetic energy requirements for successful interaction. The minimum kinetic energy required by a system of particles for chemical reactions to occur among them is called the *activation energy* and is a characteristic of all reacting chemical systems. If activation energy is below the minimum required, reaction is slow or does not occur at all; above minimum requirements it results in rapid reactions. From the fact that the velocities of particles in any given system are likely to vary according to a form of distribution curve (see p. 737), some fast, some slow, with most around a mean value, only some of them are likely to possess activation energy. Only those that do will interact chemically with others, hence such interactions can only be thought of in terms of chance or probability.

A possible hindrance to interaction between particles may appear if the particles have some special geometric structure confining the reactive portion of the particle to one particular place on it. Such particles would have to collide so that their reactive parts were properly opposed and this would be particularly important in the cases of very large molecules such as those found in living systems.

All the factors that are known to influence the rate of chemical reactions can be considered to do so by affecting the frequency at which effective collisions between the atoms and molecules occur. Among the more important factors are the physical nature or degree of subdivision of the reactant particles, the molecular concentration of the reactants, the temperature and pressure of the system and the presence of catalysts.

Effective collisions between particles are more likely to occur if greater numbers of them are exposed by fine separation. Hence the tendency for greater speed of reaction when the reactants are in solution where the solvent particles break up the solute into separate interacting molecules or ions. In some biological situations certain substances are specifically effective in breaking up larger groups of particles; the action of bile juice on fat globules is an example.

The Law of Mass Action proposes that "the velocity of a chemical reaction is proportional to the active masses of the reacting substances" where active mass means the molecular concentration or the number of moles per unit volume. In dilute solutions and gaseous systems at low pressures this is given with sufficient accuracy by the concentration of the substance in moles per litre (mol dm^{-3}) raised to the power of the number of molecules of that substance known to react according to the chemical equation describing the reaction involved.

If the molarity of two substances A and B are respectively a/V and b/V and they react within a volume V of a reacting system according to the equation—$xA + yB = products$, then the velocity of the reaction, v, is given by

$$v = K \left(\frac{a}{V}\right)^x . \left(\frac{b}{V}\right)^y$$

where K = the "*velocity constant*" of the reaction. Note that the value of K is also affected by other factors that affect reaction velocity. The effect of increasing the number of interacting particles of either or both of the reactants can be understood on the basis of the collision theory if it is the case that the greater number of particles present, the greater the likelihood of successful collisions, but it should be remembered that it will be the concentration of reactant in least supply that limits the overall rate and, of course, the total quantity of the product. It should also be remembered that whereas in the case of irreversible reactions the quantity of product has no effect on reaction rate, in the case of reversible reactions the concentration of the product will have a very marked effect.

There is no straightforward relationship between variation in temperature and rate of chemical reaction but it is generally the case that the higher the temperature, the greater the speed of reaction. But this general rule cannot be applied to biological systems over wide ranges of temperature since the enzyme catalysts are affected by temperature in a vastly different way (*see* p. 138). For ordinary chemical reactions, an increase of 10°C in the temperature of the system generally results in a doubling of the reaction rate. This can be written

$$Q_{10} = \frac{\text{Rate of reaction at } (x + 10)°C}{\text{Rate of reaction at } x°C} = 2$$

Again, the collision theory will help to explain this temperature effect. An increase in temperature increases particle velocity (*see* p. 967), giving greater kinetic energy, hence tending to increase successful collisions between the more highly activated particles.

Pressure increase has a similar effect to temperature increase on reaction rate by increasing the density of the system. The closer the particles are forced together the greater the chances of successful collisions between them, but this is applicable in general only to gaseous systems. Hence pressure variation will have little or no effect on biochemical reactions.

In many chemical systems the presence of a suitable catalyst can increase the rates at which reactions occur. In living systems the chief catalysts are the enzymes and it is thought that they play their part not so much by increasing the overall number of collisions between reactants but by ensuring that when a collision does occur between two large molecules, or between a large and a small, the large molecules are so oriented that their active portions are presented precisely to the colliding particle so ensuring some chemical activity.

Oxidation and Reduction

The term *oxidation* was originally applied to the class of reactions in which the various elements, particularly metals, combined with oxygen, implying thereby the addition of oxygen to a substance as, for example, when sulphur or carbon is burnt in air. Sometimes the oxidation of an element or compound by the addition of oxygen involves the removal of the oxygen from some other compound. For example, when magnesium is caused to burn in carbon dioxide, the gas loses oxygen and is said to be *reduced* to carbon, whilst the magnesium is said to be *oxidized* to magnesium oxide: $2 \, Mg + CO_2 = 2 \, MgO + C$.

Since then the use of the terms, oxidation and reduction, has been extended to include *all cases of electron transfer from one particle to another*. When any substance combines with oxygen that substance is giving up electrons to the oxygen and the process described above therefore comes within the extended meaning but there are other cases of electron transfer that do not involve oxygen at all.

The common factor in all oxidations is the gaining of a positive charge by loss of electrons and in reductions the loss of a positive charge by the gain of electrons. An atom that loses electrons is therefore oxidized and one that gains them is reduced. This concept of oxidation and reduction is a very useful means of describing certain chemical processes and it is possible to write the following general equation for them—

$$\text{Oxidized state} + \text{Electrons} \rightleftharpoons \text{Reduced state}$$

or

$$\text{Reduced state} - \text{Electrons} \rightleftharpoons \text{Oxidized state}$$

The formation of ionic chemical bonds between substances is always an oxidation–reduction reaction. For example, when sodium and chlorine combine to form sodium chloride, the sodium atoms each change from neutral to charge $+1$ by losing an electron to each of the chlorine atoms which change from neutral to charge -1 by gaining an electron. The sodium is oxidized and the chlorine reduced. The sodium acts as a reducing agent by contributing electrons and the chlorine acts as an oxidizing agent by accepting them.

Oxidation can also be brought about by the removal of hydrogen from a molecule. For example, in the reaction $H_2S + O = S + H_2O$, the hydrogen sulphide is oxidized to sulphur and the oxygen reduced to water. In terms of electron transfer, when the hydrogen atoms leave the hydrogen sulphide each shares an electron with the oxygen to form water. Hydrogen in the atomic state or *nascent* hydrogen is a very strong reducing agent by virtue of its affinity for oxygen and other electronegative substances. In biological systems, oxidation is generally achieved by the transfer of hydrogen from the substance to be oxidized through a series of hydrogen acceptors until finally it is passed to oxygen. A complete sequence may be: Co-enzymes 1 and 2 → Flavo-proteins → Cytochromes → Oxygen, but these acceptors are not all necessarily required in every such oxidation. The acceptors and the sequence in which they are used in any given oxidation depends upon what are known as *redox potentials* or *oxidation–reduction potentials*.

As has been stated, the essential feature of an oxidation process is a transfer of electrons from the oxidized to the reduced substance and whenever there is an oxidation there must be the complementary reduction. The readiness with which a substance can be oxidized will therefore depend upon how easily it can be made to release electrons, or its *fugacity*, as this property is sometimes called. The greater the fugacity of a substance, the more easily can it be oxidized. The redox potential, E_h, of any dissolved substance is a measure of this electron fugacity and can be found in practice. This is done by measuring the negative charge produced on a platinum electrode dipped into the solution by the electrons tending to escape. The size of this charge is proportional to the electron fugacity. A solution of an oxidizing substance will oxidize other substances whose solutions possess a more negative E_h by taking electrons from them and conversely, a reducing substance will reduce other substances whose solutions possess a more positive E_h by donating electrons to them. The further apart are the redox potentials of the two systems, the more completely will that system with the more positive (less negative) potential oxidize the other system. The values of E_h are most commonly quoted for the mid-point of an oxidation–reduction (50 per cent reduction) and at some specified pH. The value of E_h under such conditions is then given as E'_0 at the given pH. Some values of redox potentials for some biochemical substances are given in the table.

VALUES OF REDOX POTENTIALS FOR SOME BIOCHEMICAL SUBSTANCES

Substance	Potential (E'_0 at pH 7) (volts)
Oxygen	+ 0·81
Cytochrome *a*	+ 0·29
Cytochrome *c*	+ 0·26
Ascorbic acid	+ 0·058
Cytochrome *b*	− 0·04
Flavoproteins or Yellow enzymes	− 0·06
Riboflavin	− 0·208
Coenzymes 1 and 2	− 0·28
Hydrogen	− 0·42

In general, any system higher in the table will oxidize any other below it and conversely, any system will reduce one above it. When pairs of oxidizing agents are mixed together, that with the less negative redox potential will accept hydrogen or electrons, i.e. oxidize the other. If more than two are present there will be a definite sequence in which the oxidation–reduction reactions occur. In a living cell the various hydrogen acceptors that are used in oxidation–reduction processes will react in an order determined by their redox potentials.

The change undergone by some ions in solution such as that by ferrous ions to become ferric ions is also an oxidation process since it involves the transfer of an electron. It may be written as

$$Fe^{++} - e^- \rightleftharpoons Fe^{+++} \quad \text{or} \quad Fe^{++} \rightleftharpoons Fe^{+++} + e^-$$

and the reverse process represents a reduction. If the electron acceptor, that is, the oxidizing agent is oxygen, the equation can be written

$$2 Fe^{++} + \tfrac{1}{2} O_2 + 2 H^+ \rightleftharpoons 2 Fe^{+++} + H_2O$$

The importance of oxidation–reduction reactions in living systems must be realized by the biologist since the chief means of energy release within them is the removal of electrons from one substance and their subsequent transfer to another.

ENERGY

The essential characteristic that distinguishes a living system from a non-living system of particles is the ability of the former to extract energy from its surroundings, transform it, store it or expand it in ways designed to maintain itself in a condition in which it can resist and indeed reverse the normal processes of disintegration. We see this expressed in its ability to synthesize complex molecules from simpler structures, these also extracted from its environment, i.e. it is able to grow and create similar systems with the same properties.

In order to understand more fully what this really means, the biologist must possess a true concept of this something that we call energy, its forms and transformations, and use the knowledge to translate the observable living processes into physical and chemical terms.

Concepts of Energy

Matter has been defined as anything that has mass and volume. There is no equivalent straightforward way, however, in which we can explain what we mean by energy since it has no apparent concrete form in our system of things. We are aware of its existence and the quantity of it, in the ordinary way, only by the effects it produces on matter. Nevertheless, by the relativity theory, energy and the mass possessed by matter have to be considered as different aspects of the same thing, an energy E being equal to a mass E/c^2 or a mass m being equal to an energy mc^2, where $c =$ the velocity of light (see p. 992). According to classical mechanics, transformations may occur between different kinds of energy, e.g. electrical energy can be converted into heat or light, etc., whilst the total amount of energy in a closed system remains constant. But on the basis of the equivalence of mass and energy, there must be taken into account the possibility that when such transformations occur, they involve the energy of rest mass (see p. 936). That is, transformations of the energy of rest mass into radiant or other forms would mean the destruction of matter, since the possession of rest mass is the characteristic property of matter. Likewise, we must consider the possibility of the reverse transformation in which matter is created. The truth concerning the equivalence of mass and energy has been tested repeatedly by nuclear physicists and always there has been confirmation of the relationship. Energy exchanges that are involved during chemical reactions must also be accompanied by corresponding changes in mass but in this comparatively low energy range, the mass changes are too small to be detected by direct measurement.

For our purposes, initially, we need not consider energy in this way but can define it in a simpler fashion, more in terms of its effects. We can say that any system "capable of doing work" possesses energy, for it is by the transfer of energy from one system to another that work is done. This implies an understanding of the meaning of *work*, a definition of which can best be derived from a study of *forces*, in the first instance.

A force exists, acting on a body, if that body changes its velocity either in speed or direction, i.e. undergoes an acceleration, positive or negative (deceleration). The unit of force in SI is the newton, that force acting on a mass of one kilogram that will impart to it an acceleration of one metre per second per second. The velocity of a body takes account of its speed, i.e. the distance it travels in unit time, and also the direction of its movement. It is thus a rate of change of displacement in a given direction. Work is given by the product of the force applied to a body and the displacement of the body in a direction parallel to that in which the force is acting. By the same units, a joule of work is done if a force of one newton acts through a distance of one metre in the direction of the force. When a man does physical work, he must transfer some of the energy of his own muscles to some other body and the amount of energy transferred is given by the same units as the work done, since this is a measure of the amount of energy transferred. If a man lifts a rock, weight one newton, to a height of one metre above its initial position, he performs one joule of work. That is, the muscle energy once available to do work has been transferred to the rock by this amount.

But movement of a body, of itself, does not imply necessarily that work is being done on it. No work is done and no energy is transferred to anything during the orbiting of the earth by the moon, for example, if it is assumed that the moon moves in a circular orbit. This is because the force acting on the moon, the earth's gravitational force, acts through the centres of the earth and moon and the moon's velocity is at right angles to this direction at every instant. Normally, in the conditions existing at the earth's surface, movement is against a resistance, such as friction or air resistance, acting in the line of movement. Thus here a force must always be applied to initiate and maintain movement of a body; work must be done on it, that is, energy must be transferred to this body from another.

The first of the natural forces to be investigated was the *gravitational pull* of the earth, by Newton, who deduced a mathematical expression by which it could be represented. Starting from the Principle of Inertia

(Newton's First Law of Motion) which asserts that bodies move with constant velocity unless a force acts on them, Newton took as evidence of the existence of a force acting on a body, a change in velocity. He then, in his Second Law of Motion, provided a means of quantifying work done on a body by a force by asserting that the rate of change in velocity produced by the force is directly proportional to the force and inversely proportional to the mass, i.e. a force is equal to *mass* × *acceleration* (rate of change in velocity), assuming that mass remains constant with time. Thus work or energy transfer becomes *mass* × *acceleration* × *distance moved by the body*. From measured rates of change in velocity of falling bodies under the earth's gravitational influence, Newton supplied an expression for gravitational attraction (force) between two particles—

$$f_{\text{gravity}} = G \frac{m_1 m_2}{d^2}$$

where m_1 and m_2 are the masses of the two particles exerting gravitational forces on each other, d is their distance apart and G is the coefficient required to express f in the correct units (constant of proportionality), or the *gravitational constant*.

But the force of gravity is not the only natural force. As far as is known at present there are four basic *interactions* or *forces* in nature governing the structure and behaviour of matter. By far the strongest of these is the attractive *nuclear interaction* which holds the nucleons (i.e. proton or neutrons) together within an atomic nucleus. This interaction is, however, of extremely short range becoming negligible at distances greater than about six nucleon diameters (of the order of 10^{-14}m).

An *electromagnetic interaction* exists between particles that have electric or magnetic properties. It is about one hundred times weaker than the nuclear force. The electromagnetic interaction is an inverse square law force, i.e. it is inversely proportional to the square of the distance between the particles concerned. Thus it has no tendency to decrease very rapidly after a certain distance and so can still be effective at quite large distances. This is in contrast to the nuclear interaction and also in contrast is the fact that the electromagnetic interaction can be either attractive or repulsive, depending on whether the charges are unlike or like in sign. It is the electromagnetic interaction that governs the structure of matter in objects varying in size from atoms to everyday objects.

A third type of interaction is called the *weak interaction*. It is about 10^{-15} weaker than the nuclear interaction. Like the latter it is of extremely short range. It is of importance in such processes as β-decay and the decay of free neutrons.

By far the weakest of the interactions is the *gravitational interaction*, which is weaker than the nuclear interaction by the fantastically small factor of 10^{-40}. It is not easy to imagine what is meant by such a factor. Some idea may be gained, however, if it is realized that it represents approximately the ratio of the size of the smallest object (diameter of a nucleon) to the largest object (diameter of the observable universe) so far measured by man.

What should be realized is that when any kind of force acts to displace matter in the direction of the force, work is done and energy thus transferred, the work done being a measure of the energy transfer.

The emphasis on forces in physical studies lasted over two hundred years and then the approach was shifted to explanations of the energy relationships of physical and chemical phenomena in terms of *quantum physics* with the emphasis on *energy quantities* (*energies*) rather than on forces. At the beginning of this century it was becoming clear that many physical phenomena, particularly those associated with radiation, could not be explained on Newtonian principles. Similarly, chemists were becoming increasingly occupied with energies of systems in their thermo-dynamical studies and were able to make statements about the energy exchanges during chemical reactions that were not directly concerned with forces. Gradually the *quantum concept of energy* has been developed and *quantum mechanics* describes the mechanisms governing phenomena that are on such small scale that they cannot satisfactorily be explained in terms of forces, only in terms of energies. Such are the movements of electrons or nuclei in atoms and molecules. The attempt by Planck to explain the forms of the curves representing energy distribution with wavelength for radiation from hot bodies was the origin of a new concept. Before this, it had been assumed that such a body radiated its energy in a continuous stream of uninterrupted waves, the frequency of which determined the nature of the radiation as heat, light, etc. Planck postulated instead a discontinuous emission of energy in which it was sent out in "squirts" or small discrete "packages", or *quanta*, and that each such quantum had a magnitude of $h\nu$ where ν is the frequency of vibration of the radiation and h is a constant having the same value for all frequencies and now known as *Planck's constant*. Later, Einstein developed and extended the quantum theory to include absorption as well as emission of radiant energy. From that time it has been convenient for many purposes to regard radiant energy as travelling through space in the form of discrete quanta of magnitude $h\nu$. The quantity h has the dimensions of *energy* $\times$ *time* and its value is given as about $6 \cdot 6 \times 10^{-34}$ joule seconds. The energy of light of a particular single wavelength, i.e. monochromatic light, is given by $E = nh\upsilon$ where n represents the number of quanta or "packets of energy" possessed by the wave. A single quantum of energy of electromagnetic radiation, $h\upsilon$, is usually referred to as a *photon*. The shorter the wavelength, i.e. the greater the frequency, of the emission, the greater the energy in one photon. In a chemical reaction that results directly from the absorption of light energy, as in photosynthesis, one quantum of the radiation activates one molecule of substance. If the energy in one quantum is $h\upsilon$, then the energy required to activate one mole of substance is $Nh\upsilon$, where N is Avogadro's number (*see* p. 995). This amount of radiation was once called an *einstein* of energy equal to $Nh\upsilon \times 10^{-7}$ J. If c/λ is substituted for υ in this value, where c is the velocity of light and λ the wavelength, this quantity of energy becomes $Nhc/\lambda \times 10^{-7}$ J. From this expression it can be seen that because the value of λ is not constant through the spectrum, there will be variable energy content per photon of light according to its wavelength. Blue light has a shorter wavelength than red light (*see* p. 992) and thus has more energy per photon (*see* p. 1009).

Bohr's first successful postulate of atomic structure involved the quantization of energy in the way visualized by Planck. His theory included the idea that atomic electrons obey quantum laws that select for them very definite energy values and require them to move in definite orbits or energy levels (*see* p. 939). The energies in this case are not multiples of a single quantum but are nevertheless functions of integers called *quantum numbers* as well as functions of Planck's constant. The application of the theory of quantum mechanics tells us that the possible orbits for the electrons to move in are defined by the expression *angular momentum* $= n.h/2\pi$ where n is an integer known as the *principal quantum number* of the particular energy level and h is Planck's constant. Thus, as the radius of the orbit increases so does the numerical value of the quantum number, and so this number may be taken as indicating the order of the various orbits, commencing with that nearest the nucleus as $n = 1$, the next outer as $n = 2$ and so on.

Bohr also explained the phenomena of *line* or *emission spectra* and *absorption spectra* on this electron energy level basis. When an electron jumps from an orbit to another of lower energy (outer to inner), the surplus energy is emitted as *electromagnetic radiation* of a definite frequency (wavelength). Since a number of different energy levels are possible in every atom there can be simultaneous transitions of electrons and hence a series of radiations of different wavelengths are possible making up a line spectrum. Conversely, when atoms are in a condition of being excited such as may be occasioned by the application of large quantities of heat or an electrical discharge, then the electrons will absorb energy. As they do so they will move from lower energy levels to higher ones (inner to outer) and these changes in position give rise to the production of absorption lines in absorption spectra, each absorption line corresponding to a particular electron movement. Elements with similar line spectra are also chemically similar and the line spectrum of an element can be regarded as a periodic property (see p. 949). Every element shows an emission spectrum peculiar to itself and can be identified by it.

The expressions for energy quantities, $E = force \times distance$ (in the direction of the force) and $E = nh\nu$ are really exactly equivalent since all kinds of energy are ultimately measured in terms of work, but whether the laws governing all physical and chemical phenomena should be stated in Newtonian terms (forces) or those of quantum mechanics is to a large extent a matter of convenience. A useful criterion is whether or not the phenomenon in question involves an *action*, that is, product of *energy* $\times$ *time*, that is large in proportion to Planck's constant. If the phenomenon is one in which a very large number of quanta are involved, the concept of a small, discrete, indivisible quantum unit has little importance. It would be like stating the value of a very rich man's fortune in terms of the smallest unit of currency or the sand in the Sahara desert in terms of numbers of grains, to use the energy of a photon as the unit of energy measurement in such a case.

In addition to an understanding of the ways in which energy can be quantitatively expressed there are some other fundamental concepts of which the biologist should be aware.

A mass is capable of transferring energy from itself to some other mass by virtue of one or both of two things, its *motion* relative to the other mass or its *position* (energy state) relative to it. We can distinguish these kinds of energy by use of the terms *kinetic* and *potential*. When a mass has velocity it is said to possess *kinetic energy* or the energy of movement. Mathematically this energy can be expressed as a quantity by *half the product of the mass* $\times$ *the square of its speed* or $\frac{1}{2}mv^2$. This expression can be derived from the initial statement of an energy transfer, $E = force \times displacement$.

But a mass that is not moving may also possess a capacity for doing work. A simple example would be that of a fully wound watch spring, that when fixed firmly in the fully wound position, is not actually performing work but when freed will drive all the working parts of the watch. Another example is that of a mass lifted above the ground against the force of gravity and then supported in a fixed position. Whilst supported, the mass is doing no work, i.e. transferring no energy, but if the support is suddenly removed, the mass will fall and immediately its potential for doing work is realized as kinetic energy. The wound watch spring and the suspended mass are said to possess *potential energy*.

It is clearly possible for a mass possessing energy to transfer it to another either relatively quickly or relatively slowly. The rate of energy transfer or rate of doing work is known as *power* and is given by the expression $P = E/t$, i.e. energy or work done divided by the time taken to do it. The unit of power is one joule per second, usually called a *watt*. The *kilowatt* is one thousand times this quantity.

From the equation $P = E/t$, clearly $E = Pt$ and thus *power × time* gives the energy transferred or work done. The *kilowatt-hour* is the amount of energy transferred by a machine working at a rate of one thousand watts for one hour. This amount of energy is 3·6 MJ.

During the last century it was established that energy appears to exist in a number of forms, that is, energy being transferred from one body to another is evidenced by a number of apparently different phenomena. Hence we talk of *mechanical, electrical, chemical, heat, light* (and other electromagnetic vibrations) and *atomic* energies.

Mechanical energy is that directly involved in moving matter; machines are devices designed to make the most convenient use of mechanical energy.

Electrical energy is a result of either the movement of electrons through a conducting substance, kinetic electrical energy—*current electricity*; or the concentrating of electrons at one place so that they would flow if given a suitable conductor to move through, potential electrical energy—*static electricity*.

Chemical energy is that possessed by chemical compounds by virtue of the binding forces between their constituent atoms and is potential energy.

Heat is the energy possessed by a body in the form of kinetic energy of motion of its particles. It can be transmitted by direct transfer from one system of particles to another, that is, by conduction or the interaction of particles possessing greater kinetic energy with those possessing less. It can be transmitted through a liquid or a gas by the actual movement of the fluid, that is, by convection. It can be transmitted as electromagnetic waves, that is, by radiation.

Light is another form of radiant energy. Together with infra-red (heat), cosmic and γ-rays, X-rays, ultra-violet rays and radio-waves it makes up a complete and continuous spectrum of electromagnetic vibrations (*see* Fig. A.10). These are wave-motions not requiring any known material medium for their propagation. They travel with a speed of $2·9978 \times 10^{10}$ centimetres per second or roughly 186 000 miles per second. As will be seen, for some purposes, radiation in the form of waves is a useful conception, but for others it is necessary to consider radiation as having particulate nature.

Atomic energy is that associated with the binding forces between the particles that constitute an atom. It is potential energy and is the energy released from an atomic nucleus at the expense of its mass (*see* p. 935).

Sound is sometimes quoted as a form of energy. It can be produced by any vibrating source and is transmitted as a longitudinal pressure wave motion through a material medium (cf. transverse wave, *see* p. 1007). As the source vibrates it meets with resistance from the medium, say air, and therefore work has to be done to make it vibrate. Energy is thus transferred to the air and travels outwards from the source with the pressure waves.

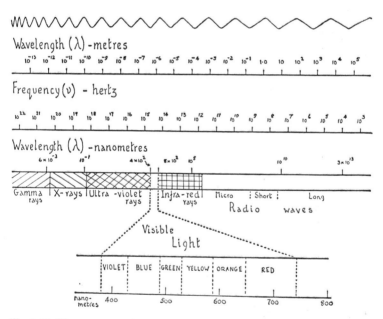

Fig. A.10. The spectrum of electromagnetic vibrations and some data relating thereto.

All these forms of energy except the last two have a very direct relationship with the normal working processes of living things and in this connexion some of them will be dealt with more fully in later sections.

It was also established during the last century that there exists what might be described as a "balance sheet" for energy transfers. Energy can neither be created nor destroyed, it can only be converted from one form to another. This is recognized as the *Principle of the Conservation of Energy*, sometimes known as the *First Law of Thermodynamics*, which may be stated as—

"Within any system of related parts, isolated from all others, the total amount of energy, in whatever form, remains unchanged."

The principle embodies the two important notions that energy may take a number of different forms and that if energy in one form increases, it does so at the expense of energy in another form in an exact, quantitative relationship. If there is any apparent deviation from it this can be explained by defining some other forms of energy. In the case of heat energy, which does not appear wholly convertible into other forms so that it can all be used in doing useful work, the concept of *entropy* fills the need of an explanation of this (*see* p. 999). This tells us that although the energy of a system is not in the overall quantity decreasing, some of it becomes non-available for general use because during the transformation of heat energy into some other recognizable form, disorder at the expense of order among the particles is being generated in the system from which the heat energy is being transferred. In moving from order to disorder some energy of the system is taken out of use or becomes non-available. This conversion of usable energy to non-usable energy is happening throughout the universe all the time and ultimately the whole system must reach the highest state of uniform disorder and all usable energy will have been made non-available. When this happens there will be no further possibility of energy exchanges or transformations and everything will come to a standstill.

One other feature of energy that needs to be mentioned is its relationship to mass. Einstein was responsible for the oft-quoted expression, $E = mc^2$, that relates energy, mass and the speed of light. What this really means is that if a body at rest absorbs an amount of energy, E, then its mass increases by an amount E/c^2 and that the mass of a body is not constant but varies according to changes of energy possessed by the body. Alternatively, it can be said that energy and mass are now regarded as different aspects of the same quantity, an energy E, being equivalent to a mass E/c^2, or a mass m to an energy mc^2. This variable quantity for mass is referred to as the *relativistic mass* of a body, to distinguish it from its *rest mass* which is constant, in the reference frame in which the body is at rest, and is the quantity normally referred to by the term "mass." Relativistic mass tends towards infinity as the speed of movement of a body approaches the speed of light.

The important consequences of this are four-fold; that *energy can be described in units of mass*, that *energy has the properties of mass*, that *matter can be transformed into energy* and that *the mass-energy of a closed system is conserved.* Some of these consequences can be demonstrated practically. For example, the mass of an electron travelling at nine-tenths the speed of light has been measured at between two and three times its mass at rest. The mass of the fragments resulting from the disintegration of uranium atoms in a nuclear reactor when they are not moving is less than the mass of the uranium atoms at rest; the rest mass of all the fragments from the complete fission of one kilogramme of uranium is 0·999 kilogram. The bending of light rays from distant stars towards the sun as they pass close to it, detectable only during eclipses of the sun, exemplifies the effect of gravitational force on radiant energy or energy without rest mass.

Before the relationship between mass and energy was discovered the Principle of the Conservation of Energy was supplemented by the Principle of the Conservation of Mass. It is now clear that the latter is only a special case of the former.

Following will be found further considerations of some of the forms of energy mentioned previously and something of their significance in the activities of living things.

Chemical Energy

In life processes, chemical energy can be considered as the fundamental energy form since the energy required in the performance of every activity is finally traceable to the chemical bonds that hold atoms and molecules together. Note, however, that this source of potential energy within the organism is not the ultimate source from which the energy of living activities comes. This is outside the organism and is almost entirely radiant energy in the form of light from the sun, but this kinetic energy is stored, as it were, during the photosynthetic processes of plants as the potential energy of chemical bonds and it is in this latter form that it later becomes the immediate internal source of energy supply to the living thing.

When a chemical bond, conceived as representing a specific amount of potential energy, is broken, this potential energy is converted into energy of movement. Such being the case, a chemical bond that unites any two atoms can be measured in units of energy and this is expressed as joules of potential energy per mole. The number of entities per mole is called *Avogadro's* constant, given most accurately as $6·02252 \pm 0·00028 \times 10^{23}$. The mole is defined on p.6. Thus the potential energy contained in various chemical compounds can be compared in terms of joules per mole.

The relative strengths of some kinds of chemical bonds have been given on p. 957 and a reference has also been made to the so-called "high-energy" phosphate bonds of ATP. Energy contents of some bonds between particular atoms, in approximate values of kilojoules per mole are

$$H—H = 437, \quad C—C = 350, \quad C—H = 416, \quad C—N = 294,$$
$$C—O = 353, \quad N—H = 390, \quad S—S = 214, \quad S—H = 340.$$

In many situations, when a chemical bond is made or broken with the accompanying transfer of energy, this energy is in the form of heat and is either absorbed from or dissipated into the surroundings according to whether the reaction is *endothermic* or *exothermic*. When elements or compounds react together very vigorously, such as phosphorus and chlorine, energy is liberated as light as well as heat.

As indicated above, chemical compounds can be classified into two groups. Those compounds that are formed from their elements with the liberation of heat such as carbon dioxide, hydrogen chloride and sulphur dioxide are called exothermic compounds whilst those compounds formed from their elements with the absorption of heat are called endothermic compounds. Examples of the latter are nitric oxide and carbon disulphide. The *heat of reaction* of a chemical change is the quantity of heat, measured in joules or kilojoules, that is liberated or absorbed when the correct proportions of particles in moles of the reactants as indicated for the reaction by the chemical equation have completely reacted. It is thus possible to write equations so as to indicate the heat changes involved, for example,

$$C + O_2 = CO_2 + 396 \cdot 7 kJ$$

The positive sign before the heat quantity indicates that the reaction is exothermic. The symbol ΔH is normally used to represent the quantity of heat exchanged in a chemical reaction. It should be noted that heats of reaction are not independent of external conditions. For example, if gases react together it is necessary to know whether they do so at constant volume or at constant pressure. At constant volume, no external work is done by or against atmospheric pressure, whereas at constant pressure, such work has to be done and this has to be taken into account by adding or subtracting the thermal equivalent of this work to or from the heat of reaction at constant volume. The heat of reaction is also affected by the physical state of the reactants and the temperature at which the reaction is conducted. In quoting a numerical value for a heat of reaction it is necessary, therefore, to be explicit about the physical state of the reactants. This is usually done by adding a suitable abbreviation to the chemical formula in the stoichiometric equation: c = crystalline; l = liquid; g = gaseous; aq = in aqueous solution, etc. Unless specified, it is assumed that a reaction commences at room temperature, the same as that to which the products of a reaction return.

The *heat of formation* of a compound is the quantity of heat, measured in joules, evolved or absorbed when one mole of the compound is produced from its component elements in their standard states. Liberation of heat is associated with the formation of exothermic compounds, e.g. carbon dioxide, whilst absorption of heat is associated with the formation of endothermic compounds, e.g. nitric oxide.

The *heat of combustion* of an element or compound is the heat liberated when one mole of the substance burns completely in oxygen or air. In such a case ΔH is generally positive and the reaction is exothermic. Exceptions are the endothermic oxides of nitrogen, NO, NO_2 and N_2O_4.

Heat is also liberated or absorbed when a substance is taken into solution by a solvent. The *heat of solution* is the quantity of heat evolved or absorbed when one mole of a substance is dissolved in such a large volume of the solvent that further addition of solvent causes no further heat change. When calculations are made involving reactions in solution it is necessary to take this heat quantity into account.

The *heat of atomization* is the heat energy required to produce one mole of atoms from the element in its standard state at standard temperature.

One further heat exchange during a chemical reaction is that which occurs when acids and bases are neutralized in aqueous solution. The *heat of neutralization* of an acid is the amount of heat evolved per mole of water formed during the neutralization of an acid by a base. For strong acid-base pairs, the heat released is fairly constant at about 57·3 kJ per mole of water produced at 25°C.

In making thermochemical calculations it is assumed that every substance possesses a definite quantity of energy inherent in its nature, that is, *internal* or *intrinsic energy*. This represents the total amount of energy stored within that substance but the absolute value of this intrinsic energy of an element or compound is not known. Some of it may be released in the form of heat if the substance takes part in a chemical reaction. The intrinsic energy of an uncombined element in its usual form is accepted as zero, thus if a compound is exothermic then its intrinsic energy must be lower than the sum of the intrinsic energies of the component elements by an amount of energy equivalent to the heat of formation of that compound. Or, the intrinsic energy of a compound is equal to its heat of formation with the sign reversed. Thus exothermic compounds have intrinsic energies that are negative and endothermic compounds have intrinsic energies with positive values. The concept of intrinsic energy follows from a consideration of the conservation of energy principle and from the fact that the formation of exothermic compounds is accompanied by the liberation of heat since it must be argued that this heat must have been produced by the conversion of an equivalent amount of some other form of energy.

If the energy-supplying reactions in living things released energy in the same way that heat of combustion is liberated, that is in one rapid conflagration, we should have to visualize their cellular activities as occurring on the same basis as a heat engine, burning fuel to raise the temperature of the system and then performing work as this heat is transferred from a region of high temperature to one of lower temperature. This would not do at all, since the temperatures involved would render the protoplasmic system "lifeless", i.e. incapable of chemical activity, by changing its physical state. The normal state of proteins, for example, would be drastically and probably irreversibly changed. In living things, the energy of chemical bonds is not released as in combustion but in relatively small amounts via a large number of successive chemical steps, each of which releases the energy in a controlled way. However, there is always some chemical energy converted to heat and in all active living systems this is noticeable as an elevation of temperature above that of the surroundings. When plant tissues are catabolically very active, for example, during respiration of germinating seeds, and heat is not allowed to escape to the surroundings as radiation or as latent heat of evaporation, their temperature often rises very much above atmospheric. In birds and mammals, where metabolic activities are usually comparatively rapid, large quantities of heat are produced from chemical reactions at rates under internal control and the body is continuously kept at a constant temperature, usually well above that of the environment.

Chemical compounds vary greatly in the quantities of potential energy that they contain and also in the ease with which this energy can be made available for work. Fats and sugars are examples of compounds containing much energy per unit volume which is also easily released and these are examples of substances used as energy stores in living things. Compounds such as water and carbon dioxide also contain potential energy but before it can be released a great deal of energy must be added to the system. Thus such substances are not normally used in living things as sources of energy.

Because living things make manifest their "life" largely by their chemical activities, entailing energy exchanges, an important consideration for the biologist is the overall balance sheet for energy shown by an organism during its active period.

Without involving the reader in some very complex considerations of the thermodynamics of chemical systems it is impossible to elaborate this as fully as is really necessary to a full understanding. The reader is therefore referred to an appropriate Textbook of Thermodynamics if he requires more than the very elementary concepts presented here.

There are two ways of representing the total energy content of a chemical system, given by the following equations—

$$U = A + TS \text{ (or } A = U - TS) \qquad . \qquad . \text{ (A1)}$$

where $U = $ *intrinsic*, *internal* or *total energy*; $A = $ the maximum amount of energy convertible to work contained in the system (sometimes called the *Helmholtz free energy*); $T = $ absolute temperature; $S = $ *entropy* or the energy not convertible into useful work, i.e. *bound energy*,

and

$$H = G + TS \text{ (or } G = H - TS) \qquad . \qquad . \text{ (A2)}$$

where $H = $ *enthalpy* or heat energy content; $G = $ energy available for useful work (sometimes called the *Gibbs free energy*); T and S as before.

In terms of changes in these quantities that may occur within the system during a reaction, (A1) and (A2) can be rewritten as

$$\Delta A = \Delta U - T\Delta S \qquad . \qquad . \qquad . \text{ (A3)}$$

and

$$\Delta G = \Delta H - T\Delta S \qquad . \qquad . \qquad . \text{ (A4)}$$

where Δ indicates the change in the quantity concerned.

Equation (A3) represents the energy changes that may occur under conditions of constant volume such as when a reaction occurs in a completely closed container and no work is done against external pressure, as would be the case if the volume of the system increased.

Equation (A4) describes the energy changes that may occur during a reversible process under conditions of constant temperature and pressure, i.e. not at constant volume.

Both quantities, A and G, are sometimes referred to as the *free energy* of a system, but it is the function G that is more important in biochemical changes and it is this quantity that is normally referred to as free energy in biological texts where, unfortunately, it is most commonly denoted by the letter F.

The relationship between ΔG and ΔA is given from a knowledge that under conditions of constant pressure,

$$\Delta H = \Delta U + P\Delta V \qquad . \qquad . \qquad . \text{ (A5)}$$

where $P = $ pressure and $\Delta V = $ change in volume. The quantity $P\Delta V$ represents the work done against external pressure during the change.

Thus from equations (A3), (A4) and (A5) we have, for a change taking place at constant pressure

$$\Delta G = \Delta A + P\Delta V \quad . \quad \quad . \quad \quad . \quad . \quad (A6)$$

Note that the free energy, F or G, is not the same as heat of combustion or the heat that may be released during any other reaction in the system. For example, the heat of combustion of glucose is $2\cdot8 \times 10^3$ kJ mol^{-1} whilst its free energy as defined above is $2\cdot88 \times 10^3$ kJ mol^{-1}.

It has been stated that reactions which require heat to be fed in for them to proceed are termed endothermic whilst those that liberate heat during their course are termed exothermic. The corresponding terms for free energy exchanges are *endergonic* and *exergonic* respectively.

If, during a chemical reaction, the system gains in free energy, i.e. is endergonic, the gain can be considered as positive and the amount of free energy gained expressed as $+\Delta F$. By contrast, in a chemical system where an exergonic reaction occurs, free energy is lost and this can be expressed as $-\Delta F$. One way of indicating whether a system during a particular reaction is gaining or losing free energy is to put the value of $+\Delta F$ or $-\Delta F$, measured in joules per mole of reactants, at the end of the equation, e.g.

$$6CO_2 + 6H_2O = C_6H_{12}O_6 + 6O_2, \quad \Delta F = +2\cdot88 \times 10^3 kJ\ mol^{-1}$$

(to represent the simplest expression of photosynthesis) and

$$C_6H_{12}O_6 + 6O_2 = 6CO_2 + 6H_2O, \quad \Delta F = -2\cdot88 \times 10^3 kJ\ mol^{-1}$$

(to represent the simplest expression of respiration).

It should be noted that whilst all exergonic reactions show an overall loss of free energy, they do not necessarily occur spontaneously. Many mixtures of potentially active reactants will never react with one another, or only very slowly, if left to themselves. Some energy must be fed in to make the reaction commence, but once started the reaction will go to its completion without the need for more energy to be added. This is explained on the basis that in such chemical systems, the reactants are characterized by having what is known as high *activation energy*, that is, unless the particles possess a high level of kinetic energy they will not readily engage with one another to initiate a reaction. The activation may take the form of a rearrangement of the atoms of the reacting molecules.

The energy exchanges that occur during chemical reactions can be expressed graphically, changes in potential energy being plotted against time. An example is shown in Fig. A.11, representing the formation of water from hydrogen and oxygen, an exergonic reaction. If hydrogen and oxygen gases are placed together in a container at standard temperature and pressure ($O°C$ and $101\cdot325 \text{ kN m}^{-2}$) nothing will happen. If, however, a small electric spark is caused to pass between two electrodes in the container, there is a minor explosion and water is formed. The electric spark supplies the activation energy (represented by h in the figure) sufficient to activate some hydrogen and oxygen particles which in their turn, as they react, activate others like a chain reaction. The whole process can be likened to a rock lying a little

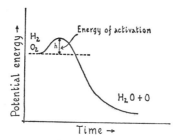

Fig. A.11. Graph representing the potential energy change that occurs when water is formed from hydrogen and oxygen.

way from the crest of a steep hill. The rock possesses high potential energy due to its position relative to its surroundings but whilst lying just off the crest none of this energy is being released. If, however, it is given a small push to the crest (absorbing energy h) it will roll rapidly down the slope, its potential energy being expended as kinetic energy of motion as it changes from the higher position to the lower.

Fig. A.12 represents the changes in free energy during the overall processes of photosynthesis and respiration, the former being endergonic and the latter exergonic. The energy stored in the carbohydrate during the synthetic processes is released during the respiratory processes. The carbohydrate at "rest" at free energy F will not proceed down the slope however without a "push" over the little hill, h.

With regard to the amount of energy of activation required for any chemical system, it should be noted that this is a characteristic of the system and it does not necessarily determine whether the reaction initiated by it is endergonic or exergonic. Neither is it the case that combination reactions of particles are always endergonic and decompositions are always exergonic; each can be either according to the particular chemical system. For instance, the decomposition of formic acid into carbon monoxide and water is endergonic.

As was pointed out previously, a change in free energy, ΔF, differs from a heat change, ΔH, in a given reaction. This difference is represented by what is called an *entropy change* in the system and the relationship between free energy and heat energy change is given by $\Delta F = \Delta H - T\Delta S$, where ΔF = free energy change, ΔH = heat energy

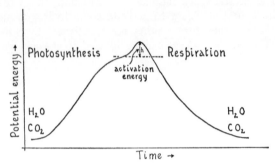

Fig. A.12. Graph representing the change in free energy during the photosynthetic and respiratory processes.

change, T = absolute temperature, and ΔS = entropy change. This concept of energy exchanges will be dealt with again in the last section of this appendix.

Heat Energy

Heat energy has its effect upon matter in changing the motion of its particles. If heat energy is added to a system the particles execute their motions of translation, rotation or vibration more rapidly and the converse is also true. Variations in the rates of motion of atoms and molecules are evident in what we call variation in *temperature*, i.e. degrees of hotness or coldness. If heat is added to a system of particles, according to their state, they move more rapidly and their temperature is said to rise; they are hotter to the touch. Vice versa, if heat is taken from a system it becomes colder as its temperature falls. We can detect changes in temperature by the physical effects that the change in heat energy content causes in a body, as well as by our own senses. Generally, when its temperature is caused to rise by the addition of heat, a body undergoes some physical change, for example, it usually increases in volume (expands); it may change its state; if it is a gas kept at constant volume its pressure increases; if it is an electrical conductor its power of conduction or resistance will change. We can compare the temperatures of bodies by comparing their effects on the physical properties of other bodies and say that one is hotter or colder than another according to a *temperature scale* arbitrarily chosen.

The commonest instrument in use for the measurement of temperature is the *liquid thermometer* which, by its design, makes easy the detection of changes in volume of a liquid as it is heated or cooled. If such an instrument is held in contact with a body of which the temperature is required to be known, the liquid of the thermometer will take up a volume in proportion to the degree of hotness it attains owing to the transfer of the energy of motion of the particles of the body to those of the liquid in the thermometer. This volume can be compared with the volume known to be occupied by the liquid at an arbitrarily fixed degree of hotness, for example, the hotness or temperature at which pure ice melts or at the temperature of the steam from boiling water, both under one standard atmosphere pressure ($101 \cdot 325 \text{ kN m}^{-2}$). These points were in fact the lower and upper fixed points of the universally adopted temperature scale (but *see* below). On what is called the *Celsius* scale the lower is designated 0°C and the upper 100°C. Other forms of temperature-measuring instruments are the constant-volume gas thermometer and the resistance thermometer.

When the bulb of a constant-volume gas thermometer is cooled below the temperature of melting ice, the pressure in the gas falls by a measured amount, approximately $P/273 \cdot 2$ for every degree fall in temperature. Thus if the gas is a perfect gas and obeys the relationship $PV = RT$, known as the *ideal gas equation* connecting pressure and volume of a quantity of gas with change in temperature, where P = pressure, V = volume, T = temperature on the absolute scale (about to be defined) and R = the molar gas constant = $8 \cdot 314\ 3 \text{ J K}^{-1} \text{ mol}^{-1}$, P will become zero at $-273 \cdot 15$°C. If temperature is measured from this *absolute zero point* instead of 0°C we call it an *absolute temperature* denoted by K, after Kelvin who first devised the scale. Thus zero K = $-273 \cdot 15$°C and 100°C = $373 \cdot 15$ K or more generally, $T = (t + 273 \cdot 15)$, where t is the temperature expressed in degrees Celsius. Note that the symbol for degree Kelvin is K, the degree sign being omitted.

Note that the earlier use of two fixed points on the temperature scale, namely, the temperature of pure ice in equilibrium with air-saturated water at one standard atmosphere pressure and the equilibrium temperature of pure water and pure steam at one standard atmosphere pressure, has been found to be unsatisfactory. This is partly because of the difficulty of achieving equilibrium between pure ice and air-saturated water, since when ice melts it tends to surround itself with pure water and thus prevents intimate contact of ice and air-saturated water and partly because the temperature of the steam point is extremely sensitive to changes in pressure.

The temperature scale in use at present is based on one fixed point only. This is the temperature at which ice, liquid water and water vapour coexist in equilibrium and is known as the *triple point of water*. To obtain best agreement between old and new scales, the temperature of the triple point is given the value of 273·16 K. The Celsius temperature scale, previously the centigrade scale, uses a degree of the same magnitude as that of the Kelvin scale, but its zero point is shifted so that the temperature on the Celsius scale of the triple point of water is 0·01° Celsius or 0·01°C Thus if t denotes the Celsius temperature, we have

$$t = T - 273·15 \text{ K}$$

where T is the absolute temperature in K. For example, the Celsius temperature, t_s, at which steam condenses at one standard atmosphere pressure is given by

$$t_s = T_s - 273·15$$

T_s is found experimentally using a constant-volume gas thermometer with the gas at vanishingly small pressure. The value of T_s is found to be 373·15, whence

$$t_s = 373·15 - 273·15$$
$$= 100·00°C$$

The significance of all this is really that, on the new scale, water at one standard atmosphere pressure freezes at approximately 0°C and boils at approximately 100°C and that these figures are no longer definitions of the temperature scale.

The true concept of an absolute zero temperature is really more complicated than that outlined above. The idea that molecular energy vanishes at absolute zero is incorrect and there is in fact considerable "zero-point energy" in all real substances. The relationship between molecular kinetic energy and temperature is undoubtedly complex and differs for different substances. Thus whilst it may be true to say that temperature is a "measure of" the molecular kinetic energy, it is not correct to assume that there is a simple correspondence between the two. In general, this energy is not proportional to the absolute temperature, nor is it zero at 0 K (nought degrees Kelvin).

With a concept of temperature in mind it is now possible to define what we mean by heat in another way; it can be said that heat energy is transferred from one body to another if, when the two are placed in contact neither changes its state, the temperature of one rises as the temperature of the other falls. No heat will flow between two bodies with the same temperature and heat can flow only in the direction from a hotter body to a colder one.

Changes in temperature of bodies can also be used to denote how much heat energy has been transferred to or from them but it must be borne in mind that whereas temperature refers only to rate of atom or molecule motion in the body, heat is a measure of the total quantity of atom or molecule motion, so that the quantity of heat in a body will depend on the kind and mass of all the particles, as well as on their rate of motion. The unit quantity of heat, as for all forms of energy, is the joule (defined on p. 7) and expresses the work that can be done by the expenditure of heat energy. The unit, calorie, is not included in SI units and is therefore not recommended for use (for conversion purposes when necessary 1 calorie = 4·18 joule, averaged over temperature range 0°C — 100°C).

The *thermal* or *heat capacity* of any body is defined as the quantity of heat required to raise its temperature 1°C or 1 K. It is proportional to the mass of the body and depends also upon the nature of the matter of which the body is composed. It is expressed as joules per Kelvin. Because of the different heat capacities per gram of different substances, different quantities of heat will be required to raise the same masses of different substances through equal ranges of temperature. The thermal capacity per kilogram of a substance is known as its *specific heat capacity* and is expressed in terms of joules per kilogram per Kelvin (J kg^{-1} K^{-1}). Water has a specific heat capacity of 4200 J kg^{-1} K^{-1}. Since values of specific heat capacity can vary considerably with temperature they are usually measured and quoted between 14·5°C and 15·5°C and at one standard atmosphere pressure.

It has been shown that heat as a quantity can be considered in terms of the kinetic energy of motion of the particles of matter as reflected by their changes in temperature and also by the work it can do. In both these cases the heat has been considered as being transferred between bodies by their contact. But it is also possible for heat to be transferred from one body to another through empty space as *electromagnetic waves*, that is, in *radiant* form. Such wave motion is a process by which kinetic energy can be transferred from one point to another without any transfer of matter between the points. Heat energy transferred in this way is usually known as *infra-red radiation* and the characteristics of this phenomenon are similar in all respects to the radiation of light energy, the two energy forms varying only in the wavelength (frequency) of the transmitted vibrations. The chief characteristics of light radiation follow in the next section.

Heat energy through its effect on the kinetic energy of atoms and molecules has a tremendous significance in the activities of living things since temperature is certainly amongst the most important of the environmental factors controlling living processes.

Light Energy

Visible light is one part of the solar radiation to reach this planet and is directly or indirectly the form of energy that supports nearly all living things.

The sun, a vast source, derives its energy from atomic transformations. Its internal temperature has been estimated at about 10^7 K and
its surface temperature at about 6×10^3 K. A good deal of the energy
radiated by the sun is absorbed by the earth's atmosphere before it
reaches the zone of living things, through the agencies of the atmospheric
gases and fine dust particles. The rate at which solar energy reaches the
outside of the earth's atmosphere has been calculated at about the
equivalent of 84 kJ per square metre per minute whereas an average
value for noon on a summer day at ground level is only about a quarter
of this. If there were not this absorption, particularly of the shorter
wavelengths, i.e. the ultra-violet rays, there would be no possibility of
life in its present form.

The true nature of solar radiation, here treated as visible light, is not
fully understood for, although it is possible to describe some of its
characteristics such as colour and how it is transmitted, reflected and
refracted, like all forms of energy, it can be fully assessed only in terms
of its effects upon matter.

To account for all the phenomena associated with light energy it is
necessary to employ two conceptions as to its nature. One is known as
the *wave theory* and the other as the *quantum theory*. The former
regards light as being propagated as electromagnetic waves through
space and the latter considers light as having some of the properties of
tiny particles, *quanta* or *photons*, possessing no rest mass but given off
by any light-emitting body and travelling in straight lines through space
until they encounter matter, when they affect it and are affected by it in
various ways. The speed of propagation in either case is constant at
about 3×10^{10} cm per second.

A somewhat crude picture of an electromagnetic wave can be derived
from consideration of electric and magnetic fields, i.e. regions of space
in which electric and magnetic forces are being exerted. A distinction
between electric and magnetic fields can be made in that the former
surround electric charges at rest whilst the latter surround electric
charges in motion (electric current). Electric lines of force radiate outwards from a stationary charge whilst magnetic lines of force surround
the path of moving charges as concentric rings. A moving charge still
possesses its electric field of force.

The conception is, that when electric charges are neither at rest nor in
uniform motion but are undergoing acceleration by vibrating back and
forth, they will produce alternating magnetic and electric fields that are
propagated outwards on the basis that a changing electric field produces
a magnetic field around it and a changing magnetic field produces an
electric field around it. A simple way of grasping this concept is to
imagine an electrically charged sphere swinging like a pendulum,

undergoing continuous change of motion. The movement of the charge constitutes a changing electric field and creates a magnetic field. As the pendulum swings its speed changes so that this magnetic field is changing. Because of this change in magnetic field an electric field is produced. Thus both types of field will be changing continuously as the electrically charged sphere vibrates, alternating with each other and each produced by the other. Maxwell first proposed this and visualized the energy of the fields spreading out into space. He calculated that its speed of propagation, assessed from electrical and magnetic measurements that he was able to make, should be about 3×10^{10} cm per second. He thus postulated that accelerating electric charges should give rise to electromagnetic waves moving through space with the speed of light, a quantity already measured by various physicists.

Waves of this nature have the properties of what are known as *transverse waves* and can be visualized by analogy to the movement of ripples on a smooth water surface. Such waves are said to be transverse because the movement of the water particles transmitting the wave is at right angles to the direction of propagation, not parallel to it as in the case of *longitudinal wave motion* such as that of sound waves.

Thus on the wave theory, light can be considered as a wave disturbance being disseminated outwards in all directions from a source and made up of periodically varying electric and magnetic fields, these being at right angles to the direction of propagation and to each other (*see* Fig. A.13). Such relationship between the components of the wave allows for the conditions that exist at any time instant, at successive points on a single light ray, to be thought of in terms of the electric field only.

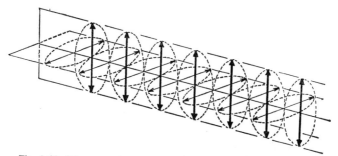

Fig. A.13. Diagrammatic representation of an electromagnetic wave.

When lines whose length represents the strength of the electric field are drawn at right angles to the ray, the ends lie on a curve as in Fig. A.14, known as the *sine curve* from the fact that if the values for the sines of all angles from 0° to 360° are graphically represented they produce the same curve. The outline forms a series of alternating crests and troughs. The *amplitude* of the wave is given by the height of a crest (or depth of a trough) from the zero value, i.e. the maximum value of the electric field in either direction, and is really a measure of the intensity of the wave-producing disturbance. *Wavelength* is the distance between any two

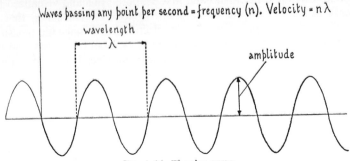

Fig. A.14. The sine curve.

corresponding points on adjacent waves. The *velocity* of the wave is the distance travelled in one second and *frequency* is given by the number of waves passing any point in unit time. Thus $c = n\lambda$, where c = velocity, n = frequency and λ = wavelength. Wavelengths of light are expressed in nanometres, one nanometre being one thousand millionth of a metre (*see* p. 21 for dimensions table).

As has been stated, a number of phenomena associated with light can be explained on the wave theory and it will be assumed that the reader has some knowledge of the basic laws relating to propagation, reflection, refraction and dispersion.

Colour, the sensation that results from the reception of visible radiation by the eye, is defined by the wavelength of the radiation. The wavelength of visible light received from the sun as a combination of many wavelengths and known as *white light*, ranges from just below 400 nm at the short, violet end of the spectrum to about 750 nm at the longer, red end. Below the shorter visible wavelengths is the ultra-violet radiation and above the longer visible wavelengths is the infra-red radiation (*see* Fig. A.10). Not much of this part of solar radiation is of interest to the biologist since radiation with wavelengths shorter than about 300 nm is rapidly lethal and that with wavelength above 1500 nm is absorbed so strongly by water that living things are virtually opaque to it. The energy in such radiation is easily transformed to heat. When the heating effect of different wavelengths of solar radiation is examined it is found that the maximum heat occurs in the orange band of the visible spectrum, not in the infra-red as is commonly believed. The energy content of the visible spectrum varies with wavelength (*see* Table 6), according to the relationship $E = hc/\lambda$ (*see* p. 990), thus blue light at about half the wavelength of red has nearly twice the energy, reference being to one photon in each case. It should be noted, however, that the numbers of red and blue photons in light are not the same so that it is not correct to say that red light contains only half the energy of blue light.

ENERGY VALUES FOR VISIBLE RADIATION

Wavelength (nm)	E (kJ mol^{-1})
400	298·95
500	239·24
600	199·37
700	170·87

Another light phenomenon that can be associated with its transmission as a wave form is the *polarization of light*. It has been said that ordinary light consists of electric and magnetic vibrations and that for each electric vibration the associated magnetic vibration takes place in a plane at right angles to it. This picture must be extended for ordinary light waves to regarding both vibrations as taking place in all possible planes containing the light ray, the vibrations themselves being at right angles to the direction of the light path. When light is transmitted through certain crystalline substances such as calcite (calcspar or Iceland spar) and tourmaline they act upon the vibrations to plane polarize them, that is, the electrical and magnetic components are confined each to one plane mutually at right angles. The plane containing the electric vibrations is called the *plane of vibration* and the plane containing the magnetic vibration is called the *plane of polarization*. Light waves in this condition are said to be *plane-polarized*.

A somewhat crude idea of the difference between ordinary and polarized light can be gained by considering a length of rope, fixed at one end but free to be oscillated at the other. If the free end is oscillated at right angles to the direction of the rope, waves will sweep along it and if the plane of oscillation is continuously rotating, the waves will occur in all planes. This simulates the transmission of transverse waves as in ordinary light. If, however, the free end of the rope is made to oscillate in one plane only, say vertically, the crests and troughs of the transverse waves sent along it will occur in that plane only. This simulates the condition of plane-polarized light. Another way of comparing ordinary light with plane-polarized can be done with reference to Fig. A.15, where (*a*) represents some of the directions of vibration of one of the fields in ordinary light and (*b*) represents the single direction of vibration in plane-polarized light. The effect of a polarizing substance to produce a single plane of vibration as in (*b*) is to eliminate completely vibrations 2–2, to transmit completely vibrations 1–1 and to transmit only those components of vibrations 3–3 that are parallel to 1–1. Thus all vibrations appear in the plane 1–1 only and the intensity of the ray is diminished owing to loss of the components parallel to 2–2. Two polarizing substances with their crystal axes set at right angles to one another will therefore absorb all the components of all the vibrations and no light will pass through the two together. Study of the effects of biological substances in causing polarization of light or in causing a rotation of the plane of polarization has been very useful in elucidating something of their structure.

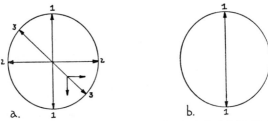

Fig. A.15. Comparison of ordinary and plane-polarized light vibrations.

Interference is another light phenomenon related to its propagation as a wave motion. If two waves are transmitted through a medium the total displacement effect at any point in the medium is the resultant of the displacements due to the separate waves. If two waves of equal wavelengths and amplitudes arrive at the same point so that they are *in phase*, that is, their crests and troughs exactly coincide, they will reinforce one another and their amplitude of vibration at the point where they meet is doubled. If, however, they are out of phase by 180°, that is, a crest of one exactly coincides with the trough of the other, they will completely annul one another's effects and the amplitude of the vibration at the meeting point will be zero (*see* Fig. A.16). Waves meeting out of phase by any value between zero and 180° will show intermediate interference effects between total reinforcement and total extinction. Such effects can be seen when light is transmitted through narrow slits and when it is reflected from thin films. The colour effects seen at a thin oil film on water or a soap film are interference effects between light rays reflected from the two sides of the film. The phase contrast microscope owes its special usefulness, in distinguishing between materials with different refractive indices, to interference effects.

Although the wave theory accounts satisfactorily for many light phenomena there are certain effects that cannot be reconciled with it. It was these that led to Planck's quantum theory, *see* p. 989. One light effect not explicable on the wave theory is the *photo-electric effect* of light on a metal surface. When such a surface is irradiated by light it releases electrons and although the numbers of electrons released depends on intensity, measurements of the velocity by which they escape show that this is dependent only on the wavelength or frequency of the light, not on the intensity or amplitude of the wave, if light is a wave motion. By the wave theory, the intensity or energy content of a wave should be proportional to the square of the amplitude and not to the frequency and the energy used in causing electron ejection would

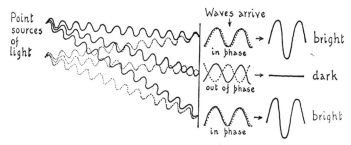

Fig. A.16. The interference effect.

need to be accumulated at the surface if this was supplied in small intensity. In fact, there is no time lag between irradiation and electron release provided a minimum threshold energy value is employed and no electrons are ever released unless the threshold value is exceeded. The only acceptable explanation of these facts is that light has some of the properties of particles, photons, containing enough energy to eject the electrons. Light of frequency v is made up of energy quanta hv. When a photo-electron of mass m is emitted with velocity v, it has kinetic energy $\frac{1}{2}mv^2$ and has also done an amount of work ϕ in coming out. Therefore, if the conservation of energy principle holds, $\frac{1}{2}mv^2 = hv - \phi$. This is Einstein's photo-electric equation. If hv is less than ϕ, the left hand side of the equation is negative and there can be no emission. If $hv_0 = \phi$ then v_0 is the *threshold frequency*.

This corpuscular conception of light and other radiation can also be used to explain a number of phenomena such as the formation of line spectra, *see* p. 990, the production of X-rays and the way in which energy and wavelength of radiation from a hot body vary with absolute temperature.

When radiant energy falls on matter, some may be reflected, some transmitted and some absorbed according to the nature of the matter and the nature of the radiation. The amount that is absorbed depends on whether or not quanta are captured by particles in the energy path and changed into some other energy form. Energy quanta of different wavelengths have different effects. In the longer infra-red region where the energy quanta are low, absorption generally results only in an increase of the vibrational energy of the absorbing particle and hence is detected as heat signified by a rise in temperature. Absorption of the shorter infra-red region, where the quanta are a bit higher in energy content, may result in both vibrational and rotational energy of particles being increased. Absorption of still higher quanta as in the visible and ultra-violet regions can cause interactions involving atomic structure and if the valency electrons are sufficiently affected, *photochemical reactions* can occur. Shorter wavelengths or still higher energy quanta can be large enough to remove electrons completely from the outer shells of atoms and cause ionization whilst the energy quanta of X-rays and γ-rays can remove inner electrons and seriously disrupt the atomic structure of the absorbing particles. Quanta of the highest energy can react with atomic nuclei. A measure of the energy of a quantum (or a sub-atomic particle) is the *electron-volt* which is the energy gained by an electron (or other particle with the same charge) in falling through a potential difference of one volt and is denoted by eV. One $eV = 1 \cdot 6 \times 10^{-19}$ J. The role of light and chlorophyll in the photosynthetic process is through the absorption of light energy quanta by the pigment and the transformation of this energy into chemical bonds in ATP.

In biological work involving the study of the effect of light on living things it is necessary to measure quantities of light energy emitted from sources and the effects of its interception. This can be done in one of two ways, *radiometrically* or *photometrically*. The former deals with the energy measurement wholly through its physical effects and as a practical technique is called *radiometry*. The latter is concerned with visual or what might be called psychophysical evaluation of light energy and is called *photometry*. The amount of energy emitted by a source is called *radiance* by the first system and *luminous flux* or *brightness* by the second. Likewise the terms *irradiation* and *illumination* mean the same thing in the two systems of measurement for describing the process of interception of radiant energy.

In radiometry the intensity of radiation from a source is measured by the rise in temperature it causes in a "perfect black body", that is, a perfect absorber. Such an absorber obeys fixed laws which relate rise in temperature to radiation absorbed. Radiometry is associated mostly with measurement of infra-red radiation.

In photometry, a source of radiation is considered to be any body that radiates the kind of energy that can be detected by the human eye. There are three quantitative aspects that must be considered:

(i) the quantity of light energy emitted by a source in all directions in unit time, known as the *total luminous flux* (Φ) or alternatively the *luminous flux emitted per unit solid angle* (*see* below) in a particular direction, known as *luminous intensity* (*I*) in that direction, and the measurement of these;

(ii) the degree of intensity of the *illumination* (*E*) of a surface at a distance from a source and the measurement of this;

(iii) the quantity of light re-emitted by an illuminated surface, that is, the *luminance* (*L*) of an illuminated surface, and its measurement.

To express the luminous flux emitted from a given point source its power to emit light energy must be compared with that of a standard emitter. This standard was originally selected as a candle of specific dimensions and composition and the standard of light-emitting power was known as the *candle-power* or *candle*. Other standards have been used, such as pentane and amyl acetate lamps working under controlled conditions of pressure and humidity. The most recently agreed standard source is that based on the light energy emitted through an aperture in a hollow enclosure kept at a temperature of the freezing point of platinum and the unit is the *candela* or "*new candle*". One square centimetre of the surface of this black body radiator, at the temperature stated, emits 60 candela. The candela, properly defined, is the luminous intensity, in the perpendicular direction, of a surface of 1/600 000 square metre of a black body at the temperature of freezing platinum at a pressure of one standard atmosphere ($101 \cdot 325$ kN m^{-2}). The luminous flux from any point source is measured in *lumens*; there is no need to use a qualifying unit of time since the speed at which the light is emitted is constant. One lumen (Lm) is the light flux emitted per unit solid angle by a uniform point source of one candela (cd). A unit solid angle (one steradian) is that having its vertex in the centre of a sphere, cuts off an area of the surface of the sphere equal to that of a square with sides of length equal to the radius of the sphere. For example, a steradian is subtended at the centre of a sphere of one metre radius by an area of one square metre at the sphere's surface. The surface area of a complete sphere is $4\pi r^2$ and therefore the solid angle subtended by a sphere at its centre is $4\pi r^2/r^2 = 4\pi$. Thus if a point source gives the same luminous intensity, *I*, in all directions, it follows that the total luminous flux, Φ, from the source is $4\pi I$, or a source of mean spherical candela power unity gives a total luminous flux of 4π lumens. *Mean spherical candela-power* refers to the fact that sources of light, particularly electric lamps, generally give different luminous intensities in different directions, and for most purposes it is necessary to specify the average luminous intensity in all directions, i.e. mean spherical candela-power, or the average intensity in a horizontal plane, i.e. the mean horizontal candela-power. The light output from an ordinary 100 watt gas-filled lamp is about 1200 lumens. Its efficiency, expressed as lumens per watt (light output per unit power input) is 12 lumens per watt, or a low efficiency. Another measure of light emitted by a source is called the *quantity of light* (Q) and is given by the product of the luminous flux of the source and the emission time in seconds, i.e. lumen seconds (1m s).

An illuminated body receives light from a luminous body or light emitter. When light falls upon a surface, the amount of light or luminous flux per unit area around a point on the surface is known as the illumination at that point, denoted by E. The unit of illumination is called a *lux* when considering a surface placed normal to the direction of the incident light at a distance of one metre from a source of one candela. From the definition of a solid angle it follows that one lux is equal to one lumen per square metre and thus illumination can be expressed in these terms.

The illumination at a surface varies inversely as the square of the distance from the source. This can be appreciated from the following reasoning. Suppose a point source, giving a total luminous flux Φ, at the centre of a hollow sphere of radius r. Since the surface area of the sphere is $4\pi r^2$, the illumination of the inner surface is given by $E = \Phi/4\pi r^2$. If I is the luminous intensity of the source and Φ, the total luminous flux is equal to $4\pi I$ (*see* above), then $E = 4\pi I/4\pi r^2 = I/r^2$.

If the incident light strikes a surface obliquely, the illumination per unit area is less than that at a surface at right angles or normal to the incident light. This is really because the light is spread over a greater area in the oblique case and the illumination is less in the ratio $1/\cos \theta$, where θ is the angle between the plane of the normal and the plane of the surface in question. The illumination of an area oblique to the source is given by $E = I/r^2 \cos \theta$.

Whereas the illumination of a surface describes the amount of light falling upon it, it is not necessarily the case that the surface will re-emit (reflect) all that light, i.e. that the luminance of a surface is equal to its illumination. Luminance of a surface is given by the product of the illumination and certain reflection factors peculiar to the surface. The unit of luminance is the candela per square metre. Unless the surface is a perfect reflector (reflection factor = 100 per cent or 1), i.e. reflecting the whole of the light falling on it, an illumination of one lumen per square metre will not give a luminance of one candela per square metre. If a surface that reflects 75 per cent of the light falling on it is illuminated to 20 lumens per square metre, its luminance will be $20 \times 75/100 = 15$ cd m^{-2}. The nearest approach to a perfect light-diffusing surface in practice is one of magnesium oxide.

Photometers are instruments used for measuring the illumination of a surface or for comparing the luminous intensities of two sources. The most convenient for use are those based on a photo-electric effect which is one involving a transfer of energy from light incident on a substance to electrons in the substance (*see* p. 1011). The reader is referred to a suitable textbook of Physics for details of such instruments.

Fluorescence and *phosphorescence* are terms applied to processes in which some substances, on being irradiated, re-emit the energy absorbed as radiation. This re-emitted radiation is always at a longer wavelength (lower frequency) than that absorbed. The substances are referred to as *luminous* and are said to be fluorescent if they re-emit the radiation within 10^{-8} to 10^{-6} second of being activated and cease to re-emit immediately the irradiation is cut off. They are described as phosphorescent if there is a lag between irradiation and re-emission. Many luminous substances are both fluorescent and phosphorescent, for example, zinc sulphide. Note that the luminescence of phosphorus, fireflies, etc. (*see* p. 71) are not due to the effects mentioned above but to chemical reactions. The red colour of a freshly extracted solution of chlorophyll is due to fluorescence consequent upon irradiation by ultra-violet light with the subsequent emission of longer wavelength visible light rays. Some luminous paints contain minute portions of radio-active substances, the emitted particles from which, falling on zinc sulphide, cause it to fluoresce. Barium sulphate is an extremely phosphorescent substance.

Energy Transformations in Living Things and the Laws of Thermodynamics

The principle of the conservation of energy has for a long time been based on a firm foundation and is embodied in the first law of thermodynamics (or heat energy transfers). The principle at its simplest states that "energy cannot be created or destroyed". The first law of thermodynamics concerns the energy from heat and in its simplest form states that when heat (energy) disappears work is done according to the relation $H = W$, where $H =$ a quantity of heat, $W =$ mechanical work done. In a more complex form this may be written, $\delta H = \delta W + \delta E$, to allow for the appearance or disappearance of other forms of energy (E) besides mechanical work, when heat disappears.

The conservation of energy principle and the first law of thermo-dynamics are really saying that the phenomena of mechanical movements, heat, light, electricity and so on, all have some common underlying identity. They are all different manifestations of something that we call energy; they are mutually interconvertible. If, for example, light can be produced in some way and this then disappears, there is left in its place the power to produce some other form of energy such as heat, electricity or mechanical movement. There is also a definite numerical relationship between the capacity to produce light energy (measured in suitable units) and the capacity to produce the heat (or other) energy that replaces it.

It must be remembered that the same laws of energy conservation apply to living things as to all other systems of particles and it is quite wrong for the reader to arrive at the conception that during its nutri-tional processes the green plant is "making" energy for itself. It is merely transforming the energy of light into another form more suitable to its internal chemical processes, namely, chemical bonds between particles.

The second law of thermodynamics involves much more complex conceptions of energy transformations. It has been stated above that the different forms of energy are equivalent in the sense that they can all be referred to a common standard and that if energy in one form disappears a corresponding quantity of it will reappear in some other form. But this fails to bring into consideration an important difference between heat and other forms of energy. It is the case that any of the energy forms other than heat can be conceived as being completely interconvertible, but there is a limitation on the power of converting heat into other forms of energy. It is possible that this apparent differ-ence is more due to the inability of man to effect a perfect transforma-tion than to a distinctive fundamental property of heat energy. Heat energy is regarded as being the total kinetic energy of the particles in a system and it may well be that man's limitation shows in his being unable to isolate these particles and direct their kinetic energy so that all of it becomes available for work. Hence he can never demonstrate that heat energy is completely convertible into other energy forms. But whether or not this is truly the case, heat energy is regarded as being of a lower grade than other forms in that, at least, there is not altogether a free choice in the use of it. It is this conception that is expressed by the second law of thermodynamics and can be stated in a number of ways, none of which may be very meaningful to the reader at this moment, but all of which in practical terms amount to the same quantitative statement of the difference between heat energy and other energy forms.

When the law is applied to systems of particles in general, the statement can be made that when there is a change of free energy within the system, this is equal to the amount of heat energy given off or absorbed by the system less an amount of energy which is given by the product of the absolute temperature at which the change occurred and the change in *entropy*. In symbols this can be written

$$\Delta G = \Delta H - T.\Delta S \text{ or } (G_1 - G_2) = (H_1 - H_2) - T(S_1 - S_2)$$

where ΔG = change in free energy $(G_1 - G_2)$, ΔH = change in heat energy $(H_1 - H_2)$, T = absolute temperature and ΔS = change in entropy $(S_1 - S_2)$.

It is the concept of entropy which should engage the interest of the biologist who wishes to compare the thermodynamics of living systems of particles with those of non-living systems. A statement attempting to define entropy could be that it is a measure of the disorder (or order) among the particles of a system, but for most of us this needs a good deal of amplification.

Picture a number of particles within an enclosure. There are an infinite number of positions of the particles relative to one another which might be conceived, ranging from all of them closely compacted together in one tiny volume of the container to the condition where all are loosely and evenly spread throughout the whole volume. The former represents a condition of maximum order or organization among the particles whilst the latter represents the minimum order or alternatively the maximum amount of disorder or disorganization. From experience we know that if a system of particles, constantly in motion and freely able to move, e.g. a gas, are enclosed in such a way, there are certain probabilities about their distribution within the container. It is highly improbable that even if they started as a compact group in a small space within the enclosure that they would remain so for very long. It is much more probable that as a result of their own random movements they would become evenly distributed throughout the whole space, there being no more or no fewer particles per unit volume in one region than another. They will tend towards a distribution at which their disorder is greatest, that is, a state of greatest entropy. To get them back into the state of highest order and to keep them in such a condition work would have to be done on them, i.e. some energy expended. Conversely, therefore, in changing from a state of high to low order they must be capable of doing some work. Thus in a system of particles some of the free energy possessed by it must be considered as being identifiable with the work necessary to reduce the particles to a state of minimum disorder. In the most probable situation where the particles are all evenly distributed with maximum disorder, this would be greatest and in this condition the entropy of the system is greatest.

To visualize the work capacity of a highly ordered system of particles in which no loss or gain of heat is involved, consider the case in which a container is separated into two compartments by a screen, impassable except for a small aperture, all the particles being contained initially on one side of the screen. Owing to their own random movements more and more of the particles must pass through the hole and spread out on the empty side, in so doing taking up the more probable distribution, i.e. creating disorder from order or increasing in entropy. If, as they pass through the hole, they strike a vane on a movable spindle they will transfer some energy to the vane and cause the spindle to rotate. The energy of their random movement which results in their disorder will be translated into work. So long as there are more particles on one side of the screen than the other there will be the tendency for more particles to be moving in one direction than the other but as soon as the particles are evenly distributed throughout the whole container the tendency will be for the spindle to remain still, being stimulated to move equally in opposite directions by the same distribution of particles on each side of the screen.

This picture of what may happen in a system of particles perhaps reinforces what has already been said and may promote a better understanding of what is meant by entropy in numerical terms. Since entropy is a measure of disorder it can be defined in mathematical terms and this definition embodies the idea that the less probable the distribution of particles in a system (i.e. the greater order) the less entropy is contained by it and conversely, the more probable the distribution (i.e. the greater the disorder), the more entropy it contains. In changing from low entropy to high entropy some energy is expended which is not heat energy. It is the case that chemical and physical reactions always tend to proceed only from states of low entropy to states of high entropy (order to disorder) unless energy is supplied. The relation between entropy values and degree of order is an inverse relation.

Returning to the equation given earlier and remembering that free energy is generally equated with the capacity to do work, it should be clear that a highly organized system is capable of exhibiting a greater change of free energy than a less well organized system. ΔH in the equation expresses the change in heat during a chemical reaction and thus the greater the amount of heat given off during a reaction, the greater the free energy exchange. If there is a large free energy change this indicates that a greater amount of potential energy has been released. Thus there is less free energy in the system after a reaction than before it with a corresponding increase in the entropy of the system.

The question may now well be asked—of what significance is all this to the biologist? The answer is that unless care is taken in considering all the facts relating to free energy in the universe as a whole it might be erroneously concluded that biological systems do not obey the Second Law of Thermodynamics. To arrive at the more tenable conclusion the reader is invited to reason as below.

There are two generalizations that can be made when considering biochemical processes in the light of the second law of thermodynamics as applicable to the universe. In the first place, for any chemical reaction to proceed there must be some force driving it along and this can be an energy change from higher to lower or an entropy change from lower to higher (order to disorder). The driving force can be envisaged as the difference in energy levels between two parts of a system, as for example, the driving force for work in the cell can be considered to be the difference in energy level between a sugar molecule and its breakdown products carbon dioxide and water.

The second generalization that follows is that all processes occurring in the universe tend towards an increase in disorder or a running down, i.e. increase in entropy. This does not mean that all the energy expended is destroyed, it is being converted into an unusable form. As the entropy of the universe increases, its free energy decreases. The energy being dissipated in the running down process, is of course, not destroyed, but with the passage of time the total amount of usable energy in the universe is decreasing.

Now, when the metabolic activities of living things, which lead to the incorporation of more material and energy into their systems, i.e. growth, are considered, it is obvious that they are actually showing an increase in free energy and a decrease in entropy. They are driving towards order from disorder. But this is true only if they are considered as isolated systems. As such, living things do build complex molecules of high free energy from substances at a lower energy level but in doing so they still rely ultimately on the energy of a non-living system, which in nearly all cases is the sun. Thus they cannot be considered as single isolated systems for the purposes of thermodynamics. Living things are able to show increases in free energy and decreases of entropy only at the expense of the free energy of the universe as a whole and the balance-sheet of energy of the whole system is showing an ever-decreasing total quantity of free energy. Momentarily, as it were, a living thing captures a small quantity of this free energy and uses it to produce and maintain its system of particles at a high level of organization and low entropy. There inevitably comes a time for each individual, however, when it is incapable of doing this any longer and its order degenerates into disorder once more. There is no doubt that the second law of thermodynamics is equally applicable to all systems of particles when all systems are considered as a whole.

Accident or Design?

Where, when and how living systems appeared on our planet are still matters of conjecture for the moment, but no doubt acceptable solutions to such questions will be put forward. Some day also may be found, somewhere in the universe, other living systems, obeying all the laws of physics and chemistry but "being alive" in totally different environmental conditions from those that we are used to. If and when they are found, it should be possible to explain their structures and functions in the same kind of biological terms as we do those living systems known to us here.

But the remaining question, that is, the why of life or the purpose of living it, wherever and in whatever form, is not a matter for biologists alone and all men should be stimulated to try to reach a conclusion. At present, there are those who clearly see a Meaning of Life, particularly for man himself, emanating from God, and there are some who see our universe, including all things that may be "alive" within it, as the result of wholly random encounters of its fundamental particles, without a Plan and without an Architect.

It would not be appropriate here to argue either case but we leave our readers to ponder the matter and to arrive at an acceptable answer for themselves.

QUESTIONS

CHAPTER 1

The Nature and Properties of Protoplasm

1. Give an account of the constitution and properties of protoplasm and indicate the importance of these properties in the activities of cells.   *L.B.*

2. Give an account of the chemical and physical nature of protoplasm. What is the rôle of the nucleus in a cell?   *C.Z.*

3. Give a general account of the structure of animal protoplasm and enumerate the chief physiological processes which take place within it.   *N.Z.*

4. What do you understand by a "colloidal system"? How does a true solution differ from such a system? Describe the properties of protoplasm, including in your description some of the evidence which suggests that protoplasm has a colloidal nature.   *C.Bi.*

CHAPTER 2

The Fine Structure of Cells

5. Describe the nature and functions of the components and inclusions in the cytoplasm of a vegetative cell.   *C.B.(S.)*

6. Give an account of the structure of nucleic acids and of their roles in protein synthesis.   *W.B.*

7. Describe the nature of DNA and discuss its role in the functioning of a cell.   *W.Bi.*

8. What is known of the structure of the mitochondrion? Explain, with reasons, which of the *two* following types of cells would have especially large numbers of mitochondria: kidney tubule cells, adipose (fatty) cells, sperm cells, cells lining the cheek.   *A.E.B.Bi.*

9. Make a clear, labelled drawing to illustrate the chief features of a plant cell and write a short account of each of the parts you have labelled.   *L.B.*

10. Give an account of the structure and function of nuclei in plant cells.
          C.B.(S.)

Chapter 3

The Relations between Cells and Their Surroundings

11. What do you understand by "osmosis" and "diffusion"? Distinguish clearly between these two processes and indicate their significance in the physiology of the plant.   *B.B.(S.)*

12. Distinguish between the osmotic pressure and the suction pressure of a cell. The osmotic pressure of the cells of a beetroot was found to be 12 atm and their suction pressure 6 atm. If thin slices of this beetroot were cut and some immersed in water and others in sucrose (cane sugar) solution of 10 atm osmotic pressure, what would happen to the cells?   *L.B.*

13. When a solution of sugar was placed inside a membrane surrounded by water it was found that the solution increased in volume and sugar passed out. Give an explanation of what happened. State what part the processes you refer to play in the absorption of water by plant roots.   *O.C.B.*

14. State what is meant by the term "osmosis." What factors control the movement of water from cell to cell in a plant?   *O.C.Bi.*

15. Explain the terms (a) osmotic pressure, (b) turgor pressure, (c) wall pressure and (d) diffusion pressure deficit (suction pressure). Demonstrate graphically how these factors will interact in a cell gradually changing from a state of incipient plasmolysis to a fully turgid condition and discuss the implications of osmosis in the physiology of higher plants.   *S.B.*

16. Give an account of the mode of entry of salts into the superficial cells of the root.   *L.B.(S.)*

17. The following data illustrate the relative absorption by carrot discs of potassium ions at different concentrations of oxygen from a potassium bromide solution at 23°C.

Conc. of O_2 (per cent) . .	2.7	12.2	20.8
Relative absorption of K .	22	96	100

What explanations can you suggest for the form of the relationship and what experiments would you make to test your suggestions?   *O.C.B.(S.)*

18. By what processes do (a) water, (b) inorganic salts enter the roots of a flowering plant? To what extent are these processes independent of one another?   *C.B.*

19. Living organisms may be considered to be clearing houses for energy exchange. Discuss this statement.   *B.Bi.(S.)*

20. What are the effects of pressure and light on the lives of animals?   *C.Z.*

Chapter 4

The Chemical Nature and Importance of the Commoner Plant and Animal Substances

21. What types of substance are included under the term carbohydrate? Describe the part played by carbohydrates in the metabolism of the flowering plant.
          O.C.Bi.

22. What are proteins? Explain their importance in the plant and indicate how they are thought to be formed from the raw materials absorbed by the plant from its environment.
N.B.(S.)

23. Discuss the importance of *either* carbohydrates *or* proteins in the metabolism and structure of plants.
B.B.(S.)

24. What do you know about the chemical nature of nucleic acids? Give an account of their known and probable functions in animals.
O.Z.(S.)

25. What is the chemical nature of fats, and what are their characteristic properties? Describe the part played by fats in the plant kingdom.
C.B.(S.)

26. What are organic compounds and why are they so called? Why is a knowledge of inorganic compounds as well as organic compounds necessary for the biologist?
C.Bi.(S.)

CHAPTER 5

The Nature and Properties of Enzymes

27. Briefly discuss the nature and role of enzymes. Explain how you would demonstrate in a given tissue the presence of enzymes which carry out an oxidation reaction.
W.Z.

28. What factors would you expect to influence the speed of an enzyme action? Describe in detail an experiment you could perform to determine quantitatively the effect of one of these factors in a *named* reaction.
O.C.B.(S.)

29. Describe the properties of plant enzymes, illustrating your answer with *three* named examples. Describe in detail how you would proceed to test for the presence of *one* of them in a potato.
O.C.Bi.(S.)

30. You have been asked to investigate whether the midgut of a newly discovered insect contains enzymes capable of digesting (*a*) sucrose, (*b*) maltose and (*c*) a simple protein such as gelatin. Write a concise account of the experiments you would perform giving an adequate explanation of the procedures.
W.Z.

31. What is an enzyme? Describe its characteristic properties. What part is played by enzymes in the germination of seeds?
W.B.

32. Outline what you know of the part played by enzymes in the metabolism of animals.
W.Z.(S.)

33. How would you proceed to show experimentally that a given solution contains a proteolytic *or* a lipolytic *or* an amylolytic enzyme?
W.Z.

34. Write an essay on "Biological catalysts."
C.Bi.

CHAPTER 6

The Nature of the Environments of Living Things: Adaptation: Homeostasis and its Maintenance

35. Discuss the special problems of higher animals and plants living in an aquatic environment.
B.Bi.(S.)

36. Compare and contrast the life conditions of freshwater algae with those of marine algae. By reference to *two* of the algae you have studied, show how freshwater and marine algae are fitted to their respective environments.
W.B.(S.)

37. What are the effects of plant and animal life on the composition of soil?
C.Bi.

38. Indicate briefly how you would determine (*a*) the water content, (*b*) the air content, (*c*) the humus content and (*d*) the pH (acidity) of a sample of soil. How may the humus content and the pH vary in different kinds of soil?
L.B.

39. Choosing *three* of the following pairs, write concise notes, illustrated where appropriate, to contrast:
(*a*) podsol and brown earth, (*b*) inorganic fertilizers and farmyard manure, (*c*) clay and sand, (*d*) peat and humus.   *N.B.*

40. Compare and contrast the properties of a sandy soil, a fertile loam and a clay soil. Discuss briefly the importance of soil in relation to the distribution of natural vegetation.   *B.B.(S.)*

41. What is humus? Explain how humus is formed and also explain its importance in the maintenance of the fertility of the soil.   *N.B.*

42. Describe those soil conditions which promote the healthy growth of plants. In what ways can man improve poor soils?   *O.C.B.(S.)*

43. Describe how the properties of a soil are related to the size of the particles composing it. How may man alter the texture of soils?   *O.C.B.*

44. What conditions are necessary for the maintenance of a healthy soil micro-flora? What contribution, if any, does the micro-flora make to food-production?   *O.C.Bi.*

45. By reference to particular physiological functions compare land and water as environments for supporting life. Most biologists believe that life originated in the sea rather than on land. Why do you think this assumption is made? *A.E.B.Bi.*

46. With reference to (*a*) availability of water, (*b*) availability of inorganic salts, (*c*) soil pH, give an account of the soil and its importance in plant growth.   *N.Bi.*

47. A farmer submits to you for analysis a sample of soil. What information do you consider should be included in your report to him? Outline the procedures you would use in carrying out the necessary tests and analyses.   *N.B.*

48. Write an account of the relationship between environmental conditions and reproduction in *either* land animals *or* marine animals *or* freshwater animals.   *N.Z.*

49. Give an account of the object and the methods used by animals, to maintain a constant internal environment.   *L.Z.(S.)*

50. List the conditions in the external environment which are necessary for the continuance of life. Explain briefly, in each case, why such conditions are essential.   *N.Bi.*

51. Give an account of the ways in which a mammal is able to maintain a steady body temperature, irrespective of its surroundings.   *N.Z.*

52. What is phytoplankton? Give an account of its occurrence and of its importance in nature.   *L.B.(S.)*

53. The sugar content of the blood of a mammal is maintained at a fairly constant level. Explain (*a*) what constituents of the diet give rise to this sugar, (*b*) how the sugar is made available from the food, and (*c*) how the blood sugar level is controlled.   *C.Z.*

Chapter 7

Water Relationships of Living Organisms

54. Explain how the aerial parts of a flowering plant lose water. How would you measure the rate of water loss from a cut shoot under different environmental conditions?   *L.B.*

55. Explain concisely why a herbaceous plant may wilt during the afternoon of a summer's day and regain its freshness by the following morning.   *L.B.*

56. Give a general account of the structure, functions and mechanism of stomata.   *W.B.(S.)*

57. Distinguish between transpiration and the transpiration stream. How would you measure the rate of transpiration from (*a*) a cut shoot, (*b*) a potted plant?   *W.B.(S.)*

58. What forces are involved in the absorption of water and its passage through plants? State briefly why large amounts of water are required by most land plants.
B.B.

59. Explain how an increase in the rate of air movement over the leaves of a plant can lead to an increase in the rate of water absorption by its roots. *L.B.*

60. State briefly what changes occur in the tissues of a leaf when it wilts. What factors external to the plant are liable to lead to wilting? *O.C.Bi.*

61. In what respects is transpiration dependent on movement of water into and out of cells? *O.B.*

62. Discuss the statement: "Transpiration in plants is a necessary evil." State what is measured by the "cobalt chloride paper method." *O.C.B.(S.)*

63. Consider the properties of water which make it so important to plants.
L.B.(S.)

64. Discuss the relative importance of water in the plant and animal kingdom.
C.Bi.(S.)

65. How are the chemical and physical properties of water made use of in living systems? *A.E.B.Bi.*

Chapter 8

Co-ordination

66. Discuss the importance of co-ordinating mechanisms in living organisms.
B.Bi.(S.)

67. Write a short account of the functions of each of the following: (a) thyroid, (b) pituitary, (c) adrenals, (d) islets of Langerhans. *N.Z.*

68. Give a list of those structures in the mammalian body *known* to have an endocrine function. What bodily functions do their secretions regulate and what disorders are known to be associated with a deficiency of endocrine secretion? *N.Z.*

69. What part is played by the hormones of the thyroid, adrenals and gonads in the development and adult life of a mammal? *O.C.Z.*

70. How would you expect young, green, terrestrial, higher plants to respond:
   (a) if given extra auxin;
   (b) if given extra gibberellin;
   (c) if auxin action were totally inhibited?
Give reasons for your answers. *O.C.Bi.*

71. Describe the part played by pituitary hormones in the co-ordination of the activities of mammals. Briefly indicate the nature of the experimental techniques utilized to investigate the action of pituitary hormones. *N.Bi.*

72. Describe the structure of a typical neuron, and show how neurons are interconnected to form reflex arcs. Give an account of the way in which a nerve impulse is initiated and transmitted, and of the processes which occur at the junctions between neurons and between a neuron and an effector. *N.Bi.*

73. Describe concisely the means by which a mammal co-ordinates its various vital activities. *C.Bi.(S.)*

74. Describe the respective roles of nervous and chemical co-ordination in the integration of the vertebrate body. *L.Z.*

Chapter 9

Translocation of Substances within the Organism

75. Xylem and phloem are tissues concerned with the translocation of inorganic and organic materials. For each tissue describe the possible translocation mechanisms involved and comment on any structural adaptations that might be specially related to translocation. *N.Bi.*

76. Some of the carbohydrate formed in photosynthesis in potato leaves is ultimately stored in the form of starch grains in the tubers. Describe the physiological processes involved in the transport and conversion of the carbohydrate.   *O.C.B.*

77. Discuss the importance of transport systems in the lives of flowering plants.   *C.Bi.(S.)*

78. Describe *two* experiments that show translocation of organic substances is taking place in a plant. What theories have been put forward to account for the mechanism of translocation.   *A.E.B.Bi.*

79. Compare the transport systems found in mammals and flowering plants.   *B.Bi.(S.)*

80. Explain how the structure and operation of the mammalian heart enable it to function as a pump in the circulatory system. What effects do the aorta, other arteries, capillaries and veins have on the flow of blood?   *L.Z.*

CHAPTER 10

Nutrition

81. What mineral elements are needed for the healthy growth of plants? Describe, with experimental details how you would show the necessity of *one* named element and indicate the part played by this element in plants.   *B.Bi.*

82. What are the essential elements obtained from the soil by a green plant? Discuss their specific roles in plant nutrition?   *A.E.B.B.*

83. What is meant by the "nitrogen cycle"? Discuss the factors affecting the supply of nitrogen to farm crops.   *O.C.Bi.*

84. Describe with full practical details an experiment to show that nitrogen is important for healthy growth in a green plant. Discuss the importance of the following processes to the life of the green plant: nitrogen fixation, nitrification.   *C.Bi.*

85. What do you understand by "trace elements?" Give examples. How would you demonstrate the importance of one of these to the plant?   *C.B.(S.)*

86. What is photosynthesis? How would you, using simple experimental methods, demonstrate the essential features of the process?   *B.Bi.*

87. What is understood by (*a*) the "end-products" of photosynthesis, and (*b*) light and dark reactions.   *A.E.B.*

88. The relationship between the rate of photosynthesis and the light intensity was determined for two species with the results given below. Comment on these results. What bearing do they have on the ecology of the species concerned?

Species	Light intensity (arbitrary units)					
	0·5	1·0	2·0	3·0	5·0	10·0
	Rate of photosynthesis (arbitrary units)					
Oxalis acetosella	0·5	1·0	2·1	2·2	2·3	2·2
Pteridium aquilinum	0·5	0·9	2·1	3·2	4·2	5·0

O.C.B.

89. Write a short account of photosynthesis. By what experiments would you show that a product of photosynthesis accumulates in the leaf during daylight and is carried away by night?
W.B.

90. Describe, with full experimental details, how you would show that carbon dioxide is necessary for photosynthesis. What factors may affect the rate of photosynthesis under natural conditions?
B.Bi.

91. Trace the path of a carbon atom into a photosynthetic cell in a leaf and its subsequent fixation assuming it becomes incorporated into carbohydrate. List the possible fates of such a carbohydrate molecule.
A.E.B.Bi.

92. Describe the experiments that you would carry out to convince a person without scientific training that the starch which accumulates in the tuber of a potato cannot be formed without a supply of carbon dioxide in the atmosphere.
L.B.

93. What do you know of the nature, distribution, and functions of chlorophyll in plants, and how would you demonstrate experimentally that it is necessary to any specific metabolic process?
W.B.

94. Give some account of protein synthesis in the green plant.
W.B.(S.)

95. Give an account of the nitrogen metabolism of a typical flowering plant.
C.B.(S.)

96. What is meant by a balanced diet? Indicate briefly what special features you would prescribe in the diets of the following individuals: a growing child; a pregnant mammal; a coal miner.
O.C.Z.

97. Compare the methods by which food is ingested in *Amoeba, Paramecium, Obelia, Planaria, Periplaneta* or *Blatta*, and *Branchiostoma (Amphioxus)*.
O.Z.

98. How are carbohydrates digested and utilized by the human body?
C.Z.

99. Describe the chemical constitution of fats, and write an essay on the digestion, absorption and utilization of these substances by a mammal.
N.Z.(S.)

100. Describe what happens to the products of digestion in a mammal from the time of their passage through the gut walls up to and including their storage, utilization or removal from the body.
B.Bi.

101. Account for the necessity in the diet of a mammal of (*a*) water; (*b*) calcium; c) iodine; (*d*) iron; (*e*) sulphur.
O.C.Bi.

102. What are amino-acids? What is their importance in metabolism? How does (*a*) a green plant, (*b*) the growing parts of a germinating seed, and (*c*) a mammal, obtain a supply of amino-acids?
C.Bi.

103. Give a concise account of the structure and properties of proteins. Summarize the functions of proteins in living organisms. Briefly indicate the means by which specific proteins are believed to be synthesized within cells.
N.Bi.

104. Give an account of the digestion, absorption and subsequent fate of protein in mammals.
L.Bi.

105. What is meant by a balanced diet? What specializations have been evolved by herbivorous animals to enable them to get the maximum nourishment out of what they eat?
N.Z.

106. The following is an outline classification of animal feeding methods:
   (*a*) Mechanisms for dealing with small particles (including filter feeders).
   (*b*) Mechanisms for dealing with large particles.
   (*c*) Mechanisms for taking in fluids.
Discuss the classification and *name* an animal from each class, giving details of its adaptation for the particular mode of feeding.
A.E.B.Bi.

107. What is the difference between intra-cellular and extra-cellular digestion? Illustrate your answer by reference to *Hydra* and a mammal.
B.Z.

CHAPTER 11

Respiration

108. Explain carefully what you understand by respiration. How is respiration influenced by external conditions? How would you examine experimentally the effect of one of these conditions on respiration?   *N.B.(S.)*

109. What is respiration and whereabouts in green plants does it occur? How could you demonstrate that respiration was proceeding in a small leafy shoot detached from a living plant?   *O.C.Bi.*

110. What aspects of aerobic respiration, as carried out by green plants, have you seen demonstrated? With the aid of annotated sketches, briefly describe three demonstrations that illustrate aspects of aerobic respiration.   *C.B.*

111. How is energy liberated in living plants?   *L.B.(S.)*

112. Give a brief account of the process of aerobic respiration in a higher plant. How would you, using simple apparatus, demonstrate the nature of the gaseous exchanges involved in respiration and measure their extent?   *B.B.*

113. Distinguish between "aerobic" and "anaerobic" respiration. Describe the apparatus you would use to measure the rates of production of carbon dioxide and of consumption of oxygen by germinating pea seeds.   *O.C.B.*

114. Describe how you would show that plants may respire in the absence of oxygen. Explain briefly why oxygen is necessary for the continued growth of most plants.   *C.B.*

115. Explain *in general terms* the stages involved in the release of energy from starch stored in a plant organ. Explain why temperature may influence the process.   *N.B.*

116. Criticize and amplify the following equation usually given to summarize the chemical reactions associated with aerobic respiration of plants

$$C_6H_{12}O_6 + 6O_2 = 6CO_2 + 6H_2O$$

L.B.(S.)

117. Describe the processes involved in the production of the energy necessary for (a) active sprout growth in a potato tuber, (b) the "budding" of a cell of yeast.   *C.B.*

118. Outline the features of respiration which are common to all living organisms.   *W.Bi.*

119. Write notes on the following to show what part each plays in the release of energy in living cells: (a) adenosine triphosphate (ATP), (b) glucose, (c) hydrogen acceptors, (d) mitochondria. (*Details* of biochemical action are not required.)   *N.Bi.*

120. Aerobic respiration involves three major steps: (a) glycolysis, (b) the Krebs cycle and (c) oxidation. Explain *in outline only* what each of these steps involves, and how energy is obtained from them. Why does anaerobic respiration of a given substrate result in a smaller amount of energy than aerobic respiration of the same substrate?   *N.Bi.*

121. Write a brief account of respiration in green plants. "Respiration in plants does not differ from that in animals." Discuss this statement.   *C.Bi.*

122. Summarize the successive stages of respiration in a mammal. What is the relation between respiration and excretion?   *C.Z.*

123. Explain how animals obtain energy.   *O.C.Z.*

124. Distinguish between respiration and breathing. Give an account of the anatomy of the structures involved in the process of breathing in a mammal and indicate briefly the physiological mechanism whereby this process is regulated.   *N.Z.*

125. Distinguish between breathing and respiration. Describe the process of oxygen uptake in one terrestrial and one aquatic animal.   *O.C.Z.*

126. What are the essential features of a respiratory surface? Show how these features are present in a frog at all stages in its life-history. *W.Z.(S.)*

127. What is meant by tissue respiration, anaerobiosis and oxygen debt?

C.Z.(S.)

Chapter 12

Locomotion

128. Give an account of the different methods of locomotion in three named unicellular organisms. *B.Bi.*

129. Describe (*a*) the gross structure, (*b*) the microscopic structure, and (*c*) the ultra-microscopic structure of striated muscle.

Give a short account of modern views concerning the way contraction occurs, in terms of molecular interactions. *N.Bi.*

130. Classify the methods of locomotion used by invertebrate animals. Describe the way in which one locomotor organ functions. *O.C.Z.*

131. There is a close correlation between the anatomy of an animal and its mode of locomotion. Illustrate this by special reference to *Amoeba, Obelia, Lumbricus* and *Scyliorhinus*. *O.Z.*

132. Outline the processes by which the chemical energy of food is converted into the mechanical energy of muscular exercise. *C.Z.(S.)*

133. Write an essay on "Animal Locomotion." *O.C.Bi.*

134. If you were provided with freshly beheaded frogs describe how you would set up and conduct laboratory experiments to investigate the responses of vertebrate striated muscle to stimulation. Make a labelled diagram (or diagrams) of the apparatus you would use and indicate the purpose of each component.

Explain what is meant by each of the following in relation to muscle contraction: summation, refractory period, tetanus, fatigue. How would you recognize their separate occurrence in your experiments? *N.Bi.*

135. Describe the histology of striated muscle and its chief physiological properties. How does the brain cause a specific muscle, for example, the biceps, to exert differences in power according to its load? *A.E.B.Bi.*

136. 'An animal can only propel itself forward by pushing backwards against its surroundings." Show how this is brought about by *Paramecium*, a planarian, a snail and a bird. *S.Z.*

137. Describe the mechanics of locomotion in the Mammalia. *C.Bi.(S.)*

138. Describe briefly the histological structure of the striated muscles of a mammal. How do these muscles bring about the locomotion of the animal? Comment in simple terms on the chemical activities that take place in these muscles when they are active. *C.Z.*

139. Write an account of the ways in which Protozoa move. Indicate wherever possible the relation between movement and feeding. *S.Bi.*

Chapter 13

Growth and Development

140. Growth has been defined as an "increase in dry weight." Criticize this definition in the light of the knowledge gained from your study of plants. *L.B.*

141. In what ways would you study the growth of pea or bean seedlings during the first four weeks from the beginning of germination? Discuss concisely the relative merits of the methods you would adopt. *L.B.*

142. What do you understand by growth in plants? Illustrate your answer by reference to growth in (*a*) a *Spirogyra* filament, (*b*) an angiosperm shoot.

W.B.(S.)

143. Describe experiments you might attempt to carry out to determine the factors that affect the opening of the buds of trees in spring.   *L.B.*

144. Describe the physiological changes in a *named* cereal grain from its dormancy to the development of the first leaf. Give a drawing to show the external appearance of the seedling at this stage.   *C.Bi.*

145. What are enzymes? Give an account of the role of enzymes in the germination of seeds.   *W.B.*

146. Describe experiments which you could perform to investigate the growth of a root, a stem and a leaf over a short period of time. State concisely the kinds of results which you would expect to get.   *S.B.*

147. Explain how the reproductive phase in flowering plants is influenced by external conditions.   *L.B.(S.)*

148. Give an account of dormancy in plants. Compare its significance with that of sleeping and hibernation in animals.   *O.C.Comb.Z. & B.III*

149. Give an account of the effects of day-length on vegetative development and reproduction in plants, distinguishing between "long-day" and "short-day" responses.   *W.B.*

150. Explain what is meant by metamorphosis. In what groups of animals does it occur?   *C.Z.*

151. Give an account of Regeneration in the animal kingdom.   *O.C.Z.*

152. It has been stated that "regenerative capacity decreases with the increased complexity of the organism." Do you agree with this statement? Give reasons for your opinion.   *W.Z.(S.)*

153. Compare the means by which each of the following may pass through a period of dormancy: (*a*) *Hydra;* (*b*) *Cystopus;* (*c*) *Taenia;* (*d*) a *named* perennial flowering plant.   *N.Bi.*

154. Give a general account of growth in plants and animals.   *B.Bi.(S.)*

155. Give an account of the gross and microscopical changes of metamorphosis in *Rana*. What factors are known to control metamorphosis in Amphibia?   *O.Z.(S.)*

155 (*a*). As part of a study of population growth an investigator introduced an equal number of flour beetles (*Tribolium confusum*) into two environments which were equal in every respect except that the first environment contained only 16 grammes of flour (food) and the second 64 grammes of flour. At intervals of time after the introduction the number of individual beetles was counted in each environment. The table below gives the results:

Days after introduction of Tribolium	Approximate number of individuals present in	
	16 g environment	64 g environment
0	20	20
7	20	20
40	200	300
60	550	800
80	560	1300
100	650	1750
120	640	1600
135	650	1900
150	645	1500

(*a*) Plot these results on a graph. (7 marks.)
(*b*) Discuss the results. (13 marks.)   *O.C.Bi.*

CHAPTER 14

Secretion and Storage

156. What do you understand by the term *secretion*? Name three examples of different organs in which secretory cells are found and describe the role of their secretions.
L.Z.

157. Write an essay on extra-cellular materials in the metazoan body. *O.C.Z.*

158. Give an account of the chief food-reserve substances present in (*a*) an onion bulb; (*b*) a potato tuber; (*c*) a carrot. Describe the essential changes these substances undergo in the life of the plant.
O.C.B.

159. What two main purposes are served by the storage of food reserves by living things? Write a short account of food storage in plants and in animals. *N.Bi.*

160. Give an account of the chemical nature and occurrence of the main reserve food materials in plants. Describe fully how you would test for the presence of any *three* of these substances.
S.Bi.

CHAPTER 15

Excretion

161. What is excretion? Compare the methods of excretion in *Amoeba*, the earthworm and a mammal.
O.C.Bi.

162. Give a labelled drawing of a complete kidney tubule, together with its blood supply.

In cases of severe bleeding no urine is formed. Why is this?

Sea water contains about 3 per cent salt, the salt content of blood is about 1 per cent and, as a maximum, the kidneys can excrete only a 2 per cent salt solution. Give a brief explanation for the serious consequences that occur after drinking a great deal of sea water.
C.Bi.

163. Explain the necessity for an excretory system in animals. Choose examples to contrast the excretory processes in aquatic and terrestrial animals. *C.Z.(S.)*

164. Write a comparative account of excretion in the invertebrates you have studied.
C.Z.(S.)

165. What is the source of "nitrogenous waste" in animal tissues? Explain how this is eliminated in *Hydra*, the earthworm and an insect. *N.Z.*

166. Why is "nitrogenous excretion" so important to animals? Give an account of this process in a mammal.
N.Z.

167. Give a diagram of a uriniferous tubule and discuss its functions. Distinguish between excretion and osmoregulation.
W.Z.

168. By reference to examples with which you are acquainted, distinguish carefully between excretion and secretion.
W.Z.

169. Excretion has been defined as the process of maintaining a constant internal environment. Outline the excretory problems associated with the marine, freshwater, and terrestrial environments, and give some examples of the ways in which different animals have overcome them.
(C.Z.(S.)

170. Describe the excretory (urinary) system of a named mammal and give a simplified account of the action of the urinary tubules. Name the principal substances excreted from the kidney.
C.Bi.

171. How does the composition of blood in the renal vein of a mammal differ from that in the renal artery? Explain the part played by the kidney in bringing about these differences.
L.Bi.

CHAPTER 16

Sensitivity and Response to Stimulation: Behaviour

172. In what ways may plants respond to external stimuli? Describe in detail the mode of any one such response. Is this response biologically advantageous?
B.B.(S.)

173. What is a tropism? How could you show experimentally that *either* a bean root is positively geotropic *or* a cereal coleoptile is positively phototropic? Comment on the biological importance of tropisms. *B.B.*

174. Illustrate, by examples, the meaning of the terms *stimulus* and *response*, when they are applied to plant movements. *List* the stimuli to which plants are known to respond. *L.B.*

175. Describe *two* experiments you would perform, *one* to demonstrate phototropism and *one* geotropism. What is known of the physiological mechanism governing these movements? *O.C.B.*

176. How would you investigate experimentally the nature and cause of the downward curvature which occurs in the radicle of a bean seedling when it is placed horizontally? *W.B.*

177. Without elaboration of detail write an account of the ways in which light affects flowering plants. *L.Bi.*

178. Give an account of experiments which have been performed to demonstrate the effect of auxins on plant growth. *W.Bi.*

179. Explain and exemplify the statement that irritability is advantageous to plants. *L.B.(S.)*

180. What influence has gravity upon the growth of the various parts of flowering plants and what is the value to them of their reactions to this environmental factor? Outline simple experiments which demonstrate the principal features of the perception of gravitational stimuli and of the response. *S.Bi.*

181. How would you demonstrate the presence of growth hormones in higher plants? What are their functions? *C.B.(S.)*

182. Explain how the reproductive phase in flowering plants is influenced by the environment. Describe briefly a series of experiments you might make to determine the conditions necessary for flowering. *C.B.(S.)*

183. Explain as far as you can three of the following observations:

(*a*) When the apical growing point of a herbaceous plant is removed, lateral buds often develop into branches;

(*b*) Many trees do not flower abundantly every year;

(*c*) A solitary pear tree frequently fails to set fruit, but can be made to do so by spraying with a hormone solution;

(*d*) sowing seeds derived from a good rose tree rarely produces "worthwhile" rose trees. *C.B.(S.)*

184. Give a detailed account, with diagrams, of the sense organs, excluding the eye and the ear, in a named mammal. *C.Bi.(S.)*

185. What is the relative importance of instinctive behaviour and of learned behaviour in the animal world. *C.Z.(S.)*

186. Describe the general sensitivity of animals to the external environment. Of what advantage is this to them? *C.Bi.(S.)*

187. Give a general account of the means by which hormones play a part in controlling and co-ordinating activities in living organisms. Illustrate your answer by reference to *two* organisms, one animal and one plant. *N.Bi.*

188. Define the term "neuron". Explain the mechanism of a "simple reflex" and a "conditioned reflex". Show how these phenomena are basic to an understanding of the behaviour of metazoan animals. *N.Z.*

189. Explain the means by which an animal (a) hears, and (b) detects, the presence of objects on its skin. *L.Bi.*

190. Balance may be described as the maintenance of position by an animal. How is balance controlled in a crayfish, a dogfish and a rabbit? Why is balance important to these animals? *L.Z.*

191. List the sensations which may be produced by external stimuli applied to the skin of a mammal. By what mechanisms does the animal become aware of, and react to, such changes in its external environment. *S.Bi.*

192. How are any animals which you have studied made aware of the external conditions in their environment? *W.Z.(S.)*

193. What are receptors and what are their functions? Classify the different types of receptor you have studied in vertebrate and invertebrate animals. Give a concise account of either (a) sound receptors or (b) receptors sensitive to position and movement. *N.Bi.*

194. From *your own observations* give an account of the behaviour of any *named* animal which you have studied. *L.Z.*

195. Write "An introduction to the study of Animal Behaviour." *O.C.Z.(S.)*

196. Give an account of the general principles involved in the production of movement in plants and animals. *B.Bi.(S.)*

CHAPTER 17

Reproduction and Sex

197. Write a general account of the ways by which plants may reproduce. Consider the biological importance and the relative success of the modes of reproduction you mention. *B.B.(S.)*

198. Outline the behaviour of the nucleus in the maturation of the germ cells. *C.Z.*

199. Give an account of the processes of spermatogenesis and oogenesis, discussing the differences between them. Comment on the significance of meiosis. *O.C.Bi.*

200. Briefly describe the structure, origin, and formation of a spermatozoon. Briefly compare and contrast the structure and formation of an egg with those of a spermatozoon and account for the *differences* shown. *L.Z.*

201. Describe the characteristics of sexual reproduction that are common to those animals you have studied and show how variations in such characteristics can be correlated with different modes of life. *O.C.Bi.*

202. Compare and contrast the advantages of sexual and asexual reproduction. Show by examples, the types of animals in which these methods of reproduction are found, and comment on the special case of parthenogenesis. *C.Z.(S.)*

203. By reference to such Protozoa as you have studied, illustrate the essential nature of sexual and asexual reproduction. *N.Z.*

204. What are the essential differences between sexual and asexual reproduction and between their possible results? *W.Z.*

205. Consider the importance of fertilization from an evolutionary point of view. *C.Bi.(S.)*

206. Discuss the process and significance of sexual reproduction. *C.Z.(S.)*

207. What are the differences between asexual and sexual reproduction and what are the advantages and disadvantages of these two processes? *A.E.B.Z.*

208. Give an account of the reproductive cycle of a named female mammal and show how this is controlled by the animal's endocrine secretions. *W.Z.*

CHAPTER 18

Variation

209. If a large number of plants of a particular species in one locality are measured for a character such as size or weight it is frequently found that the observations follow the type of curve shown in the accompanying graph. Suggest explanations of this result and state what further experiments you could carry out to test the truth of your suggestions.

Graph: normal distribution

O.C.Bi.(S.)

210. In what ways may plants belonging to one and the same species differ from one another? What is the biological significance of the kinds of variation you mention? C.B.(S.)

211. Give an account of the nature of the variation found in plants with special reference to its importance to the survival of (a) the individual, and (b) the species, over long periods of time. O.B.(S.)

212. What do you understand by the term *variation* applied to animals? What is the significance of variation in evolutionary theory? O.Z.(S.)

213. Some species of flowering plant are found to have different variants, each characteristic of a different type of habitat. Describe the experiments you would do in order to investigate this phenomenon. C.B.(S.)

214. What types of variation increase the chances of survival of an individual? Illustrate your answer by reference to specific examples. C.Bi.(S.)

215. De Vries determined the number of ray florets on the topmost "flower" of the main stem of 97 *Chrysanthemum segetum* derived from a mixed seed sample with the following results:

No. of ray florets	12	13	14	15	16	17	18	19	20	21	22
Frequency	1	14	13	4	6	9	7	10	12	20	1

Seeds from the 12- and 13-rayed plants were sown and comparable determinations were made on the resulting plants with the following results:

No. of ray florets	8	9	10	11	12	13	14	15	16	17	18	19	20	21
Frequency	2	1	0	7	13	94	25	7	7	1	2	0	3	0

Comment on the experiment. What conclusions would you draw from the results? O.C.B.(S.)

216. During a short walk straight across a small area of woodland a collection was made of fallen acorns (fruits of oak trees) from a strip two or three yards wide. The lengths of fifty acorns are given below.

Use the data given, together with your own botanical knowledge, to answer the following:

(a) Draw a graph or histograms to show variation in length.

(b) What conclusions can you reach?

(c) What other steps or information would be necessary to confirm these conclusions?

Lengths (cm)

1·8	2·0	2·4	2·3	2·2	2·5	2·3	1·9	2·0	1·5
1·9	2·3	2·0	1·8	1·7	2·0	2·2	2·0	2·3	2·2
2·4	1·7	2·3	1·9	1·9	2·3	1·8	2·1	2·4	2·0
2·1	2·5	1·9	2·2	2·3	2·1	2·4	1·9	2·2	2·3
2·0	2·2	2·3	2·6	2·0	1·9	2·0	2·2	1·9	2·1

C.B.

216 (a). The number of carpels per flower was determined for a hundred plants chosen at random from each of two spatially-isolated populations of the white water lily (*Nymphaea alba*) with the results given below. Comment on the variation observed and on the taxonomic difficulties presented by such variation. How would you attempt to determine the causes of the intra- and inter-population differences?

Number of carpels per flower	No. of plants	
	Population I	Population II
9	10	—
10	32	—
11	45	—
12	9	5
13	2	9
14	2	10
15	—	9
16	—	21
17	—	20
18	—	11
19	—	10
20	—	5

O.C.B.

CHAPTER 19

Genetics

217. What chief facts were discovered by Mendel in his study of inheritance? Illustrate your answer by reference to two pure lines which differ from each other in *two* characters. Consider the result of crossing these pure lines and the production of the subsequent F_2 generation by self-fertilization of the offspring.   *B.Bi.*

218. Explain the essentials of Mendelian inheritance. Why was Mendel more successful than his predecessors in his investigations concerning heredity?   *L.B.*

219. Give a detailed and illustrated account of the prophase of meiosis showing how genetic material becomes interchanged between members of homologous chromosomes. How can you relate this knowledge to the fact that hybrids between closely related species (e.g. the Tigon from the lion × tiger cross) are normally sterile?
A.E.B.Bi.

220. State Mendel's laws and give an example to illustrate each law. In what way did the subsequent discovery of linkage necessitate a modification of the laws?
O.C.B.

221. State the Laws of Inheritance as postulated by Mendel. How far has the validity of these laws been affected by modern studies of the mechanism of nuclear division?
W.B.(S.)

222. How far have Mendel's conclusions been borne out by subsequent discoveries?
C.B.

223. Give a brief statement of the principles of Mendelian segregation. Explain what is meant by linkage and crossing over and show how the anticipated results of a cross may be affected by these phenomena.
O.B.

224. Explain concisely the following terms: gene, heterozygote, diploid, hybrid vigour, pure line.
L.B.

225. *Either*— describe the contributions made by *three* of the following to genetics: De Vries, Bateson, Johannsen, Morgan *or* explain concisely *five* of the following terms: pure line, backcross, outbreeding, hybrid vigour, lethal character, mutation, allelomorph (allele).
L.B.

226. Define and give examples of each of the following genetical terms: (*a*) gene mutation; (*b*) dominance; (*c*) linkage. In what ways other than by gene mutation may other heritable differences arise?
O.C.B.

227. Write briefly on four of the following: heterozygote; allelomorph; haploid; linkage; independent assortment.
O.C.Bi.

228. Write short notes on (*a*) chromosome, (*b*) dominant character, (*c*) phenotype, (*d*) mutation, (*e*) allelomorph.
B.Z.

229. Distinguish, with illustrative examples, between the terms in each of the following pairs: (*a*) sperm and ovum; (*b*) chromosome and chromatid; (*c*) homozygote and heterozygote; (*d*) dominant and recessive; (*e*) genotype and phenotype.
W.Z.

230. How does our knowledge of the process of nuclear division help to explain the principles of Mendelian inheritance?
L.B.

231. What is the evidence that genes are (*a*) located on the chromosomes; and (*b*) arranged in a linear order along the chromosomes?
O.C.B.

232. What is crossing-over and how may it be detected (*a*) in breeding experiments, (*b*) in stained preparations of cell division? What is its importance in connexion with sexual reproduction?
O.C.B.(S.)

233. What is the evidence for the view that the nucleus is of great importance in the study of inheritance in plants?
C.B.(S.)

234. What are chromosomes and why are they important?
C.Z.

235. Explain how the germ cells in a metazoan afford a connexion between previous and subsequent generations.
W.Z.(S.)

236. How may a new character arise and spread in a wild population? To what extent does the size of the population concerned affect the rate of spread of the new character?
A.E.B.Bi.

237. What is the mechanism of sex determination in mammals? Account for the phenomenon of sex-linked inheritance.
S.Z.

238. Describe in detail experiments that you could perform without assistance to discover the mode of inheritance of flower colour in a species of annual plant displaying two colours.
O.C.Bi.

239. How far do you consider a species to be a distinct genetical unit? Discuss how new species may arise. *S.B.*

240. Give an account of the mode of inheritance of a sex-linked character (e.g. "white eye" in *Drosophila*). *N.Z.*

241. What is meant by linkage? How is linkage detected? A homozygous plant containing genes *A, B, C*, was crossed with another homozygous plant with corresponding but different genes *a, b, c*. The offspring produced gametes in the following proportions.

ABC	20 per cent	*AbC*	5 per cent
abc	20 per cent	*Abc*	5 per cent
ABc	20 per cent	*aBC*	5 per cent
abC	20 per cent	*aBc*	5 per cent

Do the data in the table indicate that genes *A, B, C*, are linked? Indicate clearly the reasoning on which you base your conclusions. Show by means of a diagram how gametes of type *AbC* and *Abc* were formed. *N.B.*

242. Flowers of a plant growing in an experimental plot were emasculated and pollinated from a dwarf member of the same species growing on a nearby gravel path. The resulting seeds were sown in the experimental plot, and all those that germinated gave rise to plants as tall as the one from which the seeds were collected. Discuss possible explanations for this result. Describe exactly how you would attempt to determine which explanation is correct. *C.B.*

243. In the rabbit, the gene for black fur is dominant to that for brown fur and the gene for short fur is dominant to that for long hair (Angora). Describe the appearance of such offspring as can be obtained by mating a rabbit with short black fur with a brown angora. Explain carefully and fully the theory underlying the results you enumerate. (It may be assumed that there is no linkage between the characters concerned, neither is sex-linkage present.) *N.Z.*

244. A tall plant with red flowers from a true-breeding line was crossed with a homozygous short plant with white flowers. One of the resulting plants was crossed with a short, red-flowered plant of unknown parentage. This cross gave 109 short white, 36 tall red, 29 tall white and 100 short red plants. Interpret these results. What was the phenotype of the plant produced by the first cross? *O.C.B.*

245. By means of clear diagrams and notes, show how a heterozygote *AaBb* can produce a single tetrad containing four types of spore, *AB, Ab, aB, ab*, when

(i) *A* and *B* are on the same chromosome, with *a* and *b* on its homologous partner;

(ii) *A* and *B* are on different, non-homologous chromosomes with *a* and *b* on their respective partners.

What affects the frequency of production of this type of tetrad in the two cases.

O.C.B.

246. What are mutagenic agents? Discuss the nature and significance of mutations. *W.B.*

247. A tomato plant with red fruits and smooth stems is crossed with another variety having yellow fruits and hairy stems. All the progeny in the F_1 generation have red fruits and hairy stems. Explain this. Show, with the aid of a diagram, the expectation of a yield of 2,000 plants in the F_2 generation arising from selfing the F_1 plants. *W.B.*

248. Explain the meaning of the terms *linkage* and *sex-linkage*. Show how haemophilia is exhibited by males but is transmitted through the female line. *W.Bi.*

249. Define phenotype and genotype. In the fruit fly, *Drosophila*, wild-type animals have red eyes and normal wings, while purple eyes and vestigial wings are

recessive mutations. In a laboratory experiment, wild-type flies were crossed with flies having purple eyes and vestigial wings and the F_1 progeny allowed to interbreed at random. Of the 784 flies obtained in the F_2 generation, 149 had red eyes and vestigial wings. Show that this result agrees closely with the theoretical value expected according to Mendel's second law.   *W.Bi.*

250. Snapdragons with red flowers of normal shape *RRNN* when crossed with those with white abnormal (peloric) flowers *rrnn*, give pink flowers. Construct a chequer board to give the genotypes of a cross between

    (*a*) the phenotypically pink normals of the F_1 generation;

    (*b*) a backcross of the F_1 generation to the white peloric parent.

Give the phenotypic appearance of each individual.   *W.Bi.*

251. What are sex chromosomes? In the fruit-fly, *Drosophila*, white eye is a sex-linked recessive mutation. In a laboratory experiment, homozygous red-eyed females were crossed with white-eyed males and the F_1 progeny allowed to interbreed at random. Explain why only one-quarter of the F_2 progeny possessed white eyes and that these flies were all of the same sex.   *W.Z.*

CHAPTER 20

Living things through the Ages

252. What major changes have plants and animals gone through during the process of evolution?   *C.Bi.S.*

CHAPTER 21

Evolution

253. What do you understand by organic evolution? State and discuss the main lines of evidence for this, illustrating your answer by reference to the plant kingdom.   *B.B.*

254. Discuss the evidence, provided by your botanical studies, which indicates that evolution has taken place.   *L.B.*

255. What is meant by the term "Organic Evolution"? Discuss the evidence in favour of a theory of Organic Evolution.   *B.Bi.*

256. Discuss the effects of geographical barriers and climate on the distribution and evolution of animals.   *L.Z.*

257. Write an account of the evidence in support of organic evolution which is derived from a study of the Vertebrata, past and present.   *N.Z.*

258. "The only certain evidence for a process of organic evolution comes from geological study." Criticize this view.   *O.Z.*

259. How would you attempt to convince a sceptical opponent that organic evolution has occurred?   *L.Z.*

260. Distinguish between homologous and analogous structures. Show how homologies are traced and discuss the value of the recognition of homologous structures in the study of evolution.   *O.C.Bi.(S.)*

261. What changes in plants, during the course of their evolution, appear to you to have made possible the change from an aquatic to a terrestrial habitat?   *W.B.(S.)*

262. Briefly state the theories of Lamarck and of Darwin concerning organic evolution. If you consider that the theories are not as acceptable today as when they were propounded, describe how they have been modified.   *L.Z.*

263. Discuss the contributions made by Darwin and Mendel to modern views on organic evolution.   *B.Bi.(S.)*

264. What is meant by adaptive radiation? Illustrate your answer by reference to the feeding habits of insects you have studied. *S.Z.*

265. Criticize the theories of organic evolution associated with the names of Lamarck and Darwin. *W.Z.(S.)*

266. Darwin's theory of natural selection is often expressed as "the survival of the fittest." What exactly is meant by "fitness," used as a biological term? Explain how Darwin's theory has had to be modified in accordance with modern knowledge. *N.Z.(S.)*

267. If "like tends to beget like," how do you correlate this with the fact that evolution has taken place? *W.Z.(S.)*

268. Write a short essay on Natural Selection. *C.Z.*

269. What is meant by the "Law of Natural Selection"? How far do you consider this to have been a factor in Evolution? *W.Z.*

270. What do you understand by the following evolutionary terms: (*a*) variation; (*b*) over-reproduction; (*c*) survival of the fittest? What roles are assigned to these processes in the theory of Natural Selection? *O.C.B.*

271. What is meant by (*a*) mutation, (*b*) natural selection? What part do these processes play in the evolution of new species? *C.B.*

272. Explain how habitat, isolation, and competition operate to bring about evolution. *O.C.Z.(S.)*

273. Explain, with examples, how the following provide evidence for evolution: (*a*) vestigial structures, (*b*) the geographical distribution of plants and animals. *C.Bi.*

274. What evidence is there from the structure of present-day animals to support the theory of organic evolution. *N.Bi.*

275. What new evidence for evolution has been brought forward since the days of Charles Darwin? Of these discoveries, which in your opinion, are the more important and why? *L.Bi.*

CHAPTER 22

Ecology

276. It has been suggested that the hedgerows between fields harbour various plant and animal pests harmful to crops and that it would therefore be sensible to substitute wire fences for hedgerows.

From your knowledge of ecology and the inter-relationships of living organisms, discuss any advantages or disadvantages which might accrue (*a*) to the farmer, and (*b*) to the naturalist, by such an action. *A.E.B.Bi.*

277. State the factors which determine the composition of any named plant community you have studied, and describe how these factors act on the vegetation. *O.C.B.(S.)*

278. Give a reasoned account of the methods you would use in an ecological investigation of an area previously unknown to you. *B.B.(S.)*

279. Describe concisely the methods you have used in the study of the vegetation of a *named* habitat. To what end did you use the methods you have described? *O.C.B.(S.)*

280. What is meant by "climatic climax vegetation"? Describe what changes you would expect to see, in the course of a few years, if rabbits and other grazing animals were excluded from an area of grassland in the neighbourhood of your school. State the nature of the soil. *O.C.Bi.(S.)*

281. Describe by reference to any named plant community you have studied, the use and meaning of any *three* of the following ecological terms; (*a*) dominance; (*b*) a transect; (*c*) biotic factors; (*d*) succession; (*e*) soil factors. *O.C.B.*

282. Describe the scheme of work you have adopted, and the methods you have used, in the study of the animals in the habitat you have studied.   *B.Z.*

283. Select an area you have studied in the field, outline its characteristics, and discuss the interrelationships between some of the plants and animals which you have found in it.   *C.Bi.*

284. What is meant by the term *succession*? Referring to habitats you have studied, discuss the evidence that succession was or was not taking place. *C.B.(S.)*

285. Give a brief account of a plant community which you have studied, and indicate to what features the success of the most abundant plant in the community may be attributed.   *C.B.(S.)*

286. Distinguish between a population, a community and an ecosystem. Describe the structure of any community you have personally studied in terms of the species present, their interactions, their relative abundance and the flow of energy between them. Briefly explain the terms *seral change* and *climax*, giving examples.   *N.Bi.*

287. Write a concise account of the most important environmental factors which appear to control the distribution of plants.   *W.B.*

288. Comment on the distribution and interrelations of the fauna in any locality you have studied.   *W.Z.*

CHAPTER 23

Man: His Success and His Problems

289. Discuss the dangers of over-exploitation of food resources in the light of the human population explosion. Suggest what steps you would take to supplement existing food supplies with a view to meeting future demands.   *L.Z.*

290. "Human interference with natural animal competition has always had un-foreseen and unfortunate repercussions." Discuss this statement.   *O.C.Z.*

291. What part is the biologist to play in the prevention of world hunger?
        O.C.Bi.(iii)

292. Give a general account of some of the major problems involved in agricul-tural food production at the present time.   *W.B.*

1. What are the functions of the nucleus? Describe nuclear division in the meristematic cells of any named plant. What significance do you attribute to this process?
O.C.B.

2. Give an account of meiosis. At what stage in the life-cycle does meiosis occur in the following plants: (a) *Fucus*, (b) a moss, (c) an angiosperm?
C.B.

3. Compare the vegetative structure and sexual reproduction of *Spirogyra, Mucor, Cystopus.*
C.B.

4. Give an account of the structure, life-history, and physiology of *Chlamydomonas*, and state briefly how far it differs from *Saccharomyces* in these respects.
W.B.

5. What is a gamete? Describe the structure and behaviour of the gametes of *Spirogyra, Chlamydomonas, Vaucheria* and *Fucus.*
W.B.

6. Summarize the features which the algae and the fungi have in common and those which justify their separation into two groups.
L.B.

7. Describe how the structure and reproduction of *Fucus* is related to its environment. What position is occupied by *Fucus* in the zonation of algae on the sea-shore?
O.C.B.(S.)

8. Describe the mature gametes and the method of fertilization in the following: *Chlamydomonas, Fucus, Mucor,* a *named* moss.
C.B.

9. What do you understand by the term "alternation of generations"? State, with reasons, whether you consider that alternation of generations is shown by *Spirogyra,* yeast, *Funaria* and *Dryopteris.*
O.C.B.

10. Give an illustrated account of the structure and reproduction of the sporophyte of a *named* fern. Why is this plant considered more primitive than an Angiosperm?
B.B.

11. Compare and contrast, in respect of vegetative structure and function, the leafy plant of *Selaginella* (or *Lycopodium*) with the leafy plant of a moss. *W.B.(S.)*

12. Describe in detail the vascular system of *Dryopteris* and state briefly how you would distinguish it from that of *Selaginella* (or *Lycopodium*), *Pinus* and a typical Dicotyledon.
W.B.(S.)

13. Give an account of the life-history of *Selaginella.* Indicate briefly how this life-history leads to an understanding of that of *Pinus* (or *Picea*). *C.B.*

14. Give an account of the range of structure exhibited by the gametophyte generation of green plants. Comment on evolutionary trends which are apparent.
B.B.(S.)

15. Distinguish, referring in your answer to named examples, between an *oogonium* and an *archegonium.* Which groups of plants possess archegonia? Indicate any biological advantages resulting from the possession of archegonia. *W.B.(S.)*

16. Give an account of the methods of spore dispersal in the plant kingdom, indicating their relative effectiveness.
N.B.(S.)

17. What do you understand by "alternation of generations"? Give a comparative account of this phenomenon in a named *Pellia, Selaginella* and *Pinus* (or *Picea*).
C.B.

18. "The gametophyte has never made a success of life on land and its dependence on external water for fertilization has restricted the Mosses and Ferns to limited

habitats." Discuss this statement with reference to mosses, ferns, conifers and flowering plants.   O.C.B.(S.)

19. Write an account of pollination and fertilization in *Pinus*. Compare these processes with those in a wind-pollinated flowering plant.   O.C.B.

20. What are the principal differences between *Pinus* and an angiosperm? L.B.

21. Where does meiosis take place in the life cycles of *Spirogyra* (or a similar haploid alga), a moss or fern and a flowering plant? What appears to be the importance of prolonging the diploid phase (*a*) genetically and (*b*) in the colonization of the land?   A.E.B.Bi.

22. Compare the structure and the importance in the life cycle of *three* of the following: (*a*) the anther of a flowering plant; (*b*) the microsporangium of *Pinus*; (*c*) the microsporangium of *Selaginella*; (*d*) the sporangium of a *named* fern; (*e*) the sporangium of *Mucor*.   C.B.

23. What is a spore? Describe *four* different types of spores, explaining (*a*) how they are formed and distributed and (*b*) the part they play in the life-cycles of the plants which produce them.   L.B.

24. What is a gametophyte? Give a comparative account of the gametophytes of *Pinus* and an angiosperm.   L.B.

25. Compare and contrast sexual reproduction in *Fucus*, *Pellia* and *Pinus*, indicating in each case how it is suited to the plant's environment.   L.B.

26. Describe how the gametes are brought together for fertilization in (*a*) *Oedogonium*, (*b*) *Vaucheria*, (*c*) *Funaria*, and (*d*) *Pinus*.   L.B.

27. "New requirements are met more commonly by the modification of an existing organ than by the evolution of a new organ." Discuss this statement with particular reference to specialized stems *or* leaves.   B.B.(S.)

28. Write concisely on the architecture of *either* vegetative *or* floral shoots. L.B.

29. Make annotated drawings of *one named* example of *four* of the following: siliqua, spathe, schizocarp, panicle, dichasial cyme.   L.B.

30. Give an illustrated account of the underground perennating organs of flowering plants.   L.B.

31. By means of large labelled diagrams illustrate the differences and resemblances between a leaf of *Pinus* and one of a *named* angiosperm.   L.B.

32. The mechanical stresses to which a plant organ is subjected are often reflected in its structure. Discuss this statement.   W.B.(S.)

33. Describe the principal differences between monocotyledons and dicotyledons.   L.B.

34. By reference to *one* of the dicotyledonous families you have studied, illustrate what is meant by the terms *species*, *variety*, *genus*, and *family*.   W.B.(S.)

35. Give reasons for classifying (*a*) *Spirogyra* and *Mucor* as Thallophyta, (*b*) *Pellia* and *Funaria* as Bryophyta, (*c*) *Ranunculus* and *Cycas* as Phanerogamia.   L.B.

36. Distinguish between parasitism, saprophytism, and symbiosis, referring to plants with which you are acquainted. Describe any *one* parasitic plant you have studied, laying stress on the special features in which it differs from a non-parasite.   W.B.

37. Give an account of the economic importance of fungi.   L.B.(S.)

38. Consider from a cytological aspect the relative merits of cross-pollination and self-pollination.   L.B.(S.)

39. Discuss the effect of wind upon plants.   L.B.(S.)

40. In what respects does the habitat of *Fucus* differ from that of *Spirogyra*? To what extent may differences in structure and in life-cycle be correlated with differences in habitat?   C.B.(S.)

41. It is generally believed that land plants have descended from aquatic ancestors. Discuss the evidence for this.   N.B.(S.)

42. What are the particular difficulties associated with plant life on land as compared to life in an aquatic environment? What evolutionary changes have occurred to make life on land possible for plants which are derived from an aquatic ancestry?   *L.B.(S.)*

43. State concisely what you consider to be the major steps which have occurred in the evolution of plants. Comment on the biological significance of these steps.
O.C.B.(S.)

44. Write concisely on one of the following: Nutrition in non-green plants; Insectivorous plants; Vegetative propagation.   *L.B.*

45. Write an essay on *either*: Cell differentiation; *or* Insectivorous plants. *B.B.*

46. Give an account of the importance in plants of *two* of the following:
   (*a*) β- indolyl-acetic acid (IAA)
   (*b*) phosphoglyceric acid (PGA)
   (*c*) adenosine triphosphate (ATP)
   (*d*) ribose nucleic acid (RNA).   *L.B.*

47. Give an account of the ways in which bacteria may affect soil fertility. *L.B.*

48. Write a short account of one of the following topics: (*a*) enzymes in the life of the plant; (*b*) spore dispersal; (*c*) the soil as an environment of plants. *O.C.B.*

49. Give a comparative account of growth as it occurs in plants having no vascular tissue.   *W.B.(S.)*

50. Write a short account of *one* of the following: photo-periodism, variation in plants, auxins, viruses.   *W.B.(S.)*

51. With the help of diagrams, discuss the external and internal structural features of the leaf of a *named* plant, that make it an efficient absorber of carbon dioxide.
C.B.

52. What are the essential differences between meiosis and mitosis, and what part may the former have played in the evolution of plants? At what stage in the life-history of an angiosperm does meiosis occur?   *W.B.*

53. Discuss the importance of water to plants.   *W.B.(S.)*

54. Write an essay on one of the following topics: (*a*) Bacteria; (*b*) Fungi; (*c*) Antibiotics.   *C.B.(S.)*

Zoology

55. Discuss the occurrence of bilateral and radial symmetry in the animal kingdom.   *B.Z.*

56. What is meant by radial and bilateral symmetry? Explain how each is advantageous in the mode of life of the animals possessing it.   *W.Z.(S.)*

57. By reference to labelled diagrams of transverse sections of a planarian, an earthworm through the intestinal region and a crayfish through the thoracic region, explain what is meant by the terms acoelomate, coelomate and haemocoele. *W.Z.*

58. Describe and compare the structure and functions of the body cavities of *Hydra*, the earthworm and the cockroach.   *N.Z.*

59. Comment on the nature and distribution in the Animal Kingdom of the following: coelenteron, coelom, haemocoel.   *W.Z.*

60. Which phyla contain segmented animals? Discuss possible advantages of segmentation in these phyla.   *O.C.Z.(S.)*

61. Describe those features of (*a*) adult anatomy, and (*b*) embryonic development, which show that vertebrates are metamerically segmented animals.   *N.Z.*

62. What are the functions of the skeleton in the invertebrate and vertebrate animals you have studied?   *C.Z.*

63. By means of illustrative diagrams show the relationship of the lower jaw to the cranium in dogfish, frog, and mammal. Comment on the mode of suspension in each case. *W.Z.(S.)*

64. The mechanism of the jaws of each animal is adapted to serve the needs of that animal. Show how this is true by a comparison of the mechanisms of the jaws of dogfish, frog, and rabbit. *N.Z.*

65. Construct labelled diagrams of the pectoral girdles and front limbs of the dogfish, frog, and rabbit. Write a short note on the functions of the front limbs of each of these animals. *N.Z.*

66. Give an account of the structure and functions of (*a*) a flame cell, (*b*) a nephridium, and (*c*) a green gland. In what animals would you expect to find these organs? *W.Z.*

67. Give some account of the sense organs of invertebrate animals. How is the position of sense organs related to the shape of these animals? *C.W.(S.)*

68. Distinguish between a neurone, a nerve fibre, and a nerve. Draw a labelled diagram to show how neurones are arranged to form a spinal reflex arc. Briefly describe the reflex actions which occur in the mammalian eye. *L.Z.*

69. Compare the structure and functions of the statocyst of a crayfish with those of the membranous labyrinth of a dogfish. *W.Z.*

70. What is a neuron? Describe how neurons are arranged to form a nervous system in *Hydra* and in a *named* chordate. *L.Z.*

71. Discuss the nervous systems in invertebrates. *B.Z.(S.)*

72. Give a comparative account of the structure of the female reproductive systems of the dogfish, frog, and mammal, and show how the reproduction of each animal is adapted to the animal's environment. *N.Z.*

73. Write brief notes on five of the following: solar plexus; Eustachian tube; lateral line organ; allantoic bladder; trophoblast; synovial joint; recurrent laryngeal nerve. *N.Z.*

74. Describe in detail one example of each of the various types of reproductive processes to be observed in invertebrates other than Protozoa. *N.Z.*

75. Write a general account of gametogenesis. Describe the ova of *Amphioxus*, of the dogfish, and of the frog, and the microscopic structure of the testis of a mammal. *N.Z.(S.)*

76. Write an essay on sexual reproduction in the Protozoa. *O.C.Z.*

77. Discuss the general principles relating to size and shape of animals in (*a*) aquatic, (*b*) terrestrial, (*c*) aerial environments. *B.Z.(S.)*

78. Give examples of the adaptation of animals to their environment. How do you imagine that such adaptations have come into existence? *C.Z.*

79. Give examples from the invertebrates of the ways in which animals remain in their characteristic environment and yet achieve dispersal. *C.Z.(S.)*

80. Show how the various physical factors of the environment assume different degrees of significance according to the size of an animal. *C.Z.(S.)*

81. What special features appear to be necessary to enable animals to live on land? *W.Z.*

82. With the aid of diagrams, compare and contrast the form and structure of the pelvic girdle and hind-limb of a dogfish with those of any *named* mammal. Suggest how each may be regarded as adapted to the mode of life of the animals possessing it. *W.Z.*

83. A whale was once described as "all adaptation." Explain what is meant by adaptation in zoology, and consider whether this remark could be made of all animals. *O.C.Z.(S.)*

84. How does oxygen reach the cells of the body in *Hydra*, the cockroach, and a named fish? *C.Z.*

85. Compare the functions of the blood in the cockroach and a mammal. How do some animals survive without a blood system?   *C.Z.*

86. Where would you expect to find *Paramecium, Euglena,* and *Monocystis*? Explain how they carry out the process of nutrition and relate their mode of feeding to their habitats.   *W.Z.*

87. Describe how locomotion and feeding in the Protozoa are interrelated.   *N.Z.*

88. Compare and contrast the skin of the earthworm with that of an insect. Show how the structure of the skin affects breathing, water conservation, and locomotion.   *N.Z.*

89. Compare the structure and function of the ectoderm of *Hydra* with the epidermis of a planarian, a cestode, an earthworm and an arthropod.   *W.Z.*

90. Give such an account of the protozoa you have studied as will show the increasing complexity of sexual reproduction in passing from simple to more complex forms.   *W.Z.(S.)*

91. Describe the phenomena of metamorphosis and the occurrence of larval forms in the animals you have studied. In each case, very briefly discuss the significance of the phenomena.   *O.C.Z.*

92. Define the terms "cell," "tissue," "organ," and "organism." On the basis of your definitions explain the constitution of (*a*) a spermatozoon, (*b*) a fertilized ovum, (*c*) a gastrula, (*d*) a three-day chick embryo.   *L.Z.(S.)*

93. Give an account of the early development of any named animal. How far do these stages provide evidence for the theory of recapitulation?   *O.C.Z(S.)*

94. What is mesoderm? Explain the structural and physiological consequences of the occurrence of this third body layer.   *L.Z.(S.)*

95. *Either:* discuss the similarities between birds and mammals, *or:* give an outline classification of the major groups of the vertebrates. Discuss the distinguishing characteristics of the groups you have mentioned.   *O.C.Z.*

96. Define the terms "cytology," "histology," "anatomy," and "physiology." Show what sort of information about living organisms is included in each of these categories, and how they are related to each other.   *O.C.Z.(S.)*

97. Write brief notes on *six* of the following: (i) erythrocyte; (ii) sternum; (iii) tail; (iv) larva; (v) lymph; (vi) corpus luteum; (vii) carnassial teeth; (viii) hormone; (ix) nematocyst.   *L.Z.*

98. Compare the division of labour in the dogfish, *Hydra,* and *Paramecium.*   *C.Z.*

99. Write a short essay on *one* of the following: (*a*) Larvae; (*b*) Parental care; (*c*) Flight in the animal kingdom.   *C.Z.*

100. Write a short essay on *one* of the following topics: (*a*) Cartilage; (*b*) the Cell Nucleus; (*c*) Homology and Analogy; (*d*) Instinct; (*e*) Animal Coloration.   *L.Z.*

101. Write a short essay on *one* of the following topics: (i) Asexual reproduction; ii) Vitamins; (iii) Charles Darwin.   *L.Z.*

102. Write an essay on *one* of the following topics: (i) Protoplasm; (ii) Biological control of pests; (iii) Hibernation; (iv) Metamorphosis; (v) The possible results of the elimination of wild rabbits from Great Britain.   *L.Z.(S.)*

103. Discuss the contribution of the zoologist to human welfare.   *L.Z.(S.)*

104. Describe the part played by carbohydrates in animal physiology. (Detailed biochemical descriptions are not required.)   *B.Z.*

105. Describe briefly the mechanisms for feeding and locomotion in the Protozoa. State why the efficiency of these mechanisms is dependent upon the size of the organisms.   *B.Z.*

106. Discuss the statement "Pests are the direct result of man's alteration of his environment."   *B.Z.*

107. In what ways is the body temperature of animals controlled? What are the advantages and disadvantages of such control? *C.Z.*

108. Write an essay on "Man is the most suitable animal for use in the study of animal function." *C.Z.*

109. Write an essay on "The place of zoology in the modern state." *W.Z.(S.)*

110. Describe the progressive steps in the probable evolution of the ear as observed in the series (a) dogfish, (b) frog, (c) mammal. Relate the structural modifications to the mode of life of each animal. *W.Z.(S.)*

111. Write an essay on *one* of the following: (a) Flying animals; (b) The excretion of substances in solution; (c) Animal coloration. *O.C.Z.*

112. Write a short essay on *one* of the following subjects: (a) Temperature control; (b) Muscular activity; (c) Kidneys. *O.C.Z.*

113. Write "An introduction to the study of Animal Behaviour." *O.C.Z.*

114. Write an essay on *one* of the following, in animals: (a) Osmoregulation; (b) Colour change; (c) Bioluminescence; (d) Electrical phenomena. *O.Z.*

115. Give an account of the structure and mode of operation of light-sensitive organs in animals. *O.Z.(S.)*

BIOLOGY

116. What characteristics distinguish living organisms from dead ones? Illustrate your answer by reference to *Amoeba* and *Spirogyra*. *L.Bi.*

117. Do you consider the division of unicellular organisms into Algae and Protozoa to be an artificial one? Discuss this. *B.Bi.(S.)*

118. "From a physiological standpoint, animals are degenerate plants." Discuss this statement by reference to named organisms. *B.Bi.(S.)*

119. Explain why some of the following organisms—*Chlamydomonas, Monocystis, Amoeba, Fucus*—are regarded as plants, and others as animals. *L.Bi.*

120. Compare and contrast the form and functions of the skeletal system of a mammal and the supporting tissues of a higher plant. *B.Bi.(S.)*

121. What are the differences between the "ovary" of a *named* flowering plant and that of a mammal? *L.Bi.*

122. "The surface layers of land plants and land animals represent a compromise between the need for protection and the interchange of materials between the organism and the environment." Discuss this statement. *N.Bi.*

123. Compare and contrast the structure and functions of the epidermis of (a) the leaf of a flowering plant, (b) an earthworm, and (c) a mammal. *L Bi.*

124. What are the essential parts of a cell? By referring to the structure of (a) a palisade cell of a leaf, (b) a cell of the piliferous (root-hair) layer, (c) an unstriped muscle fibre, and (d) a ciliated epithelial cell, point out the differences between plant and animal cells. *L.Bi.*

125. Write illustrated notes on five of the following: cambium, sclerenchyma, guard cell, spongy mesophyll, tracheid, endodermis, piliferous layer. *B.Bi.*

126. Carefully describe (a) the prothallus, (b) the sexual organs, and (c) the gametes of a fern. What stages in the life-history of a flowering plant are comparable with (a) and (c)? *L.Bi.*

127. What is meant by "alternation of generations"? What contribution, in your opinion, is made to our understanding of evolution by a study of plant life-cycles? *O.C.Bi.(S.)*

128. Explain what is meant by the terms sporophyte and gametophyte. Illustrate your answer by reference to *three* named plants belonging to different major subdivisions of the plant kingdom. *O.C.Bi.*

129. Give an account of reproduction which does not involve fusion of gametes, in animals and in non-vascular plants. *L.Bi.*

130. What is a spore? Discuss the origin, form and significance of spores with special reference to protozoa, fungi, and *Selaginella*. *L.Bi.*

131. Classify and describe the nutritional methods used by living organisms, citing *one* organism as an example of each method. *O.C.Bi.*

132. Consider with named examples, the range of types of nutrition found in unicellular organisms. *B.Bi.(S.)*

133. What is a parasite? What features of the structure and life-history of dodder and of the tapeworm may be correlated with the parasitic mode of life? *L.Bi.*

134. By reference to the life-histories of the malaria parasite and of the potato blight, show how a knowledge of a parasite and of its host may be used to combat disease. *L.Bi.*

135. By reference to one *named* plant parasite and one *named* animal parasite, discuss the characteristic features of parasites under the headings of (*a*) structure, (*b*) nutrition, and (*c*) reproduction. *L.Bi.*

136. Describe and discuss the biological advantages and disadvantages of a parasitic mode of life. *B.Bi.(S.)*

137. Explain the terms holophytic, parasitic, and saprophytic, illustrating your answer by reference to a named plant for each mode of nutrition. *O.C.Bi.*

138. *Hydra* and *Fucus* are both sedentary organisms living in water. Point out those features of the two organisms which appear to fit them for their mode of life. *L.Bi.*

139. What biological problems are involved in a change from an aquatic to a terrestrial mode of life? Discuss how these problems have been overcome by plants and animals. *B.Bi.(S.)*

140. Discuss the protective mechanisms adopted by the animals with which you are familiar. *O.C.Bi.(S.)*

141. Give one example each of disease caused by bacteria, viruses, and protozoa. Discuss the preventive and curative measures taken to control such disease. *O.C.Bi.*

142. Give an account of the importance to human society of the study of Zoology. *O.C.Bi.(S.)*

143. Write a short essay on *one* of the following subjects: (*a*) The application of physical methods to Biology; (*b*) The application of chemical methods to Biology; (*c*) The application of mathematical methods to Biology. *O.C.Z.*

144. Criticize and enlarge on the statement "All living organisms require oxygen". *L.Bi.*

145. Explain the biological basis for four of the following practices, and in each case show its value to man:

   (*a*) artificial insemination;
   (*b*) immunization against infectious disease;
   (*c*) composting garden and household refuse;
   (*d*) winter spraying of fruit trees;
   (*e*) incorporating leguminous plants in crop rotation;
   (*f*) restricting the use of chlorinated hydrocarbons as insecticides;
   (*g*) determination of a patient's blood group before giving him a transfusion.
   N.Bi.

146. Write short accounts of the use of any four of the following in biological investigations: (*a*) chromatography, (*b*) radioactive tracers, (*c*) electrophoresis, (*d*) statistics, (*e*) controlled environment, (*f*) stains and histochemical reagents. *N.Bi.*

147. What is meant when it is said that a substance is in the colloidal state? Write a short account of the principal properties of colloids, and show why it is believed that protoplasm is colloidal. *N.Bi.*

148. Discuss food chains in the sea or in large lakes. The concentration of algal cells in a lake in early spring was 100 cells/litre. Each cell multiplies by dividing into two. Prepare two tables showing the concentration of cells after each division during 6 successive cycles of division, assuming

(a) that there was no loss of cells and

(b) that one cell in ten was eaten in the interval between divisions.

State the final concentration of algal cells and the total number of cells eaten. What do you find interesting about these results?   *N.Bi.*

149. Argue the case for and against the following proposition:

Man as a species and men as individuals are no longer subject to natural selection.   *S.Z.*

150. Describe in detail how you would attempt to obtain living specimens of any three of the following: (a) *Mucor*, (b) *Cystopus*, (c) *Lactobacillus*, (d) *Monocystis*. Describe how you would isolate one of these types into pure culture, and maintain it as a permanent culture in the laboratory.   *S.Bi.*

151. Name one disease of the human body caused by each of the following conditions:

(a) infection by a bacterium;

(b) lack of a vitamin (name it);

(c) an infective organism carried to the body by an insect vector (name organism and vector);

(d) cigarette smoking;

(e) lack of a mineral ion (name it);

(f) infection by a virus.

In each case state what can be done to (i) prevent the disease and (ii) cure it. Outline the roles of (a) the individual citizen and (b) the government in the prevention of disease.   *S.Bi.*

FURTHER READING

The following books, grouped according to main content, are suggested as being among those suitable for further reading or reference. They are not arranged in any special order of preference or suitability. P = Programmed Text.

CELL STRUCTURE AND PHYSIOLOGY

The Ultrastructure of Cells, Bessis (Sandoz).
A Guide to Sub-cellular Botany, Stace (Longmans).
The Microstructure of Cells—Introduction for Sixth Forms, Hurry (Murray).
Methods in Cell Physiology, Ed. Prescott (Academic Press).
An Atlas of Biological Ultrastructure, Dodge (Arnold).
Structure and Function of the Cell, Parker, Reynolds and Reynolds (Pitman). P

BIOCHEMISTRY

Principles of Biochemistry, Tracey (Pitman).
Comparative Biochemistry, various authors (Academic Press).
Plant Biochemistry, Davies, Giovanelli, Rees (Blackwell).
The Chemistry of Micro-organisms, Bracken (Pitman).
Textbook of Physiology and Biochemistry, Bell, Davidson, Scarborough (Livingstone).
The Nature of Biochemistry, Baldwin (C.U.P.).
Cole's Practical Physiological Chemistry, Baldwin, Bell (C.U.P.).
Molecular Biology: A Structural Approach, Smith (Faber and Faber).

PLANT PHYSIOLOGY

The Growth of Plants, Fogg (Penguin).
An Introduction to Plant Physiology, James (Clarendon).
Plants at Work, Steward (Addison–Wesley).
The Reproductive Capacity of Plants, Salisbury (Bell).
The Living Plant, Brook (Edinburgh U.P.).
An Introduction to the Biology of Yeasts, Ingram (Pitman).
The Chemistry and Biology of Yeasts, Cook (Academic Press).
Microbes at Work, Selsam (Chatto and Windus).
Organization in Plants, Baron (Arnold).
Nitrogen Fixation in Plants, Stewart (Athlone Press).
The Physiological Aspects of Photosynthesis, Heath (Heinemann).
Rhythmic Phenomena in Plants, Sweeney (Academic Press).

ANIMAL PHYSIOLOGY

Animal Body Fluids and their Regulation, Lockwood (Heinemann).
On Growth and Form, Thompson (C.U.P.).
How Animals Move, Grey (C.U.P.).

Insect Physiology, Wigglesworth (Methuen).
Animal Structure and Function, Griffin (Holt, Reinhart, Winston).
An Introduction to General and Comparative Endocrinology, Barrington (Clarendon).
Animal Physiology, Scheer (Wiley).
The Life of Insects, Wigglesworth (Weidenfeld and Nicolson).
The Physiology of Mammals and Other Vertebrates, Marshall, Hughes (C.U.P.).
The Life of Mammals, Young (Clarendon).
An Introduction to Animal Physiology, Yapp (Oxford).
A Functional Bestiary, Vogel and Wainwright (Addison-Wesley).
The Physiology of Sense Organs, De Forest Mellon (Oliver and Boyd).
Hormones, Cells and Organisms, A. G. and P. C. Clegg (Heinemann).
A Primer of General Physiology, Coxon and Kay (Butterworth).
The Growth Process in Animals, Needham (Pitman).
Eye and Brain, Gregory (Weidenfeld and Nicolson).
Animal Locomotion, Gray (Weidenfeld and Nicolson).

ANIMAL BEHAVIOUR

An Introduction to the Behaviour of Invertebrates, Carthy (Allen and Unwin).
Bird Migrants, Simms (Cleaver–Hume).
The Study of Instinct, Tinbergen (O.U.P.).
Learning, Hill (Methuen).
Social Behaviour in Animals, Tinbergen (Methuen).
Learning and Instinct in Animals, Thorpe (Methuen).
An Introduction to Animal Behaviour, Manning (Arnold).
Genetic Analysis of Behaviour, Parsons (Methuen).

ENVIRONMENTS AND ECOLOGY

Science Out of Doors, Nature Conservancy (Longmans).
The World of the Soil, Russell (Collins).
A Guide to Field Biology, Sankey (Longmans).
Animal Ecology—Aims and Methods, MacFadyen (Pitman).
Ecology, Odum (Holt, Reinhart, Winston).
The Young Botanist, Prime (Nelson).
Patterns of Life, Dale (Heinemann).
Freshwater Biology, Macan (Longmans).
Animal Life in Fresh Water, Mellanby (Methuen).
Some Aspects of Life in Fresh Water, Popham (Heinemann).
The Marine and Freshwater Plankton, Davis (Constable).
The Soil, MacBean (Faber and Faber).
Soil Fungi and Soil Fertility, Garrett (Pergamon).
The Sea, Carson (McGibbon and Kee).
Wild Life of Wood and Forest, Edlin (Hutchinson).
Behavioural Aspects of Ecology, Klopfer (Prentice-Hall).
Britain's Green Mantle, Tansley (Allen and Unwin).
Elements of Ecology, Clarke (Wiley).
The Biology of Rocky Shores, Lewis (E.U.P.).

The Geography of Living Things, Anderson (E.U.P.).
Soil Animals, Kevan (Witherby).
Micro-organisms in the Soil, Burges (Hutchinson).
Land Invertebrates, Cloudsley-Thompson, Sankey (Methuen).
Weeds and Aliens, Salisbury (Collins).
Geology and Ourselves, Edmunds (Hutchinson).
Flora of the British Isles, Clapham, Tutin, Warburg (C.U.P.).
Dictionary of the Flowering Plants and Ferns, Willis (C.U.P.).
Connexions: Fit to Live In? Allsop (Penguin).
The Soil Ecosystem, Sheals (The Systematics Association, London).
Elements of Marine Ecology, Tait (Butterworth).
Quantitative Plant Ecology, Greig-Smith (Butterworth).
Biology of Marine Animals, Nicol (Pitman).
World Soils, Bridges (C.U.P.).
The Biosphere, Scientific American Book (Freeman).

GENETICS

Heredity: An Introduction to Genetics, Winchester (Harrap).
Genetic Effects of Radiation, Purdom (Newnes).
Molecular Genetics, Taylor (Academic Press).
Teaching Genetics, Darlington, Bradshaw (Oliver and Boyd).
Biology for the Modern World, Waddington (Harrap).
Experiments in Plant Hybridization, Mendel (Oliver and Boyd).
Experiments in Genetics with Drosophila, Strickberger (Wiley).
Fungal Genetics, Fincham, Day (Blackwell).
Classic Papers in Genetics, Peters (Prentice-Hall).
Principles of Human Genetics, Stern (Freeman).
Ecological Genetics, Ford (Methuen).
Outline of Human Genetics, Penrose (Heinemann).
Genetics of Garden Plants, Crane, Lawrence (Macmillan).
The Organization of Heredity, Lewis and John (Arnold).
The Code of Life, Borek (Columbia University Press).
DNA—Key to Life, Parker, Reynolds and Reynolds (Pitman). **P**
Genes and Populations, Geisert (Pitman). **P**
Mitosis and Meiosis, Parker, Reynolds and Reynolds (Pitman). **P**
Facets of Genetics, Scientific American Book (Freeman).

EVOLUTION

Evolution, Savage (Holt, Reinhart, Winston).
The Mechanism of Evolution, Dowdeswell (Heinemann).
A Century of Darwin, Barnett (Heinemann).
Evolution and its Implications, Kelly (Mills and Boon).
The Fossil Evidence for Human Evolution, Le Gros Clark (Chicago U.P.).
The Process of Evolution, Ehrlich, Hohn (McGraw-Hill).
Organization and Evolution in Plants, Wardlaw (Longmans).
The Origin of Species, Darwin (Dent).
Embryos and Ancestors, De Beer (O.U.P.).

Extinct and Vanishing Animals, Ziswiler (Longmans/Springer).
Plant Variation and Evolution, Briggs and Walters (Weidenfeld and Nicolson).

MAN

A Biology of Man, Hogg (Heinemann).
Smoking and Health, Royal College of Physicians (Pitman).
Man, the Peculiar Mammal, Harrison (Pelican).
The Natural History of Man in Britain, Fleure (Collins).
The Human Use of the Earth, Wagner (Free Press of Glencoe).
Human Fertility: The Modern Dilemma, Cook (Sloan; New York).
Four Thousand Million Mouths, Le Gros Clark, Pirie (O.U.P.).
The Estate of Man, Roberts (Faber and Faber).
World Population: Past Growth and Present Trends, Carr Saunders (Clarendon).
The Rape of the Earth, Jacks, Whyte (Faber and Faber).
Man and Environment, Arvill (Penguin).
Man's Ancestors, Tattersall (Murray).

STATISTICS

Statistical Methods in Biology, Bailey (E.U.P.).
Facts from Figures, Moroney (Pelican).
Statistics for Biology, Bishop (Longmans).

PHYSICAL SCIENCES RELATED TO BIOLOGY

Molecules Today and Tomorrow, Hyde (Harrap).
Energy, Life and Animal Organization, Riegel (E.U.P.).
Biophysical Principles of Structure and Function, Snell and others (Addison-Wesley).
Introduction to Thermodynamics, Spanner (Academic Press).
Physics in Botany, Richardson (Pitman).
An Introduction to Energetics with Applications to Biology, Linford (Butterworth).
A Biologist's Physical Chemistry, Morris (Arnold).
Energy, Organization and Life, Panares (Pitman). P
The Biology of Work, Edholm (Weidenfeld and Nicolson).
The Architecture of Molecules, Pauling and Hayward (Freeman).

BIOLOGICAL TECHNIQUES

Elementary Microtechnique, Peacock (Arnold).
Chromatography and Electrophoresis on Paper, Feinberg, Smith (Shandon).
The Handling of Chromosomes, Darlington, La Cour (Allen and Unwin).
Practical Microscopy, Duddington (Pitman).
Laboratory Techniques in Botany, Purvis, Collier, Walls (Butterworth).
Physical Methods in Physiology, Catton (Pitman).

GENERAL TOPICS

The Development of Modern Biology, Marshall (Pergamon).
Biology in the Modern World, Comber (Thames and Hudson).

The Origin of Life, Bernal (Weidenfeld and Nicolson).
Mathematical Ideas in Biology, Smith (C.U.P.).
Problems in Quantitative Biology, Eggleston (E.U.P.).
Studies in Biology Series (Institute of Biology, various authors) (Arnold).

INDEX

List of genera referred to in the text. Fossil forms shown separately (p. 1091)